T0135329

Advances in Intelligent Systems and Computing

Volume 1002

The series "Advances in Intelligent Systems and Computing" contains publications on theory, applications, and design methods of Intelligent Systems and Intelligent Computing. Virtually all disciplines such as engineering, natural sciences, computer and information science, ICT, economics, business, e-commerce, environment, healthcare, life science are covered. The list of topics spans all the areas of modern intelligent systems and computing such as: computational intelligence, soft computing including neural networks, fuzzy systems, evolutionary computing and the fusion of these paradigms, social intelligence, ambient intelligence, computational neuroscience, artificial life, virtual worlds and society, cognitive science and systems, Perception and Vision, DNA and immune based systems, self-organizing and adaptive systems, e-Learning and teaching, human-centered and human-centric computing, recommender systems, intelligent control, robotics and mechatronics including human-machine teaming, knowledge-based paradigms, learning paradigms, machine ethics, intelligent data analysis, knowledge management, intelligent agents, intelligent decision making and support, intelligent network security, trust management, interactive entertainment, Web intelligence and multimedia.

The publications within "Advances in Intelligent Systems and Computing" are primarily proceedings of important conferences, symposia and congresses. They cover significant recent developments in the field, both of a foundational and applicable character. An important characteristic feature of the series is the short publication time and world-wide distribution. This permits a rapid and broad dissemination of research results.

** Indexing: The books of this series are submitted to ISI Proceedings, EI-Compendex, DBLP, SCOPUS, Google Scholar and Springerlink **

More information about this series at http://www.springer.com/series/11156

Jiuping Xu · Syed Ejaz Ahmed ·
Fang Lee Cooke · Gheorghe Duca
Editors

Proceedings of the Thirteenth International Conference on Management Science and Engineering Management

Volume 2

 Springer

Editors
Jiuping Xu
Business School
Sichuan University
Chengdu, China

Syed Ejaz Ahmed
Department of Mathematics and Statistics
Brock University on McMaster
Hamilton, ON, Canada

Fang Lee Cooke
Department of Management
Monash University
Melbourne, VIC, Australia

Gheorghe Duca
Academy of Sciences of Moldova
Chisinau, Moldova

ISSN 2194-5357 ISSN 2194-5365 (electronic)
Advances in Intelligent Systems and Computing
ISBN 978-3-030-21254-4 ISBN 978-3-030-21255-1 (eBook)
https://doi.org/10.1007/978-3-030-21255-1

This Springer imprint is published by the registered company Springer Nature Switzerland AG
The registered company address is: Gewerbestrasse 11, 6330 Cham, Switzerland

Contents

Advancements in Project Management, Supply Chain Management, and Organizational Strategy at the Thirteenth ICMSEM Proceedings 1
Jiuping Xu

Part I   Project Management

Research on the Container Optimization of Communication Products 13
Jiahui Duan, Jiancheng Xu, and Jianguo Zheng

Research on the Improvement of IPD Management Mode of Engineering Project Under Network Environment 26
Ling Wan and Xiaozhong Yu

A Research of FAHP Approach in Evaluating Online Training System Alternatives 40
Liping Li and Dan Li

The Impact of Sponsorship Following on Brand Evaluation-A Test of Mediation 49
Ge Ke, Hong Wang, Hua Zhang, Wei Li, Shoujiang Zhou, and Shan Li

Study on the Relationship between Modernized Economic System Construction and Transfer and Transformation of Scientific and Technological Achievements 61
Qian Fang and Yi Sheng

A Study on the Subjects and Network of Technology Transfer: Evidence from Chinese Universities Patent Rights Transfer 75
Qiang Li and Xin Gu

The Effect of Stock Liquidity on Default Risk: An Empirical Study of China's Capital Market 93
Kun Li, Qinqin Yan, and Yu Lei

Research on the Motivation of Government Intervention in Executive Compensation of State-Owned Enterprises-An Empirical Study Based on the State-Owned Listed Enterprises 105
Sheng Ma, Shuang Li, ChuanBo Chen, and Rui Wang

Research on the Influence of User Interaction of Tourism Virtual Community on Purchase Intention 119
Jingdong Chen and Jingwen Qiao

Does the Tax Stickiness Exist? 135
Wei Yang, Shengdao Gan, Hong Wang, and Xinyun Liu

Research on the Generation and Development of Sharing Manufacturing based on the Long Tail Theory 149
Dong Chen and Sheng Yi

Evaluation and Comparative Study of Provincial Resilience in China 162
Chao Huang, Jihui Zhong, Jing Wu, and Jiaqi Fan

The Influence of Enterprise Internationalization Level and R&D Input on Enterprise Performance under the Background of "One Belt and One Road" 177
Jingjing Lu and Hongchang Mei

Do Customers Really Get Tired of Double Eleven Global Carnival? An Exploration of Negative Influences on Consumer Attitudes toward Online Shopping Website 189
Ruo Yang, Yongzhong Yang, Mohsin Shafi, and Xiaoting Song

Herd Behavior of Chinese Firm's Outward FDI Along One Belt One Road- Analysis on Chinese Listed Manufacturing Sector 201
Haiyue Liu, Shiyi Liu, Ling Huang, and Yile Wang

ETOS-based Research on Earthquake Tourism in the Longmenshan Fault Zone: Take Wenchuan for Example 221
Xu Zu, Xinyi Diao, and Zhiyi Meng

A Framework for BIM-based Quality Supervision Model in Project Management 234
Jun Gang, Chun Feng, and Wei Shu

"Element-Chain-Cluster" Industrial Project Planning Model in Poverty-Stricken Areas 243
Mingzhou Tu, Hongjiang Zhang, Yujie Wang, Chengyan Zhan, Linlin Liu, and Yi Lu

Media Attention, Investor Attention and Corporate Innovation Investment: Empirical Study Based on China GEM Listed Companies   255
Yuzhu Wei and Hongchang Mei

Analysis of Sustainable Net Cage and Fluid Flows Through the Review   270
Mengyuan Zhu, Shuijin Li, Tingting Liu, Rongwei Sun, and Jingqi Dai

Review of Energy Finance and Corresponding Policies in Promoting Renewable Energy Sustainable Development in China   279
Yanfei Deng, Lei Xu, Yuan Yuan, and Karen Mancl

Part II   Supply Chain Management

A Robust Weighted Goal Programming Approach for Supplier Selection Problem with Inventory Management and Vehicle Allocation in Uncertain Environment   295
Lishuai Wang and Jun Li

Pre-sale Pricing Strategy for Fresh Agricultural Products Under O2O   310
Wenjing Wu and Chunxiang Guo

Climatic Changes and Surface Water Quality on Republic of Moldovas Territory   325
Gheorghe Duca, Maria Nedealcov, Viorica Gladchi, and Serghei Travin

The Impact of "Internet +" on the Business Models Transformation of Traditional Enterprises   335
Jinjiang Yan, Yongyi Wang, Zhen Liu, Yong Huang, and Xinhui Wang

Modeling the Barriers of Sustainable Supply Chain Practices: A Pakistani Perspective   348
Muhammad Nazam, Muhammad Hashim, Mahmood Ahmad Randhawa, and Asif Maqbool

Impact of Supplier's Lead Time on Strategic Assembler-Supplier Relationship in Pricing Sensitive Market   365
Jiayi Wei

Effects of Customer Referral Programs on Mobile App Stickiness: Evidence from China Fresh E-Commerce   376
Qian Wang, Yufan Jiang, Chengcheng Liao, and Yang Yang

Impact of Supply Chain Management Practices on Organizational Performance and Moderating Role of Innovation Culture: A Case of Pakistan Textile Industry 390
Muhammad Hashim, Sajjad Ahmad Baig, Fiza Amjad, Muhammad Nazam, and Muhammad Umair Akram

A Performance Assessment Framework for Baijiu Sustainable Supply Chain in China 402
Xianglan Jiang, Yinping Mu, and Jiarong Luo

Spatial Evolution, Driving Factors and Comprehensive Development on Urban Agglomeration–A Case Study of Sichuan Province 415
Jialing Zhu, Quan Quan, Ming You, Sichen Xu, and Yunqiang Liu

Relationship Between Institutional Pressures, Green Supply Chain Management Practices and Business Performance: An Empirical Research on Automobile Industry 430
Jinsong Zhang, Xiaoqian Zhang, Qinyun Wang, and Zixin Ma

Non-cooperative Game Based Carbon Emission Reduction for Supply Chain Enterprises with a Cap and Trade Mechanism 450
Min Wang, Shuhua Hou, and Rui Qiu

A Comparative Study of Waste Classification Laws and Policies: Lessons from the United States, Japan and China 460
Lai Wei and Yi Lu

Design and Implementation of Sustainable Supply Chain Model with Various Distribution Channels 469
YoungSu Yun, Anudari Chuluunsukh, and Mitsuo Gen

A Multiple Decision-Maker Model for Construction Material Supply Problem Based on Costs-Carbon Equilibrium 483
Rongwei Sun, Shuhua Hou, and Rui Qiu

Do Firms Experience Enhanced Productivity After Cross-Border M&As? 492
Zihan Zhou, Lei Zhang, and Dongmei He

Future of India-China Relations: A Key Role to the Global Economy 510
Nitin Kumar and Sita Shah

Effects of Autocracy and Democracy on FDI's Inflows 523
Tahir Yousaf, Qurat ul Ain, and Yasmeen Akhtar

Part III Oganizational Strategy

Strategic Management Model for Academic Libraries – The Case
Study of Ilma University, Karachi 537
Asif Kamran, Farhana Shoukat, Nadeem A. Syed, and Sheeraz Ali

Research on the Factors Affecting the Delisting of Chinese Listed
Companies 546
Yanyan Zhang

Customer Relationship and Efficiency Analysis of the Listed
Air Companies in China...................................... 557
Ying Li, Zelin Jin, Hongyi Cen, Yung-ho Chiu, and Jian Jiao

How Social Factors Drive Electronic Word-of-Mouth on Social
Networking Sites?... 574
Muhammad Sohaib, Peng Hui, Umair Akram, Abdul Majeed,
and Anum Tariq

Online Impulse Buying of Organic Food: Moderating Role of Social
Appeal and Media Richness 586
Anum Tariq, Changfeng Wang, Umair Akram, Yasir Tanveer,
and Muhammad Sohaib

Research on the Advertising Diffusion Effectiveness on Microblog
and the Influence of Opinion Leaders 600
Dan Zhang, Chuanpeng Xu, Malian Shuai, Wenyu Xiong, Wen Jiang,
Dong Xu, Yue He, and Weiping Yu

What Affects the Innovative Behavior of Civil Servants? Survey
Evidence from China 616
Rongrong He and Shuaifeng Li

The Study on the Influence of Online Interactivity on Purchase
Intention on B2C Websites: The Interference Moderating Role
of Website Reputation 628
Rongjia Su and Dianjie Liang

Turnover: Organizational Politics or Alternate Job Offer? 641
Aimon Iqbal, Abdullah Khan, and Shariq Ahmed

Research on the Influence Mechanism of eWOM on Selection
of Tourist Destinations—The Intermediary Role of Psychological
Contract 654
Mo Chen, Jingdong Chen, and Weixian Xue

A Study on the Impact of Customer Engagement on Continued
Purchase Intention for Online Video Websites VIP Service 668
Jingdong Chen and Wenxin Xu

**A Study of Museum Experience Evaluation from the Perspective
of Visitors' Behavior** 683
Shuangji Liu, Yongzhong Yang, and Mohsin Shafi

**Changing Preference Aspects from Traditional Stores to Modern
Stores** 694
Abdullah Khan, Shariq Ahmed, and Farhan Arshad

**The Link Between Heterogeneity in Employment Arrangements,
Team Cohesion and Team Organizational Citizenship Behavior:
A Moderated Mediation Model** 705
Xuan Wang, Yanglinfeng Zheng, and Xiaoye Qian

**Administrative Resilience and Adaptive Capacity of Administrative
System: A Critical Conceptual Review** 717
Md Nazirul Islam Sarker, Min Wu, Roger C. Shouse, and Chenwei Ma

**Can Manager's Environmentally Specific Transformational
Leadership Improve Environmental Performance?** 730
Xuhong Liu and Xiaowen Jie

**Financial Indicators and Stock Price Movements: The Evidence
from the Finance of China** 743
Qiang Jiang, Xin Wang, Yi Li, Dong Wang, and Qing Huang

**Empirically Analyzing the Future Intentions of Pakistani Students
to Stay or Leave: Evidence from China** 759
Kashif Iqbal, Hui Peng, Muhammad Hafeez, Khurshaid, and Israr Khan

**Study on Fluctuation and Regulation of Potato Market Price in China:
Based on the View of Stable Crop for the Potato** 770
Qianyou Zhang, Xinxin Xu, Yuanling Zhang, Yuerong Zheng,
and Jinqiu Tian

Author Index 781

Advancements in Project Management, Supply Chain Management, and Organizational Strategy at the Thirteenth ICMSEM Proceedings

Jiuping Xu[✉]

Uncertainty Decision-Making Laboratory, Sichuan University, Chengdu 610065,
People's Republic of China
xujiuping@scu.edu.cn

Abstract. Management Science and Engineering Management (MSEM) has significantly contributed to management and control process developments in both the economy and the society. In this paper, we first describe the basic concepts covered in the 13th ICMSEM proceedings Volume II, after which we conduct a review of engineering management (EM) research to identify the key areas, the most widely discussed research areas for which have been project management, supply chain management and organizational strategy. After an analysis of the key research achievements in the three areas, the related research studies in Proceedings Volume II are described. The research trends from both MSEM journals and the ICMSEM are then summarized using CiteSpace. As always, we strive to provide researchers with an international forum through ICMSEM for academic exchanges and communication.

Keywords: Project management · Supply chain management · Organizational strategy

1 Introduction

Management science and engineering management focus on the theories, methods and engineering practices related to complex management decisions that employ information technology to solve social, economic, engineering and other management issues. By emphasizing analysis, decision making, strategic execution and innovation, MSEM research and development over the past few decades have brought new vitality to management and engineering practice. Because of the continuous development of innovative management tools, MSEM has been playing an important role in promoting global economic development and scientific management awareness in various industries. MSEM is a complex synthesis of advanced management ideas, methods, organization, and technology that employs mathematical and computer models to analyze, determine and solve problems in operations, organizational, and technology management areas. This

J. Xu et al. (Eds.): ICMSEM2019 2019, AISC 1002, pp. 1–9, 2020.
https://doi.org/10.1007/978-3-030-21255-1_1

cross-functional, multidisciplinary research improves management efficiency and assists in conserving resources.

Fig. 1. The research on MSEM

As shown in Fig. 1, MSEM research is being applied to resolve complex engineering management problems. Proceedings Volume I focuses on management science (MS) and its future development trends, and Volume II focuses on Engineering management (EM). Kocaolgu defined EM as an engineering management check that uses an integrated approach to achieve optimal engineering management results, improve organizational structures, and conserve resources with the aim of optimizing manufacturing, construction, design engineering, and industrial engineering systems [6]. Volume II of the ICMSEM meeting record focuses on three key emerging market areas: project management, supply chain management and organizational strategy.

In this paper, Sect. 2 provides a literature review on the three key areas, Sect. 3 presents the central issues presented in Proceedings Volume II, Sect. 4 analyzes EM and ICMSEM development trends, and Sect. 5 gives the conclusion and the future research trends.

2   Literature Review

To identify the most pertinent research fields and possible research directions, the most recent EM research was reviewed, from which it was found that project management, supply chain management and organizational strategy had been the most widely studied. In this section, the related literature from these three areas is reviewed to identify the specific development tracks. As Fig. 2 shows, the analysis process used different classification criteria to select and confirm the databases, after which CiteSpace was employed to analyze the database and elucidate the EM development trends.

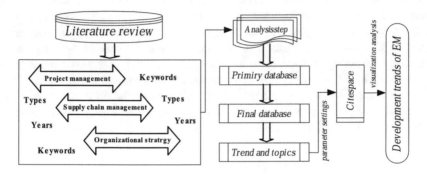

Fig. 2. The steps of analysis

2.1   Project Management

Project management competencies are the knowledge, personal attitudes, skills and relevant experience that ensure project success [1], have been regarded as one of the most important factors affecting project success [5]. Due to the increasing complexity of business activities, project-based organizations have become increasingly common [5]. Torbjrn Bjorvatn et al. studied the relationships between project complexity and project management performance and found that there were clear, direct and positive statistical correlations between project complexity and delays and overruns, indicating that effective project management begins with the integration of processes and people [2]. Because effective project management begins with the integration of processes and people in the construction project. Sevilay Demirkesen et al. investigated the impact of various integrated management components on building project management performance and quantified the relationships to achieve effective integration throughout the project life cycle [3]. In short, project management can assist managers effectively control the complete project cycle.

4   J. Xu

2.2 Supply Chain Management

Supply chains include a core enterprise, intermediate product and final product producers, retailer and consumers, and basically connects suppliers, manufacturers, distributors and end users into a functional network chain. Esteban Koberg et al. examined the key elements of sustainable supply chain management in global supply chains and found that supply chain management allocation and governance mechanisms were the key supply chain management elements [9]. Hans-Christian Pfohl et al. explored the supply chain funds flow and proposed a conceptual "supply chain finance" framework and mathematical model [13]. Good supply chain management can reduce business operating costs. Robert D.Klassen et al. examined the relationships between social management capabilities and social responsibility, risk, opportunity and performance in a supply chain [4]. Matthias Kalverkamp et al. examined closed-loop supply chains and suggested that the use of this term in academia and business was too limited from a sustainability perspective, and proposed a more adaptive management method to reverse the supply chain for end-of-life products [8]. Antonella Moretto presented a model that considered the supplier financial indicators and buyer operational assessments and used stakeholder theory to study the benefits and challenges in this model for all stakeholders [11].

2.3 Organizational Strategy

Organization structural strategies are related to internal long-term development enterprise planning models that are focused on business strategy requirements, the business environment, government policy requirements, and the necessary inter organizational relationships. Robert C. Ford et al. developed organizational strategies to ensure customers had successful service experiences [7], and Ruekert examined the relationship between market orientation and organizational processes, personal attitudes, and long-term financial performance using business unit analysis to determine overall organizational strategies, the results from which could assist managers develop and maintain market-orientations [14]. Susan E. Lynch et al. examined how formal structural changes affected the ability of senior managers to maintain intra-organizational networks, and found that the characteristics of particular network connections determined whether they were affected by formal structural changes [10]. JonWelty Peachey et al.examined the challenges faced by SDP organizations when forming and maintaining inter-organizational partnerships when participants had different backgrounds and partnership types, and identified the strategies to overcome these challenges [12].

3 Central Issues in Proceedings Volume II

The above brief analysis highlights the key areas in Proceedings Volume II associated with project management, supply chain management, and organizational strategy problems related to green and pro-environmental concepts.

Different ideas have also been presented for project management applications. Li et al. proposed a structured mathematical model to evaluate online training system alternatives using a fuzzy analytical hierarchy process (FAHP), with the primary aim of demonstrating the viability of FAHP models in solving these types of decisions in practice. Wan further refined the Integrated Project Delivery (IPD) model for core delivery using BIM technology and "Internet +" related technologies and network organization structures. To explore user intentions to continuously purchase online video website VIP services, Chen et al. used SEM to empirically study the driving influences for customer willingness to purchase based on a multi-dimensional customer cognitive, emotional and behavioral participation perspective, and then employed linear regression analysis to explore the impact of promotions on consumer negative attitudes, from which it was found that: (1) the negative impact of double eleven sales promotions had no significant effect on consumer attitudes towards the website; and (2) the AW (attitude to the website) and SS (sales satisfaction) had a significant impact on AP (attitude to promotions).

In research on supply chain management, based on the daily life fresh produce sales, Wu et al. examined optimal supplier and retailer pricing strategies in a dual-channel business model in different periods to reduce demand uncertainty and maximize supply chain profit. To achieve better protection and adequate management of water resources in vulnerable areas, Duca et al. proposed an integrated knowledge-based approach for adapting to climate change effects in the broad context of natural and social systems that incorporated scientific, legal, economic and ecological measures. Yan et al. found that the key to business model transformations due to internet impacts was to optimize and adjust internal company structures and increase the external influence of the internet. Hashim et al. investigated the impact of supply chain management (SCM) practices on organizational performance through the moderating role of innovation culture. Wei examined the impact of supplier lead-time on the assemblers strategic choice of supplier partnerships. Wang demonstrated that the emissions reduction effect of the government carbon quota per unit of product allocations was more obvious than direct total amount restriction and was more beneficial for supply chain system sustainability.

A wider range of industries have been paying attention to organizational strategies. Tariq et al. found that there was significant moderation in social appeal because of the influence of social communities and forums on both consumer cognition and affect and that websites with high media richness could increase impulse decisions and convert intentions into buying. As major domestic airlines face fierce competition from domestic and international counterparts, it is important to establish good customer relationships. Li et al. used a two-stage dynamic DEA approach to determine whether customer complaints positively and significantly impacted airline operational efficiency. Zhang et al. found that opinion leaders had a greater influence on the breadth of the heat and had a positive influence on other carriers and audiences. Iqbal et al. found that turnover had a significantly positive relationship with perceived organizational politics and the finding of alternate jobs with higher pay.

4   Evaluation of EM and ICMSEM Development Trends

In this section, the EM and ICMSEM development trends are evaluated. The CiteSpace information visualization technology developed by Chen enables researchers to analyze relevant research, terminology, and institutional and country affiliations from scientific publications. This technology includes a collaborative network that supports mixed node networks and mixed link types such as common references, co-occurrences, and directed reference links.

Citespace can be used to identify and visually represent the research trends in a specific research area. To do this, an advanced search function with engineering management as the identifier was applied to the Web of Science database with the timespan set from 1990 to 2019. Initially, 56,473 published articles were identified and after a final screening using TI = (engineering management) as the search string, 3365 articles were finally extracted and input to Citespace.

4.1   The Development Trends of EM

The data from the 3365 articles were saved and converted into CiteSpace, which transformed the data into a format that could be identified by the software for parameter selection. The time span was from 1990 to 2019 and the time slice was set at one year, with the theme selection being based on the titles, abstract subject words, identifiers, and keywords for node selection. Then, each zone with the highest keyword records was clustered and analyzed, from which a map was drawn using minimum spanning tree.

The cluster analysis results are shown in Fig. 3, from which the key node theoretical and research innovation foci can be seen; knowledge, management, development, and technology. With the increasing emphasis on integrated development, management innovations are now focused on software and tools, and there has also been increased research on the main EM subject areas; project management, supply chain management, and organizational strategy.

Figure 3 gives a timeline view of the EM research and shows the EM developments and related keywords in recent years, from which the current research scopes and directions can be understood. Many research keywords also appeared in the early days, suggesting that these topics have been key research areas for a long time. For example, project management and supply chain management are included in a wide range of research areas such as forecasting, optimization, collaboration, resource cost savings and resource mobilization.

The frequent keyword analyses also highlighted that engineering, knowledge and development research is a popular current research area. Therefore, the papers presented in this ICMSEM conference proceedings Volume II, 2019 closely reflect the most relevant engineering management research areas; project management (full-cycle project management, project training systems, integrated

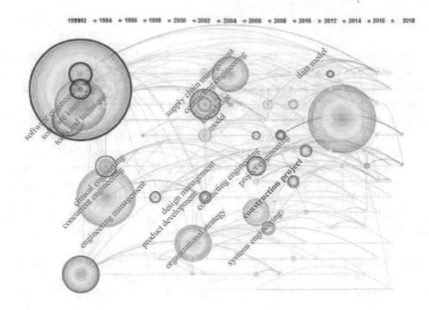

Fig. 3. The timeline view of research on EM

project delivery, internet commerce), supply chain management (agricultural trading models, water conservation, carbon emissions), and organizational strategy (airline service strategy and employee management).

4.2   Future Development Predictions

As mentioned, to further understand the engineering management field, high frequency research was included in the timeline view, from which it was found that these words have been associated with the key research areas for many years. Through the analysis of the most common words, project management, supply chain management and organizational strategy remain the most popular EM research areas.

The effective management of engineering projects and technicians requires training and experience in both management and specific engineering disciplines. Therefore, we believe that EM research needs to focus on specific EM issues while at the same time popularizing the wider use of MS knowledge. Sound academic exploration could promote the global development of engineering management; however, the future development of EM requires a research focus on practical theories, effective methods, and wider applications. To achieve this goal, it is necessary for all MSEM scholars to promote EM knowledge. In the future, greater attention needs to be paid to low carbon emissions, EM knowledge research, periodic management applications, energy use, cost control, and other popular EM issues.

5   Conclusion

Engineering management is the application of management to engineering projects; therefore, it is a discipline that combines technical organization, management knowledge, and planning capabilities. EM techniques are being widely used to monitor the operational performances of engineering-driven companies and project-operated companies. Citespace was used as the main analysis tool in this short introduction to analyze the EM and ICMSEM trends and identify the search terms and keywords. The results of these visualizations highlighted the current EM trends and the strengths of the various EM research arms. We hope the research analyses in this volume provide some direction and inspiration for researchers. EM research is constantly evolving and new themes are emerging every year; however, more EM research is needed to provide more active research discussions.

Acknowledgements. The author gratefully acknowledges Tingting Liu and Jingqi Dai's efforts on the paper collection and classification, Zongmin Li and Mengyuan Zhu's efforts on data collation and analysis, and Rongwei Sun and Yawen Deng's efforts on the chart drawing.

References

1. Association I.P.M. et al.: Icb-ipma competence baseline version 3.0. International Project Management Association, Nijkerk (2006)
2. Bjorvatn, T., Wald, A.: Project complexity and team-level absorptive capacity as drivers of project management performance. Int. J. Proj. Manag. **36**(6), 876–888 (2018)
3. Demirkesen, S., Ozorhon, B.: Impact of integration management on construction project management performance. Int. J. Proj. Manag. **35**(8), 1639–1654 (2017)
4. DKlassena, R., Ann, V.: Social issues in supply chains: Capabilities link responsibility, risk (opportunity), and performance. Int. J. Prod. Econ. **140**, 103–115 (2012)
5. Ekrot, B., Kock, A., Gemünden, H.G.: Retaining project management competencełantecedents and consequences. Int. J. Proj. Manag. **34**(2), 145–157 (2016)
6. Farr, J.V., Buede, D.M.: Systems engineering and engineering management: Keys to the efficient development of products and services. Eng. Manag. J. **15**(3), 3–9 (2003)
7. Ford, R.C., McColl-Kennedy, J.R.: Organizational strategies for filling the customer can-do/must-do gap. Bus. Horiz. **58**(4), 459–468 (2015)
8. Kalverkamp, M., Young, S.B.: In support of open-loop supply chains: Expanding the scope of environmental sustainability in reverse supply chains. J. Clean. Prod. **214**, 573–582 (2019)
9. Koberg, E., Longoni, A.: A systematic review of sustainable supply chain management in global supply chains. J. Clean. Prod. (2018)
10. Lynch, S.E., Mors, M.L.: Strategy implementation and organizational change: How formal reorganization affects professional networks. Long Range Plan. (2018)
11. Moretto, A., Grassi, L. et al.: Supply chain finance: From traditional to supply chain credit rating. J. Purch. Supply Manag. (2018)

12. Peachey, J.W., Cohen, A., et al.: Challenges and strategies of building and sustaining inter-organizational partnerships in sport for development and peace. Sport. Manag. Rev. **21**(2), 160–175 (2018)
13. Pfohl, H.C., Gomm, M.: Supply chain finance: Optimizing financial flows in supply chains. Logist. Res. **1**(3–4), 149–161 (2009)
14. Ruekert, R.W.: Developing a market orientation: An organizational strategy perspective. Int. J. Res. Mark. **9**(3), 225–245 (1992)

Part I
Project Management

Research on the Container Optimization of Communication Products

Jiahui Duan, Jiancheng Xu, and Jianguo Zheng[(⊠)]

Business School of Sichuan University, Chengdu, People's Republic of China
zhengjianguo@scu.edu.cn

Abstract. The transportation cost of a company is very high. Firstly, this paper applies the classical ideas and tools of industrial engineering such as "5W1H" and fishbone diagram to analyze the container problem; uses the system engineering ideas and methods to establish the mathematical model of goods loading and placement, and obtain the optimal solution of the highest utilization rate of the container. Secondly, Flexsim simulation software is applied to work out container load time and the number of cartons loaded in every single container. The optimization scheme has achieved the purpose of improving the efficiency of container loading, improving the capacity utilization ratio and reducing the transportation cost.

Keywords: Container loading · Optimization model · Transport optimization · Effect

1 The Research Background

F Company's communication products are mainly exported; Due to the large demand, its monthly average reaches 45 containers; The products are shipped by sea to CNSBG overseas factory; Freight charges are paid in unit of container.

1.1 Product Freight

The transportation information of F Company from January to June 2017 is shown in Table 1 below.

The formula [5] used for transportation cost calculation is:

$$\text{Transport cost} = \text{container number} \times \text{single container freight} \qquad (1)$$

Use Table 1 data and formula (1) to calculate the actual freight of the single case.

$$\text{Single case actual freight} = \text{total transportation cost}/\text{number of carton} \qquad (2)$$

© Springer Nature Switzerland AG 2020
J. Xu et al. (Eds.): ICMSEM2019 2019, AISC 1002, pp. 13–25, 2020.
https://doi.org/10.1007/978-3-030-21255-1_2

Table 1. Transport in January to June 2017 information table

Month	January	February	March	April	May	June
Container	17	40	31	85	31	51
Carton	12852	30240	23436	64260	23436	38556
Single container charge	4.8	4.8	4.8	4.8	4.8	4.8
Total shipping cost	81.6	192	148.8	408	148.8	244.8

Substitute into the formula (2), and get:

The actual transportation cost of the single case $= 81600 \div 12852 = 6.35(\$)$
$$192000 \div 30240 = 6.35(\$)$$
$$148800 \div 23436 = 6.35(\$)$$
$$408000 \div 64260 = 6.35(\$)$$
$$148800 \div 23436 = 6.35(\$)$$
$$244800 \div 38556 = 6.35(\$)$$

The actual shipping cost of the single case is shown in Table 2 below.

Table 2. Actual transportation expense list of a single case

Month	January	February	March	April	May	June
Container	17	40	31	85	31	51
Carton	12852	30240	23436	64260	23436	38556
Single container charge	4.8	4.8	4.8	4.8	4.8	4.8
Total shipping cost	81.6	192	148.8	408	148.8	244.8
Single carton charge	6.35	6.35	6.35	6.35	6.35	6.35

In order to facilitate the comparison of the actual transportation costs and theoretical calculation expenses, the theoretical transportation costs of the single-case are shown in Table 3 below:

As can be seen from Tables 2 and 3, the actual freight of a single carton is far greater than the theoretical freight of it.

The transportation cost is greatly increased. Therefore, it is urgent to optimize the packing and transportation operation to reduce the transportation cost.

1.2 Container Loading Process and Container Volume Utilization

(1) Container loading process
Through investigation and research on the whole process of shipping container, the loading process is as follows.

Table 3. Theoretical transportation expense list of a single case

Month	January	February	March	April	May	June
Container?	17	40	31	85	31	51
Carton	12852	30240	23436	64260	23436	38556
Theoretical single container freight	2.8	2.8	2.8	2.8	2.8	2.8
Theoretical total shipping cost	47.6	112	86.8	238	86.8	142.8
Theoretical single carton charge	3.7	3.7	3.7	3.7	3.7	3.7

(1) stack the boxes on the pallet first;
(2) manual handling to 1.5 floors;
(3) after packing, it can be temporarily placed in the 1.5 storehouse for handling;
(4) through manual handling, the packing boxes and pallets will be transported to the first floor stacking place, where the PE film packaging shall be practiced;
(5) When the above operation is finished, the loading operation is carried out by forklift truck.

The loading process is shown in Fig. 1 below.

Based on the data in Table 2, it is concluded that 756 cases can be loaded in at present. Two columns are carried out at the time of loading. One column is placed vertically and nine pallets can be placed. The other column is placed horizontally, and 12 pallets can be placed, as shown in Fig. 2.

For the column of stack plate placed vertically, loading operations can be done by manual forklifts;

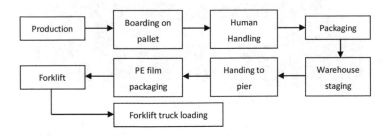

Fig. 1. The process of loading

For the horizontal column for the stack board, loading operations can be done by electric forklifts.

(2) Container volume utilization.

In the existing container mode, the relevant data of container are collected, thus the volume table of assembly equipment (Table 4) is obtained.

In order to confirm the capacity utilization rate of container loading, the loading operation process, and the loading action are reasonable, the volume utilization rate of a single cabinet is calculated by using Table 4 and formula 2.

Table 4. Volume table of assembly equipment

The volume of container m³	The volume of pallet m³	The volume of single carton m³	The number of cartons in container
65.75	3.37	0.05	756

The capacity utilization rate of a single container = (the volume of a pallet+ the volume of single carton × the number of cartons in a container)/ the volume of container

$$(3)$$

Substitute into the formula (3), and get:

The capacity utilization rate of a single container = $(3.37 + 0.05 \times 756) \div 65.75 = 62.61\%$

Because the loading operation is mainly done by forklifts, pallets are needed. The above calculated container capacity utilization rate is only 62.61%.

The reasons of low capacity utilization are:

The gap between the pallets and the height of the pallets reduces the stack height;

A large part of the space is not utilized effectively, which causes a great waste of space.

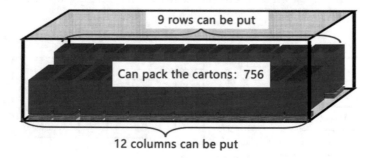

Fig. 2. Schematic diagram of the original number of carton boxes

Therefore, it is necessary to optimize the pallet size, stacking mode, operation space and operation action, so as to improve the container capacity utilization and packing efficiency, and save costs.

2   Reason Analysis

2.1   Fishbone Diagram Analysis

By observing the handling and packing operations, recording the site situation, the fishbone [4] diagram is applied to summarize the reasons for the low capacity utilization of the container.

(1) Packaging mode

The space wasting between pallets in the container is about 20.24%, and the space wasting between the pallet and the top of the container is about 19.36%, thus the packaging mode needs to be optimized.

(2) Insufficient stack layers

The container space utilization rate is low in that only six layers of carton can be placed on each pallet due to the limitation of packaging materials.

(3) Thinking set

Communicate with the site management personnel to learn about:

Some operators sometimes do not follow the loading, unloading and handling procedures;

Safety accidents due to rough handling;

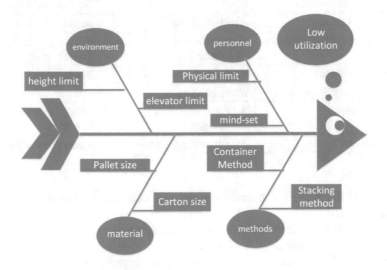

Fig. 3. Fishbone diagram

Sometimes insufficient packing;

Handling time is too long;

Handling efficiency is too low.

(4) The pallet size does not match the container.

The size of the pallet is different from the size of the carton. It must be packed by the PE film, in order to be assembled on the pallet and transported by forklift trucks.

(5) Height limits heap height.

The height of the container is 2.36 m. If there are seven layers of cartons, some workers will have difficulty in manually assembling the cartons into the container.

(6) Physical strength limits heap height.

Because the manual assembly is labor-intensive, workers are fatigable in the assembly process, which causes safety accidents.

(7) The height of the elevator limits the height of the heap.

Due to the limitation of packing, it is necessary to pack on the 1.5 floor, and move to the 1st floor to assemble PE film, and then move through the elevator, but the elevator height cannot accommodate 7 layers of cartons and pallets.

2.2 5W1H Analysis

Using the "5W1H" analysis method to analyze the low utilization rate of single cabinet, the statistical analysis results are shown in Table 5.

Table 5. 5w1h analysis table

	The situation	Why	How to optimize
What	Container	Wasting of space	Improved utilization
Why	Low space utilization	The collection mode is not reasonable.	Propose solution and Modeling solution
Where	Container	Stack height restriction	Stack seven cartons
When	Packaging time	Wasting of time	Manual assembly
Who	Porters	Height restriction, physical limitations.	Using ergonomics to Optimize
How	Forklift	Poor handling accuracy	Manual handling

By 5W1H analysis, the reasons for the low utilization of single cabinet can be summarized:

(1) The lifting height of the forklift is not high enough to stow 7 layers of cartons.

(2) Cartons and pallets need to be transported from the 2nd floor to the 1.5 floor and then the 1st floor. The process is too complex, limited and inefficient.

(3) Due to height and physical strength restrictions, it is difficult to realize manual handling.

3 Optimization Process

3.1 Element Analysis of Transport System and Scheme Design

The system engineering method can be used to analyze and solve the problem as a system.

Determine the factors of cargo handling system and their relationships.

(1) Factor determination

This system mainly includes the following 10 factors: container, pallet, pile board, carton, elevator, products being transported, forklift, porters, management personnel and customers.

(2) Relationship between factors

According to the Fig. 4, the relationship between the elements of the handling system is analyzed.

(1) The change of stack size should be taken into account when using pallets as the unit of the assembly.

(2) When using pallets as the unit of the assembly, there will be a process of packaging and wrapping PEM membrane.

(3) When the pallet size changes, the carton size will change.

(4) Customer's requirements on the assembly unit and carton size should be met.

(5) In the case of carton packing, the goods should be stowed on the board manually.

(6) The forklift operation is adopted for ease of stowing the pallets.

Yet, only 6 layers can be piled up. If it is higher than 6 layers, the forklift cannot reach the height of 7 layers.

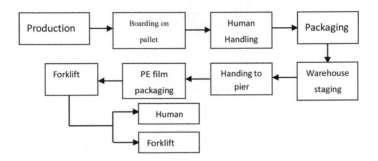

Fig. 4. Flow chart of handling operation

(3) Clarify the problem

In the previous analysis, the main reason for the slack in container utilization has been determined.

On this basis, the problem to be solved can be further clarified.

By "brain storm", the causes of underutilized container are summarized: unreasonable pallet size, unreasonable carton size, unreasonable display, unreasonable handling unit, thus decision-making flow chart is shown in Fig. 5.

(4) Determine the target

According to the previous analysis, the goal is set to improve the container utilization rate, and to design an appropriate plan that can meet the requirements of the operation and customer's needs.

(5) The project design

Based on the previous analysis and target, the decision tree is drawn and optimized according to the decision-making process in Fig. 5.

Decision tree consists of decision point and state point. Firstly, the size of the assembly unit is selected which can meet customer's requirements on the basis of the decision-making object. Secondly, the height (layer) of the stack is set which

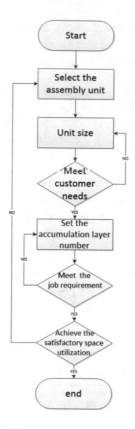

Fig. 5. The decision-making process

must meet the requirements of working precision, such as prevent forklifts from transporting single carton. Finally, the planar stacking method is determined, in order to maximize the space utilization on the basis of a certain height.

   Then get the scheme decision tree in Fig. 6.

(6) Choice of solutions

Based on Fig. 6, the solution branch is filtrated which can meet requirements of customers and working precision, and reject the solution branch of low space utilization. Obviously, with the same planar laying method, the space utilization can be improved with higher packing method. Figure 7 illustrates the final choice:

(1) Packed in cartons

(2) Maintain the original size

(3) Pack seven layers

(4) Manual handling

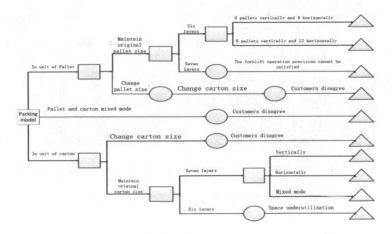

Fig. 6. Scheme decision tree

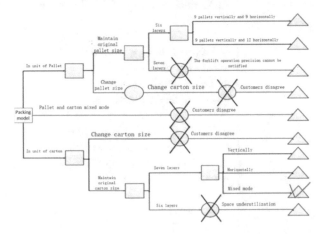

Fig. 7. Scheme selection

3.2   Layout Optimization

Use "L to L" and "W to L" (Fig. 8) pattern to pack; calculate the number of possible placement patterns using linear programming [3]; leave interstice in the middle, and change the original mode of stacking to improve the space utilization.

Container size (mm): $12060L * 2330W * 2360H$

Carton size (mm): $445L * 388W * 297H$

(1) Linear programming

Set Xi equal to the number of carton as illustrated in Fig. 8.

m, n, p, q mean the method of packing, in terms of variables 0, 1; 0 stands for unadopted placement method, and 1 stands for adopted placement method.

$$m, n, p, q = \begin{cases} 1 \\ 0 \end{cases} \tag{4}$$

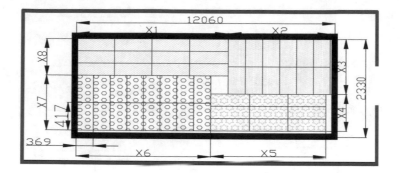

Fig. 8. Preliminary design drawing of the layout

Objective function:

$$Maxz = X_1 * X_1 X_8 + X_2 * X_3 + X_4 * X_5 + X_6 * X_7$$

Constraints [1]:

$$S.t.: \begin{cases} 445X_1 + 388X_2 <= 12060 \\ 445X_3 + 388X_4 <= 2330 \\ 445X_5 + 388X_6 <= 12060 \\ 445X_7 + 388X_8 <= 2330 \\ p(445X_4 + 388X_8) <= 2330 \\ q(445X_3 + 388X_7) <= 2330 \\ m(445X_5 + 388X_6) <= 12060 \\ n(445X_7 + 388X_8) <= 12060 \\ p + q >= 1 \\ m + n >= 1 \\ q + m >= 1 \\ p + n >= 1 \\ p, q, m, n, \in \{0, 1\} \\ x_i >= o i = 1, 2, \cdots, 8 \\ x_i \in 7 i = 1, 2, \cdots, 8 \end{cases}$$

(2) Solve with LINGO

$$X = (270000006)$$
$$Y = 162$$

The best packing method is that all the cartons are placed as 'L to L'. In this way, cartons can be packed 27 columns and 6 rows (as Fig. 9 illustrates).

3.3   Flexism Simulation

Flexism [2] software is applied to simulate with the parameters including time, layers, size, and the method of packing. The time of packing is as follow:
(1) 1:53:35 for packing a single container.
(2) 1132 cartons for a container.

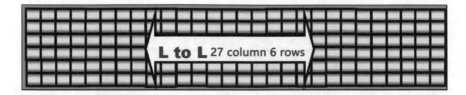

Fig. 9. Design drawing of the layout

4   Optimization Result

Single container capacity utilization and transportation cost optimization results are significant.

(1) Single container capacity utilization optimization

As shown in Fig. 10, capacity utilization has been improved by about 30%.

As shown in Fig. 11, the cost of single carton has been reduced by some 21%.

(2) Economical result

Through data from Table 6, after changing packing method, although we have to pay more human resource cost, the material cost and transportation cost can still be reduced a lot. The cost saving of 7272000 RMB/year can be finally achieved.

As shown in Table 7, after optimization, the number of cartons in a container can be improved a lot, making cost of single carton reducing from 6.35 to 4.24.

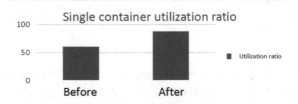

Fig. 10. Single container utilization ratio comparison diagram

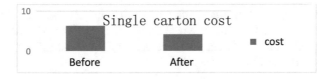

Fig. 11. Single carton cost comparison diagram

Table 6. Cost comparison table (RMB/year)

	Material cost	Transportation cost	Human resource cost
Before optimization	108000	1728000	0
After optimization	0	115400	76000
Cost saving	108000	574000	−76000
Total cost saving	606000		
	7,272,000 RMB/year		

Table 7. The comparison of single container packing number, capacity utilization ratio and freight

	Number of cartons	Capacity utilization	Cost of single carton
Before optimization	756	60.39%	6.35 RMB
After optimization	1132	87.88%	4.24 RMB

5   Conclusion and Prospect

We build optimization model and find the solution to reduce the cost of transportation, which can help the company save money.

In the future, we will use other knowledge in IE such as Human Factors Engineering to improve transportation and handling procedure. We can also use the skill of planning and scheduling to optimize the work schedule and maximize the resource utilization.

Innovation Points

Comprehensive application systems engineering, operations research, economics, industrial engineering, computer simulation and other professional knowledge optimization solutions, innovative research methods.

Author Contributions

The research is designed and performed by Jian Guo Zheng. The data was collected and analyzed by Jian Cheng Xu. The paper is written by Jiahui Duan and Jianguo Zheng, and finally checked and revised by Jian Guo Zheng. All authors read and approved the final manuscript.

Data Availability

No data were used to support this study.

Conflicts of Interest

The authors declare no conflict of interest.

References

1. Hou, D., Zou, L.: Constraint satisfaction technology for container stacking problem in container yards. Ind. Eng. Manag. (2012). (in Chinese)
2. Qin, T., Zhou, X.: Practical System Simulation Modeling and Analysis with Flexsim. Tsinghua University Press, Beijing (2013). (in Chinese)
3. Xu, J., Hu, Z., Wang, R.: Oper. Res. Science Press, Beijing (2018). (in Chinese)
4. Yi, S., Guo, F.: Fundament of Industrial Engineering. China Machine Press, Beijing (2013). (in Chinese)
5. Zeng, Q., Yue, A., Sun, H., Chen, C.: A pricing model for the container liner shipping based on price matching policies. Syst. Eng.-Theory Pract. $37(9)$, 2366–2372 (2017)

Research on the Improvement of IPD Management Mode of Engineering Project Under Network Environment

Ling Wan[1(✉)] and Xiaozhong Yu[2]

[1] Guangdong ocean university Cunjin college, Zhanjiang, Guangdong 524000, People's Republic of China
2543120345@qq.com
[2] Economics and Management School of Southwest Petroleum University, Chengdu, Sichuan 610500, People's Republic of China

Abstract. In recent years, with the advent of the era of "Internet +", the development trend of informatization and digitization exist all walks of life, BIM (building information model) as the construction industry the most advanced information and digital technology obtained the promotion and application of the government and the construction industry in our country, the application of new technology will bring a series of change, so the IPD(integrated project delivery) model appeared based on BIM technology, based on the full consideration under the network environment, it is necessary to make further improvement on the IPD model, that is the formation of the IPD mode as the core delivery, With BIM technology, "Internet +" related technology and network organization structure to support the improvement of the IPD management mode, and to analyze the basic principle and framework, operation mechanism and implementation process of the improvement of IPD model, then contrast analysis on the IPD mode before and after improve, finally to prove the improvement of IPD mode is feasible and scientific by the practical application of simulation.

Keywords: Network environment · IPD (integrated project delivery) mode · Management mode · Improvement

1 Introduction

Nowadays, the construction industry occupies an important position in the global economy. 40% of the world's energy and raw materials are used in construction production [3]. A large amount of energy and raw materials are invested in the construction industry, but the whole construction industry has been faced with the problem of low production efficiency. In the engineering construction industry, these problems are mainly reflected in rework and serious waste of resources [10]. According to the data, 57% of the time, work and material investment of the construction industry has not been converted into the final product value

© Springer Nature Switzerland AG 2020
J. Xu et al. (Eds.): ICMSEM2019 2019, AISC 1002, pp. 26–39, 2020.
https://doi.org/10.1007/978-3-030-21255-1_3

[9]. On the other hand, in the engineering construction industry, according to statistics, the actual cost of most projects in the current period is twice as much as the required cost, 72% of the projects are over budget, and 70% of the project construction period lags behind [8]. In the final analysis, the results mainly due to the traditional inefficient project management mode, the project management mode studied in this paper specifically refers to the project delivery mode. These traditional project delivery models do not reflect the advantages of the network environment, do not reflect the advantages of the integration value chain. Such as the DDB model, DB model, CM model, etc., and the DDB model is widely used in China at present stage of a project management mode, in this traditional mode, the owner signs the contract with the designer and the construction of the contractor respectively, they have their own independent contract relationship, meanwhile, in the process of the project implementation, the sequence of the project is design, then bidding, and then construction. this natural process makes it is easy to appear in the process of project engineering change, so increasing the cost, at the same time, a stage to enter the next stage after completing a job, make the progress of the projects seriously slow down. As each participant of the project is in an independent state, it is not so much to consider the maximization of the overall interests when working on the same project, but to consider the maximization of its own interests. As this traditional project management mode is difficult to achieve the state of close cooperation and collaboration between all participants involved in the project. Therefore, the project itself is not an optimal situation. With the continuous promotion and development of BIM technology in China, the traditional project management mode has been innovated and the IPD mode has been born. According to the statistics of foreign scholars, among the IPD projects investigated, 70.3% of the projects have saved costs and 59.4% have shortened the construction period [5]. Although the IPD mode is still in the initial stage of development in China, but the IPD mode is bound to become the general trend of the future development of the construction industry. Especially under the network environment, the "Internet +"and the advent of the era of big data speed up the connection of the whole society, there are changes in the whole social network structure, under this background, the construction industry is also facing unprecedented changes, this paper research on the improvement of IPD mode under the network environment, in order to make the improvement of IPD model better in combination with the information technology and network organization under the network environment, so as to improve the efficiency of the engineering project management. The main innovation in this paper is define the connotation of network environment, research on the influence of network environment on IPD mode, and propose the basic connotation of the improved IPD mode under network environment, builds a new architecture of IPD mode in the network environment, and draws the architecture diagram and the implementation flow chart of the improved IPD mode, then explains the operation mechanism of the improved IPD mode, through the comparative analysis of IPD mode before and after the improvement, to illustrate the advantages of the improved IPD mode, finally, the

simulation research on the operation and application of the improved IPD mode is conducted and a detailed SWOT analysis of the specific example is made to improve the feasibility and scientific of the improved IPD mode.

2 Development Process of Project Management Model

2.1 The Evolution of Project Management Model

For a long time, the evolution of engineering project management mode has experienced from the same unit for construction and management to professional contract, then to the gradual integration of general process [7], it is already have very mature structure of the system architecture, and it is impossible to create a new model, only under the new circumstances, take advantage of new technology or a new idea to update the existing engineering project management mode or improve the existing engineering project management mode for further improvement and innovation to improve the efficiency of management, so as to produce new model [16]. these engineering project management models are constantly updated with the progress of the times, the occurrence of each kind of model and update all have certain social background, so, in the present network environment, how to get the information network and the network convergence to the engineering project management mode is worthy to discuss, this article is based on the further improvement of the IPD mode under the network environment, It enables the improved IPD mode to exert greater advantages and further improve the efficiency of engineering project management.

2.2 The Generation of IPD Mode

IPD model is based on the building information model technology, in recent years, BIM technology is a new technology which has been actively promoted and applied by Chinese government in the construction industry in recent years, building information model is based on 3D digital technology, integration of construction engineering project engineering data of all relevant information, is an entity project facilities and function of the digital expression. Based on this appeared a new project management patterns–IPD mode, the American association of architects is defined the IPD mode as a project delivery method, it is the personnel, systems, integrated into a business and practice process, all participants make full use of the wisdom and experience, in all stages of project optimization, improve the construction process, by reducing waste to add value to projects, maximize the efficiency of the whole project and the value [1]. As a subversive project delivery model, the application of IPD model in China's construction engineering field is still in the exploratory stage [15]. At present, there are no successful cases of actual application of IPD mode in China, but IPD mode has been completed in some developed countries and achieved good results. Different from traditional delivery mode, the IPD mode is a kind of concentrated delivery mode, The IPD model is implemented under a set of standard contracts to be followed by all participants. under the constraint of the

IPD mode by the team, project team members to break the traditional model, on the basis of mutual respect and mutual benefit in an early step in the project, fully open communication, sharing, cooperation, risk-sharing, starting from the value of the project, all participants to the overall interests, the interests of the whole project participants consistent with the current of a high concentration of all-round cooperation team. In the IPD mode, project participants should collaborate closely on the basis of BIM technology, and the degree of information collaboration is very high. The information exchange between all participants involved in the project is more frequent and the accuracy of the information is higher. In addition to using BIM technology, under the network environment, it can further optimize and improve the IPD mode, improve the efficiency of engineering project management and improve the efficiency of engineering project management.

3   Improved IPD Management Mode Under the Network Environment

3.1   The Impact of Network Environment on IPD Mode

When referring to the network environment, most people will understand it as the internet network environment. In this paper, in addition to the information network under the internet background, the network environment also includes the network organization, namely the organization network. Information network mainly refers to information network technology, which is a technical means to collect, process, transmit, digest, feedback and other information data processing of information [4] in the Internet age, information network technology is an integral part of enterprise production technology, the use of information technology to speed up the IPD team members' communication and exchanges, enhance cooperate project participants, can bring technical support for the IPD mode, better guarantee the project participation of all participants to cooperate, information sharing, responsibility Shared purpose. And network organization mainly refers to the network of enterprise organizational structure. Enterprises make full use of highly developed Internet, highly developed market and highly developed social division of labor to establish a three-dimensional, integrated organizational structure [13]. Information technology lays the foundation for networking of organizational structure, information technology has accelerated the speed of the enterprise internal information exchange and transmission within the project participants, makes the enterprise internal management efficiency improved, at the same time, all the project participants in IPD mode have established a network organization structure through BIM technology, realizes the organization structure of the network, thus accelerate the enterprise external information transmission and information communication efficiency, from network organization level, improved the operational efficiency of the IPD mode. Therefore, both information technology and organizational networking speed up the management efficiency of IPD mode. Based on this network environment, this paper further improves the IPD mode and builds a more effective IPD management mode.

3.2 The Basic Principle and Architecture of the Improved IPD Mode

With the coming of information age, information technology, digital speed up the development of all walks of life, deepening the ties in all walks of life, information technology has brought about technological progress, but it also requires the innovation of traditional organizational forms, Therefore, it drives the innovation of traditional industries and promotes the transformation and upgrading of industries, for the construction industry, with the development and innovation of BIM technology, the IPD mode has been promoted, compared with the traditional project management mode, the IPD mode is advanced, However, the IPD model does not fully integrate the concept of information network technology and organizational networking into it, so under the network environment, IPD mode can be improved by means of information technology support and organizational structure networking into IPD mode, that is to achieve IPD mode as the core delivery, with the BIM technology, the "Internet +" related technology and the organization structure network to support the IPD project management pattern, makes the improved IPD model can further improve the efficiency of project management. The specific architecture diagram of the improved IPD mode is shown in Fig. 1.

3.3 The Operation Mechanism of the Improved IPD Mode

First of all, the most important thing is to establish the IPD collaborative platform based on BIM, which is the foundation to improve the operation of IPD mode. This platform serves as the information management platform and project management platform of the whole project. on the other hand, the owners, supervisors, subcontractors, general contractors, survey and design participants and suppliers of different units involved in the project form a flat organizational structure and gradually develop into a networked organization. it has gradually developed into a network organization. With the support of network organization, it has changed the traditional way of information transmission. all participants involved in the project directly obtain the required information on this platform, instead of the information being transmitted from one party to the other party, which greatly speeds up the traditional speed and accuracy of the information at the same time, each participating units can also be used in the process of project construction of the platform for professional information modification, supplement constantly improve, make the whole project management process and optimize constantly fusion, improve the efficiency of engineering project management, make the project involved in mutual win-win cooperation, the parties made the contribution of the parties to ensure the project benefit maximization [11].

Then, through the mobile terminal equipment, the cloud service platform and the Internet of things terminal the information technology, collect the information required by the project and bring it to the platform, store and accumulate

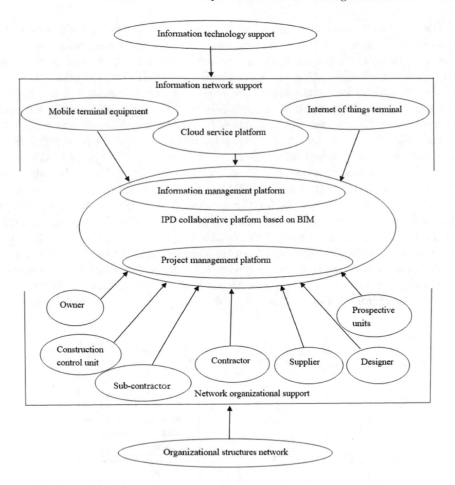

Fig. 1. Architecture of the improved IPD mode in a network environment

and big data for all information in the process of project implementation, by analyzing the data, project participants can obtain the required information from the platform at any time through certain permission, so as to grasp the current status of the project, all the IPD team members with authority can obtain the required data and BIM model on this platform, and upload the change the information or data to this platform, and realize the information management function through this platform.

3.4   The Implementation Process of the Improved IPD Mode

In the IPD mode, the implementation process of the project is generally divided into seven stages: conceptual design, standard design, detailed design inspection, construction and delivery [2], during the implementation of the project, the project participants participated in the whole project at an early stage.

During the design stage, all parties involved in the project jointly completed the design, making the design scheme more accurate and reasonable, reducing the occurrence of design changes and construction rework, shortening the construction period and saving the cost. the improved IPD mode makes full use of the support of information network technology and organizational network to further improve the work of each process of IPD mode. Information network technology accelerates the solution of problems encountered in each process, and organizational network speeds up the speed of information exchange and communication among all parties involved in the project. The specific implementation flow chart is shown in Fig. 2. As can be seen from Fig. 2, the improved IPD mode is still centered on the IPD collaborative platform based on BIM. in the implementation process, from the conceptual design, preliminary design, detailed design, construction and delivery stages, each stage is closely related to the platform, gather relevant information of all project participants through the platform, with the support of information network technology, the access to data information is faster, more convenient and accurate through the mobile terminals, the Internet of things terminals and the cloud service platforms in the process of information acquisition, which is conducive to the improvement of

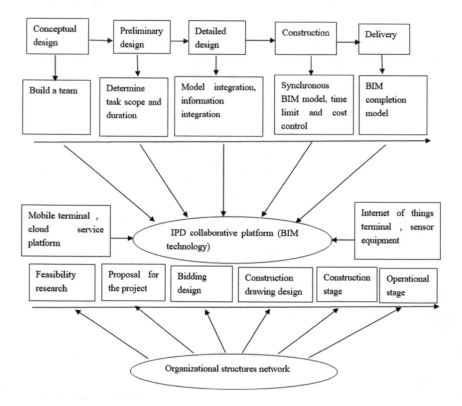

Fig. 2. Implementation flow chart of improved IPD mode in network environment

information communication efficiency in the process of project implementation. Meanwhile, in the process of the project, from the stage of feasibility research, project proposal, bidding and tendering design, construction drawing design and construction to the stage of operation and maintenance, all parties of the project form a closed organizational structure network with the help of organizational structure networking on the basis of IPD collaborative platform, which doubles the speed of information transmission. All parties involved in the project can use the platform to better communicate and traditional information, complete the whole project smoothly through the BIM model, and integrate the information network technology and organizational structure into the network at all stages of the project, so as to complete the project faster and better [14].

3.5 Comparative Analysis Before and After the Improvement of IPD Mode

The improved IPD mode is improved on the basis of IPD mode. In order to better illustrate the difference between the improved IPD mode and IPD mode, a comparative analysis is made between the improved IPD mode and IPD mode, in order to illustrate the advantages of the improved IPD mode. The specific analysis is shown in Table 1. The Table 1 shows that the improved IPD mode is based on the IPD model, on the basis of the introduction of information network technology and the organization network support, makes the improved IPD mode has solved the IPD mode on the problem of high degree of information and data requirements, he basic features of the improved IPD mode and the IPD mode is roughly same, but the improved IPD mode has a new feature, is the introduction of information network and organization network, to further improve the IPD mode of data acquisition, collection, storage and analysis ability, and improve the application efficiency of the IPD mode, on the one hand, through the information network technology, Such as mobile terminals, Internet technologies, cloud platform to support the collection and transmission of information, on the other hand, through the form of organization network to accelerate the velocity of circulation of information and data, makes the data through the IPD collaborative platform based on BIM rapid flow between the participants of the project, to make the project management efficiency have a further improvement.

4 Application Simulation Analysis on the Improved IPD Mode

4.1 Introduction of the Application Simulation Project

The project of Lize business district is located in the core area of Lize business district, Fengtai district, Beijing. It is composed of three super high-rise office buildings are respectively 150 m in height (31 floors) of F02-1, 120m in height (25 floors) of F02-2 and 199.9 m in height (42 floors) of F03, the commercial

Table 1. The comparative table of the improved IPD model and the IPD model

Type \ Nature	Feature	Advantage	Existing problem
IPD model	1. Early intervention and full participation	1. Participate in collaborative work of all parties to improve project quality	1. The parties have low trust, unclear responsibilities and uneven risk sharing
	2. Fully open exchanges and cooperation	2. Optimize the process shorten the construction period and save cost	2. The BIM technology is not mature enough and needs support from the BIM platform
	3. Information sharing, promoting and accelerating the process		3. High requirements for the contract
	4. Electronic, BIM and multi-dimensional		
	5. The overall interests of the project determine the interests of all parties		
The improved IPD model	1. intervention and full participation	1. In addition to the advantages of IPD mode, with the help of organizational networking and information technology, the efficiency of information communication can be improved to acquire complex information and data faster and better	1. High requirements for the contract, no unified standard form of the contract, some difficulties in implementation
	2. Fully open exchanges and cooperation	2. Can better process information and data and improve the efficiency of project management	2. High requirements for BIM-based IPD platform
	3. Information sharing, promoting and accelerating the process		
	4. Electronic, BIM and multi-dimensional		
	5. he overall interests of the project determine the interests of all parties		
	6. Supported by information network technology and network organization		

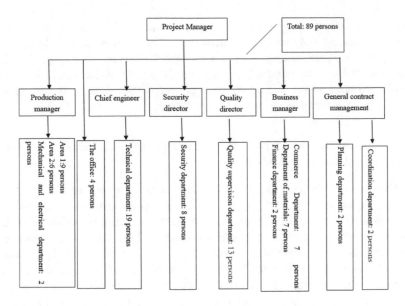

Fig. 3. Organization chart of the project in Lize business district

skirt room of 5 storeys high and 4 storeys commercial and car park basement. A total construction area of 417187 m², of which the building area, 309900 m², underground building area, 107287 m² the development organization is Beijing TianChengYongTai property co., LTD., Beijing TianChengYongYuan property co., LTD., Beijing ShouChuangChaoYang real estate development co., LTD., Fengtai branch; the design unit is Beijing architectural design and research institute; the construction control unit is ZhongZi engineering construction supervision company; the construction organization is China construction second engineering bureau co., LTD. the organization chart of this project is shown in Fig. 3.

4.2   SWOT Analysis on the Simulated Project

This project was completed based on BIM technology, but it did not use IPD project management mode, nor did it fully realize the support of information network technology and organizational network. the following is a SWOT analysis for this project, and the strategic selection for this project is analyzed. the specific analysis is as follows in Table 2.

As can be seen from the SWOT analysis in Table 2, this project has four different strategic choices based on its advantages (S), weaknesses (W), opportunities (O) and threats (T). among them, different strategies come from cross-combination according to the characteristics of different elements in the strategy. these strategic choices can be concluded as the final reasonable and scientific concrete measures and application methods by means of strengthening and complementing each other, internal and external combination, competition and cooperation. the project is located in downtown Beijing, in the network environment, corresponding SO strategy should be adopted. and select an advanced project management mode, only when the "Internet +" technology and BIM technology are integrated and the organizational structure of the engineering project is networked, can it be more conducive to the construction of the project, therefore, once this project chooses to adopt the improved IPD model, it is very favorable for the long-term development of the enterprise.

4.3   Analysis on the Implementation Process and Effect of the Application Simulation

Once this project adopts the improved IPD mode, it can accelerate the efficiency of project management. All the project participants can realize networked organization structure, strengthen the information communication of each unit and form an organizational structure centered on BIM platform, thus realizing the networking of organizational structure. As shown in Fig. 4. the networked organizational structure in the Fig. 4 above accelerate the speed of information transmission and reduces the error of information transmission. In the process of project implementation, the loss caused by the error of information transmission can be effectively avoided. at the same time, this networked organizational structure can greatly reduce the organization personnel and save costs. Secondly, under the framework of improving IPD mode, the participants sign the contract agreements. to cooperate on the basis of the contract agreement, with the support of the improved IPD mode information network technology and the IPD mode management framework based on BIM technology, the IPD team was established. from the feasible research stage of the project, the bidding design stage, the detailed construction stage, the construction stage and the operation and maintenance stage, information integration is realized through BIM technology in the whole process [12], at the same time, internet technology is used to obtain information from the internet of things terminals and the mobile terminals. for example, in the field construction stage, the mobile phone can be used to directly upload the field pictures to the IPD collaborative platform to achieve data acquisition, data transmission, information integration and data processing. Refer to the flow chart shown in Fig. 2 for implementation.

Finally, the platform project participants can use the data and information form the IPD collaborative platform, from planning and design, project construction and operations which a series of process based on the BIM technology, the IPD project members can cross all stages involved in the project, using the advantage of the characteristics of network organization structure, the IPD project members can interevent in different stages, such as the construction unit of China construction second engineering co., LTD., can be involved in the early design stage, it can discuss with the Beijing architectural design institute about the design scheme, collaborative design whether meet the requirements of construction technology and construction process, All participants involved in the project can use different BIM model data at different stages to extract the data required for the implementation of the project, such as design and

Table 2. SWOT analysis on the project in Lize business district

Internal ability / External factors	S	W
	1. The project is located in a favorable geographical position	1. No experience in IPD mode project management
	2. Talent advantage	2. Imperfect infrastructure and construction of software and hardware need to be strengthened
	3. Industrial integration advantages	
O	SO	WO
1. Information technology continues to mature, and changes brought about by "Internet +" have led to constant development of mobile technology, sensing technology and visualization technology	To advance 1. Closely follow national policies, vigorously develop BIM technology and establish a BIM team	Reinforcing 1. Use information network technology to strengthen platform construction
2. "One Belt and One Road" policy stimulus	2. Innovation of traditional management mode and use of information network technology	2. Improve basic software and hardware construction
3. National policies to promote BIM	3. Make organizational changes to meet the development needs of the new era	3. Experience in learning IPD mode
T	ST	WT
1. The application of the new model is faced with a series of uncertainties and great risks	steady 1. Closely follow national policies and vigorously develop BIM technology	prevent 1. Innovation and creation
2. High requirements and huge investment for platform construction	2. Make use of information network technology to strengthen intelligent and information construction	2. Strict recruitment and selection of talents
3. There is no standard IPD contract agreement in China		

implementation model, design collaboration model, construction coordination management model, operation period model, progress model and cost model, etc., to give play to the value of BIM model in different periods [6].

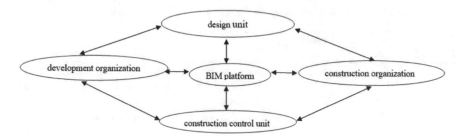

Fig. 4. The network diagram of the organizational structure of the project under the improved IPD mode

5   Conclusion

Research shows that through operating of the improved IPD mode, the final results are as follows: organization can simplify network organization, stream-line personnel, make the original project organization structure simplified at least one-third of 89 people, at the same time, it can strengthen the information exchange and collaboration among enterprises, the use of BIM platform to achieve the project in different stages can be extracted model and information, strengthen the efficiency and quality of communication, can greatly shorten the construction period, and the save time can reach more than 80%, thus greatly save the cost, improve the efficiency of management, through the simulation analysis shows that the improved IPD model is scientific and feasible, it has certain advancement.

Acknowledgements. This research topic by Science and technology special project in Zhanjiang city subsidy (NO:2018B01010). This paper was completed under the guidance of master tutor Xiaozhong Yu. I am appreciated for the guidance of professor Yu; His rigorous academic attitude and scientific research spirit have inspired me a lot. In addition, I would like to thank all the scholars in this field, which is the basis of this paper. Thank you for everyone's contribution!

References

1. https://info.aia.org/siteobjects/files/ipd-guide-2007. Technical report, AIA (2017)
2. Integrated project delivery: A guide. Technical report, AIA [2017-03-25] (2017)
3. Asmar, M.E., Hanna, A.S., Loh, W.Y.: Quantifying performance for the integrated project delivery system as compared to established delivery systems. J. Constr. Eng. Manag. **139**(11), 04013012 (2013)
4. Peffers, K., Rothenberrger, M., Tuunanen, T. et al.: Design science research in information systems. Adv. Theory Pract. **398** (2012)
5. Kent, D.C., Becerikgerber, B.: Understanding construction industry experience and attitudes toward integrated project delivery. J. Constr. Eng. Manag. **136**(8), 815–825 (2010)

6. Lahdenpera, P.: Making sense of the multi-party contractual arrangements of project partnering, project alliancing and integrated project delivery. Constr. Manag. Econ. **30**(1), 57–79 (2012)
7. Lee, S., Liang, S.: Key core project management of construction industry to study. Open J. Soc. Sci. **2**(03), 48 (2014)
8. Miller, R., Strombom, D., Iammarino, M. et al.: The Commercial Real Estate Revolution, vol. 328. Hoboken, Wiley (2009)
9. Nasfa, X., Appa, A., Agc, A.I.A.: Integrated project delivery for public and private owners (2010). Accessed 12 Dec 2012
10. Pena-Mora, F., Li, M.: Dynamic planning and control methodology for design/build fast-track construction projects. J. Constr. Eng. Manag. **127**(1), 1–17 (2001)
11. Sacks, R., et al.: Requirements for building information modeling based lean production management systems for construction. Autom. Constr. **19**(5), 641–655 (2010)
12. Thomson, J.D.: Managing integrated project delivery. In: Construction Management Association of America (CMAA), vol. 105. McLean, VA (2009)
13. Tran, Q., Tian, Y.: Organizational structure: Influencing factors and impact on a firm. Am. J. Ind. Bus. Manag. **3**(2), 229 (2013)
14. Wang, X., Xie, B., Lu, W.: Research overview of project management mode decision-making. J. Eng. Manag. **67** (2013(4)). (in Chinese)
15. Ma, Z., Ma, J.: The application of ipd and blm technology in it. Inf. Technol. Civ. Constr. Eng. 36–41 (2011). (in Chinese)
16. Ma, Z., Li, S.: The commercial real estate revolution environment. J. Tongji Univ. **991** (2018) (in Chinese)

A Research of FAHP Approach in Evaluating Online Training System Alternatives

Liping Li[1,2](✉) and Dan Li[1]

[1]School of Public Management, Sichuan University, Chengdu 610065,
People's Republic of China
248185060@qq.com
[2]School of Management, Sichuan University Jinjiang College, Pengshan 620860,
People's Republic of China

Abstract. With the coming of "Internet Plus" Era, knowledge economy forces people to keep learning and then flexible online self-study pattern becomes more and more popular. Also the selecting process of an alternative vendor has been very important issue for online training companies. This paper proposes a structured math model for evaluating online training system alternatives using the fuzzy analytical hierarchy process (FAHP), aims to demonstrate how FAHP model can help in solving such decisions in practice. After summarizing related literature review and brief overview of fuzzy AHP, the effectiveness of FAHP model was illustrated by using a case company in China. Hope to make some theoretical and practical contributions on evaluating online training system alternatives.

Keywords: Analytical hierarchy process (AHP) · Fuzzy set theory · Pair wise comparison · Alternatives selection

1 Introduction

With the coming of "Internet Plus" Era, the rise of sharing economy brought a new flexible online self-study pattern based on network platform. Computer is widely used, and online training mode has become an important part in modern E-learning. Modern online training system sends courses through audio, video (live or video) to learners. Some new trends such as MOOCS become more and more popular, with real-time and non real-time computer technology. Online training has the advantage that the traditional training has not, such as lower cost, shared resource, more flexible time, and so on. Therefore, online training is prevalent all over the world, especially has a large part of users in China. According to the research of the status quo about China's online education industry and the market outlook forecast report (2017–2023) by China Market Research Online, many enterprises including Amazon, Google, Baidu, Alibaba

J. Xu et al. (Eds.): ICMSEM2019 2019, AISC 1002, pp. 40–48, 2020.
https://doi.org/10.1007/978-3-030-21255-1_4

and Tencent have entered the online education market. On average, there are about 26 enterprises enter this industry every day in 2015, and the number of online education users is 779.69 million, which will continue to grow by more than 15% in the next few years. At present, only 10% of the employees have received online vocational training. From 2014 to 2017, the online market size of vocational certification training in China was 22.4 billion Yuan and 28.7 billion Yuan respectively, with a year-on-year growth of 281%. In General, online training is a teaching pattern that the training institution selects suppliers to provide internet technique. For an online training center, there are many alternatives to consider while hope to select the best one by using some quantitative measurement or scientific method. Former researchers have made some exploration and achievement. However, the research on appraising online training alternatives is relatively scarce in China and overseas, and few researchers pay attention on FAHP method to evaluate these alternatives. Thus, with the main purpose of solving this problem as well as enrich related research, our paper discussed how to select the best online training alternative with the method of FAHP. After doing this introduction, we organized this paper by summarizing related literature review at Sect. 2 and overview of fuzzy AHP at Sect. 3. Then the paper illustrated the effectiveness of the FAHP model by using a case company in China at Sect. 4, with conclusion at final section.

2   Literature Review of Related Research

A lot of researches are related to online training evaluation. Tal Soffer et al. examined the effectiveness of 3 online courses compared with the same 3 courses in a face-to-face (F2F) format, which had the same characteristics [12]. SUN Lu-lu et al. presented a method of evaluating the complexity of the electromagnetic environment based on the grey analytic hierarchy process (Grey-AHP) due to the effects of the complex electromagnetic environment on radar countermeasure training, which unifies subjectivity and objectivity. Practice proves that the method is of certain reference value [13]. WANG Qin-zhao et al. established the training performance evaluating index system based on AHP to solve the current problem (current methods are human-oriented and they are not objective and systematic) through analysis of the requirement and characteristics of the armored unit simulation training system [15]. Rochelle Irene Lucas et al. proposes an evaluation framework based on the Analytic Hierarchy Process (AHP) to systematically evaluate the emergence of information and communication technology (ICT) workshops designed for teachers [8]. Pan Jialiang et al. used Fuzzy analytic hierarchy process to evaluate the outboard active decoy, and this algorithm uses multiple levels and indicators to improve the comprehensiveness of outboard active decoy assessment [9].

Also a mass of publications have appeared on supply chain management, particularly in the alternative selection problem. Chan and Chung aimed at the distribution network problem in supply chain management, a multi-criteria genetic optimization method is proposed. By combining the analytic hierarchy process (AHP) with genetic algorithm (GA), the ability of multi-criteria decision making is obtained and the computing time is reduced [2]. Vaidya and Kumar introduced the literature review and the application of analytic hierarchy process (AHP) [14]. Handeld et al. integrate environmental issues into supplier evaluation decisions with the help of AHP [4]. A data envelopment analysis method for multi-target supplier selection process is proposed by Liu et al. [7] and Weber et al. [16]. Zadeh pioneered that the Fuzzy set theory is a mathematical theory aiming to model the fuzziness or imprecision of human cognitive process, which is basically a class theory without clear boundaries [17]. It is important to realize that any clear theory can be fuzzified by extending the concept of a set in that theory to the concept of a fuzzy set. Several authors include George J. Klir and Bo Yuan 1995 reviewed the theory soon afterward [5].

3   Overview of FAHP

A fuzzy number is a special fuzzy set $F= \{(x, \mu_F(x)), \in R\}$, where x takes its value on the real line $R_1 : -\infty < x <= +\infty$ and $\mu_F(x)$ is a continuous mapping from R_1 to the close interval $[0, 1]$. A triangular fuzzy number can be denoted as $M = (l, m, u)$, its membership function $\mu_M(x) : R \to [0, 1]$ is equal to:

$$\mu_M(x) = \begin{cases} \frac{1}{m-l}x - \frac{l}{m-l}, x \in [l, m] \\ \frac{1}{m-u}x - \frac{u}{m-u}, x \in [m, u] \\ 0, \qquad otherwise, \end{cases}$$

where $1 \leq m \leq u$, and a stand for the lower and upper value of the support of M respectively, and m for the modal (mid) value. When $1 = m = u$, it is a non-fuzzy number by convention [3]. The main operational laws for two triangular fuzzy numbers M_1 and M_2 are as follows (Kaufmann [1]):

$$M_1 + M_2 = (l_1 + l_2, m_1 + m_2, u_1 + u_2)$$
$$M_1 \oplus M_2 = (l_1 l_2, m_1 m_2, u_1 u_2)$$
$$\lambda \oplus M_1 = (\lambda l_1, \lambda m_1, \lambda u_1) \quad \lambda > 0, \lambda \in R$$
$$M_1^{-1} = (1/l_1, 1/m_1, 1/u_1).$$

3.1   Fuzzy Representation in A Pair Wise Comparison

In order to consider the fuzziness of evaluation when comparing selection criteria in pairwise, triangular fuzzy numbers M_1, M_3, M_5, M_7, M_9 are used to represent the assessment from "equally preferred (M_1), moderately preferred (M_3), strongly preferred (M_5), very strongly preferred (M_7), extremely preferred (M_9)" and M_2, M_4, M_6, M_8 are middle values. Fig. 1 shows the triangular numbers $M_t = (l_t, m_t, u_t)$ where t=1, 2, 3, ..., 9, l_t, u_t are lower and upper value of the fuzzy number M_t respectively and m_t is the middle value of the fuzzy number M_t. The δ is used to represent a degree of judgment where $u_t - l_t = l_t - u_t = \delta$. A larger value of δ implies a higher fuzzy degree of judgment. When $\delta = 0$, the judgment is a non-fuzzy number. Zhu et al. reported that δ should be larger than or equal to one-half. In this article, the value δ was set to be one-half [18].

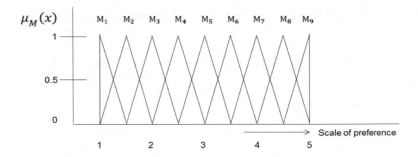

Fig. 1. The membership function of the triangular fuzzy numbers

3.2   The Steps of FAHP Research

The analytic hierarchy process (AHP) method is also known as eigenvector method. It is shown that the eigenvector corresponding to the maximum eigenvalue of the pairwise comparison matrix provides the relative priority of each factor and retains the order preference among alternatives. This indicates that if one alternative is superior to another, its eigenvector component is greater than the other. The weight vector obtained by pairwise comparison matrix reflects the relative performance of each factor. This algorithm improves the scale scheme of judgment matrix by using triangle fuzzy number, and uses interval algorithm to solve fuzzy feature vector. The four-step-procedure of this approach is given as follows:

Step 1. Comparing the performance score: Triangular fuzzy numbers $(\tilde{1}, \tilde{3}, \tilde{5}, \tilde{7}, \tilde{9})$ are used to indicate the relative strength of each pair of elements in the same hierarchy.

Step 2. Constructing the fuzzy comparison matrix: By using triangular fuzzy numbers, via pairwise comparison, the fuzzy judgment matrix $\tilde{A}(a_{ij})$ is constructed as given below;

$$
\tilde{A} = \begin{bmatrix} 1 & \widetilde{a_{12}} & \ldots & \widetilde{a_{1n}} \\ \widetilde{a_{21}} & 1 & \ldots & \widetilde{a_{2n}} \\ .. & .. & \ldots & .. \\ .. & .. & \ldots & .. \\ \widetilde{a_{n1}} & \widetilde{a_{n2}} & \ldots & 1 \end{bmatrix},
$$

where $\tilde{a}_{ij}^{\alpha} = 1$, if i is equal to j, an $\tilde{a}_{ij}^{\alpha} = \tilde{1}, \tilde{3}, \tilde{5}, \tilde{7}, \tilde{9}$ or $\tilde{1}^{-1}, \tilde{3}, ^{-1}\tilde{5}^{-1}, \tilde{7}^{-1}, \tilde{9}^{-1}$, if i is not equal to j.

Step 3. Solving fuzzy eigenvalue: A fuzzy eigenvalue, $\tilde{\lambda}$ is a fuzzy number solution to

$$
\tilde{A}\tilde{x} = \tilde{\lambda}\tilde{x}, \tag{1}
$$

where nxn fuzzy matrix containing fuzzy numbers $\widetilde{a_{ij}}$ and $\tilde{x}$ is a non-zero nxl, fuzzy vector containing fuzzy number ($\tilde{x}_i$). To perform fuzzy multiplications and additions by using the interval arithmetic and $\alpha-cut$, the equation $\tilde{A}\tilde{x} = \tilde{\lambda}\tilde{x}$ is equivalent to

$$[a_{i1l}^{\alpha}x_{1l}^{\alpha}, a_{i1u}^{\alpha}x_{1u}^{\alpha}] \oplus \cdots \oplus [a_{inl}^{\alpha}x_{nl}^{\alpha}, a_{inu}^{\alpha}x_{nu}^{\alpha}] = [\lambda x_{il}^{\alpha}, \lambda x_{iu}^{\alpha}],$$

where

$$\tilde{A} = [\tilde{a}_{ij}], \tilde{x}^t = (\widetilde{x_1}, \ldots, \widetilde{x_n})$$
$$\tilde{a}_{ij}^{\alpha} = \left[a_{ijl}^{\alpha}, a_{iju}^{\alpha}\right], \tilde{x}_i^{\alpha} = [x_{il}^{\alpha}, x_{iu}^{\alpha}], \tilde{\lambda}^{\alpha} = [\lambda_l^{\alpha}, \lambda_u^{\alpha}] \qquad (2)$$
$$For\ 0 < \alpha \leq 1,\ \text{and where}\ i = 1,\ 2,\ n;\ j-1,\ 2,\ n,$$

$\alpha-cut$ is known to incorporate the experts or decision-makers confidence over his/her preference or the judgments. Degree of satisfaction for the judgment matrix $\tilde{A}$ is estimated by the index of optimism μ. The larger value of index μ indicates the higher degree of optimism. The index of optimism is a linear convex combination (Lee, 1999) [6] defined as

$$\tilde{a}_{ij}^{\alpha} = \mu a_{iju}^{\alpha} + (1-\mu) a_{ijl}^{\alpha}, \forall\ \mu \in [0,1] \qquad (3)$$

While α is fixed, the following matrix can be obtained after setting the index of optimism μ in order to estimate the degree of satisfaction.

$$\tilde{A} = \begin{bmatrix} 1 & \widetilde{a_{12}^{\alpha}} & \cdots & \widetilde{a_{1n}^{\alpha}} \\ \widetilde{a_{21}^{\alpha}} & 1 & \cdots & \widetilde{a_{2n}^{\alpha}} \\ \cdots & \cdots & \cdots & \cdots \\ \cdots & \cdots & \cdots & \cdots \\ \widetilde{a_{n1}^{\alpha}} & \widetilde{a_{n2}^{\alpha}} & \cdots & 1 \end{bmatrix}.$$

The eigenvector is calculated by fixing the μ value and identifying the maximal eigenvalue. $\alpha-cut$: It will yield an interval set of values from a fuzzy number. Normalization of both the matrix of paired comparisons and calculation of priority weights (approx. attribute weights), and the matrices and priority weights for alternatives are also done before calculating max a_{ij}. In order to control the result of the method, the consistency ratio for each of the matrices and overall inconsistency for the hierarchy calculated. The deviations from consistency are expressed by the following equation CI, and the measure of inconsistency is called the CI,

$$CI = \frac{\lambda_{max} - n}{n - 1}. \qquad (4)$$

The consistency ratio (CR) is used to estimate directly the consistency of pairwise comparisons. The CR is computed by dividing the CI by a value obtained from a table of Random Consistency Index (RI) (Table 1);

$$CR = \frac{CI}{RI}. \qquad (5)$$

Table 1. Average random index (RI) based on matrix size (adapted from Saaty [11])

N	1	2	3	4	5	6	7	8	9	10
RCI	0	0	0.52	0.89	1.11	1.25	1.35	1.4	1.45	1.49

If the CR less than 0.10, the comparisons are acceptable, otherwise not. RI is the average index for randomly generated weights (Saaty, 1981) [10].

Step 4. The priority weight of each alternative can be obtained by multiplying the matrix of evaluation ratings by the vector of attribute weights and summing over all attributes. Expresses in conventional mathematical notation;

$$\text{Weighted evaluation for alternative k} = \sum_{i=1}^{t} (attribute\ weight_i \times evaluation\ rating_{ik})$$

For i = 1, 2, t (t : total number of attributes).

$$(6)$$

After calculating the weight of each alternative, the overall consistency index is calculated to make sure that it is smaller than 0.10 for consistency on judgments.

4   Application of the Model to the Case Study

As case study, a decision of a vocational education server of valuating online training in YIGAO Company was taken into consideration. The proposed approach was carried out by using the FAHP approach. The data entered by the user for the analysis are given in Table 2. The main attributes and attributes for YIGAO called from vocational education server are given in Table 3. In addition, Table 5 shows a diagram of the main attributes with their attributes used for this vocational education server.

First, the fuzzy comparison matrix of pairwise comparisons for the attributes using triangular fuzzy numbers $(\tilde{1}, \tilde{3}, \tilde{5}, \tilde{7}, \tilde{9})$ is given in Table 4. And the fuzzy comparison matrix of alternatives of the main attribute productivity is shown in Table 5. The lower limit and upper limit of the fuzzy numbers with respect to the were defined as follows by applying Eq. 2;

Table 2. Data from the YIGAO co

TeamViewer server type: YIGAO for general use

- Names of alternatives: Tarui (P1); Jingchuang (P2); Zhongou (P3)
- Index of optimism (μ): 0.5 (default value: 0.5, $0 < \mu < 1$)
- Confidence level (α): 0.5 (default value: 0.5, $0 < \alpha < 1$)

Table 3. List of main attributes for valuating online training alternatives selection

		Main attributes
1	Productivity	C1
2	Technology	C2
3	Service	C3
4	Flexibility	C4
5	Safety and environment	C5

Table 4. The fuzzy comparison matrix of pairwise comparisons

	C1	C2	C3	C4	C5
C1	1	$\tilde{5}$	$\tilde{7}$	$\tilde{9}$	$\tilde{3}$
C2	$\tilde{5}^{-1}$	1	$\tilde{3}$	$\tilde{5}$	$\tilde{3}^{-1}$
C3	$\tilde{7}^{-1}$	$\tilde{3}^{-1}$	1	$\tilde{3}$	$\tilde{5}^{-1}$
C4	$\tilde{9}^{-1}$	$\tilde{5}^{-1}$	$\tilde{3}^{-1}$	1	$\tilde{7}^{-1}$
C5	$\tilde{3}^{-1}$	$\tilde{3}$	$\tilde{5}$	$\tilde{7}$	1

Table 5. The main attribute productivity

C1	P1	P2	P3
P1	1	$\tilde{1}$	$\tilde{5}$
P2	$\tilde{1}^{-1}$	1	$\tilde{3}$
P3	$\tilde{5}^{-1}$	$\tilde{3}^{-1}$	1

Table 6. The final ranking of online training alternatives

Alternative attribute		Alternatives		
		P1	P2	P3
C1	0.5073	0.528	0.354	0.118
C2	0.1309	0.154	0.737	0.109
C3	0.0644	0.063	0.273	0.664
C4	0.0338	0.144	0.762	0.094
C5	0.2636	0.057	0.396	0.547
Overall E-vector (%)		31.3	42.1	26.6

As the same principle, we build C2-P, C3-P, C4-P, C5-P matrixes. Calculating the matrixes according to the Eq. 1–6, we get the following results (Table 6):

At last, we get our results. From the Table 7, we can conclude that Jingchuang is our best choice.

Table 7. B/C ratio analysis for online training alternatives

Alternative	FAHP score (%)	Purchasing cost (1000RMB)	B/C ratio
Tarui	31.3	10.6	2.953
Jingchuang	42.1	12.8	3.289
Zhongou	26.6	9.0	2.956

5 Conclusion

With the advent of knowledge economy, science and technology are more and more developed, and knowledge is being updated more and more quickly. People need to rely on computers and intelligent devices to keep learning, which brought the prosperity of online training institute. These online training institutes also have to consider many factors in the selection of network vendors. This paper describes a Fussy AHP method, also cites specific case company in the choosing alternatives of online training network vendors. By the introduction of FAHP in selecting vendors, we can avoid the defect of traditional method of 1–9 scores, and reduce the human factors in scoring. This method is easier, more systematic and scientific, which helps us to find a shortcut in selecting alternatives of online training network vendor. However, there are still some deficiencies in this paper. Firstly, we did not analyze the present literature comprehensively. Secondly, the representativeness of case company not been illustrated thoroughly. As for future research direction, the FAHP could be used to evaluate similar vendor alternatives in order to enlarge its application scope and further test its effectiveness. Also other method such as data envelopment analysis (DEA) or neural network model could be used to testify and evaluate the online training vendor alternatives.

References

1. Kaufmann, A.: Introduction to Fuzzy Arithmetic Theory and Application, p. 574. Van Nostrand, New York (1991)
2. Chan, F., Chung, S.: Multi-criteria genetic optimization for distribution network problems. Int. J. Adv. Manuf. Technol. **24**(7–8), 517–532 (2004)
3. Chang, D.Y.: Applications of the extent analysis method on fuzzy ahp. Eur. J. Oper. Res. **95**(3), 649–655 (1996)

4. Handfield, R., Walton, S.V., et al.: Applying environmental criteria to supplier assessment: a study in the application of the analytical hierarchy process. Eur. J. Oper. Res. **141**(1), 70–87 (2002)
5. Klir, G.J., Yuan, B.: Fuzzy sets and Fuzzy Logic: Theory and Applications, vol. 574. Prentice Hall PTR NJ (1995)
6. Lee, A.R.: Application of modified fuzzy ahp method to analyze bolting sequence of structural joints. UMI Dissertation Service. A Bell & Howell Company (1995)
7. Liu, J., Ding, F.Y., Lall, V.: Using data envelopment analysis to compare suppliers for supplier selection and performance improvement. Supply Chain Manag. Int. J. **5**(3), 143–150 (2000)
8. Lucas, R.I., Promentilla, M.A., et al.: An ahp-based evaluation method for teacher training workshop on information and communication technology. Eval. Program Plan. **63**, 93–100 (2017)
9. Pan, J., Xue, F., Fan, Q., Fang, J.: Research on the evaluation method of outboard active decoy based on fuzzy-ahp. Aerosp. Electron. Warf. **02**, 45–52 (2018)
10. Saaty, T.L.: The Analytical Hierarchy Process. McGraw Hill, New York (1981)
11. Saaty, T.L.: Fundamentals of Decision Making and Priority Theory with the Analytic Hierarchy Process, vol. 6. RWS Publications (2000)
12. Soffer, T., Nachmias, R.: Effectiveness of learning in online academic courses compared with face-to-face courses in higher education. J. Comput. Assist. Learn. **34**(5), 534–543 (2018)
13. Li, S., Li, H., Wl, Huang: The evaluation method of electromagnetic environment complexity in radar countermeasure training based on grey-ahp. Radar & ECM **3**, 35–42 (2011)
14. Vaidya, O.S., Kumar, S.: Analytic hierarchy process: an overview of applications. Eur. J. Oper. Res. **169**(1), 1–29 (2006)
15. Wang, Q.Z., Guo, A.B., et al.: Performance evaluation method of armored unit simulation training based on AHP. Comput. Simul. **10**, 56–63 (2015)
16. Weber, C.A., Current, J., Desai, A.: An optimization approach to determining the number of vendors to employ. Supply Chain Manag. **5**(2), 90–98 (2000)
17. Zadeh, L.A.: Fuzzy logic, neural networks, and soft computing. Commun. ACM **37**(3), 77–85 (1994)
18. Zhu, K.J., Jing, Y., Chang, D.Y.: A discussion on extent analysis method and applications of fuzzy ahp. Eur. J. Oper. Res. **116**(2), 450–456 (1999)

The Impact of Sponsorship Following on Brand Evaluation-A Test of Mediation

Ge Ke, Hong Wang$^{(\boxtimes)}$, Hua Zhang, Wei Li, Shoujiang Zhou, and Shan Li

Business School, Sichuan University, Chengdu 610065, People's Republic of China
781617414@qq.com

Abstract. Sponsorship following refers to the late-movers sponsorship that follows the first-movers sponsorship. Findings suggest that: The high primary effect improves the brand evaluation of following brand more than the primacy effect; Compared with the low-recency effect, the high-recency effect improves the following brand t evaluation more significantly; The assimilation effect and contrast effect play intermediary roles in the following brand evaluation which the primary effect and the recency effect affect. This paper enriches the evaluation of sponsorship from the perspective of competition, and expands the area of sponsorship research.

Keywords: Sponsorship following · Mediation effects · Assimilation effects · Contrast effects

1 Introduction

Sponsorship means a sponsor invests resources in a specific activity in the form of money or material, and receives an opportunity from the Sponsored activity in return. With the gradual rise of the tertiary industry in China, especially the cultural and sports industries, more and more enterprises are paying more attention to the sponsorship. According to data released from IEG, corporate-sponsored spending has been continually growing for 13 consecutive years. In 2016, global corporate-sponsored spending reached the US $ 60.1 billion, which demonstrated an increase of 4.5% over the same period of last year. It is estimated that by 2017, it would reach 62.8 billion dollars. In Asia, sponsorship spending in 2017 will reach 15.7 billion dollars, showing a year-on-year increase of 5.8% [5]. From the above, it can be seen that sponsorship is receiving intense attention from many enterprises, and more and more enterprises are investing in the fierce sponsorship competition. Therefore, as an important way of sponsorship competition, sponsorship following has been valued by more and more enterprises. Sponsorship following refers to the late-movers sponsorship that follows the first-movers sponsorship. In addition, the aim of sponsorship following is for late-movers to compete with first-movers in sponsorship. For example, RIO sponsored following Breezer in 2014. Breezer sponsored a TV series named

© Springer Nature Switzerland AG 2020
J. Xu et al. (Eds.): ICMSEM2019 2019, AISC 1002, pp. 49–60, 2020.
https://doi.org/10.1007/978-3-030-21255-1_5

"Love Apartment" for the third season and then followed by RIO as the sponsor of "Love Apartment" for Season Four. According to the survey, after the fourth season of "Love Apartment finished, 64% of consumers still considered Breezer to be the sponsor of Love Apartment. Even 70% of the viewers said they didnt know that RIO sponsored "love apartment. From above, we have to ask what determines the effects of late-movers sponsorship? What is the psychological mechanism that sponsorship following impact the consumer evaluation? These are the practicable starting points of this study. From a theoretical point of view, the current research on sponsorship following is not deep enough. First, the current research fails to guide the effects of sponsorship following. Some scholars think that there are differences between the sponsor's image, the sponsor's sales volume and media coverage differences [13]. Some scholars hold the opinions that the sponsorship assessment is divided into the ex-ante assessment and ex-post facto assessment. Some scholars think that the images and qualities of the sponsors could have a great influence on the sponsorship [3]. In general, previous studies on the evaluation of sponsorship have focused on how the sponsorship itself could affect the sponsors. Researches have not been conducted from the perspective of sponsorship competition. In other words, it is worth noting that how the late-movers sponsorship will affect the following corporate sponsorship effect. Second, the existing research cannot clarify the mechanism of following. As an important marketing concept, following is formally defined in Philip Kotler's book Marketing Management and is followed by an emphasis on "proper distance" [18]. In addition to the market following, there are also scholars who have researched following strategies and brand following [6]. In general, this paper finds that following is an important concept in the field of marketing, and its research focuses on the level of enterprise competitive strategy and market behavior. However, there are no researches on the following from the perspective of the consumer psychology and behavior. From the perspective of consumer psychology, the principle of following is the "contradictory game" between the primacy effect and the recency effect. As a contradictory concept, the core of following is to see whether the late-movers brand evaluation is similar to the first-movers primacy effect or the late-movers recency effect. More importantly, this can prove that the path of following from the perspective of consumer psychology is the assimilation effect and contrast effect [10]. Based on the analysis above, on one hand, this paper explores the sponsorship following strategies that late-movers should take; on the other hand, this paper explores core of following from the perspective of consumer psychology based on the theory of assimilation and contrast effects [7,8]. This paper has two innovations. The first is theoretical innovation. From the perspective of consumer psychology and behavior, this paper explores the core of following, which is the "contradictory game" between the primacy effect and the recency effect. In addition, as for the practical innovation, this paper focuses on how to utilize sponsoring following strategies to increase the "sponsor leverage".

2   Literature Review and Research Hypotheses

2.1   The Related Research of Following Theory

In the early period, scholars mainly studied the following theory from three aspects: market following, strategy following, and brand following. The first is market following. As an important marketing concept, following is formally defined in Philip Kotler's book Marketing Management and is followed by emphasis on "proper distance". Followers mainly have two following strategies. First, following closely. The strategy emphasizes on keeping "low profile" in stimulating the market and avoiding direct conflict with the leader. Second, following with distance. The strategy focuses on "keeping the distance right", which means to keep a few differences with the leader to create a clear distance. Not only does it pose no threat to the leader, but it also survives because the followers own a small share of the market. The second is strategy following. According to Redondo [15], depending on the market entry strategy (trailblazers, latecomers to innovation, and latecomers to non-innovation), innovative latecomers, it could enjoy higher market potential and higher repeat purchase rates . The third is brand following. Reijmersdal examined the stage of the brand's incoming product life cycle, believing that those in the growing stage were able to reach their asymptotic sales [16]. In conclusion, the paper summarizes the following connotations, as shown in Table 1.

Thus this important concept in the field of marketing researches on following focuses more on the enterprise competitive strategy and market behavior. In addition, there are no studies analyzing following from the perspective of consumer behavior. Thus, this paper tries to analyze following from the perspective of the consumer psychology and get some conclusions related to sponsorship following.

2.2   Primary Effect

Primary effect, or the first effect, or first impression effect, was first proposed by the American psychologist Lorens and referred to the influence of the first impression, that is, the effect of "preconceived" [12]. The primary effect emphasizes the effect of the first temporal one-sided impression superimposition [1]. The first impression is the strongest and longest feeling. The primacy effect is generated by the brain's pattern of information processing. The human brain processes information in a two-system model. One system is a rational system, and one system is a perceptual system [11]. The first-arrival information leaves traces in the perceptual system.

According to the accessible-diagnosable theory, consumers may transfer the brand evaluation to another if similar events or attributes occur according to the similar events or attributes [2], and this may result in a brand spillover effect. According to the accessible-diagnosable theory, the brand spillover effect will occur at this time. The brand image and evaluation of the followed brand will be transferred to the following brand. When the primary effect of the sponsorship

Table 1. The following content summary

Document source	Connotation elements	Meaning
Kotler, (Philip 1965)	Time element	Following means that one happened after another happened, and the two cannot occur at the same time
	Related Elements	There is a certain relationship between following events, which may be causal or related. The followed events to a certain extent determine the time, direction, degree, and size of following events
	Distance Element	Followers must maintain a certain market distance with the leader. The purpose of followers is not to go beyond the leader, but just to keep up with the state
	Share Element	Market leaders shares should be greater than the followers market share
(Lei et al. 2008)	Rules element	Followers follow the market share they deserve, not trying to change industry rules
(Janakiraman et al. 2009)	Strategic element	The follower's strategies should stay ahead of the leader and beware of the challenger's attack

of the competing brand is higher, the image and evaluation of the brand that transfers to the followed brand are relatively higher. Therefore, in the process of sponsorship following, because of the effect of brand following, the brand enhances the role of evaluation. Accordingly, this paper concludes that H1:

H1: The high primary effect improves the brand evaluation of following brand more than the primacy effect.

2.3 Recency Effect

The recency effect refers to the phenomenon that the brain can memorize more recent things better. The longer the information span, the more significant the recency effect would be for consumers. This is due to the fact that previous information has been obscured in memory due to the passage of time so that short-term information becomes more prominent in memory [9].

For starters, the cause of the recency effect is related to human thinking memory mode. Human memory mode of thinking focuses on storing things close to time and blurring things far away. The second is related to the way people deal with information. The brain system that processes rational information places more emphasis on the chronological order of information processing than the systems that process emotions. In the case of information from the Rational

Information Processing System, it will pay more attention to the time process and form stronger proximity. Third, the order in which information is presented has an impact on social cognition. Information presented first has a greater influence on the information presented [14].

Recency effect refers to the phenomenon that when the consumer registers a series of things, the memory effect of the last part of the item is better than that of the middle part [4]. This phenomenon is due to the effect of the recency effect. The longer the interval before and after information, the more obvious the recency effect could be. The reason is that the previous information is gradually blurred in memory, making the recent information clearer in short-term memory. As the recency effect become stronger, the clearer the memory of following brand sponsorship would be and the easier it is for consumers to form higher brand evaluations. Accordingly, this paper proposes that H2:

H2: Compared with the low-recency effect, the high-recency effect improves the following brand t evaluation more significantly.

2.4   Assimilation Effect on the Contrast Effect of the Intermediary Role

The assimilation effect refers to consumers' unconsciously biased attitudes towards certain things to other things, especially when two things are similar. Contrast refers to the perception of a stimulus and judgments deviate from the background information. The model of Inclusion/Exclusion Model (IEM) holds the point that the essence of the contrast effect and assimilation effect is at ease of information processing. When the two sets of comparative information are more available and the information consistency is higher, the consistency of information is higher. The lower the intensity of the information before and after comparison, the less difficult it is for the consumer to deal with the two sets of comparative information. Thus the consumer can easily start the assimilation mode and make a similar evaluation of the two pieces of information. Therefore, during the process of sponsor following, consumers will move the evaluation of following brand to the evaluation of competitive brand through assimilation effect, and the assimilation effect plays an intermediary role in the following of sponsorship. When the comparative information is less available, the weaker the coherence of information, the higher intensity of information comparison between before and after would be. And then it is more difficult for consumers to deal with the two pieces of comparative information. Consumers are more likely to start the comparison mode with the two pieces of information to the contrary evaluation [11]. Therefore, in the process of sponsorship following, the evaluation of following brand by contrasting effect deviates from that of competitive brand, and the contrast effect plays an intermediary role in the following of sponsorship.

H3a: The assimilation effect plays an intermediary role in the following brand evaluation which the primary effect affects;

H3b: The contrast effect plays an intermediary role in the following brand evaluation which the primary effect affects;

H3c: The assimilation effect plays an intermediary role in the following brand evaluation which the recency effect affects;

H3d: The contrast effect plays an intermediary role in the following brand evaluation which the recency effect affects.

In summary, the model of this paper is shown in Fig. 1.

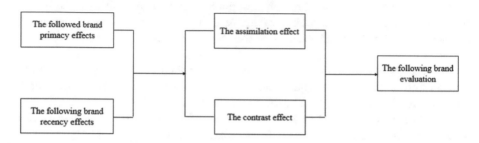

Fig. 1. Research Model

3   Experimental Design and Hypothesis Testing

The survey design of 2 (primary effect: high vs. low) × 2 (recency effect: high vs. low) was used to test the research hypothesis.

3.1   Stimulus Design

The design principles of stimulants are the following. High primacy effects stimuli are: Tide (Blue Moon) sponsors "I am a singer". "I am a singer" is a program of China Hunan TV that generates from Korea MBC introduction of singing live, created by the Hongtao team. In each program, it invites seven famous singers to compete. Specifically, low recency effects stimulus is that "brand A is a sports brand with a long history of professional production and sales of basketball supplies and it often sponsors basketball events to expand the market (It is reported that brand A sponsored the men's basketball held Asian Cup not long ago)".

The low primacy effects stimulus for effect: Tide (Blue Moon) sponsors the Anhui TV production. The design of high and low recency effects stimuli is the same as that of high and low primacy effects stimuli.

3.2   Variable Measurement

For the measurement of assimilation effect, refer to the study of Shi [17], the measurement item is "I am very concerned about the similarities between the two sponsorships." "I will follow the Blue Moon (Tide) sponsorship and the

Tide (Blue moon) evaluation. After these two sponsorship events, my assessment of Tide (blue moon) will be biased in favor of the Blue Moon (Tide) evaluation.The measurement items are "I am concerned about the difference between two sponsorship events", "I will evaluate the blue moon according to the difference between the blue moon sponsorship event and the blue moon sponsorship event", "After these two sponsorship events, I will evaluate the blue moon with me." The Moon's assessment is the opposite.

3.3   Investigation Procedure

The questionnaires were distributed to the sample through the Internet. The samples of the participants were completely random and not affected by the demographic factors. Therefore, the samples could be considered as meeting the selection criteria. The investigation procedure is as follows: First, a preliminary survey is conducted to verify the validity of the stimulus; then, a formal investigation is conducted to test the research hypothesis. The official text includes the following sections: Firstly, it is due to primacy effects and recency effects such as stimulus description; secondly, variable measurement; thirdly, it is the demographic characteristics of the item.

3.4   Pre-investigation

The first is to pre-survey the sample. This paper invited a total of 80 people involved in the re-investigation. They are divided into four groups, of which 40 people belong to the low head due to high proximity while the other 40 belong to the high head due to low proximity. Among them, there are 50 males and 30 females; however, gender has no significant effect on other variables.

In order to test whether sex significantly affects research variables and manipulated variables, sponsor involvement is measured only once, as the first and recency stimuli in this survey are all Entertainment-sponsored brands of washing products. In this paper, through one-way ANOVA test, the results showed that gender differences do not have a significant impact on the variables ($P > 0.05$).

An analysis of the qualifications of the pre-surveyed sample shows that there are 8 students in high school and below, 29 in undergraduate level, and 34 in master's degree and above. In this paper, by one-way analysis of variance test, the results show that academic differences do not significantly affect each variable ($p> 0.05$).

Secondly, the measurements of quality are as followed. the Cronbach's Alpha of the first match due to sponsorship is 0.888; the Cronbach's Alpha of first due to sponsorship attitude is 0.809; the Cronbach's Alpha of sponsored event involvement is 0.822; the primary effect is 0.804; the Cronbach's Alpha of proximity match due to sponsorship was 0.788; the Cronbach's Alpha of recency sponsorship attitude was 0.798; the Cronbach's Alpha of asymmetric brand relation was 0.824, the Cronbach's Alpha of recency effect was 0.847; the Cronbach's Alpha of following was 0.841; the Cronbach's Alpha of assimilation effect is 0.799, the Cronbach's Alpha of contrast effect is 0.858; the Cronbach's Alpha of brand sponsor evaluation is 0.888; the Cronbach's Alpha of assimilation effect is 0.759; the Cronbach's Alpha of contrast is 0.788; the Cronbach's Alpha of overall scale reliability is 0.845. Pre-survey shows that the reliability of this scale is high.

Finally, the results of pre-survey are as followed. There were no significant differences in the primacy effects match between the different types $M_{lowprimacy,highrecency}$ = 4.88, $M_{highprimacy,lowrecency}$ = 4.71, F (1,78) = 0.214, $p = 0.886 > 0.05$). There were no significant differences ($P < 0.05$) between the type of recency sponsorship match ($M_{lowprimacy,highrecency}$ = 4.46, $M_{highprimacy,lowrecency}$ = 4.61, F (1, 78) = 0.197. The first is due to sponsorship attitude in different types of no significant differences ($M_{lowprimacyandhighrecency}$ = 4.52, $M_{highprimacy,lowrecency}$ = 4.67, F (1,78) = 10.367, $p = 0.070 > 0.05$). There was no significant difference in the recency sponsorship between different types ($M_{highprimacy,lowrecency}$ = 4.30, $M_{highprimacy,lowrecency}$ = 4.07, F (1,78) = 0.225, $p = 0.879 > 0.05$). As the primary effect and the recency effect stimulants are detergent brands sponsor entertainment programs, the sponsor involvement measured only once and the data analysis found that the sponsor involvement in different types of no significant difference ($M_{lowprimacy,highrecency}$ = 4.52, $M_{highprimacy,lowrecency}$ = 4.82; F (1,78) = 0.112, $p = 0.953 > 0.05$). There was a significant difference between the top and bottom causes ($M_{Highprimacy}$ = 4.87, $M_{Lowprimacy}$ = 4.46; F (1,78) = 2.307, $p = 0.033 < 0.05$). The high and low near-effect effects were significantly different among different types ($M_{highrecency}$ = 4.63, $M_{lowrecency}$ = 4.20; F (1,78) = 0.022, $p = 0.041 < 0.05$). In summary, the pre-investigation shows the success of stimulus design and therefore reliable scale can conduct a formal investigation test hypothesis.

3.5   Formal Investigation

The first is the sample of formal survey. In this paper, a total of 371 people were invited to participate in the formal investigation. They were divided into four groups, of which 94 were in the causal group of low primacy and high recency effects, 92 were in the causal group of high primacy and low recency, 95 were in the group of high primacy and high recency and 95 were in the group of low primacy and low recency effects.

Among them, there were 223 males and 148 females, and gender has been proved to have no significant effect on other variables. To test whether gender can significantly affect the research variables and manipulated variables, this study adopts the one-way analysis of variance (ANOVA) to test whether gender differences have significant effects on various variables ($P > 0.05$). In order to test whether sex can significantly affect the research variables and manipulated variables, we used the one-way analysis of variance test, and finally, the results showed that academic differences did not have a significant impact on various variables ($P > 0.05$).

The following is manipulation test. The first due to sponsorship match in different types there is no significant difference ($M_{low primacy,high recency}$ = 4.77, $M_{highprimacy,low\ recency}$ = 4.74, $M_{high primacy,high recency}$ = 4.58,

$M_{low\ primacy, Low\ recency}$ = 4.95; F (3,368) = 1.984, $p = 0.059 > 0.05$). Proximity due to sponsorship match between the different types of no significant difference ($M_{low\ primacy, high\ recency}$ = 4.79, $M_{low\ primacy, high\ recency}$ = 4.74, $M_{high\ primacy, high\ recency}$ = 4.85, $M_{low\ primacy, Low\ recency}$ = 4.65; F (3,368) = 5.366, $p = 0.075 > 0.05$). The first due to sponsorship attitude in different types there is no significant difference ($M_{low\ primacy, high\ recency}$ = 4.83, $M_{low\ primacy, high\ recency}$ = 4.67, $M_{high\ primacy, high\ recency}$ = 4.75, $M_{low\ primacy, Low\ recency}$ = 4.17; F (3,368) = 5.492, $p = 0.091 > 0.05$).

Proximity sponsorship attitude in different types of no significant difference ($M_{lowprimacy, highrecency}$ = 4.69, $M_{low\ primacy, high\ recency}$ = 4.70, $M_{high\ primacy, high\ recency}$ = 4.67, $M_{low\ primacy, Low\ recency}$ = 4.32; F (3,368) = 13.597, $p = 0.114 > 0.05$). Because the primary effect and therecency effect stimulants are detergent sponsor entertainment programs, so the sponsor involvement measures only once. The data analysis found that there was no significant difference in the types of sponsorship among the different types ($M_{low\ primacy, high\ recency}$ = 4.56, $M_{lowprimacy, high\ recency}$ = 4.74, $M_{high\ primacy, high\ recency}$ = 4.93, $M_{low\ primacy, low\ recency}$ = 4.86; $M_{low\ primacy, high\ recency}$ = 4.74, $M_{high\ primacy, high\ recency}$ = 4.93, $M_{low\ primacy, low\ recency}$ = 4.86; F(3,368) = 1.248, $p = 0.254 > 0.05$). The primary effect was significantly different between different types ($M_{high\ primacy}$ = 4.56, $M_{low\ primacy}$ = 3.93; F (1,370) = 9.441, $p = 0.002 < 0.05$). Proximal cause effects were significantly different among different types ($M_{high\ recency}$ = 4.94, $M_{low\ recency}$ = 3.94; F (1,370) = 26.342, $p = 0.000 < 0.05$). Therefore, the stimulus manipulation is successful and we can make a hypothesis test.

The last is hypothesis testing. There was a significant difference between the first brand and the first brand due to the high first factor effect ($M_{Highprimacy}$ = 4.15, $M_{Lowprimacy}$ = 3.82; F (1,369) = 19.751, $p = 0.000 < 0.05$). Therefore, suppose H1 gets the third verification. The brand evaluation of the following brands shows significant differences in the type of near-effect due to the higher recency effect according to the brand's brand evaluation ($M_{highrecency}$ = 3.99, $M_{lowrecency}$ = 3.53, F (1,369) = 5.438, $p = 0.020 < 0.05$). Therefore, we assume that H2 is the third verification.

The brand evaluation of the following brand was higher in the top causal group with significant differences ($M_{highprimacy}$ = 4.15, $M_{highprimacy, highrecency}$ = 3.92, F (1,307) = 4.09, $p = 0.045 < 0.05$). Assume H3a is validated for the second time. The brands following the brand were rated as high and low, respectively. The low-to-near brands were significantly higher ($M_{Highprimacy}$ = 4.15, $M_{highprimacy, Lowrecency}$ = 4.79, F (1,433) = 3.633, $p = 0.029 < 0.05$), assuming that H3b is validated for the second time. The brand evaluation of the following brands in the low-head and high-close groups was higher and there were significant differences ($M_{highprimacy, lowrecency}$ = 4.79, $M_{lowprimacy, highrecency}$ = 5.09, F (1,247) = 10.558, $P = 0.002 < 0.05$), assuming H3c was validated for the second time.

In order to test the mediated effect of the assimilation effect and the comparative effect in the process of the first-cause effect on the brand evaluation, the paper uses the step-by-step regression method to test. The first step is to construct a regression equation from independent variable to dependent variable with a significant regression equation (F = 19.751, $p < 0.05$) and a coefficient of first-cause to brand following brand (r = 0.225, $p < 0.05$). The second step is to construct the regression equation of the independent variable to the intermediate variable. The regression equation for the first-cause-type to assimilation effect was significant (F = 0.039, $p < 0.05$), and the first-cause to assimilation effect was significant (r = 0.225, $p < 0.05$). The third step is to build the regression equation of the primacy effects type to the contrast effect. The regression equation for the first-cause-type to the comparative effect was significant (F − 0.001, $p < 0.05$), and the first-cause type to the comparative effect coefficient was significant (r = 0.225, $p < 0.05$). Finally, we test the influence of independent variable and mediating variable on dependent variable, constructing the regression equation of the primacy effects type and assimilation effect to the brand evaluation of the following brand. The regression equation between the primacy effects and the brand evaluation of following brand was significantly ($P < 0.05$) due to the first type and assimilation effects (F = 10.922, $p < 0.05$). The regression of brand evaluation was also significant ($p < 0.05$). Therefore, the assimilation effect played an incomplete mediation effect when the effect of first-cause effect followed the brand evaluation of the brand, assuming H4a was partially established.

We construct the regression equation of the primacy effects type and the contrast effect to brand evaluation of the following brand. The first regression equation of the first type to the brand following the brand was significant (F = 11.173, $p < 0.05$). The regression of the first type to the brand following the brand was significantly ($p < 0.05$). The regression of brand evaluation was also significant ($p < 0.05$). Therefore, the comparative effect played an incomplete mediating effect when the first-cause effect influenced the brand evaluation of the brand, assuming that the H4b part was established.

Similarly, the method for testing the effect of assimilation and contrast effects is mentioned in this paper. The influence of stepwise regression recency effect on the brand following played an intermediary role in the evaluation process, showing that the impact on the comparative effects of assimilation in effect recency effect on the brand following played intermediary roles in the evaluation process, assuming H4c and H4d are partially valid.

4   Conclusion

4.1   Research Conclusions

This paper finds out that the high primary effect improves the brand evaluation of following brand more than the primacy effect. In addition, compared with the low-recency effect, the high-recency effect improves the following brand t evaluation more significantly; The assimilation effect and contrast effect play intermediary roles in the following brand evaluation which the primary effect and the recency effect affect.

4.2   Theoretical Contribution

First of all, this paper answers the essence of follow-up from the psychological and behavioral level of consumers, and proves that the principle of following is the "contradictory game" between the primacy effect and the recency effect, while the core of the "contradictory game" is the assimilation effect and the contrasting effect.

Second, this paper studies sponsorship from the competitive angle. Previous researches studies sponsorship mainly in still ways like sponsorship fit and sponsorship objects [3,7,8,16]. This paper studies sponsorship following and find that contrast effects are important, and this finding prove that sponsorship competition could impact sponsorship effects. Therefore, this paper expands the area of sponsorship research from the perspective of sponsorship competition.

4.3   Management Implication

First of all, the following brand should focus the sponsoring events that happened before. The best sponsorship following strategy is to follow the high primacy sponsorship effects events, because high primacy effects sponsorship events always mean that the first-movers sponsorship resources are with high brand reputations and high brand values. If late-mover follow the first-movers sponsorship that is of high primacy effects, the high brand reputations and high brand values are easy to be transferred to late-movers.

Secondly, when following the brand to follow the sponsorship event at any time, using low proximity strategy is better than using high proximity strategy, and the sponsorship cost is lower. The empirical study finds out that, compared with the high primacy and the low recency effect, the low primacy and high recency effect can enhance the following brand evaluation at utmost. It indicates that the sponsoring effect is affected by competition.

Acknowledgements. We are grateful for the financial support from Natural Science Foundation of China(No.71702119) ; Humanity and Social Science Youth Foundation of Ministry of Education of China (No. 17YJC630065); 2018-2019 Sichuan University Business School Teacher-Student Joint Innovation Project; China Postdoctoral Science Foundation (No. 2018M640027).

References

1. Aydinli, B.M.A., Lambrecht, A.: Price promotion for emotional impact. J. Mark. **78**(4), 80–96 (2014)
2. Buse, J.B., Defronzo, R.A., et al.: The primary glucose-lowering effect of metformin resides in the gut, not the circulation: results from short-term pharmacokinetic and 12-week dose-ranging studies. Diabetes Care **39**(2), 198–205 (2016)
3. Chen, Y., Ganesan, S., Yong, L.: Does a firm's product-recall strategy affect its financial value? an examination of strategic alternatives during product-harm crises. J. Mark. **73**(6), 214–226 (2009)

4. Cordery, C.J., Baskerville, R.F.: Cash, sinkholes and sources. How are Community Sport and Recreation Organisations Funded and What are the Implications for Their Future Viability? Research Report 2: Football clubs. Social Science Electronic Publishing (2018)
5. Dan, N.: Ieg predicts global sponsorship: spending \$ 62.8 billion, China leads the asia-pacific mania (2017). http://mini.eastday.com/mobile/170202173803777.html
6. Davis, S.: Brand asset management2: how businesses can profit from the power of brand. J. Consum. Mark. **19**(4), 351–358 (2002)
7. Davison, M.L.: On a metric, unidimensional unfolding model for attitudinal and developmental data. Psychometrika **42**(4), 523–548 (1977)
8. Dickson, G., O'Reilly, N., Walker, M.: Conceptualizing the dissolution of a social marketing sponsorship. J Glob. Sport Manag. pp 1–24 (2018)
9. Fang, X., Kleef, G.A.V., Sauter, D.A.: Person perception from changing emotional expressions: primacy, recency, or averaging effect? Cogn. Emot. 1–14 (2018)
10. Garretson, J.A., Dan, F., Burton, S.: Antecedents of private label attitude and national brand promotion attitude: similarities and differences. J. Retail. **78**(2), 91–99 (2002)
11. Göde, S., Turhal, G., et al.: Primary sinonasal malignant melanoma: effect of clinical and histopathologic prognostic factors on survival. Balk. Med. J. **34**(3), 255–262 (2017)
12. Holbrook, C.M.B.: The chain of effects from brand trust and brand affect to brand performance: the role of brand loyalty. J. Mark. **65**(2), 81–93 (2001)
13. Jensen, J.A., Turner, B.A.: Event History Analysis of Longitudinal Data: a Methodological Application to Sport Sponsorship, pp 1–18. Social Science Electronic Publishing (2018)
14. Plonsky, O., Erev, I.: Learning in settings with partial feedback and the wavy recency effect of rare events. Cogn. Psychol. **93**, 18–43 (2017)
15. Redondo, I.: The effectiveness of casual advergames on adolescents' brand attitudes. Eur. J. Mark. **46**(11–12), 1671–1688 (2012)
16. Reijmersdal, E.A.V., Jansz, J., et al.: The effects of interactive brand placements in online games on childrens cognitive, affective, and conative brand responses. Comput. Hum. Behav. **26**(6), 1787–1794 (2010)
17. Shi, X., Zhu, Y.: Urbanization and risk preference in china: a decomposition of self-selection and assimilation effects. China Econ. Rev. **49** (2018)
18. Tsordia, C., Papadimitriou, D., Parganas, P.: The influence of sport sponsorship on brand equity and purchase behavior. J. Strateg. Mark. **26**(1), 85–105 (2018)

Study on the Relationship between Modernized Economic System Construction and Transfer and Transformation of Scientific and Technological Achievements

Qian Fang$^{(\boxtimes)}$ and Yi Sheng

Sichuan Academy of Social Sciences, Chengdu 610072, People's Republic of China
lily_lily009@163.com

Abstract. From the perspective of the construction of modernized economic system, this paper ponders over the transfer and transformation of scientific and technological achievements, which is based upon systemic theory. In this paper, a structural modeling combining the construction of modernized economic system and the transfer and transformation of scientific and technological achievements is established on the foundation of Interpretative Structural Modeling Method. It also conducts analysis on the relationship and differences between the construction of modernized economic system and the transfer and transformation of scientific and technological achievements to make their operational mechanism clarified. It puts forward a way to consolidate the construction of modernized economic system through the transfer and transformation of scientific and technological achievements, which is combined with the current situation of transfer of scientific and technological achievements at the national and regional levels.

Keywords: Modernized economic system · Scientific and technological achievements · Interpretative Structural Model

1 Introduction

As the primary power for development, innovation is also the strategic support for the construction of modernized economic system. For a long time, our country focuses on increment of innovations in science and technology but not the amount of achievement transformation. However, the weakness of Made in China is revealed in the increasingly intensive China-US trade war and the sanction to ZTE by the America. China has increasingly enhanced global voice on the aspect of applications of high technologies; however, China still has prominent defects in fields such as chips. It was emphasized by General Secretary Xi Jinping at the Opening Meeting for the 19th Meeting for Academicians of Chinese Academy of

© Springer Nature Switzerland AG 2020
J. Xu et al. (Eds.): ICMSEM2019 2019, AISC 1002, pp. 61–74, 2020.
https://doi.org/10.1007/978-3-030-21255-1_6

Sciences and the 14th Meeting for Academicians of Chinese Academy of Engineering on May 28, 2018 to realize that innovation is the primary power, and it is needed to provide sciences and technologies with high quality to support modernized economic system construction. With the macroscopic new historical background, it is needed for scientific & technological personnel and institutions in our country to observe the core orientation of modernized economic system construction as well as inspect the important roles of transfer and transformation of scientific & technological achievements in national economic development. In addition, it is needed to take transfer and transformation of scientific & technological achievements as the emphasis to drive coordinative development in sciences, technologies and economics.

2 Literature Review

The modernized economic system as well as the transfer and transformation of scientific & technological achievements are two hot topics in the academic circle, with great differences on starting points of research, performance of topic heat and discussion fields.

The discussion on the topic of transfer and transformation of scientific & technological achievements started early, which has been discussed since 1980s by scholars. The topic has favorable quantity and quality of published articles as well as concentration conditions. More than 14400 articles can be retrieved in the CNKI database with titles or key words of "transfer of scientific & technological achievements" and "technological transformation". With respect to quantity of published articles, scholars have paid increasingly more attention to this field. The annual quantity of published articles of this topic was less than 200 during 1979–2000; it was 500 during 2001–2010; it achieved more than 660 during 2011–2018. With respect to quality of published articles, more than 3200 articles are published in Chinese core periodicals, occupying more than one fifth of the total quantity of published articles, with relatively high quality in achievements.

The "construction of the modernized economic system" is the new topic proposed by the 19th National Congress of the Communist Party of China. More than 750 literatures can be retrieved with the title of key word of "modernized economic system", with main publishing time of 2017 and 2018. Scholars reflections on the modernized economic system are mainly concentrated on concepts, connotations and significances, illustrated on aspects of main social conflicts [8], constitutive reforms, new development ideas, high-quality development [5], socialism with Chinese characteristics and the new ear. The "system construction" is discussed on topics of regional development, modern industrial system [7], income distribution system, supply-side structural reforms, substantial economy and technological innovation. Considering that the concept of modernized economic system has not been proposed for long, the discussion is conducted mainly on connotations and significances. Although certain explorations have been conducted in realization paths, they are not profound or concrete. The researches are not closely related to practical conditions, with unfavorable references for practical works.

In conclusion, regarding of discussion on "achievement transformation" and "economy", scholars concentrate on the following three points: Firstly, if scientific & technological innovation can support economic development. It is pointed by Pang et al. [11] that "generally speaking, scientific & technological innovations fail to play a favorable supporting role in economic development in various provinces", with the main reason of failure in effective transformation of achievements [11]. Secondly, it involves in contributions made on economics by achievement transformation. Zhu et al. [16] construct the economic concentration degree indicators for achievement transformation, to propose that the domestic achievement transformation level shows a degressive pattern in a way of "east C middle - west" parts [16]. Thirdly, it involves in the transformation of scientific & technological achievements to realistic productivity, to become the "true gold" promoting economics, i.e., method and path. The reflections on this topic during early stage concentrated on the method to promote integration between sciences & technologies and regional economy, finance and industry. For example, Gong [9] points that it is needed to strengthen connection between scientific & technological achievements and regional economy [9]. It is proposed by Long and Li [14] to avoid risks at the same time of strengthening combination between science & technology and finance [14]. Scholars have developed their ideas in a systematic way since the 19th National Congress of the Communist Party of China, and the construction of the transfer and transformation system for achievements has been a common expression for scientific & technological departments as well as economic department, and the emphasis of transformation strategies has been transferred from scientific & technological departments and scientific research institutions to enterprises, markets, intermediary institutions and scientific research personnel. For example, it is proposed by Wang [10] to promote the approaching transactions of scientific & technological plan achievements [10].

In summary, only small numbers of scholars include "construction of modernized economic system" and "transfer and transformation of scientific & technological achievements" into the same analysis angle in their researches, with the following two reasons: Firstly, it is assumed to take technological progress as internal factors for economic growth, and "good achievement transformation will inevitably bring in good economy" becomes an unanswerable generally known truth in need of no verification. The well-known affirmation makes scholars pay less attention to the topic of "the way of scientific & technological achievement transformation to achieve better economy"; however, they lay emphasis on "how to realize scientific & technological achievement transformation in a better way" (the phenomenon is known as the "problem preposition" according to the author). Secondly, "transfer and transformation of scientific & technological achievements" is a specific problem, and it is difficult to bring this problem into the macroscopic economic field and abstract space, with great difficulties in connections of theories and practice. With respect to reality, there are great differences between scientific & technological conditions and economic conditions, which always exist, with less change. The most common problem is more invalid achievements and less transformation resources. In the new era, the economic development method of China varies from extensive framing to intensive and meticulous farming. With this background, the scientific & technological achievement transformation is applied to "modernized economic system construction", to discuss on relationship between them, so as to clarify the main channel for achievement transformation to promote economic development theoretically. In this way, it lays emphasis on the main path of achievement transformation efficiency in practice, playing an especially important role in promoting the modernized system construction in our country as well as improving transfer and transformation efficiency of scientific & technological achievements.

3   Build the Structural Model

There are a lot of factors giving rise to influences on modernized economic system construction as well as transfer and transformation of scientific & technological achievements (hereinafter referred to as "achievement transformation") in the regional economic and social system. Such factors serve for the modernized economic system and the achievement transformation. It is needed to master such factors as well as direct or indirect relations among such factors, which is especially important for us to further understand the relations between modernized economic system construction and achievement transformation. Specifically speaking, it is needed to analyze statuses of various factors in the system on the aspect of structure; i.e., it is necessary to establish the structural model.

The problem assumed in the research is to determine the factors of achievement transformation during the modernized economic system construction and the relations among such factors. Lay emphasis on the key factors of "modernized economic system construction" and "achievement transformation" based on transcendental knowledge. Arrange the 10 factors of modernized economic system according to the report of the 19the National Congress of the Communist Party of China and the important statement of General Secretary Xi Jinping.

It was pointed by General Secretary Xi Jinping in the third collective learning of the Political Bureau of the Central Committee of the Communist Political Bureau in Jan. 2018 that: "the modernized economic system is an organic whole composed by mutual relations and internal relations of various processes, aspects and fields of social economic activities." Therefore, the overall framework of the modernized economic system includes 6 sub-systems: (1) the industrial system with compact entities and overall upgrading; (2) the unified and open market system with orderly competitions; (3) the fairness-promoting distribution system with emphasis of efficiency; (4) the coordinative and optimized regional development system; (5) the green development system with saving resources and friendly environment; (6) the multiply-balanced, safe and efficient overall open system. In addition to the 6 sub-systems, 4 additional factors including substantial economy, scientific & technological innovation, modern finance and human resource are extracted from the industry system. Theoretically, the 6 sub-systems are corresponding to the four processes of social production, i.e., production, exchange, distribution and consumption. The 14 factors include sub-systems of the modernized economic system and social production processes; besides, they are extracted to abstract theoretical level from realistic functions. In addition, 18 factors related to achievement transformation have been extracted from existing researches, with the factor literature source (Table 1) and factor detailed table of Table 2.

Table 1. Key elements and sources

Element	Source
Colleges and universities	Yong et al. [13]
Scientific research institutions	Deng et al. [4]
Enterprises	Qiao Bing et al. (2016)
Scientific research personnel	Han Fenzhang et al. (2016)
Entrepreneurship	Zhu [15]
Scientific & technological innovation	Sun Yuhong et al. (2018)
Scientific & technological achievements	Chao Qinghui (2017)
Intermediary agencies	Gray and Boardman [6]
Technical market	Arora and Fosfur [1]
Information platform	Liu Qijia et al. (2018)
Intellectual property	Bing [2]
Products	Jian Hui (2015)
Benefits	Jian Hui (2015)
Demands	Liu Dayong et al. (2017)
Demand information	Liu Shuqing et al. (2011)
Technology	Xu Yuanlin et al. (2018)
Environment	Chi [3]
Capital	Werthamer et al. [12]

The upper triangular matrix is written according to the detailed statement of factors and factor relations, to establish the adjacent matrix A. The relationship of the modeling factor S_i and S_j $(i, j = 1, 2, 3 \ldots, 32)$ is researched, to establish the upper triangular matrix with Formula (1).

$$\begin{cases} (1)\, S_i \times S_j \text{ i.e. } S_i \text{ and } S_j \text{ are mutually correlated;} \\ (2)\, S_i O S_j \text{ i.e. } S_i \text{ and } S_j \text{ are not mutually correlated;} \\ (3)\, S_i \wedge S_j \text{ i.e. } S_i \text{ is correlated with } S_j, \text{but } S_j \text{ is not correlated with } S_i; \\ (4)\, S_i \vee S_j \text{ i.e. } S_j \text{ is correlated with } S_i, \text{but } S_i \text{ is not correlated with } S_j; \end{cases}$$

The computer is utilized to calculate the sum of the adjacent matrix and the unit matrix, and Boolean algebra principle is observed as the calculation rule.

Table 2. Key elements and sources

Code	Name of factor	Code	Name of factor
S1	Production	S17	Scientific research institutions
S2	Exchange	S18	Colleges and universities
S3	Distribution	S19	Enterprises
S4	Consumption	S20	Scientific research personnel
S5	Industrial system	S21	Intermediary agencies
S6	Substantial economy	S22	Intellectual property
S7	Market system	S23	Information platform
S8	Open system	S24	Scientific & technological innovation
S9	Distribution system	S25	Technical market
S10	Regional development system	S26	Products
S11	Green development system	S27	Demands
S12	Modern finance	S28	Demand information
S13	Public finance	S29	Entrepreneurship
S14	Capital	S30	Scientific & technological achievements
S15	Technology	S31	Benefits
S16	Human resources	S32	Environment

The following formula can be acquired based on the calculation:

$$A_1 \neq A_2 \neq A_3 \neq \cdots A_{10} = A_{11}$$

The reachable matrix R can be acquired based on the calculation. The reachable matrix $R = A_{10} = A_{11}$, indicating the reachability of the access with passing length among various nodes of no more than 10.

The lowest level factor T is acquired from the calculation of the reachable set and the advanced set, to judge the connectivity of factors. In condition that the intersection of the reachable sets of factors is vacant, they belong to different connected domains. In condition that the intersection of the reachable sets of factors is not vacant, they belong to the same connected domain.

The lowest level factor T can be acquired based on calculation.

Considering that $R(S_5) \cap R(S_6) = \{S_2\} \neq \varPhi$, S_5 and S_6 belong to the same connected domain.

The reachable matrix is taken as the criterion, and 32 factors in the system are divided into different levels (layers), to acquire the factors of different layers as follows:

$$T = \{S_5, S_6, S_7, S_8, S_9, S_{10}, S_{12}, S_{13}, S_{15}, S_{16}, S_{32}\}; L_1 = \{S_1, S_2, S_{11}, S_{14}\};$$
$$L_2 = \{S_{18}, S_{19}\}; L_3 = \{S_{17}\}; L_4 = \{S_{20}\}; L_5 = \{S_{29}\}; L_6 = \{S_{24}\}; L_7 = \{S_{30}\};$$
$$L_8 = \{S_{21}, S_{22}, S_{23}, S_{25}\}; L_9 = \{S_{26}\}; L_{10} = \{S_{31}\}; L_{11} = \{S_3\}; L_{12} = \{S_4\};$$
$$L_{13} = \{S_{27}\}; L_{14} = \{S_{28}\}$$

The hierarchical structure model is drawn according to the hierarchical results (omitted), and then the "structural model for the modernized economic system construction and the transfer & transformation of scientific & technological achievements" according to the hierarchical structure model, "modernized economic system construction and transfer & transformation of scientific & technological achievements-ISM" for short, as shown in Fig. 1. Considering that the system has reflection mechanism during the investment to output process, the feedback lines are added into Fig. 1. Such feedback lines are "products → industrial system (green development system), "demand information → production", "demand information → green development system" and "distribution → scientific research personnel (distribution system)".

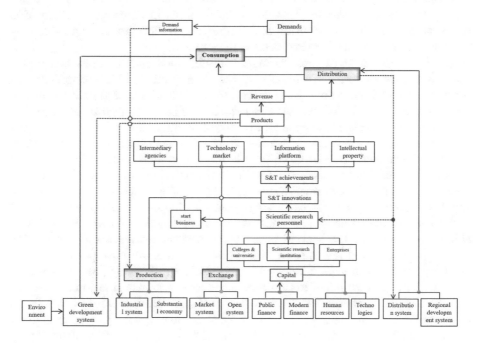

Fig. 1. Modernized economic system construction and transfer & transformation of scientific & Technological Achievements-ISM

4   Analysis Relationship Between the Elements

4.1   See Their Relationship from the Perspective of Social Production

The relations between modernized economic system and achievement transformation can be analyzed on the four processes of social production, as follows:

(1) In the production process: "Substantial economy" and "industrial system" are carriers and power to promote modernized economic system construction. "Substantial economy" is reflected on three innovative subjects in achievement transformation (enterprises, colleges & universities and scientific research institutions), and the creativity of innovative subjects originates from "scientific research personnel". Therefore, it is needed to motivate scientific research personnel to act as the driving force for accelerating achievement transformation. It should be noted that "entrepreneurship" is also the path to motivate innovations, which is a work widely mobilized in our country, with favorable implementation conditions at present.

In the production process, the logic line of the relations between them is listed as follows: "modernized economic system → substantial economy → enterprises, colleges & universities and scientific research institutions → innovation and entrepreneurship → scientific & technological achievements → products → industrial system → modernized economic system".

(2) In the exchange process: "Market system" and "open system" are important contents of modernized economic system construction, which ensure efficient configuration of state economic resources, to meet demands of globalization and establishment of the macroscopic market view. The open system enlarges the market scale, the exchange space and the product summary, which is the extensive expansion to exchanges. "Market system" is constituted by four factors during achievement transformation, including intermediary agencies, technical market, information platform and intellectual properties. In order to transform scientific & technological achievements to "products" with market values, to acquire and compensate for early stage cost and increase subsequent power, it is needed for governments and markets to work together to provide a favorable exchange scene. Intellectual property protection, achievement information issuing platforms and intermediary institution services are important means to promote effective connection between supply and requisitioning parties as well as fields with key improvement in domestic achievement transformation at present.

The logic line between them during the exchange process is listed as follows: modernized economic system → market system (open system) → intermediary institutions, technical markets, information platforms, intellectual property → products → revenue → market system → modernized economic system.

(3) In the distribution process: "Distribution system" and "regional development system" are important contents for modernized economic system, promoting social equity and balanced development of regions. In achievement transformation, "distribution system" is mainly embodied as scientific distribution of "revenue"; i.e., if scientific research personnel (scientific research teams) can be provided with revenue keeping enthusiasm for innovation. By comparison, achievement transformation leads to less influence on the "big distribution system" on the intermediary and macroscopic layers, or even no influence on "regional development system". However, distribution of achievement transformation gives rise to direct influences on power of innovation of scientific research teams and personnel, playing a critical role.

The logic line between them during the distribution process is listed as follows: modernized economic system → distribution system (regional development system) → revenue → distribution → scientific & technological personnel (distribution system) → modernized economic system.

(4) In the consumption process: The modernized economic system is not sufficiently embodied in the consumption process, mainly in "green development system". "Green" is one of high-quality development characteristics, which is a general requirement on production and living by the public. "Green development system" shall be taken as the orientation for the process and results of achievement transformation. It is required to improve environments with good inventive ideas and products with green awareness, so as to improve and perfect the green development system.

The logic line between them in the consumption process is listed as follows: modernized economic system → green development system → consumption → demands → demand information → production → green development system → modernized economic system.

4.2 See Their Relationship from the Perspective of System Intersection

(1) The system intersection during the production process is embodied in substantial economy and factors of production (capital, modern finance, human resource and public finance). The production process of the modernized economic system is corresponding to "substantial economy". The production process of achievement transformation is corresponding to carriers of enterprises, scientific research institutions and colleges & universities, as well as the innovation process of substantial economy. The "factors of production" during the production process are key of the modernized economic system and achievement transformation; lack of factors, unfavorable structure and unbalanced distribution are general problems for them two.

(2) The system intersection during the exchange process is embodied in "market system". The exchange process of modernized economic system is corresponding to market system. The exchange process of achievement transformation is embodied on markets ensuring achievement transformation, such as "intermediary agencies", "technical market", "information platform" and "intellectual property", and related factors.

(3) The system intersection during the distribution process is embodied in "distribution". The exchange process of the modernized economic system is corresponding to the "distribution system", and the distribution process of achievement transformation is corresponding to "distribution". The difference is that the distribution of the modernized economic system tends to distribution of resident income, and the distribution of achievement transformation tends to distribution of revenue after transformation.

(4) The system intersection during the consumption process is embodied in "demands". The green development system is adopted to meet demands of people to beautiful environments; at the same time, achievement transformation

is conducted to improve life of people based on product updating, so as to realize transformation to environments.

4.3   See Their Differences from the Perspective of System Characteristics

They two have the relationship of whole and local. Achievement transformation is a constituting part of the modernized economic system construction, which is embedded in the system as the important path promoting modernized economic system construction. The achievement transformation operation contributes to "production process" and "exchange process" of the modernized economic system as well as substantial economy, industrial system, market system and open system construction.

They have different emphasis layers. Modernized economic system construction is considered on the intermediary and macroscopic layer; however, the carriers of achievement transformation are considered on the microscopic layer, with different decision-making and management departments. Considering that the modernized economic system construction is proposed by the 19th National Congress of the Communist Party of China, the connotations of which are enriched by addressed from state leaders, especially those from finance and economics field. In addition to laws and regulations in our country such as "Law of Promoting Transformation of Scientific & Technological Achievements of PRC", most policies related to achievement transformation are implementation plans or supplementary policies of regional governments and scientific & technological departments.

They have different emphasis points. Modernized economic system construction lays emphasis on cooperation awareness while achievement transformation pays more attention to the problem awareness. With respect to modernized economic system construction, coordinative and cooperative development is needed for the 6 sub-systems, with no "buckets effect". It is needed to have global awareness for modernized economic system construction by taking balanced development of regions, groups and economic society. With respect to achievement transformation, the emphasis is concentrated in processes of production, exchange and distribution, especially production and exchange. In order to keep advanced technologies, it is needed to rely on market feedback as well as innovation enthusiasm and imagination of scientific research personnel. Therefore, it is needed to select aiming at the realistic "short slabs" for the emphasis of achievement transformation.

5   Suggestions for Improvement

Based on the above analysis, in combination of current conditions and problems of scientific & technological achievement transformation in our country as well as national and regional policies, it is needed to lay emphasis on the following five aspects in promoting of carrier construction, improving of market functions and insisting on demand orientation to realize modernized economic system construction with achievement transformation.

5.1 Promote Carrier Construction and Construct Enriched Innovation Polar Nucleus with Three Types of Factors

The achievement transformation process mainly includes the demander subjects represented by enterprises, the supplier subjects represented by colleges, universities and scientific research institutions as well as regulating subjects represented by governments. Therefore, it is needed to build carriers with differential functions around the three types of subjects and form innovative polar nucleus. Firstly, it is needed to promote the construction of innovative carriers with enriched enterprise resources. By taking high-class high-tech zones, economic development zones and industrial parks as the objects, it is needed to give full play to advantages of enriched enterprise resources and perfect functions of zones and parks around industrial orientation, so as to promote research, development and sharing of common technologies. In addition, it is needed to strengthen the pilot plant test function of parks and encourage related enterprises to create national industry innovation centers and enterprise technology centers and engineering technology centers. Secondly, it is needed to promote the construction of innovation carriers with enriched intelligence resources. It is needed to improve scientific & technological incubation and transformation functions of university science parks, and guide to establish achievement transformation funds with multiplied subjects. Thirdly, it is needed to promote the construction of the innovation carriers with enriched policy resources, to gather innovation resources with better policy environments, so as to construct more energetic linked mechanism among governments, industry, education and research.

5.2 Perfect Market Functions, and Improve Scientific and Technological Service Supply Power in Public Domains

Fragmented supply and demand information, insufficient detection and realization of prices and lack of professional services are important reasons leading to unfavorable support for scientific & technological achievement transformation by market mechanism. Considering that such problems must be solved with public and external efforts, which cannot be solved only by self-construction by suppliers and demanders or natural adjusting by markets, efforts from governments are needed. Firstly, it is needed to establish professional institutions for scientific & technological achievement transformation. It is necessary to encourage colleges, universities and scientific research institutions to establish professional institutions such as technology transfer offices and technology concept verification centers, to improve self-transformation ability. Secondly, it is needed to promote concentrated and open transactions of scientific & technological achievements. It is needed to give full play to information driving, to adopt multiple methods such as technology bid invitation, listed transaction and achievement entrustment, in combination of online and offline policy guidance, so as to realize concentrated transactions with carriers of information platforms, to promote further gathering of supply and demand information. Fourthly, it is needed to cultivate intermediary service institutions for science and technology and a batch of transformation managers with high professional quality, to strengthen talent team construction of technology transfer managers, technology brokers and technology managers.

5.3 Insist on Demand-Orientation and Connect Among Industry, Science and Technology and Regional Development Tasks

It is needed to find out correct emphasized demands and insist on demand-orientation with the viewpoint of regional development, so as to lay solid foundation for modernized economic system construction, to form the transfer and transformation mode of driving supply with demand. Firstly, it is needed to lay emphasis on key processes of industry development in combination of demands on regional industry development, to promote coordination between technology policies and the tertiary industry policy, in order to form policy resultant force. Secondly, it is necessary to pay attention to leading role played by financial funds, to promote interaction and combination of scientific & technological finance and industry finance. Thirdly, take development demands of medium-sized and small enterprises into consideration. Explore colleges, universities and scientific research institutions with annual research and development expenditure higher than stipulated amount, and subsidize the joint research and development of medium, small and micro-sized enterprises in key fields or achievement transformation with certain proportion of funds.

5.4 Strengthen Obligations on Transformation and Improve Assessment and Incentive Systems for Scientific Research Personnel

Firstly, it is needed to clarify obligations of scientific & technological achievement transformation. It is needed to strengthen obligations and responsibilities of achievement transformation of colleges, universities and scientific research institutions, to include achievement transformation into performance assessment of colleges, departments and individuals. Secondly, it is needed to improve the assessment system for scientific research personnel. Colleges, universities and scientific research institutions under the state and local governments are required to establish assessment systems in accordance with rules and work characteristics of scientific & technological achievement transformation. It is needed to take conditions of scientific & technological achievement transformation as basis for work performance and title review of scientific research personnel during entrepreneurship or part-time job. Thirdly, it is needed to improve incentive system for scientific research personnel. It is necessary for colleges, universities and scientific research institutions to connect with local policies and reach agreements with scientific research personnel about rights and obligations of part-time job and demission related to transformation, as well as formulate administrative rules for scientific research personnel (such as professors) with par-time job.

5.5 Implement Reform Measures and Construct the Institutional System Taking Fairness and Efficiency into Consideration

Firstly, it is needed to conduct institutional design by taking achievement executors as the center, to change the institutional design idea by taking title scientific achievement holders as the center, in which benefits of achievement executors are ignored. It is needed to motivate enthusiasm of achievement executors based on achieving balance between achievement executors and achievement holders. Secondly, construct the right system for achievement executors. It is necessary to formulate "Implementation Methods for Award and Remuneration of Scientific & Technological Achievement Transformation" and detail rules for remuneration and award of scientific & technological achievements. It is needed to distinguish natures of institutions based on "Regulations for Implementation of Promoting Transformation of Scientific & Technological Achievements", to conduct classified design for transformation implementation methods.

Acknowledgements. This paper is supported by Soft Science Research Projects of Sichuan Department of Science and Technology (2018ZR0377); Soft Science Research Project of Sichuan Science and Technology Department((2018ZR0377) and Key Projects for Talent Introduction: Research on Industrial Development Strategy of Tianfu New Area in Sichuan for 2025.

References

1. Arora, A., Fosfuri, A.: Licensing the market for technology. J. Econ. Behav. Organ. **52**(2), 277–295 (1999)
2. Bing, X.: The characteristics and evaluation of interest allocation policies in scientific and technological achievements. Forum on Science & Technology in China (2014)
3. Chi, K., Wu, C.: Innovation environment, r & d linkages and technology development in hong kong. Reg. Stud. **29**(6), 533–546 (1995)
4. Deng, L., Li, Y., et al.: Effect of media promotion on industrialized transformation of agricultural scientific and technological achievementsǁcase study in guangxi academy of agricultural sciences. J. South. Agric. **42**(7), 821–824 (2011)
5. Chi, F.: Construction of the modernized economic system with the core objective of high quality development (in Chinese). Adm. Reform **12**, 4–13 (2017)
6. Gray, D., Boardman, C.: The new science and engineering management: cooperative research centers as intermediary organizations for government policies and industry strategies. J. Technol. Transf. **35**(5), 445–459 (2013)
7. Huang, H.: Construction of modernized industrial system, the highlight of the modernized economic system construction (in Chinese). Rev. Econ. Res. **63**, 3–5 (2017)
8. Jia, K.: Construction of the modernized economic system in the new era - opinions on taking supply-side structural reform as the principal line from the perspective of social principal contradiction transformation in our country (in Chinese). People's Forum Acad. Pioneer **05**, 52–54 (2018)
9. Gong, L.: Connection strategies between scientific research achievements in colleges & universities and regional economy (in Chinese). Chin. Univ. Technol. Transf. **06**, 46–47 (2010)

10. Wang, L.: Reflection on transaction policies of scientific & technological plan achievements (in Chinese). Cooperative Econ. Technol. **21**, 82–84 (2018)
11. Pang, R., Fan, Y., Yang, L.: Do scientific & technological innovations in china support economic development? (in Chinese). Quant. Tech. Econ. **10**, 37–52 (2014)
12. Werthamer, N.R., Raymond, S.U.: Technology and finance: the electronic markets. Technol. Forecast. Soc. Change **55**(1), 39–53 (1997)
13. Yong, C., Zu, S., Yang, Z.: The obstacles and countermeasures of the transformation of achievements in sci. & tech. to practical productive of higher instituion. Explor. Nat. (1995)
14. Long, Y., Li, Z.: A study on breakthroughs of the mechanism integrated with science & technology and finance (in Chinese). Sci. Manag. Res. **01**, 109–112 (2012)
15. Zhu, J.: The industry-university-research cooperation of chinese medium and small technological enterprises taking scientific papers supported by the tif between 2000–2010 for sample. Adv. Eng. Forum **6–7**, 149–155 (2012)
16. Zhu, X., Xu, Z., Xing, Z.: Provincial comparative study on scientific & technological transformation levels in our country (in Chinese). Sci. Manag. Res. **04**, 21–24 (2018)

A Study on the Subjects and Network of Technology Transfer: Evidence from Chinese Universities Patent Rights Transfer

Qiang Li[1,2(✉)] and Xin Gu[2]

[1] Department of Port and Shipping Management, Guangzhou Maritime Institute,
Guangzhou 510725, People's Republic of China
[2] Sichuan University, Chengdu 610064, People's Republic of China
2332948208@qq.com

Abstract. The patent rights transfer has increasingly become an important means to measure the ability of technology transfer of universities, but a whole recognition on the transfer of patent rights in academia is still scarce, especially about the subjects involved in these activities and the networks among different regions. Based on the patent rights transfer of granted invention between 2007 and 2015, this study attempts to reveal the situation of technology transfer of universities from transferors, transferees and the focal patents, as well as the network characteristics of technology transfer at the provincial level through combining statistics and social network analysis. Our findings indicate that, firstly, the number of transfer patents of universities has been increased year by year, but the percentage is still low. Secondly, the breadth and depth of technology transfer of universities are constrained. Thirdly, most of patents are transferred into enterprises. Finally, the characteristics of patent rights transfer in provinces demonstrates that the spatial and network distributions gradually spread from central regions to edge regions. Among provinces, the coastal developed regions such as Shanghai, Beijing, Guangdong play the role of central hub in technology transfer network, and they also play an important role in information integration.

Keywords: Technology transfer · Patent rights · Network · Chinese university

1 Introduction

Innovative technology plays a vital role in promoting economic development and is an important driving force for regional and national economic growth as well as the establishment of competitive advantage [13,16]. Technology transfer is one of the most important means for each innovation subject to acquire external innovative knowledge and technology [8,19]. Meanwhile, in order to overcome the imbalanced distribution of technical resources and ability to innovate in

© Springer Nature Switzerland AG 2020
J. Xu et al. (Eds.): ICMSEM2019 2019, AISC 1002, pp. 75–92, 2020.
https://doi.org/10.1007/978-3-030-21255-1_7

each region, make up for the shortcomings of technological development in various regions, promote balanced and coordinated development in all regions, in the Outline of the National Strategy for Innovation-Driven Development, the Chinese government has stressed the establishment of a network for technology transfer among different regions so as to realize the flow of innovative resources between the east, the central and the west in China.

As the main source of technological innovation, universities are an important part of Chinas national innovation system and regional innovation system, and play a supporting role in the development of technological innovation of enterprises. In addition, universities also undertake most national, local or enterprise scientific research projects. They have obtained rich research results, published a large number of papers and applied for many patents every year. In recent years, with the development of technological innovation, the number of patent applications, authorization, implementation and revenue of colleges and universities are on the rise [20]. At present, various universities have produced and accumulated a large number of patented technology. The transfer of patent rights has become an important way for universities to transfer innovative technology and implement patented technology.

This study contributes the theoretical and practical development of technology transfer in some ways. First, we offer more abundant materials and analysis results to a comprehensive understanding on technology transfer of Chinese universities. Second, this study is contributed to technology suppliers, technology demander and government departments who are responsible for policy formulation, by providing them valuable reference.

The remainder of this paper is organized as follows. Section 2 reviews the relevant literature. Section 3 introduces the data source and methodologies. Section 4 presents the results of the analysis. Finally in Sect. 5, we provide in-depth discussions, including some implications and the limitations of this study which is also the direction of future studies.

2   Literature Review

The existing research on universities patent rights transfer has not attracted wide attention in academic fields yet, although some scholars have gradually begun to notice that universities play an important role in the transfer of patent rights and discussed this phenomenon generally. For example, Miller [12] pointed out that the modes of patent transfer of universities can be divided into direct cooperation between universities and enterprises, patent technology incubation, intermediary platform service and high-tech enterprise entrepreneurship. These modes are divided into three phases, including universities transfer technology directly by themselves, universities transfer technology through intermediary institutions, and universities establish enterprises and entrepreneurial universities. While Wang and Li [18] divided the transfer modes of universities patents into three modes, which are local university mode, team enterprise mode and science and technology intermediary mode. Wang et al. [17] thought that the ability of patent transfer in universities is growing, but it is still very weak. Zhou work also showed that the efficiency of technology transfer in Chinese universities is low [21]. Fan et al. [2] analyzed and found that the technology transfer efficiency of universities in China is influenced by the promotion of universities themselves, the relationship between enterprises and universities, together with the development level of GDP in different regions. Lee [7] state the influence of relevant policies of universities and the government on patent transfer of universities. In addition, the scholars also discussed the influences of R&D investment, the number of R&D personnel, the scale of technology transfer institutions, enterprise characteristics, economic environment and regional environment on the patent technology transfer of universities [12,15]. However, these researches mainly focus on the mode, efficiency and influencing factors of patent transfer. There is no systematic discussion on the overall characteristics of patent rights transfer of universities, that is, the transferors, the transferees, the transferred patented technology and the characteristics of the regional network formed by technology transfer. At present, the understanding on the overall characteristics of patent rights transfer in universities is not comprehensive and deep enough. Therefore, research on patent transfer of universities still needs to be further deepened.

As an important provider of innovative technology, universities have strong technical support for small and medium-sized enterprises at the present stage of weak technological innovation ability in China. Enterprises can acquire technologies that they need for development through the transfer of patent rights of universities, so as to improve their innovation ability, enhance market competitiveness and obtain financial performance. Meanwhile, for the universities, they can realize the social and economic value of their technologies. Thus, this study provides decision-making information for strengthening the technology transfer ability of universities by offering a deeper understanding of the overall characteristics of patent technology transfer universities, individual technology transfer ability and opportunities brought by the provinces and cities where universities are located. Thus, this could be a valuable reference for universities, enterprises, and governments.

3   Method

In the following contents, the data sources, data processing methods and research methods used in this paper will be mainly described.

3.1   Subsection Heading

The sample data used in this paper are patent rights transfer of invention patents which published by National Intellectual Property Administration, PRC (CNIPA) from 2007 to 2015. According to Chinese patent law, the patent whose patent rights is transferred needs to be put on record in CNIPA, namely the assignors (transferors) and the assignees (transferees) should conclude a contract in written form for the patent to be transferred, which shall be published on the official website of the CNIPA. These announcements mainly include basic information such as the patent number, patent name, application date, transfer date, contract number, the name and address of the transferors and transferees of the transferred patent. This study uses the granted invention patents. The main reason is that invention patent is granted after patent inspectors rigorous examination on its novelty, inventiveness and practical applicability. Compared to the other two types of patents, utility models and designs, invention patents have a high technical content, which can better reflect the ability of patent transfer of universities. Furthermore, the information about the transfer of patent rights can be obtained from the official website of CNIPA. Based on the data of invention patents transferred by universities, we first exclude the records of only the name or address change of the transferor rather than the patent rights change. Then, we remove the records missing the name of applicants, and the records which applicants belong to Hong Kong, Macao and Taiwan. Finally, there are 6,514 invention patents transferred by Chinese universities from 2007 to 2015 are collected and used.

3.2   Methodology

Aiming at the characteristics of transferors, transferees and transferred patents, we analyze the sample data by using mathematical statistics method, in which, the geographical distribution of transferors and transferees is drawn in ARCGIS software. As for the characteristics of transferors and transferees formed at the provincial level, we use the social network analysis method, which has been widely applied to knowledge flow and technology transfer in academic fields [4,9]. For example, Drivas [1] use patent citation to prove that social proximity between inventor cooperative networks is the basic driving force of knowledge flow. Zhan et al. [6]. Use social network based on patent citations to describe the flow and spillover of knowledge in China. In this study, we build the patent transfer network of provinces where the transferors and the transferees are located, using the patent transfer data cleaned up and screened and 31 provinces in China. The network matrix is shown in formula (1)

$$M = \begin{bmatrix} m_{1,1} & m_{1,2} & \cdots & m_{1,30} & m_{1,31} \\ m_{2,1} & m_{2,2} & \cdots & m_{2,30} & m_{2,31} \\ \vdots & \vdots & \ddots & \vdots & \vdots \\ m_{30,1} & m_{30,2} & \cdots & m_{30,30} & m_{30,31} \\ m_{31,1} & m_{31,2} & \cdots & m_{31,30} & m_{31,31} \end{bmatrix}. \tag{1}$$

In this network of patent rights transfer, the region where the transferors and the transferees are located is the associated node, with a total of 31 network nodes. $m_{i,j}$ represents the number of transferred patents from the transferors in province i to the transferees in province j. It presents that the patent technology is transferred from the region where the supplier is located to the region where the demander is located.

The main indexes of network characteristics to be analyzed in this paper are degree centrality, betweenness centrality and structure hole [3,5,10]. Among them, degree centrality represents the direct spatial distance and close relationship between one node in the network and other nodes, which in this study refers to the spatial distance and close relationship between technology demander and technology supplier. The closer the spatial distance of two nodes is, the closer the relationship of two involvers is. That is, the node at the center of the network or with a high degree of centrality plays an important role in the network, for example, it has more advantages to gain and take more resources and development opportunities [14]. Degree centrality can be divided into two types, which are Indegree and Outdegree, indicating the frequency of input and output technology of a node, respectively. The calculation methods are as follows in formula (2) and (3), where $a_{i,j}$ represents the number of patents transferred from node i to node j , on the contrary, $a_{j,i}$ represents the number of patents transferred from node j to node i .

$$\text{OutDegree}_i = \sum_{j=1}^{n} a_{i,j} \tag{2}$$

$$\text{InDegree}_i = \sum_{j=1}^{n} a_{j,i} \tag{3}$$

Betweenness centrality represents the shortest distance between two nodes and usually acts as an intermediary. The higher the value of this index, the more this node controls other nodes connected to it, and the more opportunities there are between these nodes. The formula is as follows. $C_{(AB_i)}$ is the sum of betweenness centrality of all nodes in the network.

$$\text{Betweenness}_i = C_{(AB_i)}/(n^2 - 3n + 2) \tag{4}$$

Structural holes are used to connect bodies that are not directly connected. The higher the value of a nodes structural hole, the more other nodes this one is closed to, the stronger control it has over other nodes, and the stronger the ability of using structural holes. The formula is as follows.

$$CO_i = \sum_{j} (p_{ij+} \sum_{q,q \neq i,q \neq j} p_{iq}p_{qj})^2 \tag{5}$$

This indicator consists of two components. Direct, $P(ij)$, when a node consumes a large proportion of a networks time and energy, and indirect, $\sum_{q,q \neq i,q \neq j} p_{iq}p_{qj}$, when a node controls other nodes, who consume a large proportion of a networks time and energy.

4   Results

In this part, we present and analyze the current situation of patent rights transfer of universities from the characteristics of transferors, transferees, patents, as well as the network formed by transferors and transferees at the provincial level.

4.1   The Characteristics of Transferors

In the data of patent transfer records from CNIPA, the types of transferors include enterprises, individuals, universities, research institutions and others. Due to their differences in the innovation capability and technology transfer ability of different patentees, the number and percentage of patent transfer are quite different.

The transfer of universities patent rights accounts for a relatively small proportion. Among the subjects of patent transfer, enterprises are the leading ones. In our sample, 5,635 transferred patents are from enterprises, holding 77.09% of the total sample. It is followed by individuals. The number of patents transferred by individuals is 381, accounting for 14.75%. The number of patents transferred by universities and research institutions are 309 and 173, accounting for 4.76% and 3.11%, respectively. While the others only account for 0.29% (see Fig. 1).

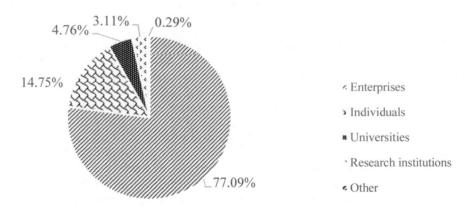

Fig. 1. The percentage of patent transfer by transferors

The number of universities transferred patents is increasing rapidly. The annual quantity distribution of universities transferred patents is increasing gradually, and the increase range is large. As is shown in Fig. 2, in 2007, there are only 18 patents are transferred by universities. This number has risen to 1,306 in 2015. In 2014, the number of patents transferred by universities peaked at 1,428. In general, from 2007 to 2015, the annual average growth rate of transferred patents of universities is 19.39%. In addition, from the perspective of the cumulative percentage growth, the annual growth rate of transferred patents of universities is the fastest from 2011 to 2015. During this period, the cumulative percentage of transferred patents of universities closes to 80%, with an annual growth rate of more than 20%.

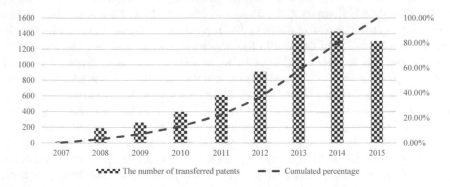

Fig. 2. The distribution of university transferred patents and its cumulated percentage

The proportion of transferred patents of comprehensive universities is high. In terms of quantity ranking of universities patents transfer, comprehensive universities are more active in patent transfer than universities of science and engineering (see Table 1). From 2007 to 2015, Shanghai Jiao Tong University has transferred 300 patents, accounting for 4.61% of the total number of patents transferred, followed by Southeast University, Tsinghua University, Peking University, Shanghai University, Wuhan University, Zhejiang University, transferring 286, 201, 161, 154, 130 and 116 patents, accounting for 4.39%, 3.09%, 2.47%, 2.36%, 2.00% and 1.78%, respectively. In the universities of science and engineering, Northwestern Polytechnical University ranks first transferring 159 patents, accounting for 3.09%, followed by Hangzhou Dianzi University and Changzhou University, which transferred 125 and 118 patents, accounting for 1.92% and 1.81%, respectively. Among the top 10 universities, comprehensive universities transferred 1,218 patents which accounts for 18.70%, while universities of science and engineering transferred 402 patents, accounting for only 6.17%.

The breadth and depth of universities patent transfer are generally insufficient. Luo et al. (2013) applied the breadth and depth of knowledge flow to measure the ability of knowledge flow of universities [11]. On the basis of existing research, this study analyzes the patent transfer abilities of different universities from two dimensions, the breadth and the depth. The breadth of patent transfer is measured as the total number of transferees when a university transfers its patents. The depth of patent transfer refers to the average number of patent transfers between universities and all transferees. Table 2 presents the top 30 universities breadth and depth of patent transfer.

The characteristics of patent transfer network for each universities are different, but the breadth and depth of patent transfer in most universities are low, which indicates that the capacity of patent technology transfer in most universities is weak. As is shown in Table 2, the largest value of the breadth is Shanghai Jiao Tong University (177), which indicates that it has transferred its patents to 177 different transferees, followed by Northwestern Polytechnical University (103), Shanghai University (85), Jiangnan University (80), Harbin Institute of Technology (59). The last one among the top 30 universities is Shanxi University, with a breadth value of 24. While the depth of patent transfer, Peking Universities is the top 1 with a value of 6.19, which means that Peking University transfers patents for each transferee on average 6 times, followed by Tsinghua University (4.9), Wuhan University of Technology (3.52), Hangzhou Dianzi University (3.21), Nanjing Normal University (3.11). The last one is Jiangsu University of Technology, with a depth value of 1.03. According to the breadth and depth of patent transfer of universities, we can classify the ability of patent technology transfer of each university. Given the average value of breadth and depth (48.96 and 2.13, respectively), Fig. 3 can be divided into four quadrants, which correspond to four types of technology transfer capabilities of universities, including low breadth and low depth, low breadth and high depth, high breadth and low depth, high breadth and high depth. In the first quadrant, universities with low breadth and low depth establish patent technology transfer relationship with fewer transferees, and the frequency of their technology transfer activities is also relatively low. This type of universities accounts for the largest proportion. However, in the fourth quadrant, universities with high breadth and high depth have established a relatively wide patent technology transfer relationship with a large number of transferees, and the frequency of their technology transfer activities is also relatively high. There are only three such universities, namely Southeast University, Wuhan University and Beijing University of Technology. This indicates that the breadth and depth of patent transfer of universities are generally insufficient in China. Universities with high breadth and low depth have established extensive patent technology transfer relations with more transferees, but not enough technology transfer activities. There are 5 universities with high breadth and low depth, including Shanghai Jiao Tong University, Northwest Polytechnic University, Shanghai University, Jiangnan University and Harbin Institute of Technology. While universities with low breadth and high depth have established patent technology transfer relationships with fewer transferees, but the times of technology transfer activities are more. There are six universities with this type, including Peking University, Tsinghua University, Wuhan University of Technology, Hangzhou Dianzi University, Nanjing University and Xian Jiaotong University. Generally speaking, the degree of technology commercialization in Chinese universities is not high.

The geographical distribution of universities with patent transfer is unbalanced. Figure 4 shows the interprovincial geographical distribution of universities as patent transferors in 2007, 2010, 2015, and the whole observation period from 2007 to 2015. The figure not only shows whether patent transfer activities of universities exist in each province, but also gives a rough impression about its proportion. These pictures illustrate that during the period from

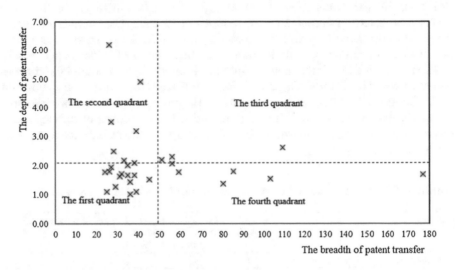

Fig. 3. The matrix of breadth and depth

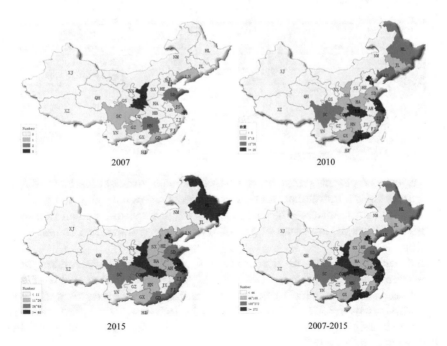

Fig. 4. The geographical distribution of transferors. *Note 7.3.*: The scale in this figure is selected according to ARCGIS natural discontinuity classification method, which can reduce the sample error

2007 to 2015, the transferors of patent rights in universities gradually shows the characteristics of decentralized distribution. In 2007, the transfer of university patents occurred only in a few regions, such as Shaanxi. In 2010 and 2015, the activities of university patent transfer mainly occurred in developed coastal areas, such as Beijing, Shanghai, Zhejiang, Guangdong and some inland areas, such as Shaanxi and Hubei. In general, from 2007 to 2015, in addition to universities in Jiangsu, Shanghai, Zhejiang, Tianjin, Beijing and other coastal developed provinces, have transferred patent rights, and a large number of universities in central and northeast China participated in the transfer of patent rights, such as Hubei and Heilongjiang.

Table 1. The top 10 universities by the number of transferred patents, 2007–2015

Universities	No. of transferred patents	Percentages of transferred patents (%)	Types of universities*
Shanghai Jiao Tong University	300	4.61	Comprehensive
Southeast University	286	4.39	Comprehensive
Tsinghua University	201	3.09	Comprehensive
Peking University	161	2.47	Comprehensive
Northwestern Polytechnical University	159	2.44	Science and engineering
Shanghai University	154	2.36	Comprehensive
Wuhan University	130	2.00	Comprehensive
Hangzhou Dianzi University	125	1.92	Science and engineering
Changzhou University	118	1.81	Science and engineering
Zhejiang University	116	1.78	Comprehensive

Note 7.1.: Collected by authors.

4.2 The Characteristics of Transferees

The subjects of patent rights transferee of Chinese universities include enterprises, universities, individuals and other subjects. By identifying the characteristics of the main transferees of patent rights of universities, we can deepen our understanding on the situation of patent technology transfer in Chinese universities.

Enterprises are the main transferees of Chinese universities patent rights transfer. The patents of Chinese universities are transferred to different types of transferees, but most of them are transferred to enterprises. Shown as Fig. 5, enterprises account for more than 86%. Parts of patents are transferred to other universities, with a proportion of 10.59%, followed by the individuals, with a proportion of 2.66%.

The scientific research and technical service industry accounts for a relatively high proportion in the enterprises transferees. The top 10 transferees of patent rights come from scientific research and technology service, manufacturing, production and supply of electric, heat power, and water, among which 6 are in

Table 2. The breadth and depth of top 30 universities, 2007–2015

No.	Universities	Breadth	Depth	No.	Universities	Breadth	Depth
1	Shanghai Jiao Tong University	177	1.69	16	Jiangsu University	36	1.44
2	Southeast University	109	2.62	17	Jiangsu University of Technology	36	1.03
3	Northwestern Polytechnical University	103	1.54	18	Suzhou University	35	1.69
4	Shanghai University	85	1.81	19	Nanjing Tech University	35	2.03
5	Jiangnan University	80	1.39	20	Xi'an Jiaotong University	33	2.18
6	Harbin Institute of Technology	59	1.78	21	Fudan University	32	1.72
7	Wuhan University	56	2.32	22	Dalian University of Technology	31	1.65
8	Zhejiang University	56	2.07	23	China Agricultural University	29	1.28
9	Beijing University of Technology	51	2.22	24	Nanjing University	28	2.50
10	Tongji University	45	1.53	25	Wuhan University of Technology	27	3.52
11	Tsinghua University	41	4.90	26	Nanjing University of Aeronautics and Astronautics	27	1.96
12	Xi'an University of Technology	39	1.10	27	Shandong University	26	1.81
13	Hangzhou Dianzi University	39	3.21	28	Peking University	26	6.19
14	Chongqing University	38	2.11	29	Guangdong University of Technology	25	1.12
15	South China University of Technology	38	1.68	30	Shanxi University	24	1.79

Note 7.2.: Top 30 means the top 30 universities in patent transfer number.

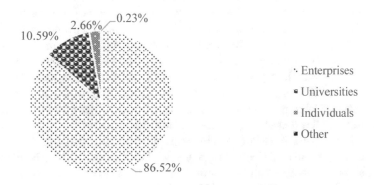

Fig. 5. The percentage of patent transfer by transferees

the industries of scientific research and technology service, with 570 transferred patents, accounting for 60%. The top 10 enterprises obtained 11.76% invention patents of the total number of our sample, in which the science and technology service enterprises accounted for 8.77%. Liyang Changda Technology Transfer Center Co., Ltd. has obtained the largest number of patents from universities, with 225, accounting for 3.45%. Followed by Changshu Nanjing Normal University Development Research Institute Co., Ltd., with 98, accounting for 1.5%. Wuxi Institute of Applied Technology of Tsinghua University obtained 84 patents, accounting for 1.29%. Zhongxin International Integrated Circuit Manufacturing (Shanghai) Co., Ltd. received only 32 patents from universities, accounting for 0.49%.[1]

The geographical distribution of transferees received patents from universities is unbalanced. Figure 6 shows the interprovincial geographical distribution of transferees in 2007, 2010, 2015, and the whole observation period from 2007 to 2015. In 2007, the transferees are mainly distributed in Shaanxi and Guangdong. In 2010 and 2015, there were more and more transferees in other provinces, indicating that the technology demander in each province introduced patented technology from universities more and more frequently. By the end of 2015, the patented technology of universities has been absorbed by most developed coastal regions (such as Guangdong, Zhejiang, Shandong, Shanghai, etc.) and some central and western regions (such as Sichuan, Hebei, etc.).

4.3   The Characteristics of Tnterprovincial Network between Patent Transferors and Transferees

The following part is an analysis of the characteristics of patent rights transfer of universities in China from the perspective of the overall interprovincial network. In order to further reveal the evolution of interprovincial network of patent rights transfer of universities in China, the networks visualization of patent transfer between provinces and cities in each year from 2007 to 2015 are carried out. The size of the line is used to indicate the patent transfer relationship and the number of transferred patents between two provinces, and the arrow is used to indicate the direction of universities patent transfer. The evolution of interprovincial network of universities patent rights transfer. The interprovincial network of patent transfer of Chinese universities has undergone a gradual evolution from simple to complex, from part to whole, and from province to other province. As is shown in Fig. 7, we selected the network of 2007, 2010, 2015 and all years for analysis. In 2007, only four links were connected to the university patent transfer network, with Shandong, Sichuan, Guangxi, Guizhou, Hunan, Guangdong and Zhejiang participating. It shows that the number of patent transfers between universities and the main subjects in other provinces in this year is small, and most of the patent transfer of universities takes place within the provinces. Subsequently, the number of connections in the patent transfer network of universities gradually increased to 65 in 2010 and 157 in 2015. It means that more and more

---

[1] The table is not listed in the context because of limited space.

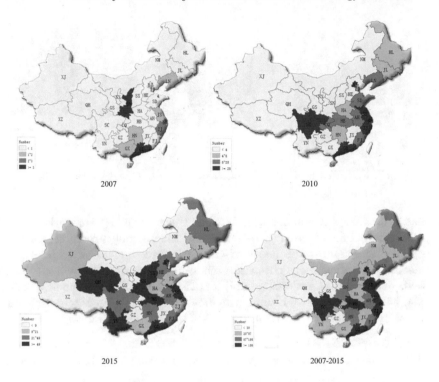

Fig. 6. The geographical distribution of transferees

universities in various provinces participate in patent transfer activities, or universities transfer more and more patents to other provinces, or more and more subjects introduced patented technologies from universities in other province. The provinces that participate in the network of university patent transfer have changed from a few scattered provinces to almost all of them. In general, from 2007 to 2015, there were 335 links in the patent transfer network of universities. The provinces with active patent transfer in the network were mainly concentrated in the economically developed eastern regions such as the Pearl River delta, the Yangtze River delta and the Beijing-Tianjin-Hebei, where universities mainly exported patents. While universities in Ningxia, Tibet and Gansu participate in the patent transfer network with absorbing patents.

The outdegree and indegree of interprovincial patent transfer network of universities. In order to reveal in more detail the number of transferred patents in interprovincial patent transfer network, the output and input number of patent transfer in each province are represented by the outdegree and indegree. Table 3 shows that Jiangsu, Beijing, Shanghai, Hubei, Zhejiang, Shaanxi and Guangdong are all active in patent output and input. However, the number of patents exported by Jiangsu is more than the number of patents imported, while the number of patents imported by Beijing, Shanghai, Hubei, Zhejiang, Shaanxi and Guangdong is more than the number of patents exported. One of

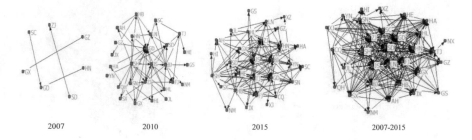

Fig. 7. Network distribution of patent transfer of Chinese universities

the reason for this phenomenon may be that there are many enterprises in Beijing, Shanghai, Hubei, Zhejiang, Shaanxi and Guangdong provinces, and local universities cannot meet the needs of enterprises in innovation technology, so they need to introduce more patented technology from other provinces to meet the needs of enterprise development and innovation. In addition, Inner Mongolia, Ningxia, Tibet, Qinghai and Hainan provinces have almost no patent technology export, all rely on other provinces to provide innovative technology.

Table 3. The outdegree and indegree of interprovincial patent transfer network

Provinces	Outdegree	Indegree	Provinces	Outdegree	Indegree	Provinces	Outdegree	Indegree
Jiangsu	1507	2233	Shandong	170	218	Gansu	24	13
Beijing	816	490	Chongqing	128	74	Yunnan	15	35
Shanghai	736	484	Hunan	108	87	Guizhou	12	17
Hubei	451	313	Shanxi	85	90	Xinjiang	12	25
Zhejiang	400	364	Henan	80	80	Inner Mongolia	9	30
Shaanxi	388	104	Anhui	70	103	Ningxia	1	2
Guangdong	272	583	Fujian	64	119	Tibet	0	4
Heilongjiang	258	125	Jilin	53	54	Qinghai	0	8
Sichuan	198	186	Hebei	46	84	Hainan	0	16
Liaoning	190	174	Guangxi	44	59			
Tianjin	181	144	Jiangxi	41	41			

The betweenness centrality of interprovincial patent transfer network of universities. Various provinces play different mediating roles in interprovincial patent transfer network. Figure 8 shows the characteristics of betweenness centrality of patent transfer network. Among them, Beijing, Jiangsu, Tianjin, Guangdong and Shanghai are among the top 5 provinces in terms of betweenness centrality, especially Beijing and Jiangsu. It indicates that these regions have the highest status in the interprovincial network of patent transfer, more connections with other regions, shorter routes and frequent technology transfer. Therefore, strong technology transfer capacity plays an indispensable and important role in obtaining resources, information and development opportunities, and can be the intermediary by providing access to technology for both parties with poor and asymmetric information. There is a lower value of betweenness centrality for Hunan, Hubei, Guangxi, Heilongjiang, and Sichuan. Whilst for Xinjiang, Guizhou, Yunnan, Gansu, Qinghai, Tibet, and Ningxia, the value of betweenness centrality is almost zero. It indicates that the central and western regions play a weak intermediary role in the interprovincial network of patent transfer, and their technological development is mainly driven by other provinces and cities.

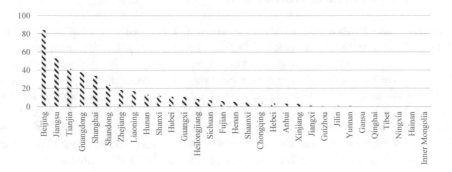

Fig. 8. The betweenness centrality of patent transfer network

The structural holes of interprovincial patent transfer network of universities. There are great differences in the structural holes of interprovincial network between universities in different provinces. The ones with more quantity and higher quality of universities have higher values of structural holes, while the ones with fewer quantity and lower quality have lower values of structural holes. As can be seen in Fig. 9, some eastern regions and inland regions, such as Beijing, Guangdong, Shandong, Tianjin, Zhejiang, Henan, Hubei and Hunan, have the highest value of structural holes. In general, these regions have more connections with surrounding provinces, frequent technical exchanges, and

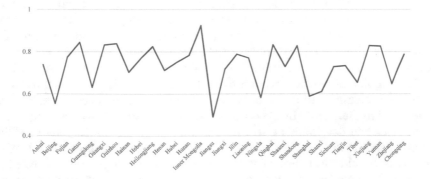

Fig. 9. The structural holes of interprovincial patent transfer network

are pioneers in the technology transfer network. In addition, they have higher independent development and initiative. On the other hand, the interprovincial network of universities in Hainan, Ningxia, Jiangxi, Tibet and other provinces has lower value structural holes, indicating that they have weak ability to use structural hole in the technology transfer network, and these regions are still in the state of technology absorption.

5   Discussion and Conclusion

The main conclusions of this study include, (1) the number of patents trans-ferred by universities is increasing rapidly every year, but their proportion in the total transferred patents is still low. Most universities have the low breadth and low depth of patent transfer, while the high breadth and low depth or low breadth and high depth account for a small proportion, let alone the bet-ter development in high breadth and high depth. (2) The main transferees of patent rights of university are enterprise. Individual and other types of trans-ferees are relatively few. (3) Patents transferred by universities gradually spread from central cities to marginal cities in geographical distribution and network distribution. (4) While exporting patents, developed coastal regions also import a large number of patents, but most of the central and western regions mainly import patents. (5) Coastal developed regions play a central role in the technol-ogy transfer network and play an important role in information integration and intermediary. Some central and western regions, such as Shaanxi, Sichuan and Hubei, are also playing an intermediary role in the patent technology transfer network. According to the research conclusion, several suggestions are put for-ward. First, in view of the weak patent technology transfer ability of universities, specific suggestions include (1) the government should perfect the legislation on the transfer of patent technology of universities, relax the policy restrictions, and create a relaxed environment for the universities to transfer their patent technologies, including tax policies, financial policies, government procurement policies and other policies. (2) Universities should change the traditional idea that only emphasize the quantity patent transfer, rather than the quality of patent transfer. They can formulate incentives for transferring patent technol-ogy, arrange special funds to ensure the smooth development of the technology transfer, and formulate detailed technology transfer rules to protect the rights and interests of all parties involved in the transfer and effectively control the possible risks. Second, suggestions on the differences in patent transfer of uni-versities in different regions include that (1) there is a big difference in patent transfer of universities in different regions. Developed provinces in the east are far more active than those in the central and western regions, which is due to the uneven economic development and innovation ability among regions in China. This unbalanced trend cannot be eliminated in a short period of time. But the government should formulate relevant policies on technology transfer, establish regional technology transfer market, promote technology transfer between uni-versities and enterprises, and play an effective regulatory role. (2) In the central and western regions where innovation knowledge is already scarce, enterprises are encouraged to actively absorb external technologies, transform and apply them in light of local development level, and stimulate their enthusiasm for innovation and development. This study provides a preliminary analysis and discussion on the ability of patent technology transfer of Chinese universities from the basic characteristics of patent transfer and its social network. However, we do not further study the process and mechanism of technology transfer, the factors that may affect universities ability of technology transfer, etc. So there are many questions remaining to be further in-depth study.

Acknowledgements. This research is funded by National Science Foundation of China (NSFC, grant No. 71571126), and Soft Science Project of Sichuan province (grant No. 2019JDR0149).

References

1. Drivas, K., Zhen, L., Wright, B.D.: Academic patent licenses: roadblocks or signposts for nonlicensee cumulative innovation? J. Econ. Behav. Organ. **137**, 282–303 (2017)
2. Fan, B., Yu, J.: Research on regional differences and influencing factors of technology transfer efficiency in colleges and universities. Stud. Sci. Sci. **33**(12), 1805–1812 (2015)
3. Guan, J., Zhang, J., Yan, Y.: The impact of multilevel networks on innovation. Res. Policy **44**(3), 545–559 (2015)
4. Jiang, J., Goel, R.K., Zhang, X.: Knowledge flows from business method software patents: influence of firms' global social networks. J. Technol. Transf. 1–27 (2017)
5. Jie, X., Zuo, L.: Characteristics and innovation performance of enterprise collaborative innovation network: a study on the mediating effect based on knowledge absorption ability. Nankai Bus. Rev. **16**(3), 47–56 (2013)
6. Kan, Z., Sun, J.: A study of the UIR in jiangsu Province based on the perspective of social network analysis. Sci. Res. Manag. **37**(10), 33–34 (2016)
7. Lee, C., Park, G., Kang, J.: The impact of convergence between science and technology on innovation. J. Technol. Transf. **43**(2), 522–544 (2018)
8. Lemley, M.A., Feldman, R.: Patent licensing, technology transfer, and innovation. Am. Econ. Rev. **106**(5), 188–192 (2016)
9. Liu, F., Na, Z., Cong, C.: An evolutionary process of global nanotechnology collaboration: a social network analysis of patents at USPTO. Scientometrics **111**(3), 1449–1465 (2017)
10. Lu, Y., Lee, J., Kim, E.: Network characteristics matter in politics on facebook: evidence from a US. national survey. Online Inf. Rev. **42**(3), 372–386 (2018)
11. Luo L (2013) Knowledge flow capability of "985 universitie" based on patent license network. Chin. J. Manag. (in Chinese)
12. Miller, K., Mcadam, R., Mcadam, M.: A systematic literature review of university technology transfer from a quadruple helix perspective: toward a research agenda. R & D Manag. **48**(1), 7–24 (2018)
13. Nordensvard, J., Yuan, Z., Xiao, Z.: Innovation core, innovation semi-periphery and technology transfer: the case of wind energy patents. Energy Policy **120**(1), 213–227 (2018)

14. Savin, I., Egbetokun, A.: Emergence of innovation networks from R&D cooperation with endogenous absorptive capacity. J. Econ. Dyn. Control **64**, 82–103 (2016)
15. Silva, D.R.D.M., Furtado, A.T., Vonortas, N.S.: University-industry R&D cooperation in Brazil: a sectoral approach. J. Technol. Transf. **43**(2), 285–315 (2018)
16. Thompson, N.C., Ziedonis, A.A., Mowery, D.C.: University licensing and the flow of scientific knowledge. Res. Policy **47**(6), 1060–1069 (2018)
17. Wang, J.: A comparative study on patent conversion ability in Chinese universities-a case study of "project 985" university. Chin. Univ. Technol. Transf. **9**, 55–57 (2015)
18. Wang, J., Li, Y.: Analysis on the effective mode of university patent transformation based on patent transfer. Forum Sci. Technol. China **4**, 70–75 (2016)
19. Wang, Y., Liying, J., Chen, J., Lu, Z.: Technology licensing in china. Sci. Public Policy **42**(3), 293–299 (2015)
20. Zhang, G., Duan, H., Zhou, J.: Investigating determinants of inter-regional technology transfer in China: a network analysis with provincial patent data. Rev. Manag. Sci. **10**(2), 345–364 (2016)
21. Zhou, Y.: Building global products and competing in innovation: the role of Chinese university spin-outs and required innovation capabilities. Int. J. Technol. Manag. **2**, 180–209 (2014)

The Effect of Stock Liquidity on Default Risk: An Empirical Study of China's Capital Market

Kun Li, Qinqin Yan[✉], and Yu Lei

School of Business, Sichuan University, Chengdu 610065, People's Republic of China
1615991270@qq.com

Abstract. This study examines a sample of listed firms in the Chinese equity market and explores the relationship between stock liquidity and firm default risk. The empirical results of this study are as follows: (1) There is a significant negative correlation between the stock liquidity and default risk of listed companies. Additionally, for companies with a high risk of default, the effect of reducing the risk of default caused by stock liquidity is more pronounced. (2) The implementation of margin trading policies significantly improves stock liquidity and reduces firm default risk.

Keywords: Stock liquidity · Corporate debt default · Margin trading policy · PSM-DID · Capital markets

1 Introduction

Default risk is a matter of great concern for company managers, creditors, and investors. Debt default is harmful to business. It leads to supply chain disruption and the loss of valued employees, which greatly affects a company's productivity, induces legal and administrative costs, and is detrimental to the interests of customers. Therefore, listed companies should limit their default risk to within a reasonable range. The academic community has explored various ways to reduce debt default, and the relationship between liquidity and default risk has received much attention [3,14]. The current research on the relationships between stock liquidity and debt default focus on three areas.

First, the liquidity of a company's stock might affect the cost of debt financing. The most important function of the capital market is the efficient allocation of resources. The ability of listed companies to obtain financing from the capital market is closely related to their stock liquidity. Butler and Wan show that listed companies with higher stock liquidity could secure better debt issuance opportunities and could also significantly reduce direct issuance costs [6]. For stocks with low liquidity, listed companies face higher financing costs in the capital market [9,11,16]. However, some researchers argue that high liquidity encourages unsuspecting investors to manipulate stock prices and push prices

© Springer Nature Switzerland AG 2020
J. Xu et al. (Eds.): ICMSEM2019 2019, AISC 1002, pp. 93–104, 2020.
https://doi.org/10.1007/978-3-030-21255-1_8

down through short selling. Researchers suggest that managers may mistakenly interpret artificially depressed stock prices as market investor disapproval of the firm's valuation thereby eliminating potential investment projects, which leads to a reduction in future cash flows and an increased risk of default [10,17].

Second, high stock liquidity might be a reflection of investors' heterogeneous beliefs. Stocks with higher liquidity can inject more market information into their stock prices. Therefore, the stocks have more information content than stocks with lower liquidity; that is, stocks with high liquidity contain more company-level information. The additional share price information reduces the total risk of the entire market, weakens the noise component of the price, and conveys more valuable information to management [13]. Changes in stock liquidity can affect the behavior of investors in the market and, ultimately, be reflected in the stock price [1,7,10].

Finally, better-governed firms are strongly associated with a lower level of default risk, and the firm's stock liquidity is what determines this association. Compared with smaller shareholders, large shareholders benefit more from higher stock liquidity because this may reduce transaction costs when trading. To increase stock liquidity, large shareholders are incentivized to encourage listed firms to improve corporate governance [2,5,15]. In terms of company value, Subrahmanyam and Titman show that stock liquidity affects a company's default risk as well as its future cash flow through its impact on corporate value. Stock prices have a strong feedback effect on corporate-investors' decision-making, which boosts the company's stock liquidity [18].

At present, domestic research on liquidity mainly focuses on the relationship between liquidity and corporate characteristics such as capital structure and corporate value. Existing literature on the relationship between liquidity and default risk is largely based on the perspective of banks and bonds, and the relationship is rarely studied in terms of listed companies. Using data on the Chinese market, this study reveals the relationship between stock liquidity and default risk and serves as a practical guide for business managers, policy makers, market investors, and market makers.

2   Research Design

Based on previous theoretical analyses, we suggest that an increase in a company's stock liquidity can reduce its default risk. With this premise, we propose the basic hypothesis of this study: stock liquidity is negatively correlated with debt default risk.

2.1   Measurement of Default Risk

According to debt default theory, option pricing model theory, and the expected default frequency (EDF) model, this study uses total debt, stock market value, stock price volatility, and other indicators to calculate the company's EDF value and measure the company's default risk.

$$DD_{i,j} = \left(\log \left(\frac{Equity_{i,t} + Debt_{i,t}}{Debt_{i,t}} \right) + \left(r_{i,t-1} - \frac{\sigma^2_{vi,t}}{2} \right) * T_{i,t} \right) / \left(\sigma_{vi,t} * \sqrt{T_{i,t}} \right)$$

(1)

where DD is the default distance of company I in time t, equity is the stock market value, debt is the total debt value, r is the yield of stock I in period t-1, e^2 is the total volatility of stock I in period t, and T is the time span. The selected quarter of this study is a basic time span, T = 0.25.

$$\sigma_{vi,t} = \frac{Equity_{i,t}}{Equity_{i,t} + Debt_{i,t}} * \sigma_{Ei,t} + \frac{Debt_{i,t}}{Equity_{i,t} + Debt_{i,t}} * (0.05 + 0.25 * \sigma_{Ei,t})$$

(2)

Total volatility is the ratio of the current market value to total assets multiplied by the stock volatility plus the ratio of the liabilities market value to total assets multiplied by the relative value. According to the simplification of the Merton model by Jonathan Brogaard, this coefficient is (0. 05. 25 * stock volatility). Finally, the default risk EDF is equal to:

$$EDF_{i,t} = N(-DD_{i,t})$$

(3)

The EDF is the default risk of a specific company. The negative default distance ensures that the EDF value obtained is positively changing along with the default risk of the company. The larger the EDF, the greater the default risk and vice versa. Table 1 shows other variables:

Table 1. Default risk

Variable	Symbol	Definition
Circulation market value	Equity	Share price * number of shares in circulation
Liabilities	Debt	Long-term liabilities + current liabilities
Volatility	Σ	Standard deviation of stock price fluctuation
Rate of return (excluding dividends)	R	Rate of return
Default distance	DD	Calculated
Default risk	EDF	Standard normal distribution Value of negative default distance

2.2 Measurement of Stock Liquidity

To measure the liquidity of stocks objectively and comprehensively, we select five indicators for regression analysis: Amihud illiquidity indicators, Amivest liquidity indicators, relative spreads, turnover ratios, and zero-return days (Table 2 shows the specific definitions). These indicators cover the dimensions of the transaction volume, price method, and volume-price combination methods to ensure the reliability and stability of our research conclusions.

Table 2. Definition of liquidity measurement

Variable	Symbol	Definition
Relative bid-ask spread	RS	(Lowest price to highest price)/average price
Amivest	Amivest	T-time turnover/Cumulative value of price change rate in T period
Amihud	Amihud	T period yield/T period trading volume
Turnover rate	hsl	Trading volume/current market value
Zero	Zero	The ratio of zero return days to trading days

2.3 Control Variables

According to the analysis of the factors affecting liquidity and default risk, and based on the multi-factor model of Bharath and Shumway [4], we take default risk as the explained variable and corporate stock liquidity as the explanatory variable in the empirical model of default risk influencing factors. Market capitalization, corporate liabilities, volatility, ROA, and excess returns are used as control variables. Table 3 shows the specific definitions:

Table 3. Control variables

Variable	Symbol	Definition
Circulation market value	Equity	Share price * number of shares in circulation
Liabilities	Debt	Long-term liabilities + current liabilities
Return on total assets	ROA	Revenue/total assets
Volatility	$\sum$	Standard deviation of stock price fluctuations
Excess returns	Er	Stock returns-index returns
Quarterly rate of return	Qtrret	Rate of return per quarter

3   Empirical Results and Analysis

3.1   Data Sources and Processing

We select a sample of A-share listed companies that were listed before 2008 and have not been delisted since 2008. We process the sample as follows:

(1) Owing to the unique structure of financial and insurance companies' assets and liabilities, most of the financial and insurance industry data are excluded.

(2) We exclude companies marked by ST from 2007Q1 to 2017Q3 because of their financial situation (the stock price limits of such companies are different from those of general company stocks).

(3) We exclude samples with missing data from 2007 to 2017.

(4) We exclude companies with abnormal data and large differences in duplicate values.

(5) To avoid the interference of outliers, we winsorized the observations of the main continuous variables distributed on the 1% and 99% quantiles.

We focus on five indicators to measure the liquidity of stocks. The original data are from the WIND and CSMAR databases. We collect specific data, financial indicators, corporate governance, and other data from the CSMAR database. We obtain the yields, volatility, and liabilities from the WIND database. Some of the missing data come from Ruisi and other financial databases. We use STATA14 and SPASS software to analyze and process the data.

After processing the samples as described above, we obtain 193 listed companies as balanced panel data between 2007 and 2017. To process the specific data, we first sort the data of each sub-indicator and then use STATA to merge and eliminate the missing and duplicate values. We reduce the sample size from the original 10,987 observations to 6,300 sample observation values.

3.2   Descriptive Statistics

This sample includes the data from the first quarter of 2007 to the third quarter of 2017, spanning almost 10 years. After the above data processing and analysis, 19 companies and approximately 8,300 data samples remain. We select 13 dependent variables, an independent variable, and a control variable, and we set 2 dummy variables. Table 4 provides descriptive statistics for each variable.

In the 10 years after 2007, the average annual default risk (EDF) per company is 0.28, the median and 75% quantile are 0.26 and 0.43, and the average volatility is 0.2. These results indicate that some enterprises face higher default risk, so the median of the overall sample is greater than the sample mean and equals 75% of the quantile. In terms of stock liquidity measurement, the maximum value of the Amihud index is 11.9, the minimum value is 0.68, and the difference is approximately 18 times. The minimum value of the Amivest liquidity index is 18, the standard deviation is 1.15, and the maximum value is 27, which is 1.5 times. To a certain extent, some listed companies in the market have poor liquidity that needs to be improved. The zero index has a mean value of 0.03, a minimum value of 0, and a maximum value of 0.31. The zero indicator reflects

Table 4. Descriptive statistics

Variable	N	mean	sd	min	p50	p75	max
Edf	8098	0.28	0.20	0.00	0.26	0.43	0.87
Amihud	8299	5.46	1.22	0.68	5.46	6.26	11.9
RS	8299	−0.04	0.02	−0.11	−0.04	−0.03	0.00
Zero	8299	0.03	0.03	0.00	0.02	0.03	0.31
Amivest	8296	22.41	1.15	18.21	22.41	23.13	27.39
Hsl	8299	4.50	0.01	3.80	4.65	5.24	6.95
Ln (equity)	8106	22.53	1.20	19.13	22.42	23.23	27.44
Ln (debt)	8098	21.64	2.00	11.26	21.55	22.56	29.33
Stv	8299	2.85	1.16	0.52	2.64	3.56	8.30
Er	8299	0.04	0.22	−0.63	0.01	0.13	2.39
ROA	8299	0.03	0.04	−0.28	0.02	0.04	0.41
Qtrret	8299	0.06	0.28	−0.64	0.01	0.18	2.85
dd	8098	0.90	1.13	−1.15	0.63	1.29	12.01

the proportion of zero-yield days, and the selected company generates stock price fluctuations on most trading days. This zero indicator shows that in the sample, the minimum number of zero-income days in a quarter for the company is 0 days, the maximum number is 18.6, the median is 1.2 days, and the P75 value is 1.8 days. According to the zero-yield days, most companies have fewer days with zero earnings, which is in line with the basic facts of the capital market.

The control variables reflect the industry characteristics and the company characteristics of listed companies. Table 4 shows that the average volatility of the listed companies selected in this study is 2.85, and the ROA is 3%, which indicates that the overall profitability of the Chinese market needs to be improved. For excess return (Er), the average value is 0.03, and the maximum value is 2.93, which preliminarily indicates that the return rate of Chinese stocks is greatly affected by the market.

In order to explore the correlation between variables, we use the correlation coefficient to determine the correlation coefficient between the selected variables, which helps explain the mutual relationship. Table 5 presents the details:

The various indicators are measured differently depending on the nature of the liquidity at the time of creation. Therefore, some of the indicators we selected are measured by price, partly by measuring the volume of trading and partly by the combination of volume and price, so that the empirical results are more comprehensively reflected. Panel A shows the correlation coefficient between various liquidity metrics. The negative correlation between the Amihud index and the Amivest index reaches 0.97, indicating a strong correlation between the two indexes. Simultaneous placement of the Amihud indicator with the Amivest indicator results in a high degree of collinearity, which will be analyzed separately

Table 5. Correlation coefficient

Panel A

	Amihud	RS	Zero	Amivest	hsl
Amihud	1				
RS	−0.1834*	1			
Zero	0.0598*	0.4359*	1		
Amivest	−0.9736*	0.1172*	−0.0936*	1	
Hsl	−0.0673*	−0.6055*	−0.3045*	0.1091*	1

Panel B

	edf	Amihud	equity1	debet1	stv	er	roa
Edf	1						
Amihud	0.1404*	1					
Ln (equity)	−0.1300*	−0.7578*	1				
Ln (debt)	0.2803*	−0.4829*	0.6432*	1			
Stv	0.7848*	0.1735*	−0.2202*	−0.1971*	1		
Er	0.1519*	−0.0003	−0.0869*	−0.0385*	0.2141*	1	
Roa	−0.2358*	−0.0711*	0.1164*	−0.2030*	−0.0399*	−0.0059	1
qtrret	−0.0001	−0.1373*	−0.1168*	−0.0284*	0.0680*	0.8000*	0.0006

in subsequent studies. The Amihud indicator reflects illiquidity, and the coefficient symbol indicates that the zero indicator is also a non-liquidity indicator. The other three indicators are liquidity indicators, which have a low correlation coefficient, indicating that there is no obvious collinearity between these indicators.

Panel B in Table 5 shows the correlation coefficient between the other control variables. The correlation coefficient between stock market value and debt is 0.643. The correlation coefficient between stock market value and the Amihud illiquidity index is negative 0.75. The quarterly rate of return and the excess return rate correlation coefficient is 0.8, the correlation coefficient between the indicators is larger, indicating a strong correlation between the variables, which conforms to the actual capital market. In addition, the correlation coefficient between other indicators is low, indicating that the independence between the variables is high, which helps to improve the accuracy of subsequent regression analysis results.

From the statistical results, we draw a preliminary conclusion: the correlation coefficient between the default risk and the illiquidity index is not zero at the 1% confidence level, and its value is 0.14, indicating that a large stock liquidity means a small default risk.

3.3   Multivariate Analysis

First, we perform linear regression according to the multi-factor model of Bharath and Shumway[4]. We select five control variables: Ln(Equity), Ln(Debt), 1/E, Excess Return, and ROE. The measured liquidity indicators (collectively referred to as liquidity indicators) are Amihud indicators, Amivest indicators, turnover, relative spreads, and zero indicators, with default risk (EDF) as a dependent variable for regression. To reduce the interference caused by endogenous problems, we lag the independent variable and the control variable by one period. Error is the variable error term. The basic regression model is as follows:

$$EDF_{i,t} = \alpha + \beta * Liquidity_{i,t-1} + \gamma 1 * Ln(Equity)_{i,t-1} + \gamma 2 * Ln(Debt)_{i,t-1}$$
$$+\gamma 3 * \frac{1}{\sigma_{Ei,t-1}} + \gamma 4 * ExcessReturn_i, t-1 + \gamma 5 * \frac{Income}{Assets_i}, t-1 + \theta Firm + \psi Year$$
$$+Error_i, t$$

$$(4)$$

In the model, Ln(Equity) is the natural logarithm of stock market capitalization, Ln(Debt) is the natural logarithm of total debt, 1/E is the reciprocal of the standard deviation of stock fluctuations, Excess Return is the excess return of the stock(Er), Income/Assets is the ROA indicator, and Table 6 shows the regression results.

Table 6 presents the results of the mixed least squares regression using the fixed effect model. The first column shows the regression result using the Amihud index with a coefficient of 0.012, indicating that an increase in each unit of the Amihud index will correspondingly lead to an improvement in the EDF index value per unit of 0.012. The second column is the regression result obtained using the turnover rate to measure liquidity. It is significantly positive, which means that the increase in turnover rate will lead to an increase in default risk. Owing to the imperfections in the construction of various indicators, the turnover rate indicator considers the impact of the number of transactions, and the results have certain limitations. The comprehensive consideration of multiple indicators leads to the best results. The third column shows the regression result of the Amivest liquidity measure. We find that the regression coefficient is negative 0.006, which is significant at the 1% confidence level, indicating that with an increase of one unit in the stock liquidity of the Amivest index value, the default risk is reduced by 0.006 units. Both have a negative correlation. The fourth column shows the regression result of relative bid-ask spread, which is significantly negatively correlated at the 1% level. The smaller the relative bid-ask spread, the lower the liquidity and the higher the risk of default. The fifth column shows the regression result of zero return days, and its regression coefficient is 0.081. The zero index also verifies the conclusion that stock liquidity is negatively correlated with default risk. Combined with a variety of indicators, the regression results initially verify the basic assumptions of this study; that is, stock liquidity is negatively correlated with debt default risk.

Table 6. Stock liquidity and debt default

	Model 1	Model 2	Model 3	Model 4	Model 5
	EDF	EDF	EDF	EDF	EDF
Amihud	0.012***				
	(0.00)				
Ln (equity)	−0.042*9**	−0.053***	−0.048***	−0.053***	−0.054***
	(0.00)	(0.00)	(0.00)	(0.00)	(0.00)
Ln (debt)	0.040***	0.042***	0.040***	0.047***	0.067***
	(0.00)	(0.00)	(0.00)	(0.00)	(0.00)
1/stv	−0.797***	−0.764***	−0.797***	−0.675***	−0.769***
	(0.01)	(0.01)	(0.01)	(0.01)	(0.01)
Er1	−0.019***	−0.035***	−0.020***	−0.043***	−0.030***
	(0.01)	(0.01)	(0.01)	(0.01)	(0.01)
ROA	−0.436***	−0.397***	−0.431***	−0.389***	−0.339***
	(0.04)	(0.04)	(0.04)	(0.04)	(0.03)
Hsl		0.000***			0.000***
		(0.00)			0
Amivest			−0.006***		
			(0.00)		
RS				−2.394***	
				(0.10)	
Zero					0.081*
					(0.05)
Cons	0.646***	0.888***	0.978***	0.649***	0.349***
	(0.08)	(0.05)	(0.05)	(0.05)	(0.02)
N	8106	8106	8103	8106	8106
r	0.666	0.669	0.664	0.688	0.732
r2- a	0.657	0.661	0.656	0.68	0.732
F	2625	2668	2609	2901	3162
Standard	*p<0.1, **p<0.05, ***p<0.01				

4 Robustness Test

The above results show that the company's default risk has a significant negative correlation with its stock liquidity. However, it is difficult to identify whether higher stock liquidity leads to lower default risk or vice versa [8]. To overcome the reverse causal problem, Galai lagged the independent variable by one period, but the results showed that the problem still existed.

Using the PSM-DID model to solve the above-mentioned reverse causal problems, we introduce the implementation of margin trading policy as a natural

Table 7. Difference-in-difference method

Variable	After matching
	EDF
After	−0.077***
	(0.01)
Treat	0.037***
	(0.01)
After*treat	−0.062***
	(0.01)
Ln (equity)	−0.049***
	(0.00)
Ln (debet)	0.055***
	(0.00)
1/stv	−0.786***
	(0.02)
Er	−0.078***
	(0.01)
ROA	0.072
	(0.06)
Cons	0.599***
	(0.05)
N	1455
F	1300
Legend:	*p<0.1; **p<0.055; ***p<0.001

experiment of exogenous shocks and test the relationship between stock liquidity and default risk when the external shock from the implementation of the margin trading policy occurs [12]. We divide the sample companies into three groups according to the degree of change in liquidity before and after the implementation of the 2010 margin trading policy. We take the group with the highest degree of change as the experimental group and the group with the lowest degree of change as the control group. Then, we focus on the liquidity indicators including Amihud, Amivest, and RS. In order to meet the preconditions of the DID model and the common development trend, we use the PSM model to perform propensity score matching, using the radius matching method and setting the radius value at 0.01. After matching, we use two methods to test whether the matched samples met the parallel trend hypothesis: the difference analysis before and after the matching and PSM propensity score description. Finally, we regress the matched samples using the DID analysis method. The basic model is as follows:

$$EDF_{i,t} = \alpha + \beta 1 * Treatment_i * After_t + \beta 2 * Treatment_i + \beta 3 * After_i + \gamma * Control_{i,t} + Error_{i,t}$$

$$(5)$$

Table 7 shows the results. The value of difference-in-difference is minus 0.07, which also verifies the basic assumption of this study that increased liquidity will reduce the risk of default.

5 Main Research Conclusions

This study explores the relationship between the stock liquidity and default risk of A-share listed companies in China. The empirical analysis shows that stock liquidity has a negative correlation with listed company default risk. In order to overcome the reverse causality and endogeneity problems, we apply the double difference method to analyze the relationship between liquidity and default risk, taking margin trading policy as an exogenous event. The results consistently show that higher stock liquidity reduces the risk of default. This study also finds that the effect is more pronounced for enterprises with higher default risk.

References

1. Admati, A.R., Pfleiderer, P.: A theory of intraday patterns: volume and price variability. Rev. Financ. Stud. **1**(1), 3–40 (1988)
2. Ali, S., Liu, B., Su, J.J.: Does corporate governance quality affect default risk? The role of growth opportunities and stock liquidity. Int. Rev. Econ. Financ. **58**(S1059056017307), 554 (2018)
3. Amihud, Y., Mendelson, H.: Asset pricing and the bid-ask spread. J. Financ. Econ. **17**(2), 223–249 (2006)
4. Bharath, S.T., Shumway, T.: Forecasting default with the merton distance to default model. Rev. Financ. Stud. **21**(3), 1339–1369 (2008)
5. Brockman, P., Chung, D.Y.: Investor protection and firm liquidity. J. Financ. **58**(2), 921–938 (2010)
6. Butler, A.W., Wan, H.: Stock market liquidity and the long-run stock performance of debt issuers. Rev. Financ. Stud. **23**(11), 3966–3995 (2010)
7. Chen, Y., Eaton, G.W., Paye, B.S.: Micro(structure) Before Macro? The Predictive Power of Aggregate Illiquidity for Stock Returns and Economic Activity. Social Science Electronic Publishing (2018)
8. Copeland, T.E., Dan, G.: Information effects on the bid-ask spread. J. Financ. **38**(5), 1457–1469 (1983)
9. Denis, D.J., Mihov, V.T.: The choice among bank debt, non-bank private debt, and public debt: evidence from new corporate borrowings. J. Financ. Econ. **70**(1), 3–28 (2002)
10. Dow, J., Goldstein, I., Guembel, A.: Incentives for information production in markets where prices affect real investment. In: Meeting Papers (2008)
11. Drehmann, M.N.K.: Funding liquidity risk: definition and measurement. J. Bank. Financ. **37**(7), 2173–2182 (2013)
12. Fang, V.W., Tian, X., Tice, S.: Does stock liquidity enhance or impede firm innovation? Soc. Sci. Electron. Publ. **69**(5), 2085–2125 (2014)

13. Kang, Q., Liu, Q.: Stock market information production and executive incentives. J. Corp. Financ. **14**(4), 484–498 (2008)
14. Lipson, M.L.: Market microstructure and corporate finance. J. Corp. Financ. **9**(4), 377–384 (2003)
15. Nadarajah, S., Ali, S., Liu, B., Haung, A.: Stock liquidity, corporate governance and leverage: new panel evidence. Pac.-Basin Financ. J. S0927538X16302669 (2016)
16. Odders-White, E.R., Ready, M.J.: Credit ratings and stock liquidity. Rev. Financ. Stud. **19**(1), 119–157 (2006)
17. Ozdenoren, E., Yuan, K.Z.: Feedback effects and asset prices. J. Financ. **63**(4), 1939–1975 (2008)
18. Subrahmanyam, A., Titman, S.: Feedback from stock prices to cash flows. J. Financ. **56**(6), 2389–2413 (2010)

Research on the Motivation of Government Intervention in Executive Compensation of State-Owned Enterprises-An Empirical Study Based on the State-Owned Listed Enterprises

Sheng Ma[1], Shuang Li[1(✉)], ChuanBo Chen[1], and Rui Wang[2]

[1] Business School, Chengdu University, Chengdu 610106, People's Republic of China
438067036@qq.com
[2] Business School, Sichuan Normal University, Chengdu 610066, People's Republic of China

Abstract. This paper makes an in-depth analysis of the characteristics, the motivation and the intervention effect of government intervention in executive compensation in state-owned enterprises (SOEs). The research shows that the local government always makes and implements differentiated compensation regulation policies according to the marketization degree, the income gap, the financial deficit level, the unemployment rate and the GDP growth rate. However, from the perspective of intervention effect, such government intervention in executive compensation weakens the compensation performance sensitivity of SOEs executives, effects the future performance of Chinese SOEs, and it has a greater negative impact on competitive SOEs. The conclusion indicates that the government should improve the pertinence and validity of government behavior and try to establish differentiated executive's compensation incentive mechanism that matches the governance behavior of modern SOEs with Chinese characteristics.

Keywords: Government intervention · SOEs · Executive compensation · Compensation regulation · Motivation

1 Introduction

In the 19th National Congress of the Communist Party of China, it pointed out it is significant to deepen state-owned enterprises (SOEs) reform and cultivate world-class enterprises with global competitiveness. The high quality development of SOEs is beneficial to perfect socialist market economy system, and the sound corporate governance and market-oriented management system are necessary to ensure the high quality development of SOEs. As the most important part of corporate governance, the rational executive's compensation incentive

© Springer Nature Switzerland AG 2020
J. Xu et al. (Eds.): ICMSEM2019 2019, AISC 1002, pp. 105–118, 2020.
https://doi.org/10.1007/978-3-030-21255-1_9

mechanism can help to stimulate the operating energy of SOEs, so the government regulation over executive compensation of SOEs runs through the relevant documents on the SOEs reform over the years. After the establishment of the State-owned Assets Supervision and Administration Commission of the State Council (SASAC) in 2003, the Chinese SOEs have gradually established compensation evaluation system based on performance. However, as the executive compensation continues to rise in SOEs, the central government has adopted a series of policies aimed at regulating the excessively high executive compensation in SOEs. At the same time, many scholars have joined the study of executive compensation in SOEs.

The relevant foreign researches on executive compensation and enterprise performance, show that performance-based compensation incentive mechanism make the executives to maximize enterprise performance and shareholder wealth when they pursue their own maximum compensation. Murphy, 1985; Lambert et al., 1987; Sloan, 1993; Core et al., 1999; Leone et al., 2006; Jackson et al., 2008 found that there is significant positive correlation between executive compensation and enterprise performance [4,6,7,10,11,14]. The researches on the relation between government intervention and executive compensation pointed out that redundancy burden caused by government intervention significantly reduces the compensation performance sensitivity of SOEs executives, increases the level of compensation stickiness and leads to more executive perks [13].

The Chinese research papers focus on discussing how the government intervention influences executives compensation incentive mechanism in SOEs. Xin QingQuan (2009) argued that market-oriented reform increases the sensitivity of executive compensation to enterprise operating performance, and SOEs located in the provinces with higher marketization degree have fewer perks [15]. Dai Xiu Li et al. (2013) found that the government intervention affects the validity of compensation incentive and equity incentive, and strengthens the invisible incentive such as on-the-job consumption and political promotion in local SOEs [16]. These literatures seldom refer to the motivation of government intervention in executive compensation in SOEs. Therefore, this paper makes an in-depth analysis of the characteristics, the motivation and the intervention effect of government intervention in executive compensation in SOEs. It enriches the theoretical literature of the motivation of government intervention in executive compensation and provides reference for the reform of SOEs.

2   Theoretical Analysis and Research Hypothesis

2.1   The Characteristics of Compensation Regulation in China

The characteristics of executive compensation regulation in China are significantly different from other countries because of the hidden rules derived from the special Chinese system background. Firstly, in terms of the motivation of compensation regulation, the specific nature of property right of the SOEs determines that the executive compensation cannot be fully priced according to the market, while Chinese cultural customs and social tradition emphasize collectivism, therefore, the government shows strong fairness preferences in the income distribution of SOEs during transformation, that is, the government has the mental of "not fearing deficiency but equalization". Meanwhile, as large shareholder of SOEs, the government always forces them to undertake multiple social responsibilities such as maintaining social fair, extending local employment, and alleviating financial deficits, correspondingly the executive compensation regulation stems from the government regulation under the background of social goal pluralism. Secondly, as for the feasibility of compensation regulation, the SOEs bare heavy policy burden [12], so the executives of SOEs should possess excellent operating and management ability and they are expected to achieve diversified social goals like officials too. The identity of "not only officials but also businessman" of SOE executives reflects that there exist government intervention in the manager market of SOEs, which makes it possible for the government to intervene in the executives compensation in SOEs. Furthermore, on the methods of compensation regulation, there are increasingly complicated trends in terms of control level, competition degree, controlling shareholding and operation target of SOEs, which makes the state-owned assets supervision department in a position of information inferiority. It is hard for government to evaluate the SOEs managers with low-cost, so the government is more willing to adopt one-size-fits-all compensation regulation measures to manage the executive compensation of SOEs. Judging from the policies on executive incentives of SOEs over the years, it is found that linking the executive compensation to the ordinary employees' compensation is the main way for the government to regulate the executive compensation of SOEs [2].

2.2   Motivation Analysis of Compensation Regulation of SOEs Executives

The State Council promulgated the "Decision on Implementing Tax-sharing Financial Management System" in 1993, which marked the formation of the intergovernmental financial decentralization system. From then on, the local government began to manage most of the SOEs, and they assumed more political responsibility for economic development, full employment, social endowment, fairness and stability, which made the relation between local government and enterprises more complex. On one hand, as an important mean to develop economy, the marketization improvement is conducive to the optimal distribution of social resources and the promotion of enterprise performance. Endowing the enterprises with self-operation to develop the local economy, the local government officials can win in the "political tournament", therefore, the local government always support the local marketization reform.On the other hand, the marketization improvement also reduce the local governments' influence on local SOEs, which is harmful to the realization of political goals in SOEs. Based on the two reasons above, the local government will balance the marketization degree according to their own interests, resulting that the marketization degree varies from region to region, and the higher the marketization degree, the more intense the competition between the enterprises. To make SOEs obtain winningly in the fierce market competition, the local government always intervene less in SOEs including executive compensation affairs, so the executives of SOEs will face weaker compensation regulation in the regions with higher marketization degree.

When evaluating the local government officials, the central government is concerned of the political performance in economic development, taxation, environmental protection, social stability, and social employment. To make the evaluation standard more intuitive, the macroscopic indexes such as financial deficit, unemployment rate, income gap and GDP growth are used as the main basis for rewarding and punishing local officials. Under the pressure of performance evaluation, the local government will pay more attention to the indexes mentioned above.

First of all, under the background of the tax sharing reform, in addition to relevant expenditures of local economic development, the local government still need to bear the financial expenditures decentralized by the central government, such as social insurance expenditure and the welfare reform expenditure of SOEs, which caused heavy financial pressure and serious financial deficit for local government. In this situation, in the regions with more serious financial deficit, the local SOEs must share policy burden with local government, and the SOEs will face higher degree of local government intervention. So the executive compensation of the local SOEs will inevitably be regulated to reduce the government expenses.

Secondly, exorbitant executive compensation will decrease the number of employees in the enterprises with certain cost of enterprise labours, meanwhile exorbitant executive compensation will arouse public doubt and anger [9], the phenomenon is more obvious in the regions with severe unemployment. Therefore, the government will intervene more in the executive compensation of SOEs in regions with serious unemployment for the social stability and fair. Thirdly, social stability and fair are the social goals that local government pay the most attention, larger income gap will lead to income polarization and public discontent, ultimately cause social problems. So the local government will implement stricter measures of compensation regulation of SOEs executive in regions with larger income gap, to avoid widening the social income gap.

Finally, as an important economic index for performance evaluation, the GDP growth rate plays a pivotal role in the political promotion of local officials. Correspondingly, the higher GDP growth rate indicates that the development of local SOEs measure up to the performance evaluation standard of local officials, so the local government seldom intervenes in the local SOEs as well as the executive compensation affairs. Conversely, if the GDP growth rate is too low to reach the performance evaluation standard of local officials, the government will intervene more in the SOEs including executive compensation affairs to punish the nonfeasance of SOEs executives. Based on the analysis above, the paper put forward hypothesis as follows:

Hypothesis one: The degree of compensation regulation in SOEs is negatively correlated with local marketization degree and GDP growth rate.

Hypothesis two: The degree of compensation regulation in SOEs is positively correlated to the local financial deficit rate, unemployment rate, and income gap.

2.3   Effect Analysis of the Compensation Regulation

The "Temporary Method of Compensation Management of the Head of the Central State Enterprise" promulgated by SASAC in 2004 emphasized that the compensation of the head of central enterprises should be composed of basic compensation, performance compensation and mid-long term incentive compensation, and the performance compensation is determined by linking to operating performance, afterwards the compensation performance sensitivity of SOEs executives, has gradually formed [8]. However, as the executive compensation continues to rise in SOEs, the government has taken a series of compensation regulation measures to limit the executive compensation of SOEs, the measures put more emphasis on the egalitarianism in the income distribution of SOEs, increasing the probability of corrupt behavior of SOEs executives. Therefore, the government compensation regulation weakens the sensitivity of executive compensation to operating performance in SOEs, indicating that the executives behavior are not for the maximization of enterprise value but for their own interests.

Moreover, the monopolistic SOEs are mostly public welfare SOEs and commercial SOEs involving national economic and social security, the enterprises must shoulder such comprehensive responsibilities as realizing economic benefits safeguarding the people's livelihood, and providing public services. Therefore, government intervention has little impact on the executive incentive effect in such enterprises. However, the competitive SOEs attach utmost importance to the operating performance and their operating mechanism is more market-oriented, the government intervention in executive compensation seriously affects the validity of executive incentive contract in competitive SOEs. According to analysis above, the paper put forward the following hypothesis:

Hypothesis three: The government compensation regulation will reduce the compensation performance sensitivity of SOEs executives, weaken the incentive effect of SOEs executives.

Hypothesis four: The competitive SOEs will face lower compensation performance sensitivity of executives and worse future performance under the background of compensation regulation.

3   Research Design

3.1   Samples and Data Sources

The SASAC promulgated the "Temporary Method of Operating Performance Assessment On the Head of the Central State Enterprise" in 2003, it marked the standardization of senior executive compensation system in the SOEs. Therefore, this paper takes all A-share listed companies from 2003 to 2014 as research samples and screened samples according to the following standards: First, due to the different accounting standards of financial and other SOEs, the data from the financial industry were removed. Second, all the ST and PT companies, the companies with missing data from 2003 to 2014 were rejected from the A-share listed companies. Third, the paper winsorized all the continuous variables at the level of 1%–99% to reduce the effect of extreme value.

Data for executive compensation, government intervention index, corporate finance, and corporate governance come from the CSMAR database built by the GuoTaiAn Enterprise. Figures of enterprise property rights were collected according to the SOEs' actual controller information and annual reports. While the indexes of government intervention were obtained by arranging and calculating data from the China Population and Employment Statistics Yearbook (2003–2014) and the Chinese Statistical Yearbook websites.

3.2   Model Construction and Variable Definition

Referring to the relevant literatures on compensation regulation and executive incentives [1,3], the following regression models were established:

1. Measurement Model of Compensation Regulation

$$Rcom_{i,t} = \alpha + \beta_1 Roa_{i,t} + \beta_2 Roa_{i,t-1} + \beta_3 Size_{i,t} + \beta_4 Lev_{i,t} + \beta_5 Dual_{i,t} \tag{1}$$
$$+ \beta_6 Scale_{i,t} + \beta_7 Idd_{i,t} + \beta_8 Excushr_{i,t} + \beta_9 Shrcr_{i,t} + \Sigma Ind + \Sigma Year + \varepsilon,$$

$$Regulation_{i,t} = predict(Rcom_{i,t}) - Rcom_{i,t}. \tag{2}$$

The Rcom in model (1) is the relative executive compensation, and the Regulation in model (2) is the degree of compensation regulation. To measure the degree of compensation regulation, the first step is to define the relative executive compensation model (1) from the aspects of operating performance, asset scale, internal and external governance effect, industry characteristics, and annual characteristics. The second step is to use the samples of non-state-owned enterprises closer to the marketization level to estimate the coefficient of model (1), and then bring the samples of SOEs to model (1), so as to estimate the expected value of the relative executive compensation in SOEs. And the difference between the expected value and actual value of relative executive compensation in model (2) is used to measure the degree of compensation regulation. The influencing factors of relative executive compensation include not only the government regulation, but also other relevant economic factors, so the difference method above mentioned was used to exclude the economic factors, and reasonably estimate the impact of government compensation regulation on the relative executive compensation.

2. Motivation Model of Compensation Regulation

$$Regulation_{i,t} = \alpha + \beta_1 Intervention_{i,t} + \beta_2 Monopoly_{i,t} + \beta_3 Roa_{i,t} + \beta_4 Roa_{i,t-1} \\ + \beta_5 Size_{i,t} + \beta_6 Lev_{i,t} + \beta_7 Dual_{i,t} + \beta_8 Scale_{i,t} + \beta_9 Idd_{i,t}. \tag{3}$$

In model (3), the Intervention represents the variables of government intervention which include the intervention factors such as marketization degree, financial deficit, unemployment rate, income gap, and GDP growth rate. The rest variables in model (3) are control variables.

3. Effect Model of Compensation Regulation

$$Com_{i,t} = \alpha + \beta_1 Roa_{i,t} + \beta_2 Regulate_{i,t} + \beta_3 Roa_{i,t} \times Regulate_{i,t} + \beta Control_{i,t} + \varepsilon, \tag{4}$$

$$Roa_{i,t+1} = \alpha + \beta_1 Regulate_{i,t} + \beta Control_{i,t} + \varepsilon. \tag{5}$$

The model (4) is used to test the impact of government compensation regulation on the compensation performance sensitivity of SOEs executives, and model (4) is used to test the impact of compensation regulation on the future performance of SOEs. Among them, the explanatory variable Com represents the executive compensation, the Roa represents the index of enterprise performance, the Regulate represents the dummy variable of the compensation regulation, which valued 1 when the degree of compensation regulation is greater than 0, and valued 0 when it is less than or equal to 0, and the Control is a set of control variables. Other variables and explanations in this paper are shown in Table 1.

4 Empirical Analysis

4.1 Descriptive Statistical Analysis

1. Determination of the prediction model coefficient of compensation regulation.

This paper firstly uses the samples of non-state-owned enterprise to estimate the coefficient of the regression model (1), and then obtains the prediction model of relative executive compensation in SOEs, the regression results of the prediction model of relative executive compensation in SOEs are shown in the Table 2. From the regression results, the relative executive compensation in SOEs is positively correlated with the enterprise performance of current and previous period, asset size, duality of CEO and chairman, and board scale at the significant level of 1%, it is positively correlated with the board independence at the significant level of 10%, and it is negatively correlated with the executive shareholding proportion and the shareholding proportion of the largest shareholder at the significant level of 1%.

2. Determination of the degree of compensation regulation.

Bring the samples of the SOEs in prediction model (1) to estimate the expected value of relative executive compensation in SOEs, then calculate the

Table 1. The variables and explanations

1. Main research variables	
Regulation	Degree of compensation regulation, represented by the difference between the expected value and actual value of relative executive compensation in SOEs
Regulate	Dummy variable of compensation regulation, valued 1 when the degree of compensation regulation is greater than 0, and valued 0 when it is less than or equal to 0
Rcom	Relative executive compensation, the average compensation of the top three executives with the highest compensation/the average compensation of employees, used to estimate the expected value of the relative executive compensation
Com	Executive compensation, the logarithm of the average compensation of the top three senior executives
Roa	Return on assets, enterprise net profit/average balance of total assets of enterprise
Market	Marketization degree, government tax/financial revenue
Deficit	Financial deficit, the absolute value of financial deficit/financial revenue
Unemploy	Unemployment rate, the registered urban unemployed/the urban employed
Gap	Income gap, the average income of the top 20% population in income ranking/the average income of the last 20% population in income ranking
GDPinc	GDP growth rate, GDP value in the current period/GDP value in the previous period
2. Control variable	
Monopoly	Monopolistic SOEs, valued 1 if the enterprises are in imperfect competitive industry, otherwise valued 0
Size	Asset size, the natural logarithm of total assets of enterprise
Lev	Asset-liability ratio, total liability/total assets of enterprise
Dual	Duality of CEO and chairman, valued 1 if the CEO doubles as chairman, otherwise valued 0
Scale	Board scale, the number of board members
Idd	Board independence, the proportion of independent directors to the total number of board members
Excushr	Executive shareholding, the proportion of the shares held by senior executives at the end of the year to the total share capital
Shrcr	Shareholding proportion of the largest shareholder, the proportion of the shares held by the largest shareholder to the total share capital

Table 2. The regression results of the prediction model of relative executive compensation in SOEs

Rcom	Coefficient	T-value
Roat	8.131***	7.64
Roat − 1	5.798***	5.97
Size	1.790***	27.24
Lev	0.098	0.32
Dual	0.976***	6.94
Scale	0.142***	3.22
Idd	2.342*	1.71
Excushr	−1.188***	−3.61
Shrcr	−1.633***	−3.72
cons	−31.930***	−19.83
Ind	Control	
Year	Control	
Adj-R^2	0.1372	
N	8621	

***, **, * refer respectively to the statistical significance at the levels of 1%, 5%, 10%

difference between the expected value and actual value of relative executive compensation in SOEs by model (2), and the degree of compensation regulation of SOEs executives can be naturally obtained. A positive difference means that executive compensation in SOEs is regulated, and the bigger the positive number is, the higher the degree of compensation regulation on SOEs executives, while a negative difference indicates that the executive compensation is not regulated in SOEs.

4.2 Empirical Results and Analysis

1. Analysis of the motivation of government intervention on executive compensation of SOEs.

According to the regression results in Table 3, there is a negative correlation between the marketization degree and the degree of compensation regulation at the significant level of 5%. The GDP growth rate is positively correlated with the degree of compensation regulation at the significant level of 5%, this is inconsistent with Hypothesis one, it is probably because the government officials pay more attention to the impact of GDP growth rate on their own political promotion in areas with higher GDP growth rates, they will intervene more in local SOEs as well as executive compensation affairs to win the "political tournament". The financial deficit and the unemployment are positively correlated with the degree of compensation regulation at the significant level of 1%. The income gap is negatively related to the degree of compensation regulation at the significant level of 1%, it is exactly the opposite of hypothesis two, it is likely because the income gap itself is a way to measure the compensation gap between executive and ordinary employees, the greater income gap means lower degree of compensation regulation on SOEs executive.

What is more, we put all regional-level motivation variables of government intervention into the same model for regression, and the regression results are still similar to the separate regression results. As for the effect of control variables on the degree of compensation regulation, the monopolistic SOEs and the SOEs with larger asset size, larger board scale, higher proportion of independent director and higher shareholding proportion of the largest shareholder will face higher degree of compensation regulation. The SOEs with better enterprise performance of current and previous period, higher asset-liability ratio and higher proportion of executive shareholding will face lower degree of compensation regulation, and the SOEs in which the CEO doubles as chairman will also face lower degree of compensation regulation.

2. Analysis of the effect of compensation regulation on compensation performance sensitivity of SOEs executives.

Table 4 shows the effect of compensation regulation on compensation performance sensitivity of SOEs executives. The regression coefficient of executive compensation and enterprise performance in the first column is 2.784, indicating the executive's compensation incentive mechanism in SOEs based on enterprise performance has been formed. The regression coefficient of Roa*Regulate in the second column is −0.709, indicating that the compensation performance sensitivity of SOEs decreases to 2.495 (3.204–0.709) after government compensation regulation, the weakening effect reaches 22%, verifying hypothesis three. All the results are significant at the level of 1%. From the third and fourth column, the regression coefficient of the Roa*Regulate in monopolistic SOEs is −0.212, which is only significant the level of 10%, while the regression coefficient of the Roa*Regulate in competitive SOEs is −0.829, which is significant at the level of 5%, so the compensation performance sensitivity of executives in competitive SOEs is more weakened, confirming hypothesis four.

3. Analysis of the effect of compensation regulation on the future performance of SOEs.

Table 5 shows the regression results of the effect of compensation regulation on the future performance of SOEs. The first column is the regression results of the complete samples, the coefficient of Regulate is negative and it is significant at the level of 1%, this indicates that the government compensation regulation damages the enterprise value, and the compensation regulation significantly weakens the validity of executive incentives in SOEs, verifying hypothesis three. In the second column and the third column, there are the subsample regression results, it can be seen that the coefficient of Regulate in monopolistic SOEs is negative, but it is not statistically significant. While the coefficient of Regulate in competitive SOEs is negative and it is significant at the level of 1%, this confirms

Table 3. The regression results of the compensation regulation and government intervention factors

Independent variables	Dependent variable: regulation$_t$					
	(1)	(2)	(3)	(4)	(5)	(6)
Market$_t$	−2.128**					−3.528***
	(−2.51)					(−2.59)
Deficit$_t$		0.401***				0.247***
		−4.07				−3.49
Unemploy$_t$			0.826***			0.885***
			(9.87)			(7.29)
Gap$_t$				−0.565***		−0.555***
				(−6.32)		(−4.86)
GDPinc$_t$					5.069**	6.708**
					(2.39)	(2.00)
Monopoly$_t$	1.960***	1.844***	1.790***	2.070***	1.894***	2.126***
	(6.16)	(6.88)	(6.72)	(7.13)	(7.07)	(6.06)
Roa$_t$	−8.993***	−7.042***	−7.349***	−6.828***	−7.161***	−3.402***
	(−5.82)	(−5.44)	(−5.71)	(−5.05)	(−5.52)	(−3.24)
Roa$_{t-1}$	−8.016***	−5.967***	−6.210***	−5.982***	−6.095***	−2.943***
	(−5.60)	(−4.92)	(−5.15)	(−4.68)	(−5.02)	(−2.90)
Size$_t$	0.561***	0.702***	0.733***	0.696***	0.675***	0.640***
	(7.33)	(10.64)	(11.23)	(10.03)	(10.30)	(7.78)
Lev$_t$	−0.609	−0.695*	−1.006***	−0.926**	−0.642*	−1.479***
	(−1.31)	(−1.83)	(−2.66)	(−2.32)	(−1.69)	(−2.95)
Dual$_t$	−0.038	−0.218	−0.168	−0.513**	−0.22	−0.261
	(−0.15)	(−1.04)	(−0.80)	(−2.27)	(−1.04)	(−0.94)
Scale$_t$	0.046	0.018	0.029	0.047	0.025	0.110**
	(1.05)	(0.53)	(0.82)	(1.22)	(0.72)	(2.28)
Idd$_t$	2.677*	1.855	2.368*	2.507*	1.943	3.509**
	(1.76)	(1.39)	(1.78)	(1.77)	(1.45)	(2.16)
Excushr$_t$	−10.170**	−11.99***	−10.16***	−12.27***	−12.64***	−6.067
	(−2.47)	(−3.17)	(−2.70)	(−2.95)	(−3.34)	(−1.33)
Shrcr$_t$	5.093***	4.615***	4.922***	4.348***	4.562***	5.449***
	(9.45)	(10.32)	(11.04)	(9.09)	(10.20)	(9.25)
Cons	−10.86***	−14.33***	−17.91***	−10.54***	−19.18***	−12.94***
	(−6.46)	(−10.01)	(−12.15)	(−6.62)	(−6.61)	(−7.19)
Ind, Year	Control	Control	Control	Control	Control	Control
Adj.R^2	0.1462	0.1309	0.1423	0.1297	0.1295	0.1164
N	4961	6650	6650	5907	6650	4258

***, **, * refer respectively to the statistical significance at the levels of 1%, 5%, 10%

Table 4. The effect of compensation regulation on compensation performance sensitivity of SOEs executives

Independent variable	Dependent variable compensation			
	(1) Complete samples regression	(2) Complete samples regression	(3) Sub-samples regression: imperfect competitive industry	(4) Sub-samples regression: perfectly competitive industry
Roa	2.784***	3.204***	2.301***	3.291***
	(22.78)	(16.05)	(5.26)	(14.75)
Regulate		−0.587***	−0.516***	−0.600***
		(−35.05)	(−13.00)	(−32.43)
Roa*Regulate		−0.709***	−0.212*	−0.829**
		(−3.14)	(−1.75)	(−2.40)
Control	Control	Control	Control	Control
Adj.R^2	0.536	0.6303	0.644	0.6374
N	7341	6650	1459	5191

***, **, * refer respectively to the statistical significance at the levels of 1%, 5%, 10%

Table 5. The effect of compensation regulation on the future performance of SOEs

Independent variable	Dependent variable: $Roa_{i,t+1}$		
	(1) Complete samples regression	(2) Sub-samples regression: imperfect competitive industry	(3) Sub-samples regression: perfectly competitive industry
$Regulate_{i,t}$	−0.000521***	−0.000339	−0.000980***
	(−3.46)	(−1.56)	(−2.78)
Control	Control	Control	Control
Adj.R^2	0.1005	0.2104	0.0813
N	6008	1307	4701

***, **, * refer respectively to the statistical significance at the levels of 1%, 5%, 10%

the hypothesis four, showing that the future performance of competitive SOEs is more negatively affected under the background of compensation regulation.

4.3   Robustness Test

In order to ensure the stability of the research conclusions, the following robustness test was performed. First, the paper replaced the measurement index of marketization degree with the marketization index compiled by Fan Gang et al. (2011) [5], the marketization degree is still negatively correlated with the degree of compensation regulation at the significant level of 1%. Second, considering that the local government intervention is likely to affect local SOEs only, the paper rejected the samples of central SOEs from the complete samples and made the empirical analysis again, the above empirical results were still valid. Third, this paper replaced this year's variables of government intervention with the last year's variables of government intervention, and the empirical analysis was performed once again, the above empirical results are still valid. Finally, the paper replaced the average compensation of the top three executives with the highest compensation with the average compensation of the top three directors with the highest compensation to calculate the relative executive compensation in SOEs, then all the hypothesis were tested again, the above conclusions are still correct.

5   Conclusion

This paper makes an in-depth analysis of the characteristics of compensation regulation, the motivation and intervention effect of government intervention in executive compensation. The main conclusions drawn are as follows: The local government makes and implements differentiated compensation regulation policies according to the marketization degree, income gap, financial deficit, unemployment rate and GDP growth rate. However, in the view of intervention effect, the government compensation regulation has reduced the compensation performance sensitivity of SOEs executives, weakened the incentive effect of SOEs executives, and it has greater negative impact on the competitive SOEs. The conclusions indicate that government should properly handle the relationship with SOEs, improve the pertinence and validity of government behavior, and the government should also try to establish differentiated executives compensation incentive mechanism and differentiated executives compensation distribution method, which matches the function orientation and operating performance of SOEs.

References

1. Brick, I.E., Palmon, O., Wald, J.K.: CEO compensation, director compensation, and firm performance: evidence of cronyism? J. Corp. Financ. **12**(3), 403–423 (2006)
2. Chen, D., Fan, Y., Shen, Y., Zhou, Y.: Employee incentives, wage rigidity, and corporate performance‖based on empirical evidence of state-owned non-listed companies. Econ. Res. **4**, 116–129 (2010). (in Chinese)
3. Conyon, M.J., He, L.: Executive compensation and corporate governance in China. J. Corp. Financ. **17**(4), 1158–1175 (2011)
4. Core, J.E., Holthausen, R.W., Larcker, D.F.: Corporate governance, chief executive officer compensation, and firm performance. J. Financ. Econ. **51**(3), 371–406 (1999)
5. Fan, G., Wang, X.L., Zhu, H.P.: Chinese marketization index: the report on the relative progress of marketization in various regions in 2011. Beijing Economic Science Press (in Chinese) (2011)

6. Holmstrom, B., Weiss, L.: Managerial incentives, investment and aggregate implications: scale effects. Rev. Econ. Stud. **52**(3), 403–425 (1985)
7. Jackson, S.B., Lopez, T.J., Reitenga, A.L.: Accounting fundamentals and CEO bonus compensation. J. Account. Public Policy **27**(5), 374–393 (2008)
8. Jiang, F.X., Zhu, B., Wang, Y.T.: Does the manager incentive contract of state-owned enterprises pay more attention to performance? Manag. World **9**, 143–159 (2014). (in Chinese)
9. Joe, J.R., Louis, H., Robinson, D.: Managers' and investors' responses to media exposure of board ineffectiveness. J. Financ. Quant. Anal. **44**(3), 579–605 (2009)
10. Lambert, R.A., Larcker, D.F.: An analysis of the use of accounting and market measures of performance in executive compensation contracts. J. Account. Res. 85–125 (1987)
11. Leone, A.J., Wu, J.S., Zimmerman, J.L.: Asymmetric sensitivity of CEO cash compensation to stock returns. J. Account. Econ. **42**(1–2), 167–192 (2006)
12. Lin, J.Y., Cai, F., Li, Z.: Competition, policy burdens, and state-owned enterprise reform. Am. Econ. Rev. **88**(2), 422–427 (1998)
13. Parthasarathy, A., Menon, K., Bhattacherjee, D.: Executive compensation, firm performance and governance: an empirical analysis. Econ. Polit. Wkly. 4139–4147 (2006)
14. Sloan, R.G.: Accounting earnings and top executive compensation. J. Account. Econ. **16**(1–3), 55–100 (1993)
15. Xin, Q., Tan, W.: Marketization reform, corporate performance, and state-owned enterprise manager compensation. Econ. Res. **11**, 68–81 (2009). (in Chinese)
16. Dai, X.L.: The research on the validity of executive compensation under government intervention. Friends Account. **13**, 22–25 (2013). (in Chinese)

Research on the Influence of User Interaction of Tourism Virtual Community on Purchase Intention

Jingdong Chen$^{(\boxtimes)}$ and Jingwen Qiao

Faculty of Economy and Management, Xi'an University of Technology, Xi'an 710054, People's Republic of China
1019352292@qq.com

Abstract. Based on the development trend of "Tourism + Internet" deepening integration, the tourism virtual community has become an important platform to interact for community members. Based on the multi-dimensional user interaction perspective of product information interaction, human-computer interaction and interpersonal interaction, this paper studies the influence of user interaction of tourism virtual community on purchase intention. The empirical results show that: (1) The user interaction of tourism virtual community plays a positive role in the purchase intention, and the influence of interpersonal interaction is the deepest; (2) Product information interaction and interpersonal interaction in user interaction have a significant positive impact on consumer trust, while the impact of interpersonal interaction is not significant; (3) Consumer trust has a significant positive impact on purchase intention. Therefore, the research results are of great significance for guiding the user interaction of tourism virtual communities, and provide the theoretical basis for enhancing consumers' intention to purchase and network marketing of tourism enterprises.

Keywords: User interaction · Consumer trust · Purchase intention · Tourism virtual community · Interaction

1 Introduction

As a product of the deep integration of "Tourism + Internet", tourism virtual community provides community users with a platform for interactive communication and sharing of tourism-related information anytime and anywhere because of its characteristics of sharing, openness and virtuality. As an emerging research area, the tourism virtual community not only meets the needs of people collecting information about tourism, but also creates business value by purchasing online travel products and services. Compared with traditional shopping methods, online consumption can make consumers out of the limitations of time and space, online to compare the price and quality of the products they need to purchase, with the advantage of convenience and create more value. In the era

© Springer Nature Switzerland AG 2020
J. Xu et al. (Eds.): ICMSEM2019 2019, AISC 1002, pp. 119–134, 2020.
https://doi.org/10.1007/978-3-030-21255-1_10

of e-commerce booming, users are looking for a unique experience to realize the value of personal experience, and interaction is a concrete way to feel the unique experience. User interaction is a common behavior of travel virtual community users, sharing product information and consumption feelings with other users through posting and replying activities in the community. In the past, the research objects selected for user interaction mostly focused on the traditional service consumption field. In recent years, online user interaction has gradually emerged, and the research in the online travel environment is rare. From the research topic, there is not much research on the psychological mechanism in the process of user interaction, and there is a lack of in-depth discussion, and there are relatively few empirical studies on the issue of online shopping trust.

There are two main contributions in this paper. First, based on the consumer-led logic, combined with consumer trust as a mediator variable, this paper constructs a relationship model between user interaction and purchase intention of tourism virtual community. Then verify the model through empirical analysis, and enriches currently existing research results of this fields. Second, provide practical guidance and advice for travel enterprises to promote user interaction through virtual communities, thereby increasing consumers intention to purchase.

The structure of this paper is as follows: Sect. 2 is a review of relevant theories and researches, including the research of tourism virtual community, user interaction and consumer trust. Section 3 is the analysis and hypotheses of the research model. Section 4 is the research design and data analysis. Section 5 is the conclusion and suggestion.

2   Literature Review

2.1   Tourism Virtual Community

The virtual community which includes blogs, virtual forums, SNS social networking sites, etc. or network collections of the above categories was firstly proposed by Rheingold [8] in 1993 and received much attention. Emotional open discussion initiated by community members on a common topic, a network of interpersonal relationships formed in virtual cyberspace through this interaction. Emotional open discussion initiated by community members on a common topic, through which interactions can form a network of relationships in virtual cyberspace. Tourism virtual community is the practical application of virtual community in tourism. It needs to be deepened in the definition of virtual community. It is defined as a group composed of netizens with common tourism interests or travel experiences and communicating and interacting through the Internet [23].

At present, the tourism virtual community has become an important way for major tourism enterprises to conduct consumer relationship management, accurately understand consumer needs, and enhance their core competitiveness. The operators of the tourism virtual community can ensure the reliability and accuracy of the shared information in the community through strict content review, provide users with a good experience through simple, stylish and friendly interface design, and improve the security of the website. Protect the privacy of users, thereby increasing the user's stickiness, and ultimately affecting the user's willingness to share and promoting the user's sharing behavior [4]. This kind of community is mostly established by online tourism enterprises, and the activities are initiated by the tourism enterprise managers, in order to finally complete the tourism transaction and promote the purchase decision of the tourism consumers.

2.2 User Interaction

Up to now, domestic and foreign research experts have not reached a unified opinion on the concept of user interaction. Wiener is the first scholar to define the meaning of interaction. He believes that interaction refers to two-way communication, which is the process by which the information receiver feedbacks the received information and then corrects the information according to the feedback generated [24]. Haeckel deemed that interaction is the exchange of information between people or between people and technology. Through this exchange of information, it will change one's knowledge and influence individual behavior [7]. Libai defined the interaction between customers from the perspective of information transfer, which was considered as the process of transferring information from one (or a group) of customers to another (or a group of customers). Such information may change customer preferences and actual purchases behavior or way of further interaction [13]. Lu Zhang [30] believed that network interaction includes communication between text users in the form of text, images, sounds, etc. It also included the acquisition and distribution of information through online media, as well as the spiritual interaction of virtual brand community stakeholders and other interactive behaviors. Junjie Zhou believes that user interaction is a process in which different virtual community users interact with each other through various information exchange activities such as communication, contact, communication and interaction. Under the traditional conditions, users are often in a state of onlookers and passives, but the emerging electronic platform can demonstrate the active participation and interaction of users. Network interaction and virtual community are closely related, and even can be said that interaction is the key and core of virtual community. The interaction in this paper mainly refers to the customer interaction in the virtual environment.

According to the dimensional analysis of online interaction by Massey and Levy [14], Wang et al. [22], this paper divides user interaction into three dimensions: product information interaction, human-computer interaction and interpersonal interaction. Product information interaction refers to the interaction between community users on the theme of travel and holiday products, air tickets, hotels, destinations, etc. Human-computer interaction refers to the process of using the network media computer as the medium, the community users to browse the content of the section, and the use and evaluation of the webpage function. Interpersonal interaction means that the community members question and answer each other questions, share experiences, etc. in the virtual community. User interaction brings a whole new social relationship and is used to achieve various social and economic goals because it has an important impact on individual consumption behavior.

2.3   Consumer Trust

Trust is an important research perspective that is of common concern to many disciplines such as sociology, psychology, and marketing. In the marketing arena, consumer trust is seen as an important determinant of consumers' intention to purchase. With the rapid development of e-commerce, more and more scholars pay attention to consumer trust in online shopping. Kramer and Tyler [11] argued that trust is the main body of the network by estimating the behavior and intentions of others. When the value of this estimate is positive, it indicates that the subject of trust has a mental state without defense which depends on others [17]. Mcknight believed that trust refers to the willingness of one party to trust and rely on the other. High level of trust can be measured by trust tendency and belief. The former is a person's willingness to rely on others'willingness in the same situation, while the latter is a person's positive prediction of others' goodwill, ability and honesty [16]. Govier's [20] study showed that trust is based on the expectation of the kindness and ability of the trusted party. This expectation comes from the honesty of the other party's attribution or assumptions. The other party is considered to be kind, even if there are risks and uncertainties, it still has good intentions and expectations for the other party's expected behavior. It is an emotional-based attitude [20]. Lee, Turban's research showed that online transaction trust refers to the buyer's positive expectation that the seller will act according to the pre-agreed trading rules, and does not care about the seller's supervisory control ability, and believes that he will not be hurt by interests from the seller [12].

In summary, different scholars have different definitions of trust, mainly in three cases. The first is based on the premise of the existence of risk; the second is based on the consumer's trustworthy expectations of the salesperson or the counterparty; the third is based on the consumer's willingness to rely on. These three statements explain trust from different angles and focuses, but the three are not contradictory and involve each other. Therefore, this paper defines the trust as the consumer trust, that is, the trust of online travel enterprises. When consumers know the existence of transaction risks in the online market environment, they still have the confidence to online suppliers who can satisfy their own consumption, release real information and keep promise.

3   Research Models and Hypotheses

3.1   Tourism Virtual Community User Interaction and Consumer Trust

Due to the tourism experience and good expectation of the enterprises behavior brought by the interaction behavior, the consumer will even increase the trust of the enterprises. Xiang and Shen [28] proposed that the premise of keeping consumers loyal to online merchants is no longer just to provide services and information, but to create a communication platform that users can communicate and interact with each other, which will help to increase Consumer trust in the enterprise. Dandan Zhao deemed that retailers can increase consumers' trust through providing high-quality information and services, and a good communication platform [32].

As mentioned above, user interaction is divided into three dimensions: product information interaction, human-computer interaction, and interpersonal interaction. Jeppesen and Molin [9] suggested that the increased frequency of product information interactions provides consumers with an opportunity to access all aspects of the product. As consumers become more aware of the product information they need to purchase, the trust in the supplier who supplies the product increases. Human-computer interaction mainly indicates that the virtual community is an important interactive channel between consumers and online suppliers. Online merchants can provide positive interactions with consumers' product service requirements, which can greatly affect consumers' trust attitude towards online suppliers. In the process of interpersonal interaction, consumers are relatively more affected by other community participants. According to Hagel and Armstrong, they create a high sense of trust through sharing their experiences and having regular interaction, because most members of the virtual community have the same hobby or a high level of interest in certain topics [2]. Therefore, it is hypothesized that.

H1a. Product information has a significant positive effect on the consumer trust.

H1b. Human-computer interaction has a significant positive effect on the consumer trust.

H1c. Interpersonal interaction has a significant positive effect on the consumer trust.

3.2   Consumer Trust and Purchase Intention

Wu and Huang [26] proposed that consumers lack trust in online shopping websites, it leads to low online shopping willingness due to the virtual nature of online shopping, but when the user's trust increases, his purchase intention increases accordingly. In the research, Si Wu proposed that the interaction of members in the community has an impact on trust, and various mechanisms can lead to the establishment of trust [27]. The Technology Acceptance Model (TAM) believes that when a user accepts a new technology, his attitude is more influential than the subjective norms of others. In turn, the consumer's trust attitude will play a positive role in its online purchase behavior. When consumers build trust in network providers, it means that consumers are confident that both parties remain honest in the transaction. The positive purchase intention is formed by the merchants to deliver the goods timely, and the achievement of their own goals is based on the behavior of both parties. Therefore, it is hypothesized that.

H2. Consumer trust has a significant positive effect on the purchase intention.

3.3   Tourism Virtual Community User Interaction and Purchase Intention

The development of virtual community and network interactions reflects the tendency of people to use the sociality of the Internet to achieve their social and economic goals [25]. Consumers interact with other members or sellers in the virtual community to obtain valid information of product before they are willing to purchase the desired item. Therefore, user interaction is an important channel for consumers to obtain the required product information when they are willing to purchase. As a platform for consumers to interact with community users or sellers, the personal networks formed by virtual communities have a more direct impact on consumers' intention to purchase.

Xiaoping Fan's earlier research showed that the virtual community's network interaction can bring consumers a variety of effects and affect consumers' intention to purchase [6]. Adjei research shows that members of online brand communities communicate with the perception of Brands and the experience of product use through the channels of online media. In this context, users interaction has a positive impact on consumers' intention to purchase [1]. Zhao and Jie [31] proposed that the collective characteristics of virtual communities, the mutual trust and frequent interactions of community members affect the purchasing behavior of community members. Jingjing Dong's research showed that online experience interaction between consumers and merchants affects online presence and online shopping willingness in the context of online shopping [5]. Therefore, it is hypothesized that.

H3a. Product information will have significant positive effect on Purchase Intention.

H3b. Human-computer interaction will have significant positive effect on Purchase Intention.

H3c. Interpersonal interaction will have significant positive effect on Purchase Intention.

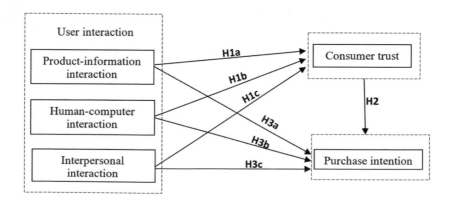

Fig. 1. Research model of this paper

Based on the research of virtual community user interaction and consumer trust, combined with the selection of research measurement dimension, the research model of tourism virtual community user interaction to purchase intention is proposed in Fig. 1.

4   Research Design and Data Analysis

4.1   Variable Operation

The user interaction in the tourism virtual community mainly measures the specific content of the interaction between the consumer and the different subjects. Based on the research scales of Wang and Ma [22] and Nambisan and Baron [18], the modification of the characteristics of the tourism virtual community is carried out. The details are shown in Table 1.

The item of consumer trust mainly refers to the result of the trust of the consumers in the online community after the interaction of the community users in the virtual community, mainly referring to the research scales of MckNight et al. [15] and Preece [19], as shown in Table 2.

The intention to purchase is the subjective intention of the consumer to purchase the tourism product or service after browsing the information and interacting with others in the tourism virtual community. This study mainly considers the pre-tourism buying behavior, mainly referring to the research scales of Zeithsml et al. [29] and Jia [10], as shown in Table 3.

4.2   Questionnaire Design and Data Collection

This questionnaire mainly includes three parts. The first part defines the concept of tourism virtual community, the second part is the scale problem, which measures user interaction, consumer trust and purchase intention, and the third part is the basic characteristics of the object under investigation. On the basis of the relevant mature scales studied by the predecessors, combined with the actual situation of the tourism virtual community, the questionnaire of this study was formed. The scoring method of the scale has been used more maturely. This study selected the Likert 5 scale. All items were measured by the Likert 5 scale: 5 means "very consistent" and 1 means "very inconsistent". The questionnaire was mainly distributed by network survey and field survey. Firstly, we use the WenJuan website to conduct online questionnaire survey, mainly through social media such as Weixin and Weibo. Secondly, in order to ensure the uniform distribution of the questionnaire samples, we select the area with a large local traffic to issue questionnaires. A total of 350 questionnaires were issued, of which 325 were returned and 278 valid questionnaires were obtained, with an effective rate of 85.54%. Through the analysis of the valid questionnaire, the descriptive statistics of the distribution of respondents are shown in Table 4.

Table 1. User interaction operational item

Measurement dimension	Operational variable	Item	Scale source
Product-information interaction	PII1	The travel virtual community contains a wealth of travel product information (air tickets, hotels, attractions tickets, etc.)	Nambisan and Baron [18], Ma and Wang [22]
	PII2	Tourism virtual community updates product information frequently and has timeliness	
	PII3	The product information provided by the tourism virtual community is in line with my interest and has credibility	
Human-computer interaction	HI1	I think the navigation, content and links of this travel virtual community website are very convenient to use	Nambisan and Baron [18], Ma and Wang [22]
	HI2	I feel that the online payment function of this travel virtual community is safe	
	HI3	I think the webpage of this travel virtual community is beautifully designed and runs smoothly	
Interpersonal interaction	II1	I will post comments, ask questions, and get responses from other community members in the travel virtual community	Nambisan and Baron [18], Ma and Wang [22]
	II2	I often participate in travel topics initiated by other members, discuss and help each other	
	II3	I can find friends with similar interests from the community to communicate and build relationships	

4.3   Reliability and Validity Analysis

In this paper, the SPSS22.0 software is used to test the reliability and validity of the sample data. The results show that the Cronbach's α coefficient of each variable is greater than 0.7, indicating that the scale has preferable internal consistency. Exploratory factor analysis was performed on the sample data after the KMO sample measure and the Bartlett sphere test. The results show that the measurement items of the same variable in the scale are distributed in the same factor, and the factor loadings are greater than 0.5, which meets the test

Table 2. Consumer trust operational item

Measurement dimension	Operational variable	Item	Scale source
Consumer trust	CT1	Merchants will try their best to help and support consumers	Mcknight [15], Preece [19]
	CT2	I think merchants post real information and will follow the promise	
	CT3	I believe that merchants have a wealth of product knowledge and the ability to meet customer needs	

Table 3. Purchase intention operational item

Measurement dimension	Operational variable	Item	Scale source
Purchase intention	PI1	If I need to buy a travel product, I will make a purchase in the travel community	Zeithsml et al. [29], Jia [10]
	PI2	I am willing to browse the travel website to find the products or services I need	
	PI3	I am very happy to buy a business from this website community	

Table 4. Descriptive analysis of sample features

Characteristics	Type	Sample size	Percentage (%)
Sex	Man	181	65.10%
	Woman	97	34.90%
Education	College	39	14.03%
	Bachelor	156	56.12%
	Master degree and above	83	29.85%
Monthly per capita income	RMB 3,000 and below	35	12.59%
	RMB 3000–5000	167	60.07%
	RMB 5000–7000	69	24.82%
	RMB 7000 and above	7	2.52%
Use travel website frequency	Less than 1 time/week	148	53.24%
	1–3 times/week	89	32.01%
	3–7 times/week	33	11.87%
	More than 8 times/week	8	2.88%

128   J. Chen and J. Qiao

Table 5. Reliability and validity analysis

Variable	Item	Factor loading	Cronbachs	KMO
Product-information interaction	PII1	0.845	0.745	0.858
	PII2	0.855		
	PII3	0.817		
Human-computer interaction	HI1	0.736	0.883	0.901
	HI2	0.843		
	HI3	0.710		
Interpersonal interaction	II1	0.743	0.830	0.883
	II2	0.857		
	III3	0.864		
Consumer trust	CT1	0.841	0.885	0.893
	CT2	0.754		
	CT3	0.777		
Purchase intention	PI1	0.777	0.872	0.832
	PI2	0.868		
	PI3	0.966		

standard level, which proves that the validity of the scale meets the standard, thus ensuring that the scale is effective, as shown in Table 5.

4.4   Model Hypotheses Test

(1) Correlation analysis
Use Pearson correlation analysis to confirm the correlation between user interactions, consumer trust and purchase intention of tourism virtual community. After research, it was found that user interaction, consumer trust and purchase intention was significantly correlated. The results are shown in Table 6.

Table 6. Correlation matrix between variables

	PII	HI	II	CT	PI
PII	1				
HI	0.145**	1			
II	0.517*	0.028	1		
CT	0.549**	0.326*	0.632**	1	0.548**
PI	0.937**	0.327*	0.501**	0.596**	1

Note **Significant at $p < 0.01$; *significant at $p < 0.05$

(2) Regression analysis

In this section, we will continue to use SPSS22.0 for multiple linear regression analysis to further determine the direction of influence and the degree of influence between variables. The results of regression analysis are shown in Table 7.

In order to verify H1a, H1b, H1c, this study using regression analysis takes tourism virtual community user interaction (product information interaction, human-computer interaction, interpersonal interaction) as an independent variable, and consumer trust as a dependent variable. As shown in the Table 7, product information interaction and interpersonal interaction have a significant positive impact on consumer trust ($\beta = 0.225, t = 3.518, p < 0.001; \beta = 0.462, t = 7.184, p < 0.001$). Therefore, H1b and H1c are established, and human-computer interaction has no significant influence on consumer trust ($\beta = 0.014, t = 0.372, p > 0.001$). H1a does not hold.

In order to verify H2, this study uses consumer trust as an independent variable and purchase intention as a dependent variable. As a result, as shown in the Table 7, consumer trust has a significant positive impact on purchase intention ($\beta = 0.684, t = 15.54, p < 0.001$), so H2 is established. In order to verify H3a, H3b, H3c, this study takes the tourism virtual community user interaction (product information interaction, human-computer interaction, interpersonal interaction) as the independent variable, purchase intention as the dependent variable, the results are shown in Table 7, product information interaction, human-computer interaction, interpersonal interaction have significant positive impacts on purchase intention ($\beta = 0.209, t = 8.941, p < 0.001; \beta = 0.161, t = 3.470, p < 0.001; \beta = 0.443, t = 2.763, p < 0.001$), so H3 is established.

Table 7. Regression analysis comprehensive result statistics

Dependent variable	Independent variable	Beta	T value	P value
PI	PII	0.209	8.941	***
	HI	0.161	3.470	***
	II	0.443	2.763	***
PI	CT	0.684	15.54	***
CT	PII	0.225	3.518	***
	HI	0.014	0.372	0.262
	II	0.462	7.184	***

Notes Beta: Standardize regression weight; ***significant at $p < 0.001$

(3) Consumer trust mediation effect test

According to Baron and Kenny [3], a variable (M) is a mediator when it meets the following conditions: (1) There is a significant correlation between X and Y; (2) A change in X can significantly affect the change of M; The change of M will significantly affect the change of Y; When M is added, the relationship between X and Y is not significantly significant (with full mediation), or the relationship between X and Y is obviously weaker but still significant (with partial mediation). In this paper, the independent variable X is the user interaction, the dependent variable Y is the purchase intention, and the mediation variable M is the consumer trust. Condition (1) (2) (3) has been verified in the previous section and is now tested for condition (4).

Since the previously verified product information interaction has no significant influence on consumer trust and does not meet the significant requirements of the regression equation, it will not be discussed further. After joining the consumer trust, the regression coefficient of product information interaction and purchase intention, interpersonal interaction and purchase intention is significant but obviously weaker ($\beta' = 0.074 < \beta = 0.209; \beta' = 0.316 < \beta = 0.443$), so consumer trust has partial mediation effects in user interaction and purchase intention, as shown in Table 8.

Table 8. Model mediation effect test

Model	Variable	Beta	P Value	Mediating Role
1	PII-PI	0.209	***	Partial mediation
	PII,CT-PI	0.074	***	
2	II-PI	0.443	***	Partial mediation
	II,CT-PI	0.316	***	

Notes Beta: Standardize regression weight; ***significant at $p < 0.001$

5 Conclusions and Discussion

5.1 Conclusions

In order to facilitate the analysis and discussion in this section, the hypotheses presented above were summarized by the table, as shown in Table 9.

This study used the tourism virtual community as a background to study the relationship between community users' interaction and purchase intention. Combined with the above empirical results, the following conclusions can be drawn:

(1) The three factors of user interaction in tourism virtual community had different degrees of significant influence on purchase intention. Among them, interpersonal interaction had the deepest influence on the purchase intention, and the influence of product information interaction and human-computer interaction on purchase intention is second. Interpersonal interaction is a way for community users to achieve emotional communication by interacting with each other and sharing experiences. This kind of online interpersonal communication makes it easier for users to generate empathy and promote mutual trust

Table 9. Hypothesis test result table

Label	Hypothesis	Test result
H1a	Product information has a significant positive effect on the consumer trust	$\checkmark$
H1b	Human-computer interaction has a significant positive effect on the consumer trust	$\times$
H1c	Interpersonal interaction has a significant positive effect on the consumer trust	$\checkmark$
H2	Consumer trust. has a significant positive effect on the purchase intention	$\checkmark$
H3a	Product information has a significant positive effect on the Purchase Intention	$\checkmark$
H3b	Human-computer interaction has a significant positive effect on the Purchase Intention	$\checkmark$
H3c	Interpersonal interaction has a significant positive effect on the Purchase Intention	$\checkmark$

Notes $\checkmark$: Support for hypothesis attained; $\times$: Lacking support for hypothesis

mechanism, thereby enhancing users' intention to purchase. The three dimensions of interaction affect consumers' intention to purchase to varying degrees. Online travel enterprises and platform managers should focus on promoting and maintaining these three interactions, and strive to build and increase the user's stickiness to online travel brands.

(2) Consumer trust had a positive impact on the intention to purchase. Consumer trust emphasizes that consumers still generate and build trust in online travel enterprises while knowing that there has a risk in the online market. When a enterprise establishes a good sense of trust among consumers, consumers are willing to actively trust the enterprise and the products it provides, so that consumers believe that the enterprise has the ability to meet their needs and safeguard their interests. Under this premise, it is bound to increase consumers' intention to purchase.

(3) Product information interaction and interpersonal interaction in the tourism virtual community had a significant impact on consumer trust. Product information interaction is for consumers to obtain travel product information in the virtual community. When the community provides comprehensive and real product information, consumers will enhance their trust to enterprise due to the more aware of the information. Interpersonal interaction is reflected in the two-way communication of members with the same hobbies in the community. They will have a high sense of trust for using high-frequency interaction to question and answer with each other and share their experiences. Human-computer interaction has not had a significant impact on consumer trust. The reason may be

that human-computer interaction only allows consumers to obtain instrumental and practical needs, but fails to cause consumers' emotional reactions.

5.2 Discussion

Based on the positive impact of tourism virtual community user interaction on purchase intention with the partial mediation effect of consumer trust, the following is the enlightenment to the operation and development of tourism virtual community.

(1) Deepen the content of community product information and enhance the level of user trust. As the manager of the virtual community, it should promptly identify and filter a large amount of complicated information, so that consumers can obtain real and effective information in a short time. In this way, establish a good and professional community image and increase the trust of community users in the community.

(2) Optimize website page module design and improve community function mechanism. By increasing the variety of communication methods of the virtual website, the layout of the web page is clear at a glance, which is convenient for consumers to use the webpage function efficiently. Through user segmentation to construct different communities, create more effective interactions within the community.

(3) Strengthen the interpersonal interaction mechanism of virtual communities, and online communication can be transformed into offline activities. Strengthening the community's interpersonal interaction mechanism platform is conducive to enhancing social identity among users and establishing long-term stable relationships. Enterprise can organize or encourage users to personalize and customize travel themes actively, and even transfer online activities to offline activities, to promote the establishment of real life interpersonal networks.

This paper also has some limitations. It takes the buyer of online transactions as the research object, and the transaction risk in the online transaction is coexisting in the buyer and seller, so the trust problem of the transaction is also mutual, but this study does not consider the seller's trust in the buyer. It needs to be improved in future research in order to obtain more objective conclusions.

Acknowledgment. This paper is supported by the Shaanxi Province Social Science Fund (No. 2016R015) and Research on Promoting to Construct Powerhouse by Cultural Coriented Tourism-Based on Customer Marketing (No. 2016R015).

References

1. Adjei, M.T., Noble, S.M., Noble, C.H.: The influence of c2c communications in online brand communities on customer purchase behavior. J. Acad. Mark. Sci. **38**(5), 634–653 (2010)
2. Armstrong, A.: The Real Value of On-line Communities. Creating Value in the Network Economy. Harvard Business School Press (1999)

3. Baron, R.M.: Kenny, D.A.: The moderator-mediator variable distinction in social psychological research: conceptual, strategic, and statistical considerations. J. Pers. Soc. Psychol. **51**(6) (1986)
4. Chen, Y., Yu, J., et al.: The relationship between user stickiness and sharing behavior in tourism virtual community: the regulating role of self-construction. Tour. Sci. **32**(03), 1–12 (2018). (in Chinese)
5. Dong, J., Xu, Z., et al.: Model construction of the influence of online experiential interaction between consumers and merchants on their willingness to buy. J. Manag. **15**(11), 1722–1730 (2018). (in Chinese)
6. Fan, X., Ma, Q.: Study on the influence of network interaction based on virtual community on network purchase intention. J. Zhejiang Univ. (Hum. Soc. Sci.) (05), 149–157 (2009). (in Chinese)
7. Haeckel, H.S.: About the nature and future of interactive marketing. J. Interact. Mark. **12**(1), 63–71 (1998)
8. Howard, R.: The Virtual Community: Finding Connection in a Computerized World. Addison-Wesley Longman Publishing Co., Boston (1993)
9. Jeppesen, L.B., Mans, J.M.: Consumers as co-developers: learning and innovation outside the firm. Technol. Anal. Strat. Manag. **15**(3), 363–383 (2003)
10. Jia, L.: The influence of Internet word-of-mouth and value creation on consumers' willingness to purchase under social network. Beijing University of Posts and Telecommunications, (2015). (in Chinese)
11. Kramer, R.M., Tyler, T.R.: Trust in organizations: frontiers of theory and research. Adm. Sci. Q. **43**(1) (1996)
12. Lee, M.K., Efraim, T.: A trust model for consumer internet shopping. Electron. Commer. **6**, 75–91 (2001)
13. Libai, B., Bolton, R., Bugel, M.S.: Customer-to-customer interactions: broadening the scope of word of mouth research. J. Serv. Res. **13**(3), 267–282 (2010)
14. Massey, B.L., Levy, M.R.: Interactivity, online journalism, and English-language web newspapers in Asia. J. Mass Commun. Q. **76**(1), 138–151 (1999)
15. McKnight, D.H., Choudhury, V., Kacmar C.: Developing and validating trust measures for e-Commerce: an integrative typology. Inf. Syst. Res. **13**(3) (2002)
16. McKnight, D.H., Cummings, L.L., Chervany, N.L.: Initial trust formation in new organizational relationships. Acad. Manag. Rev. (1998)
17. Moreland, K.R., Tyler, R.T.: Trust in organizations: frontiers of theory and research. Adm. Sci. Q. **43**(1) (1996)
18. Nambisan, S., Baron, R.A.: Virtual customer environments: testing a model of voluntary participation in value co-creation activities. J. Prod. Innov. Manage. **26**(4), 388–406 (2009)
19. Preece, J.: Sociability and usability in online communities: determining and measuring success. Beha. Inf. Technol. **20**(5), 347–356 (2001)
20. Trudy, G.: Social Trust and Human Communities (1997)
21. Wang, Y., Ma, S.: An empirical study on the driving factors of customer interaction in virtual brand community and its impact on customer satisfaction. J. Manage. **10**(09), 1375–1383 (2013). (in Chinese)
22. Wang, Y., Ma, S., Sun, B.: Intermediary role of self-determination in customer interaction and community satisfaction:an empirical study based on sor theory and self-determination theory. J. Shanxi Univ. Financ. Econ. **34**(08), 99–107 (2012). (in Chinese)
23. Wang, Y., Yu, Q., Fesenmaier, D.R.: Defining the virtual tourist community: implications for tourism marketing. Tour. Manag. **23**(4), 407–417 (2002). (in Chinese)

24. Wiener, N.D.: Subtle and obvious keys for the minnesota multiphasic personality inventory. J. Consult. Psychol. **12**(3), 164–170 (1948)
25. Wind, Y., Mahajan, V.: Digital Marketing: Global Strategies from the World's Leading Experts. Wiley (2002)
26. Wu, P., Huang, Y.: Research on the factors of website consumption intention. Manag. Rev. **2006**(11), 18–25+63 (2006). (in Chinese)
27. Wu, S., Ling, Y., Wang, W.: Study on the relationship between interaction, trust and willingness to participate in virtual brand community. J. Inf. **30**(10), 100–115 (2011). (in Chinese)
28. Xiang, H., Shen, Z.: Virtual community knowledge sharing activities bring business opportunities to e-commerce. Inf. Theory Pract. **05**, 472–474+461 (2004). (in Chinese)
29. Zeithaml, V.A., Berry, L.L., Parasuraman, A.: The behavioral consequences of service quality. J. Mark. **60**(2), 31–46 (1996)
30. Zhang, L.: Research on the Influence of Network Interaction Based on Virtual Community on the Image Perception of Tourism Destinations[D]. Zhejiang University (2011)
31. Zhao, X., Jie, J.: Study on the Influence of Social Characteristics of Virtual Community on Online Purchase Behavior. Shopping Modernization, pp. 19–21 (2010). (in Chinese)
32. Zhao, D.: "Buy" or "Abandon": research on online retail channel conversion behavior. Manag. World **34**(06), 184–185 (2018). (in Chinese)

Does the Tax Stickiness Exist?

Wei Yang[1(✉)], Shengdao Gan[1], Hong Wang[1], and Xinyun Liu[2]

[1] Business School, Sichuan University, Chengdu 610064, People's Republic of China
yangwei0627@yeah.net
[2] Beijing Normal University-Hong Kong Baptist University United International
College, Zhuhai 519000, People's Republic of China

Abstract. In the macro economy, the growth rate of tax revenue is consistently higher than the GDP growth rate, which is the thought source of this paper. Is the tax payment performance of micro businesses consistent with this macroeconomic phenomenon? For this reason, this article puts forward the concept of corporate tax stickiness, while exploring its existence, analyzing its theoretical basis from different angles. Then we analyze the existence of tax stickiness, its time trend and influencing factors, and put forward assumptions and use the A-share listed companies' annual financial data from 2008 to 2015 to launch a series of empirical tests. It is found that the tax stickiness of listed companies in China does exist and tends to weaken with time extension. The macro economic growth effect was not significant, while the enterprise's capital intensity has an enhanced influence on tax stickiness, and asset-liability ratio can relief the tax stickiness. However, in different length of time span, the influence of each factor is different, and the effect of comprehensive on tax stickiness is unstable. This paper is only a preliminary study, there are some limitations and a ground of research space for further exploring and improvement.

Keywords: Corporate tax · Stickiness · Listed companies · Financial data processing

1 Introduction

In August 2016, the State Council issued the Work Program for Reducing the Costs of Enterprises in the Real Economy, which fully launched the campaign for reducing the costs of enterprises in the Real Economy. However, there are many entrepreneurs still complaining that they are suffering from a heavy tax burden. Why? This triggers our curiosity greatly. At the same time, we find a micro-economic phenomenon called tax elasticity, which shows that domestic taxation still maintains high growth when our current economic growth slows. Since corporate tax is the main source of fiscal revenue, if we want to know the real reason of that, we have to explore the micro tax turned over by unique companies. As long as we extend the concept of tax elasticity into the sensitivity of corporate tax (the GDP is placed by business income), some similarities can be

© Springer Nature Switzerland AG 2020
J. Xu et al. (Eds.): ICMSEM2019 2019, AISC 1002, pp. 135–148, 2020.
https://doi.org/10.1007/978-3-030-21255-1_11

found between those two concepts. A company's tax, meanwhile, will be unlikely to rise always. When the business income change is in decline, will the change of tax decrease by the same degree or a less degree? The latter is more possible to occur. Therefore, we can speculate that the change in corporate tax in different directions is "asymmetric"—a characteristic of stickiness (Shown as Fig. 1).

Tax stickiness in certain companies hasn't received much attention, but it has great practical significance. Firstly, tax stickiness is a microscopic source of macro-economic phenomenon of tax elasticity. Secondly, as tax is a necessary cost of business operation, understanding its marginal changes, especially the corresponding relationship with business performance, can promote the rational cost planning and tax planning of enterprises. Further and even more importantly, tax stickiness reveals the fact that when the operating situation deteriorates, the reduction of tax is limited, which actually increases the burden on the company in disguise, and indicates that there may be some unfair income distribution between the governments and corporates.

In conclusion, exploring and researching tax stickiness is in line with the needs of macro and micro economic operation, which is beneficial to government tax collection, management and reform, along with corporate tax planning and operation improving.

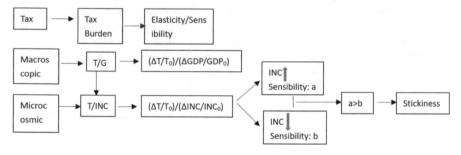

Fig. 1. The deduction process of tax stickiness

2   Theoretical Analysis and Research Hypothesis

2.1   Basic Concepts

Stickiness generally refers to the "asymmetry" of a certain kind of cost when the business income increases or decreases. Anderson et al. [9] called this phenomenon as cost stickiness and set an empirical test. They found that the change of American listed companies' sales, management and operating costs are asymmetrical during the period from 1979 to 1998: revenue increased by 1%, these costs increased by 0.55%; while revenue decreased by 1%, expenses only decreased by 0.35% [9]. Now, the study on cost stickiness has been extended to various specific cost subjects, including business cost stickiness [12], salary stickiness [3], and R&D cost stickiness [20]. Similar to other expenditure, enterprise tax is essentially deducted from business income to cover the costs incurred in the process of production and operation. Will there be similar characteristics in its changes? Some scholars focus their research on the relative changes of income tax expenses and pre-tax profits. They discussed the stickiness of income tax expenses from the perspectives of tax-accounting differences, earnings management, tax collection and management, and so on. Through empirical analysis, they drew a conclusion that stickiness of income tax expenses does exist widely in Listed Companies in China [14,18]. It is definite that income tax expenses cannot fully reflect the real tax amount of government collection and enterprises turning over as it is only a certain part of the whole corporate tax. These studies are not enough to explore the changing rule of the general corporate tax. From the overall tax perspective, Cong Yi and Zhou Yijun [21] pointed out that the tax burden of enterprises in China has a rigid feature, which means that the corporate tax burden does not decrease but rises when the profit level falls [21]. However, since the research object is different from this paper, the conclusions drawn are also different. In fact, it has been pointed out that the growth rate of total tax is higher than the growth rate of business income [15]. It is only because the changes of tax in different directions are not distinguished that the asymmetric sticky characteristic of corporate tax were was not been realized. By way of conclusion, even though there is no unified conclusion about the research of tax stickiness, the possibility of the existence of tax stickiness can also be found from the existing literatures.

With regard to the corporate tax, the majority of conventional researches were based on the perspectives of absolute tax burden and actual rate of income tax expenses, such as ETRs [5,11,16,17,19]. These measurements ignore China's real situation of tax structure—the income tax expense is not the major of the total tax but the turnover tax indeed is. Therefore, using this way to study Chinese companies' tax expenditure may be unreasonable. Other measuring method is to measure the overall tax expenditure of an enterprise by the "taxes and fees paid" or the net tax expenditure: "all taxes and fees paid—tax and fee returns received" in cash flow statements [1,7,10,13]. This paper holds that the later measurements can reflect the overall tax of enterprises more comprehensively. They will be used in the principle regression and robustness checks respectively.

2.2 The Existence of Tax Stickiness

From the references outlined above, it is evident that tax stickiness has some theoretic basis. From the perspective of cost theory, tax is a kind of cost of an enterprises' operation. To decompose the structure of tax, some parts are related to the scale of enterprises and inherent capital, such as property tax, land use tax, vehicle and ship tax, stamp tax. Others are related to business volume, such as value added tax, consumption tax, resource tax, enterprise income tax, etc. When the business income of an enterprise continues to rise, managers usually expand their business scale and increase their investment to meet the demand of production. At this point, all tax revenue related to the business income and the scale will rise rapidly. But when the business income of an enterprise declines, the tax will not reduce correspondingly due to the scale and fixed capital. That is to say, taxes, like other costs of enterprises, cannot be adjusted in time according to changes in the business, and cause stickiness.

Scholars also have pointed out that the instability of tax collection and management is an important factor for the stickiness of enterprise income tax expenditure [18]. Our tax work is to levy according to the target plan [8]. Under the pressure of fulfilling the task, the collection and management departments will take certain measures, such as levying "excessive tax", "prepaid tax", "first levy and then return tax", "tax inspection and supplementary tax", etc. To increase fiscal revenue, the government will raise the rate of enterprise tax management, which will make enterprise tax easy to increase and difficult to reduce in the year of income decline [4]. These circumstances will lead to the reduction of corporate tax in the year of business income decline is limited, showing sticky characteristics. In addition, tax stickiness may also be related to the enterprise's own tax policy. For example, active tax policy would be adopted by enterprises, they pay attention to the cost-reducing effect of tax and fee by having good control of them, and they also make tax planning, tax evasion and seek tax preferences when business income declines, so the reduction of tax may be more obvious; but in reality, most enterprises still adopt negative tax revenue, paying less attention on taxes. Then, they will not reduce taxes by means of tax planning, reasonable tax avoidance, tax negotiation and so on, resulting in limited reduction of tax costs. In conclusion, we promote the first and basic hypothesis:

H1: Tax stickiness exists in Chinese listed companies.

2.3   The Time Trend of Tax Stickiness

According to Sunzheng's research, cost stickiness would decrease with the period extension [22]. Tax will have the same characteristic? Even though corporate tax will change in line with the change of business income in theory, there is still a certain time lag in reality for the following reasons. Given the theory of contract, the changes of both tax laws made by government and tax plans conducted by companies lag behind the change of overall economic development and business operation. Nonetheless, this lag should be limited, that is, when the trend of development becomes clear, policy makers and business managers have time to adjust the tax correspondingly, so that the tax stickiness will decrease with time extension.

H2: Tax stickiness of Listed Companies in China tends to diminish over time.

2.4   The Influencing Factors of Tax Stickiness

Is the tax stickiness affected by macro-economy? In good macro-economic periods, it is unlikely that the government will use tax to intervene the market. Tax "neutrality" will play a role [2]. It will not affect the distribution of resources among different departments, having no obvious impact on the tax stickiness. However, at the micro-level, the optimism of business managers on the situation will lead to increased investment and expansion of enterprises. Scale promotes the increase of tax revenue related to stock, thus enhancing the tax stickiness. So the hypothesis is put forward:

H3a: In good macroeconomic times, the tax stickiness will increase.

From the perspective of cost theory, the stickiness mainly comes from the tax related to scale and fixed capital. Therefore, the high proportion of fixed capital will limit the decrease of tax when the business income is in decline. The degree of capital intensity of enterprises, meanwhile, directly affects the amount of tax and the degree of tax stickiness.

H3b: The greater the capital intensity, the stronger the tax stickiness.

According to the revised MM theory, due to the existence of tax, the capital structure of the company will affect the value of the enterprise because of the tax deduction effect of interest on debt. Interest is generated by debt. The more interest you pay, the bigger the tax shield, and the more obvious the decline in the final tax. Therefore, the degree of debt directly affects the extent of tax expenditures and tax changes. We use the asset-liability ratio as a proxy index for debt levels and make the following assumptions:

H3c: The greater the asset-liability ratio, the weaker the tax stickiness.

3   Research Design and Sample Selection

3.1   Research Design

The Existence of Tax Stickiness Using Anderson's method [9], we use logarithmic model to test the existence of tax stickiness of Listed Companies in China, and propose the Model I:

$$\ln(\frac{\text{Tax}_{i,t}}{\text{Tax}_{i,t-1}}) = \beta_0 + \beta_1 \ln(\frac{\text{Revenue}_{i,t}}{\text{Revenue}_{i,t-1}}) + \beta_2 \times D_{i,t} \times \ln(\frac{\text{Revenue}_{i,t}}{\text{Revenue}_{i,t-1}}) + \varepsilon_{i,t}$$

(1)

The relationship between tax changes and income changes can be found by regression. $D_{i,t}$ is fictitious variable. The specific relationship is as follows:

When $Revenue_{i,t} > Revenue_{i,t-1}$, $D_{i,t} = 0$, business income increased by 1%, tax expenditure increased by $\beta 1\%$.

When $Revenue_{i,t} < D_{i,t} = 0$, $D_{i,t} = 1$, business income decreased by 1%, tax expenditure decreased by $(\beta 1\ 1+\beta 2)\%$.

Obviously, when $\beta 1 > \beta 1 + \beta 2$, $\beta 2 < 0$, the existence of the tax stickiness—the change of tax when business income increases are larger than that when the business income decreases-can be verified. When the absolute value of $\beta 2$ is larger, meanwhile, the tax stickiness is stronger.

The Time Trend of Tax Stickiness For the time trend of tax stickiness, we use the extended model of model I to enlarge the time span from one year to three years, considering the different changes of tax and business income in one year, two years and three years, then to compare the degree of tax stickiness in order to verify hypothesis 2, and building the Model II:

$$\ln(\frac{\text{Tax}_{i,t}}{\text{Tax}_{i,t-n}}) = \alpha_{0,n} + \alpha_{1,n} \ln(\frac{\text{Revenue}_{i,t}}{\text{Revenue}_{i,t-n}}) + \alpha_{2,n} \times D_{i,t} \times \ln(\frac{\text{Revenue}_{i,t}}{\text{Revenue}_{i,t-n}}) + \varepsilon_{i,t} \quad (2)$$

The Influencing Factors of Tax Stickiness According to Liang Shangkun [6], and considering the influence of various factors on the coefficient $\beta2$, model III is constructed:

$$\ln(\frac{\text{Tax}_{i,t}}{\text{Tax}_{i,t-1}}) = \beta_0 + \beta_1\ln(\frac{\text{Revenue}_{i,t}}{\text{Revenue}_{i,t-1}}) + \beta_2 \times D_{i,t} \times \ln(\frac{\text{Revenue}_{i,t}}{\text{Revenue}_{i,t-1}}) + \beta_3 \times \text{Growth}_t \times D_{i,t}$$
$$\times\ln(\frac{\text{Revenue}_{i,t}}{\text{Revenue}_{i,t-1}}) + \beta_4 \times \ln\frac{\text{Assets}_{i,t}}{\text{Revenue}_{i,t}} \times D_{i,t} \times \ln(\frac{\text{Revenue}_{i,t}}{\text{Revenue}_{i,t-1}}) +$$
$$\beta_5 \times \ln\frac{\text{Debts}_{i,t}}{\text{Assets}_{i,t}} \times D_{i,t} \times \ln(\frac{\text{Revenue}_{i,t}}{\text{Revenue}_{i,t-1}}) + \varepsilon_{i,t}$$

$$(3)$$

$Growth_t$ represents the growth rate of China's GDP; $Assets_{i,t}/Revenue_{i,t}$ and $Debts_{i,t}/Assets_{i,t}$ represent the capital-intensive degree and asset-liability ratio respectively.

3.2   Sample Selection

This paper chose 2008–2015 as the time window and takes all A-share listed companies as initial samples to ensure the comprehensiveness and representativeness of the samples. Then we excluded them according to the following criteria. They are: ST companies, companies in financial industry and samples with missing data. Besides, companies with less than five years' existence are also be excluded to meet the research needs by considering the time span involved in the model design. In the end, 16143 companies-annual samples were obtained. The data comes from CSMAR database, and the statistical software is stata13.0.

4   Empirical Results and Analysis

Firstly, the mixed OLS model is selected by F-value test and B-P test, as it is more suitable for regression model of panel data in this paper. At the same time, cluster regression is used to report t-value adjusted by heteroscedasticity (Robust t) in order to improve the robustness of regression results.

4.1   Testing the Existence and Time Trend of Tax Stickiness (H1 and H2)

Table 1 reports the descriptive statistics of the main variables. It can be seen that for the sample companies, the median of annual tax expenditure and annual business income are 0.82 billion yuan and 1.446 billion yuan; the median ratio of tax and business income is 5.94%, indicating that the company will pay 5.94 yuan for every 100 yuan of income. Further, the observations in the column of 75% quantile show that the growth rates of tax in one year, two years and three years are 37.2%, 69.3% and 104.9%, while the growth rates of business income are 27.3%, 57.5% and 89.5%. The growth rate of tax is higher than that of income. This may be an intuitive manifestation of the tax stickiness. Judging from the standard deviation, there are great differences among different samples, so this characteristic is not reflected in the average, 25% and median values. However, the gap between the current tax growth rate and the income growth rate gradually narrows until it exceeds the 75% quantile.

Table 1. Definition and descriptive statistics of main variables in model I and II

Variable name	Variable definition	Sample	Mean value	Standard deviation	25% quantile	50% quantile	75% quantile
Tax	Tax	16143	34322.37	89193.87	3350.46	8182.91	23243.9
Revenue	Business income	16143	570139.42	1416797.13	60349.76	144595.41	384844.61
TaxRe	Tax/Business income	16143	0.0772	0.0614	0.0353	0.0594	0.101
Tax1	1-year tax change rate	13913	3.393	124.8	0.886	1.104	1.372
Revenue1	1-year income change rate	13913	12.53	1148	0.961	1.106	1.273
Tax2	2 year tax change rate	11697	6.123	173.5	0.897	1.227	1.693
Revenue2	2-year income change rate	11697	18.05	1430	1	1.239	1.575
Tax3	3-year tax change rate	9483	8.222	202.9	0.937	1.375	2.049
Revenue3	3-year income change rate	9483	24.44	1380	1.046	1.389	1.895

Unit Ten thousand yuan
Note Continuous variables have been winsorized at the level of 1% and 99%.

The regression coefficients of each variable in model I and model II are shown in Table 2 below. $beta2$ is -0.130 and significant at 1% level, which means that when the company's business income increases by 1%, the growth rate of tax is 0.715%, but when the business income decreases by 1%, the change rate of tax is 0.585%, the tax stickiness does exist. H1 is true. By analogy, when the time span is extended to 2-year and 3-year periods, the levels of $\alpha2$, 2 and 2, 3 are -0.129 and -0.115 respectively, significant at 1% level. This shows that tax stickiness still exists in the process of time extension. Because of $|-0.130| > |-0.129| > |-0.115|$, it can be seen that the degree of tax stickiness decreases with time, thus the hypothesis H2 is verified.

4.2 Testing the Influencing Factors of Tax Stickiness (H3a, H3b, H3c)

By regressing model III, the results are shown in Table 3. $\beta3$, $\beta4$, $\beta5$ indicate respectively the influence of macro-economy, capital intensity and asset-liability ratio on tax stickiness.

In the 1-year model, $\beta3$ is -0.599, which is in line with the expected negative value, but because its t value is not significant, it cannot be statistically recognized that macroeconomic growth has an enhanced effect on tax stickiness; $\beta4$ is -0.051, which is in line with expectations, and t value is statistically significant at the level of 5%, indicating that capital intensity contributes to tax stickiness, H3b is verified. $\beta5$ is 0.080, which is in line with the expected positive value, significant at the level of 5%. This shows that every 1% increase in asset-liability ratio will lead to a 0.08% decrease in tax stickiness, and H3c hypothesis has been verified. However, the value of 2 is positive and not statistically significant,

Table 2. Regression results of model I and model II

Variable coefficient	Model I	Model II	
		$n = 2$	$n = 3$
$\beta 1$	0.715***		
	(31.92)		
$\beta 1$	−0.130***		
	(−3.30)		
$\alpha 1, 2$		0.819***	
		(40.66)	
$\alpha 2, 2$		−0.129***	
		(−3.36)	
$\alpha 1, 3$			0.854***
			(37.56)
$\alpha 2, 3$			−0.115***
			(−2.60)
$\beta 0, \alpha 0, 2, \alpha 0, 3$	0.016***	0.013*	0.018
	(3.37)	(1.73)	(1.53)
Observations	13,912	11,696	9,482
R-squared	0.231	0.361	0.438
Adj_R2	0.230	0.361	0.438
F	657.3	1036	914.4

Note ***, **, * mean the coefficient is significant at 1%, 5%, and 10%, () is t-value.

indicating that the comprehensive effect of various factors on tax stickiness is not yet clear.

In the 2-year and 3-year models, the values of $\beta 3$ and $\beta 4$ are positive, contrary to the expected symbols, but not statistically significant. Therefore, it is impossible to draw a conclusion that macroeconomic and capital intensity have no enhanced effect on tax stickiness. The values of $\beta 5$ are 0.042 and −0.008 respectively, indicating that the 2-year model is consistent with the prediction, while the 3-year model is inconsistent with the prediction. This shows that the mitigation effect of $\beta 5$ on tax stickiness is weakening, but because t-test fails, the influence cannot be supported in statistics.

5   Robustness Checks

5.1   Substituting Variables

In order to check the robustness of the above conclusions, this paper uses Li Linmu and Wang Chong [10]'s the net tax expenditure in cash flow statement (all taxes paid—tax returns received) as an alternative of corporate tax. Given that tax policies like government subsidies, export rebate, and other methods

Table 3. Regression results of model III

Variable coefficient	Model III		
	1-year	2-year	2-year
$\beta 0$	0.017***	0.007	0.015
	(3.74)	(0.96)	(1.33)
$\beta 1$	0.714***	0.829***	0.857***
	(31.15)	(39.75)	(36.77)
$\beta 2$	0.075	−0.291***	−0.241**
	(0.84)	(−2.86)	(−2.00)
$\beta 3$	−0.599	0.527	0.266
	(−1.08)	(1.5)	(0.99)
$\beta 4$	−0.051**	0.031	0.007
	(−2.12)	(0.91)	(0.2)
$\beta 5$	0.080**	0.042	−0.008
	(1.98)	(1.13)	(−0.15)
Observations	13,911	11,687	9,471
R-squared	0.233	0.36	0.436
Adj_R2	0.232	0.36	0.436
F	286.1	421.8	367.5

Note ***, **, * mean the coefficient is significant at 1%, 5%, and 10%, () is t-value

may make enterprises receive some tax, using this index also has certain practical meaning and remain the results stable. Repeating the above models, the results are shown in Tables 4 and 5.

It can be seen that the results in Table 4 are not significantly different from those of the original models I and II. The negative values of $\beta 2$, $\alpha 2, 2$, $\alpha 2, 3$ and their t values are significant, then, the hypothesis that tax stickiness exists and the stickiness effect tends to weaken with time is verified again. The results of $\beta 3$, $\beta 4$ and $\beta 5$ in Table 5 are not significantly diffcrent from those in Table 3 of the original model III regression.

5.2 Using the Moving Average of Variables

The moving average of macroeconomic growth rate, capital intensity and asset-liability ratio can be set in the model to control the dynamic attributes of each influencing factor. Repeat model III regression. The results are as follows Table 6, the results are not significantly different from those in Table 3.

5.3 Using Manufacturing Samples

It is generally believed that the manufacturing industry has the most complete supply chain. Actually, there are 68.49% samples belonging to the manufacturing

Table 4. The Existence of net tax stickiness

Variable coefficient	Model I	Model II	
		n=2	n=3
$\beta 1$	0.877***		
	(22.62)		
$\beta 2$	−0.206***		
	(−3.48)		
$\alpha 1,2$		0.909***	
		(27.25)	
$\alpha 2,2$		0.102***	
		(−3.65)	
$\alpha 1,3$			0.898***
			(26.01)
$\alpha 2,3$			−0.127**
			(−2.14)
$\beta 0, \alpha 0,2, \alpha 0,3$	−0.007	−0.021*	−0.015
	(−0.95)	(−1.87)	(−0.90)
Observations	13,382	11,146	8,978
R-squared	0.156	0.252	0.316
Adj_R2	0.156	0.252	0.316
F	351.2	487.1	487.5

Note ***, **, * mean the coefficient is significant at 1%, 5%, and 10%, () is t-value

Table 5. Influencing factors of net tax stickiness

Variable coefficient	Model III		
	1-year	2-year	2-year
$\beta 0$	−0.01	−0.031***	−0.021
	(−1.29)	(−2.64)	(−1.23)
$\beta 1$	0.886***	0.926***	0.910***
	(21.93)	(26.79)	(25.98)
$\beta 2$	−0.004	−0.381***	−0.386**
	(−0.03)	(−3.08)	(−2.18)
$\beta 3$	−0.732	0.577	0.499
	(−1.02)	(1.53)	(1.45)
$\beta 4$	−0.005	0.053	0.031
	(−0.16)	(1.22)	(0.76)
$\beta 5$	0.167***	0.078	−0.015
	(2.74)	(1.41)	(−0.21)
Observations	13,381	11,137	8,967
R-squared	0.158	0.252	0.317
Adj_R2	0.158	0.252	0.316
F	151.2	214.7	200.7

Note ***, **, * mean the coefficient is significant at 1%, 5%, and 10%, () is t-value

Table 6. The average influencing factors and tax stickiness

Variable coefficient	Model III		
	1-year	2-year	2-year
$\beta 0$	0.023***	0.032***	0.040***
	(4.21)	(3.77)	(3.10)
$\beta 1$	0.695***	0.810***	0.842***
	(26.75)	(34.96)	(31.37)
$\beta 2$	0.087	−0.625*	−0.089
	(0.16)	(1.89)	(−0.29)
$\beta 3$	−0.271	0.621	−0.108
	(−0.31)	(1.39)	(−0.31)
$\beta 4$	−0.044	0.025	0.034
	(−0.99)	(0.60)	(0.80)
$\beta 5$	0.03	−0.008	−0.047
	(0.52)	(−0.15)	(−0.72)
Observations	11,687	9,468	7,251
R-squared	0.219	0.358	0.429
Adj_R2	0.219	0.357	0.429
F	225.7	330.1	281.4

Note ***, **, * mean the coefficient is significant at 1%, 5%, and 10%, () is t-value

industry. Thus, using the manufacturing samples to test the above assumptions can represent the whole listed companies. The results of H1 and H2 are as Table 7. The results of H3 are shown in Table 8.

In Table 7, $\beta 2$ is −0.073 and significant at the level of 10%. H1 is verified again. The value of $\alpha 2,2$ is −0.095, which is significant at 5% level. $\alpha 2,3$ is −0.065, whose absolute value is less than $\alpha 2,2$, though it is not significant in statistics. We can conclude that the hypothesis of H2 still has been partially validated in manufacturing samples.

The values of $\beta 5$ in Table 8 are 0.088 and 0.055 in the 1-year and 2-year models, significant in statistical, which means that H3c has been verified again, but H3a and H3b haven't been verified. In conclusion, there is no significant difference between the overall and the initial regression results.

6   Conclusions and Limitations

This paper explores the existence of corporate tax stickiness, which is a relatively innovative concept. At the same time, the authors analyze the reasons for the tax stickiness from various perspectives, along with citing the latest and related references. Then the corresponding research hypothesis and statistical models are put forward. Based on the data of A-share companies listed in China from 2008 to

Table 7. Tax stickiness in manufacturing samples

Variable coefficient	Model I	Model II	
		n=2	n=3
$\beta1$	0.798***		
	(31.71)		
$\beta2$	−0.073*		
	(−1.67)		
$\alpha1,2$		0.861***	
		(33.63)	
$\alpha2,2$		−0.095**	
		(−2.10)	
$\alpha1,3$			0.877***
			(30.30)
$\alpha2,3$			−0.065
			(−1.27)
$\beta0$, $\alpha0,2$, $\alpha0,3$	0.010**	0	0.002
	(2.00)	(0.03)	(0.13)
Observations	8,789	7,331	5,890
R-squared	0.267	0.373	0.453
Adj_R2	0.267	0.372	0.453
F	669.5	772.3	667.2

Note ***, **, * mean the coefficient is significant at 1%, 5%, and 10%, () is t-value

2015, this paper empirically examines the existence, time trend and influencing factors of tax stickiness, the basic conclusions can be drew as the following: (1) Tax stickiness does exist in Chinese listed companies; (2) Tax stickiness of listed companies tends to weaken with the time extension; The existence of stickiness is due to the time lag of enterprise management and policy formulation-with the passage of time, the situation becomes clear, and the space of stickiness will become smaller. (3) Macroeconomic growth rate has no significant effect on tax stickiness; the greater the capital intensity of enterprises in that year, the stronger the tax stickiness; the asset-liability ratio has a mitigating effect on tax stickiness. However, the impact of factors in different length of periods are different, yet their comprehensive effect on tax stickiness is unstable. In view of the fact that this paper is only a preliminary exploration of tax stickiness of listed companies in China and relevant researches are limited, there are still some limitations: (1) Besides the factors highlighted above, we believe that the influencing factors of tax stickiness are comprehensive and complex, including tax structure, tax policy, enterprise's own characteristics, such as property rights, industry attributes, tax planning and so on. (2) The impact of tax stickiness on enterprise performance is still unclear. Tax stickiness should be considered as an

Table 8. Factors of tax stickiness in manufacturing samples

Variable coefficient	Model III		
	1-year	2-year	2-year
$\beta 0$	0.003	−0.013	−0.004
	(0.57)	(−1.41)	(−0.26)
$\beta 1$	0.820***	0.886***	0.885***
	(31.45)	(33.94)	(29.34)
$\beta 2$	−0.205	−0.442***	−0.353**
	(−1.46)	(−3.07)	(−2.13)
$\beta 3$	1.061	1.046***	0.610*
	(1.17)	(2.71)	(1.87)
$\beta 4$	0.032	0.072*	0.054
	(1.01)	(1.75)	(1.10)
$\beta 5$	0.088**	0.055*	0.011
	(2.22)	(1.70)	(0.19)
Observations	8,788	7,327	5,886
R-squared	0.268	0.375	0.452
Adj_R2	0.268	0.374	0.451
F	275.6	320.8	271.3

Note ***, **, * mean the coefficient is significant at 1%, 5%, and 10%, () is t-value

effective way to influence companies' operation. Then, what is the moderation of corporate tax stickiness? What is the relationship between tax stickiness and firm performance? They are questions worth exploring and need further arguments and analyses.

Acknowledgment. This paper is supported by China National Social Science Foundation Project (17BJY186): Research on the linkage and coordination of tax system reform, tax distribution system and tax burden pressure of enterprises after replacing business tax with value-added tax.

References

1. China Development Publishing House (2017) Performance evaluation report of chinese listed companies in 2017
2. Deng, Z.J., Deng, L.P.: Tax neutrality, tax regulation and industrial policy. Fisc. Res. **29**(9), 32–35 (1995)
3. Fang, J.X.: Is there stickiness in executive compensation of listed companies in china? Econ. Res. **2009**(3), 110–124 (2009)
4. Gao, P.Y.: The mystery of sustainable and high-speed tax growth in china. Econ. Res. **2006**(12), 13–23 (2006)
5. Graham, J.R., Mills, L.F.: Using tax return data to simulate corporate marginal tax rates. J. Account. Econ. **46**(2), 366–388 (2007)

6. Liang, S.K.: Media attention, information environment and corporate cost stickiness. China's Ind. Econ. **2017**(2), 154–173 (2017)
7. Liu, J., Feng, L.: Fiscal centralization, government control and corporate tax burden: Evidence from china. China J. Account. Stud. **1**(3–4), 168–189 (2013)
8. Lu, B.Y.: Guo QW (2011) The source of rapid tax growth in china: Explanation under the framework of tax capacity and tax effort. Chin. Soc. Sci. **2**, 76–90 (2011)
9. Anderson, M., Banker, R.: Are selling, general, and distribution costs sticky? J. Account. Res. **41**(1), 47–63 (2003)
10. Li, L., Wang, C.: Tax burden, innovation ability and enterprise upgrading: Empirical evidence from new third board listed companies. Econ. Res. **11**, 119–133 (2017)
11. Porcano, T.M.: Corporate tax rates: Progressive, proportional, or regressive. J. Am. Tax. Assoc. **1986**(7), 17–31 (1986)
12. Kong, Y.: Cost stickiness research: Empirical evidence from chinese listed companies. Account. Res. **2017**(11), 58–65 (2017)
13. Pan, X.Z.: New enterprise income tax law and enterprise tax and fee burden: Based on the microscopic perspective of listed companies. Financ. Trade Res. **24**(5), 113–119 (2013)
14. Qin, H.N., Cheng, H.W., Peng, Q.: Accounting-tax differences and firm tax burden stickness. Bus. Account. **3**, 9–12 (2018)
15. Ren, Y.: The relationship between tax revenue and business income: Based on the annual report data of listed companies. J. Grad. Stud. Cent. South Univ. Financ., Econ. Law **2011**(6), 43–48 (2011)
16. Shevlin, T.: Taxes and off-balance-sheet financing: Research and development limited partnerships. Account. Rev. **62**(3), 480–509 (1987)
17. Stickney, C.P., Mcgee, V.E.: Effective corporate tax rates the effect of size, capital intensity, leverage, and other factors. J. Account. Public Policy **1**(2), 125–152 (1982)
18. Wang, C.B., Sun, Jiang, G.: Viscosity study of enterprise tax expenditure: From the perspective of government collection and management. Accounting Research **5**, 28–35 (2018)
19. Wilkie, P.: Corporate average effective tax rates and inferences about relative tax preferences. J. Am. Tax. Assoc. **10**(1), 75–88 (1988)
20. Xu, L.: Research on the viscous behavior of R&D expenditure stickiness of enterprises. Southeast University, Technical report (2014)
21. Cong, Y., Zhou, Y. J.: The characteristics, effects and policy suggestions of "tax burden rigidity" in china's current tax system: an empirical analysis based on the data of listed manufacturing companies from 2013 to 2016. South. Econ. **2017**(6), 53–63 (2017)
22. Zheng, S., L, Hao: Cost sticky behavior of chinese listed companies. Econ. Res. **12**, 26–34 (2004)

Research on the Generation and Development of Sharing Manufacturing based on the Long Tail Theory

Dong Chen[✉] and Sheng Yi

Sichuan Academy of Social Sciences, Chengdu 610071, People's Republic of China
1135382297@qq.com

Abstract. Based on the connotation and causes of sharing manufacturing, this paper introduces the long tail theory into sharing manufacturing, and analyses the realization path of sharing manufacturing. By building a sharing manufacturing platform to realize the sharing of manufacturing resources and production capacity, the long tail portion of the market demand curve in the manufacturing sector is continuously extended, widened, and "short head" and "long tail" are integrated to form a flat market demand and realize benefit sharing. At the same time, through the analysis of two typical cases of sharing platform "Tao Factory" and the intelligent machine tool platform i5, the commercial operation mode of sharing manufacturing platform is analyzed. In a conclusion, with the expansion and development of the sharing manufacturing platform, the long tail effect will gradually appear.

Keywords: Sharing manufacturing · Long tail theory · Development path · Operation model · Case analysis

1 Introduction

Germany proposed the concept of "Industry 4.0" at the Hannover Messe in 2013, the United States proposed "Industrial Internet" in 2014, China's "Made in China 2025" program is proposed in 2015, and countries such as Britain, France and Japan also put forward their own manufacturing development strategies. Under the background of the emergence and development of new technologies such as Internet of Things, Big Data and Artificial Intelligence, all countries are seeking the transformation and upgrading of manufacturing industry. In order to keep the competitive advantage and the technological leadership of manufacturing industry, the breakthrough of all countries will be intelligent manufacturing represented by "Industry 4.0". With the rapid development and expansion of the "sharing economy", the new business model of "not seeking ownership, but seeking utilization" has improved the efficiency of resource allocation and made full use of idle resource elements. In the manufacturing sector, the sharing economy has increased the utilization rate of idle manufacturing resources, enabling

© Springer Nature Switzerland AG 2020
J. Xu et al. (Eds.): ICMSEM2019 2019, AISC 1002, pp. 149–161, 2020.
https://doi.org/10.1007/978-3-030-21255-1_12

both supply and demand to profit in this process and promoting the transformation and upgrading of the manufacturing industry. The concept of "sharing manufacturing" has also arose. This paper is based on the connotation of sharing manufacturing, focusing on the sharing economic operation mode in the manufacturing field, and introducing the long tail theory into the sharing manufacturing, analyzing the realization approach and providing theoretical basis for the development of sharing manufacturing.

2   Literature Review

The sharing economy is one of a research hotspots right now, scholars at home and abroad often use many well-known original concepts such as "collaborative consumption", "peer-to-peer network economy", "on-demand economy", "right-of-use economy" and "platform economy" to explain the new concept of sharing economy, and try to start their own discussion in a clearer "domain" [10]. Researchers also try to use some theories as the theoretical basis for the study of sharing economy, such as property rights theory, transaction cost theory, Coase theorem, marginal cost, marginal utility and externality, information technology and platform theory, etc [2,6,9,13]. In terms of the research content of the sharing economy, the main research abroad includes the macroscopic sustainable development dimension, the evolution of micro-organizational behavior, the technology of sharing economic realization, the operation management and regulation of the sharing economy, etc [8]. The main problems of China's sharing economic research focus on three major research themes: what is sharing economy, how does the sharing economy develop, and how does the government implement shared economic operations management and regulation. Besides, after combing the research literature at home and abroad, we find that the current research on sharing economy is mainly concentrated in the field of consumer goods, and the research on idle capacity and equipment of enterprises in the manufacturing field have begun to appear sporadically [19].

Stephen and Mille put forward that the sharing economy is emerged in the service industry such as transportation and accommodation, and gradually expanded to the manufacturing sector in the later stage [18]. Shi takes the "i5" intelligent sharing platform of IMTCL (Shenyang Machine Tool Group) as a case, and focus on the intelligent production sharing business model, and believes that the intelligent production sharing mode design should take the customer value proposition as the starting point [21]. Chen's team carried out a series of studies based on the integration and development of the shared economy and traditional industries, and deeply demonstrated the important role of the sharing economy in industrial innovation. In particular, they focused on innovative and integrated research in the sharing economy and manufacturing industry, and formed a new research field based on the "intellectual sharing model" of manufacturing. Among them, Bao studied the integration of shared economy and intelligent manufacturing enterprises from the perspective of intelligent manufacturing sharing and user experience [22]. In addition, research on sharing manufacturing also includes sharing strategy selection, sharing platform construction, related technical issues, and manufacturing enterprises' knowledge sharing research and so on. For example, Pan studied the two-level supply chain of manufacturing capacity sharing problem formed by a single cloud platform and a manufacturer by means of differential game, and found that the sharing of cooperation can improve the overall profit of the supply chain. Under the certain condition, the cost-sharing contract can realize the coordination of the supply chain [20]. Chen analyzes the advantages and disadvantages of China's knowledge-sharing development by analyzing China's two major manufacturing smart phone manufacturing and intelligent network service manufacturing industrics [4].

In general, many studies have gradually focused on the sharing economy in the manufacturing sector, but the holistic research is still fragmented, especially the introduction of long tail theory on sharing manufacturing. In the business model of shared economy, the focus is often on the long tail customers which are not paid attention to by the traditional business model. With the advancement of technology, the long tail portion may be indefinitely extended or the curve is moved upwards, so that the long tail area is increased, and niche commodities will become an important profit growth point for enterprises under the sharing economy [14].

3   The Connotation of Sharing Manufacturing

At present, there is no clear and consistent definition of sharing economy in academic circles [17], the same as sharing manufacturing. Yu believes that the sharing manufacturing is based on a new generation of information technology, and enterprises or individuals can easily share all idle and redundant resources (including software and hardware resources), and capabilities, demanders can conveniently obtain these resources and capabilities at any time and on demand, so that both supply and demand sides of resources and capabilities can achieve a win-win situation [7]. The State Information Center (Administration Center of China E-government Network) also gives a relevant definition:

The capacity sharing of manufacturing mainly refers to a new economic form based on the Internet platform and characterized by the sharing of usage rights, integrating and disposing of distributed manufacturing resources and manufacturing capabilities around the manufacturing process to maximize manufacturing efficiency. Combining the above points, the understanding of sharing manufacturing can be seen separately from two aspects of "sharing" and "manufacturing": sharing refers to a short-term transfer of their products to others, in which the ownership of the product does not change [11]; manufacturing includes manufacturing resources such as equipment, technology, talent, and knowledge. Manufacturing capabilities are design, production, management. The purpose of sharing idle manufacturing resources and capabilities is to obtain income which can be in monetary or non-monetary form [3]. At the same time, sharing manufacturing is a new business phenomenon with the development of Internet.

Therefore, sharing manufacturing can be defined as a new economic model based on the Internet platform, in which manufacturing enterprises lease the idle manufacturing resources such as equipment, technology, talent, knowledge and the manufacturing capacity such as design, production and management for a short time to gain profits.This sharing manufacturing model not only maximizes manufacturing productivity, but also meets the needs of consumers for personalization and customization.

4   The Cause of Sharing Manufacturing

The shift of the shared economy model to manufacturing industry has already begun in the practice of enterprises, but the relevant academic research lags far behind the practice of enterprises [12]. According to the estimates of the State Information Center, China's manufacturing capacity-sharing market in 2017 is about 412 billion yuan, and more than 200,000 enterprises provide services through the platform, especially the small and micro enterprises. Internet companies have begun to enter the manufacturing sector and set up third-party platforms to meet manufacturing supply and demand; In order to seek opportunities for transformation and development, traditional manufacturing companies have also begun to use the sharing economy to open up resources and capacity. This has generated quite a number of sharing manufacturing platforms, such as "Tao Factory", "CASICloud", "i5" Intelligent Production Platform, HCH and so on.

Table 1. Overview of major capacity sharing platforms in China's manufacturing industry

Sharing platform	Affiliated company	Establishment time	Nature
HCH	Haier Group	May-14	Sharing platform for manufacturing enterprises
Midea Open Innovation Platform	Midea Group	Sep-15	
CASICloud	China Aerospace Science & Industry Corp	Jun-15	
"i5" Intelligent Production Platform	SMTCL	2014	
Ming Jiang Inteligent System	Shanghai Ming Jiang	2015	
Tao Factory	Alibaba Group	2013	Internet enterprises enter manufacturing industry
IngDan	Cogobuy Group	2014	

Data sources: online data collection and collation

The reason why sharing manufacturing develop rapidly and become a new business model are as follows:

First, due to the rapid development of shared economy, the shared economy has accelerated in all walks of life and various fields. The key areas include knowledge and skills, life services, housing and accommodation, transportation, medical sharing, shared finance and production capacity. According to the data released by the National Information Center, in 2017, the transaction volume of China's shared economic market is about 492.5 billion yuan, which is an increase of 47.2% over the previous year. And the share of total market transactions in non-financial sharing area increased to 42.6%. Shared economic structure is still improving, and the market transactions in manufacturing capacity sharing area are also in a fast rising stage.

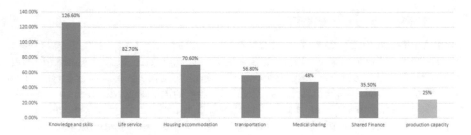

Fig. 1. Growth of market turnover in key areas of China's shared economy in 2017

Second, the need for structural reform on the supply side. As China's economy has shifted from high-speed growth to high-quality development, the "troika" of consumption, investment and export, which drives economic growth, has begun to show signs of weakness. China's economic growth mainly depends on investment, and the demand-side pulls to the supply-side. Shared economy, as a new business model, accurately matches the supply and demand sides through Internet technology, improves the utilization rate of idle resources, and plays a significant role in reducing production capacity and expanding effective supply.

Third, the supporting role of innovative technologies such as cloud computing, big data, artificial intelligence, and the Internet of Things. The premise of manufacturing capacity sharing is the application of Internet technology and intelligent production. Only when the related operation processes and manufacturing services of manufacturing achieves linking, management and information docking in the network cloud, the shared economy can achieve control. The development of these innovative technologies not only reduces the transaction costs such as search costs, but also improves the transaction matching rate by relying on information mining and integration matching.

Fourth, the global manufacturing industry is in the critical period of the alternating role of group technological breakthroughs and fundamental institutional innovation, and the phenomenon of information civilization has emerged. Under the pressure of global homogenization overcapacity, on the one hand, the trend of "customized design, single-piece small-batch production, individualized consumption" promotes the rapid development of intelligent, collaborative and cloud-based manufacturing technology. On the other hand, in order to meet the needs of "standardized design, mass production and homogeneous consumption", the bureaucratic "modern enterprise system" is becoming increasingly difficult to adapt to the requirements of intelligent, collaborative and cloud-based manufacturing technology for the transformation of production mode. Shared economy, as a new economic model, can better promote the transformation and the upgrading of manufacturing industry in line with the development of the times.

5   Sharing Manufacturing from the Perspective of Long Tail Theory

5.1   Important Information

Chris Anderson firstly puts forward the long tail theory [16], he believes that if the storage and circulation channels are large enough, the market share of products with low demand or poor sales can be comparable to or even larger than that of those with few hot sellers [1]. The long tail theory divides the market into two parts: "head" and "tail". The head demand curve is steep and towering, representing most of the market share, the tail market demand curve is flat and outreached. The types of commodities covered can be extended indefinitely, but the sales volume is small [15]. The basic principle of the long tail theory is that many small markets are concentrated into market energy which can compete with the mainstream big market, and in essence is connected to the scope economy. The long tail theory focuses not on the "head" of the demand curve with Pareto distribution in traditional economy, but on the "tail" with a large variety and a small number of products. With the development of technology and social economy, the improvement of the quality of life makes people more and more individualized.The further development of the Internet meets people's personalized needs on a large scale, and it is possible to aggregate a large number of niche markets. The long tail portion may be infinitely extended or the curve moves upwards, so the long tail area is increased, and niche commodities will become an important profit growth point for the enterprise in the future.

5.2   The Realization Path of Shared Manufacturing from the Perspective of Long Tail Theory

In the traditional manufacturing system, manufacturing enterprises with large-scale production capacity focus on the needs of many consumers, providing a large number of related mainstream products, while the needs of small consumers are often ignored. Mainstream products occupy the market and limit people's choice. At the same time, large enterprises have large-scale production capacity and core processing technology and equipment. Due to the lack of these soft and hard conditions, the majority of small and medium-sized enterprises or individual users cannot provide the individualized and diversified needs of the market. As shown in the Fig. 2, the "short head" part has a small variety of manufactured products and a high output, which is the large-scale standardized production; the "long tail" part of the manufacturing products have many varieties type and a low output, mainly meet the need of small and medium-sized enterprises and individual users. It belongs to customized production.

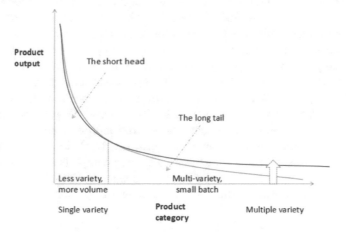

Fig. 2. The long tail theory in manufacturing

According to the long tail theory, once the production channels are widened, on the one hand, a large number of multi-variety and small-volume orders are matched to the appropriate large and medium-sized enterprises through the sharing manufacturing platform, releasing the idle manufacturing resources and capabilities of the manufacturing enterprises. On the other hand, small and medium-sized enterprises with insufficient resources, capital and R&D investment can rent equipment and manufacturing capabilities of large enterprises and core technology enterprises through shared manufacturing platforms for customized production. As a result, the long tail of the manufacturing sector shows a tendency to prolong and widen (as shown in Fig. 2), and the market share of long tail products will continue to expand. Sharing manufacturing platform plays a key role, which can achieve manufacturing supply and demand matching, agglomeration and integration of the online and offline resources. Due to the sharing economic platform, the characteristics of openness, sharing, and network externality, etc. the production channels become large enough, and transaction costs are also greatly reduced. It raises the conditions for realizing the long tail effect.

The rapid development of shared manufacturing platforms will bring two major changes to the market demand curve in the manufacturing sector. On the one hand, with the development of new generation information technologies such as cloud computing, big data, artificial intelligence, and the Internet of

Things, the marginal cost of sharing manufacturing platform channels, circulation, marketing, etc. tends to zero. By matching various types of manufacturing companies and satisfying the diversified and personalized manufacturing needs of users, it is possible to improve the users' scale and platform performance and achieve a range economy. The long tail part continues to extend and widen, and the market demand curve will flatten [5]. On the other hand, sharing manufacturing platforms have the typical characteristics of bilateral markets and network externalities. The more companies on the platform provide idle manufacturing resources and production capacity, the more companies or users with manufacturing needs will be attracted. The more demand-side users on the platform, the more the manufacturing supplier is willing to provide manufacturing resources and production capacity on the platform. Participants at both ends of the platform need complementary and interdependent. The utility of the demand side and the supply side increases as the number of users in the other market increases, reflecting positive cross-network externalities. The shared manufacturing platform attracts and retains demanders and suppliers through experience, mutual subsidies, and the provision of rich and practical cloud products. A large number of high-quality users and manufacturing companies exist on the platform to generate demand and supply for each other. As the scale of the platform expands, the externalities of the network continue to emerge and achieve a virtuous circle.

When the market curve in the manufacturing sector tends to be flat and stable, it shows that the long tail effect begins to exert an influence. The market share of direct matching of niche manufacturing products on the sharing manufacturing platform is sufficient to compete with the market share of manufactured products under the traditional manufacturing model. The "head" and "tail" are beginning to mix up, and users are more able to get diverse and personalized products. Through the new business model of sharing manufacturing, the idle capacity of manufacturing enterprises can be fully utilized. It can not only remove excess capacity, but also improve effective supply and meet diversified and personalized needs. The overall output of society will also improve.

5.3   Case Analysis

"Tao Factory" Tao Factory is founded in 2013. At first, Ali set up a matching platform between individual fashion sellers and factories in 1688. Now it is a platform to link Taobao sellers and factories, and solve the problems of finding factories, trying orders, turning over orders and developing new models for Taobao sellers. It is a solid bridge between sellers and high-quality factories. From the demand side, "Tao Factory" is a third-party trading platform with credit certificates and secured transactions agreements, and aggregated a large number of factory resources (currently it has about 15,000 factories). It can provide Taobao sellers a high-quality supply chain platform. On the supply side, "Tao Factory" can quickly match hundreds of billions of processing orders (more than 150 million registered enterprise users) in Taobao market through cloud computing, big data and artificial intelligence technology, which helping factories realize e-commerce transformation. Massive factory resources and 100 billion processing orders have enabled the expansion of niche products produced on the "Tao Factory" platform, and clearly have the conditions to achieve long tail effect in the clothing processing and manufacturing industry.

As a third-party platform of clothing production industry with Two-sided market characteristics, the operation mode of "Tao Factory" is similar to the platforms such as Didi and Uber. In the demand market (side 1 market), through the introduction and station of a large number of buyers, and then according to the sales volume and capacity of buyers, buyers are graded hierarchically to form a high-quality buyers pool; In the supply market (side 2 market), the sellers pool is formed by introducing investment and setting up the entrance threshold of the manufacturer, and classified according to the business attributes and quantity level of the factory. The main types of factory include women's clothing, men's clothing, children's clothing, bags, accessories, packaging and so on. The relationship between buyers and sellers on the platform is interdependent and mutually reinforcing. More and more users on "Tao Factory" attracts factories to participate in this platform. Conversely, the more factories the platform has, the more users will be attracted to stay, repeatedly, the network externalities are formed. After that, the platform matches the demand of both suppliers precisely through cloud computing, big data and artificial intelligence technology, cultivates the strategic win-win cooperative supply relationship between buyers and sellers, and borrows loans, credit and other business based on the credit system of buyers and sellers.

Generally speaking, "Tao Factory" has the following characteristics:

The first one is capacity sharing. The factory on the platform has been opened for a 30-days idle period. The idle schedule indicates the willingness of the factory to take orders. Taobao sellers can quickly search the matching factory on the platform. If the factory has no idle schedule, the search will be filtered out by default.

The second one is that factories with high flexibility will be recommended priority. Flexibility refers to the factory with large elasticity of production, covering the minimum order quantity, proofing cycle, production cycle, etc.

The third one is customized according to users' needs. The platform requires Tao factories to be stationed to provide free proofing, quotation and deadline for users, and the start-up standard is 30 pieces, production within seven days, and secured transactions in credit documents. Users can try small batches of test orders and quickly turn them over.

The fourth one is that the platform provides services such as transaction matching, security and so on. The platform solves the problems of lack of funds and security of funds in transactions by means of financial credit and secured transactions. At the same time, through multi-dimensional data analysis, the factories are classified to promote rapid and accurate information docking between supply and demand sides.

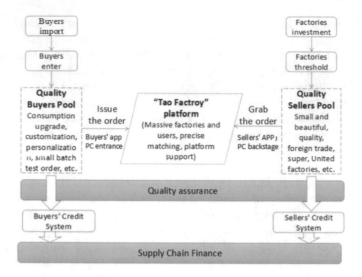

Fig. 3. "Tao Factory" business operation model

"i5" Intelligent Production Platform From the launch of i5 intelligent machine tools in 2014 to the construction of smart factories, intelligent manufacturing valleys and iSESOL cloud platforms, IMTCL (Shenyang Machine Tool Group) has initially built a new ecosystem of i5 intelligent manufacturing, and the company has accelerated its transformation from a single manufacturer to an integrated service provider. By the end of 2017, iSESOL cloud platform had about 10,620 online machine tools, and successfully connected to more than 500 enterprises and institutions, more than 2,000 network factory customers, distributed in more than 30 cities and regions across the country, with more than 2360 thousand service hours, and online order volume exceeded 5,500.

The iSESOL cloud platform is essentially a link of "cloud service + IoT + smart terminal", which realizes the "Internet + intelligent manufacturing" operation mode, and connects the value-added network of stakeholders through the layout of intelligent terminal devices. The operation of the sharing platform is roughly as shown in Fig. 4.

It can be seen from the figure that before the implementation of the sharing manufacturing business model, IMTCL used the traditional production model from machine tool manufacturing to sales, and products, technology, capital, logistics and other basic one-time transactions were completed. After implementing the business model of sharing manufacturing, the intellectualization of operation, service, diagnosis and programming has been set up. In order to hook more potential users, including small and medium-sized enterprises, enjoy the sharing platform at the same time, IMTCL uses lease management, resource recovery, and fee-based commissioning to transfer smart manufacturing technology at a low cost, making it a compatible platform for business model innovation. The basic form of intelligent manufacturing sharing mode is B2B: IMTCL, as

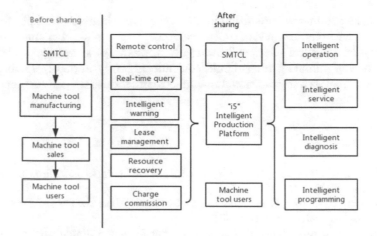

Fig. 4. iSESOL cloud platform operating model [22]

the supplier of i5 intelligent manufacturing sharing, uses the "low price, zero price" strategy to transfer the right to use the shared platform at low cost, so that the majority of manufacturing enterprises can enjoy industrial manufacturing services. In this way, small and medium-sized enterprises with insufficient resources, capital and R&D investment can rent IMTCL equipment and manufacturing capacity through "i5" sharing manufacturing to customize the production. Manufacturing long tails will also show a tendency to lengthen and widen.

6   Conclusion and Enlightenment

From the above analysis, we can see that sharing manufacturing is a new economic model based on the Internet platform. Manufacturing enterprises lease idle manufacturing resources such as equipment, technology, talent, knowledge and manufacturing capacity such as design, production and management for a short time to gain profits. The sharing manufacturing model not only maximizes manufacturing productivity, removes excess capacity and increases effective supply, but also meets the need of consumer of individualization and customization. In the context of the continuous development of new generation information technologies such as cloud computing, big data, artificial intelligence, and the Internet of Things, this paper introduces the long tail theory into sharing manufacturing, and analyses the realization path of sharing manufacturing. By building a shared manufacturing platform to realize the sharing of manufacturing resources and production capacity, the long tail portion of the market demand curve in the manufacturing sector is continuously extended, widened, and "short head" and "long tail" are integrated to form a flat market demand and realize benefit sharing. At the same time, through the analysis of two typical cases of "Tao Factory" sharing platform and "i5" intelligent machine tool platform, we analyzed the commercial operation mode of sharing manufacturing platform. We found that with the expansion and development of the sharing manufacturing platform, the long tail effect will gradually appear. As a new type of business model, sharing manufacturing should continue to study in depth. In the future, Internet companies and traditional manufacturing industries can further develop in this direction.

References

1. Anderson, C.: The long tail—why the future of business is selling less of more. Hyperion **24**(3), 274–276 (2006)
2. Bardhi, F., Eckhardt, G.M.: Access-based consumption: the case of car sharing. J. Consum. Res. **39**(4), 881–898 (2012)
3. Belk, R.: You are what you can access: sharing and collaborative consumption online. J. Bus. Res. **67**(8), 1595–1600 (2014)
4. Chen, W.: Research on knowledge sharing in intelligent manufacturing enterprises in the vision of open innovation. Reform **2018**(10), 102–110 (2018). (in Chinese)
5. Chen, X.: Research on shared finance development strategy based on long tail theory. Mod. Manag. **37**(3), 5–7 (2017). (in Chinese)
6. Costantini, F.: The "peer-to-peer" economy and social ontology: legal issues and theoretical perspectives. Int. Semant. Web Conf. **2015**, 15–22 (2015)
7. Costantini, F.: Research on the Planning System in Sharing Manufacturing. Zhejiang University (2016). (in Chinese)
8. Dai, K., Chen, W., Li, X.: Sharing economic research context and its development tren. Econ. Perspect. **2017**(11), 126–140 (2017). (in Chinese)
9. Evans, D.S., Schmalensee, R.: Multi-sided platforms. In: International Journal of Industrial Organization. The New Palgrave Dictionary of Economics. Palgrave Macmillan UK (2017)
10. Frenken, K.: Political economies and environmental futures for the sharing economy. Philos. Trans. **375**(2095), 23–25 (2017)
11. Habibi, M.R., Davidson, A., Laroche, M.: What managers should know about the sharing economy. Bus. Horiz. **60**(1), 113–121 (2016)
12. Heinrichs, H.: Sharing economy: a potential new pathway to sustainability. GAIA-Ecol. Perspect. Sci. Soc. **22**(4), 228–231 (2013)
13. Hellwig, K., Morhart, F., et al.: Exploring different types of sharing: a proposed segmentation of the market for "sharing" businesses. Psychol. Mark. **32**(9), 891–906 (2015)
14. Hjorth-Andersen, C.: Chris anderson, the long tail: how endless choice is creating unlimited demand. The new economics of culture and commerce. J. Cult. Econ. **31**(3), 235–237 (2007)
15. Lan, H., Liu, L., Liu, X.: Analysis on the generation and development of sharing retailing based on the long tail theory. China Bus. Mark. **2018**(7), 12–19 (2018). (in Chinese)
16. Marshall, W., Consedine, T.: The long tail. Mon. Labor Rev. **130**(3), 69–70 (2007)
17. Martin, C.J.: The sharing economy: a pathway to sustainability or a nightmarish form of neoliberal capitalism? Ecol. Econ. **121**, 149–159 (2016)
18. Miller, S.R.: First Principles for Regulating the Sharing Economy, pp. 147–202. Social Science Electronic Publishing (2015)

19. Ndubisi, O.N., Ehret, M., Wirtz, J.: Relational governance mechanisms and uncertainties in nonownership services. Psychol. Mark. **33**(4), 250–266 (2016)
20. Pan, X., University, T.: Analysis of dynamic sharing strategies of manufacturing capacity under cloud manufacturing environment. J. Ind. Technol. Econ. **35**(05), 16–29 (2016). (in Chinese)
21. Shi, Z., Cai, R., Zhu, X.: Study of business model innovation for intelligent production sharing. China Soft Sci. **2017**(6), 130–139 (2017). (in Chinese)
22. Bao, S., Cai, R.: Intelligent production sharing and user experience: a case study of shenyang machine tool. Ind. Eng. Manag. **22**(03), 77–82 (2017). (in Chinese)

Evaluation and Comparative Study of Provincial Resilience in China

Chao Huang[✉], Jihui Zhong, Jing Wu, and Jiaqi Fan

School of Public Administration, Sichuan University, Chengdu 610065,
People's Republic of China
danceofriver@gmail.com

Abstract. With the growth of population, urbanization, industrialization and other social processes, China has entered a period of high incidence of all kinds of emergencies. The problem of hazard prevention and mitigation has become increasingly prominent. Thus more and more attention has been paid to the concept of resilience. In this paper, we evaluated the provincial resilience of China with the entropy method. The evaluation was based on the BRIC model proposed by Carter, which consists of five aspects: resilience of the economy, the resilience of society, resilience of environment, resilience of infrastructure, and resilience of organization. The data of the 31 provinces originated from the National Statistical Yearbook, China Environmental Statistical Yearbook, and China Energy Yearbook et al. The results of evaluation and comparison showed that the general level of provincial resilience is quite low. Meanwhile, the differences among provinces are relatively large. The resilience has a significant correlation with the economic development level of provinces, infrastructure construction, and ecological environment.

Keywords: Provincial resilience · Evaluation of resilience · Comparative study · Entropy value method

1 Introduction

In recent years, there have been a series of catastrophes of different types around the world, such as the 2008 Wenchuan earthquake in China, the 2010 oil spill in the Gulf of Mexico, the 2011 earthquake in eastern Japan and the resulting tsunami. These great disasters remind us that the world today is facing the threat of terrorism, public health events, cyclical economic crises, natural disasters, and technological disasters. Faced with kinds of uncertainties, the traditional disaster prevention concept can no longer meet the future development needs of the city. To figure out how to deal with these uncertain disturbances and impacts, scholars and policymakers have conducted relevant researches and put forward countermeasures. Theconcept of resilience, which has become a specific issue studied by scholars, provides new ideas for disaster response from the perspective of dynamics, adaptation, and development.

© Springer Nature Switzerland AG 2020
J. Xu et al. (Eds.): ICMSEM2019 2019, AISC 1002, pp. 162–176, 2020.
https://doi.org/10.1007/978-3-030-21255-1_13

The study of resilience has spread from theoretical operationalization to the empirical application of the system. The scholars in China have also made some progress in the research on resilience, but most of the research is qualitative study. In this paper, we reviewed the resilience of the related research, expounded the basic concept of resilience, and combined with the actual situation in China to construct an evaluation index system for provincial resilience. Based on the index, we evaluated the resilience of 31 provinces in China (except Hong Kong, Macao, and Taiwan). At last, some relevant conclusions and suggestions are put forward on China's resilience construction through comparison.

2   Literature Review

"Resilience" is derived from the Latin word "resilio", which is originally meant to "recover to the original state". In the 1970s, Holling, a Canadian ecologist, introduced the idea of "resilience" into the field of systems ecology for the first time, and proposed "ecosystem resilience" to describe the persistence of ecosystem when faced of changes caused by natural or human-made factors [11]. Since the 1990s, "resilience" has been gradually applied to engineering, social and economic fields, and its connotation has been constantly enriched [19]. In the 21st century, along with the deepening of people's awareness of natural and human-made disasters and the improvement of their awareness of coping, "resilience" began to be widely used in disaster management to study the response and adaptation of urban systems to future uncertain and unpredictable impacts [1,9,21,25]. "Resilience" has become an important part of disaster research. UNISDR defines "resilience" as the ability of systems, communities or society to resist, absorb, adapt and recover from the impacts of attacks in a timely and effective manner, including protecting and restoring its necessary infrastructure and functions [18]. "Resilient city" has three characteristics: firstly, it can sustain substantial changes, absorb disasters and maintain essential functions and structures; secondly, it can adapt itself to disasters, resist and recover from them; thirdly, it can learn through self-organization to more severe disasters in the future [2,4,7,10,30]. How to scientifically quantify the resilience evaluation can help to apply the resilience concept to the resilience construction practice and improve the ability of the complex social ecosystem to cope with various uncertainties in the future [27].

At present, there are two types of resilience evaluation systems. The first category is the resilience evaluation study for a specific region (e.g., coastal city [24] or a disaster (e.g., climate change [28], earthquake [6,22], floods and typhoons [26] and other disasters). The second type is the universal, comprehensive resilience evaluation research [3,13,15–17,23,29]. Mayunga regards community capital as the resource reserve that can be mobilized when it encounters uncertain disturbance and quantifies community resilience through five capital forms, namely, society, economy, material, human and nature [3]. Burton measured "geographical county-level resilience" from social, economic, institutional, infrastructural and environmental dimensions, and used Hurricane Katrina and the restoration of the Mississippi bay coast as case studies [7]. Cutter et al. established a resilience evaluation system in 2008 from six aspects: society, economy, system, infrastructure, ecology and community [15,16]. Moreover, they (2010) took into account the particularity of ecosystems in different regions, which would have different impacts on resilience, and therefore could not be compared among regions, so they the removed the ecological part [17]. In 2014, Cutter further put forward the Baseline Resilience Indicators for Communities (BRIC), which quantifies the Resilience of different regions of the United States from six dimensions including society, economy, community, system, infrastructure, and the environment by using public data of government or research institutions [29].

As in China, Meng Lingjun et al. analyzed the resilience level of a community in Tianjin from three aspects, including social composition, community space, and management system, by using the resilience evaluation system of RATA [20]; Yun Yingxia, Xu Manchen et al. discussed the evaluation index system of community disaster resilience based on fire safety [31]; Liu Jiangyan and Zeng Zhongping comprehensively studied domestic and foreign studies, drew lessons from the evaluation index system of sustainable urban development, and took Wuhan as the research area, they established the evaluation index system of urban resilience from four dimensions including ecological resilience, economic resilience, engineering resilience and social resilience [14]. In the second category, Li Ya and Zhai Guofang constructed the urban disaster resilience evaluation system of China from six aspects: economic resilience, social resilience, environmental resilience, community resilience, infrastructure resilience and organizational resilience [19].

From the existing research, the scholars out of China have preliminarily constructed a complete research system including theoretical operationalization, index system construction and the empirical application of the system. Cutter's BRIC model can be used for horizontal and vertical comparison of urban resilience, which is highly practical and replicable and has been adopted by many scholars and research institutions [19]. The research in China mainly focuses on the evaluation of resilience in specific areas, which is of low reproducibility to popularize. Although Li Ya and Zhai proposed the evaluation index system at the municipal level, the index system could not be widely applied in the whole country due to data acquisition and other reasons. Therefore, while drawing on the evaluation framework of Cutters BRIC model, this paper takes into account the differences between China and western countries, and follows the principles of the establishment of the indicator system (consistent with the tenacity implication principle, the ability comparison principle, the representation principle and the feasibility principle) [17], and localize its specific indicators.

3   Data and Methodology

Based on the existing research results, we formed the index system according to the concept of resilience, as shown in Table 1.

The Data came from the national statistical yearbook, China environment statistical yearbook, China law statistical yearbook, China energy yearbook,

Table 1. Indexes and index weight

First-class indexes	Index weight	Index weight	Second-class indexes	Data source
Economic resilience	0.181577	0.003890	Average annual GDP growth rate %	China statistical yearbook 2017
		0.042384	GDP Per capita /yuan	China statistical yearbook 2017
		0.039926	The proportion of tertiary industry in GDP %	National Bureau of statistics annual data 2017
		0.009762	Energy consumption per unit of GDP /ton of SCE/10000 yuan	China energy yearbook 2017
		0.085615	Number of active inventor patents of industrial enterprises above the scale	National Bureau of statistics annual data 2017
Social resilience	0.256823	0.014604	Natural population growth rate %	China statistical yearbook 2017
		0.015532	The population of age 14 and under in total %	China statistical yearbook 2017
		0.018034	Total dependency ratio %	China statistical yearbook 2017
		0.019306	The proportion of the population over 64 years old %	China statistical yearbook 2017
		0.012223	Urban population density/ person/sq.km	China statistical yearbook 2017
		0.034850	Per capita annual disposable income/yuan	China statistical yearbook 2017
		0.015297	Average number of students in colleges and universities per 100,000 population	China statistical yearbook 2017
		0.003510	the proportion of illiteracy in the population aged 15 and over %	China statistical yearbook 2017
		0.009336	Social services composite index	China civil affairs statistical yearbook 2017
		0.010531	The proportion of Social services account in GDP %	China civil affairs statistical yearbook 2017
		0.057758	Unemployment insurance rate %	China labor statistics yearbook 2017
		0.013116	Number of cases handled by courts per 10,000 population	China law yearbook 2017
		0.032727	Unemployment rate registered in urban areas %	China statistical yearbook 2017

continued

Table 1. continued

First-class indexes	Index weight	Index weight	Second-class indexes	Data source
The environment of resilience	0.199159	0.015646	Per capita area of parks/sq.m	China statistical yearbook 2017
		0.012455	The coverage rate of afforestation in the developed area %	China statistical yearbook 2017
		0.027760	Per capita grain output/kg	China statistical yearbook 2017
		0.068519	Forest coverage rate %	National Bureau of statistics annual data 2017
		0.027491	Innocuous disposal rate of living garbage (%)	China statistical yearbook 2017
		0.005830	Percentage of sewage treatment %	Chinese city statistical yearbook 2017
		0.007425	The rate of comprehensive utilization of industrial waste residue %	China statistical yearbook 2017
		0.007317	The proportion of pollution control accounts in GDP %	National Bureau of statistics annual data 2017
		0.006763	The annual per capita electricity consumption /10000 kwh	China statistical yearbook 2017
		0.003890	The annual per capita water consumption /m^3/ person	National Bureau of statistics annual data 2017
		0.010129	The proportion of the direct economic loss caused by natural disasters accounts in GDP %	National Bureau of statistics annual data 2017
		0.005934	Number of environmental emergencies	China environment yearbook 2017
Infrastructure resilience	0.186817	0.008860	Per capita area of parks/sq.m	China statistical yearbook 2017
		0.023172	Public transport vehicles per 10000 persons	China statistical yearbook 2017
		0.025446	Internet penetration rate %	National bureau of statistics annual data 2017
		0.029358	Access to telephones (include mobile phones) (set/100 persons)	National Bureau of statistics annual data 2017
		0.042424	Turnover of freight traffic (billionton - km)	National Bureau of statistics annual data 2017
		0.017055	Number of medical institutions per 10000 people	National Bureau of statistics annual data 2017
		0.019226	Number of health technical personnel per 10000	China statistical yearbook 2017
		0.021276	Number of beds in medical institutions per 10000 population	China statistical yearbook 2017

continued

Table 1. continued

First-class indexes	Index weight	Index weight	Second-class indexes	Data source
Organizational resilience	0.175623	0.064072	Per capita General public budget revenue /10000 yuan	China statistical yearbook 2017
		0.050444	Per capita General public budget expenditure /10000 yuan	China statistical yearbook 2017
		0.026664	Number of social services and facilities per 10000 people	China civil affairs statistical yearbook 2017
		0.017641	Number of social service workers per 10000 person	China civil affairs statistical yearbook 2017
		0.016802	Number of social organizations per 100000 people	China civil affairs statistical yearbook 2017

China civil administration statistical yearbook, China social service statistical data, China city statistical yearbook, and China bureau website, all are public data set of government.

In order to eliminate the influence of different index dimension, we advocate using extreme value method for non-dimension of indicators. The positive indicator is that the larger the original data are, the better it is for resilience. The treatment formula is shown in Eq. 1. The negative indicator is that the smaller the original data are, the better it is for resilience. The treatment formula is shown in Eq. 2.

$$Y_{ij} = \frac{X_{ij} - Min(X_{ij})}{Max(X_{ij}) - Min(X_{ij})} \tag{1}$$

$$Y_{ij} = \frac{Max(X_{ij}) - X_{ij}}{Max(X_{ij}) - Min(X_{ij})} \tag{2}$$

X_{ij} represents the original data of the evaluation index, Y_{ij} is the standardized data, represents the first-level index ($i = 1, 2, 3, 4, 5$), and j represents the second-level index corresponding to the first-level index. For example, X_{34} refers to the area of cultivated land per capita in environmental resilience.

In this study, there are 12 negative indicators: ten thousand yuan GDP energy consumption, the proportion of the population under 14 years old and, 65 years old and overpopulation, total dependency ratio and density of urban population, the proportion of illiterate population accounted for more than 15 years of age and population, court cases per ten thousand people, urban registered unemployment rate, power consumption, water consumption per capita, per capita natural disasters caused by the direct economic loss of a share of GDP, the number of emergency environmental accidents. The remaining 31 indicators are all positive indicators. Standardized data were obtained after the positive or negative processing of different indicators.

Index weight can reflect the importance of each index in the index system. In order to allocate weight more objectively and scientifically, the entropy method is adopted in this study. Entropy method is a mathematical method that gives actual weight by taking into the amount of information provided by each index comprehensively [12]. According to the information entropy theory, information entropy is a measure of information uncertainty. The more significant the amount of information is, the smaller the uncertainty will be, and the same as the information entropy. If the entropy value of an indicator is smaller, the higher the role it plays, it should be given a higher weight [5,8].

Firstly, calculate the proportion P_{ij} of the secondary index j in the corresponding first-level index i.

$$P_{ij} = \frac{X_{ij}}{\sum\limits_{i=1}^{n} X_{ij}} \tag{3}$$

Secondly, calculate the information entropy e^j of the index j as shown in Eq. 4. (We use the ratio of the actual entropy to the maximum entropy $\ln n$, and the value range of information entropy is revised between 0 and 1, that is, if there is no difference between the secondary indicators, $P_{ij} = \frac{1}{n}$, $e^j = 1$, if there is only one secondary index, all the other regions are 0, $P_{ij} = 1$, $e^j = 0$).

$$e^j = -\frac{1}{\ln n} \sum\limits_{i=1}^{n} (P_{ij} \ln P_{ij})(i = 1, 2, 3, \ldots, n; j = 1, 2, 3, \ldots, m) \tag{4}$$

Thirdly, use entropy e^j to determine the weight w_j as shown in Eq. 5.

$$w_j = \frac{1 - e^j}{\sum\limits_{j=1}^{m} (1 - e^j)}(j = 1, 2, 3, \ldots, m) \tag{5}$$

Finally, use linear weighting to get the score Z_{ij} of resilience in Table 2.

$$Z_{ij} = \sum\limits_{j=1}^{m} w_j \cdot X_{ij}(i = 1, 2, 3, \ldots, n; j = 1, 2, 3, \ldots, m) \tag{6}$$

4   Evaluation and Comparison

4.1   Overall Characteristics of Provincial Resilience in China

Table 2 is obtained after descending order of resilience score of each province. The mean value of provincial resilience in China is 33.32, and the region with the highest resilience is Beijing, with a score of 63.12, followed by Zhejiang (55.48), Tianjin (50.88), Shanghai (49.97), Guangdong (49.69), Jiangsu (41.51) and Tibet (41.42). Jiangxi gets the lowest resilience score at 20.70. Scores of Yunnan (21.76), Guangxi (21.91), Henan (22.13) and Guizhou (22.65) were also

Table 2. Consumer trust operational item

Region	Score
Beijing	63.41617
Zhejiang	55.48176
Tianjin	50.87912
Shanghai	49.97185
Guangdong	49.69441
Jiangsu	41.51371
Tibet	41.41827
Heilongjiang	38.44775
Sichuan	36.56430
Inner Mongolia autonomous region	35.10801
Shandong	34.73167
Hubei	34.21470
Liaoning	31.18884
Fujian	30.74120
Shaanxi	30.62300
Ningxia	30.39004
Xinjiang	30.14439
Chongqing	29.24414
Qinghai	29.06036
Hainan	27.92343
Shanxi	27.69199
Gansu	26.57008
Hebei	25.88788
Jilin	24.71228
Anhui	24.64367
Hunan	23.50467
Guizhou	22.64581
Henan	22.13288
Guangxi	21.91220
Yunnan	21.76004
Jiangxi	20.70030

relatively low. The highest value is triple that of the lowest score of provincial resilience, which indicates that there is a significant difference in resilience between different provinces. The scores are divided into five intervals: 70–60, 60–50, 50–40, 40–30, and 30–20. The corresponding toughness strength was higher resilience, high resilience, medium toughness, low toughness, and lower resilience.

(The resilience score are shown in Fig. 1, which is made by the "map hui" web-page). The ratio of cities with different resilience from high to low is 3.23%, 6.45%, 12.90%, 32.26%, and 45.16%. Among them, the number of regions with higher resilience and high resilience accounts for 9.68% of the total, while the number of regions with lower resilience and low resilience accounts for 77.42% of the total. Therefore, it can be seen that the overall of provincial resilience in China is low, and these regions are the inadequate inability of coping with various uncertainties and recovering from damage.

4.2   Analysis of First-Class Indexes

Figures 2, 3, 4, 5, 6 show the scores of different dimensions of resilience in different provinces. It can be seen that the regional difference of economic resilience is enormous, which is consistent with the spatial distribution of overall resilience score. The scores of social resilience are relatively high, and the eastern coastal cities have more advantages. The scores of environmental resilience are generally low, and the differences are not evident except Sichuan and Tibet. Most of the areas with low scores on infrastructure resilience are in the southwest. The scores of organizational resilience are generally low, among which organizational resilience is relatively high in south China and Qinghai-Tibet plateau.

According to the scores of different provinces at different dimensions of resilience, we draw the radar charts as shown in Fig. 7.

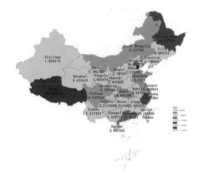

Fig. 1. Provincial resilience   **Fig. 2.** Economic resilience

4.3   Analysis of Second-Class Indexes

We get the top ten secondary indexes with the highest weight in our study as shown in Table 1: industrial enterprises above designated size active num-ber of invention patents, the forest coverage rate, general budget revenue per capita, unemployment insurance rate, general budget expenditure per capita, goods turnover amount, per capita GDP, industrial structure (the proportion of

Fig. 3. Social resilience

Fig. 4. Environmental resilience

Fig. 5. Infrastructure resilience

Fig. 6. Organizational resilience

the third industry), the per capita disposable income, urban registered unemployment rate. Then we can summarize the factors that have a significant impact on regional resilience into four aspects: levels of economic development, infrastructure construction, organization, and social management, and ecological environment.

Levels of economic development mainly include the scale and mode of economic development. First of all, the scale of economic development represents the existing economic strength of a region and is an important factor influencing the resilience. It goes through and supports the whole process of urban construction and development, directly affects local infrastructure construction, and directly determines the level of government organization and management as well as the investment in ecological and environmental protection. Secondly, the mode of economic development represents the economy sustainable development ability. The development of technology and the third industry can drive the sustainable development of the economy. Regional resilience requires not only muscular economic strength but also the optimization of industrial structure and the sustainability of economic development. Infrastructure is the essential condition for

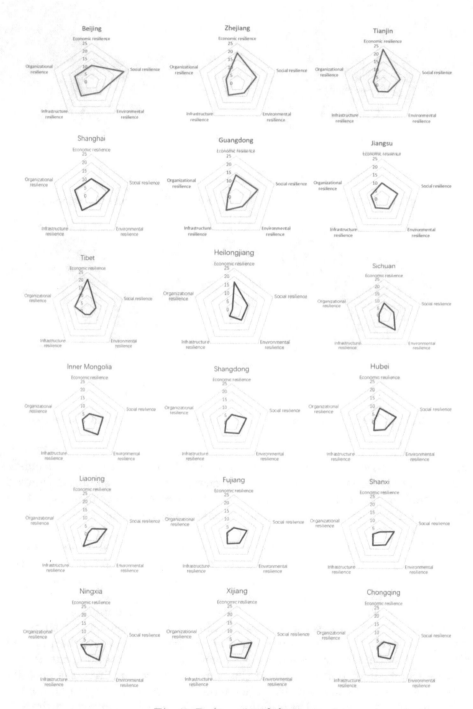

Fig. 7. Radar map of the score

ensuring the functions of the region, as well as the disaster body for the region to resist various kinds of disasters. In addition to surviving the disaster to maintain functions of the region, it also includes covering losses caused by disasters, providing relief and participating in post-disaster reconstruction and development. Infrastructure resilience affects disaster resilience in two aspects: disaster resistance and post-disaster relief. Organization and social management are essential parts of regional resilience software construction. They emphasize the role of human beings, indicates not only satisfying material needs and spiritual expression of human beings but also the attention and input of the government, creating a stable social environment and improve people's ability to cope with various emergencies. Ecosystems are capable of breaking down and digesting waste from human production and life. Therefore, we should keep development and construction activities within the legal carrying capacity of the natural environment and maximize the protection of the natural ecological environment.

5   Conclusions and Recommendations

Based on Cutter's BRIC model and the actual situation in China, we build the provincial resilience evaluation indexes covering five aspects, including economic resilience, social resilience, environmental resilience, infrastructure resilience, and organizational resilience, confirm index weight by entropy method, and then we evaluate and compare the resilience of 31 provinces in China. Overall, the provincial resilience level of China is generally low, with substantial regional differences. The overall resilience score is consistent with the regional distribution of economic resilience score. From each dimension of resilience, the level of provincial social resilience is relatively high, and the eastern coastal areas have more advantages than other areas. The level of environmental resilience is generally low in all provinces except Sichuan and Tibet. The infrastructure resilience of southwest region is generally low. The level of organizational resilience is also generally low, especially in south China and the Qinghai-Tibet plateau. According to the second-class indexes, the main factors affecting the provincial resilience include the scale and mode of economic development, organization and social governance, infrastructure construction level and ecological and environmental protection.

Given the above findings, we propose the following suggestions:

Firstly, adjust the industrial structure and increase investment in high-tech industries to enhance economic strength and small regional differences. The development of high and new technology industries meets the dual requirements of saving resources and protecting the environment. Therefore, all provinces should actively promote industrial transformation and upgrading, guide the development of high-tech, low-energy, and low-pollution industry. The provincial comprehensive resilience is also closely related to economic resilience. Therefore, narrowing the regional economic development gap is conducive to narrowing the regional resilience gap. Secondly, strengthen infrastructure, especially in southwest China. As the material carrier of social production and residents' life, infrastructure is the basic condition to guarantee the smooth progress of people's production and life. However, in the face of disasters, the infrastructure, as the disaster-bearing body, will be most directly impacted. Therefore, the government should strengthen the completeness, robustness and of infrastructure, and pay special attention to the redundancy of infrastructure, which plays an important role in maintaining the normal functions of the region when it is damaged. Thirdly, pay attention to organization and social management. Organization and social management, as the software condition of urban resilience, mainly include government management and social security. Fiscal revenue and expenditure can measure the government's ability to mobilize resources for resilience building. The social security level is mainly reflected in creating a stable social environment, providing residents with medical care, employment and other security, which helps reduce social risks and enhance residents' ability to resist disasters. Therefore, the government should pay attention to organization and social management, gradually improve the social security system, explore the security system in line with regional characteristics, and improve the ability of residents to resist disasters. Fourthly, strengthen environmental protection and promote conservation culture. The ecosystem can decompose and digest human waste and provide abundant resources for human production and life. To strengthen environmental protection, on the one hand, it is necessary to protect the natural ecological environment and maintain the natural purification capacity of the ecosystem itself. On the other hand, negative impacts on the environment should be reduced, such as energy conservation and emission reduction, and economic development mode transformation.

Acknowledgements. This work was financially supported by the Sichuan Social Science Planning Project, "Research on the Theory and Method of Statistical Measurement of Provincial Resilience in China" (SC18TJ017), Chengdu Social Science Planning Project, "Comparative Study on the Sustainable Development of Chengdu and World Cities Based on Resilience Theory" (2018A12), and the NSFC project "Study on Risk Assessment and Resilience Construction Strategy of Urban Lifeline System in Xiong' an New District", (71741038).

References

1. Adger, W.: Social and ecological resilience: are they related. Prog. Hum. Geogr. **24**(3), 347–364 (2000)
2. Alliance, R.: Resilience (2017). http://www.resalliance.org/576.php
3. Alshehri, S.A., Rezgui, Y., Li, H.: Delphi-based consensus study into a framework of community resilience to disaster. Nat. Hazards **75**(3), 2221–2245 (2015)
4. Barnett, J.: Adapting to climate change in pacific island countries: the problem of uncertainty. World Dev. **29**(6), 977–993 (2001)
5. Boer, P.T.D., Kroese, D.P., et al.: A tutorial on the cross-entropy method. Ann. Oper. Res. **134**(1), 19–67 (2005)
6. Bruneau, M., Reinhorn, A.: Exploring the concept of seismic resilience for acute care facilities. Earthqu. Spectra **23**(1), 41–62 (2007)

7. Burton, C.G.: A validation of metrics for community resilience to natural hazards and disasters using the recovery from hurricane katrina as a case study. Ann. Assoc. Am. Geogr. **105**(1), 67–86 (2015)
8. Zhu, F.X., Chen, Y.H.: The method to determine the weight of interval number decision matrix attribute—Entropy value method. J. Anhui Univ. Sci. **30**(5), 4–6 (2005)
9. Fishwick, M.W.: The resilient city: how modern cities recover from disaster. J. Am. Cult. **28**(4), 456–456 (2010)
10. Gorczyca, B.: Searching for safety. State J. (2006)
11. Holling, C.S.: Resilience and stability of ecological systems. Ann. Rev. Ecol. Syst. **4**(4), 1–23 (1973)
12. Hu, J., Yan, Y., et al.: Similarity and entropy measures for hesitant fuzzy sets. Int. Trans. Oper. Res. (8) (2017)
13. Chen, J.C.: Who has been observing and discussing the disaster response capacity of the coastal areas in Hong Tai area (2013)
14. Liu, J.Y., Zeng, Z.P.: Construction of elastic city evaluation index system and its empirical study. E-Gov. **03**, 82–88 (2014)
15. Cutter, S.L., Barnes, L., et al.: Community and regional resilience: perspective from hazard, disasters, and emergency management. Commun. Reg. Resil. Initiat. **1**, 383–398 (2008a)
16. Cutter, S.L., Barnes, L., et al.: A place-based model for understanding community capability to natural disasters. Glob Env. Chang. **10**, 598–606 (2008b)
17. Cutter, S.L., Burton, C., Emrich, C.T.: Disaster resilience indicators for benchmarking baseline continuations. J. Homel. Secur. Emerg. Manag. **1**, 1–12 (2010)
18. Li, T., Niu, P., Gu, C.: Overview of research framework of elastic city. J. Urban Plan. **5**, 23–31 (2014)
19. Li, Y., Zhai, G.F.: Study on urban disaster resilience evaluation and improvement strategies in china. Planners **33**(08), 5–11 (2007)
20. Meng, L.J.: Based on the rata resilience evaluation system of community already promotion strategy—royal disaster in tianjin hedong district cooperation way both communities, for example Urban planning society of China, Shenyang municipal people's government. In: Planning for 60 Years: achievements and Challenges, Proceedings of 2016 Annual Meeting of China's Urban Planning. The urban planning society of China, Shenyang municipal people's government: 2016–12 (in Chinese) (2016)
21. Maru, Y.T.: Resilient regions: clarity of concepts and challenges to nay measurement. CSIRO Sustainable graphics (2010)
22. Mishra, A., Ghate, R., et al.: Building ex ante resilience of disaster-exposed mountain communities: drawing insights from the nepal earthquake recovery. Int. J. Disaster Risk Reduct. **22**, 167–178 (2017)
23. Noy, I., Yonson, R.: Economic vulnerability and resilience to natural hazards: a survey of concepts and measurements. Sustainability **10**(8), 2850 (2018)
24. Orencio, P., Fujii, M.: An analytic hierarchy process (AHP) in an analytic hierarchy process (AHP). Int. J. Disaster Risk Reduct. (2016)
25. Pickett, S.T.A., Cadenasso, M.L., Grove, J.M.: Resilient cities: meaning, models, and metaphor for integrating the ecological, socio-economic, and planning realms. Landsc. Urban Plan. **69**(4), 369–384 (2004)
26. Sensier, M., Bristow, G., Healy, A.: Measuring regional economic resilience across europe: operationalizing a complex concept. Spat. Econ. Anal. **11**, 1–24 (2016)
27. Shao, Y., Jiang, X.: Understanding urban resilience: a conceptual analysis based on integrated international literature review. Urban Plan. Int. (2015)

28. Shaw, R., Team, I.: Climate disaster resilience: focus on coastal urban citizen asia. Asian J. Environ. Disaster Manag. **1**, 101–116 (2009)
29. Cutter, S.L., Ash, K.D., Emrich, C.T.: The geographies of community disaster resilience. Glob. Environ. Chang. **29**(29), 65–77 (2014)
30. Walker, B., Holling, C.S., et al.: The resilience, the adaptability and transformability in social-ecological systems (2017)
31. Yun, Y.X.: Discussion on the evaluation index of community disaster resilience from the perspective of fire safety. (2017)

The Influence of Enterprise Internationalization Level and R&D Input on Enterprise Performance under the Background of "One Belt and One Road"

Jingjing Lu and Hongchang Mei[✉]

Chongqing Technology and Business University, Chongqing 600000, People's Republic of China
m_hchang68@163.com

Abstract. The "One Belt, One Road" strategy is an important opportunity for Chinese companies to carry out internationalization. This paper selects the panel data of 98 Belt and Road countries along the route in 2013–2017, uses the OLS hybrid regression method to study the impact of corporate internationalization level on corporate performance, and analyzes the adjustment effect of R&D investment to the relationship between internationalization level and corporate performance. The results show that: (1) The relationship between internationalization level and corporate performance is S-type related; (2) R&D investment has a positive adjustment effect on the relationship between internationalization level and enterprise performance. The introduction of R&D investment instead of the past industry and location advantages has broadened the research perspective. According to the conclusions, several suggestions for the Chinese company were proposed: (1) Enterprises should learn from foreign experience, identify the positioning of the company in the market and try to avoid excessive internationalization. (2) Enterprises should increase investment in research to improve product competitiveness.

Keywords: Belt and road · Internationalization level · Company performance · R&D investment · Regression method

1 Introduction

The world economy is unevenly developed due to the deep-seated impact of the international financial crisis in the world today continues to manifest itself. China proposed the "Belt and Road" strategy in order to conform to the trend of economic globalization in 2013. This strategy aims to explore the path of economic growth and create new types of cooperation in the region. On the basis of the "one belt and one way" strategy, the nineteen party proposed that China support qualified enterprises to "go out". According to data from the Ministry

© Springer Nature Switzerland AG 2020
J. Xu et al. (Eds.): ICMSEM2019 2019, AISC 1002, pp. 177–188, 2020.
https://doi.org/10.1007/978-3-030-21255-1_14

of Commerce, Chinese enterprises had added a total of 4.67 billion US dollars in trade investment to the countries along the "Belt and Road" from January to April 2018.In the past five years of the "Belt and Road" strategy, Chinese companies along the line have continued to develop internationally. In this process, the relationship between the degree of corporate internationalization and corporate performance is unknown. So the author chooses to do research from this perspective. This paper selects the Chinese enterprises along the "Belt and Road" as a research object, studies the relationship between the degree of internationalization and corporate performance through the mixed OLS method, and observes the adjustment effect of R&D investment between the two variables. The study concluded that the internationalization level of listed companies in the selected samples is significantly related to corporate performance. The contribution of this research lies in the attempt to answer the question of whether the internationalization of Chinese enterprises along the "Belt and Road" can promote the development of enterprises in the long run, and how to transfer and acquire the advantages of R&D in the process of internationalization. Let the decision-making level of the enterprise develop relevant strategies to solve the phased problems in a targeted manner on the basis of grasping the trend.

The framework of this paper has five parts. The first part is the introduction of the article. This part mainly summarizes the selected economic background, the main purpose and contribution of the research, the research methods adopted, and the organizational structure of the article. The second part of the literature review is based on the specific concerns of this paper, and comprehensively combs and analyzes the theory and research related to enterprise performance and international business management, research and development, and gives a comment. The third part is the research design in the core part, and proposes relevant research hypotheses such as the relationship between internationalization and performance, the adjustment effect of R&D intensity, selecting the sample and explaining the data source, and designing and constructing the relationship between the dependent variables of each independent variable in the model. Give the model formula, the list shows the meaning of each variable. The fourth part is empirically tested with the software Stata 14.0, and the regression results are analyzed. Descriptive statistics, correlation analysis and regression analysis are carried out. The fifth part summarizes the research of the full text and puts forward relevant suggestions.

2   The Theoretical Analysis and the Research Hypothesis

2.1   Internationalization and Corporate Performance

By studying the relationship between the degree of internationalization and corporate performance from different perspectives, domestic and foreign scholars have come up with five types: positive correlation, negative correlation, U-shaped, inverted U-shaped and S-shaped. Grant [9], Kim and Lyn [13], and

Dunning [8] pointed out that there is a linear positive correlation between internationalization and firm performance by using empirical research; however, Harveston et al. [10], Siddharthan and Contractor [6], Denis [7], Jianting and Yong [12] found that they have a linear negative correlation; other scholars draw different conclusions from the empirical perspective through optimization models and measurement methods: Song [16], Ruigrok and Chang [19], Lin Zhihong [3], Chen et al. [4] proved the U-type relationship is a more reasonable explanation between internationalization and firm performance; Brock et al. [1], Hitt et al. [11], Aleson, Chen et al. [5] and so on have obtained the inverted U-shaped relationship between them; Moon [17], Jianting and Yong [12], Chen and Tan [2] believe that there is an S-type relationship between the two.Based on the five conclusions, Zeng et al. [20] put forward different views. They used the panel data of 140 domestic auto industry SMEs from 2005 to 2009 to draw the degree of enterprise internationalization and corporate performance. The conclusion of the M-curve relationship.

Scholars who have a significant positive correlation with the degree of corporate internationalization believe that the increased internationalization of multinational corporations will increase the profitability of the company's operations. Its specific advantages can be more competitive in the transaction and achieve economies of scale. Dunning [8] and other believe that the specific advantages of the company's research and development capabilities, marketing capabilities, etc. can be transformed into intermediate products within the enterprise transfer, in the process of enterprise internationalization can bring positive effects to the enterprise. Scholars with a linear negative correlation view believe that they will lead to additional transaction costs and increase the risk of foreign investment when companies attach importance to overseas investment and increase investment in overseas resources.

Scholars who believe the two are U-shaped hold that multinational corporations will suffer short-term losses due to the burden of newcomers and the disadvantages of outsiders at the beginning of internationalization, and then gradually adapt to the local organizational mechanism and give play to the advantages of outsiders. Douglas selected 386 large Mexican companies as research objects, and found that in the early stage of internationalization, the cost of learning local business operation mechanism will be greater than the internationalization income. However, after gaining market experience, the early learning cost will be transformed into later stage. Profit support. The view of the inverted U-type relationship is derived by the scholars who combined the Uppsala model. The model believes that in the process of internationalization location selection, enterprises generally choose markets that are close to the geographical distance of the home country or have high cultural similarity, and gradually spread to locations with farther psychological distances. Therefore, the slight difference in the previous environment will reduce the management cost. As the degree of internationalization deepens, the uncertainty and learning cost increase, and the enterprise performance drops sharply after reaching the inflection point. Aleson, who takes the entropy index as the measure of internationalization degree concluded that there is an inverted U-shaped relationship between the degree of internationalization and corporate performance by studying 103 manufacturing industries in Spain.

Scholars with an S-type view divide the degree of internationalization of enterprises into three stages. In the early stage of the internationalization process, the increase of learning costs will weaken the performance of enterprises. With the improvement of adaptability, the scope economy and economies of scale will gradually show advantages, the performance of enterprises will be significantly improved, and the stage of high internationalization will achieve high profits for multinational corporations. Coverage of the rate region, only the secondary market with low potential profit remaining, and the loose operation mode will lead to a significant increase in the coordination cost of the enterprise, and the income will once again enter the downward turning point. Overall, it shows the evolution trend of "decline, rise, and decline". Bbillo and zturriage thought that internationalization and corporate performance are S-type relationship which is not affected by the internal and external environment of the enterprise through the study of 1,500 large enterprises in Western Europe from the perspective of corporate strategy.

Based on the above research, this paper proposes the following hypothesis based on the background of the Belt and Road:

Hypothesis 1: The relationship between the degree of internationalization of Chinese companies along the line and corporate performance is S-type related.

Hypothesis 2: The relationship between the degree of internationalization of Chinese companies along the line and corporate performance is U-type related.

2.2   The Role Of R&D Investment in the Adjustment of Internationalization and Corporate Performance

In the process of internationalization, Rugman and Verbeke [15] and Yuan [18] believe that the company's research and development capabilities are embedded in the organization regardless of location restrictions. They can reach Global economies of scale without too much adjustment when crossing the international market. Moreno [14] believes that developing countries generally carry out technical accumulation and may even be improved and innovated in the process of learning to form some difficult to be developed on the basis of "special learning experience" by multinational corporations in developed countries due to the limitations of the technical economy. After 40 years of reform and opening up, China has also seen a large number of enterprises with strong R&D capabilities, independent intellectual property rights. In this sense, the R&D investment of Chinese enterprises can brought good performance in the process of internationalization. So the assumptions are as follows:

Hypothesis 3: China's "Belt and Road" enterprise R&D intensity is positively regulating the relationship between internationalization level and corporate performance.

3   Research Design

3.1   Sample Selection

This article selects the "One Belt, One Road" Chinese listed companies in the "One Belt, One Road" column of the Wind Database for research in 2013–2017 which lists the list of Chinese A-share listed companies that have business dealings with companies along the "Belt and Road" countries. According to the annual report of the company, the enterprises with incomplete data were supplemented, and the listed companies with ST, *ST, PT and data were excluded, and the characteristic panel data of 44 (two cross-section) Chinese enterprises were finally obtained.

3.2   Variable Design

(1) Dependent variable.
This paper selects the performance of listed companies as the dependent variable. Common indicators for measuring firm performance include profit margins, stock prices, ROA, ROE and so on. ROA and ROE are used the most.The author believes that the value of ROA is more comprehensive and stable than ROE, which can fully reflect the impact of corporate structure on performance. At present, most scholars use the total return on assets of enterprises to measure corporate performance.

(2) Independent variables.
According to the World Investment Report 2000—Transnational M&A and Development published by the United Nations Conference on Trade and Development, the measure of the degree of internationalization of multinational corporations is the average number of three ratios (the ratio of overseas sales to total sales and the proportion of overseas assets to total assets). That is, the ratio of FATA, FSTS, and the number of foreign employees to the total number of employees, called the transnationalization index. Therefore, this paper uses the transnationalization index as an independent variable.

(3) Adjustable Variables.
This paper uses R&D intensity to measure the intensity of R&D investment as a measure of technological superiority take into account the impact of technological innovation on the road to internationalization of enterprises. The R&D capability is measured by the ratio of R&D expenses of the company to total operating income.

(4) Control variables.
This study selected five levels of control variables. The company size (SIZE) is expressed by the company's total assets. The debt level (DEBT) is expressed by the asset-liability ratio; The shareholding structure (CON) is expressed by the equity concentration indicator. Other indicators are government subsidies (GOV) and number of employees (PER). The relevant variables are described in Table 1.

Table 1. Variable metric

	Variable	Symbol	Measure
Dependent variable	Business performance	ROA	Net profit / total assets
Independent variable	Degree of internationalization	TNI	(foreign assets/total assets + foreign sales / total sales + foreign employees / total employees) / 3 * 100%
Adjustable variables	R&D intensity	R&D	Research and development expenses / total operating income
	Business scale	SIZE	Total assets of the enterprise
	Equity concentration	CON	Number of shares held by major shareholders/total number of shares
Control variables	Debt level	DEBT	Total liabilities / total assets
	Government subsidies	GOV	Government subsidy / operating income * 100
	Number of workers	PER	Number of workers

3.3   Construction of Model

Domestic and foreign scholars mostly establish regression equations when studying the impact of internationalization on firm performance. In this article, the model is set as:

$$ROA_{it} = \beta_0 + \beta \sum X_{it} + \varepsilon_{it} \tag{1}$$

$$ROA_{it} = \beta_0 + \beta_1 TNI_{it} + \beta \sum X_{it} + \varepsilon_{it} \tag{2}$$

Model 1 is the regression analysis of firm performance and control variables, and model 2 is the relationship between firm internationalization level and company performance in the short term. The dependent variable ROAit is the performance of i company in year t; the independent variable TNIit is the internationalization level of i company in year t; $\sum X_{it}$ is the various control variables of i company in year t; ε_{it} is the error of i company in year t.

$$ROA_{it} = \beta_0 + \beta_1 TNI_{it} + \beta_2 TNI_{it}*TNI_{it} + \beta \sum X_{it} + \varepsilon_{it} \tag{3}$$

$$ROA_{it} = \beta_0 + \beta_1 TNI_{it} + \beta_2 TNI_{it}*TNI_{it} + \beta_3 TNI_{it}*TNI_{it}*TNI_{it} + \beta \sum X_{it} + \varepsilon_{it} \tag{4}$$

Internationalization is the company's long-term development strategy. Therefore, the internationalization of secondary and tertiary items in Model 3 and Model 4 reflects the long-term development process of internationalization.

According to the internationalization theory of enterprises, if β_1 is significant, it shows that the internationalization level of "One Belt, One Road" Chinese enterprises has a significant linear relationship with company performance. If β_1 and β_2 are significant and β_3 is not significant, the two are U-shaped or inverted U-shaped. If β_1, β_2 and β_3 are both significant, it indicates that the internationalization level has a significant S-type relationship or an inverted S-type relationship with enterprise performance. Model 5 adds the level of regulation variable development based on Model 2.

$$ROA_{it} = \beta_0 + \beta_1 TNI_{it} + \beta_4 RD_{it} + \beta_5 TNI_{it}*\text{RD}_{it} + \beta \sum X_{it} + \varepsilon_{it} \qquad (5)$$

4 The Empirical Test and Analysis

Based on the existing literature adjustment test method, stata 14.0 was selected as the data analysis software, and the mixed OLS regression was used for empirical research according to the research hypothesis of this paper. Before the regression, the data was standardized.

4.1 Descriptive Statistical Analysis

Descriptive statistics for the main variables are summarized in Table 2. It can be seen that the standard deviation of the listed company's internationalization level (TNI) and R&D input (RD) is larger than the average value, and the difference between the maximum value and the minimum value is large. It shows that the internationalization level and R&D investment of different listed companies are very different. It is demonstrated that the internationalization of different listed companies is a process of constant change and long-term development. Descriptive statistics for other variables are normal.

Table 2. Descriptive statistical variables involved in model

Variable	Number of samples	Mean	S.D.	Min	Max
roa	220	0.52	3.12	−0.19	31.35
tni	220	6.07	21.23	0.01	189.66
rd	220	1.87	2.02	0	8.12
gov	220	16.23	4.87	0	21.29
size	220	24.12	1.88	20.66	28.51
debt	220	23.66	2.07	18.76	27.73
con	220	3.71	0.4	2.45	4.35
per	220	8.93	2.05	4.86	13.21

4.2   Correlation Test of Variables

It can be seen from Table 3 that the correlation coefficient between the explanatory variables in the same equation is low, and most of the absolute values are
less than or equal to 0.45. It means that there is no serious correlation between
the variables. The degree of correlation between the size of the enterprise and
the number of employees can be explained in practical terms. The scale of the
enterprise is large and the number of employees required is large. Therefore,
it will not affect the overall study. The variance expansion factor (VIF) test
results show that the variance expansion factor between the explanatory variables in the same equation is mostly below Table 3, indicating that there is no
multi-collinearity. Due to space limitations, the results of the multi-collinearity
test are not listed.

Table 3. Variable correlation matrix

	roa	tni	rd	gov	size	debt	con	per
roa	1							
tni	0.43***	1						
rd	−0.08	−0.12	1					
gov	0.07	0.03	0.32***	1				
size	−0.24***	−0.32***	0.09	0.05	1			
debt	0.26***	0.28***	−0.07	0.23***	−0.04	1		
con	0.16**	0.24***	−0.1	0.20***	−0.22***	0.37***	1	
per	−0.14*	−0.20***	−0.01	0.08***	0.45***	0.14*	−0.20***	1

Note: ***, **, and * in the table indicate significant levels at 1%, 5%, and 10%,
respectively, and regression values for t-values in parentheses

4.3   Regression Analysis of Degree of Internationalization and Firm Performance

The mixed OLS regression results are shown in Table 4. Model 1 is a regression analysis of firm performance and control variables. The results show that
government subsidies greatly enhances corporate performances as a manifestation of direct state-to-business support; the larger the firm, the worse the firm's
performance may be showed by the table. This could be caused by lacking of
the characteristics of flexible and changeable small-scale enterprises in the big
company. It may be that large-scale enterprises are not flexible and changeable
due to their large size, and they face uncertainties in the early stage of internationalization, which makes large enterprises unfavorable to the improvement of
corporate performance; other control variables have a weak impact on corporate
performance.

In model 2, the degree of internationalization is only passed at the level of significance of 0.05, and model 3 is added to the quadratic term of degree of internationalization based on model 2. The coefficient of internationalization of model 3 is significantly negative. The squared (long-term) coefficient of internationalization is significantly positive, the primary and secondary coefficients are passed at the significance level of 0.01 and 0.05. Respectively, the internationalization degree of the model 3 (0.1534) is greater than the model 2 (0.1104), which means the curve model is somewhat better than the linear model. The primary, quadratic, and cubic terms of the model 4 are significant. The short-term coefficient of the internationalization level is negative, if β_1, β_2, β_3 are both significant according to the model setting and the theory of enterprise internationalization. Simultaneously, the coefficient of the square term is positive and the coefficient of the cubic term is negative, that means Business Performance are following the trend of "fall, rise, fall again". The result indicates that the internationalization level of listed companies in the selected sample is significantly related to the performance of the enterprise, and Hypothesis 1 is verified. Model 5 introduces interaction items between R&D intensity and internationalization level. The interaction term passes at the significance level of 0.05, and the coefficient is 0.0234. The R&D intensity positively adjusts the relationship between internationalization level and enterprise performance. Hypothesis 3 is verified.

5   Conclusion and Suggestion

This study mainly explores the relationship between the degree of internationalization of enterprises and corporate performance along the "Belt and Road", and also examines the role of R&D investment in regulating the relationship between the two. The study found that:

(1) On the whole, the level of internationalization has an S-type relationship with corporate performance. In the initial internationalization stage, companies clearly feel the burden of globalization. The benefits of globalization at this stage are lower than the cost of learning. In the middle stage of internationalization, the scope of economic advantages gradually emerges;enterprises will face the problem of excessive internationalization later. The cost of business management has increased dramatically. Corporate performance decreases as internationalization increases.

(2) The research results show that increasing R&D investment in the process of internationalization can bring good performance to the enterprise. The company's technical capabilities enable companies to create products with core competencies.

According to the research conclusions, relevant suggestions are put forward:

(1) In the first stage, enterprises should actively absorb the advanced experience of foreign companies in service awareness and business philosophy, optimize resource allocation and establish a business model that conforms to the international development of enterprises.

Table 4. Mixed OLS regression empirical results

Variable	Model 1	Model 2	Model 3	Model 4	Model 5
TNI		0.1104**	0.1533***	-0.2024***	0.1262***
		-5.33	-6.13	-7.48	-8.03
TNI*TNI			-0.0053**	0.0477***	
			-1.98	-2.52	
TNI*TNI2				-0.0198***	
				1.81	
RD					-0.0079*
					-0.72
TNI*RD					0.0234**
					-0.83
GOV	0.6162***	0.7231***	0.6147**	0.5869*	0.6243*
	-3.23	-3.3	-2.96	-2.28	-2.99
SIZE	-0.2389*	-0.0595*	0.0142	-0.1662**	-0.2064 -1.02
	-1.24	-0.35	-0.08	-2.91	
DEBT	0.3164**	0.0769**	0.0682*	0.0543*	0.0823**1.01
	-3.49	-0.86	-0.77	-0.65	
CON	0.2431**	-0.3679**	-0.5114*	0.1417	0.360774
	-0.38	-0.83	-1.15	-0.96	
PER	0.1716	0.0377**	0.0160**	0.0302**[1]	0.0417
	-0.76	-0.24	-0.29	-1.72	-1.41
N	220	220	220	220	220
R^2	0.1233	0.5838	0.5934	0.7559	0.5873
Adj R^2	0.0974	0.569	0.5763	0.7538	0.5674
F	2.76	6.28	7.81	9.41	7.17

Note: ***, **, and * in the table indicate significant levels at 1%, 5%, and 10%, respectively, and regression values for t-values in parentheses

(2) After the company enters the stage of internationalization and maturity, the company's performance technology is improved. Managers should fully realize that this stage is the best stage for the international operation of the company, and seize the opportunity to develop business-related businesses.

(3) The S-type theory enlightens the manager that blindly increasing the degree of internationalization will bring huge burdens to the enterprise. Enterprises can't take care of this and lose sight of it. Focusing on the domestic market, overseas layout is the best positioning for the internationalization of enterprises. Excessive internationalization will bring the enterprise into the abyss.

(4) Chinese companies need to have a comprehensive and in-depth understanding of the domestic and international competition situation. Enterprises

must always strengthen their independent research and development capabilities in the process of transnational operations. Increase investment in product research and development to form core technology and high-quality brand image. In this way, Chinese companies can better meet the needs of the international market.

References

1. Brock, D.M., Yaffe, T., Dembovsky, M.: International diversification and performance: A study of global law firms. Journal of International Management **12**(4), 473–489 (2006)
2. Chen, S., Hao, T.: Region effects in the internationalizationcperformance relationship in chinese firms. Journal of World Business **47**(1), 73–80 (2012)
3. Chen Y, Xu RY (2013) A study on impacts of corporate internationalization speed on performancel analyses on the listed chinese manufacturers 2006c2010. In: International Conference on Management Science & Engineering
4. Chen, L., Liu, J., Zhang, S.: The influence of imitation isomorphism on enterprise internationalization-performance relationship Han empirical study based on the justification of institutional theory (in chinese). Chinese industrial economy **09**, 127–143 (2016)
5. Chen, Y., Jiang, Y., Wang, R.: Product diversification strategy, enterprise resource heterogeneity and internationalization performance: An empirical test of china's listed companies in manufacturing industry from 2008 to 2011 (in chinese). Management Review **26**(12), 131–141 (2014)
6. Contractor, F.J.: Why do multinational firms exist? a theory note about the effect of multinational expansion on performance and recent methodological critiques. Global Strategy Journal **2**(4), 318–331 (2012)
7. Denis, D.J., Denis, D.K., Yost, K.: Global diversification, industrial diversification, and firm value. The journal of Finance **57**(5), 1951–1979 (2002)
8. Dunning J (1977) Trade location of economic activities and the mne: A search for an eclectic approach: The international allocation of economic activities. In: Proceedings of a Nobel Symposium Held in Stockholm
9. Grant, R.M.: Multinationality and performance among british manufacturing companies. Journal of International Business Studies **18**(3), 79–89 (1987)
10. Harveston, P.D., Kedia, B.L., Francis, J.D.: Mne's dependence on foreign operations and performance: a study of mnes from the "triad" regions. International Business Review **8**(3), 293–307 (2011)
11. Hitt, M.A., Tihanyi, L., Miller, T., et al.: International diversification: Antecedents, outcomes, and moderators. Social Science Electronic Publishing **32**(6), 3–4 (2009)
12. Jianting, F., Yong, L.: Differences between chinese and foreign enterprises in the relationship between degree of internationalization and performance: Empirical evidence from top 500 enterprises (in chinese). Journal of Management Sciences **21**(06), 110–126 (2018)
13. Kim, W.S., Lyn, E.O.: Excess market value, the multinational corporation, and tobin's q-ratio. Journal of International Business Studies **17**(1), 119–125 (1986)
14. Moreno, L.: The determinants of spanish industrial exports to the european union. Applied Economics **29**(6), 723–732 (1997)
15. Rugman, A.M., Verbeke, A.: Subsidiary-specific advantages in multinational enterprises. Strategic Management Journal **22**(3), 237–250 (2010)

16. Song, Y., Li, Y., Wang, Y.: Enterprise resources, nature of ownership and degree of internationalization ll evidence from chinese manufacturing listed companies (in chinese). Management Review **23**(2), 53–59 (2011)
17. Xiao, S.S., Jeong, I., Moon, J.J., et al.: Internationalization and performance of firms in china: Moderating effects of governance structure and the degree of centralized control. Journal of International Management **19**(2), 118–137 (2013)
18. Yuan X, Dai W (2014) The impact of r&d investment on corporate financial performance (in Chinese). Science Management Research (3)
19. Chang Y (2011) Dynamic characteristics of overseas direct investment performance of chinese cntorprises: An empirical analysis based on large state-owned enterprises. Finance & Trade Economics
20. Zeng, D., Ya, S., Wan, W.: Research on the relationship between the degree of internationalization and firm performance: m-curve. Science and Science and Technology Management **37**(04), 25–34 (2016)

Do Customers Really Get Tired of Double Eleven Global Carnival? An Exploration of Negative Influences on Consumer Attitudes toward Online Shopping Website

Ruo Yang, Yongzhong Yang[(⊠)], Mohsin Shafi, and Xiaoting Song

Business School, Sichuan University, Chengdu 610065,
People's Republic of China
yangyongzhong116@163.com

Abstract. The total turnover of Double Eleven Global Carnival on Tmall is RMB 213.5 billion on November 11th in 2018. Behind the great success, various sales promotions are significant. However, not all sales promotions have only good effects, some of them will have negative influences on consumers, resulting in some consumers' dislike to online shopping website. Based on the grounded theory, we conducted a qualitative analysis of the viewpoints of 123 consumers who dislike the double 11 shopping carnival. After coding, we found that the complex rules of promotion, fake discounts, compelling marketing, disadvantages of goods/services might have caused consumers' aversion to shopping festival. But this negative situation is not common. The study used linear regression to analyze data collected from 156 Chinese respondents, exploring the impact of sales promotions' negative factors on consumer attitudes. We found that (1) negative influences of the double eleven sales promotion have no significant impact on consumers' attitudes towards the website; (2) AW (attitude toward website) as well as SS (sales promotion satisfaction) have significant impact on AS (attitude toward sales promotion).

Keywords: E-commerce · Double eleven global carnival · Consumer attitudes

1 Introduction

In 2009, Alibaba's Taobao Mall (now renamed Tmall) made a seemingly ordinary attempt to hold promotion activities on November 11 every year to take advantage of the gap between National Day golden week and Christmas promotion season to expand the brand influence of Taobao Mall. However, people did not expect that since then, with JD, Dangdang and other e-commerce companies

© Springer Nature Switzerland AG 2020
J. Xu et al. (Eds.): ICMSEM2019 2019, AISC 1002, pp. 189–200, 2020.
https://doi.org/10.1007/978-3-030-21255-1_15

joining in one after another, "double eleven" has gradually become the largest commercial promotion carnival on the Internet in China. [13]

This study examines attitudes of consumers in China toward online shopping website under the influences of various sales promotions of Double Eleven Global Carnival.

Negative attitudes toward websites have been identified as critical factors that impact purchase intention in an online context [15], so even if merchants hope that sales promotions can improve current sales, they should be careful not to hurt consumers' attitudes towards the website. The more consistent the sales promotion is with the benefits the consumer expects to receive, the more favorable their attitudes towards the sales promotion will be [12], but now the increasingly complex promotion activities and preferential rules of singles' day have exhausted consumers.

With more and more negative comments about double eleven sales promotion on the Internet in recent years, two questions have been raised: Does double eleven sales promotion really have negative impacts on customers? Would they hate the website because of negative impacts? Therefore, this paper focus on those sales promotions which have negative influences on consumer attitudes. We find sales promotions do make some customers dislike or even hate Double Eleven Global Carnival through qualitative analysis, but it is not a universal phenomenon cause through quantitative research we find the negative aspects of promotion did not have a significant impact on consumer attitudes. This is probably because consumers can tolerate the negative factors of promotion within a certain limit when they get the discount.

The main structure of this paper is as follows: (1) introduction, we listed questions we want to study under the research background; (2) literature review, we reviewed papers related to consumers' attitude and sales promotion; (3) qualitative analysis, since there are few papers studying the negative impact of the double eleven sales promotion, we mainly analyzed the views of consumers who dislike/hate double eleven sales promotion and extract the negative factors of the sales promotion by coding; (4) quantitative analysis, we designed questionnaire according to qualitative analysis, collected data, and analyzed the impact of negative factors on consumer attitudes with linear regression; (5) conclusion, we summarized our findings. (6) discussion, we discussed possible reasons for the conclusion.

2   Literature Review

Consumers' attitude

Attitude, which defined as a person's overall evaluation of a concept, have two types: attitudes toward objects/toward behaviors [1].Consumers' attitude towards online shopping websites is a strong predictor to online shopping behavior and satisfaction, which has been explained by technology acceptance model, theory of reasoned action, and theory of planned behavior [12]. Customer attitude towards the online store affects the customers' behavioural intentions to repurchase from the online store [9].

Sales promotion

Free gift offers, price offs, extra product offers, exchange offers, buy-more-and-save offers, contests and sweepstakes are include in different promotions [6]. Some studies compare different types of promotion, for example, retailers and manufacturers often rely on non-price promotion techniques, such as premium promotions, where consumers receive a free gift with the purchase of a product, but impact of premiums on purchase behavior is systematically lower than that of equivalent price cuts [4]. Some studies explored negative effects of conditional sales promotion on consumers' feelings of unfairness [8]. But there are few studies with systematic analysis of sales promotion's negative influence on consumer.

3 Qualitative Analysis

Qualitative research methods are applicable to some fields without any or without complete historical information and data [16], so we use qualitative analysis as one of the methods in the study. Since there are not many previous studies on the specific topic studied in this paper for reference, we use qualitative analysis as a method to explore what kind of bad influence the Double Eleven Global Carnival has on customer, using online communication as the data. And online communication [14], which include online posts [11], was proved that it can be used as a data source for research.

According to the viewpoint of Grounded Theory proposed by Barney G. Glaser and Anselm L., their "strategy of comparative analysis for generating theory puts a high emphasis on theory as process; that is, theory as an ever-developing entity, not a perfect product." [5] So we discussed a lot during every stage of the research and added the necessary additional investigations and pilot study to clarify the questions raised in the study, which is why our research always adds something new to the next stage during the process. The paper also shows the necessity of this and its important role in drawing the conclusion.

3.1 Data Collection

We used baidu to search for the expression of dislike "double eleven" and got initial data. In order to avoid the interference of other people's summary on qualitative research, we did not use news, WeChat articles for analysis but took netizens' direct answers/ comments to relevant questions or discussion online as the main analysis materials. The explanation of the analyzed materials is as following: 99 effective answers of the question "Why did you start to hate the double 11?" on Zhihu, 8 effective answers which related to "dislike double 11" of 8 different questions on Zhihu, 2 answers of the question "Is everyone disgusted with the current e-commerce double eleven?" on Zhihu, 4 answers of the question "Is there anyone like me who hates double eleven now?" on Baiduzhidao, 10 answers of the discussion "Double eleven doing things like this really make people dislike them" on Jiangyou forum, so the sample of the qualitative analysis is 123 answers related to the study online.

3.2   Coding

To analyze the data, constant comparative approach [5,10] and Atlas.ti 8.0 software are used by two independent coders. Results of coding (Fig.1) helps us to construct the initial questionnaire served at pilot study. Then 8 categories are formed: complex rules, fake discounts, compelling marketing, disadvantages of goods/services, disturb life, time-wasting, negative emotions, irrational consumption.

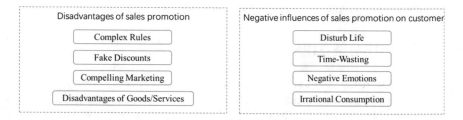

Fig. 1. Categories of qualitative analysis results

(1) Complex rules in this study mainly refers to the rules of promotion are complicated and difficult to be understood, some coupons are difficult to get and the rules of calculating prices are complicated as well. For example, netizen a*******g said: There is no consideration for customers. I can't tell what coupon can be used and what can't be used. I can't easily find the entrance to get the coupon.

(2) Fake discounts in this study mainly refers to the discount is not true or not as good as those stores claimed, including the phenomenon such as the businessman show the discount after they raises the price of product. For example, netizen l****y said: I can only say that I don't believe in sales promotion, discount any more. Several times the original price marked up. Then they make a small discount, make your heart a little comfortable.

(3) Compelling marketing in this study mainly refers to marketing activities make consumers feel compelled. For example, when you want to play computer games, the advertisements for promotion are overwhelming and surround you; you are "forced" to go to the promotion website thumb up (click supporting button) for your friends because of promotion; or you receive a lot of junk promotional messages. For example, netizen h***r said: These years Tmall/ Taobao/ JD advertising are more and more and more fancy. I just opened the bottom right corner of the computer and open search engine to search information, the ads of Taobao out and accounted for four-fifths of the entire page!

(4) Disadvantages of goods/services in this study mainly refers to bad points/short coming of goods (like poor quality) or services (like express delivery is too slow, Online customer service ignore customer and do not answer question). For example, netizen z**v said: I remember when I bought something on Double 11, It was really slow to deliver, which I really didn't like.

(5) Disturb life in this study mainly refers to that promotional activities disturb the life of consumers, for example, the purchase of goods during the activity has disrupted the daily life arrangements, occupied the time to do the main work. For example, when answer the reason why you hate Double Eleven Global Carnival, netizen z****u said: Mainly (because it) disturb life.

(6) Time-wasting in this study mainly refers that the sales promotion waste customer's time. For example, netizen h***e said: Activities hold by Taobao, the game, to send red packets are very small not worthy to waste time. At the beginning two years of the double 11 sales promotion I can also grab red envelopes with hundreds of yuan (RMB), now these also usually have ten or twenty yuan, a waste of time and emotion, really not worth (to take part in).

(7) Negative emotions in this study mainly refers to negative emotions caused by sales promotion, like worry or afraid to miss the discount. For example, netizen b*****g said: To tell the truth, singles day makes me feel very tired. I have to start browsing the things I want to buy for a long time. I am afraid of missing the discount.

(8) Irrational consumption in this study mainly refers that consumers made unreasonable consumption decisions during the promotion period. For example, netizen a*c said: People blindly follow the trend and buy things they don't need.

In the process of data analysis, we found that poor or feeling/recognizing poverty is one of the reasons why the netizens in the samples hate the double 11 sales promotion. The word "poor" was mentioned 18 times by 15 netizens. However, we did not include it into the category, the main reasons are as follows: firstly, it is not a problem which the promotion activity itself can change or improve; secondly, strictly speaking, such negative effects are not caused by sales promotion. People are reminded that they are poor simply because of sales promotion.

4   Quantitative Analysis

In order to find out the influences of "disadvantages of sales promotion" and "negative influences of sales promotion on customer" on consumer attitudes toward online shopping website, the quantitative analysis was carried out.

4.1   Pilot Study

According to the results of qualitative analysis, we took 8 categories as variables and designed items based on the content of online communication in qualitative analysis. In addition, we added two variables concerning consumer attitude, namely, consumers' attitude towards the website and consumers' attitude towards the stores on the website, so as to find the answer of our research question. In pilot study, 97 respondents were approached to fill the questionnaires and 84 valid questionnaires were returned (response rate 86.6%). Through the factor analysis we found that :(1) Consumers' attitude towards the website and consumers' attitude towards the store on the website are detected to be one factor.

We speculate that this is due to the particularity of Taobao, it has few Self-operated Business, mainly connects consumers and merchants, so consumers' attitude towards the website and consumers' attitude towards the stores on the website are overlapping. Therefore, we only used one variable in the main study, namely, consumers' attitude towards the website. (2) "Negative emotions" was a part of "disturb life. Although at the very beginning we want to separate the "disturb life" and the emotional changes caused by the promotion, respondents appeared to have generated negative emotions are not much, and the types of negative emotions mentioned in the qualitative analysis data were not endorsed by the respondents. So we incorporated the negative emotions into the disturb life as an item. To disturb also means to make sb. worry, we thought it was acceptable to do this. Through linear regression, we found that when consumers' attitude towards the website or consumers' attitude towards the stores on the website are taken as dependent variables and the rest of the variables (complex rules, fake discounts, compelling marketing, disadvantages of goods/services, disturb life, time-wasting, negative emotions, irrational consumption) are taken as independent variables (method: Enter, condition: In cases where probability of F to be input $< = .050$, probability of F to be removed $> = .100$), all the independent variables are excluded. This may be because consumers are simply dissatisfied with the promotion and not angry with the website. Therefore, Attitude toward Sales Promotion (AS) was added as a variable for the main study analysis. In addition, lot of customers do not agree that the promotion is Time-wasting. In questionnaire used in pilot study, we have a variable indicates that the sales promotion of Double Eleven is a waste of time. And one item of it is "Compared with the time spent, the benefit I got is not worth mention." Many respondents disagree with this item, which maybe because people think that although Double Eleven Global Carnival is time-wasting, the return is really worth it. After all, the more consistent the sales promotion is with the benefits the consumer expects to receive, the more favorable their attitudes towards the sales promotion will be [3]. Besides, savings is one of utilitarian benefits which promotions provide [2]. Since saving money should be a thing which consumer expects to receive, whether it saves money should be the key point that we need to consider and we add Sales Promotion Satisfaction as a variable, which include the item "It saved me a lot of money", into the questionnaire.

4.2 Main Study

We made some modifications according to the result of pilot study, using 9 variables in questionnaire (Appendix 1).

(1) **Samples**

The research object of this paper is the person who knows double eleven global carnival. Snowball sampling technique is adopted for data collection. We invited people to fill out questionnaires round their opinions to double eleven sales promotion on Taobao/Tmall. Strictly speaking, double eleven Double Eleven Global Carnival was run for Tmall, but now it also benefits stores on Taobao, and people can search stores of Tmall directly on Taobao. So the study did not strictly differentiate between Taobao and Tmall. 156 respondents were approached to fill the questionnaires and 151 valid questionnaires were returned (response rate 96.8%). Descriptive Statistics of samples are shown in Table 1 as follows.

Table 1. Research sample demographic profile

Characteristic	N	%
Gender		
Male	53	35.1
Female	98	64.9
Education		
Primary school	1	0.7
Junior high school	5	3.3
High school	27	17.9
Bachelor	92	60.9
Master	11	7.3
Doctor or above	15	9.9
Monthly income (RMB)		
<2000	57	37.7
2000–4000	2	1.3
2001 4000	47	31.1
4001 6000	28	18.5
6001 8000	9	6.0
>8000	8	5.3
Age		
<18	1	0.7
18 25	61	40.4
26 30	20	13.2
31 40	20	13.2
41 50	11	7.3
51 60	32	21.2
>60	6	4.0

(2) Reliability and Validity Test

The exploratory factor analysis technology was used by SPSS to measure the reliability and validity of questionnaire before empirical analysis. Table 2 shows the internal consistency of the scale is with good validity and it is suitable for further analysis with Cronbach's alpha of all the variables above 0.7 and KMO of all the variables above 0.6.

Table 2. Reliability statistics

Variables	Number of items	Cronbach's Alpha	KMO
Complex Rules (CR)	3	0.863	0.722
Compelling Marketing (CM)	3	0.849	0.704
Disadvantages of Goods/Services (DGS)	3	0.708	0.648
Disturb Life (DL)	3	0.880	0.721
Fake Discounts (FD)	3	0.923	0.760
Irrational Consumption (IC)	3	0.924	0.748
Attitude toward Sales Promotion (AS)	3	0.849	0.709
Sales Promotion Satisfaction (SS)	3	0.847	0.698
Attitude toward Website (AW)	3	0.824	0.701
Total	27	0.853	0.855

(3) Model Setup and Results
Model (1)

In order to verify whether the negative influences of the "Double Eleven Global Carnival" which discussed above have an impact on consumers' attitudes towards websites, this paper constructs regression Model as follows:

$$\text{Model}(1): \text{AW} = \alpha_1 + \beta_{11}\text{CR} + \beta_{12}\text{CM} + \beta_{13}\text{DGS} + \beta_{14}\text{DL} + \beta_{15}\text{FD} + \beta_{16}\text{IC} \quad (1)$$

Results of Model (1)

The linear regression results of Model (1) by SPSS are as follows (method: Enter, condition: probability of F to be input $< = .050$, probability of F to be removed $> = .100$): goodness of fit of Model (1) is bad (adjusted $R^2 = .003$) and the all the independent variables are excluded. These results show that negative influences and shortcomings of the double eleven sales promotion have no significant impact on consumers' attitudes towards the website.

Model (2)

Next, this study explores which factors may influence consumers' attitudes towards the sales promotion (AS). In addition to the variables which summarized in the qualitative analysis and retained after the first round of questionnaire as independent variables (CR, CM, DGS, DL, FD, IC), we also added two independent variables (SS, AW). According to the qualitative analysis and the results of the first round of questionnaire, we found that if people could save money through sales promotion, many negative effects would not be important, so we added SS as an independent variable to the analysis of the mode (2). In addition, the analysis of correlation in the second round of the questionnaire revealed that the AW and consumers' attitudes to promotional activities significantly correlated, so we add the AW as the independent variable to the model analysis as well, explore that whether people will like sales promotions hold by Taobao, just as "Love me, Love my dog". The paper constructs regression Model as follows:

$$Model(2) : AS = \alpha_2 + \beta_{21}CR + \beta_{22}CM + \beta_{23}DGS + \beta_{24}DL + \beta_{25}FD \quad (2)$$
$$+ \beta_{26}IC + \beta_{27}SS + \beta_{28}AW$$

Results of Model (2)

The linear regression results of Model (2) by SPSS are shown in Table 3. CR, CM, DGS, DL, FD, IC do not have significant impact on AS; AW ($\beta_{28} = 0.578$, $p < 0.01$) as well as SS($\beta_{27} = 0.233$, $p < 0.01$) have significant impact on AS, and AW's impact on AS is bigger than SS's on it, which indicates that if consumers like the website, then it is easy for them to like the sales promotion on the website.

Table 3. Linear regression results of Model (2)

Adjusted R^2	Dependent variable	Independent variable	Unstandardized beta	Sig.
.515	AS	AW	.578	.000
		SS	.233	.001

Method: Stepwise, condition: probability of F to be input $< = .050$, probability of F to be removed $> = .100$
Prediction variables in the model: constant, AW, and SS
Excluded variables: CR, CM, DGS, DL, FD, IC

5   Conclusion

Qualitative analysis revealed that the sales promotion of Double Eleven Global Carnival had four shortcomings: complex rules, fake discounts, compelling marketing, disadvantages of goods/services. And its negative influences on customer mainly included four kinds: sales promotion of Double Eleven Global Carnival disturbed customer life, wasted their time, made their emotions negative and led to irrational consumption. Pilot study of quantitative analysis indicated that when it comes to Taobao, consumers' attitude towards the website and consumers' attitude towards the stores on the website did not differ significantly. Many customers did not find the promotion time-wasting. In other words, they thought it's worthwhile to spend time on joining Double Eleven Global Carnival. Model (1) showed that negative influences and shortcomings of the double eleven sales promotion had no significant impact on consumers' attitudes towards the website. Model (2) indicated that Attitude toward Sales Promotion was better when Attitude toward Website or Sales Promotion Satisfaction was better. The main contributions of this paper are: We found the disadvantages of the double eleven sales promotion on Taobao/Tmall and the negative effects they might have on customers. But these negative influences had no significant impact on consumers' attitudes towards the website. Attitude toward the website, as well as sales promotion satisfaction, had significant impact on consumers' attitudes toward sales promotion. Future research may focus on how to avoid the negative impact of sales promotion, how to improve people's satisfaction with sales promotions and how to promote positive attitudes towards shopping websites.

6   Discussion

Although the results of qualitative studies and results of quantitative analysis seems different, but it is understandable. Even we can see so many negative impacts of Double Eleven Global Carnival on customers, but compared with the shopping convenience provided by Taobao for 365 days, annual Carnival is still difficult to shake people's impression of Taobao websites. However, in qualitative research, we can find that problems like fake discounts, complex rules can still make people dislike the Carnival, posing a threat to people's evaluation of websites. Therefore, it is necessary for e-commerce platforms to clean up the phenomenon of fake discounts, make rules easier.

Of course, strictly speaking, the effects should be mutual. The reason why consumers like this website may be that its sales promotion enables them to enjoy the benefits, meanwhile, the reason why consumers are willing to participate in the sales promotion of the website may be that the website makes them have a good impression, so they are willing to shopping on this website. On the other hand, a website which can be loved by consumers should also have the capability to run a sales promotion which consumers love.

We can see a "paradox" through this study. But when we think about it, we can find something interesting. The paradox between qualitative analysis and quantitative analysis is like this: according to qualitative analysis, people will hate double eleven global carnival because of its negative influence; according to quantitative analysis, complex rules, compelling marketing, disadvantages of goods/services, disturb life, fake discounts, irrational consumption have no significant impact on consumers' attitudes towards the promotion activity of double eleven global carnival on Taobao. This paradox between them may explained by reasons as follows:

(1) Qualitative analysis materials focus on the opinions of netizens who are disgusted with "Double Eleven". However, in daily life, the proportion of consumers who are disgusted with "Double Eleven" is not large, and more people like it.

(2) As long as the sales promotion satisfies the core point of saving money for consumers, other disadvantages can be ignored.

(3) Even though there are so many disadvantages of the "Double Eleven" sales promotion, it is still excellent compared with other promotional activities which consumers can participate in, so it can still be favored by consumers (Table 4).

Table 4. Appendix 1

Variables	Items	Measurement items	References
Complex Rules (CR)	CR1	I find it difficult to calculate the actual price of the goods	
	CR2	I find the rules of discount complicated and difficult to understand	
	CR3	Coupon collection method is complex and difficult to operate	
Compelling Marketing (CM)	CM1	Forced to give a like to friends for "double eleven"	
	CM2	It was bombarded by advertisements of "double eleven"	
	CM3	Too many promotional text messages	
Disadvantages of Goods/Services (DGS)	DGS1	The express delivery is slow, and it took a long time to receive the goods	
	DGS2	Customer service does not answer	Based on qualitative analysis
	DGS3	Quality of goods is not good.	
Disturb Life (DL)	DL1	"Double eleven" gives me negative emotions	
	DL2	Shopping for "Double eleven" disrupted the schedule	
	DL3	It takes up time to do important things	
Fake Discounts (FD)	FD1	Claim preferential price, did not reduce price actually	
	FD2	Raise the price before reducing the price, exaggerate the degree of discount	
	FD3	The price is not lower than usual	
Irrational Consumption (IC)	IC1	Buy things I don't need	
	IC2	Impulse buying something, don't use it later	
	IC3	Stock up on goods cause they are cheap, but can't use up	
Attitude toward Sales Promotion (AS)	AS1	I like this sales promotion	
	AS2	I would recommend others to take part in the sales promotion	
	AS3	The sales promotion is good	[7]
Sales Promotion Satisfaction (SS)	SS1	I'm satisfied that the discount is a substantial one	
	SS2	It saved me a lot of money	Based on pilot study
	SS3	It offers bigger discounts than other websites	
Attitude toward Website (AW)	AW1	I like this website	
	AW2	Compared to other shopping websites, I'll look at it first	
	AW3	It is a good website	[7]

Acknowledgements. This study was supported by the Key Project of the National Social Science Fund(18AGL024).

References

1. Al-Debei, M.M., Akroush, M.N., Ashouri, M.I.: Consumer attitudes towards online shopping: the effects of trust, perceived benefits, and perceived web quality. Internet Res. **25**(5), 707–733 (2015)
2. Chandon, P., Wansink, B., Laurent, G.: A benefit congruency framework of sales promotion effectiveness. J. Mark. **64**(4), 65–81 (2000)
3. Crespo-Almendros, E., Del Barrio-García, S.: Online airline ticket purchasing: influence of online sales promotion type and internet experience. J. Air Transp. Manag. **53**, 23–34 (2016)
4. Foubert, B., Breugelmans, E., et al.: Something free or something off? a comparative study of the purchase effects of premiums and price cuts. J. Retail. **94**(1), 5–20 (2018)
5. Glaser, B.G., Strauss, A.L., Strutzel, E.: The discovery of grounded theory; strategies for qualitative research. Nurs. Res. **17**(4), 364 (1968)
6. Jallow, H., Dastane, O.: Effect of sales promotion schemes on purchase quantity: a study of Malaysian consumers (2016)
7. Kwahk, K.Y., Kim, B.: Effects of social media on consumers' purchase decisions: evidence from taobao. Serv. Bus. **11**(4), 803–829 (2017)
8. Lee, W.H.: Effects of conditional sales promotion tactics on consumers' feelings of unfairness. J. Promot. Manag. **22**(3), 301–320 (2016)
9. Mpinganjira, M.: An investigation of customer attitude towards online stores. Afr. J. Sci., Technol., Innov. Dev. **8**(5–6), 447–456 (2016)
10. Riley, L., Mili, S., et al.: Using qualitative methods to understand physical activity and weight management among Bangladeshis in New York city, 2013. Prev. Chronic Dis. **13**, E87 (2016)
11. Rodda, S.N., Hing, N., et al.: Behaviour change strategies for problem gambling: an analysis of online posts. Int. Gambl. Stud. **18**(3), 420–438 (2018)
12. Sarkar, S., Khare, A.: Moderating effect of price perception on factors affecting attitude towards online shopping. J. Mark. Anal. **5**(2), 68–80 (2017)
13. Weiwei, T.: When online shopping becomes a habit - a brief analysis of the survey report on online shopping users in China in 2015. China Stat. **2**, 15–17 (2016)
14. Wittenberg-Lyles, E., Washington, K., et al.: "It is the 'starting over' part that is so hard": Using an online group to support hospice bereavement. Palliat. Support. Care **13**(2), 351–357 (2015)
15. Wu, W.Y., Quyen, P.T.P., Rivas, A.A.A.: How e-servicescapes affect customer online shopping intention: the moderating effects of gender and online purchasing experience. Inf. Syst. E-Bus. Manag. **15**(3), 689–715 (2017)
16. Yang, Y.: An introduction to creative management. Econ. Manag. Publ. House (2018)

Herd Behavior of Chinese Firm's Outward FDI Along One Belt One Road- Analysis on Chinese Listed Manufacturing Sector

Haiyue Liu[✉], Shiyi Liu, Ling Huang, and Yile Wang

Business School of Sichuan University, Chengdu, Sichuan,
People's Republic of China
seamoon@scu.edu.cn

Abstract. This paper takes the listed firms of China's manufacturing industry as the research object. FGLS model and Probit model are adopted to analyze 930 OFDI (Outward Foreign Direct Investment) events by listed manufacturing firms from 2004 to 2015. It is found that there exists obvious herd behavior in Chinese manufacturing firms' OFDI, which is more significant in the investment to developed countries. The BRI (Belt and Road Initiative) launched in 2013 significantly enhanced the herding effect, enabling firms to allocate large amounts of money to countries along OBOR (One Belt and One Road). Firms' OFDI herding behavior is more obvious in countries with lower GDP growth rate and infrastructure level. This study supports the location selection theory regarding Chinese firms' OFDI, and provides a reference for optimizing OFDI location selection of Chinese manufacturing firms and improving the investment efficiency.

Keywords: Herd effect · Outward FDI · One belt and one road ·
Location choice · Manufacturing industry

1 Introduction

With the adoption of the "going global" strategy, more and more Chinese firms began to regard OFDI as a method to break down trade barriers, open overseas markets and integrate global resources. As a result, Chinese OFDI has increased significantly. According to the *Chinese OFDI Statistic Report*, in 2017, China's net flow of OFDI was 158.29 billion US dollars, ranked as the third largest among all countries. The launch of Belt and Road Initiative (BRI)[1] in 2013 triggered a

---

[1] In September and October 2013, Chinese President Xi Jinping proposed the cooperative initiatives of building the "New Silk Road Economic Belt" and the "Marine Silk Road in the 21st Century", namely the BRI (The Belt and Road Initiative), aiming at holding high the banner of peaceful development and actively developing economic partnership with the countries along OBOR by borrowing the historical symbols of the ancient Silk Road.

© Springer Nature Switzerland AG 2020
J. Xu et al. (Eds.): ICMSEM2019 2019, AISC 1002, pp. 201–220, 2020.
https://doi.org/10.1007/978-3-030-21255-1_16

new stage in the development of Chinese firms' overseas investment with more firms considering the OBOR area as a destination for investment. As a result, outward foreign direct investment has become an important research topic and the supporting body of literature has expanded to reflect this.

Existing literature suggests that the motives of OFDI include promoting trade, seeking resources, acquiring technology and human capital, opening markets, seeking government supports and broadening financing channels [4,9,23]. Such a diverse array of motivating factors mean that the investment decision-making process is influenced by many factors, including the heterogeneity of host countries and differences in firms' own conditions [26]. Dunning summed up the three most basic factors that determined international direct investment and the behavior of international firms with the OLI model, representing the ownership advantage (Ownership), the location advantage (Location) and the market internalization advantage (Internalization) [7]. Subsequently, many scholars began to study OFDI from the perspective of motivation and location choices.

In China, it's also a widely-discussed topic. As a large developing country, the geographic distribution of Chinese OFDI presents an obvious point of interest. The investment experience of earlier peers greatly influences the OFDI decision of late-comers, and lead to similar behavior of firms even with different characteristics such as productivity and the market position. Due to information asymmetry, it is difficult for investors to make reasonable expectations on the future market. Thus, they tend to absorb information through observing other firms' behavior in the same industry, expecting to reduce the risk and cost of entering the market. In the process of continuous information transmission, different messages reinforce each other and become increasingly similar with each other, resulting in herd behavior. The decision of one firm encourages more firms to follow and eventually OFDI herding effect is formed [11]. The effect of BRI is highly correlated with herding behavior. According to the data displayed by the *Chinese OFDI Statistic Report 2017*, the amount of Chinese OFDI along the OBOR area was 14.6 billion US dollars in 2016 and 20.17 billion US dollars in 2017, showing a substantial increase compared to before 2013. During this process, the investment decisions of pioneer firms provide a great reference for following firms.

The herding effect has significant impacts on investment efficiency, as well as the competitiveness and long-term operating performances of OFDI firms. There are few studies on firms' herd behavior in outward investment decision. Relevant articles are mainly theoretical analyses on managers' decision-making behavior. Typical examples include the "information speculation theory"[2] put forward by Bkhhchandani et al. [1], the theory of "shirking responsibility"[3] put forward by

---

[2] "Information speculation theory": After observing predecessors' behavior, they follow without considering their own information, then information cascade occurs, and then local consistency and group behavior of information are formed.

[3] "Shirking responsibility theory": When a manager makes a decision beforehand, he or she will consider how to shirk his or her personal responsibility and protect his

Shiller and the theory of "remuneration structure"[4] proposed by Aug, E. and N. Naik [17].

Based on the exiting studies, this paper selects data on listed manufacturing firms to study the herding effect in OFDI. As a pillar industry of the national economy, the manufacturing industry influences the overall economic situation of China and its national strength. According to Chinese OFDI Statistic Report 2017, among all industries, the manufacturing industry ranked second in investment flow with an investment volume of \$29.51 billion. In 2017 there was an annual increase rate of 1.6%, and the overall investment accounted for 18.6% of the total annual flow (Table 1). Under this trend, the herding effect and investment agglomeration phenomenon is becoming increasingly obvious. It is of great significance to study the development trend of OFDI in the manufacturing industry and explore efficient investment strategies, so as to enhance China's overall OFDI efficiency.

This paper collects details of 930 cases of listed Chinese manufacturing firms' outward foreign direct investment from 2004 to 2015, aiming to explore whether there is herding effect in firms' OFDI behavior. Host countries are divided into developed countries and developing countries to see the difference of herding effect within these two subsamples. Then we study whether there is a relationship between BRI and herding effect. Finally, the host country determinants of herding behavior are analyzed. The main innovations of this paper are as follows. Firstly, in contrast to existing OFDI literature, this research is carried out based on the herd mentality and follower behavior; the aim is to investigate the impacts of other firms' investment experience on firms' location choice in OFDI. Secondly, most literature uses OFDI stock of the host country to define herding behavior. Country-level data might generate biased result when investigating firm's herding behavior. This paper collects data at firm level which comes from the collation and matching of three databases, including national, industrial and firm levels so we can explore herding behavior in a more detailed way. Finally, this paper integrates the investment frequency and the investment volume into the evaluation system to constructs a more complete set of agglomeration indices to ensure the robustness of the result. This study provides a new angle for the OFDI location selection and structural optimization of Chinese manufacturing firms. We believe this study could shed light on the sustainable development of firms' OFDI, improve the investment efficiency in China's manufacturing industry, as well as providing a reference for guiding firms to invest rationally and implementing OFDI policies.

---

or her professional reputation after the failure of the investment, so as to adopt the strategy of following others' actions.

[4] "Remuneration structure theory": When managers' compensation is related to the performance of other investment managers, managers tend to herd to avoid underperformance.

Table 1. Industry subjects of Chinese OFDI flow in 2016–2017

Industry	2016			Industry	2017		
	Investment flow	Year-on-year(%)	Proportion(%)		Investment flow	Year-on-year(%)	Proportion(%)
Leasing and business services	65.78	81.4	33.5	Leasing and business services	54.27	-17.5	34.3
Manufacturing	29.05	45.3	14.8	Manufacturing	29.51	1.60%	18.6
Wholesale and retail	20.89	8.7	10.7	Wholesale and retail	26.31	25.9	16.6
Information service	18.67	173.6	9.5	Finance	18.79	25.9	11.90%
Real estate	15.25	95.8	7.8	Real estate	68	55.1	4.3

2   Literature Review

When making investment decisions, firms usually focus on investigating the determinants related to the host country's heterogeneity. Host countries' economic, political and legal factors, geographic distances as well as cultural differences have been studied and existing literature have revealed the connections between them and Chinese OFDI location decisions [3,5,20,25]. Some literature has taken OBOR area as a focus and found that BRI significantly promoted Chinese firms' investment in related countries [16]. Thus, domestic policy becomes another variable of interest for researchers whose focus is on studying the motivation behind Chinese OFDI. Not much of the existing literature related to Chinese OFDI explores the herding effect in location decisions despite the tendency of Chinese firms to follow others when investing in high risk countries the majority of countries along the OBOR are found to be of high relative political and financial instability [14].

The spatial agglomeration of OFDI provides a basic pre-condition for firms' herding behavior. Unlike developed countries, investment firms from developing countries with relatively backward home markets find it difficult to avoid "outsider disadvantage", which includes additional costs and competitive disadvantages, institutional differences, cultural differences, geographical coordination, etc. The phenomenon of regional agglomeration appears as a result. Then, the self-reinforcement effect" is constantly produced; the accumulative effect" of production and operation factors in the agglomeration area emerges. For example, the gathering of OFDI firms makes the local business service facilities construct in the direction that is beneficial to the development of foreign firms, thus reducing the operational cost. The agglomeration of OFDI promotes the legalization of location to a certain extent, and entry into the region is conducive to the legitimacy of foreign firms [2]. In conclusion, the positive externalities generated by the geographical concentration of economic activities have a gravitational effect on the investment decision-making of later firms, prompting firms to follow their predecessors to invest in the agglomeration areas [14], resulting in herding behavior. Research has shown that China's OFDI has an obvious agglomeration effect in the host countries [21] which did play an important positive role in OFDI location choice [15]. The phenomenon is more significant in high-income host countries [22]. Some scholars have explored this issue from the perspective of experience. They have examined the impacts of investment experience on firms' location choice. The experience often comes from historical investment cases of the same firm [6,27]. But in China, more than 80% of OFDI firms have no historical OFDI. As Qi et al. found in her study in 2015, most of these impacts come from other firms' investment experience (especially those in the same industry) [19].

Previous studies often employed dummy variables and the absolute value of industry investment to measure herd behavior. In contrast, this paper introduces continuous variables of firms' investment volume and builds a more comprehensive set of agglomeration indices using both investment volume and frequency. The rest of this paper is structured as follows: the third part describes main indicators and variables; the fourth part introduces the model and analyzes empirical results; the fifth part draws conclusions and puts forward to policy recommendations.

3   Variable Descriptions

3.1   Sample Selection

Research objects of this paper are 930 cases of listed Chinese manufacturing firms' outward foreign direct investment from 2004 to 2015, of which 306 cases occurred in the area of OBOR and 628 cases occurred in developed countries.[5] The sample covers 45 host countries,[6] and includes 8 country-level heterogeneity indicators. All data comes from the three databases which have been matched and integrated. The first one is the directory of overseas investment firms published by the Ministry of Commerce; it includes information about domestic investors, names of overseas firms, the scope of business, the places of investment inflow and the time of investment. The second one is Chinese OFDI Statistic Report published by the Ministry of Commerce. The dataset includes information about the flow and stock of China's FDI to each host country. The third is information and financial data about all listed firms that have persisted in Shanghai and Shenzhen stock markets from 2004 to 2015 according to the Wind database and the CSMAR database. Host country indicators have been collected from the source of World Bank, IFS database. More information about the data sources is shown in Table 2.

3.2   Variables

Explained Variable The explained variable of this study is *Lcap* which represents the natural logarithm of OFDI case i's volume to the host country j in year of t. The data are collected from the directory of overseas investment firms and the annual reports of firms.

Explanatory Variables (1) OFDI agglomeration indices[7]
    The indices are used study the impacts of OFDI flow in year $t - 1$ on the investment volume of Chinese manufacturing firms in year t, and to verify herding effect. In order to improve the robustness of results, this study is carried out

---

[5] The World Bank classifies the world's economies into four income groups: high income (OECD), medium to upper income (UMC), medium to lower income (LMC) and low income (LIC). This paper refers to the World Bank's fiscal year 2018 classification of country income groups, and regards the latter three categories as developing countries, and OECD as developed countries.

[6] 20 countries are in the area of BRI while 25 countries are out of BRI related area, see Appendix Table E.1 for details. 22 countries are developed countries while 23 are developing countries, see Appendix Table E.2 for details.

[7] Appendix Table E.3 shows annual average of agglomeration indices in 45 countries.

Table 2. Variable definition and data sources

Variables	Data sources	Definition
PE^a	International Country Risk Guide	It measures the host country's political environment. It is one of the completely independent systems on comprehensive political risk prediction and is widely used in the field of OFDI research [16,18]. Specific contents are shown in Appendix Table E.3
$GDPgr$	World Bank	It measures host country's GDP growth rate, which reflects host country's market potential or absorptive capacity. [19]
$RSRC$	World Bank	It measures the natural resources of the host country, which calculated by the sum of fuel exports (% of merchandise exports) and ores and metals exports (% of merchandise exports). [19]
$Tech$	The United Nations	It measures the host country's level of technological development. It is the ratio measured by the host country's high technology exports to nominal GDP (current U.S. dollar in a given year). [10]
$Infra$	World Development Indicators	It measures the host country's infrastructural facilities, which uses the natural logarithm of telephone lines per 100 people as a proxy [12]
$Meanr$	Authors' calculation. IFS Database	The first moment (average) of monthly real exchange rates around year t (defined to include all monthly observations during year t-1 and year t) for each host country. It measures the exchange rate level of host country's currency. An increase means host country's currency depreciation, vice versa. [16] Formula $Meanr = \Sigma_{k=1}^{2} 4X_k$ (X_k is monthly observation in t and $t-1$)
$Open$	World Bank	It measures the openness (measured by the ratio of the sum of imports and exports to GDP) of host country in constant prices. It reflects economic and trade links between the host country and China. [16]
$Finfree$	The Heritage Foundation	It measures the market openness of the host country. The index is evaluated through the following five sub indicators: the degree of government regulation of financial services; the extent to which the state interfered with the banks or other financial firms through direct and indirect ownership; the government affects the credit allocation; the degree of financial and capital market development; the opening of foreign competitors. [8]

[a]The smaller the PE, the greater the host country's risk. Detailed variables included in PE can be seen in Appendix Table E.3

from two perspectives: investment frequency and volume. Variables are natural logarithms in the regression analysis.

Frequency agglomeration index on country-industrial level: the ratio of the frequency of Chinese manufacturing firms' investment in the host country j in the year of t to the total frequency of all Chinese firms' investment in that country that year.

$$Aggnum_{t,j} = \frac{N_{t,j}}{\Sigma N_{t,j}} \tag{1}$$

Volume agglomeration index on country-industrial level: the ratio of the volume of Chinese manufacturing firms' investment in the host country j in the year of t to the total volume of all Chinese firms' investment in that country that year.

$$Aggvol_{t,j} = \frac{V_{t,j}}{\Sigma V_{t,j}} \tag{2}$$

(2) Host country heterogeneity indicator

This paper takes a set of comprehensive host-country indicators as control variables to investigate whether political environment, market potential, natural resources endowment, the technology level, the infrastructure level, openness etc. can influence a firm's OFDI. Detailed variable definitions and supported literature are listed in Table 2. Table 3 shows a summary of statistics.

Table 3. Summary statistics

	Observation	Mean	Std. dev.	Min	Max
Lcap	930	18.87071	4.00324	5.03449	26.76114
Agnum	861	0.877807	0.167123	0.2	1
Agvol	861	0.916157	0.210456	0.00179	1
PE	866	4.60554	0.88172	1.33333	5.73485
GDPgr	927	2.56634	3.27935	−7.82089	21.6726
LRSRC	897	0.25413	2.65869	−7.89643	4.08266
Ltech	900	23.78591	2.21446	10.55198	26.12143
Linfra	926	3.28619	0.90424	−3.9095	4.17983
Lmeanr	893	−4.89039	2.78649	−7.22181	3.48707
Lopen	922	4.15131	0.87433	0.27154	6.0458
Finfree	893	66.84211	17.56777	10	90

[a]LRSRC, natural logarithm of opening period for host country during t period; Lopen, natural logarithm of opening degree of host country during t period; Ltech, natural logarithm of technology development level in host country during t period; Linfra, natural logarithm of infrastructure level in host country during t period; Lmean, natural logarithm of exchange rate of host country during t period

3.3   Analysis of Spatial Structure of OFDI in Manufacture Sector

We use Hirschman-Herfindahl index (HHI) to quantitatively analyze the spatial distribution structure of OFDI of China's listed manufacturing firms in various sub industries. HHI is a commonly used index to measure the market concentration. It has been widely used by scholars to analyze the location distribution structure of China's export commodities and other regional differences. Similarly, this index is also applicable to the analysis of OFDI spatial distribution structure in this paper.

The formula of HHI:

$$HHI = s_1^2 + s_2^2 + ...s_n^2 = \Sigma_{j=1}^N s_j^2 \tag{3}$$

$$s_j = \frac{ofdi_j}{\Sigma_{j=1}^N ofdi_j} \tag{4}$$

In the formula, s_j indicates that the amount of OFDI of firms in the host country j accounts for the share of the world, N represents the total number of host countries, $ofdi_j$ represents the amount of OFDI in the host country j, and $\Sigma_{j=1}^N ofdi_j$ represents the total amount of OFDI of Chinese listed firms in

Table 4. Spatial distribution structure of OFDI in manufacturing

Sub industry[b]	HHI (yearly average)[a]	Investment frequency	Years of investment
C28	0.9856 (1)	4	3
C27	0.7961 (2)	30	12
C33	0.7497 (3)	191	11
C32	0.6662 (4)	13	7
C29	0.6429 (5)	32	12
C34	0.6387 (6)	38	9
C38	0.6186 (7)	122	12
C22	0.5556 (8)	5	3
C36	0.5366 (9)	54	9
C26	0.5209 (10)	11	7
C35	0.5200 (11)	127	11
C30	0.4706 (12)	14	9
C39	0.3803 (13)	268	12

[a]Numbers in brackets represent the order of spatial structure agglomeration degree reflected by each index, from 1 to 22, agglomeration degree decreases in turn
[b] C28: Chemical Fiber Manufacturing Industry; C27: Pharmaceutical Manufacturing Industry; C33: Metal Products Industry; C32: Non-ferrous Metal Smelting and Rolling Industry; C29: Rubber and Plastic Products Industry; C34: General Equipment Manufacturing Industry; C38: Electrical machinery and equipment manufacturing industry; C22: Paper and Paper Products Industry; C36: Automobile Manufacturing Industry; C26: Manufacturing of Chemical Materials and Chemicals; C35: Manufacturing of Special Equipment; C30: Non-metallic Mineral Products Industry; Manufacturing of Computers, Communications and Other Electronic Equipment

the world. The value of HHI index ranges from 0 to 1. The closer the HHI index approaches to 1, the higher the degree of spatial concentration; the HHI index equals 1, indicating that the industry only invested in one host country in that year; the smaller the HHI index, the lower the degree of concentration, that is, the more average the OFDI location distribution of the industry.

Results are shown in Table 4.

Table 4 shows the average HHI index of each manufacturing sub industry in China over the years, the total number of investments and the number of years of OFDI events in each industry from 2004 to 2015. Generally speaking, excluding the industry where only one firm carries out OFDI in one year, the average annual HHI of most sub-industries exceeds 0.5, and the spatial distribution structure shows significant clustering characteristics.

4   Empirical Analysis

4.1   Model Design

Firstly, B-P test,[8] LM test[9] and VIF test[10] are carried out on the chosen sample. Results show that heteroscedasticity exits between groups while autocorrelation exits within each group; variable collinearity does not exist. Like many previous OFDI studies from the country-level, the FGLS (Feasible Generalized Least Squares) model with control panel heteroscedasticity is adopted in this study. The model allows estimation in cases of autocorrelation in the panel and correlation in the cross-section. Regression results are used to find out whether Chinese manufacturing firms invest more in OFDI agglomeration countries, which indicates the existence of herding effect. The basic model is as follows:

$$Cap_{ij,t} = \alpha_0 + \alpha_1 Agg_{ij,t-1} + \Sigma_m \alpha_m Country_{ij,t} + \mu_{ij,t} \qquad (5)$$

In the formula, $Cap_{ij,t}$ is the dependent variable, which represents the natural logarithm of the amount of OFDI case i to the host country j in the year of t. $Agg_{ij,t-1}$ is the main explanatory variable, representing the natural logarithm of the agglomeration index for the last year in the host country. Among them, $i, j = 1, 2, 3...n$. $Country_{ij,t}$ denotes host country heterogeneity, including political risk, exchange rate volatility, openness degree, infrastructure level, technological advantage, financial freedom and other variables. $\mu_{ij,t}$ is the residual term.

---

[8] B-P test is used to test whether there is heteroscedasticity in the model. If the results significantly reject the original hypothesis, there is heteroscedasticity.

[9] LM test (Lagrange multiplier) is used to test whether there is high-order autocorrelation in the residual sequence of regression equation. If the result significantly rejects the original hypothesis, there is sequence correlation.

[10] VIF test is used to test the existence of multiple collinearity between variables. It is generally believed that there is no obvious multicollinearity when a single VIF does not exceed 10 and the average VIF does not exceed 5.

4.2   Is There a Herding Effect in the OFDI of Chinese Listed Manufacturing Firms?

If new Chinese firms tend to investment more when the host countries have more Chinese OFDI agglomeration in the previous year, we can identify firm's herd behavior and claim that the host country has experienced a herding effect with regards to new Chinese investments. FGLS regression was used on the whole sample. The empirical results in Table 4 show that the regression coefficients of aggregation indices relative to the new OFDI volume of Chinese manufacturing firms are significantly positive at the level of 1%, which proves that new entrants tend to invest in countries with high OFDI flows in the previous year. It means when making investment decisions, they will pay attention to the investment trends of other firms in the same industry. The gathering of OFDI firms actually conveys a signal that the cultural environment and institutional policies of the region are suitable for foreign firms to operate, thus reducing the risk and cost of firms entering the market. Obvious herding behavior can be found, which is robust under the two measurement methods of the quantity index and the volume index.

In order to provide robustness to the results, and explore whether there is a difference of herding effect between developed and developing countries, we divide the sample into two sub-samples. The results are shown in Table 4. Compared with developing countries, firms investing in developed countries exhibit more significant herding behavior. The reason is that there are much more previous investment cases in developed countries than in developing countries (see Numbers of Obs. in Table 5). It makes the reference and follow-up to a higher degree, thus providing a more convenient condition for a firm's herding behavior.

4.3   BRI and Herding Effect

After BRI was launched, Chinese firms' investment in the areas along the OBOR has increased significantly [13, 24]. Spatial distribution of OFDI in the whole area shows the obvious feature of concentration [13]. Therefore, herding effect plays an important role in investment of Chinese firms in related area. In order to further explore how the herding effect affects the investment of China's manufacturing firms in countries along The Belt and Road", we divide host countries into two sub samples. One sub sample includes countries along the OBOR and the other sub sample contains countries outside OBOR. A dummy variable *Initiative* is set as the proxy variables of BRI effect. OFDI cases occurred after 2013 takes the value of 1; OFDI events occurred before 2013 (including 2013) takes the value of 0.

At the same time, dummy variables of Agnum_dum and Agvol_dum are introduced to divide host countries into strong-agglomeration countries and weak-agglomeration countries. If the agglomeration index of host country j is greater than the median of the sample in the year of t, the country will be defined as a strong-agglomeration country and take the value of 1. Otherwise, the country will be defined as a weak-agglomeration country and take the value of 0. The values are delayed by one year in regression. Results are as follows[11]:

As shown in Table 6, the agglomeration indices of two sub samples are significantly positive under the perspective of both frequency and volume, indicating that the existence of herding effect is robust. In OBOR sub sample, the regression coefficients of cross items Agnum_dum Init and Agvol_dum Init are significantly positive, indicating that the joint action of BRI and herding effect will increase the investment in OBOR area; the herding effect can enhance the response degree of Chinese manufacturing firms to BRI to a certain extent. More over, they tend to follow other firms and allocate a large amount of money to countries which have attracted lots of OFDI from China's manufacturing industry. These decisions further generate cumulative effect and attract even more corporations to invest in the same country. Previous studies also confirm that the spatial agglomeration of China's OFDI does exist in OBOR area. After the agglomeration reaches a certain scale, it may even generate a spatial spillover effect and then spreads to neighboring countries and regions [13].

4.4   Host Countries' Determinants of Firm's Herd Behavior

In order to explore factors influencing firms' herding behavior, this paper further establishes a Probit model. *Herdnum* and *Herdvol* are introduced as proxy variables of herding behavior on country-industry level. Variable assignment goes as follows: In the year of t, if OFDI case i's host country j is a strong-agglomeration country, the firm's decision is identified as herding behavior with the variable value of 1. In the year of t, if OFDI i's host country j is a weak-agglomeration country, the firm's decision is identified as non-herding behavior and the variable value is 0.

The specific model is stated below

$$Herd_{ij,t} = \alpha_0 + \Sigma_m \alpha_m Country_{ij,t} + \mu_{ij,t} \tag{6}$$

where $Herd_{ij,t}$ represents herd behavior proxy variable, $Country_{ij,t}$ means host heterogeneity, and $\mu_{ij,t}$ is the residual item.

Results are shown in Table 7.

It is found that poor economic development and low infrastructure level are more likely to stimulate the herding behavior of Chinese manufacturing firms to some extent for both sub-samples. Possible reasons include, the information acquisition of host countries with poor conditions (lower *GDPgr* and *Infra*) is difficult, while the investment risk is relatively high. Therefore, firms need to learn from predecessors' experience to reduce the likelihood of investment failure. At the same time, the shirking responsibility theory" shows that when investment fails, herding behavior help managers to diffuse responsibility. To sum up, the cost of following-up is lower than seeking information on their own when investing in such countries.

---

[11] To avoid the influence of collinearity, the cross variables are centralization.

Table 5. Estimation results of herding effect identification

Variables	Frequency index			Volume Index		
	Total sample	Developed countries	Developing countries	Total sample	Developed countries	Developing countries
LAgnum	0.876***	3.153***	−2.176***	0.257***	1.268***	0.185***
	(0.22)	(0.136)	(0.507)	(0.0027)	(0.133)	(0.0242)
LAgvol						
PE	−0.542***	−1.708***	−1.265***	−0.522***	−1.859***	−0.864***
	(0.0443)	(0.166)	(0.299)	(0.0329)	(0.152)	(0.264)
GDPgr	0.0983***	0.213***	−0.0285	0.132***	0.0491***	0.0758*
	(0.0126)	(0.0151)	(0.0471)	(0.00842)	(0.0127)	(0.0459)
LRSRc—	0.103***	0.262***	−0.687**	0.173***	0.307***	−0.477*
	(0.0133)	(0.0206)	(0.296)	(0.00684)	(0.0208)	(0.267)
Ltech	0.127***	0.157***	0.512***	0.170***	−0.0556**	0.556***
	(0.0133)	(0.0206)	(0.0458)	(0.0102)	(0.0255)	(0.0811)
LInfra	−0.659***	−0.478*	−0.606***	−0.478***	−0.306	−0.650***
	(0.0854)	(0.248)	(0.189)	(0.0569)	(0.192)	(0.223)
Lmeanr	−0.0640***	−0.425***	0.300***	−0.0869***	−0.395***	0.277***
	(0.0132)	(0.013)	(0.0373)	(0.00605)	(0.0115)	(0.0302)
Lopen	0.0997***	0.464***	−0.536***	0.199***	0.719***	−0.510***
	(0.0288)	(0.0448)	(0.0942)	(0.0197)	(0.0498)	(0.108)
Finfree	0.0997***	−0.000724	0.0271**	0.199***	−0.0148***	0.0452***
	(0.0288)	(0.00642)	(0.0118)	(0.0197)	(0.00479)	(0.00962)
Constant	17.66***	21.98***	16.98***	15.42***	27.48***	12.84***
	(0.519)	(1.577)	(1.364)	(0.363)	(1.282)	(1.748)
Year dummy	Controlled	Controlled	Controlled	Controlled	Controlled	Controlled
Industry dummy	Controlled	Controlled	Controlled	Controlled	Controlled	Controlled
Number of Obs	726	491	235	726	491	235

Notes: Robust standard errors in parentheses, ***p¡<0.01, **p¡<0.05, *p¡<0.1 (the same as in the following table)

Table 6. Estimation results of herding effect in OBOR

Variables	Frequency index		Volume index	
	OBOR	Non-OBOR	OBOR	Non-OBOR
Agnum_dum	1.605***	1.290***		
	−0.194	−0.0959		
Agvol_dum			0.701***	0.800***
			−0.169	−0.0736
Initiative	−2.008***	1.980***	−1.679***	1.704***
	−0.571	−0.322	−0.614	−0.279
Agnum_dumInit	1.983***	−1.030***		
	−0.242	−0.265		
Agvol_dumInit			1.342***	−1.025***
			−0.242	−0.271
PE	0.499***	−1.682***	0.155	−1.345***
	−0.152	−0.104	−0.248	−0.132
GDPgr	−0.0553	0.204***	−0.0847**	0.126***
	−0.041	−0.0277	−0.0347	−0.0241
LRSRC	0.0698**	0.323***	0.0456	0.270***
	−0.0351	−0.0413	−0.0549	−0.055
Ltech	0.230***	−0.0185	0.182***	−0.0356
	−0.0325	−0.0383	−0.0443	−0.0372
LInfra	−1.406***	3.781***	−1.409***	2.852***
	−0.081	−0.251	−0.0909	−0.204
Lmeanr	0.449***	−0.412***	0.391***	−0.416***
	−0.0242	−0.0185	−0.0339	−0.022
Lopen	−0.106*	0.450***	−0.00902	0.511***
	−0.0576	−0.0408	−0.068	−0.053
Finfree	0.113***	−0.00407	0.0921***	−0.011
	−0.00508	−0.00474	−0.00783	−0.00678
Constant	10.51***	7.684***	14.58***	11.07***
	−1.339	−1.213	−1.936	−1.008
Year dummy	Controlled	Controlled	Controlled	Controlled
Industry dummy	Controlled	Controlled	Controlled	Controlled
Number of Obs	226	500	226	500

The results also show divergent responses of firms' herding behavior with related to other host-country characteristics. In detail, firms' herd behavior quantified by investment frequency and volume occurs more in OBOR host-countries with rich natural resources and less financial openness, it means the investment

Table 7. Estimation results of factors affecting herding behavior

Variables	Herdnum		Herdvol	
	OBOR	Non-OBOR	OBOR	Non-OBOR
PE	1.464***	0.314	−0.428	−2.277**
	−0.548	−0.455	−0.312	−1.119
GDPgr	−1.037***	−0.791***	−0.304***	−0.151*
	−0.233	−0.157	−0.105	−0.0807
LRSRC	1.314***	−2.557***	0.405***	−2.963***
	−0.298	−0.343	−0.0926	−0.599
Ltech	−0.127	−0.874***	0.025	−1.484***
	−0.115	−0.143	−0.0902	−0.335
LInfra	−0.137	−10.41***	−0.390*	−6.962***
	−0.331	−1.305	−0.216	−0.89
Lmeanr	−0.115	−0.128	−0.133**	−0.658**
	−0.0985	−0.127	−0.0581	−0.269
Lopen	−0.608*	0.685***	−0.142	0.185
	−0.321	−0.175	−0.229	−0.166
Finfree	−0.117***	0.0483**	−0.0314***	0.0855***
	−0.0246	−0.0197	−0.0113	−0.028
Constant	15.33***	57.14***	7.302***	65.57***
	−2.598	−5.841	−1.899	−13.29
Year dummy	Controlled	Controlled	Controlled	Controlled
Industry dummy	Controlled	Controlled	Controlled	Controlled
Adj.	0.7103	0.8026	0.5651	0.6922
Number of Obs	226	500	226	500

motivation of obtaining natural resources is considered as a very important factor to form herding behavior, abundant resources of host country will cause firms to herd there along OBOR. When investing in countries with low financial freedom, herding can help reduce entry cost to the market [2]. However, in non-OBOR areas, it's the resource scarcity and highly open financial market that promotes the herding behavior. Compared with seeking natural resources, firms investing in non-OBOR areas are more concerned about the good financial environment, thus forming investment agglomeration and improving the possibility of herding behavior.

5   Conclusions and Suggestions

The paper investigates 930 OFDI events by listed Chinese manufacturing firms from 2004 to 2015. Two agglomeration indicators are set from the country level and industrial level considering both investment frequency and volume. The results show that a herding effect does exist in Chinese manufacturing firms' OFDI and it's more obvious when investing in developed countries. The herding effect is also significant in countries along OBOR and amplifies effect of BRI. Probit regression shows that the herding behavior in OFDI can be significantly enhanced in countries with lower GDP growth rate and infrastructure construction level for the whole sample, while responding differently to other host-country indicators like natural resources endowment and financial environment in OBOR and non-OBOR sub samples.

The research implies that, to improve the allocation of resources and optimize the investment location, we need the joint efforts of firms and the government. On one hand, Chinese manufacturing firms should enhance their ability to integrate information and firm's long-term strategy. Firms can learn from other firms' investment experience to help decision-making, but on the basis of that, they also need to maintain rationality and try to make the best decision through comprehensively considering all relevant factors, including predecessors' experience, domestic and foreign policies, the conditions of the firm itself and the environment of the host country. In addition, firms also need to improve their operational strength and production efficiency, so as to better cope with possible impacts of investment agglomeration and institutional environment changes in the host country. Furthermore, the government and relevant departments should build effective information dissemination channels to further promote the smooth flow of investment information among firms, provide better host-country risk evaluation and consultancy, lead capital to allocate to countries and industries that could generate better performances, and build up think tank and information exchange platform to ensure firms are better prepared for international competition.

Acknowledgements. This work was supported by Humanities and Social Sciences Fund of Ministry of Education, the Empirical Study on "Herd Effect" of Chinese Firms' OFDI (17YJC790094).

Appendix

See Tables E.1, E.2 and E.3

Table E.1. List of countries along One Belt One Road

OBOR		Non-OBOR	
Coutrycode	Name	Coutrycode	Name
ARE	The United Arab Emirates	ARG	The Republic of Argentina
BRN	Negara Brunei Darussalam	AUS	Commonwealth of Australia
HUN	People's Republic of Hungary	AUT	The Republic of Austria
IDN	Republic of Indonesia	BEL	The Kingdom Of Belgium
IND	Republic of India	BRA	The Federative Republic of Brazil
IRN	Islamic Republic of Iran	CAN	Canada
JOR	The Hashemite Kingdom of Jordan	COD	The Republic of Congo
JPA	Japan	DEU	The Federal Republic of Germany
KAZ	The Republic of Kazakhstan	DNK	The Kingdom of Denmark
KHM	The Kingdom of Cambodia	ESP	The Kingdom of Spain
MNG	Mongolia	FRA	French Republic
MYS	Malaysia	GBR	United Kingdom of Great Britain and Northern Ireland
PAK	The Republic of Panama	GHA	The Republic of Ghana
PHL	Republic of the Philippine	ITA	The Republic of Italy
POL	The Republic Of Poland	KOR	Republic of Korea
RUS	The Russian Federation	LUX	The Grand Duchy of Luxembourg
SGP	Republic of Singapore	MAC	Macao
THA	Kingdom of Thailand	MAR	The Kingdom of Morocco
UZB	The Republic of Uzbekistan	MEX	The United States of Mexico
VNM	Socialist Republic of Vietnam	NLD	The Kingdom of the Netherlands
		PNG	Independent State of Papua New Guinea
		SWE	The Kingdom of Sweden
		TUN	The Republic of Tunisia
		USA	United States of America
		ZAF	The Republic of South Africa

Table E.2. List of developed and developing countries

Devloped countries		Developing countries	
Coutrycode	Name	Coutrycode	Name
ARE	United Arab Emirates	BRA	Brazil
ARG	Argentina	COD	Congo, Dem. Rep
AUS	Australia	GHA	Ghana
AUT	Austria	IDN	Indonesia
BEL	Belgium	IND	India
BRN	Brunei Darussalam	IRN	Iran, Islamic Rep
CAN	Canada	JOR	Jordan
DEU	Germany	JPN	Japan
DNK	Denmark	KAZ	Kazakhstan
ESP	Spain	KHM	Cambodia
FRA	France	MAR	Morocco
GBR	United Kingdom	MEX	Mexico
HUN	Hungary	MNG	Mongolia
ITA	Italy	MYS	Malaysia
KOR	Korea, Rep.	PAK	Pakistan
LUX	Luxembourg	PHL	Philippines
MAC	Macao SAR, China	PNG	Papua New Guinea
NLD	Netherlands	RUS	Russian Federation
POL	Poland	THA	Thailand
SGP	Singapore	TUN	Tunisia
SWE	Sweden	UZB	Uzbekistan
USA	United States	VNM	Vietnam
		ZAF	South Africa

Table E.3. Variables incorporated into PE (Political environment)

Grade 1 index	Grade 2 index
Voice and accountability	Military in politics
	Democratic accountability
Globalization	Social globalization
	Political globalization
	Overall globalization
Freedom	Business freedom
	Fiscal freedom
	Labor freedom
	Monetary freedom
	Investment freedom
	Financial freedom
Control of corruption	Corruption perceptions of transparent
	Freedom from corruption
Law and rights	Rule of law
	Property rights
	Political rights

References

1. Bikhchandani, S., Hirshleifer, D., Welch, I.: A theory of fads, fashion, custom, and cultural change as informational cascades. J. Polit. Econ. **100**(5), 992–1026 (1992)
2. Chang, S.J., Park, S.: Types of firms generating network externalities and mncs' co-location decisions. Strat. Manag. J. **26**(7), 595–615 (2005)
3. Chen, S., Liu, H.: The effects of governance of host countries on chinese outward foreign direct investmentĦan empirical study based on panel data. Res. Econ. Manag. **6**, 71–78 (2012)
4. Cheung, Y.W., Qian, X.: Empirics of china's outward direct investment. Pac. Econ. Rev. **14**(3), 312–341 (2009)
5. Chi, J.Y., Fang, Y.: Institutional constraints for chinese ofdi's location choices. Int. Econ. Trade Res. **01**, 81–91 (2014)
6. Chung, W., Song, J.: Sequential investment, firm motives, and agglomeration of japanese electronics firms in the united states. J. Econ. Manag. Strat. **13**(3), 539–560 (2004)
7. Dunning, J.H.: Toward an eclectic theory of international production: Some empirical tests. J. Int. Bus. Stud. **11**(1), 9–31 (1980)
8. He, Y., Xu, K.: Study on the effect of the economic institutions of countries along the "belt and road" on chinese outward foreign direct investment. Journal of International Trade **1**, 92–100 (2018)
9. Holtbrügge, D., Kreppel, H.: Determinants of outward foreign direct investment from bric countries: An explorative study. Int. J. Emerg. Mark. **7**(1), 4–30 (2012)

10. Hou, W., Su, J.: Research on the factors influencing china's ofdi location choice. Commer. Res. (1) (2018)
11. Jun-xiong, F.: Corporate investment decision-making convergence in china: Herd behavior or wave phenomenon? J. Financ. Econ. **11**, 93–103 (2012)
12. Kamal, M.A., Li, Z., et al.: What determines china's fdi inflow to south asia? Mediterr.Ean J. Soc. Sci. **5**(23), 254 (2014)
13. Li, Q., Xu, W.: China's location selection of "one belt and one road" global ofdi: Based on the perspective of spatial effects. Macroeconomics (8), 3–18+102 (2017)
14. Li, Y., Wu, C.: Agglomeration on effects and outward fdi from developing countries: The case of china. Q. J. Manag. (z1), 24–46 (2016)
15. Liu, H., Jiang, J., et al.: Ofdi agglomeration and chinese firm location decisions under the belt and road initiative. Sustainability **10**(11), 4060 (2018)
16. Liu, H.Y., Tang, Y.K., et al.: The determinants of chinese outward fdi in countries along one belt one road. Emerg. Mark. Financ. Trade **53**(6), 1374–1387 (2017)
17. Maug, E., Naik, N.: Herding and delegated portfolio management: The impact of relative performance evaluation on asset allocation. Q. J. Financ. **1**(02), 265–292 (2011)
18. Pang, H.E., Wang, J., School, E.: A study on the political risks of china's ofdi in brics countries and precautions. J. Xidian Univ. (1) (2018)
19. Qi, J., Liu, H.: Is previous experience relevant for location choice of sequential investment: Evidence from china's industrial enterprises. Econ. Theory Bus. Manag. **V35**(10), 100–112 (2015)
20. Song, W., Xu, H.: The research of factors affecting location choice of foreign direct investment. Res. Financ. Econ. Issues **10**, 40–50 (2012)
21. Sun, C., Shao, Y.: The effect of economic cooperation on china's outward foreign direct investmentła spatial panel data approach. Emerg. Mark. Financ. Trade **53**(9), 2001–2019 (2017)
22. Wang, F., Liu, J., Cong, S.: Outward foreign direct investment and export performance in china. Can. Public Policy **43**(S2), 1–16 (2017)
23. Wang, J.Y.: What drives China's growing role in Africa? pp. 7–211, Social Science Electronic (2007)
24. Xiong, B., Wang, M.: Study on the influencing factors of chinese ofdi in countries along the belt and road: Based on spatial perspective. J. Int. Trade (2) (2018)
25. Yu, G.: Host countries' economic risk and binary growth location choice of china's foreign direct investment: An empirical study based on panel data threshold effect model. J. Cent.L Univ. Financ. Econ. **6**, 74–81 (2017)
26. Zhang, Y.: Human development, political risk in host country and chinese enterprises' ofdiła panel tobit model analysis. Rev. Invest. Stud. **4**, 103–117 (2017)
27. Zhu, A.: Essays on the discrete choice model: Application and extension. Ph.D. thesis (2014)

ETOS-based Research on Earthquake Tourism in the Longmenshan Fault Zone: Take Wenchuan for Example

Xu Zu[1], Xinyi Diao[2], and Zhiyi Meng[3(✉)]

[1] Business School, Sichuan Agricultural University, Chengdu 611830,
People's Republic of China
[2] College of Economics, Sichuan Agricultural University, Chengdu 611130,
People's Republic of China
[3] Business School, Sichuan University, Chengdu 610065, People's Republic of China
zhiyimeng@scu.edu.cn

Abstract. Since the distinguishing characteristics- having the most widely affected area, involving the most geological units, and having the richest and best preserved geological relics landscape, especially the new earthquake disaster relics landscape and lake landscape system-Longmenshan Fault Zone (LFZ) becomes an ideal area for research into earthquake relics and tourism landscapes. Understanding the opportunities for earthquake tourism within any environment involves recognizing the carrying capacity of the development areas, and how to develop and resolve the contradiction between development and preservation reasonably. Therefore in this article, a management tool and an earthquake tourism system are proposed considering the actual situation in the LFZ.

Keywords: Earthquake disaster relics · Longmenshan fault zone · Tourism landscapes · ETOS

1 Introduction

Since both the Wenchuan and Lushan earthquakes occurred in the LFZ (Longmenshan Fault Zone), the post-earthquake tourism research is of great significance in China's post-disaster reconstruction [18]. In this article, a management tool and an earthquake tourism system are proposed considering the actual situation in the LFZ.

General information about the disaster-stricken areas, e.g., the disaster situation, biodiversity, geological conditions, environmental functions and so forth were studied through document analysis, field investigations, on-site interviews, telephone interviews, and participatory observations. Compared to other researchers, we have a unique and advantageous position. First, our institute has a unique geographical advantage as it is located in Chengdu, the capital of earthquake-ravaged Sichuan province [19]. Chengdu is near the earthquake-hit

J. Xu et al. (Eds.): ICMSEM2019 2019, AISC 1002, pp. 221–233, 2020.
https://doi.org/10.1007/978-3-030-21255-1_17

regions, so it was convenient to conduct the field studies. Wenchuan lies in the southeast part of the Tibetan-Qiang Autonomous Prefecture of Aba, 146 km to the northwest of Chengdu [20]. Lushan is a hilly area about 156 km from Chengdu along the LFZ heavily impacted by the 2008 Wenchuan earthquake. Most disaster relief materials including quilts, disinfectants, tents and rice were shipped from Chengdu. Second, the Institute of Emergency Management and Post-disaster Reconstruction has a scientific research advantage as it is professionally engaged in the post-disaster reconstruction study. Our team has suggested many reconstruction strategies to the authorities and a series of studies have been made of the post-disaster reconstruction and emergency management in both earthquakes [16,17,20,22]. On that basis, previous studies about earthquake tourism are compacted, and then a classification management method for an earthquake management system is proposed.

2   Material and Method

2.1   Study Area

The topography of the LFZ is varied and complicated [15], where live the Han, Tibetan, Yi, Qiang, Lisu and other nationalities. Their different cultures and customs provide inexhaustible tourism resources [20]. This paper outlines a new framework, based on the practical situation of the LFZ within which opportunities for earthquake tourism may be realized.

Firsthand data were collected immediately by relief volunteers of our team in the hard-hit areas (e.g., Dujiangyan, Wenchuan, Beichuan, Lushan) through field observations. In particular, Wenchuan earthquake data was collected from May 12 to May 30, 2008, and Lushan earthquake from April 20 to May 10, 2013 [20]. The interviewees were randomly selected to gather thoughts about the restoration and reconstruction with the victims, rescuers, officials and representatives of nongovernmental organizations. In addition, firsthand data were confirmed and complemented through telephone interviews with the previous interviewees. Also, secondary facts, data, and evidence were collected through document analysis.

2.2   ETOS

Development of the ETOS framework is set against the study of many key problems. Understanding the opportunities for earthquake tourism within any environment involves recognizing the carrying capacity of the development areas, and how to develop and resolve the contradiction between development and preservation reasonably.

Physical characteristics
Altitude: Altitude is an important determinant for shaping earthquake tourism development. The altitude of the LFZ varies from 200 m in the Sichuan Basin to 7556 m at the peak of Mount Minya Konka. The altitude is distinguished by

high-altitude tourism characteristics: low; moderate; high. Every 1000 m increase in altitude reduces air density by 11% [3].

Landform: A landform is a feature on the Earths surface that is part of the terrain, including four major types: Mountains, hills, plateaus, and plains. Minor landforms include buttes, canyons, valleys, and basins [14]. The LFZ can be divided into types: (a) basin and plain; (b) hills and mountains; (c) mountains and plateaus.

Climate: Climate is a measure of the average pattern of variation in temperature, humidity, atmospheric pressure, wind, and other meteorological variables in a given region over long periods [3]. The climate of the LFZ is complex and varied because of the complicated geography and various landforms. Therefore, the climates are classified into three types: mild; moderate; diverse.

Cultural background

Cultural background constitutes the ethnic, religious and other socioeconomic factors that shape an individual's upbringing. It is an important way to define an individual's identity [6]. The comfortable working and living environment of Chengdu brings worldwide people of different cultural background to interact with each other. However, tourists visiting the Longmenshan National Geo-park have a natural tourism experience because no people are living in the park, so tourists only enjoy the unique scenery, the rich natural resources and the primeval forest. Therefore, cultural background diminishes gradually from the modern areas to the primitive areas: large; moderate; small.

Access

Transportation: The transportation categories we use were adopted from TOS: (A) non-motorized conveyance; (B) mixed non-motorized and motorized; (C) many traffic modes with few non-motorized. Explorers and researchers are expected to prefer non-motorized forms of transportation, such as canoes or foot travel to limit the impact on the environment. Motorized forms of transportation would be the preference for most tourists because of convenience and efficiency.

Marketplace: The marketplace would also differ between the types of tourists: individual; retailers; wholesalers [5]. Explorers and scientists would prefer to travel alone. In contrast, older people and students prefer to travel as part of an organized tour. As a result, the market is diverse, and not as general as for mass tourism. Other types may be developed from the marketplace created by local tourism operators who own accommodations within the earthquake tourism destination areas. Information system: In general, an information system consists of all the ways that people communicate with others [11]. An earthquake tourism information system is a specific type of information system which consists of all information channels used in a business or community to promote itself. These channels can be divided into three types: word of mouth; social source advisory; commercial.

Other Resources Use

The presence of other resources and their relationship with earthquake tourism is treated as an important factor within the ETOS framework. Compatibility is the possible goal which would be dependent on the nature and extent of the earthquake tourism activity within a region and the nature of the other uses. In this paper, the degree of compatibility can be divided into three types: high; moderate; minimal.

Attractions

The diversity of a saved and constructed earthquake landscape is a major factor in the ETOS framework. The LFZ Earthquake Landscape has unique geological landscapes and diverse national cultures, thereby containing the double properties of natural and cultural values. The ETOS attractions were analyzed and three points were derived: (N) primitive quake relics only; (M) mixed quake landscapes with modern and primitive; (E) modern quake landscapes with some learning about the primitive environment.

Skill level

The skill level factor in ETOS is based on TOS [4], where similar levels were adopted to indicate the level of skill and prior knowledge needed to participate in the earthquake tourism experience: extensive; medium; minimal.

Job opportunities

Job placement is an important part of the post-earthquake reconstruction task. To help the affected people out of the psychological shadow and to start a new life, the government has invested in massive projects to promote economic recovery and job creation. The job opportunities from Earthquake Modern areas to Earthquake Primitive areas decrease with the economy's needs and environmental protection.

Social interaction

The appropriate amount of social interaction is an important characteristic of different tourism opportunities [13]. Social interaction beyond expected encounter levels may result in changes in the experience obtained within a region, which may impact the opportunities the region may present to tourists.

Facilities

An improved facility-classify is presented in the ETOS based on the ROS: (H) many comforts and conveniences; (I) moderate comforts and conveniences; (J) few or no comforts or conveniences [4].

Ecological ethics

Eco-Ethics ideology refers to the ethical relationship between nature and humans and how the environment is treated. Therefore, under the constraint conditions of the limited earthquake tourism resources in the LFZ, tourism must adhere to sustainable development. The attraction of earthquake tourism from the modern to the primitive setting is rather diverse, we therefore divide three levels of ecological ethics: very high; high; general.

Life ethics

Life ethics ecology should be incorporated into earthquake tourism development as it is necessary to consider the emotion and opinions of the victims and those left behind. Post-disaster rebuilding projects need to emphasize care for the people. The classification of life ethics is based on the disaster situation and reconstruction scale: very high; high; general.

Acceptance of visitor impacts

The acceptable impact level for earthquake tourism is a concern to both managers and tourists. Managers must be concerned about maintaining opportunities for quality tourism as well as protecting other tourism resources. The average tourist may be aware of the impact from the Earthquake Modern experience and be sympathetic and thus be willing to accept the strict control over the number of groups permitted, their size and allowable activities.

Acceptance of management regime

Butler and Waldbrook [4] proposed TOS in an attempt to control tourism development and define responsibility. As a new form of tourism, earthquake tourism needs an adaptable managerial regime. Earthquake tourism management involves different departments and agencies. However, repeated information and data make matters so complex that a professional organization should be established and lawmakers need to define the responsibilities and the boundaries. Specifics on what type of organization should be involved, its structure, and level of responsibility are open to much debate by the stakeholders involved with earthquake tourism and the various groups that may be involved in the overall decision-making process [2].

An approach for combining all factors is shown in Fig. 1. All the conditions represented by the thirteen factors are arrayed along a modern to primitive opportunity continuum. For any generic type of opportunity, a band of acceptable combinations can be described in area management plans through the use of objectives and standards.

3   Results and Analyses

The central notion of ETOS is to offer tourists alternative settings in which they can have a variety of tourism experiences. In Fig. 1, the range of conditions that a factor can have represents the relative rather than the absolute limits of what is acceptable and appropriate along the ETOS. When the framework is applied, specific criteria must be developed. Our objective, however, is to focus on the process by which the ETOS factors can be managed to achieve the desired objectives. After the tourists understand the approach, then more specific values for each factor can be estimated.

Because the specific parameter and threshold data include confidential information, detailed data would not be made public. In this discussion, we assume that all factors have equal weight owing to the specificity and sensitivity of earthquake tourism and all influence earthquake tourism behavior in a similar way.

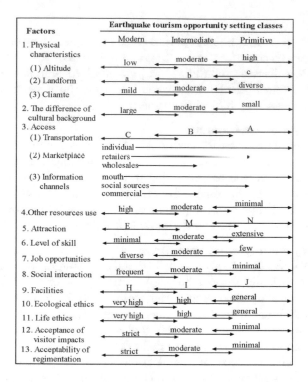

Fig. 1. Earthquake tourism opportunity setting classes

Though in reality, this is probably not the case, such individual differences may balance out when tourism choices are considered in general. Further research is required to determine under what conditions these assumptions are appropriate. The study areas in the LFZ were divided into three types of earthquake tourism through an expert evaluation method (Fig. 2).

It is evident from Fig. 2 that the Earthquake Primitive opportunities are concentrated in the hardest-hit areas, such as Qingchuan, Pingwu, Beichuan, Wenchuan, Dujiangyan, Pengzhou, and Lushan. These places theoretically belong to the forbidden development zone, so the relics need to be retained in the museum to commemorate the deceased, conduct scientific research and remember the history. Because of complicated geological and economic conditions, some harder-hit areas such as Jiange, Yanting, Zhongjiang, Songpan, Maoxian, and Lixian are in a restricted development zone, so are suitable for developing Earthquake Intermediate type tourism. The Earthquake Modern type is most suitable for relatively developed urban areas in Chengdu, Ya'an, Deyang and Mianyang. The study areas are divided into the different tourism opportunities types, which allow for the development of local earthquake tourism ideas and also provide diverse earthquake landscapes for tourist routes.

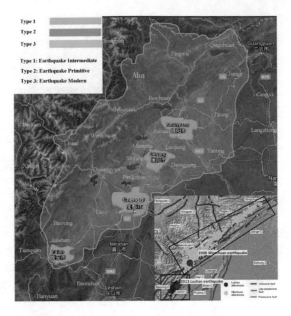

Fig. 2. Three types of earthquake tourism

4   Suggestion on Optimization

Earthquake tourism is adopting new organizational and managerial principles including quality, flexibility, customization, innovation, diagonal integration and environmental soundness [12]. This paper explores the issues associated with the establishment of the management mechanism and especially considers the management of the different periods of earthquake tourism.

Management architecture
ETOS is a tool with universal appeal that could be applied in the world's most earthquake-prone countries; China, Japan, the United States, Mexico, Chile and Indonesia. ETOS improves our understanding of the complexity of earthquake tourism management, strengthens sound professional judgment, and enables a manager to make better decisions. Earthquake tourism management involves several stakeholders, multiple periods and different areas, as seen from the ETOS factors. This paper proposes a three-dimensional management architecture (time, space and object) for earthquake tourism (Fig. 3).

   The "Time dimension" requires respecting the evolutionary rules of earthquake disasters and obeying the process from the beginning, through development, to the changing and the dying down. The development of earthquake tourism should be launched at the right time with a proper plan. The "Time dimension" is divided into three periods; pre-earthquake, earthquake-process and post-earthquake. The "Space dimension" consists of three types; forbidden development zones, restricted development zones and key development zones. The "Space dimension" requires a response and interaction with the concept

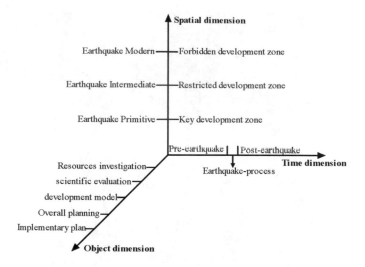

Fig. 3. Three-dimensional management architecture for earthquake tourism

of "Man-land relationship". The "Object dimension" refers to the achievement of the sustainable development of earthquake tourism through resources investigations, scientific evaluations, development models, overall planning, implementing planning, among other measures [10]. The theoretical basis for the integrated management system includes ekistics, geology, sociology, catastrophology, tourism, environics, ethics, engineering, management and other areas.

Integrated system

An earthquake disaster is one of the most destructive natural disasters, causing casualties, economic damage and social turbulence, but simultaneously also creating a landscape possessing inspiration, visual impact, various spaces and a concise structure [12]. The rational development and utilization of these resources in the LFZ not only enriches the earthquake tourism products and tourist sources but also protects the unique "historical mark" through its exploitation [9]. Under the guidance of "Three-dimensional management architecture", this paper proposes an "Integrated management system for earthquake tourism", including several interacting subsystems, such as social activities, the regional economy, the ecological environment, resource management, and policies and regulations.

The scientific management of earthquake tourism is a systematic project, which should be "Government instructs, Enterprise acts, Society participates, Market operates" (Fig. 4). An effective crisis response should be viewed as a living system within an organization [9]. There are five stakeholders in the management system-the government, a competent authority, a tourism enterprise, the community, and an expert group-who interact to develop a balance of interests for the protection and exploration of earthquake relics. Considering local development, central and local governments coordinate the interests of all parties. The Ministry of Land and resources applies Laws, regulations, policies and

Government instructs- Enterprise acts- Society participates- Market operates

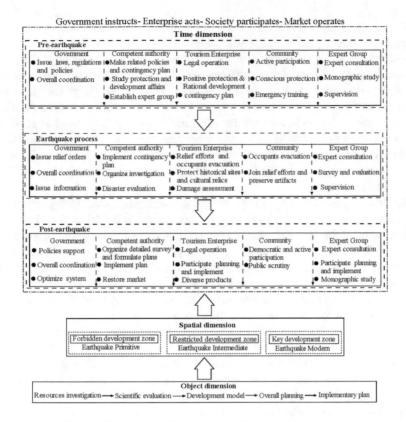

Fig. 4. The system framework of earthquake tourism

other means, the seismological bureau, the tourism bureau and other authorities to clear property rights, and develop the principles and plans for the care of the earthquake relics. A market-oriented earthquake tourism operation can advance management efficiency and reduce protection costs while ensuring a high-quality tourism experience. The community can be regarded as the witness to the earthquake disaster, which includes both public and social organizations in the disaster-hit areas. The communities should take an active part in the development and management of earthquake tourism and fulfill their responsibilities. Tourism management needs experts to participate in every aspect from the design of the plan to operations management. The participation of experts who offer intellectual support can effectively assist in development and management.

In the pre-earthquake period, strengthening defense activities against natural disasters is the focus [8]. Meanwhile, an emergency reserve of earthquake relics should be established to protect and maintain them. The primary task of the government is to enact the related laws, standards and policies and to coordinate the relationship among departments [1]. The competent authorities coordinate and handle the specific administration, such as putting together expert groups and studying the development and protection of earthquake relics. Tourism enterprises should obey related laws and regulations in their daily work. Additionally, they need to develop relevant contingency plans that ensure the implementation of these planning measures. The public should actively participate in the decision-making, protection and development of local tourism resources [7]. Furthermore, under the guidance of the relevant departments, training in disaster prevention and alleviation is also necessary. The experts can offer consulting services for other participants. There remains a still more important task for experts to carry out: monographic studies in earthquake tourism under the auspices of other stakeholders.

The "earthquake-process" is the period about a week after the quake hit. The main tasks at this time are to support earthquake relief efforts and deal with continuous aftershocks and secondary disasters. The governments issue relief orders the first time, and uniformly plan, deploy and coordinate the relief work. Competent authorities are responsible for the concrete implementation of the contingency plan and the mobilization of the relief efforts for tourists and major artifacts. These authorities should organize an investigation of the earthquake tourism resources and conduct an assessment of the situation [9]. The information obtained will not only help decision-makers in the relief efforts, but also offer suggestions for the next phase work for reconstruction. For tourism enterprises, the tourists should be evacuated as quickly as possible and the injured hospitalized. The community residents also need to be evacuated to safe zones as quickly as possible and can help the injured and preserve artifacts to assist the relevant departments. The important tasks of the expert group are to provide timely and effective solutions in emergencies and supervise the practical working of the related departments [10]. The experts are also needed to participate in the investigation and hazard assessment organized by the competent authorities.

The post-earthquake period is important for the protection and development of earthquake tourism resources. In this period, governments need to take further measures to promote post-disaster reconstruction. Governments should play a dominant role in coordinating different departments and checking and monitoring the progress of each department. The emergency management system needs to be further optimized based on practical experience [20]. The authorities conduct detailed surveys, formulate regional plans which concentrate on monographic studies in earthquake tourism, and put forward development plans by clarifying priorities. Some measures should be taken to restore market confidence in conjunction with the plan implementation [1]. Enterprises should operate following the law, avoid risks, prevent vicious competition, and safeguard their overall interests and good image. Further, enterprises need to have a clear understanding of the implementation plans, so they can follow their usual workflow but also provide active coordination and assistance for the related departments [12]. Customer segmentation refers to a process that divides customer sets into one or many classes using customer attribute data. By customer segmentation, enterprises could provide a variety of products for different types of visitors.

Mass participation in the protection and development of earthquake tourism is evidence of a judiciary democracy. Democratic participation can achieve a just, fair society and also ensure that arbitrary restrictions prevent any abuse of power [21]. People should actively take part in plan implementation and supervise the work of other departments. The experts should develop their knowledge and influence and offer advice for planning and implementation. The monographic studies should be taken and written up as news stories.

As an important tourism resource, increasing attention has been paid to research on earthquake tourism in the LFZ. In the post-Wenchuan earthquake era, owing to the complex and diverse geological conditions and cultural differences in the LFZ, there may be inner resistance to the development of earthquake tourism. The principle of "three-dimensional architecture" was applied to the research on earthquake relics, and multiple subsystems were designed, coordinated and controlled to identify the key tasks in the different periods from a "time, spatial and object dimension" perspective. On that basis, the framework for an integrated management system for earthquake tourism in different periods has been established through an extensive literature survey and field investigations.

The key hypothesis of this management system is that every geological disaster will bring a new circulation for this model in the time dimension. For example, Donghekou is a village in Qingchuan County, Guangyuan City. Before the Wenchuan earthquake, Donghekou was in an active defense village. After it, the related departments actively made efforts to conserve and protect the earthquake relics. In the post-earthquake period, Donghekou Quake Relic Park was built to commemorate the deceased. However, a flood on August 19, 2010, damaged the park and damaged this post-earthquake period. The procedure moved back to the period "earthquake-process".

5   Conclusions

LFZ is an earthquake relic and a world-class special tourism destination, with the potential to present both the natural and cultural relics to the world. The distinguishing characteristics of LFZ, such as the most widely affected area, involving the most geological units, and the richest and best preserved geological relics landscape, especially the new earthquake disaster relics landscape and lake landscape system, make LFZ an ideal area for research into earthquake relics and tourism landscapes.

Post-disaster reconstruction is a systematic project and is worthy of considering how these tragedies can be developed into earthquake tourism to make people pay attention to the forces of nature, keep in mind the human catastrophe, and feel the preciousness of life. Therefore, how to maintain harmony and ensure the sustainable development of earthquake tourism calls for deeper exploration due to its complex natural and social environment.

This paper proposed an ETOS to manage the earthquake tourism experience, which would cover both the hard and soft ranges of the experiences being sought.

The practical situation of the LFZ is adequately considered in the ETOS which incorporated ideas from the ROS and TOS. The overarching goal of ETOS is to provide planners and managers with a framework and procedure for making better decisions to exploit high-quality, diverse earthquake tourism opportunities. However, technology, population structures and economic changes often produce impacts beyond the ability of managers to fully anticipate or control. Such changes can produce dramatic shifts in the type and intensity of demand for earthquake tourism.

Although the future can be only imperfectly predicted, the ETOS does provide a framework for accommodating these shifting demands, as well as estimating the kinds of impacts associated with these changes. There are multi-factors, multi-periods, multi-regions and multi-steps involved in earthquake tourism management, so the proposed integrated management system was based on a three-dimensional management architecture. The protection and the development of earthquake relics is a complex project that calls for more concern from society. The general applicability of the ETOS and the management system needs to be tested and perfected further by follow-up studies and application practices.

Acknowledgments. This research was supported by the Humanities and Social Sciences Foundation of the Ministry of Education of China (Grant Nos. 16YJC630089). It was also supported by Soft Science Program of Sichuan Province (Grant No. 2019JDR0155).

References

1. Barton, L.: Crisis management: Preparing for and managing disasters. Cornell Hotel. Restaur. Adm. Q. **35**(2), 59–65 (1994)
2. Boyd, S.W., Butler, R.W., et al.: Identifying areas for ecotourism in northern ontario: Application of a geographical information system methodology. J. Appl. Recreat. Res. **19**(1), 41–66 (1994)
3. Breiling, M., Charamza, P.: The impact of global warming on winter tourism and skiing: A regionalised model for austrian snow conditions. Reg. Environ. Chang. **1**(1), 4–14 (1999)
4. Butler, R., Waldbrook, L.: A new planning tool: The tourism opportunity spectrum. J. Tour. Stud. **2**(1), 2–14 (1991)
5. Clark, R.N., Stankey, G.H.: The recreation opportunity spectrum: a framework for planning, management, and research. USDA Forest Service, General Technical Report, p. 98. (PNW-98) (1979)
6. Hystad, P.W., Keller, P.C.: Towards a destination tourism disaster management framework: Long-term lessons from a forest fire disaster. Tour. Manag. **29**(1), 151–162 (2008)
7. Manyara, G., Jones, E.: Community-based tourism enterprises development in kenya: An exploration of their potential as avenues of poverty reduction. J. Sustain. Tour. **15**(6), 628–644 (2007)
8. Mistilis, N., Sheldon, P.: Knowledge management for tourism crises and disasters. Tour. Rev. Int. **10**(1–2), 39–46 (2006)

9. Paraskevas, A.: Crisis management or crisis response system? A complexity science approach to organizational crises. Manag. Decis. **44**(7), 892–907 (2006)
10. Pender, L., Sharpley, R.: The Management of Tourism. Sage, London (2004)
11. Poon, A.: Tourism, Technology and Competitive Strategies. CAB International (1993)
12. De Sausmarez, N.: Crisis management for the tourism sector: Preliminary considerations in policy development. Tour. Hosp. Plan. Dev. **1**(2), 157–172 (2004)
13. Smith, V.L.: Hosts and Guests: The Anthropology of Tourism. University of Pennsylvania Press (2012)
14. Tsai, C.H., Chen, C.W.: An earthquake disaster management mechanism based on risk assessment information for the tourism industry-a case study from the island of taiwan. Tour. Manag. **31**(4), 470–481 (2010)
15. Wang, H., Li, H., et al.: Paleoseismic slip records and uplift of the longmen shan, Eastern Tibetan plateau. Tectonics **38**(1), 354–373 (2019)
16. Wu, Z., Xu, J., He, L.: Psychological consequences and associated risk factors among adult survivors of the 2008 wenchuan earthquake. BMC Psychiatry **14**(1), 126 (2014)
17. Xu, D., Hazeltine, B. et al.: Public participation in ngo-oriented communities for disaster prevention and mitigation (n-cdpm) in the longmen shan fault area during the wenchuan and lushan earthquake periods. Environ. Hazards **17**(4), 371–395 (2018a)
18. Xu, J., Liao, Q.: Prevalence and predictors of posttraumatic growth among adult survivors one year following 2008 sichuan earthquake. J. Affect. Disord. **133**(1–2), 274–280 (2011)
19. Xu, J., Lu, Y.: Meta-synthesis pattern of post-disaster recovery and reconstruction: based on actual investigation on 2008 wenchuan earthquake. Nat. Hazards **60**(2), 199–222 (2012)
20. Xu, J., Lu, Y.: A comparative study on the national counterpart aid model for post-disaster recovery and reconstruction: 2008 wenchuan earthquake as a case. Disaster Prev. Manag. **22**(1), 75–93 (2013)
21. Xu, J., Wang, P.: Social support and level of survivors' psychological stress after the wenchuan earthquake. Soc. Behav. Pers.: Int. J. **40**(10), 1625–1631 (2012)
22. Xu, J., Wang, Q. et al.: Types of community-focused organisations for disaster risk reduction in the longmen shan fault area. Environ. Hazards **17**(3), 181–199 (2018b)

A Framework for BIM-based Quality Supervision Model in Project Management

Jun Gang[1]([✉]), Chun Feng[1], and Wei Shu[2]

[1] Sichuan Institute of Building Research, Chengdu 610081,
People's Republic of China
gangjun@aliyun.com
[2] Chengdu Raincloud Technology Co., Ltd., Chengdu 610081,
People's Republic of China

Abstract. The building industry plays an important role in the economic development in China. It is more and more unrealistic to rely on large-scale and extensive production mode to promote the development of construction industry. Building information model (BIM), a new technology, has been applied in various fields of construction industry. First, this paper explores advantages and characteristics of BIM in quality supervision. Furthermore, a framework for BIM-based quality supervision model in project management is developed. Finally, an example is implemented to illustrate the proposed framework and research conclusions and future directions are provided.

Keywords: Building information model · Quality supervision ·
Project management

1 Introduction

With the rapid development of economy and the acceleration of urbanization, the living standards and quality requirements of people have also been improved. Accordingly, people have higher and higher requirements on the appearance, function and comfort of buildings, which makes the structure of buildings more complex, and how to ensure the quality of building products has become significant. As a pillar of the country, the construction industry has been facing many problems, such as serious environmental pollution, excessive waste of resources, and intensive labor force. These problems may lead to inefficient project management and poor communication between participants. Under this background, Building Information Modeling (BIM) came into being.

BIM integrate big data, cloud computing and Internet of Things in the field of construction so as to enhance data utilization and information level, which change production methods and management modes of construction industry. Specially, scholars have paid more attention in quality management. Chen and

© Springer Nature Switzerland AG 2020
J. Xu et al. (Eds.): ICMSEM2019 2019, AISC 1002, pp. 234–242, 2020.
https://doi.org/10.1007/978-3-030-21255-1_18

Luo (2014) investigated the advantages of 4D BIM for a quality management based on construction codes, and a case study was given to validate the effectiveness of the proposed 4D BIM application for quality control during the construction phase [2]. In terms of defect management, considering the relationship of defect information flow, Park presented a conceptual system framework for construction defect management, which combined augmented reality (AR) with BIM [1]. Besides, a defect management system was developed including an image-matching system to enable quality inspection and a mobile app, which indicated the potential applicability of BIM, image-matching, and AR technologies in construction defect management [5]. Kim et al. (2015) proposed an end-to-end framework for dimensional quality assessment of elements based on BIM and 3D laser scanning, where the proposed framework was composed of four parts: the inspection checklists, the inspection procedure, the selection of an optimal scanner and the inspection data storage [4]. In order to implement real-time quality control and early defects detection, an integrated system of BIM and Light Detection and Ranging (LiDAR). Based on the same idea, Wang et al. (2015) developed a real-time onsite quality information collecting and processing for quality control [7]. In addition, some other researches on BIM have been developed such as expanding from 3D to computable nD [3], re-engineering processes for cloud-based BIM [6] and risk management [8]. From these researches, we can see that a BIM-based framework is a significant and practical link for quality management. Therefore, we focus on establishing a framework for BIM-based quality supervision model in project management, where the establishing the BIM model and applying the BIM model phase are discussed in detail.

The paper is structured as follows: Sect. 2 introduces the advantages and characteristics of BIM for quality supervision. In Sect. 3, the framework for BIM- based quality supervision model is developed, where designer, government, constructor, builder, supplier and supervisor are integrated. 4D simulation (i.e. dynamic site management, dynamic resource management, construction process simulation and simulated tracking) and some quality management tools are considered in this framework. Furthermore, an example is implemented to illustrate the effectiveness of the proposed model in Sect. 4. Finally, Sect. 5 provides conclusions and future research directions.

2   The Advantages and Characteristics of BIM for Quality Supervision

In this section, we will introduce the difference between traditional quality supervision and the one based on BIM, which will derive the advantages and characteristics of BIM for quality supervision.

The framework of traditional quality supervision and BIM-based quality supervision is shown in Fig. 1.

From the Fig. 1, we can observe that the traditional quality supervision model can connect part participants, where information communication also happen in part participants. On the contrary, BIM-based quality supervision can

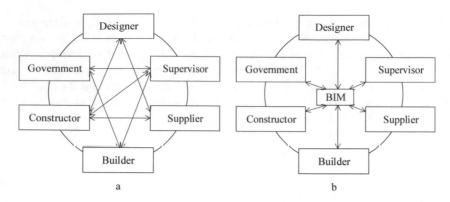

a   b

Fig. 1. The difference between traditional quality supervision and BIM-based quality supervision

Table 1. The advantages and characteristics of BIM for quality supervision

Categories	The advantages of BIM for quality supervision
Working ways	Understand and judge easily the project situation, especially in joint review of drawings from 2D to 3D
	Grasp the key points of construction and reduce the operation errors
Mechanical management	Simulate the whole process of construction using the existing machinery
	Mechanical equipment layout
Material management	Integrating information about materials and equipment, such as manufacturers, suppliers, sizes, factory certificates, quality assurance certificates, etc.
	BIM model can also automatically count the supply plan of building materials and equipment in any period of time
Measures	The virtual reality can preview different construction technologies, compare different schemes, optimize the process and sequence of operation, and select the best construction scheme
	Based on BIM model, managers can trace, record and count the quality of construction products through advanced measuring tools and mobile terminal equipment
Environment	The combination of BIM and GIS technology can conveniently make managers understand the natural environment around the construction project
	Through the simulation of the working environment of construction, the scientific and reasonable construction site layout can be realized
	In terms of management environment, participants cooperate on the same platform, which can achieve efficient communication and information sharing

connect all participants based on the BIM platform. Furthermore, BIM lead to the phenomenon where whole process' quality supervision and prior control are paid more attention. In detail, advantages and characteristics of BIM for quality supervision include the following five aspects shown in Table 1.

From Table 1, we can draw the conclusions that the advantages and characteristics of BIM for quality supervision lies in working ways, mechanical management, material management, measures and environment. In terms of working ways, the BIM can be integrated with other new technologies, which can improve the working efficiency and enhance the ability of quality management. For mechanical management, the plan of construction machinery can be optimized and adjusted so as to achieve accurate planning of the entry and exit routes and operation scope. As for material management, constructer can assess the quality of materials, equipment and service of manufacturers to form the internal material and equipment procurement database and provide basis for selecting high-quality suppliers. In the aspects of measures, BIM and other technologies can monitor and master the quality status of construction products at any time and avoid unknown quality risks. In terms of environment, nature environment, operation environment and management environment can be improved to implement better quality control.

3   The Framework for BIM- based Quality Supervision Model

At present, the quality supervision of construction projects mainly depends on the inspection and record of field managers. The information level of quality control is low. There are many documents in quality inspection records, which cause many difficulties in quality management. Therefore, it is necessary to establish a framework of BIM-based quality supervision system shown in Fig. 2.

From Fig. 2, we can observe that the framework of BIM-based quality supervision system includes three parts, establishing the BIM model, applying the BIM model, and establishing the BIM information platform. This framework integrates the designer, government, constructor, builder, supplier and supervisor. By using BIM-based model, a more comprehensive and detailed quality plan can be established. By comparing the inspection results with the quality plan at any time, regulator can find out the quality deviation and formulate reasonable corrective measures. Based on existing quality control theories, BIM covers all relevant information in the whole life cycle of construction projects. Therefore, BIM is benefit to the quality control and runs through all links of quality planning. Based on BIM information platform, combined with quality control tools, the quality control process can be improved and the project quality objectives can be achieved.

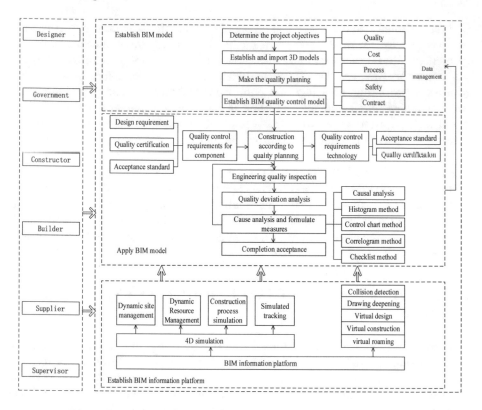

Fig. 2. The framework of BIM-based quality supervision system

4   An Applicable Example

Yibin east station is located in Yibin City, the southern margin of Sichuan
Basin. It is belonging to the junction of Sichuan, Yunnan and Guizhou provinces.
Because Jinsha River and Minjiang River converge here to form a river, it is
known as "the first city of the Yangtze River". The Yibin east station is located
in Yibin Market Street Office with good traffic conditions. Yibin East Station is
a medium-sized railway passenger station. The design scale is 3 platforms and 7
lines, where includes 2 main lines. The station is designed based on the maximum
numbers of people gathered. The basic platform is $450 \times 12 \times 1.25$, the second
platform is $450 \times 12 \times 1.25$, and third platform is $50 \times 12 \times 1.25$, with equal-
length and equal-width reinforced concrete canopy. The building area of Yibin
east station is $1193\,\mathrm{m}^2$, and the center mileage of the station is DK143+817. The
life of the main structure of the station building is 50 years. The safety grade of
the station building is grade 2, similarly, the safety grade of the platform canopy
is grade 1. The durability of the concrete structure is 50 years, and the design
grade of the foundation is grade C. The lower part of the station building is a
reinforced concrete frame structure system, and the roof is a steel grid structure.

The seismic grade of the station building is grade 2 and the rain shed is grade 3.

By using a series of BIM technology, this project has good quality control during life cycle. The detail process is shown as follows:

The 3D model of Yibin East Station is shown in Fig. 3, the BIM-based reinforcement and prestressing model of main beam shown in Fig. 4, the3D construction coordinate extraction by 3D model shown in Fig. 5. Through REVIT software, the three-dimensional construction coordinate extraction of any point can be inquired in real time, and the plane coordinate and elevation of the lofting point can be checked. Collision inspection of design drawings arc shown in Fig. 6. Collision checking of design drawings by using REVIT can realize entity reading of structure and reduce unnecessary rework, which can help find collision problems in advance so as to implement better quality management. Simulation of pavement construction is described in Fig. 7 and traction installation on beam is shown in Fig. 8. Before the project construction, the BIM model is established to

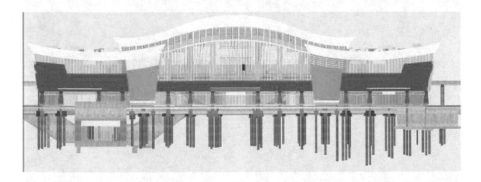

Fig. 3. 3D model of Yibin east station

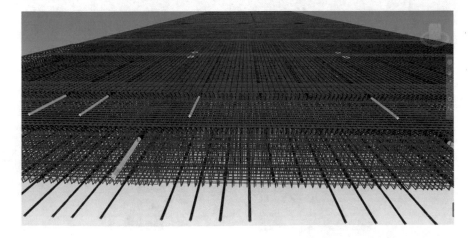

Fig. 4. The BIM-based reinforcement and prestressing model of main beam

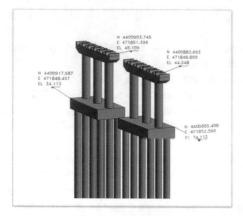
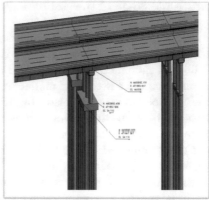

Fig. 5. 3D construction coordinate extraction by REVIT

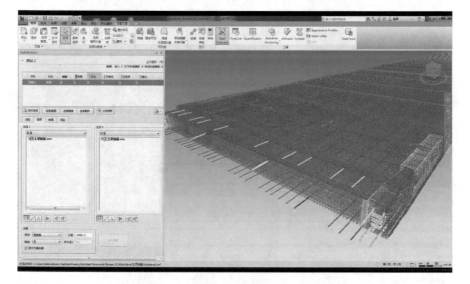

Fig. 6. Collision inspection of design drawings

Fig. 7. Simulation of pavement construction

Fig. 8. Traction installation on beam

incorporate the construction plan, site environment and scheme into the model, and the BIM software is used to simulate the construction. The on-site video detection can be carried out on the construction site, which greatly reduces the quality and safety problems of the building, rework and renovation.

5 Conclusions

In this paper, the differences between traditional quality supervision and BIM-based quality supervision are first presented. Furthermore, the advantages and characteristics of BIM for quality supervision are discussed in terms of working ways, mechanical management, material management, measures and environment. The BIM model can effectively conduct the quality control. Based on the advantages and characteristics of BIM, a framework for BIM-based quality supervision model is developed, where some quality management tools are adopted such as causal analysis, histogram method, control chart method, correlogram method and checklist method. In addition, in order to verify the effectiveness of proposed framework, an example concerning on Yibin East Station is carried out in terms of 3D construction coordinate extraction by REVIT, Collision inspection of design drawings and simulation construction. This example demonstrates that the BIM-based quality supervision model is beneficial to quality control.

This proposed framework is developed from the perspective of management demonstration, which need more quantitative tools of mathematical statistics for quality control. The application point analysis of BIM quality management system in each stage is not comprehensive enough, which needs further improvement in the future research. In addition, BIM can be combined with cloud computing, large data management and other modern technologies in the future.

Acknowledgements. This research is supported by Funding for Science and Technology Achievement Transformation of Scientific Research Institutes in Sichuan Province (Grant No. 2017YSZH0022).

References

1. Chan-Sik, P., Do-Yeop, L., Oh-Seong, K., Wang, X.: A framework for proactive construction defect management using BIM, augmented reality and ontology-based data collection template. Autom. Constr. **33**(8), 61–71 (2013)
2. Chen, L.J., Luo, H.: A BIM-based construction quality management model and its applications. Autom. Constr. **46**(10), 64–73 (2014)
3. Ding, L., Zhou, Y., Akinci, B.: Building information modeling (BIM) application framework: the process of expanding from 3D to computable nD. Autom. Constr. **46**, 82–93 (2014)
4. Kim, M.K., Cheng, J.C.P., Sohn, H., Chang, C.C.: A framework for dimensional and surface quality assessment of precast concrete elements using bim and 3D laser scanning. Autom. Constr. **49**, 225–238 (2015)
5. Kwon, O.S., Park, C.S., Lim, C.R.: A defect management system for reinforced concrete work utilizing BIM, image-matching and augmented reality. Autom. Constr. **46**(10), 74–81 (2014)
6. Matthews, J., Love, P., et al.: Real time progress management: Re-engineering processes for cloud-based BIM in construction. Autom. Constr. **58**, 38–47 (2015)
7. Wang, J., Sun, W., et al.: Integrating BIM and LiDAR for real-time construction quality control. J. Intell. Robot. Syst. **79**(3–4), 417–432 (2015)
8. Yang, Z., Kiviniemi, A., Jones, S.W.: A review of risk management through BIM and BIM-related technologies. Saf. Sci. **97**, 88–98 (2017)

"Element-Chain-Cluster" Industrial Project Planning Model in Poverty-Stricken Areas

Mingzhou Tu, Hongjiang Zhang, Yujie Wang, Chengyan Zhan, Linlin Liu, and Yi Lu(✉)

Business School, Sichuan University, Chengdu 610065, People's Republic of China
luyiscu@163.com

Abstract. Poverty has become one of the most acute social problems in modern society, and poverty reduction has always been high on the agenda of governments. At present, significant progress has been made in China, however, as its poverty alleviation work entered a crucial stage, problems arise: the lack of scientific nature and sustainability, lack of systematicness, and disconnections between projects. This paper summed up the previous literature and brought up the innovative method: "Element-Chain-Cluster" (ECC) poverty alleviation project planning model, and carries out detailed project planning: the formation of three major industrial clusters: ecological agriculture, green industry and cultural tourism in Ganluo County, a typical poverty-stricken county in China. The SAF model is then used to analyze the feasibility of introducing and popularizing ECC model so as to ensure general guidance can be provided for the development plan of poverty-stricken areas home and abroad.

Keywords: Planning model · Poverty-stricken areas ·
Industrial poverty alleviation · Targeted poverty alleviation

1 Introduction

In a world haunted by severe poverty, alleviation practices are carried out worldwide. However, governments encounter many bottlenecks in the process of industrial project planning: lack of systematicness and the disconnections between projects [13]. On top of that, project planning tends to focus on the development of backboned industry and rarely promotes the coordinated development of three sectors (industry, agriculture and tourism), which strictly limit the sustainable development of economy in poverty-stricken areas.

In modern society, the poverty status varies from country to country, on the basis of which, a number of innovative and effective poverty alleviation models were proposed [9]. Kusumastuti [10] developed a poverty alleviation decision-making model tailored to the condition of poverty-stricken counties in the eastern provinces of Indonesia, so as to enhance the selection criteria of the most

© Springer Nature Switzerland AG 2020
J. Xu et al. (Eds.): ICMSEM2019 2019, AISC 1002, pp. 243–254, 2020.
https://doi.org/10.1007/978-3-030-21255-1_19

efficient processed products. Kala State Agricultural University (KAU) [6] effectively promoted mechanization of paddy agriculture by mobilizing the educated unemployed youths in Keral. In order to protect forest resources and alleviate the overall level of poverty in poor areas, Islam [7] proposed a participatory agroforestry program (PAP) based on the geological conditions of Bangladesh. After conducting researches on the local tourism resources and poverty level, Guoqing and Yang [4] proposed a poverty alleviation model through tourism development based on the four-quadrant method. Cheng S H [3] proposed that international policy discourse and conservation and development investments were paying a lot of attention to utilize forest ecosystems and forest-based resources to poverty reduction. Based on the panel data of nine national poverty-stricken counties in Chongqing in the Three Gorges Reservoir Region from 1998 to 2015, Lu H [11] proposed that the national policy for targeted poverty alleviation has a significant positive effect on poverty reduction.

In general, most of the poverty alleviation models proposed by the existing research institutes only target a single industry, i.e. concentrate on the development of the backboned industry in poverty-stricken areas, and rarely involve the coordinated development of three sectors of industry, agriculture and tourism. In addition, most of the current researches are tailored to the local characteristics, which means there is no systematic and universally applicable industrial project planning model for poverty alleviation, which will otherwise provide an important reference and guiding value. This paper attempts to put forward a poverty alleviation model that can be universally applied to the industrial project planning in poverty-stricken areas. Moreover, this model could be integrated with the local characteristics and then give birth to a targeted and specific planning scheme which excellently meets the local needs. Therefore, in this way this model lays the foundation of the systematic project planning in poverty-stricken areas, which would definitely save lots of trouble.

Currently, some of the areas in poverty-stricken areas in China encountered problems such as the lack of systematicness, the unbalanced development of industries. Ganluo county is a typical one. In recent years, with the ongoing process of poverty alleviation work in Ganluo county, many high-quality projects have been planned and implemented initially. In the process of poverty alleviation planning, problems arose, for example the disconnections between different projects, irrational planning, unscientific selection of support projects and unbalanced development of three industries. At the invitation of the Investment Promotion Bureau of Ganluo County, the project team went to the county to conduct in-depth research and participate in the project planning.

After several months of field research and literature research, the team summarized a set of theoretical frameworks, on the basis of which, ECC industrial project planning model is proposed. Furthermore, the researchers have applied this model to the industrial planning of Ganluo county based on its current situation, and then analyzed the feasibility of introducing and popularizing ECC model in order to ensure it could serves as a guide or foundation rather than another individual case.

2   Design Framework of ECC Model

The "Element-Chain-Cluster" model is of great scientific nature and sustainability. Design framework of the ECC poverty alleviation model is shown in Fig. 1.

(1) The project begins with the theoretical analysis on the feasibility, risk assessment and benefit evaluation of each existing project, on basis of which the those with good prospects are chosen as an independent "project element". As the resources are less accessible in poverty-stricken areas, the efficiency of resource is of vital importance as it directly concerns the local development. Therefore, only feasible and beneficial project should be given priority.

(2) With each project as an element, a "project chain" can be created on the basis of the connection between them, limited resources should then be preferentially allocated to key elements such as, three-dimensional agricultural service center which promotes most of the elements, fresh food e-commerce (downstream sales channels) that ensures there are distribution channels for the products, etc. In this way, key elements could be optimized and well explored, thus better serving as a pillar for other elements. Synergy effect would then be achieved [5].

(3) After the formation of "project chain", comprehensive consideration should be given to the connections between all the chains in the industry, for example their overall risk-return, degree of urgency and positive effects on poverty alleviation work, on the basis of which, chains can be sequenced. The government should then focus on building the core project chain, and reinvest the profits into the follow-up projects to construct chains in stages, and finally forms the industrial project cluster.

(4) In the long-term, the government should dig into the connections between three major industries, and prioritize the allocation of resources to supportive projects, such as infrastructure construction, downstream logistics system construction, sales system construction, thus ensuring the close integration of three major industries. The probability of project failure arising from weak agricultural or industrial base should be minimized [12].

(5) Finally, with the combination of the internal project elements, chains of different industries and different project clusters, a closed industrial loop will form as three major industries join together as an organic, ecological and sustainable unity, fulfilling the purpose of targeted poverty alleviation.

To summarize, the industrial loop is formed from element to chain, then to cluster, during the period of which unsustainable projects screened out, and the rest of those merge organically with local characteristics. Chains are formed after exploring the correlation between elements, and clusters are built by coordinating the three sectors. As a result, projects are complementary to each other and the optimal economic development cycle come into existence.

3   ECC Model's Application to Ganluo

Ganluo county, which is a prominent representative of poverty-stricken areas in China, is chosen as research subject to study the ECC model for a patch of reasons.

Fig. 1. Design framework of the "Element-Chain-Cluster" poverty alleviation model

First, Ganluo County is one of the most severely poverty-stricken counties in Liangshan Prefecture. It shares the typical characteristics of a poor county, such as limited financial resources, poor infrastructure, seriously irrational industrial structure, and weak agricultural and industrial bases etc. However, in a positive light, the county has abundant natural resources (including mineral resources, water, plant, medicinal resources, animal resources, etc.), as well as unique ethnic cultural tourism resources, thus having great potential of economic development. Moreover, with the support of local government and national policies such as the Village Revitalization Plan, a bunch of poverty alleviation projects have swarm into Ganluo.

Second, there is no such development process that is smooth sailing all the time, and the rule applies to Ganluo County. During implementation of poverty alleviation projects, the government encounters many obstacles: the instability of the national supportive policies, the inability to develop steadily due to its internal industrial structure; chaotic development glossed over by specious planning language, the difficulty in investment promotion; incoordination of the development of the tertiary industry deriving from the lack of propaganda and poor infrastructures etc. Among them, investment attraction is a key part of targeted poverty alleviation. In the process of project planning, there are always problems like disconnection between industries, and the blindness of project implementation, resulting in the lack of scientificity and sustainability in the overall planning. Investment promotion is both an opportunity and a challenge. It entails rational planning and the ECC poverty alleviation model is undoubtedly a better resort at present.

Third, the project team participated in the local investment promotion project planning at the invitation from Investment Bureau of Ganluo County in Liangshan Yi Autonomous Pre fecture of Sichuan Province. After carrying out researches and analysis on Ganluo County, the team embarked on project planning under the guidance of ECC. Three clusters (agriculture, industry and tourism) emerge harmoniously as an organic whole, which leads to the coordinated development of economic and social undertakings in Ganluo County. This article will take Ganluo County as an example to elaborate on the application of ECC model in industrial project planning.

In the project cluster of ecological agriculture, there are mainly two subclusters: poultry and livestock breeding, processing, selling and agricultural orchards and vegetable gardens, Chinese herbs, and basic industrial sub-clusters. The

Structure of Industrial Planning under ECC Model in Ganluo is shown in Fig. 2. Ecological Agricultural Project Cluster is illustrated in Fig. 3. Besides Fig. 4 is a typical project chain.

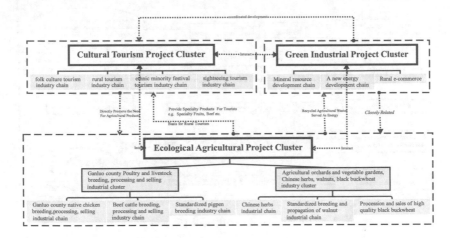

Fig. 2. The structure of industrial planning under ECC model in Ganluo

(1) Poultry and livestock breeding, processing and selling.

The cluster can be subdivided into three categories: native chicken breeding, processing and selling industry chain; standardized pigpen breeding industry chain and beef cattle breeding, processing and selling industry chain. The three industrial chains can all be further divided into upstream (breeding, cultivation), midstream (slaughtering, processing), and downstream (wholesaling, retailing). This model can also be applied in general poultry livestock breeding projects.

The upstream breeding can be elaborated mainly from three dimensions: original species protection, livestock reproduction and professional breeding, the crux of which is standardization, specialization and industrialization. In midstream, the introduction of deep processing enables the development of Ganluo County to step out of the original comfort zone, make bold move and cooperate with high-tech chemical companies so as to enhance the added value of products fundamentally. Although currently there are many bottlenecks, if fostering strengths and circumventing weaknesses, high-quality leading enterprises can also be attracted with the combination of standardization of specialized breeding processing zones and lower production costs, downstream Chinese e-commerce sales are of most importance, as it acts as the juncture of Ganluo County's agricultural development. Answering to the national-level project of popularizing e-commerce, the county should firmly grasp the advantages of the e-commerce platform, apply business models (B2C, O2O) and resolve traffic occlusion of Ganluo.

(2) Agricultural orchards and vegetable gardens, Chinese herbs, walnuts, black buckwheat industry clusters.

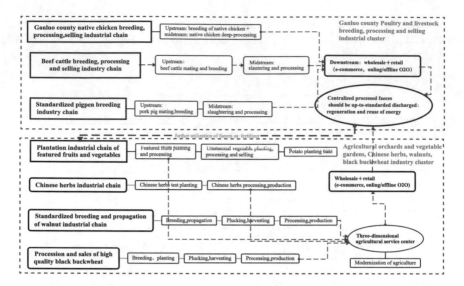

Fig. 3. Ecological agricultural project cluster

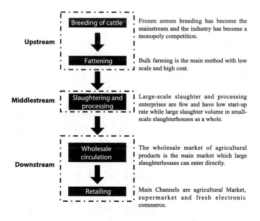

Fig. 4. A typical project chain C beef breeding, fattening, processing and retailing

The cluster is subdivided into plantation of featured fruits and vegetables, Chinese herbal medicine industrial chain; standardized breeding and propagation of walnut industrial chain; planting, processing and selling of high-quality black buckwheat. Among which the three-dimensional agricultural service center is the core of the industrial cluster, playing the role of resource integration and allocation.

Similar to the poultry livestock breeding program, in the crop industry cluster, monitoring of the whole processes is also needed ranging from planting, picking, processing to transferring and selling. Among them, assembling and transferring is key to the value-added process of the industry chain, and government needs to switch from putting blind faith in the scattered-planting model to organically aggregating various fruits and vegetables, Chinese herbal medicines and other products. Attention should therefore be focused on rational planning of suppliers thus strengthening the systematic and coordinated development of industrial clusters.

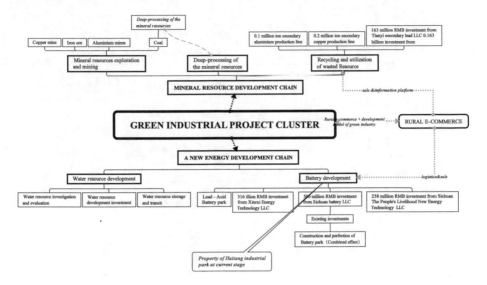

Fig. 5. Green industrial project cluster

The green industrial project cluster is divided into a mineral resource development chain and a new energy development chain. The former can then be divided into several project elements: exploration, mining, deep processing and waste recycling; in order to better explore resources, the procedures above should be chained into an industrial loop and the core is combining lead–zinc ore with waste incineration to form a circular economy industrial chain ("collection and incinerating – deep processing – regeneration – reprocessing"). Green industrial project cluster is shown in Fig. 5.

The new energy development chain consists of hydropower, wind energy, photovoltaic development and solar energy development. At present, the development of hydropower resources is relatively mature, and its development concepts covers two parts: the in-place conversion of hydropower resources and long-range hydropower transmission. From the perspective of in-place conversion, the government should focus on utilizing electricity to attract investment and exploring two following modes: direct power supply for companies and subscribing shares of corporations to promote cooperation. On top of that, officials should focus on the introduction of energy-intensive industries with high value-added products, especially those highly related to demand of power, meanwhile promoting the in-place conversion of hydropower resources and extending the hydropower industry chain.

In the entire mineral resource and new energy development chain, e-commerce is integrated into selling in order to create an online information platform, break the information barriers of the suppliers and consumers and make the entire green industrial cluster an interrelated and inseparable recyclable industrial cluster.

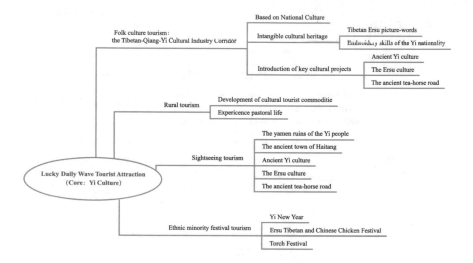

Fig. 6. Cultural tourism project cluster

The cultural tourism project cluster is divided into four major industrial chains: ethnic minority festival tourism industry chain, sightseeing tourism industry chain, folk culture and ethnographic tourism industry chain and rural tourism industry chain. Cultural Tourism Project Cluster is shown in Fig. 6.

In terms of festival tourism, it is vital to make full use of the geographical advantages of Ganluo county. Ganluo is well placed as it is located at the border of Liangshan yi nationality autonomous District, while at the same time adjacent to Ya'an, Meishan, Chengdu and other areas. Furthermore, there is various traditional festivals of the ethnic minorities around Ganluo such as Yi New Year, Torch Festival, Ersu Tibetan and Chinese Chicken Festival.

As for sightseeing tourism, there are three places of interest in Ganluo county, including the ancient town of Haitang, the yamen ruins of the Yi people and the ancient roads of Qingxi. The government should fully exploit their historical and cultural characteristics and high-quality landscape resources to stimulate the development of tourism.

This paper also brings forward a scheme to develop folk culture tourism by inheriting the heritage of folk custom, refining rural culture, and exploring folk intangible cultural heritage. First of all, the government should focus on the Tibetan–Qiang–Yi Cultural Industry Corridor and launch cultural tourism projects on the basis of ancient Yi culture, the Ersu culture, and the ancient tea-horse road. Then, it should enable visitors to truly experience the local folk culture, which in turn promotes its interactive development with cultural value chain.

With respect to rural tourism, the tourism souvenir processing industry around the Yi culture will be developed. The rural tourism business model should also be introduced, which combines tourism with orchard, vegetable garden, cash crops, poultry and livestock breeding, in which case farmers can provide tourists with green and pollution-free food, vegetables and poultry, meanwhile promote the sales of agricultural and sideline products to encourage the consumption of food and beverage.

4   Discussion

4.1   The Advantages of ECC Model

Compared with the current poverty alleviation models which continually lead to chaos, disorder and inefficient use of limited resources [2], the ECC poverty alleviation model seems to be a better option for the following reasons:

(1) Unnecessary cost is saved via screening out projects that are not in line with long-term strategy of poverty alleviation.

In many poverty-stricken areas, resources are often blindly invested in low-return projects, or projects which do not meet the need of the current situation, leaving projects half done due to the poor screening mechanism. Apart from the resources waste, it also leads to 'crowding-out effect' on high-quality investment projects, which has seriously hindered the progress of further poverty alleviation [14].

(2) The projects will develop in the form of a series projects joint together to form a closed industrial chain.

In "project chain", the ECC model fully explores local resource advantage and connects the core poverty-alleviation projects, which aims to realize the vertical integration of the industrial chain, thus achieving the economies of combined operations and improving the internal control and co-ordination. Further, it also saves transaction costs, making the development of the industry more stable [1].

(3) Maximize short-term returns while recognizing the importance of long-term benefits.

In the process of applying ECC model, the project chains developed in stages. Priorities are given to the basic and key project chains during the allocation of resources. In order to maximize short-term benefits while ensuring sustainability in the long-term, it is important to stimulate subsequent development through reinvestment, thus maximizing the output of existing projects.

(4) Optimize infrastructure for long-term interests.

The ECC model gives priority to basic infrastructures, which not only meets the needs of the current status, but also lays a solid foundation for the ecological project clusters. Jerome (2011) once mentioned that although the work may not make remarkable achievements in the short term, the infrastructure's complementary value to other projects will gradually emerge in the synergy of clusters [8].

(5) Formation of industrial clusters and utilization of synergistic effect.

By forming three industrial clusters in stages, synergistic effect can be fully utilized. Take agricultural development as an example, it provides material and cultural basis for tourism; the tourism boom can then stimulate the development of other tertiary industries while indirectly stimulate that of the primary and secondary industries.

4.2 Feasibility Analysis Applying the SAF Model

(1) Suitability analysis

The ECC model responds to the call of targeted Poverty Alleviation Policy. The implementation of targeted poverty alleviation mechanisms in poverty-stricken counties is an important part of the future poverty alleviation work. Targeted poverty alleviation requires sustainable investment promotion, and efficiency of management are worth close attention. The ECC poverty alleviation model closely adheres to the principle of targeted poverty alleviation, and can be generally applied to the poverty-stricken counties that encounter obstacles in the early stage of poverty alleviation.

The ECC model meets the condition of the current poverty alleviation. With the influx of a series of investment projects, obstacles of the project planning arise. In most cases, projects cannot be implemented simultaneously due to the financial situation in certain poor counties. Therefore, it is indispensable to have a rational arrangements of existing investment projects, and the ECC model provides solutions for the problems above.

(2) Acceptability analysis

From the perspective of profitability, the application of the ECC model enhances the economy efficiency from five aspects: saving unnecessary costs, forming a closed industrial chain, maximizing long-term returns, optimizing infrastructure, and synergy among the three industrial clusters.

From the perspective of risk, the ECC poverty alleviation model has relatively limited practical experience at home and abroad to draw on, however, it is a guiding concept based on existing projects, the application of which will by all means exerts effect on resource optimization, thus enhancing the disordered layout in poverty-stricken areas. Besides, counties' cooperation with colleges means nominal direct cost. Therefore, the risk of the ECC poverty alleviation model is at acceptable level.

From the perspective of stakeholders' acceptability, the ECC poverty alleviation model is closely related to people, thus being widely accepted. On top of that it provides theoretical guidance in the planning process of investment promotion projects in Ganluo County, which is in desperate need at present, thus it is further increasing its popularity. Besides, the ECC poverty alleviation model has made a reasonable design for the current poverty alleviation problem, if further improvement and promotion are carried out, the model will be of great social significance, thus improving the acceptability on the national and social levels.

(3) Feasibility analysis

As for financial support, considering the lack of poverty alleviation funds, the ECC poverty alleviation model attaches great importance to resource optimization. Although financial support provided by the local government is limited, taking the financial support of the state, the government, and all sectors of society into consideration, the ECC poverty alleviation model can effectively use existing funds to bring county more sustainable income.

As for development resources, there are not only a series of policy supports like targeted poverty alleviation, rural revitalization project, but also inherent self-driving factors like rural industrial economic development, industrial structure upgrading and supports from local industry. Besides, currently, government are actively promoting the participation of universities in the poverty alleviation work of surrounding poverty-stricken areas, laying great foundations for the stable source of the talents.

To summarize: the ECC model to some extent fills the vacancy in research of industrial project planning in poverty-stricken areas. It provides the most needed resources for thousands of poverty-stricken areas in China at a relatively low cost-a guiding ideology and a model for reference.

Acknowledgements. This research is supported by the Humanity and Social Science Youth Foundation of Ministry of Education of China (Grant No. 17YJC630096).

References

1. Argyres, N., Mahoney, J.T., Nickerson, J.: Strategic responses to shocks: comparative adjustment costs, transaction costs, and opportunity costs. Strateg. Manag. J. **40**(3), 357–376 (2019)
2. Cheng, H., Dong, S., et al.: A circular economy system for breaking the development dilemma of 'ecological fragility-economic poverty' vicious circle: a CEEPS-SD analysis. J. Clean. Prod. **212**, 381–392 (2019)
3. Cheng, S.H., MacLeod, K., et al.: A systematic map of evidence on the contribution of forests to poverty alleviation. Environ. Evid. **8**(1), 3 (2019)
4. Guoqing, H., Yang, Z.: Research on tourism poverty alleviation model of reservoir region. Information and Business Intelligence, pp. 603–608. Springer, Berlin (2012)
5. Haken, H.: Synergeticsłan interdisciplinary approach to phenomena of self-organization. Geoforum **16**(2), 205–211 (1985)
6. Indira Devi, P., Hema, M., Jaikumaran, U.: Value chain in poverty alleviation-a model for institutional initiatives for organizing and capacity building of farm work force. Agric. Econ. Res. Rev. **23**(2010), 523–526 (2010)
7. Islam, K., Hoogstra, M., et al.: Economic contribution of participatory agroforestry program to poverty alleviation: a case from Sal forests, Bangladesh. J. For. Res. **23**(2), 323–332 (2012)
8. Jerome, A.: Infrastructure, economic growth and poverty reduction in Africa. J. Infrastruct. Dev. **3**(2), 127–151 (2011)

9. Kaidi, N., Mensi, S., Amor, M.B.: Financial development, institutional quality and poverty reduction: worldwide evidence. Soc. Indic. Res. 1–26 (2018)
10. Kusumastuti, R.D., Martdianty, F., et al.: A decision-making model for community-based food industry to support poverty alleviation in East Indonesia: case study of East Sumba regency. Asia-Pac. Manag. Bus. Appl. 5(1), 1–15 (2017)
11. Lu, H., Chen, N.: A study on how to eliminate spiritual poverty to achieve the accurate poverty alleviation in a real sense. Asian Agric. Res. 10 (2018)
12. Maulidi, C., Sitanggang, R.M., et al.: Sustainable tourism development review: steeping poverty on Samosir cultural landscape. Adv. Sci. Lett. 24(12), 9441–9445 (2018)
13. Mazzetto, S.: A practical, multidisciplinary approach for assessing leadership in project management education. J. Appl. Res. High. Educ. 11(1), 50–65 (2019)
14. Saidjada, K.M., Jahan, S.I.: Public and private investment nexus in Bangladesh: crowding-in or out? J. Dev. Areas 52(4), 115–127 (2018)

Media Attention, Investor Attention and Corporate Innovation Investment: Empirical Study Based on China GEM Listed Companies

Yuzhu Wei and Hongchang Mei[✉]

Chongqing Technology and Business University, Chongqing 600000,
People's Republic of China
m_hchang68@163.com

Abstract. In the information age, media attention and investor attention have become two important external governance factors affecting corporate innovation investment, and the specific impact mechanism on corporate innovation investment is still unclear. Taking the China GEM listed companies from 2013 to 2017 as the research samples, this paper investigates the direct and interaction effects of media attention and investor attention on innovation investment. The results show that both media attention and investor attention have a significant positive impact on corporate innovation investment; and the interaction between the two also has a significant positive influence on corporate innovation investment, that is, there is a complementary effect between media attention and investor attention. The study will help to improve the external governance mechanism of corporate innovation investment, and provides a new way to solve the problem of insufficient innovation investment of Chinese enterprises.

Keywords: Innovation investment · Media attention · Investor attention · Empirical studies

1 Introduction

Under the background of innovation-driven national development strategy, innovation investment has become an important force to improve the market competitiveness of enterprises and promote the sustainable and efficient development of China's economy. However, compared with other developed countries, the corporate innovation investment of China is relatively insufficient. Therefore, exploring which factors affect the innovation investment is not only the concern of enterprises, but also the focus of the state in promoting economic transformation. Existing studies have found that both internal governance characteristics such as executive incentive [3,32], political connection [37], internal control [33], board structure [24], equity structure [13], and external governance

© Springer Nature Switzerland AG 2020
J. Xu et al. (Eds.): ICMSEM2019 2019, AISC 1002, pp. 255–269, 2020.
https://doi.org/10.1007/978-3-030-21255-1_20

factors such as government subsidies [26], the protection of intellectual property right [19], marketization [5], institutional environment [12] can affect corporate innovation investment. The study on internal governance is becoming more and more perfect, and the impact mechanism of external environment on innovation investment needs to be further explored. With the deepening of social informatization, media attention and investor attention, as two important external governance factors for corporate innovation investment, have played an important role in alleviating information asymmetry between companies and their stakeholders [6,25], restraining the behavior of companies and their executives [27,36]. At the same time, the impact of investor attention on innovation investment depends on the media attention to reduce information asymmetry and avoid investment risk. Due to the investors limited attention and information asymmetry with enterprises, investors rely on media attention to obtain enterprise information. Conversely, the impact of media attention on innovation investment also relies on investors' investment behavior generated by receiving information. Because without the attention of investors, it is difficult for media attention to exert an influence on innovation investment. So, we can draw a conclusion that media attention and investor attention have a direct impact on corporate innovation investment, meanwhile, the interaction between the two also affects corporate innovation investment. However, most of the existing studies discussed the impact of the two on stock market from the aspects of stock liquidity and stock return rate, there is still a lack of further in-depth theoretical analysis and empirical research on the direct impact and interaction of the two on corporate innovation investment. In view of this, based on investors limited attention, information asymmetry and principal-agent theory, this paper deeply analyses the complex relationship between media attention, investor attention and corporate innovation investment, and empirically tests the direct and interaction of the two on innovation investment through multiple regression analysis.

This paper attempts to make progress in the following two aspects through theoretical explanation and empirical test: (1) Exploring the direct impact of media attention and investor attention on corporate innovation investment examining the information mediation effect of the media, the active search behavior of investors and the external supervision role played by reputation mechanism, which provide a reference for effectively reducing the degree of information asymmetry, alleviating the agency problem of the company, and improving the external governance of innovation investment. (2) Investigating the interaction effect of the two on innovation investment, so as to fully explain the influence mechanism of the two on innovation investment, and provide new ideas and methods for improving the corporate innovation investment degree and promoting the sustainable and healthy growth of enterprises.

2   The Theoretical Analysis and the Research Hypothesis

2.1   Media Attention and Corporate Innovation Investment

With the rapid development of global Internet technology, the media, as the "fourth power" of the western captain market, has gradually shown its function of information intermediary and supervision [8]. Its impact on political activities, capital market and corporate governance has been widely concerned. In the Chinese capital market, most of the studies empirically test the external corporate governance effects played by the media from the aspects of minority shareholder protection [28], earnings management [7] and administrative governance [17]. Although some studies have shown that the media coverage has self-interested behavior of disseminating false news and chasing the sensationalism, the authenticity of media reports will affect the reputation of the media. Because the media with lower reputation may be punished by investors in the way of voting-by-feet [29], and only the media reporting real information can obtain sustained and long-term attention of the public and investors [34]. Reputation has become an important factor for the continued development of the media in the fierce market competition. Therefore, this paper mainly considers the positive governance effect of the media attention on the corporate innovation investment through the information intermediary effect and reputation restraint mechanism. First, the media attention affects corporate innovation investment through exerting the "information mediation" effect to alleviate the information asymmetry between enterprises and external stakeholders [4]. Media attention can effectively alleviate agency problems [14] and improve the information environment of investors [28] through reducing the degree of information asymmetry between management or major shareholders and stakeholders. At the same time, media attention, as an important channel for enterprises to understand the information of relevant innovative investment projects, can provide more valuable information about corporate innovation investment. Not only it can reduce the cost of collecting information for innovative projects, but also can improve the decision-making efficiency and quality of corporate innovation investment. In addition, banks are an important financing channel for corporate innovation activities. The media attention plays an active role in alleviating the information asymmetry between banks and enterprises and helping enterprises to obtain the credit loans from banks [16]. Second, the media attention exerts an impact on corporate innovation investment through influencing the reputation of the company or its managers [8]. The media attention supervises the violations and self-interest behaviors of companies and its managers through the reputation mechanism, encourages them to do the meaningful investment activities which promoting the sustainable development of enterprises to build a good reputation. Meanwhile, the market pressure hypothesis believes that the strong market pressure brought by media attention prompts executives to make reasonable and innovative strategic decisions [30], the higher media attention the more likely it is to reduce the possibility of managers manipulating innovation investment projects thus reducing agency costs. Based on the above analysis, the following assumption is proposed:

Hypothesis 1: Media attention has a positive effect on enterprise innovation investment, that is, the higher the media attention, the higher the corporate innovation investment.

2.2 Investor Attention and Corporate Innovation Investment

Investor attention is the core concept in the field of behavioral economics. It refers to the investors' response to the stocks due to the specific and attracting events of the enterprises [9], which is reflected in buying or selling the target stocks. Kahneman (1973) said that attention is a scarce cognitive resource, and the ability of individuals to input and process information is restricted by attention [15]. On this basis, Merton (1987) put forward the "investor cognitive hypothesis", pointing out that investor attention may be related to stock pricing and liquidity. Because the information is incomplete, investors can not understand all stock information [21], they often pay attention to limited information and adjust their investment judgment and decision-making behavior accordingly [1]. Barber (2008) argued that a sharp increase in the search for target stocks by investors with limited time and attention will make them more inclined to "buy" rather than "sell" stocks [2]. The existing studies mainly discussed the impact of investor attention on stock trading behavior, information interpretation and information disclosure behavior, and less on corporate innovation investment. However, the investment capital brought by investor attention is an important external financing channel for enterprises. Therefore, based on the review of the existing literature, this paper believes that the investor attention affects corporate innovation investment mainly through the active search behavior of investors to reducing information asymmetry between investors and enterprises, and the market pressure mechanism introducing external reputation constraint mechanism. On the one hand, the investor attention reduces the degree of information asymmetry, increases investors' willingness to invest in stocks and provides important external financial support for corporate innovation investment through improving the efficiency of information interpretation. This paper focuses on the investors actively searching for stock information behavior of target companies before investing. It believes that investor attention plays an important role in alleviating the information asymmetry, reducing the investment risk of investors, and increasing investors' trust and investment possibility in corporate innovation investment [20]. Capital has become an important channel for listed companies to innovate, and investors' stock investment behavior provides sufficient financial support for enterprises to continue innovative investment. On the other hand, the investor attention introduces external reputation supervision through market pressure mechanism to alleviate the second kind of agency problem, so as to effectively avoid the short-sighted behavior of ultimate shareholders and their managers, and encourage other major shareholders to actively participate in the corporate innovation investment decision-making activities. At present, due to the imperfect legal system of China's capital market, the agency problem caused by information asymmetry is more prominent, and because of the high risk high uncertainty long return cycle of innovation investment the ultimate controlling shareholders and company management who are deeply involved in the second kind of agency problem may lack the motivation and willingness to sustain high-level R&D investment [22]. Based on attention-driven trading behavior theory, it is believed that higher investor attention improves the liquidity of stocks and promotes other major shareholders to participate actively in corporate governance. Good equity balance will help reduce the agency costs associated with the second type of agency questions, and promote enterprises to invest more in R&D. In addition, the investor attention improves the level of enterprises' initiative to disclose R&D information through market pressure mechanism, reduces the level of manipulative earnings announcement effect and management self-interest behavior through effective restrictions on management earnings announcement timing behavior [36], and promotes managers' innovative investment for the sustainable and healthy development of enterprises. Therefore, this paper proposes the following assumptions:

Hypothesis 2: Investor attention has a positive effect on corporate innovation investment, that is, the higher the investor attention, the higher the corporate innovation investment.

2.3   Interaction Between Media Attention and Investor Attention

Media attention and investor attention as two important external governance factors for corporate innovation investment have played a role in reducing information asymmetry and constraining executive short-sight behavior. Studies have shown that the interaction between the two affects stock returns [18], but whether the interaction effect of the two has complementary or substitutive effects on corporate innovation investment has not been verified. Therefore, based on the intimate relationship between media attention and investor attention, this paper further considers the impact of the interaction between the two on corporate innovation investment. On the one hand, the impact of investor attention on innovation investment depends on the information intermediary effect of media attention to effectively alleviate the information asymmetry between investors and enterprises, avoid investment risks, and encourage investors to invest in R&D. Due to the limited attention of investors and the information asymmetry between investors and enterprises, media attention as a key link in information dissemination positively strengthens investors' attention to target enterprise information [10], which can not only effectively reduce the degree of information asymmetry between investors and enterprises, improve the information environment of investors. but also reduce the investors' investment risk and protect investor interests through affecting investor emotions [11]. And studies have shown that the more the media reports on listed companies, the more attention attracting investors. So, we can see that media attention may expand the impact of the investor attention on corporate innovation investment. On the other hand, the influence of media attention on innovation investment depends on the investment behavior generated after investors receiving information. For media attention, investor behavior results from investor's attention to relevant media coverage. No matter how high the quality is, it is difficult for media attention to have an impact on corporate innovation investment without the investor attention. And for media companies they also tend to disclose the listed companies with high investor investment preferences. The higher investor attention indicates that the company has strong market competitiveness and good development potential. The media that reports such high-quality enterprises with good development prospects, not only can get more user traffic, but also promote further innovation investment to maintain market competitive advantage and obtain continuous attention of investors. Therefore, Investor attention may expand the impact of media attention on enterprise innovation investment. To sum up, media attention and investor attention may have complementary effect, that is, media attention can expand the impact of investor attention on enterprise innovation investment, conversely, the marginal promotion effect of media attention on innovation investment presents an increase trend with the increase of investor attention. So this paper proposes the following assumption:

Hypothesis 3: The interaction between media attention and investor attention has a positive impact on corporate innovation investment, that is, media attention and investor attention have a complementary effect on corporate innovation investment.

3   Research Design

3.1   Variable Definition

1. Dependent variables. In this paper, the proportion of R&D expenditure to total assets is used as a measure of corporate innovation investment (RD) [31].

2. Independent variables. (1) Media Attention (Media). There are two main methods for measuring media attention in the existing studies: the number of items in the news search engine and the number of media reports in the main newspaper. Due to the rapid development of the Internet, online media has become the main medium for people to understand the information of listed companies. Therefore, we use the number of Baidu news reports as a measure of media attention, specifically by substituting the listed companies into Baidu news for time-phase search. Get the total number of news reports of the company in each year, and then add 1 to take the natural logarithm, which is the agent variable of media attention. (2) Investor Attention (IA). When testing theories of attention, empiricists face a substantial challenge that we do not have direct measures of investor attention. We have indirect proxies for investor attention such as stock trading volume, turnover rate and advertising expenditure. However, the above proxy variables are the transaction characteristics and price behavior of the financial assets themselves, DA (2011) proposed a novel and direct measure of investor attention using the weekly search volume index(SVI) of Google search engine [35]. Baidu as a main search engine of China, using Baidu search index as a proxy variable for investor attention, which has certain

Validity. Considering that some company abbreviations not only represent its enterprises themselves, it is easy to cause result errors. This paper takes the natural logarithm of the company's overall daily search volume in the Baidu search index platform as its proxy variable.

4   Control Variables

In this paper, the control variables include Size, Lev, Growth, Indep and Top1, etc. In addition, this paper also introduces Year and Industry as Virtual variables. Specific variables and their measurements involved in this paper are shown in Table 1.

Table 1. Definitions of main variables

Types	Name	Symbols	Variables' definitions
Dependent variable	Innovation investment	RD	R&D Expenditure/Total Assets.
Independent variables	Media attention	Media	Log (each listed companys total annual news of Baidu news + 1).
	Investor attention	IA	Log (enterprise code annual Baidu index search volume).
Control variables	Company size	Size	Ln (total assets).
	Asset-liability ratio	Lev	Total liabilities/Total assets.
	Company growth	Growth	Increase rate of business income.
	The proportion of independent directors	Indep	Number of independent directors/ Number of Board of Directors
	Ownership concentration	Top1	The shareholding ratio of the largest shareholder.
	Industry	Indu	According to the Guidelines for Industry Classification of Listed Companies, GEM is divided into 13 industry categories, with manufacturing enterprises as the reference frame, setting 12 virtual variables.
	Year	Year	With 2013 as the reference, four virtual variables were set up in 2014, 2015, 2016 and 2017

4.1   Regression Model

Firstly, this paper builds the following Eq. (1) to test hypothesis 1:

$$RD_{i,t} = \beta_0 + \beta_1 Media_{i,t} + \beta_2 Lev_{i,t} + \beta_3 Growth_{i,t} + \beta_4 Size_{i,t} + \beta_5 Indep_{i,t}$$
$$+ \beta_6 Top1_{i,t} + \sum Year_{i,t} + \sum Indu_{i,t} + \varepsilon_{i,t}$$

$$(1)$$

In Eq. (1), i and t represent enterprises and years respectively, the independent variable is media attention (Media)and the other control variables are shown in Table 1; β_0 is the constant term, β_1 is the regression coefficient of independent variable, β_2 β_6 are regression coefficient of the control variables, and ε is the random perturbation term.

Secondly, this paper builds the following Eq. (2) to test hypothesis 2:

In Eq. (2), the independent variable is investor attention (IA) and the remaining control variables and the regression coefficient are consistent with Eq. (2).

$$RD_{i,t} = \beta_0 + \beta_1 Media_{i,t} + \beta_2 Lev_{i,t} + \beta_3 Growth_{i,t} + \beta_4 Size_{i,t} + \beta_5 Indep_{i,t}$$
$$+ \beta_6 Top1_{i,t} + \sum Year_{i,t} + \sum Indu_{i,t} + \varepsilon_{i,t}$$

$$(2)$$

Thirdly, this paper builds the following Eq. (3) to test hypothesis 3:

$$RD_{i,t} = \beta_0 + \beta_1 Media_{i,t} + \beta_2 IA_{i,t} + \beta_3 Media_{i,t} \times IA_{i,t} + \beta_4 Lev_{i,t} + \beta_5 Growth_{i,t}$$
$$+ \beta_6 Size_{i,t} + \beta_7 Indep_{i,t} + \beta_8 Top1_{i,t} + \sum Year_{i,t} + \sum Indu_{i,t} + \varepsilon_{i,t}$$

$$(3)$$

In Eq. (3), media attention (IA), investor attention (IA) and their interaction (Media*IA) are added at the same time, and the remaining variables are also consistent with Eqs. (1) and (2).

5   The Empirical Study

5.1   Data Resource and Sample Selection

The research sample of this paper is selected from the GEM of Shenzhen Stock Exchange. There are two main reasons why we choose China GEM listed companies as our research object. On the one hand, GEM listed companies are mostly high-tech small and medium-sized enterprises (high-tech SMEs), which have high growth potential and are important subjects for promoting national sustainable innovation. Their innovation investments are easy to get the sustained attention of the media and investors. On the other hand, compared with the larger and multi-business main-board market, Baidu News and Baidu Index are more representative in choosing GEM listed companies as the research object, which can effectively avoid the disruptive data generated by non-investor attention and non-media attention. Therefore, this paper selects companies listed before December 31, 2013, and takes the data of China GEM listed companies for five consecutive years from 2013 to 2017 as the initial research sample. The data sources of this paper are as follows: First, we obtained the data of the proportion of independent directors (Indep), the control variable, from CSMAR database, and the dependent variable(innovation investment) and control variables, such as ownership concentration (Top1), company size (Size), asset-liability ratio (Lev), company growth (Growth) from WIND database. Second, we collected the independent variables data through manual data collection, the data of Baidu news and Baidu search index are used as the substitution variables of media attention (Media) and investor attention (IA) respectively. After obtaining the initial sample, the following principles are adopted to screen the data to ensure the validity of it: (1) eliminating listed companies with missing innovation investment data; (2) eliminating listed companies with incomplete financial data; (3) eliminating listed companies with abnormal data of independent variables. Besides, in order to avoid the influence of extreme values of independent variables, the variables less than 5% and more than 95% were processed by Winsonrize. Finally, a total of 1,450 observational samples of 290 GEM listed companies were selected as valid research samples. In practical research, data filtering and basic operations are processed by Excel software, and descriptive statistical analysis, regression analysis and robust test are processed by Stata14.0.

5.2 Descriptive Statistical Analysis

The descriptive statistical analysis results of the main research variables in this paper are shown in Table 2. We can see that the average value of innovation investment (RD) is 0.027, accounting for 2.7% of the total assets of enterprises, the minimum and maximum values of innovation investment are 0 and 0.122 respectively, indicating that compared with developed countries, the overall innovation investment of GEM enterprises in China is relatively insufficient and there is a big difference in the innovation investment degree among different enterprises in China's GEM. The average values of media attention and investor attention are 1.499 and 2.519 respectively, which not only shows that investor attention

Table 2. Descriptive analysis of variables

Variable	N	Mean	SD	Min	Max
RD	1450	0.027	0.019	0.000	0.122
Media	1450	1.499	0.445	0.699	2.386
IA	1450	2.519	0.163	2.267	2.851
Lev	1450	0.298	0.167	0.011	1.037
Growth	1450	0.289	0.470	−0.700	6.431
Size	1450	9.283	0.340	8.525	10.659
Indep	1450	0.383	0.056	0.250	0.600
Top1	1450	0.302	0.125	0.044	0.689

is higher than media attention, but also media attention is relatively less than that of the main board enterprises; the minimum and maximum values of media attention are 0.699 and 2.386 respectively, and the minimum and maximum values of investor attention are 2.267 and 2.851 respectively, this indicates that for different listed companies, media attention is quite different, while the distribution of investor attention is relatively uniform.

5.3   Regression Analysis

In order to test the hypotheses, this paper conducted a multivariate regression analysis. The results of the regression analysis are shown in Table 3. We have firstly estimated model (1) as a control group and analyzed the impact of the control variables on innovation investment (adj-R^2 = 0.2077). In model (2), the media attention (Media) is added as the independent variable, and the inde-

Table 3. The Regression results for the Impact of media attention, investor attention and their interaction on innovation investment

Variable	−1	−2	−3	−4
Media		0.00548***		0.00390***
		4.04		2.78
IA			0.0140***	0.0100***
			3.85	2.68
IA*Media				0.0132***
				3.02
Lev	−0.00678**	−0.00708**	−0.00679*	−0.00703**
	(−2.14)	(−2.25)	(−2.16)	(−2.25)
Growth	0.000801	0.000475	0.000891	0.000614*
	0.81	0.48	0.9	0.62
Size	−0.00735***	−0.00952***	−0.00948***	−0.0111***
	(−4.34)	(−5.38)	(−5.34)	(−6.15)
Indep	0.0268***	0.0258***	0.0277***	0.0265***
	3.3	3.2	3.44	3.3
Top1	−0.00802**	−0.00678*	−0.00579	−0.00588
	(−2.13)	(−1.80)	(−1.52)	(−1.55)
_cons	0.0886***	0.103***	0.0727***	0.0933***
	5.52	6.28	4.4	5.44
INDU&YEAR	YES	YES	YES	YES
F-statistic	19.08	19.16	19.07	18.5
Adj R-squared	0.2077	0.2161	0.2153	0.2247
N	1450	1450	1450	1450

*at 10%significant level; **at 5% significant level; ***at 1%significant level.

pendent variable coefficient is significantly positive at 1% significance level(β is 0.00548 and $p < 0.01$), indicating that the higher the media attention, the higher the level of innovation investment. This result proves the hypothesis 1 and the model explanatory power is strengthened (adj-$R^2 = 0.2161$). In model (3), investor attention (IA) is added as the independent variable, and the regression coefficient is also significantly positive at 1% significance level (β is 0.014 and $p < 0.01$), which indicates that investor attention has a significant positive impact on enterprise innovation investment, thus the hypothesis 2 is verified. In addition, from the regression coefficients of Model 2 and Model 3, we think that the impact of media attention corporate on innovation investment is weaker than investor attention. In model (4), We add media attention, investor attention and the intersection of the two as independent variables to test the interaction effect of the two. In order to avoid the multi-collinearity problems caused by the interaction of media attention and investment attention, the two variables are centralized and the interaction terms are calculated and brought into the

Table 4. Robust test results

Variable	1	2	3	4
Media		0.0180***		0.0129***
		3.75		2.59
IA			0.0523***	0.0408***
			4.05	3.06
IA*Media				0.0279*
				1.8
Lev	−0.0928***	−0.0938***	−0.0929***	−0.0936***
	(−8.28)	(−8.40)	(−8.33)	(−8.42)
Growth	−0.0111***	−0.0122***	−0.0108**	−0.0117***
	(−3.15)	(−3.46)	(−3.07)	(−3.32)
Size	0.000334	−0.0068	−0.00759	−0.0125
	0.06	(−1.08)	(−1.21)	(−1.94)
Indep	0.0572**	0.054*	0.0608**	0.0571*
	1.99	1.89	2.12	2
Top1	−0.0519***	−0.0478***	−0.0436**	−0.0432**
	(−3.88)	(−3.58)	(−3.23)	(−3.21)
_Cons	0.0898	0.136**	0.0303	0.0892
	1.58	2.34	0.52	1.46
INDU	YEAR	YES	YES	YES
F-statistic	18.74	18.69	18.82	17.84
Adj R-squared	0.2045	0.2117	0.213	0.2181
N	1450	1450	1450	1450

*at 10%significant level; **at 5% significant level; ***at 1%significant level.

regression equation. The empirical result of Model 4 shows that media attention, investor attention, and their intersection are significantly positive at the
1% significance level, and the model is further optimized (adj-R^2 = 0.2247). That
means there is a complementary effect between the two, and the hypothesis 3 is
verified.

5.4   Robust Test

This paper replaces the calculation method of the dependent variable and takes
the ratio of R&D expenditure to business income as the proxy variable [23] to
verify the stability of the empirical results in this paper. the robust test results
are shown in Table 4. We can see that after changing the calculation method of
the innovation investment, the direct influence of media attention and investment
attention on the innovation investment are still significantly positive and their
interaction coefficient is also significantly positive. The results of robust test are
basically consistent with those of the previous regression analysis, that means
the regression model of this paper is stable.

6   The Conclusions and Discussion

Based on the relevant theoretical analysis, this paper selects the China GEM
listed companies from 2013 to 2017 as the research sample, and uses the multiple regression analysis to empirically test the direct impact and interaction of
media attention and investor attention on corporate innovation investment. The
research results show that: (1) Both media attention and investor attention have
positive effects on corporate innovation investment, while media attention has a
weaker impact on innovation investment than investor attention. (2) The interaction between the two has a significant positive impact on corporate innovation
investment, that is, there is a complementary effect between media attention
and investor attention to promote the innovation investment.

In order to effectively use the positive external effects of media attention
and investor attention on corporate innovation investment, enhance the level of
innovation investment and achieve the sustainable growth of enterprises, this
paper puts forward the following suggestions from perspective of government,
enterprises and investors. Government should strengthen media supervision, continue to promote the industrialization reform of the media industry to ensure
the health development of the media industry and improve the authenticity and
authority of media coverage, so as to better play the role of media attention in
external governance; At the same time, government should urge the legislature
to improve the legal system of investor protection to provide legal protection
for the interests of investors. For enterprises, they need to actively cooperate
with the media, timely disclose R&D activities information to investors through
the information intermediary effect of the media attention. It can maintain the
good image and reputation of enterprises and attract investors' attention, thus

increasing the trust and investment possibility of investors in corporate innovation investment, and obtaining external financing for corporate innovation investment. For investors, it is necessary to pay attention to media coverage and other investor attention for the enterprises based on Baidu search platform, and choose the enterprises with high media and investor attention to invest; In addition, they should strengthen their information discrimination ability through their active search behaviors to seek out as much information as possible about the innovation investment of target enterprises under the condition of information asymmetry, thereby reducing investment risk, protecting their own interests, and promoting investors continuous attention to the corporate innovation investment.

References

1. Aboody, D., Lehavy, R., Trueman, B.: Limited attention and the earnings announcement returns of past stock market winners. Soc. Sci. Electron. Publ. **15**(2), 317–344 (2010)
2. Barber, M., Terrance, O.: All that glitters: the effect of attention and news on the buying behavior of individual and institutional investors. Rev. Financ. Stud. **21**(2), 785–818 (2008)
3. Balkin, D.B., Markman, G.D., Gomezmejia, L.R.: Is CEO pay in high-technology firms related to innovation? Acad. Manag. J. **43**(6), 1118–1129 (2000)
4. Bushee, B.J., Core, J.E., Guay, W., et al.: The role of the business press as an information intermediary. J. Account. Res. **48**(1), 1–19 (2010)
5. Chen, L., Wu, B.: Marketization, education and family business's R&D investment (in Chinese). Sci. Res. Manag. **35**(7), 44–50 (2014)
6. Cen, W., Tong, N., Yue, L.: Information advantage, investor attention and insider trading gains. Chin. Rev. Financ. Stud. **7**(02), 28–42 (2015). (in Chinese)
7. Chen, K.: Media supervision, rule of law and earnings management of listed companies (in Chinese). Manag. Rev. **29**(07), 3–18 (2017)
8. Dyck, A., Volchkova, N., Zingales, L.: The corporate governance role of the media: evidence from Russia (in Chinese). J. Financ. **63**(3), 1093–1135 (2008)
9. Engelberg, J., Manski, C.F., Williams, J.: Comparing the point predictions and subjective probability distributions of professional forecasters. J. Bus. Econ. Stat. **27**(1), 30–41 (2009)
10. Engelberg, J.E., Parsons, C.A.: The causal impact of media in financial markets. J. Financ. **66**(1), 67–97 (2011)
11. Fang, L.H., Peress, J., Zheng, L.: Does media coverage of stocks affect mutual funds' trading and performance? Soc. Sci. Electron. Publ. **27**(12), 3441–3466 (2014)
12. Fang, L., Zheng, Y., Xi, Y.: Institutional environment, tax incentives and enterprise innovation investment (in Chinese). Manag. Rev. **28**(2), 61–73 (2016)
13. Geletkanycz, M.A., Boyd, B.K.: CEO outside directorships and firm performance: a reconciliation of agency and embeddedness views. Acad. Manag. J. **54**(2), 335–352 (2011)
14. Jin, L.: Corporate governance role of media coverage: from the perspective of dual agency cost. J. Financ. Res. **10**, 153–166 (2012)
15. Kahneman, D.: Attention and Effort. Prentice-Hall, Englewood Cliffs (1973)
16. Lai, L., Ma, Y., Xia, X.: Media coverage and credit access. J. World Econ. **2016**(09), 124–148 (2016)

17. Li, P., Shen, Y.: The corporate governance role of media: empirical evidence from China. Econ. Res. J. **2010**(04), 14–27 (2010)
18. Liu, F., Ye, Q., et al.: Impacts of interactions between news attention and investor attention on stock returns: empirical investigation on financial shares in China (in Chinese). J. Manag. Sci. China **17**(01), 72–85 (2014)
19. Li, W., Yu, X., Cai, L.: Study on the relationship between government R&D investment, intellectual property right protection and enterprises R&D investment (in Chinese). Stud. Sci. Sci. **34**(3), 357–365 (2016)
20. Li, X., Zhu, C.: Investors' limited attention and information interpretation (in Chinese). J. Financ. Res. **2011**(8), 128–142 (2011)
21. Merton, R.C.: A simple model of capital market equilibrium with incomplete information. J. Financ. **42**(3), 483–510 (1987)
22. Tang, Y., Zhuo, J.: Ownership property, blockholders governance and corporate innovation. J. Financ. Res. **2014**(6), 177–192 (2014)
23. Tong, L., Yin, D.: Corporate governance and innovation: differences among industry categories. Econ. Res. J. **49**(01), 115–128 (2014)
24. Pugliese, A., Bezemer, P.J., Zattoni, A.: Boards of directors' s contribution to strategy: a literature review and research agenda. Corp. Gov. Int. Rev. **17**(3), 292–306 (2010)
25. Tetlock, P.C.: Does public financial news resolve asymmetric information? Rev. Financ. Stud. **23**(9), 3520–3557 (2010)
26. Wu, J., Tian, Z., et al.: The impact of government subsidies on corporate innovation in strategic emerging industries (in Chinese). Stud. Sci. Sci. **2018**(01), 158–166 (2018)
27. Xiong, X., Wu, S.: Investor attention, accrual mispricing and earnings manipulation (in Chinese). Account. Res. **2012**(6), 46–53 (2012)
28. Xu, L.: Media governance and minority shareholder protection (in Chinese). Nankai Bus. Rev. **2011**(6), 36–47 (2011)
29. Xiong, Y., Li, C., Wei, Z.: Media sensatinoalism: transmission mechanism, economic consequences and reputation punishment: a case study based on overlord incident. Manag. World **2011**(10), 125–140 (2011). (in Chinese)
30. Yu, X., Junke, F., Fei, Y.: Media coverage, effectiveness of internal control and innovation performance. Collect. Essays Financ. Econ. **228**(12), 88–96 (2017)
31. Yu, W., Ning, B.: Cross-listing, investor attention and corporate innovation: empirical study based on Chinese a-share listed companies. Foreign Econ. Manag. **2018**(1), 50–62 (2018). (in Chinese)
32. Yi, M., Sheng, L., Li, W.: Executive incentive, innovation input and corporate performance: an empirical study based on endogeneity and industry categories (in Chinese). Nankai Bus. Rev. **21**(01), 109–117 (2018)
33. Zhang, J., Huang, Z.: Internal control, innovation and firm performance: empirical evidence from Chinese manufacturing companies. Econ. Manag. J. **38**(09), 120–134 (2016)
34. Zheng, Z.: Corporate governance role of extra-legal system: a literature review (in Chinse). Manag. World **2007**(9), 136–147 (2007). (in Chinese)
35. Zhi, D.A., Engelberg, J., Gao, P.: In search of attention. J. Financ. **66**(5), 1461–1499 (2011)
36. Zhou, K., Ying, Q., Chang, Z.: Can media coverage improve corporate governance? evidence from fraud by listed firms in China. J. Financ. Res. **2016**(06), 193–206 (2016)

37. Zhou, M., Zhang, Q.: "Face Project" or "True Talent" a research based on the relationship between political promotion incentive and innovation activity in state-owned listed companies. Manag. World **2016**(12), 116–132 (2016). (in Chinese)

Analysis of Sustainable Net Cage and Fluid Flows Through the Review

Mengyuan Zhu[1], Shuijin Li[2], Tingting Liu[1], Rongwei Sun[1], and Jingqi Dai[1(✉)]

[1] Business School of Sichuan University, Chengdu 610065,
People's Republic of China
dajingqiscu@163.com
[2] School of Science and Engineering, University of Dundee, Dundee, United Kingdom

Abstract. More and more aquaculture inshore farms will move offshore, as there are more benefits to work offshore for these industries. The research on the fluid flows through a net cage has a lot of positive significance, which can improve the utilization rate of feed to improve the fish production and understand the diffusion law of the pollution source in order to protect the marine environment. Computational simulation has a lower cost compared to laboratory experiment. Computer simulations were used to study the changes in the fluid regime as the fluid passed through the net cage, as well as changes in the concentration of contaminants. This review and relevant discussions on the softwares and application can provide some suggestions for the future research on aquaculture and net cage.

Keywords: CFD · Aquaculture · Net cage · Review

1 Introduction

1.1 Background

With the development of society, the global population has increased rapidly. How to feed 9.8 billion people in 2050 under many tremendous challenges of climate change, economic and financial uncertainty, and growing competition for natural resources has become one of the biggest challenge in the world [7]. According to the survey [9], the vigorous development of the aquaculture industry is a good solution to the food problem as the annual output growth of the aquaculture industry has increased greatly. And on the other hand, in recent years, the proportion of fish in the human diet has also increased.

In order to supply fish steadily, human beings consider changing their approach to fetching fish food from capture fishery production to aquaculture, since overfish can lead to ecosystem collapse [14]. Since modern time, especially from the late 1980s, the capture fishery production has become relatively stable, while aquaculture has been mainly a thriving trend [9]. Although aquaculture

© Springer Nature Switzerland AG 2020
J. Xu et al. (Eds.): ICMSEM2019 2019, AISC 1002, pp. 270–278, 2020.
https://doi.org/10.1007/978-3-030-21255-1_21

accounted for only 7% of fish for human consumption in 1974, it reached 26% by 1994 and 39% by 2004 surprisingly. Among them, China has played a major role in these growths since the aquaculture production in China accounts for 60% of the world's aquaculture production. Nevertheless, the rest part of the world, excluding China, also benefited from the booming development of aquaculture since 1995 [9]. Compared with fish capture industry, the aquaculture is more friendly to the environment although it still has some impact on surrounding environment [6]. On the other hand, the rate of growth in global fish supply over the recent past 50 years, with an annual rate of growth of 3.2% over the period 1961 C 2013, is double than the growth rate of human population, which makes per capita availability increased [9].

This paper mainly discusses some simulation methods in the ocean industry which are applied to detact the change of tide and its relevant influence on the fish industry. Specifically, a comparative discussion among various simulation software and literature review can provide some new management paradigm on the culture of fish.The remainder of this paper is organized as follows. Section 1.2 demeritson aquaculture. In Sect. 2, literature review is given as an abstract of the current research condition, Sect. 2.1 proposed the concept of fluid around a cage, which highlights the importance of improving the level of aquaculture in fluid research, and Sect. 2.2 summarize the important researches to this research field. The conclusion and future research suggestions are given in Sect. 3.

1.2 Simulation Softwares

Nowadays, there are many powerful computer simulation softwares on the market, such as ANSYS, OpenFOAM, Bentley, Revit and so on. Other specialized software (such as Numeca, Flotherm, etc.) firmly occupy the market relying on their respective fields with their professionalism, which has good properties on pre and post processing, numerous physical models, high efficiency and reliability. For computational fluid dynamics (CFD) in aquaculture, the softwares that are widely used is ANSYS Fluent and OpenFOAM.

The advantages and disadvantages of the two CFD softwares are shown in Fig. 1.

ANSYS Fluent is commercial software, which has an expensive price. ANSYS FLUENT has some advantages such as powerful, easy to learn, fast, and has rich learning materials and friendly graphical user interface (GUI). ANSYS provided a student version for student free to use, this version was limited in some ability to complete some complex problems. Besides, the core of the software is a black box. The internal working code of the commercial software is not open to users. Users will not have any way to understand some mathematical-physical model in processing details. Although commercial software can be easy to use, the cover of details would cause problems for users in debugging. Since the code for commercial software is closed, it is hard to extend the functionality of the commercial software for the users. However, OpenFOAM is a open-source free CFD software, based on the Linux platform, which makes up for the shortcomings of ANSYS. Users can modify any code in OpenFOAM, and to build a new

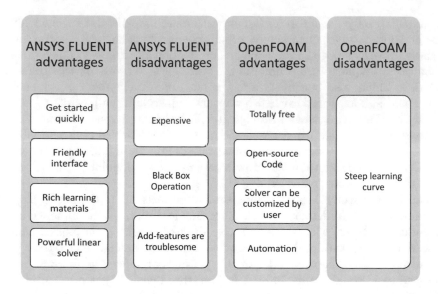

Fig. 1. The advantages and disadvantages of ANSYS Fluent and OpenFOAM

user-defined solver for a new algorithm. Unlike ANSYS, OpenFOAM doesn't have a friendly GUI, this makes user learn slower

2   Literature Review

In this part, an overview for the research in the past decades years are given with software Citespace, and concrete topic of cage aquaculture and fluid around cage are also analysed.

Figure 2 shows the results of cluster analysis using the citations data downloaded from the web of science from 2010 to 2019. It can be seen that there are respective 8 obvious clusters in the past literature, which means they are the main research directions in net cage and aquaculture. Among them, the sustainable aquaculture management, even though only originated from 2010 to 2012 or so, is the newest field and hottest issue currently according to the time-row analysis. The research overview demonstrates that the sustainable aquaculture management have been concerned by experts throughout the world, therefore, to solve an effective and efficient method to continue the fishery but at the same time, to keep the quantity of ocean fish is the main point in today's sustainable aquaculture management. Research data illustrates the fluid have a significant relationship with the activities of ocean fishes. Based on these given points, two areas are explained in detail in the following literature review.

As is illustrated in Fig. 3, the period sequence of relevant research are visualized based on the origin and development time. For example, the first opinion and also continue be concerned by experts in this field from our literature these years is hydrodynamic behavior, which is labeled on the right first. Then is the

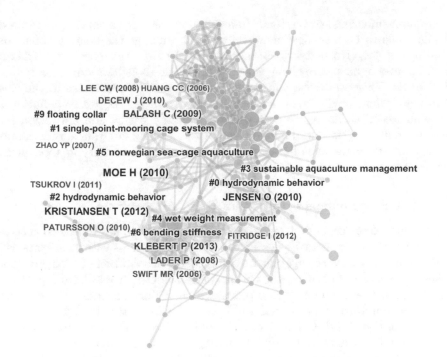

Fig. 2. The cluster results of literature from 2010 to 2019

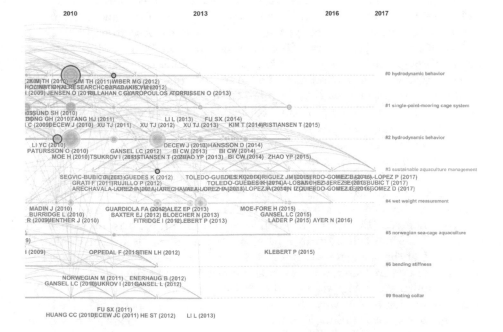

Fig. 3. The cluster results of literature from 2010 to 2019

single-point-mooring cage system. However, it is also shown that, some research are not able to be investigated by about 2013, such as the floating collar and bending stiffness (No. 6 and No. 9 respectively). In 2017, the most attracted issue is sustainable aquaculture management (No.3) even though its start mainly from 2010 or so. This can be explained that in the past ten years, due to the raise of human beings environment awareness and the increasing water pollution in aquaculture. In addition, the topics of wet weight measure and norwegian sea-cage aquaculture (No. 4 and No. 5) also triggered many interests and discussion these years, they are some branches in developing sustainability in aquaculture and fishery.

2.1   The Environmental Impact of Cage Aquaculture

The use of offshore cages, or pens, to culture marine finfish has been used success-fully to culture a number of species for many years. Success with cage systems is common in sea area with cage culture being used since the 1950s in some regions. Whether onshore or offshore, the culture of fish in cage would have a potential impact on the surrounding environment which mainly depends on the hydro-logical surroundings, type of feed, mode of cultivation, the density of stocking and type of fish [5, 12]. Other than a limited ability to influence the quality of the surrounding water by modifying feeding and other management practices, open, offshore grow out systems are inherently dependent on natural conditions to provide a suitable growing environment.

Generally speaking, The cage wastes were incorporated into the food web and support the wild fishes in the vicinity of the fish farm and most of the feed that are put into culture system would be wasted. According to statistics [26], about 85% of phosphorus, 80–88% of carbon and 52–95% of nitrogen are wasted through feed, excretion of fish, production of faeces, breathing and other forms to release to the environment. China is the largest aquaculture country in the world now with accounting for more than 60% of the total aquaculture of the world. However, some problems will inevitably occur in the rapid develop-ment of China's aquaculture, which includes overdevelopment of some seawater, eutrophication in some freshwater lakes and marine area, distortion of the local ecosystem, aesthetic deterioration in the coastal area [9, 12, 25]. Among these problems, eutrophication is the most serious environmental problem caused by aquaculture, which occurs in many parts of China, such as in Dianchi Lake, Tai Lake Bohai Bay and so on, leading to the outbreak of cyanobacteria and some other phenomena [15]. Eutrophication is a difficult problem to be solved. For example, a shrimp culture with a good management on ecosystem would lose about 30% of the residual bait, and 12.8% of nitrogen and 4% of phosphorus are wasted into culture water, and eventually discharged into the sea.

Cage culture is a high-density high-feeding open-ended aquaculture ecosys-tem. The major reason why the growth potential of properly fed fish is not fully realized is that stocking density has exceeded the carrying capacity of the system. Most of the nutrition sources in this ecosystem are mainly small trash fish with a feed coefficient of 8–10 (using small trash fish as feed are seriously do damage to resources in aquaculture), whereas only 20% of the protein in the feed is used for fish growth while the remaining 80% of the protein is excreted in the sediment of the cage bottom or sea bottom in the form of excrement and residual bait. Then, the waste through the decomposition of heterotrophic bacterial to release nitrogen and phosphorus into the water. In short, the contaminant of the aquaculture has a significant negative impact on the coastal water [25].

Wastewater after chemical flooding can not be easily treated by conventional physicochemical and biological processes, which can be regarded as a method of hard path. Therefore, innovative management methods are demanded from the soft path. We need to assess environmental capacity scientifically with respect to the negative environmental impact produced by cage culture. In order to evaluate the environmental capacity, it is necessary to accurately calculate the concentration field of a culture pollution source and determine response relationship of the concentration between the control point of water quality and the pollution source. The transference and diffusion of material would be influenced by the change of hydrodynamic conditions of contaminant caused by choked flow and damping wave due to the presence of cages [16]. Additionally, the layout of aquaculture farms has shown a tendency to extend from inland bay with a gentle flow to the offshore with a grand wave stream due to several reasons as continued input of contaminant from land exacerbates deterioration of seawater quality, other developments in the coastal zone (such as tourism), deep-water cage technology has developed rapidly and maturely in recent years. Along with this trend, it becomes more meaningful and urgent to study the influence of cages on material transport under complex hydrodynamic conditions [16]. Besides, the negative influence of carbon and nutrient enrichment from intensive marine cage culture can also endanger the fishery.

Whether onshore or offshore, cage aquaculture can have a potential impact on the surrounding environment, depending on the hydrological environment, feed types, planting methods, stocking densities and fish species [5,12]. According to a statistic summarized by [26], it is illustrated that there are about 85% of phosphorus, 80–88% of carbon and 52–95% of nitrogen are wasted through several approaches such as feed, excretion of fish, production of faeces, breathing and other forms to release the nitrogen to the environment. One of the severe problems caused by aquaculture is eutrophication, which is difficult to solve.

Therefore, in order to reduce the impact to the environment caused by aquaculture, it is high time to scientifically assess the environmental capacity of the target sea in response to the negative environmental impacts of cage aquaculture. On the other hand, to achieve this goal, we must first accurately calculate the field of fluid concentration formed by a certain aquaculture pollution source and determine the concentration relationship between the fluid flow regime and the pollution source [18].

The control system of the prototype was a digital, software-based approach. The advantages to using software-based digital control are especially applicable to research aquaculture in deep sea. In this way, computational simulation in a digital control system can be done with software. Ease of reconfiguration is essential, since the system will not be easily accessible. Furthermore, a computaional simulation digital control system allows expandability. Examples of expandability could include additional temperature, GPS, or other sensor data to be part of the control system. Moreover, computational simulation is a good way to achieve this research goal. Compare with laboratory experiments, computational simulation has the advantages of cheapness and high error-tolerance.

2.2   The Research of Fluid Around a Cage

The research of fluid around a cage has importance in today's aquaculture. Since the expansion of near-shore aquaculture is becoming more difficult because of multi-use issues and environmental impact concerns, the feasibility of moving aquaculture into the open ocean is being studied. Marine aquaculture has been predominantly experimented in protecting near-shore waters. With the conflicts between utilization and environmental conflicts being irreconcilable as the near-shore aquaculture industry continues to grow with an obvious tendency. The feasibility of extending operations into the energetic open ocean has recently been addressed. Such a transition is not trivial because suitable species and technologies for this environment are not yet developed to an economically viable point. Many scholars have studied the flow of fluid around the cages, thus studying the extent and range of environmental pollutant in cage culture [13]. In the research of simulation of the flow field through net cage, [2,20] supported the theoretical foundation. Besides, [1] conducted a series of experiments to study the velocity distribution in the cage system and developed a deceleration formula for net cage. [10] used field measurements to study the flow velocity in a cage in the open sea and found that the velocity was reduced by about 10%. [17] conducted a series of experiments to the cage under uniform flow and measured that there was an average 20% velocity reduction in the cage. [8,22,23] conducted a large number of on-site tests on the effect of mass diffusion on the currents in different ways of aquaculture such as mussel farms, fish farms, marine aquaculture and floating raft. They found that these aquatic facilities can hinder the flow of the water, resulting in the flow of the sea to make the process of mass diffusion changed, and thus affect the production of aquaculture.

These efforts mentioned above are the contribution to the physical model tests and field measurements. In recent years, with the development of the computer, simulation began to be applied to study the flow through net cages. [21] began to use the porous media approach to simulate and verify the feasibility of this method. [11] study the flow field inside and around the porous cylinders by using particle image velocimetry (PIV) approach, which provides a benchmark to the following numerical models for predicting the fluid flow regime through the porous media.[3,27] used the 3D numerical porous model to simulate a current fluid flow regime through 4 different cages and also prove that the porous model is good enough to make the simulation. A series of experiments were carried out by [18,24] on the lab to study the effects of the grids with different densenesses on the distribution of convection field. The experimental flume uses a flow control system to produce a constant uniform flow, followed by a fluorescent dye and a continuous laser to track the flow of particles in the water. The results of the experiments coincide with [4, 19].

3   Conclusion

The food problem has always been accompanied by the development of human society. The proportion of fish in the human diet is growing, and the development of aquaculture has become one of the hot spots today. Aquaculture is more environmentally friendly and has more production in the capture industry. Cage culture is one of the patterns in aquaculture, which is a high-density high-feeding open-ended aquaculture ecosystem. However, the common feeding method would waste 80% of the feed's nutrition, which would cause the environmental impacts and the feeding efficiency. Assess the environmental capacity of the target sea in response to the negative environmental impacts of cage aquaculture can effectively reduce environmental impact. On the other hand, to study the effects of flows in the net cage can help the industry improve the bait utilization, increase productivity, reduce water eutrophication, and control the environmental impacts. In this field, the theoretical foundation, such as the interaction between flows and net cage, the velocity reduction when the fluid flows through the net cage, the concentration fluid around the net cage, has been supported. Then, several experiments, using field measurements, on-site tests, were conducted. In recent years, the computational simulation is applied to the field. Porous media has been proven to be used in place of cage models to simulate. Also, particle image velocimetry approach was provided for predicting the fluid flow regime through the porous media. However, more computational analysis is still needed to study the effects of the fluid flows through the net cage to provide a means to increase aquaculture productivity.

References

1. Aarsnes, J., Rudi, H., Løland, G., et al.: Current forces on cage, net deflection. In: Engineering for Offshore Fish Farming. Proceedings of a conference organised by the Institution of Civil Engineers, Glasgow, UK, 17–18 October 1990, Thomas Telford, pp 137–152 (1990)
2. Bear, J.: Dynamics of Fluids in Porous Media. Courier Corporation, North Chelmsford (2013)
3. Bi, C., Dong, G., Liu, X., Zhao, Y., et al.: Numerical simulation of the flow field around fishing net under current. In: The Twenty-first International Offshore and Polar Engineering Conference, International Society of Offshore and Polar Engineers (2011)
4. Bi, C.W., Zhao, Y.P., Dong, G.H., et al.: Experimental investigation of the reduction in flow velocity downstream from a fishing net. Aquac. Eng. **57**, 71–81 (2013)
5. Black, K.D.: Environmental Impacts of Aquaculture, vol 5. Taylor & Francis, US (2001)
6. Cao, L., Wang, W., Yang, Y., et al.: Environmental impact of aquaculture and countermeasures to aquaculture pollution in China. Environ. Sci. Pollut. Res.-Int. **14**(7), 452–462 (2007)

7. DESA, U.: World population prospects: The 2015 revision, key findings and advance tables. United Nations Department of Economic and Social Affairs. Population Division Working Paper no ESA/P/WP 241 (2015)
8. Drapeau, A., Comeau, L., Landry, T., Stryhn, H., Davidson, J.: Association between longline design and mussel productivity in prince Edward Island, Canada. Aquaculture **261**(3), 879–889 (2006)
9. Food and Agriculture Organization of United Nations.: The State of World Fisheries and Aquaculture 2016. Contributing to Food Security and Nutrition for All, p. 200. Rome (2016)
10. Fredriksson, P.G.: How pollution taxes may increase pollution and reduce net revenues. Public Choice **107**(1–2), 65–85 (2001)
11. Gansel, L.C., McClimans, T.A., Myrhaug, D.: Average flow inside and around fish cages with and without fouling in a uniform flow. J. Offshore Mech. Arct. Eng. **134**(4), 041,201 (2012)
12. Hall, P., Holby, O.: Environmental impact of a marine fish cage culture. In: ICES EM, pp. 1–14 (1986)
13. Holmer, M.: Environmental issues of fish farming in offshore waters: Perspectives, concerns and research needs. Aquac. Environ. Interact. **1**(1), 57–70 (2010)
14. Jackson, J.B., Kirby, M.X., Berger, W.H. et al.: Historical overfishing and the recent collapse of coastal ecosystems. Science **293**(5530), 629–637 (2001)
15. Jin, X., Liu, H., Tu, Q.: Eutrophication of Chinese Lakes. China Environmental Science Publisher, Beijing (1990)
16. Klebert, P., Lader, P., Gansel, L., Oppedal, F.: Hydrodynamic interactions on net panel and aquaculture fish cages: A review. Ocean Eng. **58**, 260–274 (2013)
17. Lader, P.F., Enerhaug, B., Fredheim, A., Krokstad, J.: Modelling of 3d net structures exposed to waves and current. In: 3rd International Conference on Hydroelasticity in Marine Technology, pp. 19–26. Department of Engineering Science, The University of Oxford Oxford, UK (2003)
18. Lei, T.: Experimental Study on Transportation and Diffusion of Pollutants in Cage Culture (2016)
19. Liu, X.: Numerical Simulation of Flow Field Around a Planar Net in Water Flow (2011)
20. Nield, D.A., Bejan, A., et al.: Convection in Porous Media, vol. 3. Springer, Berlin (2006)
21. Patursson, Ø., Swift, M.R., Tsukrov, I., et al.: Development of a porous media model with application to flow through and around a net panel. Ocean Eng. **37**(2–3), 314–324 (2010)
22. Plew, D.R., Stevens, C.L., Spigel, R.H., Hartstein, N.D.: Hydrodynamic implications of large offshore mussel farms. IEEE J. Ocean Eng. **30**(1), 95–108 (2005)
23. Pusceddu, A., Fraschetti, S., et al.: Effects of intensive mariculture on sediment biochemistry. Ecol. Appl. **17**(5), 1366–1378 (2007)
24. Shao, D.: Experimental study on wake and mass transfer process of planar mesh in a constant uniform flow. In: The 14th National Conference of Environmental Mechanics (CEM-2018) (2018)
25. Wang, D., Gao, J.: he development and current status of marine aquaculture in china. Chin. Aquac. **4**, 39–42 (2015)
26. Wu, R.: The environmental impact of marine fish culture: Towards a sustainable future. Mar. Pollut. Bull. **31**(4–12), 159–166 (1995)
27. Zhao, Y.P., Bi, C.W., et al.: Numerical simulation of the flow field inside and around gravity cages. Aquac. Eng. **52**, 1–13 (2013)

Review of Energy Finance and Corresponding Policies in Promoting Renewable Energy Sustainable Development in China

Yanfei Deng[1], Lei Xu[2]([✉]), Yuan Yuan[3], and Karen Mancl[4]

[1] School of Management, Southwest University for Nationalities, Chengdu 610041, People's Republic of China
[2] School of Economics, Xihua Univeristy, Chengdu 610039, People's Republic of China
leihsu@163.com
[3] China Power Construction Corporation, Sichuan Electric Power Design & Consulting Limited Company, Chengdu 610016, People's Republic of China
[4] Soil Environment Technology Learning Lab, The Ohio State University, Columbus, OH 43210, USA

Abstract. As global climate warming and the continued use of fossil energy are unsustainable, it has become imperative to accelerate the development of renewable energy. As one of the largest developing countries, this is especially true for future sustainable development in China; therefore, focusing on renewable energy can significantly reduce its dependence on imported fossil fuels. With policy support, China's renewable energy industries have had rapid development and are now at the international forefront level in some fields. However, the renewable energy industries are facing challenges, particularly in terms of energy finance. Basically, the renewable energy industry in China is facing a contradictory dilemma because of funding shortages and inefficient investment strategies, due in part to government-centered renewable energy investment and financing. In view of this, the development status, modes and renewable energy investment characteristics, financing and the corresponding polices in China are discussed in detail in this paper. As certain constraints are hampering the promotion of sustainable renewable energy development, based on the analysis of the key issues, optimizing countermeasures and feasible proposals are presented. All in all, this study is of significance for energy finance and for the development of corresponding policies for sustainable renewable energy development in China.

Keywords: Renewable energy · Energy finance policy · Sustainable development

© Springer Nature Switzerland AG 2020
J. Xu et al. (Eds.): ICMSEM2019 2019, AISC 1002, pp. 279–292, 2020.
https://doi.org/10.1007/978-3-030-21255-1_22

1   Introduction

Renewable energy (RE) has become the fundamental direction and core content of the global energy transformation. As renewable energy is environmentally friendly, low-carbon energy is the main force behind the technological revolution in energy and has become a major emerging industry growth point. China enacted a series of polices to promote sustainable RE development to increase investment in RE industries and to gradually eliminate the obstacles to RE sustainable development. These government actions have improved the RE subsidy mechanism and incentives to ensure that renewable energy generation maintains a steady upward development trend. As China is only in the middle stages of industrialization and urbanization, energy demands continue to grow. The average annual RE installed capacity is expected to reach 2600 kW or more every year in the future. There is a need, for example, for new investment in the solar RE industry of around 190 billion CNY to meet the goals of annual new installed capacity of 20 million kW. Wind and solar power development together needs future annual investments of at least 400 billion CNY, with a total RE investment of 2 trillion CNY required for the "13th Five-Year" period [30].

International experts have now categorized RE into traditional and new, with the former referring to giant hydropower and directly burnt biomass and the latter referring to small hydropower, solar energy, wind energy, biomass energy, geothermal energy and others [2]. The renewable energies discussed in this article fall into the latter category. There are abundant RE resources in China; therefore, to make good use of these RE resources, China needs to promote the development of large new energy resources. Seeking to increase RE use and deployment to encourage a move towards sustainable development, governments have developed and implemented RE support policies. However, as the renewable energy sector grows, it is facing increasingly more complex investment, financing and policy challenges. Because of the initial high costs of renewable energy projects, market development requires focused policy support. If RE financing mechanisms are incomplete or poorly focused, renewable energy project financing has higher costs and attracts less investment; therefore, innovative investment and financing mechanisms are needed to break through the RE financing bottleneck and the continued reliance on government guidance and incentives and to encourage financial market innovations and policy incentives. Existing studies have found that renewable energy laws and regulations as well as incentive policies are an important driving force in the rapid RE development in China [1,7,9,10,22,25,29] such as incentive policies, financial subsidy policies, tax rebates and exemption policies, preferential feed-in tariff policies and technical support policies. China's current renewable energy policies include assistance with development plans, industrial guidance, technical support, legal responsibility and popularization [3,11,13,20,31,35]. Based on a questionnaire, Lam et al. found that the most important drivers of RE development in China were government financial assistance, technical support, feed-in-tariff policies and tax incentives.

Strategically, it is important for China to develop domestic renewable energy manufacturing and promote RE applications for several reasons. First, the promotion and development of RE can ensure that China achieves its goal of 15% non-fossil energy by 2020 and 20% by 2030. Second, RE gives developing countries an important opportunity to catch up with developed countries in terms of RE production capacity and to improve their energy structures [25]. Therefore, with these goals in mind, it is important to assess China's renewable energy financing and policies issues to understand how such large scale investments in renewable energies have been achieved, the major renewable power generation investment and financing channels, the types and effectiveness of Chinese government financing incentive mechanisms, the characteristics of China's renewable energy and whether the current policy mechanisms are suitable for the future development of renewable energy. These issues are fully discussed in this paper. This paper first reviews the development status of China's new energy industries in terms of the RE status and industry investment and financing then, based on the status analysis, a policy analysis is conducted to examined the RE incentives that are driving China's renewable energy development. Finally, a discussion on the countermeasures is presented to solve the identified constraints.

2   Overview of Renewable Energy Development in China

2.1   Renewable Energy Tendencies

China's renewable energy sector has experienced rapid development in recent years. By 2015 China's RE total supply production had reached 3.9×108 tones of standard coal. If solar energy, wind energy, biomass energy, geothermal energy and small hydropower energy are all taken into account, the total supply production in 2015 was as high as 4.4×108 tonnes of standard coal, or over 10% of total energy consumer demand. As of 2015, the total RE power generation capacity was 199GW, for which hydropower, wind power, solar power, biomass power, geothermal and ocean energy power generation accounted respectively for 79.87%, 13.46%, 2.86%, 3.76% and 0.04% [30], as shown in Fig. 1. Renewable energy power generation increased from 761 million kW in 2010 to 1377 million kW by the end of 2015, an almost doubling of generating capacity. China's RE generating capacity is shown in Fig. 2. China's renewable energy sector development is critical to China's aims of achieving 15% renewable energy by the end of 2020 and 20% by 2030. With continuing economic growth and rising pressure from the growing industrial and residential demands, there is a need to guarantee that future energy needs can be sustainably met. Therefore, China needs to move away from its reliance on fossil fuels and promote the coordinated development of hydropower, wind power, solar power, biomass power, geothermal power and other new energy resources.

2.2   Renewable Energy Resources in China

The Chinese government has attached great importance to the development of RE. At present, wind, solar, biomass, and geothermal are relatively mature

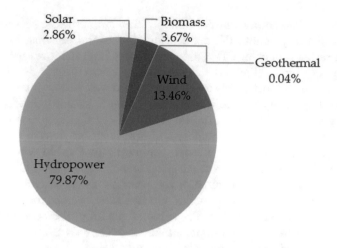

Fig. 1. Composition and proportion of China's RE power generation in 2015

technologies and already have preliminary industrial bases. The National Bureau of Statistics in China identified the main types of renewable energy as solar, wind, geothermal, small hydropower and biomass [2,25]. These renewable energy resources are introduced in this section.

The Chinese Academy of Meteorological Sciences estimated the total wind energy resources in 2014 at 2.49×109 kW, with the main distributions being in the southeast, northeast, north, southeastern coastal areas and inland areas, including the "Sanbei Region" (Hebei, Qinghai, Inner Mongolia, Xinjiang and Hexi Corridor in Gansu Province), as shown in Fig. 3a.

China is blessed with abundant solar energy resources, with its land surfaces receiving annual solar radiant energy of 1.7×1012 tones of standard coal. Two-thirds of the land areas in China have more than 2200 h of sunshine per year and the annual solar radiation has been calculated at 3340C8400 MJ/m^2 across the country. Areas with better conditions have solar energy of 5852 MJ/m^2, with a daily average radiation of $4 \text{ kWh}/ \left(m^2/\text{day} \right)$; however, the resources vary widely, ranging from less than $2 \text{ kWh}/ \left(m^2/\text{day} \right)$ in parts of the southeast to more than $9 \text{ kWh}/ \left(m^2/\text{day} \right)$ in parts of the west. The most abundant solar energy resources are on the Tibetan Plateau [24,33]. Solar energy resources are shown in Fig. 3b.

Geothermal energy resources in China are mainly of low temperature but are around 8% of available resources in the world. The distribution of geothermal energy in China is shown in Fig. 3c. Geothermal energy comes from deep layers of the earth. China has 3.69 million kW of power capacity and 12.6 billion kW of geothermal potential as of October 2011. The National Ministry of Science and Technology estimated the geothermal resources at a depth of 2000 meters to be equivalent to 2500 trillion tones of standard coal, with the initial development equivalent to 3.6 million tones of standard coal [20].

Biomass resources have been widely used in China. More than two-thirds of China's biomass resources are distributed in Inner Mongolia, Sichuan, Henan,

Shandong, Anhui, Hebei, Jiangsu, with biomass generation in these regions accounting for about 70% of the total in China. The employable biomass energy in China can be divided into straw in terms of crop stalks and agricultural process residue; dung from livestock and poultry manure; industrial organic waste residue, waste water and municipal solid waste; and forests such as fire wood residue. The biomass resources distribution is shown in Fig. 3d. It is estimated that the annual exploitable straw resource is around 440 Mtec, the total exploitable fire wood residue resource is equivalent to 200 Mtec, at least 48 Mtec could be generated from biogas produced by industrial organic waste residue and waste water livestock and poultry manure, and around 12 Mtec could be generated from solid waste [14].

2.3   The Driving Forces behind RE Development Financing

Seeking to satisfy the growing energy needs and to reduce greenhouse gas emissions, it is important that more RE resources be harnessed. However, the wider consumption of RE resources is closely related to financial support. Government financial support mechanisms and appropriate finance policies are important to enhance RE sustainable development. It has been widely recognized that renewable energy policies, such as incentive policies financial subsidy policies, tax rebates and exemption policies, preferential feed-in tariff policies, technical support policies, development plans, industrial guidance and technical support, price incentives, legal responsibility and popularization, are essential for the promotion of renewable energy development [3,31,35]. The renewable energy market share has been expanding and in recent years, China's renewable energy has had significant development because of the implementation of the Renewable Energy Law and related policies. Existing studies have shown that incentive policies have been an important driving force behind the rapid RE development in China [21].

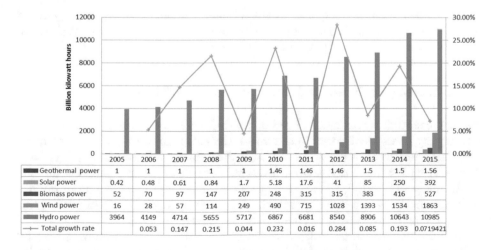

	2005	2006	2007	2008	2009	2010	2011	2012	2013	2014	2015
Geothermal power	1	1	1	1	1	1.46	1.46	1.46	1.5	1.5	1.56
Solar power	0.42	0.48	0.61	0.84	1.7	5.18	17.6	41	85	250	392
Biomass power	52	70	97	147	207	248	315	315	383	416	527
Wind power	16	28	57	114	249	490	715	1028	1393	1534	1863
Hydro power	3964	4149	4714	5655	5717	6867	6681	8540	8906	10643	10985
Total growth rate		0.053	0.147	0.215	0.044	0.232	0.016	0.284	0.085	0.193	0.0719421

Fig. 2. RE generation in China from 2005–2015

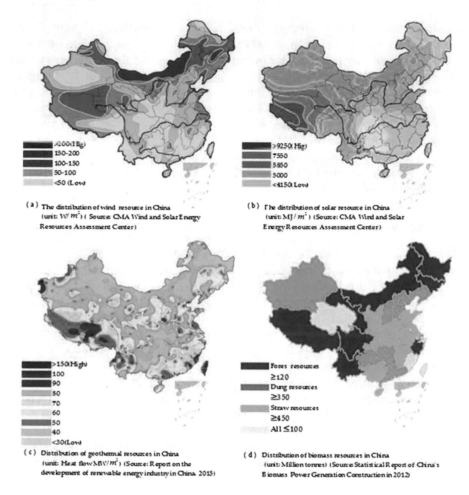

Fig. 3. Distribution of the four main RE resources in China. **a** The distribution of wind resource in China; **b** The distribution of solar resource in China; **c** Distribution of geothermal resources in China; **d** Distribution of biomass resources in China

3   Renewable Energy Investment Situation

As new energy investment has been increasing every year, China has become a new energy investment power in the world. China's renewable energy investment has been steadily growing since 2011. To date, around $124.4 billion has been invested, with an average annual investment of $24.9 billion [23]. During the "12th five-year" period, the new energy investment began to grow more rapidly with the total renewable energy investment reaching 110.5 billion in 2015, at an average annual investment of 75.8 billion, as shown in Fig. 4.

The renewable energy "13th Five-Year" development plan proposed that by 2020 non fossil energy is expected to account for 15% of total energy consumption with a 20% goal by 2030. Europe's RE investment was down 44% from 2012, and, for the first time ever, China invested more in renewable energy than all of Europe combined at 56.3 billion USD (including R&D) of new investment in renewable energy, down 6% from 2012. Despite the overall investment decline, China's investment in additional renewable power capacity surpassed fossil fuel capacity additions in 2013 for the first time [12] stated that in 2015, global renewable energy investment reached a new record, with China's investment reaching 110.5 billion U.S. dollars, which was 36% of the total global RE investment, making the total RE investment in developing countries greater than in developed countries for the first time.

China has invested the most in RE, consolidating its position as the world leader in RE deployment and manufacturing. In the near future, the average annual new installed capacity is expected to be more than 26 million kW, and the 2020 PV investment planning goal is 190 billion CNY. In the next few years, the cumulative installed RE capacity is expected to be more than 4000 kW, with an annual new installed capacity of 20 million kW and an annual investment of more than 180 billion CNY. Wind power and solar energy future annual investment in the "13th Five-Year" period was around 400 billion yuan, with a total investment of nearly 2 trillion CNY [6].

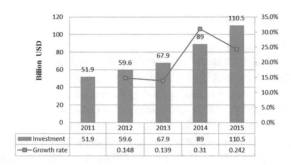

Fig. 4. Annual investment from 2011–2015 in China

Therefore, the demand for renewable energy funds is significant and because of the high costs of renewable energy projects, financing and policy support is needed for comprehensive market development. Unfortunately, because of these high costs, investment in the renewable energy sector is not attractive to investors. In the following section this paper analyzes financing channels and financing means.

4   Policies for Renewable Energy Development

As in the developed world, government support is the key to the development of renewable energy in China because the renewable energy sector is typically policy-oriented [28]. Due to the "Renewable Energy Law" and other related policies, China's renewable energy sector has achieved significant development over the past decade and a half [4, 5, 17–19], and has significantly contributed to the achievement of China's emissions reduction targets. However, China's RE development still requires strong government input such as funding and policy support. China's rapid economic development has continued now for several decades and has provided a strong economic foundation for the support and development of the renewable energy sector [34].

4.1 Laws and Regulations

The existence of a renewable energy act may help reduce policy uncertainty, legally protect the stakeholder interests and boost investor confidence [27]. The Chinese government has attached importance to the development of the RE sector and in line with this has promulgated a series of associated laws, plans and policies to promote RE sector development. Since the implementation of "Renewable Energy Law of the People's Republic of China" on February 2, 2005, a series of specific laws, regulations, policies and industrial plans for hydropower, wind, solar/PV, biomass and geothermal development have been proposed and put into effect [16].

4.2 Incentive Approaches

(1) Financial Subsidies

Financial subsidies, such as investment subsidies, product subsidies and user subsidies, are designed to encourage industrial development in specific areas. Investment subsidies are direct subsidies given to renewable energy program developers and investors [28]. Except for hydropower, the cost of renewable energy power generation remains relatively high in China, making it difficult to compete with conventional energy supplies. Therefore, the Chinese government provides financial subsidies to support RE development. Different subsidy standards have been formulated for the different renewable energy types, as shown in Table 1. These subsidies not only encourage equipment manufacturers to produce large-scale, highly efficient equipment, but also contribute to energy saving and emissions reductions [10].

(2) Tax Support

Tax support policies have been used around the world to encourage RE investment and development; however, these have been only narrowly applied in China. The most widely applied tax policies have been value added tax and income tax. The Renewable Energy Law stipulated favorable tax breaks for hydro, wind, solar, biomass and geothermal power generation projects. The preferential tax policy covers value-added tax (VAT), income tax and import duties. These preferential tax policies shown in Table 2 clearly demonstrate the government's commitment to the development of the RE industry. Compared to the standard VAT rate of 17%, the preferential RE development VAT policy significantly reduces the tax burden on RE enterprises. Similarly, as the income tax rate is 25% in

Table 1. RE subsidy policies

RE	Subsidy policies	Core objective
Wind power	Adjustment of Import Tax Preferential Policies for Large-scale Wind Energy Electricity Generator Equipment, Key components and Raw Materials. (2008). As of the end of 2014, The subsidies of wind power was rewarded at the standard of 0.2CNY/kW	Stimulate wind power generation
PV power	Tentative Procedures for the Management of Financial Subsidy Funds on Solar Photovoltaic Construction and Application: Stipulate the subsidy conditions, standards and usage of the funds (2009). Notice on Improving Solar PV Power Generated Electricity Price Policy. Notice on Grid Power Tariff Adjustment. (2011). The subsidy standard in 2013 has raised to 0.6CNY/kWh, which can be appropriately increased if the project is combined with smart grid or micro grid. Announcement of Value-added Policies for PV Generated Electricity. Reduction of tax of solar power (2013)	Stimulate solar power generation
Biomass power	Implementation Opinions of Financial Supporting Policies to Develop Biomass Energy and Biochemical Engineering and emphasized elastic deficit subsidy for biomass energy. The biomass power generation projects can enjoy a subsidy of 0.25 CNY/kWh for 15 years after commercial operation. Since 2010, the subsidy of newly approved projects dwindles by 2% per year	Strengthen biomass generation

China, large income tax concessions provide a strong stimulus to enterprises to invest in renewable energy production projects [10,36].

(3) Tariff Incentives

(4) R&D Incentives

Technical research and development (R&D) and R&D fund support to promote renewable energy has been important in many countries. An important incentive for the promotion of RE R&D is providing support for the establishment and continuance of R&D institutions. In China, such institutions have significantly contributed to the technological improvements in the Chinese RE industry [8] as the Chinese government has long attached importance to technical RE R&D and has provided major R&D funding support through the Renewable Energy Development Special Fund. Relevant laws and regulations also stipulate that some of these special funds are to be used to support the R&D of new products as well as the industrialization of key technologies in the solar power industry [12, 32]. The stability and continuity of R&D funding to support renewable energy including resource exploration, standard-setting, and project demonstration is as important as market stability to the success of RE policies (Table 3).

Table 2. Tax policies and regulations to support RE sustainable development in China

Type	Contents
VAT	(1) The VAT rate for methane biomass, small hydro and wind power projects is 13%, 6% and 8.5% respectively. (2) From October 1, 2013 to December 31, 2015, the VAT for solar energy power products can enjoy a tax refund of 50%
Income tax	The wind power, solar power and geothermal power projects can enjoy an income tax concession, which consists of income tax exemption in the first 3 years and a 50% reduction of the income tax from the 4th to the 6th year, starting from the year when the projects firstly obtained income
Tax policies	Regulation on the Implementation of the Enterprise Income Tax Law. Opinions on the Implementation of Financial and Tax Policy to Support the Development of Biomass Energy and Biological Chemical: Regulate the financial and tax policy to support the development of biomass energy and biological chemical

5 Issues and Countermeasures

5.1 Issues of Renewable Energy Finance

This paper has closely examined optimization and innovation in RE finance and policy issues to ensure RE sustainable development in China. From this close examination, there are several outstanding issues that need to be resolved to guarantee the continued long-term sustainable success of RE in China. The main issues are as follows:

(1) Incomplete financing systems for RE projects. Generally, investment options have focused on the industrialization of RE technologies only rather than on innovation. Second, domestic enterprises are not investing enough and have not been sufficiently integrated into RE development chains. However, for the RE true sustainable development, investment chains for renewable industry sectors and investment integration is necessary to ensure that the investments contribute to overall long-term sustainability.

(2) Lack of policy coordination and consistency. As there appear to be differences in the goals of each of the RE sectors, it has been difficult to ensure consistency in policy frameworks making it difficult to develop a long-term and effective policy system to support renewable energy's sustainable development. Further, as energy issues encompass federal, state and local governments and several different departments such as energy, science and technology and agriculture, there has been inconsistency and a lack of coordination in policy development across these boundaries, ultimately eroding state macro-control and resulting in wasted investments due to duplication.

(3) Weakness Financial subsidies for renewable energy investment projects are much lower than for other sectors and China's support for renewable energy through taxation methods is fairly weak, with the actual taxation benefits for

Table 3. Feed-in tariff for RE developments in China

Type	Contents
Wind power	(1) China's territory is divided into four types of wind resource regions where different feed-in tariffs are defined. (2) Offshore wind power is divided into coastal and intertidal wind power. The feed-in tariff of coastal and intertidal wind power projects is 0.85 CNY/kWh and 0.75 CNY/kWh respectively before 2017
Solar PV power	In 2009, the feed-in tariff for the first batch of concession projects was 1.09 RMB/kWh, while the temporary feed-in tariff for solar photovoltaic power plant project in Ningxia was 1.15 CNY/kWh in April, 2010. The benchmark tariffs are 0.9, 0.95 and 1 CNY/kWh respectively
Geothermal power	The feed-in tariff of geothermal power projects is determined considering their actual construction and operation costs as well as reasonable profits
Biomass power	(1) The "Improving Policies on the Feed-in Tariff of agriculture and forestry biomass power" issued in 2011 set at the feed-in tariff of the agriculture and forestry biomass power projects was 0.75 CNY/kWh (2) The feed-in tariff of waste incineration power projects is 0.65 CNY/kWh
Small hydropower	(1) The feed-in tariff of inter-provincial hydropower stations is determined by negotiation between the electricity supplier and buyer; Other hydropower stations follow benchmark tariff policy, and the price is formulated by taking into consideration the hydropower exploitation cost, the provincial average electricity purchase price, and the demand and supply of the market. (2) Provinces with large amount of hydropower adopt time-of-use electricity price, which varies according to three different periods of electricity supply: flow period, wet period and dry period

most renewable energy projects being similar to or of less benefit than conventional energy tax benefits.

(4) Inadequate investment in technical R&D and weak independent RE development innovation. At the moment, China lacks a powerful platform to support technological research. What is needed is a clear long-term continuous and rolling R&D development investment plan, with guaranteed funding to support R&D. Most core technologies in most domestic enterprises come from abroad, which will ultimately affect the sustainable development of the RE sector.

5.2   Countermeasures

The following countermeasures are proposed to address these issues:

(1) Develop a market-oriented investment and financing system. China needs to actively promote RE industry marketization reform and give full play to the basic market role of resource allocation. Constructing a diversified system to

support investment and financing, encouraging policy institutions to support the renewable energy industry and broadening financing methods could lead to improved RE investment structures.

(2) Enlarge investment on research and development to improve technical innovation. The Chinese government needs to subsidize RE industry R&D activities through government-sponsored laboratories, or by direct funding to the private sector. To achieve better RE regulation development, RE regulation systems need to be reformed by coordinated policy innovations which encourage and promote independent innovation.

(3) Improve the coordination of policies and enhance policy innovation. Related departments need to continue to prioritize power system reform so as to establish an RE power system operating mechanism that allows for RE integration and coordination to promote local micro-grid applications. RE industry sector development must take into account the central, provincial and local needs, so there needs to be an established coordination mechanism across all governmental levels to encourage information sharing. Related authorities also need to make full use of the tariff and price subsidy market mechanisms when allocating resources to support policy mechanisms.

6   Conclusions

This paper systematically reviewed the investment status, financing, and corresponding policies for sustainable development of the renewable industry sector in China. Investment, financing and corresponding policies have played a significant role in sustainable RE development. Investment, financing, finance means and related laws, regulations and major incentive mechanisms in China have progressed significantly in recent years. Various incentive mechanisms such as the RE preferential fiscal policies and R&D fund support for the commercialization of technologies has gone some way to decreasing the development costs of renewable energy projects. In the future, joint financing and policies and support mechanisms for renewable power generation need to be further strengthened. Fiscal and tax incentives have significantly relieved the financial burdens on renewable energy power generation enterprises and preferential tax policies for VAT, income tax and import duties for the renewable energy power generation industry as well as strong financial subsidies have provided invaluable support for renewable power projects. Tariff incentives, in particular, have ensured reasonable profits for renewable energy power generation enterprises. However, investment and financing renewable energy policies need to be strengthened further as China's RE financing methods and development policies have some inherent disadvantages. For the future, it is necessary to develop a range of coordinated strategies to ensure long-term efficient developments for the RE industry sector. In brief, this paper provides a comprehensive analysis of China's RE sustainable development. The major financing methods and identified incentive strategies and approaches can assist researchers and industry practitioners gain a better understanding of how RE sustainable development can be promoted by the

Chinese government. The lessons learned from this China study can also be compared with or used to formulate innovative financing methods and incentive polices for RE sustainable development in other countries.

Acknowledgements. The work was supported by the Central Universities Project under Grant (No. 2016NZYQN13). The authors are indebted to the editors and reviewers for their valuable comments and suggestions.

References

1. Abolhosseini, S., Heshmati, A.: The main support mechanisms to finance renewable energy development. Renew. Sustain. Energy Rev. **40**, 876–885 (2014)
2. Cao, X., Kleit, A., Liu, C.: Why invest in wind energy? Career incentives and chinese renewable energy politics. Energy Policy **99**, 120–131 (2016)
3. Cui, H., Wu, R.: Feasibility analysis of biomass power generation in China. Energy Procedia **16**, 45–52 (2012)
4. Development, N., Reform Commission, C.: The 12th Five Year Plan for National Economic and Social Development of the People's Republic of China 2011–2015 (2011)
5. Development, N., Reform Commission, C.: Notice on Adjusting Benchmark Tariff of Onshore Wind Power Generation Project (2015)
6. Development, N., Reform Commission, C.: The 13th Five Year Plan for National Economic and Social Development of the People's Republic of China 2016–2020 (2016)
7. Geng, W., Ming, Z., Lilin, P., et al.: China's new energy development: Status, constraints and reforms. Renew. Sustain. Energy Rev. **53**, 885–896 (2016)
8. He, Y., Pang, Y., Zhang, J., Xia, T., Zhang, T.: Feed-in tariff mechanisms for large-scale wind power in China. Renew. Sustain. Energy Rev. **51**, 9–17 (2015)
9. He, Y., Xu, Y., Pang, Y., Tian, H., Wu, R.: A regulatory policy to promote renewable energy consumption in China: Review and future evolutionary path. Renew. Energy **89**, 695–705 (2016)
10. Hu, Z., Wang, J., Byrne, J., Kurdgelashvili, L.: Review of wind power tariff policies in China. Energy Policy **53**, 41–50 (2013)
11. Kang, J., Yuan, J., Hu, Z., Xu, Y.: Review on wind power development and relevant policies in China during the 11th five-year-plan period. Renew. Sustain. Energy Rev. **16**(4), 1907–1915 (2012)
12. Koseoglu, N.M., van den Bergh, J.C., Lacerda, J.S.: Allocating subsidies to R&D or to market applications of renewable energy? Balance and geographical relevance. Energy Sustain. Dev. **17**(5), 536–545 (2013)
13. Lam, J., Woo, C.K., Kahrl, F., Yu, W.: What moves wind energy development in China? Show me the money!. Appl. Energy **105**, 423–429 (2013)
14. Liu, J., Wang, S., Wei, Q., Yan, S.: Present situation, problems and solutions of China's biomass power generation industry. Energy Policy **70**, 144–151 (2014)
15. Zeng, M., Song, X., Ma, M.J., Zhu, X.L.: New energy bases and sustainable development in China: A review. Renew. Sustain. Energy Rev. **20**, 169–185 (2013)
16. Zeng, M., Liu, X.M., Li, Y.L., Peng, L.L.: Review of renewable energy investment and financing in China: Status, mode, issues and countermeasures. Renew. Sustain. Energy Rev. **31**, 23–37 (2014)

17. Standing Committee of the National People's Congress, C.: Standing Committee of the National People's Congress. Renewable Energy Law of the People's Republic of China (2006)
18. Standing Committee of the National People's Congress, C.: Standing Committee of the National People's Congress. Energy Conservation Law of the People's Republic of China (2007)
19. Standing Committee of the National People's Congress, C.: Atmospheric Pollution Prevention and Control Law of the People's Republic of China (2015)
20. National Renewable Energy Center, C.: The National Renewable Energy Development Report (2015)
21. Ng, T.H., Tao, J.Y.: Bond financing for renewable energy in Asia. Energy Policy **95**, 509–517 (2016)
22. Ouyang, X., Lin, B.: Levelized cost of electricity (lCOE) of renewable energies and required subsidies in China. Energy Policy **70**, 64–73 (2014)
23. Zhang, P.D., Yang, Y.L., et al.: Opportunities and challenges for renewable energy policy in China. Renew. Sustain. Energy Rev. **13**(2), 439–449 (2009)
24. Pinkse, J., Van den Buuse, D.: The development and commercialization of solar PV technology in the oil industry. Energy Policy **40**, 11–20 (2012)
25. Shen, J., Luo, C.: Overall review of renewable energy subsidy policies in China-contradictions of intentions and effects. Renew. Sustain. Energy Rev. **41**, 1478–1488 (2015)
26. Song, A., Lu, L., Liu, Z., Wong, M.: A study of incentive policies for building-integrated photovoltaic technology in Hong Kong. Sustainability **8**(8), 769 (2016)
27. The National People's Congress, C.: Renewable Energy Law of the People's Republic of China (2009)
28. Wang, N., Chang, Y.C.: The development of policy instruments in supporting low-carbon governance in China. Renew. Sustain. Energy Rev. **35**, 126–135 (2014)
29. Wang, Z., Qin, H., Lewis, J.I.: China's wind power industry: Policy support, technological achievements, and emerging challenges. Energy Policy **51**, 80–88 (2012)
30. Yang, X.J., Hu, H., Tan, T., Li, J.: China's renewable energy goals by 2050. Environ. Dev. **20**, 83–90 (2016)
31. Zeng, M., Li, C., Zhou, L.: Progress and prospective on the police system of renewable energy in China. Renew. Sustain. Energy Rev. **20**, 36–44 (2013)
32. Zhang, H., Li, L., Zhou, D., Zhou, P.: Political connections, government subsidies and firm financial performance: Evidence from renewable energy manufacturing in China. Renew. Energy **63**, 330–336 (2014)
33. Zhang, S., He, Y.: Analysis on the development and policy of solar pv power in China. Renew. Sustain. Energy Rev. **21**, 393–401 (2013)
34. Zhang, X., Shen, L., Chan, S.Y.: The diffusion of solar energy use in HK: What are the barriers? Energy Policy **41**, 241–249 (2012)
35. Zhao, Z.Y., Zuo, J., Fan, L.L., Zillante, G.: Impacts of renewable energy regulations on the structure of power generation in China-a critical analysis. Renew. Energy **36**(1), 24–30 (2011)
36. Zhao, Z.Y., Chen, Y.L., Chang, R.D.: How to stimulate renewable energy power generation effectively?—China's incentive approaches and lessons. Renew. Energy **92**, 147–156 (2016)

Part II
Supply Chain Management

A Robust Weighted Goal Programming Approach for Supplier Selection Problem with Inventory Management and Vehicle Allocation in Uncertain Environment

Lishuai Wang and Jun Li[✉]

School of Economics and Management, University of Electronic Science and
Technology of China, Chengdu 611731, People's Republic of China
fengliujianyun@163.com

Abstract. This research work deals with the multi-product multi-period supplier selection problem with inventory management and vehicle allocation in uncertain environment. In this paper, a robust weighted goal programming approach is developed to solve the multi-product multi-period supplier selection problem. The purpose is to consider parameter uncertainty by controlling the impact of estimation errors on the procurement strategy performance. In the proposed model, total costs, rejected items and late-delivered items are considered as three objectives that have to be minimized over the decision horizon. To illustrate the feasibility of the proposed model, a numerical example is presented.

Keywords: Supplier selection · Robust weighted goal programming · Multi-objective optimization

1 Introduction

In the past few decades, the supplier selection problem as the first chain of supply chain has attracted great attention in both the theoretical and applied literature. Supplier selection problem can find best suppliers to significantly reduce costs and maintain the continuity of the supply chain. Pan proposed a mathematical programming model to determine the suppliers and quantities [21]. After that, the multi-product single-period supplier selection model is widely used in production and operations management [12,20]. It is easy to confirm that the value of scheduling suppliers over the multi-period horizon along with the supplier selection can be significantly higher than planning over a single period. Therefore, some researchers focus on the multi-product multi-period supplier selection problem [4,14].

It is difficult for decision makers to obtain complete information for most of the input parameters. Therefore, the input parameters cannon be accurately estimated, which means that parameters include uncertainty. The parameter

© Springer Nature Switzerland AG 2020
J. Xu et al. (Eds.): ICMSEM2019 2019, AISC 1002, pp. 295–309, 2020.
https://doi.org/10.1007/978-3-030-21255-1_23

uncertainty problem may lead to an infeasible or meaningless procurement strategy because the solutions obtained from the mentioned model are sensitive to the input parameters. Robust optimization (RO) is introduced in this research to address the parameter uncertainty problem. Unlike stochastic programming and fuzzy optimization, robust optimization does not require full information such as probability distribution of the uncertain parameters, which are occasionally excessively strong and criticized for their validity. Based on the previous researches [5,22], robust optimization has achieved a great progress, and applied to many fields such as combinatorial optimization and supply chain. Other latest articles can be found in Ben-Tal et al. [6], Babrel et al. [8], Chassein and Goerigk [11], etc.

This paper deals with a multi-product multi-period supplier selection problem with inventory management and vehicle allocation, where demand and quality are considered as uncertain data. To obtain robust solutions, a robust weighted goal programming approach for supplier selection problem is developed. In addition, the budget uncertainty set is considered in order to guarantee the linear formulation.

The remainder of the paper is organized as follows. A brief review of the literature provided in Sect. 2. A robust weighted goal programing model for multi-product multi-period supplier selection problem with inventory management and vehicle allocation in an uncertain environment is presented in Sect. 3. A numerical example is given to illustrate the proposed approach in Sect. 4. The conclusions are drawn in Sect. 5.

2   Literature Review

In today's competitive environment, companies are forced to take advantage of every opportunity to optimize their business processes. Supplier selection and order allocation are effective ways to reduce costs. The first research is traced back to the 1950s [23]. Since then, subsequent works were conducted [12,21]. Though single supplier can foster better collaboration and partnership, many firms have adopted multiple suppliers to decrease the risk of supply disruption. Some studies consider single product multi-period supplier selection problem with deterministic demand [25] and multi-product multi-period supplier selection problem with deterministic demand [1,4]. However, the parameter uncertainty problem is involved in the input data of supplier selection problem. To against the parameter uncertainty problem, we introduce the robust optimization to this study. Unlike stochastic programming [19] and fuzzy method [26], robust optimization does not require full information such as probability distribution of the uncertain parameter, which is excessively strong and criticized for their validity. Furthermore, robust optimization has been widely used in the field of supply chain. Aouam and Brahimi developed a model that integrates production planning and order acceptance decisions while taking into account demand uncertainty [2]. Thorsen and Yao proposed a general methodology based on robust optimization for an inventory control problem with demand and lead-time uncertainty [24]. Aras and Bilge developed a multi-period supply chain network design model with multi-products problem of a firm in the food sector and adopted the minimax regret approach to against demand uncertainty [3]. To obtain robust and sustainable biofuel supply chains, Hombach et al. proposed a robust multi-objective approach to solve the multi-objective supply chain problem with uncertain data [16]. Joonrak et al. developed a robust closed-loop supply chain model to respond to uncertainty from reverse logistics [17].

Robust goal programming is an efficient method in dealing with multi-objective programming problem with uncertain parameter. Charnes et al. first introduced goal programming (GP) in 1955 [10]. However, the traditional GP cannot deal with the parameter uncertainty problem. To against the parameter uncertainty problem, Kuchta studied robust goal programming that constructs interval-based uncertainty sets [18]. Subsequently, Ghahtarani and Najafi applied robust goal programming approach to portfolio selection problem [13]. More researches on robust goal programming can be found in Hanks et al. [15] and Bojd et al. [9]. This research investigates a robust weighted goal programming approach for supplier selection problem to against parameter uncertainty problem. To keep the continuity of the supply chain, order allocation, inventory management and vehicle allocation are also considered.

3   Proposed Model

We consider a multi-product multi-period supplier selection problem in this paper. We assume that: (1) Products are provided by a set of suppliers. (2) In the process of production operation, the enterprise should avoid shortages. (3) The product demand and rate of rejected items are uncertain parameters. (4) Transportation costs are included as an important link in the supply chain.

3.1   Notations

Notations are as follows.
Indices
$\quad i = 1, 2, \cdots, I$ index for products;
$\quad j = 1, 2, \cdots, J$ index for suppliers;
$\quad t = 1, 2, \cdots, T$ index for periods;
$\quad k = 1, 2, \cdots, K$ index for vehicles;
$\quad m = 1, 2, \cdots, t.$
Parameters
$\quad P_{ijt}$: purchasing price of product i from supplier j in period t;
$\quad O_{ijt}$: ordering cost of product i from supplier j in period t;
$\quad P_k$: fixed transportation cost for vehicle k;
$\quad V_{ijk}$: unit transportation cost for product i from supplier j by the vehicle t;
$\quad H_i$: holding cost for product i per period;
$\quad D_{it}$: demand of product i in period t;
$\quad C_{ijt}$: capacity of product i from supplier j in period t;
$\quad V_i$: unit volume of product i;

W_i: unit weight of product i;

V_k: bulk capacity of vehicle i;

W_k': weight capacity of vehicle i;

M_{jt}: maximum kinds of products supplier i can provide in period t;

N_{it}^{min}: minimum quantity of suppliers is selected to meet the demand of product i in period t;

N_{it}^{max}: maximum quantity of suppliers is selected to meet the demand of product i in period t;

f_{ijt}: percentage of rejected items of product i delivered by supplier j in period t;

g_{ijt}: percentage of late delivered items of product i delivered by supplier j in period t.

Variables

X_{ijt}: number of units of product i that buyer procures from supplier j in period t;

X_{ijtk}: number of units of product i that buyer procures from supplier j by using vehicle k in period t;

Y_{ijt}: 1 if supplier j is selected to provide product i in period t, 0 otherwise;

X_{ijtk}': the number of vehicles allocated to supplier j to transport product i in period t.

3.2   Mathematical Formulation

A multi-objective mixed integer programming for supplier selection problem (SSP) is given as follows:

$$\min \quad Z1 = \sum_{i=1}^{I}\sum_{j=1}^{J}\sum_{t=1}^{T} P_{ijt}X_{ijt} + \sum_{i=1}^{I}\sum_{j=1}^{J}\sum_{t=1}^{T}\sum_{k=1}^{K}(P_k X_{ijtk'} + V_{ijk}X_{ijtk}) + \sum_{i=1}^{I}\sum_{j=1}^{J}\sum_{t=1}^{T} O_{ijt}Y_{ijt}$$

$$+ \sum_{i=1}^{I}\sum_{t=1}^{T} H_i(\sum_{m=1}^{t}\sum_{j=1}^{J}(X_{ijm} + X_{ij(m-1)}g_{ij(m-1)} - X_{ijm}g_{ijm}) - \sum_{m=1}^{t} D_{im}) \tag{1}$$

$$\min \quad Z2 = \sum_{i=1}^{I}\sum_{j=1}^{J}\sum_{t=1}^{T} f_{ijt}X_{ijt} \tag{2}$$

$$\min \quad Z3 = \sum_{i=1}^{I}\sum_{j=1}^{J}\sum_{t=1}^{T} g_{ijt}X_{ijt} \tag{3}$$

$$s.t. \quad X_{ijt} = \sum_{k=1}^{K} X_{ijtk}, \forall i, \forall j, \forall t, \tag{4}$$

$$X_{ijtk}V_i \le X_{ijtk}'V_k', \forall i, \forall j, \forall t, \forall k, \tag{5}$$

$$X_{ijtk}V_i \le X_{ijtk}'V_k', \forall i, \forall j, \forall t, \forall k, \tag{6}$$

$$\sum_{m=1}^{t}\sum_{j=1}^{J}(X_{ijm} + X_{ij(m-1)}g_{ij(m-1)} - X_{ijm}g_{ijm}) - \sum_{m=1}^{t} D_{im} \ge 0, \forall i, \forall t, \tag{7}$$

$$X_{ijt} \le C_{ijt}Y_{ijt}, \forall i, \forall j, \forall t, \tag{8}$$

$$\sum_{i=1}^{I} Y_{ijt} \le M_{jt}, \forall i, \forall t, \tag{9}$$

$$N_{it}^{min} \le \sum_{j=1}^{J} Y_{ijt} \le N_{it}^{max}, \forall i, \forall t, \tag{10}$$

$$X_{ijt}, X_{ijtk}, X'_{ijtk} \ge 0 \text{ and Integer}, \forall i, \forall j, \forall t, \forall k, \tag{11}$$

$$Y_{ijt} \in \{0, 1\}, \forall i, \forall j, \forall l. \tag{12}$$

The mixed integer multi-objective linear programming model consists of three objective functions. The objective function (1) is to minimize the total costs which include: (i) purchasing costs associated with the suppliers, (ii) ordering costs associated with the selected supplier, (iii) transportation costs from the selected suppliers, and (iv) the inventory holding costs. In the former researches, transportation costs were usually assumed to be a constant. It is worth noting that the transportation costs are composed of the fixed costs determined by the type of vehicles and variable costs associated with the suppliers, vehicles and products. The objective function (1) refers to the rejected items of the suppliers. The objective function (3) minimizes the late delivered items of the suppliers.

Constraints (4)–(12) associated with the vehicle allocation problem constrain quantity, volume and weight. Constraint (4) represents that the quantity of product delivered by supplier in period are equal to the quantity transported by different type of vehicles. Constraints (5) and (6) indicate that the volume and weight of the procured products cannot exceed the upper limit, respectively. Constraint (7) ensures that the demand should be met regardless of which scenario eventually occurs. Constraint (8) shows that each supplier cannot beyond its production capacity. Constraint (9) constrains the type of product by each supplier. Constraint (10) represents that the number of suppliers of a product are available under some restrictions. Constraint (11) defines the non-negativity and integer variables, while constraint (12) defines the binary variables.

3.3   Solution Method

In order to solve (SSP) with uncertain parameter, we propose a robust weighted goal programming approach. Then a robust weighted goal programming approach for supplier selection model is developed.

Robust Weighted Goal Programming Goal Programming (GP) is a widely used approach in multi-objective programming [2]. As a (GP) variant, the weighted goal programming (WGP) allows for direct trade-offs between all unwanted deviational variables by placing them in a weighted. In (WGP), the values of both goal and weight coefficient are set by decision maker. However,

most of parameters are uncertain because these parameters are mostly estimated by the historical data. A robust weighted goal programming model that
integrates weighted goal programming with robust optimization is developed to
tackle with uncertain parameters. In the robust optimization framework, one of
the most important tasks is to build an uncertainty set. In order to flexibly adjust
to the conservatism of the robust solutions and obtain a linear formulation, we
follow the approach proposed by Bertsimas and Sim [7].

Considering the following multi-objective linear programming:

$$
\min \left\{ c_i^T x, i = 1, 2, \cdots, p \right\} \\
\text{s.t. } Ax \le b.
$$

(13)

where $x \in R^n$ is the decision vector, and $c_i = (c_{i1}, c_{i2}, \cdots, c_{in})^T, i = 1, 2, \cdots, p$,
is the vector for the ith objective function, $A \in R^{m \times n}$ is a matrix of elements
a_{kj} for $k = 1, 2, \cdots, m$ and $j = 1, 2, \cdots, n$, $b \in R^m$ is the right-hand vector and
$b = (b_1, b_2, \cdots, b_m)$. Let $C \in R^{p \times n}$, $C = (c_1, c_2, \cdots, c_p)^T$. Here C, A and b are
deterministic coefficients.

We assume that g_i is the specified target values for ith linear objective functions, the positive (negative) deviations between the achievement goals and their
specified target values are denoted as d_i^+ (d_i^-), and the corresponding weight
coefficients based on the relative importance of the objectives are represented as
w_i^+ (w_i^-). Thus, the weighted goal programming model is given as follows:

$$
\min \sum_{i=1}^{p} \left(w_i^- d_i^- + w_i^+ d_i^+ \right) \\
\text{s.t. } c_i^T x + d_i^- - d_i^+ = g_i, \forall i, \\
Ax \le b, \\
d_i^- \ge 0, d_i^+ \ge 0, \forall i.
$$

(14)

Bertsimas and Sim introduced an adjusted non-negative integer to control
the level of conservatism of robust solutions [7]. In this study, we assume that
the c and A are uncertain parameters and take value in the following budget
uncertainty sets $\mathscr{U}_C$ and $\mathscr{U}_A$, respectively.

$$
\mathscr{U}_C = \left\{ C \in R^{p \times n} : \bar{c}_{ij} - \hat{c}_{ij} \eta_{ij} \le c_{ij} \le \bar{c}_{ij} + \hat{c}_{ij} \eta_{ij}, \sum_{j=1}^{n} \eta_{ij} \le \Gamma_i, 0 \le \eta_{ij} \le 1, \forall i, \forall j \right\},
$$

$$
\mathscr{U}_A = \left\{ A \in R^{m \times n} : \bar{a}_{kj} - \hat{a}_{kj} \zeta_{kj} \le a_{kj} \le \bar{a}_{kj} + \hat{a}_{kj} \zeta_{kj}, \sum_{j=1}^{n} \zeta_{kj} \le \Gamma'_k, 0 \le \zeta_{kj} \le 1, \right. \\
\left. \forall k, \forall j \vphantom{\sum_{j=1}^{n}} \right\},
$$

where, Γ_i takes value in the interval $[0, |J_i|]$, and Γ'_k takes value in the inter-
val $[0, |J'_k|]$, J_i and J'_i are the sets of coefficients c_{ij} and a_{kj}, respectively. c_{ij},
$j \in J_i$ takes value in the interval $[\bar{c}_{ij} - \hat{c}_{ij}, \bar{c}_{ij} + \hat{c}_{ij}]$ and a_{kj}, $j \in J'_k$ takes
value in the interval $[\bar{a}_{kj} - \hat{a}_{kj}, \bar{a}_{kj} + \hat{a}_{kj}]$. The uncertainty set $\mathscr{U}_C$ shows that
up to $\lfloor \Gamma_i \rfloor$ of these coefficients are allowed to change, and one coefficient c_{it}
changes by $(\Gamma_i - \lfloor \Gamma_i \rfloor) \hat{c}_{it}$. For the chosen Γ_i, we consider a subset satisfying
the conditions $S_i \subseteq J_i$ and $|S_i| = \lfloor \Gamma_i \rfloor$. The uncertainty set $\mathscr{U}_A$ indicates that

up to $\lfloor \Gamma'_k \rfloor$ of these coefficients are allowed to change, and one coefficient a_{kt} changes by $(\Gamma'_k - \lfloor \Gamma'_k \rfloor)\hat{a}_{kt'}$. For the chosen Γ'_k, we consider a subset satisfying the conditions $S'_k \subseteq J'$ and $|S'_k| = \lfloor \Gamma'_k \rfloor$.

Based on the above analysis, we propose a robust weighted goal programming approach (RWGP) as given bellow:

$$\min \sum_{i=1}^{p} (w_i^- d_i^- + w_i^+ d_i^+)$$

$$s.t. \sum_{j=1}^{n} c_{ij}x_j + \max_{\{S_i \cup t_i | S_i \subseteq J_i, |S_i| = \lfloor \Gamma_i \rfloor, t_i \in J_i \setminus S_i\}} \left\{ \sum_{j \in S_i} \hat{c}_{ij}|x_j| + (\Gamma_i - \lfloor \Gamma_i \rfloor)\hat{c}_{it_i}|x_{t_i}| \right\}$$

(15)

$$+ d_i^- - d_i^+ = g_i, \forall i,$$

$$\sum_{j=1}^{n} \bar{a}_{kj}x_j + \max_{\{S'_k \cup t'_k | S'_k \subseteq J'_k, |S'_k| = \lfloor \Gamma'_k \rfloor, t'_k \in J'_k \setminus S'_k\}} \left\{ \sum_{k \in S'_k} \hat{a}_{kj}|x_j| + (\Gamma'_k - \lfloor \Gamma'_k \rfloor)\hat{a}_{kt'_k} x_{t_k}| \right\}$$

$$\leq b_k, \forall k,$$

(16)

$$d_i^- \geq 0, d_i^+ \geq 0, \forall i.$$

If Γ_i is chosen as an integer, the ith constraint is protected by $\beta_i(x, \Gamma_i) = \max_{\{S_i | S_i \subseteq J_i, |S_i| = \lfloor \Gamma_i \rfloor\}} \left\{ \sum_{j \in S_i} \hat{c}_{ij}|x_j| \right\}$. Note that when $\Gamma_i = 0$, the constraints are equivalent to that of the nominal problem. By varying $\Gamma_i \in [0, |J_i|]$, we have the flexibility of adjusting the robustness of the method against the level of conservatism of the solution. Let

$$\beta_i(x^*, \Gamma_i) = \max_{\{S_i \cup t_i | S_i \subseteq J_i, |S_i| = \lfloor \Gamma_i \rfloor, t_i \in J_i \setminus S_i\}} \left\{ \sum_{j \in S_i} \hat{c}_{ij}|x_j| + (\Gamma_i - \lfloor \Gamma_i \rfloor)\hat{c}_{it_i}|x_t| \right\}.$$

(17)

Model (17) is equivalent to the following linear optimization problem:

$$\min \sum_{j \in J_i} \hat{c}_{ij}|x_j^*|\eta_{ij}$$
$$s.t. 0 \leq \eta_{ij} \leq 1, \forall j \in J_i$$
$$\sum_{j \in J_i} \eta_{ij} \leq \Gamma_i, \forall j \in J_i$$

(18)

The dual problem of the primal problem (18) is

$$\min \sum_{j \in J_i} p_{ij} + \Gamma_i z_i$$
$$s.t. z_i + p_{ij} \geq \hat{c}_{ij}|x_j^*|, p_{ij} \geq 0, \forall j \in J_i,$$
$$z_i \geq 0.$$

(19)

Since problem (18) is feasible and bounded for $\Gamma_i \in [0, |J_i|]$, it follows from strong duality theorem that the dual problem (19) is also feasible and bounded and the optimal objective value of problem (19) is equal to the optimal objective value of primal problem (18). Thus, constraint (15) is equivalent to

$$
\sum_{j=1}^{n} \bar{c}_{ij} x_j + z_i \Gamma_i + \sum_{j \in J_i} p_{ij} + d_i^- - d_i^+ = g_i, z_i \geq 0, \forall i,
$$
$$
z_i + p_{ij} \geq \hat{c}_{ij} y_j, p_{ij} \geq 0, \forall i, \forall j \in J_i, \tag{20}
$$
$$
-y_j \leq x_j \leq y_j, \forall j \in J_i.
$$

Similarly, the constraint (16) is equivalent to the below:

$$
\sum_{j=1}^{n} \bar{a}_{kj} x_j + \Gamma'_k z'_k + \sum_{j \in J'_k} p'_{kj} \leq b_k, z'_k \geq 0, \forall k,
$$
$$
z'_k + p'_{kj} \geq \hat{a}_{kj} y_j, p'_{kj} \geq 0, \forall k, \forall j \in J'_k. \tag{21}
$$

Therefore, (RWGP) is reformulated as follows:

$$
\min \sum_{i=1}^{p} (w_i^- d_i^- + w_i^+ d_i^+)
$$
$$
s.t. \sum_{j=1}^{n} \bar{c}_{ij} x_j + z_i \Gamma_i + \sum_{j \in J_i} p_{ij} + d_i^- - d_i^+ = g_i, \forall i,
$$
$$
\sum_{j=1}^{n} \bar{a}_{kj} x_j + \Gamma'_k z'_k + \sum_{j \in J'_k} p'_{kj} \leq b_k, \forall k,
$$
$$
z_i + p_{ij} \geq \hat{c}_{ij} y_j, p_{ij} \geq 0, \forall i, \forall j \in J_i, \tag{22}
$$
$$
z'_k + p'_{kj} \geq \hat{a}_{kj} y_j, p'_{kj} \geq 0, \forall k, \forall j \in J'_k,
$$
$$
-y_j \leq x_j \leq y_j, \forall j,
$$
$$
z_i \geq 0, d_i^- \geq 0, d_i^+ \geq 0, \forall i,
$$
$$
z'_k \geq 0, d'^-_k \geq 0, d'^+_k \geq 0, \forall k.
$$

A Robust Weighted Goal Programming Approach for Supplier Selection Problem In this study, we assume that the demand and the percentage of rejected items are uncertain data. The target value for each objective function is denoted by f_i^* ($i = 1, 2, 3$). Thus, we propose the following robust weighted goal programming for supplier selection model (RWGP_SSP):

$$
\min Z = w_1^+ d_1^+ + w_2^+ d_2^+ + w_3^+ d_3^+
$$
$$
s.t. \sum_{i=1}^{I} \sum_{j=1}^{J} \sum_{t=1}^{T} P_{ijt} X_{ijt} + \sum_{i=1}^{I} \sum_{j=1}^{J} \sum_{t=1}^{T} \sum_{k=1}^{K} (P_k X'_{jtk} + V_{ijk} X_{ijtk}) + \sum_{i=1}^{I} \sum_{j=1}^{J} \sum_{t=1}^{T} O_{ijt} Y_{ijt}
$$
$$
\tag{23}
$$
$$
+ \sum_{i=1}^{I} \sum_{t=1}^{T} H_i \left(\sum_{m=1}^{t} \sum_{j=1}^{J} (X_{ijm} + X_{ij(m-1)} g_{ij(m-1)} - X_{ijm} g_{ijm}) - \sum_{m=1}^{t} D_{im} \right) + d_1^- - d_1^+
$$
$$
= f_1^*,
$$
$$
\sum_{i=1}^{I} \sum_{j=1}^{J} \sum_{t=1}^{T} f_{ijt} X_{ijt} + d_2^- - d_2^+ = f_2^*, \tag{24}
$$

$$\sum_{i=1}^{I}\sum_{j=1}^{J}\sum_{t=1}^{T} g_{ijt}X_{ijt} + d_3^- - d_3^+ = f_3^*, \tag{25}$$

$$d_i^-, d_i^+ \geq 0, i = 1, 2, 3 \tag{26}$$

Conatraints (4)–(24) are also included where, d_i^+ and d_i^- ($i = 1, 2, 3$) are the deviation variables representing the distances between the attained objective value and the target goal.

Let D_{it} be the uncertain demand of product i in period t, $\bar{D}_{it}$ be the nominal value, and $\hat{D}_{it}$ be the maximum deviation from nominal value. Thus, the uncertain demand D_{it} takes value in the interval . For products i, $i = 1, \cdots, I$, we assume that the uncertainty level in period t is chosen as Γ_{it} which takes value in the interval $[0, \lfloor J_{it} \rfloor]$. Let J_1' is the set of coefficients f_{ijt}, $(i, j, t) \in J_1'$ that are subject to parameter uncertainty; i.e., f_{ijt}, $(i, j, t) \in J_1'$ takes value according to a symmetric distribution with mean equal the nominal value f_{ijt} in the interval $[\bar{f}_{ijt} - \hat{f}_{ijt}, \bar{f}_{ijt} + \hat{f}_{ijt}]$.

In the model (RWGP_SSP), the uncertain parameters appear in the constraints (7) and (23)–(24). To simplify the complexity of the model (RWGP_SSP), we discuss the robust equivalent formulation of the constraints (7) and (23)–(24).

For constraint (7), we consider the following formulation:

$$\sum_{m=1}^{t}\sum_{j=1}^{J} (X_{ijt} + X_{ij(m-1)}g_{ij(m-1)} - X_{ijm}g_{ijm}) - \max_{\{S_{it}\cup n_{it}|S_{it}\subseteq J_{it}, |S_{it}|=\lfloor \Gamma_{it}\rfloor, n_{it}=J_{it}\setminus S_{it}\}}$$
$$\left\{ \sum_{m\in S_{it}} \hat{D}_{im} + (\Gamma_{it} - \lfloor \Gamma_{it} \rfloor)\hat{D}_{in_{it}} \right\} \geq 0 \tag{27}$$

However, constraint (27) still has a nonlinear term. According to the robust weighted goal programming approach, it is easy to obtain that the constraint (27) is equivalent to the following:

$$\sum_{m=1}^{t}\sum_{j=1}^{J} (X_{ijt} + X_{ij(m-1)}g_{ij(m-1)} - X_{ijm}g_{ijm}) - (\Gamma_{it}Z_{it} + \sum_{m\in J_{it}} p_{im}) \geq 0, \tag{28}$$

$$Z_{it} + p_{im} \geq \hat{D}_{im}, \forall m \in J_{it}, \tag{29}$$

$$Z_{it} \geq 0, \tag{30}$$

$$p_{im} \geq 0, \forall m \in J_{it}. \tag{31}$$

Similarly, we can equivalently transform the constraint (23) into the given below:

$$\sum_{i=1}^{I}\sum_{j=1}^{J}\sum_{t=1}^{T}P_{ijt}X_{ijt} + \sum_{i=1}^{I}\sum_{j=1}^{J}\sum_{t=1}^{T}\sum_{k=1}^{K}(P_k X'_{jtk} + V_{ijk}X_{ijtk}) + \sum_{i=1}^{I}\sum_{j=1}^{J}\sum_{t=1}^{T}O_{ijt}Y_{ijt}$$

$$+\sum_{i=1}^{I}\sum_{t=1}^{T}H_i\left(\sum_{m=1}^{t}\sum_{j=1}^{J}(X_{ijm} + X_{ij(m-1)}g_{ij(m-1)} - X_{ijm}g_{ijm}) - (\Gamma_{it}Z_{it} + \sum_{m\in J_{it}}p_{im}))\right)$$

$$+ d_1^- - d_1^+ = f_1^*, \tag{32}$$

Constraints (23.20) (23.31) is also included $\hspace{3cm}$ (33)

The formulation of constraint (24) is considered as follows:

$$\sum_{i=1}^{I}\sum_{j=1}^{J}\sum_{t=1}^{T}\bar{f}_{ijt}x_{ijt} + d_1^- - d_1^+ +$$

$$\max_{\{S_1\cup n_1|S_1\subseteq J', |S_1|=\lfloor\Gamma_1\rfloor, (i_1,j_1,t_1)\in n_1\subseteq J'\setminus S_1\}}\left\{\sum_{(i,j,t)\in S_1}\hat{f}_{ijt}x_{ijt} + (\Gamma' - \lfloor\Gamma_1\rfloor)\hat{f}_{i_1 j_1 t_1}x_{i_1 j_1 t_1}\right\} = f_1^* \tag{34}$$

According to the proposed model, we obtain that the constraint (34) is equivalent to the following:

$$\sum_{i=1}^{I}\sum_{j=1}^{J}\sum_{t=1}^{T}\bar{f}_{ijt}x_{ijt} + \sum_{(i,j,t)\in J_1}p_{ijt} + \Gamma'Z' + d_1^- - d_1^+ = f_1^*$$

$$Z' + p_{ijt} \geq \hat{f}_{ijt}x_{ijt}, \ \forall(i,j,t)\in J'_1, \tag{35}$$

$$p_{ijt} \geq 0, \ \forall(i,j,t)\in J'_1,$$

$$Z' \geq 0.$$

Therefore, (RWGP_SSP) can be equivalently transformed to the following:

$$\min Z = w_1^+ d_1^+ + w_2^+ d_2^+ + w_3^+ d_3^+$$

$$\sum_{i=1}^{I}\sum_{j=1}^{J}\sum_{t=1}^{T}P_{ijt}X_{ijt} + \sum_{i=1}^{I}\sum_{j=1}^{J}\sum_{t=1}^{T}\sum_{k=1}^{K}(P_k X'_{ijtk} + V_{ijk}X_{ijtk}) + \sum_{i=1}^{I}\sum_{j=1}^{J}\sum_{t=1}^{T}O_{ijt}Y_{ijt}+$$

$$\sum_{i=1}^{I}\sum_{t=1}^{T}H_i\left(\sum_{m=1}^{t}\sum_{j=1}^{J}(X_{ijm} + X_{ij(m-1)}g_{ij(m-1)} - X_{ijm}g_{ijm}) - \sum_{m=1}^{t}\bar{D}_{im} - \Gamma_{it}Z_{it} - \sum_{m=1}^{t}p_{im}\right)$$

$$+ d_1^- - d_1^+ = f_1^*,$$

$$\sum_{i=1}^{I}\sum_{j=1}^{J}\sum_{t=1}^{T}\bar{f}_{ijt}x_{ijt} + \sum_{i=1}^{I}\sum_{j=1}^{J}\sum_{t=1}^{T}p_{ijt} + \Gamma'Z'$$

$$+ d_2^- - d_2^+ = f_2^*,$$

$$\sum_{i=1}^{I}\sum_{j=1}^{J}\sum_{t=1}^{T}g_{ijt}X_{ijt} + d_3^- - d_3^+ = f_3^*,$$

$$\sum_{m=1}^{t}\sum_{j=1}^{J}(X_{ijm} + X_{ij(m-1)}g_{ij(m-1)} - X_{ijm}g_{ijm}) - \sum_{m=1}^{t}\bar{D}_{im} - \Gamma_{it}Z_{it} - \sum_{m=1}^{t}p_{im} \geq 0, \forall i, \forall t,$$

$$p_{im} \geq 0, \forall i, \forall m,$$

$$Z_{it}, p_{ijt}, d_i^-, d_i^+ \geq 0, \forall i, \forall j, \forall t,$$

$$Z_1 \geq 0.$$

Constraints (23.4)−(23.6) and (23.8)−(23.12) are also included.

$$\tag{36}$$

4   A Numerical Example

This section studies a multi-product and multi-period supplier selection problem. Table 1 shows the data including purchasing price, ordering cost, capacity, rate of defected items and rate of late delivered items. The maximum kinds of products are also included in Table 1. Table 2 presents the volume, weight, unit inventory cost of each product and the demand for each product in each period. The intervals that appear in Table 1 and Table 2 illustrate the range of the parameter. For instance, [25] represents that the nominal value, the best case value and the worst case value are 20, 12 and 28, respectively. Table 3 shows the data about vehicles.

Under the deterministic environment, we get the optimal solution when one objective is considered, and the objective values are $f_1^{min} = 37159.5$, $f_2^{min} = 101.8$, $f_3^{min} = 33.2$. We assume that the goal values $f_1^* = 38000$, $f_2^* = 110$, $f_3^* = 35$. There are two kinds of uncertain parameters including the rate of rejected items and product demand. From the above analysis, we can obtain that Γ' takes value in the interval $[0, |J'|]$, where $|J'|$ is 48 in this numerical example. The role of the parameter Γ_{it} and Γ' are to adjust the robustness of the proposed model against the level of conservatism of the solution. In order to compare the performance of the results, we assume that Γ' are 0, 2, 4 and

Table 1. Data of products

j	M_{jt}	p_{ijt}			O_{ijt}			C_{ijt}			f_{ijt}			g_{ijt}		
		$i=1$	$i=2$	$i=3$	$i=1$	$i=2$	$i=3$	$i=1$	$i=2$	$i=3$	$i=1$	$i=2$	$i=3$	$i=1$	$i=2$	$i=3$
1	2	20	50	70	100	300	400	100	50	30	[12, 28]	[9, 21]	[8, 16]	5	6	4
2	2	19	48	70	100	300	400	100	60	40	[12, 24]	[8, 18]	[9, 15]	7	5	4
3	2	18	48	68	50	200	500	80	60	40	[13, 23]	[7, 17]	[7, 13]	7	8	5

Table 2. Data of suppliers

I	V_i	W_i	H_i	N_{it}^{min}	N_{it}^{max}	D_{it}			
						$t=1$	$t=2$	$t=3$	$t=4$
1	10	5	2	1	3	[90, 110]	[75, 85]	[90, 110]	[75, 85]
2	20	10	5	1	3	[40, 60]	[55, 65]	[35, 45]	[48, 52]
3	50	20	10	1	3	[25, 35]	[28, 32]	[38, 42]	[28, 32]

Table 3. Data of vehicles

K	P_k	K_p	V_k	$V_{ijk}, i=1$			$V_{ijk}, i=2$			$V_{ijk}, i=3$		
				$t=1$	$t=2$	$t=3$	$t=1$	$t=2$	$t=3$	$t=1$	$t=2$	$t=3$
1	200	200	80	4	5	5	10	11	11	15	14	13
2	300	600	200	3	4	4	8	8	9	14	14	12
3	500	1000	500	2	3	3	7	7	6	12	13	11

16. We define that $\Gamma = (\Gamma_{11}, \Gamma_{12}, \Gamma_{13}, \Gamma_{14}, \Gamma_{21}, \Gamma_{22}, \Gamma_{23}, \Gamma_{24}, \Gamma_{31}, \Gamma_{32}, \Gamma_{33}, \Gamma_{34})$. The value of Γ_{it} should be determined by decision maker. To present the solution under different level of conservatism, we define three uncertain levels as $\Gamma^1 = (0,0,0,0,0,0,0,0,0), = (0.5, 1, 1.5, 0.5, 1, 1.5, 0.5, 1, 1.5)$, and $\Gamma^3 = (1, 2, 3, 1, 2, 3, 1, 2, 3)$.

We applied the model (RWGP_SSP) to the multi-product multi-period supplier selection problem considering inventory management and vehicle allocation problem in uncertain environment. The robust solutions are shown in Table 4. The results are plotted with different levels of uncertain demand on the horizontal and the goal value on the vertical axis is shown in Fig. 1. Using the proposed model, the goal value and ordering strategy are various in different uncertain conditions. From Fig. 1, we can draw the following conclusions. First, when Γ' is fixed, with the increasing of the uncertain level of demand, the objective value is increasing. Second, when Γ_{it} is fixed, the objective value is increasing with the increase of the uncertain level of product quality. Moreover, with the increase of the level of uncertainty, the larger degree of the objective value deviates from the target value.

Table 4. Solutions of the problem

Γ'	Γ_{it}	Solution	Objective value	Achieved value			Deviation value			Sat		
				Z1	Z2	Z3	Z1	Z2	Z3	Z1	Z2	Z3
0	G1	Global	0.036	41472.1	111.8	35	3472	1.8	0	No	No	Yes
	G2	Global	0.092	43138.9	121.9	36.1	5138.9	11.9	1.1	No	No	No
	G3	Global	0.125	44829.9	124.9	37.1	6829.9	14.9	2.1	No	No	No
2	G1	Global	0.079	41969.7	124.5	35	3969.7	14.5	0	No	No	Yes
	G2	Global	0.139	43241.4	135.3	36.7	5241.4	25.3	1.1	No	No	No
	G3	Global	0.172	44837.5	138.9	37.6	6837.5	28.9	2.1	No	No	No
4	G1	Global	0.101	42220.2	123.5	37.4	4220.2	13.5	2.4	No	No	No
	G2	Global	0.165	44039.7	128.9	40.7	6039.7	18.9	5.7	No	No	No
	G3	Global	0.198	45708.3	132.2	41.7	7708.3	22.2	6.7	No	No	No
16	G1	Global	0.16	41478.2	152.7	35	3478.2	42.7	0	No	No	Yes
	G2	Global	0.223	43122.5	162.2	37.1	5122.5	52.2	2.1	No	No	No
	G3	Global	0.262	44684.4	166.4	38.4	6684.4	56.4	3.4	No	No	No

Considering the existing problems, some suggestions are given. Firstly, the decision makers should carefully consider the uncertain level of the uncertain data. The uncertain level of the data has a great influence on the decision strategy, which means that an improper uncertain level may cause infeasible or meaningless decisions. Secondly, the decision makers need to be aware of the weight coefficients. The decision makers can flexible adjust the weight of each objectives based on production requirements. Thirdly, a high quality computer should be

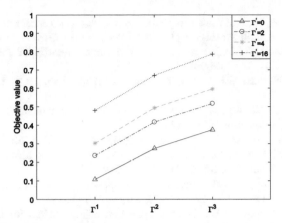

Fig. 1. Comparison of the objective value under different level of uncertain data

equipped for calculation because we established a complex multi-objective mixed integer programming which contains lots of integer variables and continuous variables.

5   Conclusion

Supplier selection problem with inventory management and vehicle allocation determines the sourcing, order allocation, and vehicle allocation over the decision horizon. Selecting the right suppliers and allocating the order to the selected suppliers become a major challenge for a decision maker. This paper presents a multi-objective mixed integer linear programming approach for supplier selection problem. In order to tackle with uncertain parameter, we introduce a robust weighted goal programming approach to obtain its robust solutions. The proposed model can determine the following problems: which suppliers should be selected, how much should be purchased from each supplier in each period, and which vehicle is suitable for delivering purchased items in an uncertain environment.

Acknowledgment. We are thankful for financial support from the National Natural Science Foundations (Grant No. 71571031).

References

1. Alfares, H.K., Turnadi, R.: Lot sizing and supplier selection with multiple items, multiple periods, quantity discounts, and backordering. Comput. Ind. Eng. **116**, 59–71 (2018)
2. Aouam, T., Brahimi, N.: Integrated production planning and order acceptance under uncertainty: A robust optimization approach. Eur. J. Oper. Res. **228**(3), 504–515 (2013)

3. Aras, N., Bilge, Ü.: Robust supply chain network design with multi-products for a company in the food sector. Appl. Math. Model. **60** (2018)
4. Basnet, C., Leung, J.M.Y.: Inventory lot-sizing with supplier selection. Comput. Oper. Res. **32**(1), 1–14 (2005)
5. Ben-Tal, A., Nemirovski, A.: Robust convex optimization. Math. Oper. Res. **23**(4), 769–805 (1998)
6. Ben-Tal, A., El Ghaoui, L., Nemirovski, A.: Robust optimization. Princeton University Press, Princeton **2**(3), 542 (2009)
7. Bertsimas, D., Sim, M.: The price of robustness. Oper. Res. **52**(1), 35–53 (2004)
8. Bertsimas, D., Brown, D.B., Caramanis, C.: Theory and applications of robust optimization. Siam Rev. **53**(3), 464–501 (2010)
9. Bojd, F.G., Koosha, H.: A robust goal programming model for the capital budgeting problem. J. Oper. Res. Soc. **69**(7), 1105–1113 (2018)
10. Charnes, A., Cooper, W.W., Ferguson, R.O.: Optimal estimation of executive compensation by linear programming. Manag. Sci. **1**(2), 138–151 (1955)
11. Chassein, A., Goerigk, M.: Compromise solutions for robust combinatorial optimization with variable-sized uncertainty. Eur. J. Oper. Res. **269**(2), 544–555 (2018)
12. Faez, F., Ghodsypour, S.H., O'Brien, C.: Vendor selection and order allocation using an integrated fuzzy case-based reasoning and mathematical programming model. Int. J. Prod. Econ. **121**(2), 395–408 (2010)
13. Ghahtarani, A., Najafi, A.A.: Robust goal programming for multi-objective portfolio selection problem. Econ. Model. **33**(33), 588–592 (2013)
14. Gorji, M.H., Setak, M., Karimi, H.: Optimizing inventory decisions in a two-level supply chain with order quantity constraints. Appl. Math. Model. **38**(3), 814–827 (2014)
15. Hanks RW, Weir JD, Lunday BJ (2017) Robust goal programming using different robustness echelons via norm-based and ellipsoidal uncertainty sets. Eur. J. Oper. Res. https://www.sciencedirect.com/science/article/abs/pii/S037722171730303X
16. Hombach, B.C.W.G.L.E.: Robust and sustainable supply chains under market uncertainties and different risk attitudes-a case study of the german biodiesel market. Eur. J. Oper. Res **269**, 302–312 (2017)
17. Kim, J., Chung, B.D., Kang, Y., Jeong, B.: Robust optimization model for closed-loop supply chain planning under reverse logistics flow and demand uncertainty. J. Clean. Prod. (2018). https://www.sciencedirect.com/science/article/pii/S0959652618318079
18. Kuchta, D.: Robust goal programming. Control. Cybern. **33**(3), 501–510 (2004)
19. Lei, L., Zabinsky, Z.B.: Incorporating uncertainty into a supplier selection problem. Int. J. Prod. Econ. **134**(2), 344–356 (2011)
20. Mansini, S.M.W.P.R.: The supplier selection problem with quantity discounts and truckload shipping. Omega **40**(4), 445–455 (2012)
21. Pan, A.C.: Allocation of order quantity among suppliers. J. Supply Chain. Manag. **25**(3), 36–39 (1989)
22. Soyster, A.L.: Technical note-convex programming with set-inclusive constraints and applications to inexact linear programming. Oper. Res. **21**(5), 1154–1157 (1973)
23. Stanley, E.D., Honig, D.P., Gainen, L.: Linear programming in bid evaluation. Nav. Res. Logist. **1**(1), 48–54 (2010)
24. Thorsen, A., Yao, T.: Robust inventory control under demand and lead time uncertainty. Ann. Oper. Res. **257**(1–2), 207–236 (2017)

25. Ventura, J.A., Valdebenito, V.A., Golany, B.: A dynamic inventory model with supplier selection in a serial supply chain structure. Eur. J. Oper. Res. **230**(2), 258–271 (2013)
26. Yua, M.C., Lin, H.C.: Fuzzy multi-objective vendor selection under lean procurement. Eur. J. Oper. Res. **219**(2), 305–311 (2012)

Pre-sale Pricing Strategy for Fresh Agricultural Products Under O2O

Wenjing Wu and Chunxiang Guo[✉]

School of Business, Sichuan University, Chengdu 610064,
People's Republic of China
gchx1@sina.com

Abstract. Fresh agricultural products are indispensable necessities in daily life. The imbalance between supply and demand leads to large price fluctuations in normal state, thus torturing producers and consumers. When fresh agricultural products are pre-sold online, we discuss the optimal pricing strategies of suppliers and retailers in different periods based on dual-channel business model. Demand uncertainty is reduced to maximize supply chain profit. Research shows that the retailer has lower optimal price under centralized strategy than that under decentralized strategy. However, the supplier has the same optimal price under the two strategies. In official and discount sales periods, the optimal pricing is closely related to online market share. Suppliers and retailers adjust market shares to achieve optimal profit of supply chain. The optimal profit of supply chain increases first, and then decreases with the increase of consumer price elasticity index. Therefore, the suppliers and retailers can adjust sales strategy to affect consumer price elasticity index, thus achieving an optimal profit of supply chain.

Keywords: O2O · Fresh agricultural products · Pre-sale · Pricing

1 Introduction

As necessaries for urban residents and important incomes for farmers, fresh agricultural products are easily decayed and damaged. At the same time, the supply chain of agricultural products has the characteristics of high supply uncertainty and high demand uncertainty. First of all, the supply chain of agricultural products is affected by natural conditions such as season and climate. The production cycle of agricultural products is relatively long, and there are more uncontrollable factors, so the uncertainty of supply is high. Secondly, the decentralization of production and consumption of agricultural products makes market information extremely dispersed, and it is difficult to grasp the comprehensive market demand, competitors and partners' information, so the uncertainty of demand is also high [16]. Sometimes, seasonal products of bumper harvest are unsalable. For fresh agricultural products, unsalable deterioration directly leads to the

© Springer Nature Switzerland AG 2020
J. Xu et al. (Eds.): ICMSEM2019 2019, AISC 1002, pp. 310–324, 2020.
https://doi.org/10.1007/978-3-030-21255-1_24

decline of farmer income and the loss of planting enthusiasm, affecting sustainable development of agriculture. With random life cycles, the demands for fresh agricultural products are affected by freshness. The loss of products (freshness) is reduced by realizing the circulation of fresh agricultural products between producers and consumers. Double uncertainty of demand and output is determined to solve production fund shortage of farmers. It is of great theoretical and practical significance to ensure the supply of urban residents, the prices, the livelihood and enthusiasm of farmers.

The rise of Internet provides a platform for rapid and effective circulation of products between producers and consumers. To meet personalized and diversified needs of consumers, the traditional supply chain changes from supply-oriented business model to demand-oriented O2O business model, such as JD's strategic investment in Fruit Day and Alibaba's investment in Yiguo. Therefore, domestic and foreign scholars focus on the supply chain innovation and design in the Internet+ era, including dual-channel circulation model.

Since Balasubramanian in 1998, the researches of dual-channel supply chain management focus on pricing decisions, channel selection, conflicts and coordination. Park and Keh [13] proved that dual-channel sales model contributes to improving profit and overall performance of supply chain and reducing retailers' profits. Chiang et al. [3] found that it is beneficial for manufacturers to adopt dual-channel sales model when customers have different preferences for online sales. In the dual-channel supply chain, channel conflicts are inevitable because of self benefit maximization. Jing et al. [5] explored the coordination contract in dual-channel supply chain when manufacturer is the leader of Stackelberg game. Webb and Kevin [17] demonstrated that channel conflicts are reduced by reasonable pricing of online direct sales channels or brand differentiation between dual channels. Meanwhile, product channel pricing is a key factor in channel conflicts. Therefore, the pricing method is an important part of dual-channel supply chain management research. Rodrìgue [15] analyzed the pricing and classification decision-making problems of supply chain when considering inventory costs in dual-channel sales model. Chen [20] discussed network and traditional channel pricing strategies for alternative products in supply chain by game theory, thus obtaining cooperation strategies between manufacturers and retailers. Li [6] studied the pricing and coordination of dual-channel supply chain with risk-averse retailers. Dual-channel supply chain coordination aims at eliminating channel conflicts and improving overall performance of dual-channel supply chain. In the context of dual-channel closed-loop supply chain, Cao [1] discussed the pricing and coordination decision of dual-channel closed-loop supply chain based on inconsistent consumer preferences for traditional retail and network direct sales channels.

With the change of consumer purchase patterns, manufacturing companies redesign the distribution channel structures, not excepting sales of agricultural products. Yu et al. [19] studied oligopoly competition of fresh agricultural product differentiation by processing the exponential time decay with arc multiplier. Zhao [21] found that a closed-loop supply chain system is built to improve value and quality, thus reducing operation cost of product. Guillermo et al. [4] discussed dynamic price competition of alternative and complementary perishable products in oligopoly market. Niu et al. [12] analyzed conditions and opportunities of contract agricultural agreement for company-peasant and farmer-cooperative-company structures. Xujin [18] compared the operation efficiencies of single farmer-supermarket and dual-channel models. It is found that the profit of cooperative increases in the dual-channel mode; the profit of supermarket decreases. Liu [7] proved that the three-stage game method has a significant integration effect on fresh agricultural products resources in dual-channel supply chain.

The above literature shows that the dual-channel circulation model contributes to breaking through regional and seasonal restrictions, increasing trading opportunities and improving the efficiency of fresh agricultural product supply chains. By solving problems such as non-circulation of commodity information and high cost of product addition, the greater value is created to generate system synergy. However, the dual-channel sales model cannot solve the imbalance between supply and demand of agricultural products, the double uncertainty of demand and output, and the shortage of funds. Meanwhile, the pre-sale strategy attracts the attention of scholars as an important mean of coordinating production operations and sales in a supply chain system. Literature on pre-sale strategies indicate certain advantages of pre-sale strategies. For example, pre-sales can predict demand in a targeted manner, obtaining partial funds in advance. Inventory risks and demand uncertainty are reduced by solving problems such as sales of agricultural products and market price stability. Indeed, dealers can predict product demand through sales data from the pre-sales process, thus reducing demand uncertainty. Especially, the consumers are attracted to participate in the discount pre-sales activities, rising dealer profits by boosting the sales.

Ashutosh [14] studied the pre-sale pricing and inventory decision-making problems of newsboy retailers. Research showed that pre-sale can reduce demand uncertainty while expanding demand. Nasiry [11] described the impact of expected regret on consumer decision-making, company profit and policy in pre-sale environment with uncertain buyer valuation. Mei [10] analyzed the effect of undisclosed pricing and lead time on the pre-sale system. Research proved that the lead time has significant effect on manufacturer's policies. Zeng [2] discussed how the compositions of experienced and under-experienced consumers affect retailers' optimal pricing strategies and profits. Yu et al. [8] explored the effect of product quality information disclosure on pre-sale strategies. Mao [9] compared joint strategy of pre-sale and repurchase with single pre-sale strategy by considering the strategy-based consumer behavior. The dealer income models under two strategies are constructed to obtain the corresponding optimal pre-sale price and optimal order quantity.

This paper applies the dual-channel sales model to the sale of fresh agricultural products, realizing the online and offline sales of fresh agricultural products. At the same time, online marketing channel can break through the geographical and time constraints, so the use of online pre-sale can not only promote the sale of agricultural products, but also effectively solve the imbalance between supply and demand of agricultural products. Freshness is an important basis for consumers to purchase agricultural products. Therefore, the work discussed the pre-sale pricing strategy of fresh products under O2O mode based on the impact of freshness on prices and demands. Dual-channel demand function was constructed according to exponential decay of freshness with time. Meanwhile, the optimal prices were discussed under decentralized and centralized decisions, respectively. Supply chain members were assisted to control demand uncertainty, thus guiding reasonable planting of farmers.

2   Problem Description

(1) The product sales are divided into three stages: the pre-sale, official and discount sales periods. The pre-sale period is carried out before the product is planted.

(2) The supply chain consists of one supplier and one retailer. The former is responsible for online sales; the latter for offline sales.

(3) Pre-sales are only carried out online. The official sales period is divided into online and offline sales. The discount sales period is only conducted offline.

(4) The supply chain only produces a fresh agricultural product, which is pre-sold during the production period and delivered at the beginning of official sales period.

(5) During the official sales period, the products ordered online by the consumers are delivered immediately after the order is placed.

(6) The logistics in the supply chain is provided by the supplier. The retailer's order lead time is 0.

(7) During the pre-sale period, the supplier and the retailer have no inventory cost. The supplier only has the production cost. During the official sales period, the supplier has inventory, loss and transportation costs. The retailer has inventory and loss costs.

As a perishable product, fresh agricultural product has the problem of freshness loss. The product should be sold during the shelf life. In Fig. 1, $t_3 - t_1$ is the length of product shelf life. As a necessity for life, fresh agricultural product has relatively stable demand in the absence of emergencies. It is related to price and freshness. Fresh agricultural product is the freshest at the end of pre-sale period. It is denoted that θ_0 is the freshness at this time; $\theta(t)$ the freshness decay function. There is an initial freshness before product transportation and sales. The freshness is considered to be 0 at the end of shelf life. Therefore, the freshness decay function is constructed as a piecewise function (Table 1).

$$\theta t = \begin{cases} \theta_0 (t = t_1) \\ \theta_0{}^{t-t_1} (t_1 < t < t_3) \\ 0 (t = t_3). \end{cases}$$

Thus, the initial freshness θ_0 also reflects the decay rate of agricultural product. The loss function is expressed as $\lambda(t) = -\ln\theta_0 \cdot \theta_0{}^{t-t_1} (t_1 \leq t \leq t_3)$.

Table 1. Model parameters and meaning

Parameter	Meaning
c	Unit product production cost
s	Unit product loss processing cost
w	Supplier's wholesale price
g	Unit product transportation cost
h	Unit inventory cost
t_1	Pre-sale period
t_2	Official sales period
t_3	Discount sales period
α	Market share of online channels during official sales period $(0 < \alpha < 1)$
θ_0	Initial freshness $(0 < \theta_0 < 1)$
$\theta(t)$	Freshness function of fresh agricultural product
$\lambda(t)$	Freshness loss function
$\iota(t)$	Retailer's inventory level at t
k	Price demand elasticity, $k > 1$
Q_P	Supplier's production
Q_s	Retailer's order quantity
I_P	Supplier's inventory
I_s	Retailer's inventory
d_j	Market size in each period $(j = 1, 2$ and 3; 1 represents the pre-sale period; 2 the official sales period; 3 the discount sales period)
D_j	Product demand in each period $(j = 1, 2$ and $3)$
Π	Total profit in supply chain
Decision variables	Meaning
P_g	Product price in each period $(g = 1, 2, 3$ and 4; 1 represents the unit price of product in pre-sale period product; 2 the unit price of product in official sales period; 3 the unit price of product in official sales period; 4 the unit price of product in discount period)

3 Model Construction

3.1 Demand Functions in Different Periods

(1) Demand function in pre-sale period

In the pre-sale period, the freshness is θ_0, and the consumer mainly considers the price factor. Based on the market demand rate function, we have the following function.

$$D_1 = \int_0^{t_1} d_1 p_1^{-k} dt \tag{1}$$

(2) Demand function in official sales period

$$D_{2u} = \int_{t_1}^{t_2} \alpha d_2 p_2^{-k} \theta_0^{t-t_1} dt \tag{2}$$

$$D_{2d} = \int_{t_1}^{t_2} (1-\alpha) d_2 p_3^{-k} \theta_0^{t-t_1} dt \tag{3}$$

(3) Demand function in discount sales period

$$D_3 = \int_{t_2}^{t_3} d_3 p_4^{-k} \theta_0^{t-t_1} dt \tag{4}$$

3.2   Retailer Correlation Functions

Retailers have the products in stock from official to discount sales period. Based on the above analysis and hypothesis, the metamorphic inventory model is used to derive the inventory model of fresh agricultural products with the loss of freshness. In Period (t_1, t_2), the change in retailer inventory can be expressed as follows.

$$\frac{dI_{s_1}(t)}{dt} - In\theta_0\theta_0^{t-t_1} I_{s_1}(t) = -(1-\alpha) d_2 p_3^{-k} \theta_0^{t-t_1}$$

The solution of the equation is:

$$I_{s_1}(t) = e^{\theta_0^{t-t_1}} \left[\alpha + (1-\alpha) d_2 p_3^{-k} \frac{e^{-\theta_0^{t-t_1}}}{In\theta_0} \right]$$

Since there is continuity between the two stages, Constant a can be calculated by $I_{s_1}(t_2) = I_{s_2}(t_2)$. In Period (t_2, t_3), the change of retailer inventory can be expressed as follows.

$$\frac{dI_{s_2}(t)}{dt} - In\theta_0\theta_0^{t-t_1} I_{s_2}(t) = -d_3 p_4^{-k} \theta_0^{t-t_1}$$

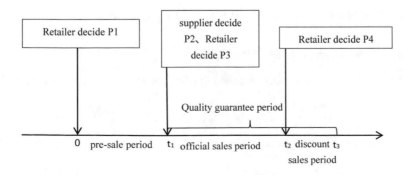

Fig. 1. Event sequence

The solution of the equation is

$$I_{s_2}(t) = e^{\theta_0^{t-t_1}}\left(b + d_3 p_4^{-k}\frac{e^{-\theta_0^{t-t_1}}}{In\theta_0}\right)$$

Constant b is calculated by $I_{s_2}(t_3)=0$.

$$b = -d_3 p_4^{-k}\frac{e^{-\theta_0^{t_3-t_1}}}{In\theta_0}$$

The final equation is expressed as follows.

$$I_{s_2}(t) = \frac{d_3 p_4^{-k}}{In\theta_0}\left(1-e^{\theta_0^{t-t_1}-\theta_0^{t_3-t_1}}\right)\quad(t_1 \le t \le t_3)$$

It is substituted into $t = t_2$ to obtain

$$I_{s_2}(t_2) = \frac{d_3 p_4^{-k}}{In\theta_0}\left(1-e^{\theta_0^{t_2-t_1}-\theta_0^{t_3-t_1}}\right)$$

Using $I_{s_1}(t_2) = I_{s_2}(t_2)$, $I_{s_1}(t)$ can be expressed as

$$I_{s_1}(t) = \frac{1}{In\theta_0}\left[(1-\alpha)d_2 p_3^{-k} + d_3 p_4^{-k}(e^{\theta_0^{t-t_1}-\theta_0^{t_2-t_1}} - e^{\theta_0^{t-t_1}-\theta_0^{t_3-t_1}})\right.$$
$$\left. -(1-\alpha)d_2 p_3^{-k}e^{\theta_0^{t-t_1}-\theta_0^{t_2-t_1}}\right]\quad(t_1 \le t \le t_2) \tag{5}$$

The order quantity of retailer is expressed as

$$Q_s = I_{s_1}(t_1) = \frac{1}{In\theta_0}\left[(1-\alpha)d_2 p_3^{-k} + d_3 p_4^{-k}(e^{1-\theta_0^{t_2-t_1}} - e^{1-\theta_0^{t_3-t_1}})\right.$$
$$\left. -(1-\alpha)d_2 p_3^{-k}e^{1-\theta_0^{t_2-t_1}}\right] \tag{6}$$

Retailer loss in Period 1 $= s\int_{t_1}^{t_2} I_{s_1}(t)\lambda(t)dt$

Retailer loss cost in Period 2 $= s\int_{t_2}^{t_3} I_{s_2}(t)\lambda(t)dt$

Retailer inventory cost in Period 1 $= \frac{h}{2}\int_{t_1}^{t_2} I_{s_1}(t)dt$

Retailer inventory cost in Period 2 $= s\int_{t_1}^{t_2} I_{s_1}(t)\lambda(t)dt$.

3.3 Supplier Correlation Functions

In Period $(t1, t2)$, the rate of change in supplier inventory is expressed as follows.

$$\frac{dI_p(t)}{dt} - In\theta_0\theta_0^{t-t_1}I_p(t) = -\alpha d_2 p_2^{-k}\theta_0^{t-t_1}$$

The solution of the equation is

$$I_p(t) = e^{\theta_0^{t-t_1}}\left(c + \alpha d_2 p_2^{-k}\frac{e^{-\theta_0^{t-t_1}}}{In\theta_0}\right)$$

The initial condition $I(0) = Q_p - D_1, I(t_2) = 0$ is substituted to obtain c. Then the solution is expressed as

$$I_p(t) = \frac{\alpha d_2 p_2^{-k}}{In\theta_0} \left(1 - e^{\theta_0^{t-t_1} - \theta_0^{t_2-t_1}}\right) \quad (t_1 \le t \le t_2) \tag{7}$$

Supplier output is

$$Q_p = I_p(t_1) + D_1 + Q_s$$

$$= \frac{\alpha d_2 p_2^{-k}}{In\theta_0} \left(1 - e^{1 - \theta_0^{t_2-t_1}}\right) + d_1 p_1^{-k} t_1 + \frac{1}{In\theta_0}$$

$$\times \left[(1-\alpha)d_2 p_3^{-k} + d_3 p_4^{-k}(e^{1-\theta_0^{t_2-t_1}} - e^{1-\theta_0^{t_3-t_1}}) - (1-\alpha)d_2 p_3^{-k} e^{1-\theta_0^{t_2-t_1}} \right] \tag{8}$$

Supplier loss cost $= s \int_{t_1}^{t_2} I_p(t)\lambda(t)dt$

Supplier inventory cost $= s \int_{t_1}^{t_2} I_p(t)\lambda(t)dt$.

3.4   Profit Functions

(1) Supplier profit function
The supplier income consists of the pre-sales, official sales and retailer purchase incomes. The expenditure consists of production, inventory, loss transportation and cost.

$$\Pi_p = D_1 p_1 + D_{2u} p_2 + Q_s w - (g+c)Q_p - s \int_{t_1}^{t_2} I_p(t)\lambda(t)dt - \frac{h}{2} \int_{t_1}^{t_2} I_p(t)dt \tag{9}$$

(2) Retailer profit function
The retailer income consists of official and discount sales incomes. The expenditure is composed of purchase, inventory and loss costs.

$$\Pi_s = D_3 p_4 + D_{2d} p_3 - s \int_{t_1}^{t_2} I_{s_1}(t)\lambda(t)dt - s \int_{t_2}^{t_3} I_{s_2}(t)\lambda(t)dt - \frac{h}{2} \int_{t_1}^{t_2} I_{s_1}(t)dt - \frac{h}{2} \int_{t_2}^{t_3} I_{s_2}(t)dt - Q_s w \tag{10}$$

(3) Supply chain profit function
In a centralized decision-making environment, the supply chain system income includes the online product sales incomes of supplier and retailer. The expenditure consists of the inventory, loss, production and transportation costs of supplier and retailer.

$$\Pi = D_1 p_1 + D_{2u} p_2 + D_{2d} p_3 + D_3 p_4 - (g+c)Q_p - s \int_{t_1}^{t_2} I_{s_1}(t)\lambda(t)dt$$

$$- s \int_{t_2}^{t_3} I_{s_2}(t)\lambda(t)dt - \frac{h}{2} \int_{t_1}^{t_2} I_{s_1}(t)dt - \frac{h}{2} \int_{t_2}^{t_3} I_{s_2}(t)dt - s \int_{t_1}^{t_2} I_p(t)\lambda(t)dt$$

$$- \frac{h}{2} \int_{t_1}^{t_2} I_p(t)dt \tag{11}$$

4 Optimal Pricing Decision

4.1 Optimal Pricing Under Decentralized Decision

(1) Optimal pricing decision of retailer

For the retailer, and are the decision variables. The first and second order partial derivatives of $\prod_s$ (about p_3 and p_4) are solved to maximize the retailer profit, respectively.

If
$$\frac{\theta_0^{t_2-t_1}-1}{In\theta_0}(k-1)p_3 + s\int_{t_1}^{t_2}(1+k)\theta_0^{t-t_1}e^{\theta_0^{t-t_1}-\theta_0^{t_2-t_1}}dt + (k+1)\frac{w}{In\theta_0}[e^{1-\theta_0^{t_2-t_1}} -$$
$$1] - (1+k)\frac{h}{2In\theta_0}[(t_2-t_1) + \int_{t_1}^{t_2}e^{\theta_0^{t-t_1}-\theta_0^{t_2-t_1}}dt] + \frac{s(k+1)(\theta_0^{t_2-t_1}-1)}{In\theta_0} < 0, \text{ then}$$
$$\frac{\partial^2\prod_s}{\partial p_3^2} < 0.$$

Let $\frac{\partial\prod_s}{\partial p_3} = 0$, and we can obtain

$$p_3^* = \frac{In\theta_0}{(\theta_0^{t_2-t_1}-1)(1-k)(1-\alpha)d_2}$$
$$\left\{ \begin{array}{l} -s\left[\frac{-k(\theta_0^{t_2-t_1}-1)(1-\alpha)d_2}{In\theta_0} + \int_{t_1}^{t_2}k(1-\alpha)\theta_0^{t-t_1}d_2e^{\theta_0^{t-t_1}-\theta_0^{t_2-t_1}}dt\right] \\ +\frac{h}{2In\theta_0}\left[-k(1-\alpha)(t_2-t_1)d_2 + k(1-\alpha)d_2\int_{t_1}^{t_2}e^{\theta_0^{t-t_1}-\theta_0^{t_2-t_1}}dt\right] \\ +\frac{w}{In\theta_0}\left[-k(1-\alpha)d_2 + k(1-\alpha)d_2e^{1-\theta_0^{t_2-t_1}}\right] \end{array} \right\} \quad (12)$$

If
$$\frac{\theta_0^{t_3-t_1}-\theta_0^{t_2-t_1}}{In\theta_0 k+1}d_3p_4^{-k-1} + \frac{h}{2In\theta_0}d_3p_4^{-k-2}\int_{t_1}^{t_2}(e^{\theta_0^{t-t_1}-\theta_0^{t_3-t_1}} - e^{\theta_0^{t-t_1}-\theta_0^{t_2-t_1}})dt +$$
$$\frac{h}{2In\theta_0}\int_{t_2}^{t_3}e^{\theta_0^{t-t_1}-\theta_0^{t_3-t_1}}dt - \frac{h}{2In\theta_0}d_3p_4^{-k-2}(t_3-t_2) - \frac{w}{In\theta_0}d_3p_4^{-k-2}\int_{t_1}^{t_2}(e^{1-\theta_0^{t_2-t_1}} -$$
$$e^{1-\theta_0^{t_3-t_1}})dt + sd_3p_4^{-k-2}\int_{t_1}^{t_2}\theta_0^{t-t_1}(e^{\theta_0^{t-t_1}-\theta_0^{t_2-t_1}} - e^{\theta_0^{t-t_1}-\theta_0^{t_3-t_1}})dt +$$
$$sd_3p_4^{-k-2}(\frac{\theta_0^{t_3-t_1}-\theta_0^{t_2-t_1}}{In\theta_0} - \int_{t_2}^{t_3}\theta_0^{t-t_1}e^{\theta_0^{t-t_1}-\theta_0^{t_3-t_1}}dt) < 0, \text{ then } \frac{\partial^2\prod_s}{\partial p_4^2} < 0.$$

Let $\frac{\partial\prod_s}{\partial p_4} = 0$, we can obtain

$$p_4^* = \frac{In\theta_0}{(\theta_0^{t_3-t_2}-\theta_0^{t_2-t_1})(1-k)d_3}$$
$$\left\{ \begin{array}{l} ks\int_{t_1}^{t_2}\theta_0^{t-t_1}d_3(e^{\theta_0^{t-t_1}-\theta_0^{t_2-t_1}} - e^{\theta_0^{t-t_1}-\theta_0^{t_3-t_1}})dt \\ +sd_3k(\frac{\theta_0^{t_3-t_1}-\theta_0^{t_2-t_1}}{In\theta_0} - \int_{t_2}^{t_3}\theta_0^{t-t_1}e^{\theta_0^{t-t_1}-\theta_0^{t_3-t_1}}dt) \\ -\frac{hk}{2In\theta_0}d_3\left[\int_{t_1}^{t_2}(e^{\theta_0^{t-t_1}-\theta_0^{t_2-t_1}} - e^{\theta_0^{t-t_1}-\theta_0^{t_3-t_1}})dt\right. \\ \left.+(t_3-t_2-\int_{t_2}^{t_3}e^{\theta_0^{t-t_1}-\theta_0^{t_3-t_1}}dt)\right] - \frac{wk}{In\theta_0}d_3 \\ (e^{1-\theta_0^{t_2-t_1}} - e^{1-\theta_0^{t_3-t_1}}) \end{array} \right\} \quad (13)$$

Proposition 1 Based on decentralized decision and certain conditions, the retailer profit function is a convex function of the unit price in official sales period and the unit price in discount sales period. The convex function achieves the maximums at the inflection points and, respectively.

(2) Optimal pricing decision of supplier

For the supplier, p_1 and p_2 are the decision variables. The first and second order partial derivatives of $\prod_p$ (about p_1 and p_2) are solved to maximize the supplier profit, respectively.

If $\frac{(\theta_0{}^{t_2-t_1}-1)(k-1)}{In\theta_0(k+1)}p_2 - \frac{(c+g)}{In\theta_0}(1 - e^{\theta_0{}^{-t_1}-\theta_0{}^{t_2-t_1}}) + \frac{h}{2In\theta_0}\int_{t_1}^{t_2} e^{\theta_0{}^{t-t_1}-\theta_0{}^{t_2-t_1}}\,dt$

$+s(\frac{\theta_0{}^{t_2-t_1}-1}{In\theta_0} - \int_{t_1}^{t_2}\theta_0{}^{t-t_1}e^{\theta_0{}^{t-t_1}-\theta_0{}^{t_2-t_1}}\,dt)$, then $\frac{\partial^2\prod_p}{\partial p_2^2} < 0$.

Let $\frac{\partial\prod_p}{\partial p_2} = 0$, and we can obtain

$$p_2^* = \frac{In\theta_0}{(\theta_0^{t_2-t_1}-1)(1-k)}$$

$$\left\{ \begin{array}{l} \frac{-k(c+g)}{In\theta_0}(1 - e^{\theta_0{}^{-t_1}-\theta_0^{t_2-t_1}}) \\ +sk(\frac{\theta_0^{t_2-t_1}-1}{In\theta_0} - \int_{t_1}^{t_2}\theta_0^{t-t_1}e^{\theta_0^{t-t_1}-\theta_0^{t_2-t_1}}\,dt) - \frac{kh}{2In\theta_0} \\ (t_2 - t_1 - \int_{t_1}^{t_2} e^{\theta_0^{t-t_1}-\theta_0^{t_2-t_1}}\,dt \end{array} \right\} \quad (14)$$

Proposition 2 Based on decentralized decision and certain conditions, the supplier profit is a convex function of the unit price p_2 in official sales period, achieving the maximum at the inflection point p_2^*.

The second order partial derivative of $\prod_p$ (about p_1) is solved to obtain the following inequality.

$$\frac{\partial^2\prod_p}{\partial p_1{}^2} = t_1 d_1 k(k-1)p_1^{-k-1} - t_1 d_1(c+g)(k+1)kp_1^{-k-2} < 0$$

It is denoted that the first order derivative is equal to 0. Then,

$$p_1^* = \frac{(c+g)k}{(k-1)} \quad (15)$$

Proposition 3 Based on decentralized decision and certain conditions, the supplier profit is a convex function of the unit price p_1 in official sales period, achieving the maximum at the inflection point p_1^*.

4.2   Optimal Pricing Under Centralized Decision

The first and second order partial derivatives of Π (about p_1, p_2, p_3 and p_4) are solved to maximize the profit of supply chain. It is denoted that the first order partial derivative is equal to 0; the second order partial derivative is less than 0.

$$p_1^* = \frac{(c+g)k}{(k-1)} \quad (16)$$

$$p_2^* = \frac{In\theta_0}{(\theta_0^{t_2-t_1}-1)(1-k)}$$

$$\left\{ \begin{array}{l} \frac{-k(c+g)}{In\theta_0}(1 - e^{\theta_0^{-t_1}-\theta_0^{t_2-t_1}}) \\ +sk(\frac{\theta_0^{t_2-t_1}-1}{In\theta_0} - \int_{t_1}^{t_2}\theta_0^{t-t_1}e^{\theta_0^{t-t_1}-\theta_0^{t_2-t_1}}\,dt) - \frac{kh}{2In\theta_0} \\ (t_2 - t_1 - \int_{t_1}^{t_2} e^{\theta_0^{t-t_1}-\theta_0^{t_2-t_1}}\,dt \end{array} \right\} \quad (17)$$

$$p_3^* = \frac{In\theta_0}{(\theta_0^{t_2-t_1}-1)(1-k)(1-\alpha)d_2}$$

$$\left\{ \begin{array}{l} -s\left[\dfrac{-k(\theta_0^{t_2-t_1}-1)(1-\alpha)d_2}{In\theta_0}\right.\\[2mm] \left.+\int_{t_1}^{t_2}k(1-\alpha)\theta_0^{t-t_1}d_2e^{\theta_0^{t-t_1}-\theta_0^{t_2-t_1}}dt\right]\\[2mm] +\dfrac{h}{2In\theta_0}\left[-k(1-\alpha)(t_2-t_1)d_2\right.\\[2mm] \left.+k(1-\alpha)d_2\int_{t_1}^{t_2}e^{\theta_0^{t-t_1}-\theta_0^{t_2-t_1}}dt\right] \end{array} \right\} \qquad (18)$$

$$p_4^* = \frac{In\theta_0}{(\theta_0^{t_3-t_2}-\theta_0^{t_2-t_1})(1-k)d_3}$$

$$\left\{ \begin{array}{l} ks\int_{t_1}^{t_2}\theta_0^{t-t_1}d_3(e^{\theta_0^{t-t_1}-\theta_0^{t_2-t_1}}-e^{\theta_0^{t-t_1}-\theta_0^{t_3-t_1}})dt\\[2mm] +sd_3k(\dfrac{\theta_0^{t_3-t_1}-\theta_0^{t_2-t_1}}{In\theta_0}-\int_{t_2}^{t_3}\theta_0^{t-t_1}e^{\theta_0^{t-t_1}-\theta_0^{t_3-t_1}}dt)\\[2mm] -\dfrac{hk}{2In\theta_0}d_3\left[\int_{t_1}^{t_2}(e^{\theta_0^{t-t_1}-\theta_0^{t_2-t_1}}-e^{\theta_0^{t-t_1}-\theta_0^{t_3-t_1}})dt\right.\\[2mm] \left.+(t_3-t_2-\int_{t_2}^{t_3}e^{\theta_0^{t-t_1}-\theta_0^{t_3-t_1}}dt)\right] \end{array} \right\} \qquad (19)$$

Proposition 4 Based on centralized decision and certain conditions, the supplier profit function is a convex function of p_1 and p_2; the retailer profit function is a convex function of p_3 and p_4. There is a unique optimal solution of price variable to maximize profit.

Demonstration:

When $\frac{\theta_0^{t_2-t_1}-1}{In\theta_0}(k-1)p_3+s\int_{t_1}^{t_2}(1+k)\theta_0^{t-t_1}e^{\theta_0^{t-t_1}-\theta_0^{t_2-t_1}}dt-(1+k)\frac{h}{2In\theta_0}[(t_2-t_1)$

$+\int_{t_1}^{t_2}e^{\theta_0^{t-t_1}-\theta_0^{t_2-t_1}}dt]+\frac{s(k+1)(\theta_0^{t_2-t_1}-1)}{In\theta_0} < 0$, then, $\frac{\partial^2\Pi_s}{\partial p_3^2} < 0$ make $\frac{\partial\Pi_s}{\partial p_3} = 0$
we can obtain p_3^*.

Similarly, the remaining conditions satisfying optimal price are available.

5   Case Analysis

p_1, p_2, p_3 and p_4 are the optimal prices of supply chain system in centralized decision-making environment; p_1 (p_1'), p_2', p_3' and p_4' the optimal prices in decentralized decision-making environment; π_1 and π_2 the profits of supply chain system in centralized and decentralized conditions.

The other parameters are set as follows: $d_1 = 5$; $d_2 = 8$; $d_3 = 2$; $t_1 = 3$; $t_2 = 4$; $t_3 = 1$; $g = 0.5$; $w = 3$. For centralized and decentralized supply chains, the optimal pricing of pre-sale product is the same as that of online product in official sales period. Therefore, the curves are merged.

In Fig. 2, the optimal price and system profit of centralized and decentralized decisions increase with the increase of initial freshness. The initial freshness θ_0 reflects the decay rate of agricultural product freshness. At the end of the pre-sale period, the product is the freshest. Therefore, p_1 (p_1') does not change with the initial freshness. The optimal prices rise with the increase of initial

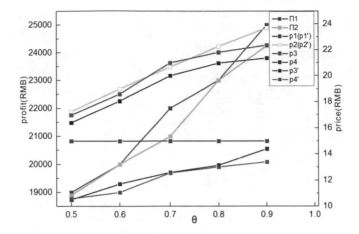

Fig. 2. Event sequence

freshness except for p_1 (p_1'). However, the upward trend gradually slows down. As the initial freshness rises, there is large investment in initial cost and rapid rise in price. In the later period, the cost increment slows down. Therefore, the price growth rate decreases. Figure 1 shows that the faster freshness decay rate leads to the lower optimal pricing of agricultural products in dual-channel mode. Meanwhile, the optimal price under decentralized decision is higher than that under centralized decision when the price fluctuates with initial freshness.

In Fig. 3, p_1 is irrelevant to the online channel market share; p_2 increases first and then stabilizes; p_3 stabilizes after decrease; p_4 shows a downward trend, with small fluctuation. Supply chain profit increases first, and then decreases with online market share. In the example, the supply chain profit reaches maximum when the online market share is 0.4. Therefore, the size of α has a great impact on supply chain profit.

In Fig. 4, the price overall decreases with the increases of price demand elasticity index. However, the decline is in the slow-fast-slow speed. When the price demand elasticity index has little fluctuation, the consumers are not sensitive to prices. Product prices cannot be reduced by supplier or retailer. As the sensitivity of consumer increases, the price largely decreases to ensure sales and interests of all parties. In the later period, the product price eventually tends to a lower limit due to production and cost constraints. In the example, the profit of the entire supply chain increases first, and then decreases with the increase of price elasticity index. The supply chain profit reaches maximum when the price elasticity index is around 1.4.

6   Conclusions

In this work, consumer demand function model was established according to the freshness and price of fresh agricultural products. When fresh agricultural products were pre-sold online, we discussed the optimal pricing strategies of suppliers

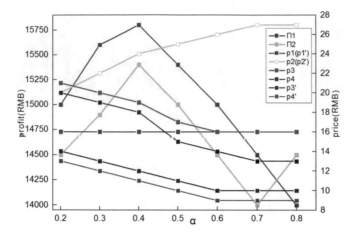

Fig. 3. Event sequence

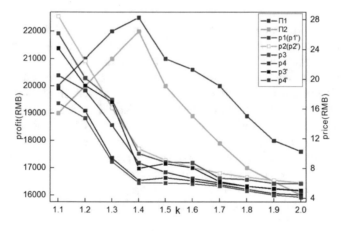

Fig. 4. Event sequence

and retailers in different periods. Research showed that the retailer had lower optimal price under centralized strategy than that under decentralized strategy. However, the supplier had the same optimal price under the two strategies. In official and discount sales periods, the optimal pricing was closely related to online market share. Suppliers and retailers adjusted market shares to achieve optimal profit of supply chain. The optimal profit of supply chain increased first, and then decreased with the increase of consumer price elasticity index. Therefore, the suppliers and retailers can adjust sales strategy to affect consumer price elasticity index, thus achieving an optimal profit of supply chain. Based on the double-channel sales of fresh agricultural products and pre-sale strategy, we can further explore the uncertainty of consumers' valuation of agricultural products in the future.

Acknowledgements. Many people have made invaluable contributions, both directly and indirectly to my research. I would like to express my warmest gratitude to Chunxiang Guo, my supervisor, for her instructive suggestions and valuable comments on the writing of this thesis. Without her invaluable help and generous encouragement, the present thesis would not have been accomplished. Besides, I wish to thank my colleague at Sichuan University, who helped me search for reference. My heart swells with gratitude to all the people who helped me.

References

1. Cao, X., Zheng, B., Wen, H.: Pricing and coordination decision of the dual channel closed-loop supply chain considering the customer preference. Chin. J. Manag. Sci. (2015)
2. Zeng, C.: Optimal advance selling strategy under price commitment. Pac. Econ. Rev. **2**, 233–258 (2013)
3. Chiang, W.Y.K., Chhajed, D., Hess, J.D.: Direct marketing, indirect profits: a strategic analysis of dual-channel supply-chain design (2003)
4. Gallego, G., Hu, M.: Dynamic pricing of perishable assets under competition (2014)
5. Jing, C., Hui, Z., Ying, S.: Implementing coordination contracts in a manufacturer Stackelberg dual-channel supply chain. Omega **40**(5), 571–583 (2012)
6. Li, B., Hou, P., Chen, P., Li, Q.: Pricing strategy and coordination in a dual channel supply chain with a risk-averse retailer. Int. J. Prod. Econ. **178**, 154–168 (2016)
7. Liu, Z., Yu, Z., Zhang, S.: Three stage game research of dual-channel supply chain of fresh agricultural products under consumer preference. Int. J. Comput. Sci. Math. **9**(1), 48–57 (2018)
8. Yu, M., Ahn, H., Kapuscinski, R.: Rationing capacity in advance selling to signal quality (2015)
9. Mao, Z., Liu, W., Hui, L.: Joint strategy of advance-selling and buy-back for seasonal perishable products. J. Manag. Sci. China (2016)
10. Mei, W., Li, D., et al.: The effects of an undisclosed regular price and a positive leadtime in a presale mechanism. Eur. J. Oper. Res. **250**(3), 1013–1025 (2016)
11. Nasiry, J., Popescu, I.: Advance selling when consumers regret. Manag. Sci. **58**(6), 1160–1177 (2012)
12. Niu, B., Jin, D., Pu, X.: Coordination of channel members efforts and utilities in contract farming operations. Eur. J. Oper. Res. **255**(3), 869–883 (2016)
13. Park, S., Keh, H.: Modelling hybrid distribution channels: a game-theoretic analysis. J. Retail. Consum. Serv. **10**(3), 155–167 (2003)
14. Prasad, A., Stecke, K., Zhao, X.: Advance selling by a newsvendor retailer. SSRN Electron. J. (2009)
15. Rodríguez, B., Aydin, G.: Pricing and assortment decisions for a manufacturer selling through dual channels. Eur. J. Oper. Res. **242**(3), 901–909 (2015)
16. Wagner, S., Bode, C.: An empirical investigation into supply chain vulnerability. J. Purch. Supply Manag. **12**(6), 301–312 (2006)
17. Webbmbabs, K.: Understanding hybrid channel conflict: a conceptual model and propositions for research. J. Bus.-to-Bus. Mark. **4**(1), 39–78 (1997)
18. Xu-jin, P., De-long, J.: The operational efficiency measurement of agro-food supply chains: the singlefarmer-supermarket direct purchase vs. dual channel. Chin. J. Manag. Sci. (2017)
19. Yu, M., Nagurney, A.: Competitive food supply chain networks with application to fresh produce. Eur. J. Oper. Res. **224**(2), 273–282 (2013)

20. Chen, Y., Fang, S., Wen, U.: Pricing policies for substitutable products in a supply chain with internet and traditional channels. Eur. J. Oper. Res. **224**, 542–551 (2013)
21. Zhao, B., Zhang, X.: Research on construction of green agriculture products closed supply chain from the perspective of model differences. Adv. Artif. Intell. (2011)

Climatic Changes and Surface Water Quality on Republic of Moldovas Territory

Gheorghe Duca[1]([✉]), Maria Nedealcov[2], Viorica Gladchi[3], and Serghei Travin[4]

[1] Department of Physical and Inorganic Chemistry, Institute of Chemistry, 2028
Chisinau, Republic of Moldova
ggduca@gmail.com
[2] Climatology and Environmental Risks Laboratory, Institute Ecology and
Geography, 2028 Chisinau, Republic of Moldova
[3] Chemistry and Chemical Technology Faculty, Moldova State University, 2071
Chisinau, Republic of Moldova
[4] Chemical Safety Problems Department, Semenov Institute of Chemical Physics of
Russian Academy of Sciences, Moscow 119991, Russian Federation

Abstract. The current state of global warming requires the expansion
of the overall knowledge as well as regional particularities of climate
change. The study finds that the Republic of Moldova is facing an accelerated pace of climate change that contributes to a more intense aridis-
ation process and heat waves, with long periods of manifestation in the
last years and argues that adaptation to the effects of climate change
must be an important element for national policies and actions. Also,
the authors establish that for Moldova climate change is a new challenge
for water quality that requires great attention and a distinct scientific
approach, which implies the participation of various domains of science.
In this context, the study proposes an integrated knowledge based approach towards adapting to climate change effects in a broad context of
natural and social systems, which incorporate scientific, legal, economic
and ecological measures. The authors propose the development of plans
and actions based on knowledge transfer and new technologies in order to
achieve a better protection and adequate management of water resources
in vulnerable areas.

Keywords: Climate change · Climate system · Natural ecosystems ·
Surface water · Water quality

1 Introduction

The issue of climate change is of a crucial importance and is one of the great
challenges of mankind, which we will have to deal with both now and in the
near future. Increasing temperatures, melting glaciers, droughts and more frequent floods are signs that climate change is really happening. It is also believed

© Springer Nature Switzerland AG 2020
J. Xu et al. (Eds.): ICMSEM2019 2019, AISC 1002, pp. 325–334, 2020.
https://doi.org/10.1007/978-3-030-21255-1_25

that, due to human activities, high concentrations of greenhouse gases in the atmosphere intensify the "greenhouse effect" naturally, thus increasing the temperature of the Earth. Greenhouse gas concentrations, especially carbon dioxide (CO_2) increased by 22% compared to 1970. In the last century, temperature in Europe rose by almost 1 °C, this occurrence is faster than the world average and is the fastest growing in the last 50 years. Climatic conditions have become variable: temperatures have reached extreme values, in northern Europe rainfall and snowfall occurrences have increased significantly, causing more floods. In contrast, in southern Europe, rainfall has declined considerably and droughts are more common. Economic losses caused by extreme weather events have increased significantly in recent decades. Due to the constant accumulation of gases in the atmosphere, these climate changes will probably continue for decades, even if emissions would stop at this point in time. The impact of climate change will progressively increase globally. Although worldwide efforts are undertaken to reduce greenhouse gas emissions, the global average temperature continues to increase, which makes the most urgent adaptation to the effects of climate change measures an absolute necessity [9].

The 5th Global Climate Change Assessment Report (AR5), prepared by the Intergovernmental Panel on Climate Change (IPCC), presents comprehensively the latest scientific results and observations on the causes of climate change and on the long-term impact in short, medium and long terms. The report contains analysis of different options for adaptation to climate change and emission reduction, including interdependencies specific to sustainable society development, taking into account the socio-economic and scientific relevance in the long term [7, 12].

2   Literature Review

The relevancy of studying future climatic conditions as well as the correlation between climate change and water resources, including their sustainability, is enforced by the opinions and studies of various scholars and practitioners.

In this respect, we can see that there are clear changes in studying local impacts on evapotranspiration, mean, low and high runoff and snow water equivalent between a 1.5, 2 and 3 °C degree warmer world. In a warmer world, the hydrological impacts of climate change are more intense and spatially more extensive [4]. Climatic projections for the nearest future years, especially in countries with heightened aridity, reveal a significant increase in temperature even for the next years [10]. Adaptation to the effects of climate change must be an important element for national policies, due to the fact that even if greenhouse gas emissions would fall over in a near-term horizon, this does not imply the mitigation of the global warming phenomenon. Moreover, long-term climatic changes connected with the changes in precipitation models, precipitations variations and temperatures are most probable to cause an increase of droughts and floods frequency. Furthermore, late onset, early cessation and prolonged dry spell periods are becoming common which adversely affect the agricultural system. It was also found that very low values of rainfall anomaly which corresponds to severe drought periods had been linked with ENSO events where they coincide or follow the episode shortly. It is, therefore, imperative to adjust the agriculture activity with the variability situation and design planned climate change adaptation strategies so as to enhance the adaptive capacity and resilience of rainfall dependent smallholder farmers [1, 11].

In the absence of an effective strategy for adaptation to the effects of climate change, there is a possibility that the Republic of Moldova will face in the future and undesirable outcome where the implementation of measures for adaption to the effects of climate change will come with higher costs and without effectiveness from economic and social point of view. It is therefore necessary that the measures for estimated effects with a high degree of certainty should be implemented in the shortest time possible.

3   Materials and Methods

Existing research [5, 8, 9] reveals the fact that after the year 1961 climate warming was more pronounced and covered almost the entire country of Moldova. A similar situation is attested at a global and regional level [1–3, 7, 11, 12]. Highlighting climate change in Moldova, the trend has been calculated for all meteorological resorts and emphasized the tendencies of increasing the average annual temperature for two reference periods 1961–1990 and 1981–2010 (the reference periods were proposed by the World Meteorological Organization). The outcome database has served as a pillar for the proposed space-time estimates. The temporal analysis was performed through the Statgraphics Centurion XVI program and digital maps were developed based on the Radial Basic method of the Surfer program.

4   Results and Discussions

Analysis of observational data for long periods of time has revealed that global warming is an ongoing phenomenon, which is also accepted by the international scientific community. Simulations using global climatic models have shown the main factors that determine this phenomenon, both natural (variations in solar radiation and volcanic activity) and anthropogenic (changes in atmospheric composition due to human activities). The cumulative effects of the two categories of factors may explain the observed changes in global mean temperature over the past 150 years. The increase in greenhouse gas concentrations in the atmosphere, especially carbon dioxide, was the main cause of heating by $0.13\,°C$ in the last 50 years of the 20th century, which is about 2 times the value of the last 100 years, as shown in AR5 of the IPCC [12].

Between the years 1880 and 2012, according to multiple sets of independent data, the average global air temperature has increased by about $0.85\,°C$ (with variations of 0.65–1.06), on average by $0.06\,C$ per decade. The overall increase

between the average of the 1850–1900s period and the period of 2003–2012 is
0.78 °C (0.72–0.85 °C) based on only one existing data set.

Europe's climate recorded a warming of around 1 C in the last century,
higher than the global average. The 21st century (2001–2013) is among the
top 15 warmest, globally, since 1880, according to the 2013 National Oceanic
and Atmospheric Administration (NOAA) report. The year 2013 is the fourth
warmest year of the last 133, being the 37th consecutive year with a medium
temperature superior to the 20th century. The years 2010, 2005 and 1998 occupy,
in that particular order, represent the top three hottest years since 1880.

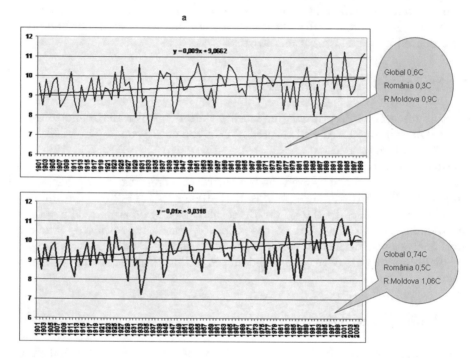

Fig. 1. Trends in air temperature change in regional and national aspect

Air temperature in Moldova has increased in the last century above the
global average and the average recorded in Europe. Thus, during the period
1901–2000, the global average temperature increase by 0.06 °C per decade. In
Romania this indicator was below the global average on 0.3 °C during century
and in the Republic of Moldova it consisted of approximately 0.5 °C above the
global average of 0.6 °C. Another comparative analysis of regional and national
data comes to confirm that there is an accelerated pace of increase of temperature
for the territory of the Republic of Moldova: in the period 1901–2006, the global
increase was 0.74 °C, in Romania 0.5 °C [2,7], and in Republic of Moldova of
1.06 °C (Fig. 1).

The tendency of air temperature increase (with 0.0122 °C/year observed in the series of instrumental observations, 1887–2017) in the country allows us to conclude undoubtedly that regional climate change is characterized by a rather accelerated rhythm. The year 2007 is among the warmest years and remains the warmest in the last 130 years, after which the years 2015, 2017 and 2016 are set with significant thermal values. So, for the last three years, we have recorded some of the highest values, which enables us to conclude that we are on the verge of substantial climate change. The year 2009 is ranked fifth, compared to the previous survey. The three-year high (2015, 2016, 2017) high thermal values show that 2012 is the temporal limit of the extreme warmth years, when the average annual temperature was 11.2 °C compared to the 9.6 °C multiannual average. We note that the year 2013, which is placed above the limit of the extreme warmth years, also recorded high values (11.1 °C).

If, according to [9], during the last two decades the manifestation of the very warm years had a repeatability of once in 2 years (Table 1), with the inclusion of the last 6 years, we find that 8 very warm years of the top 10 (1887–2017) belong to the period 2000–2017 (2007, 2015, 2016, 2017, 2009, 2008, 2000, 2012). The 11th year is 2013, which in the previous study period (1887–2016) ranked 10th in the top of the hottest years. We note the significant share of the last three years in estimating the climate warming trend at regional level. Thus, only with the inclusion of the year 2017, the trend values increase by 0.0006 °C, i.e. from 0.0123 °C/year (1887–2016) to 0.0129 °C/year (1887–2017), for the whole series of instrumental observations.

Table 1. Top of the coolest and warmest years recorded in 1887–2017

1887–2010 [9]				1887–2017			
1933	7,2	**2007**	**12,1**	1933	7,2	**2007**	**12,1**
1929	7,9	**2009**	11,4	1929	7,9	**2015**	**12,0**
1934	8,0	1990	11,3	1934	8,0	**2017**	**12,0**
1985	8,0	1994	11,3	1985	8,0	**2016**	**12,0**
1912	8,1	**2008**	11,3	1912	8,1	**2009**	**11,4**
1940	8,1	**2000**	11,2	1940	8,1	1990	11,3
1987	8,1	1999	11,0	1987	8,1	1994	11,3
1888	8,3	1966	10,9	1888	8,3	**2008**	**11,3**
1976	8,3	1989	10,9	1976	8,3	**2000**	**11,2**
1980	8,3	**2002**	10,8	1980	8,3	**2012**	**11,2**

The above-mentioned facts once again demonstrate that climate change persists with a pronounced warming trend. The climate of the Republic of Moldova is influenced by its position on the globe (in the northern part of the northern parallel - 45°N), as well as its geographic position on the continent (including its meridional extent), which gives the climate a temperate continental character, persisting in greater differentiation between the south and north of the country. Thus, the annual average temperature in the south of the country amounts to about 10.5–11 °C, while in the north, at comparable altitudes, the values of this parameter are lower by about 1.5–2.0 °C (Fig. 2). At the same time, during the evolution of the regional climate there are periods when the principle of zoning is not observed and inverse situations are observed, when in the north one can observe a positive trend and in the southern parts, on the contrary, one can observe the tendency of the annual temperature decrease.

We should mention that the south, south-eastern and central parts of the country register a negative trend, i.e. a decrease of the air temperature during the period of 1961–1990 (Fig. 2a), and in the north and north-east an increasing trend by 0.3 °C was observed. Temperature increase is registered in the entire area during 1981–2010, with its significant footprint in the southern part of the country (Fig. 2b). So if the trend is 1.5 ... 1.6 °C in the north, then the temperatures increase faster by 0.7 °C, summing up to 2.3 °C in the southern extremity.

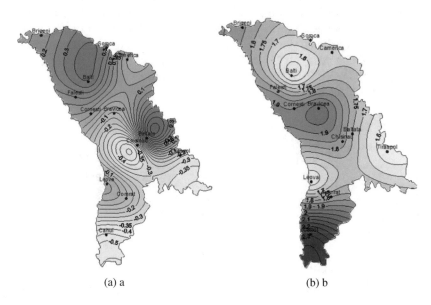

(a) a   (b) b

Fig. 2. This is lot of figures arranged side by side in matrix form with captions for each and a main caption

The trend of temperature increase has also intensified the phenomenon of dryness and heat, the latter being evaluated for the summer season (June–August).

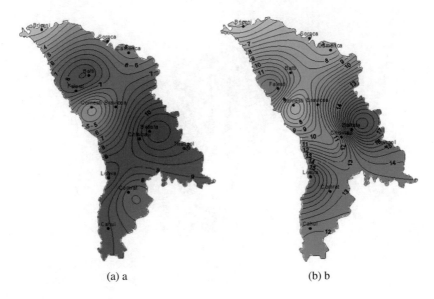

(a) a   (b) b

Fig. 3. The phenomenon of dryness and heat during two reference periods (**a**: 1961–1990; **b**: 1981–2010)

The trends of this phenomenon at national level compared to the periods 1961–1990 and 1981–2010, respectively are shown in Fig. 3.

As it can be seen, between 1961 and 1990, the phenomenon of "dryness and heat" is manifested with a reduced intensity (3–7 days), for the most part of the country, which means that agricultural crops are not generally affected by the heat stress generated by air temperatures above 25 °C and relative air humidity below 30% frequently. Locally, in the southern part of the country, the phenomenon of "heat" shows a higher intensity (10 days). Between 1981 and 2010, the area affected by the heat is expanding, showing a "pole" in the center, in the west of the country and in the southwest of the territory (Fig. 3a, b), determined largely by the magnitude of the synoptic processes which generate these phenomena.

Thus, during certain years (2015), the phenomenon of dryness and drought can exceed the multi-year average values by 7–8 times (Fig. 4), while having different trajectories, i.e. spatial distributions. This phenomenon can contribute to the double pollution of the standing waters (Lake Manta, Lake Beleu) in the southern part of the country, due to the significant increase of evaporability over the multiannual average values [6]. At the same time, the surface graph analysis of the Water Pollution Index (on the example of Lake Beleu) demonstrates its direct connection with atmospheric precipitations and evaporability observed in certain years (Fig. 5).

As a conclusion we can state that at a national level, the accelerated pace of climate change contributes to a more intense desertification process and heat waves, with long periods of manifestation in the last years (2009, 2012, 2015, etc.).

Fig. 4. The phenomenon of dryness and heat in some concrete years (2015)

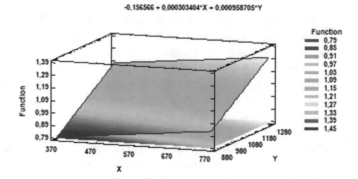

Fig. 5. Surface graph of IPA dependence on atmospheric precipitation and evaporability

At the same time, the observed spatial differentiation demonstrates the presence of a multitude of current climate conditions caused by climate change. We underline that it is important to know them depending on the local physical and geographical factors, as well as on the possible climate changes expected in the coming years in the south of the country, where there is also a pronounced deficit of water resources. The results obtained fit into the country's needs for an adequate and qualitative use of water resources, in particular of surface waters.

Generalizing the findings, we can underline that climate change has contributed and continues to contribute, along with other phenomena, to the increase of natural pollution of waters, especially in the south and south-eastern part of the country, but also to the occurrence of flood events in the same areas. In this context, we consider it appropriate to develop plans to protect the southern areas, to carry out preliminary actions to promote projects and transfer knowledge and technologies in the field of protection and management of vulnerable areas. To achieve the abovementioned objectives additional studies on mapping, modelling and the division of vulnerable areas into sub-sectors is required as well as a simulation of the future trend to modify the river bed in the event of floods and/or droughts. The results of the studies will become valuable documentary and technical support for the continuation of the rehabilitation works and the protection of the vulnerable zones to climate change and associated risks, as well as for the identification of vulnerable zones to the effects of climate desertification. The identification of vulnerable areas for water supply is directly related to the development of Hydrographic Basin Management Plans and their ongoing updating, which needs to take climate change into consideration. More efficient water management would imply: preservation in extremely dry periods through the rehabilitation of transport and distribution networks and also through improvements, the construction of special exploitation accumulations that would increase the availability of water resources in very dry periods, the application of a set of special solutions that would allow storage and exploitation of rainwater especially during very dry periods. Developing maps on the risk of desertification at hydrographic basin level, including them in regional development plans, linking spatial development plans with spatial strategy and hydrological drought risk management plans that could essentially contribute to the efficient use of water resources. Developing a river management strategy with an ecosystem approach, starting from the statement, that rivers are complex ecosystems that depend on the watercourse regime where flow, sediment transport, water temperature and other variables have a well-defined role, contributes to the effective use of scarce water resources. In case of changes of these variables compared to the naturally occurring values, the ecological balance is affected, leading to a restructuring of the biocenosis, namely the loss of species, the replacement of valuable species with other less valuable ones. As a result, the arrangement of the rivers through hydro-technical work must have the objective of maintaining in time and space the integrity and the ecological balance of the aquatic ecosystems, respectively of the water courses. Instead of crossing rivers between digs, a solution usually adopted so far, the current strategy of the European Union resides in the new concept of "more river space" which supports the need to floodplain and drain floods accordingly, while also keeping intact the biocenosis in the event of dryness phenomena. Therefore, the sustainable quantitative and qualitative management of water, the management of natural disasters caused by excess presence or lack of water, the preservation of biodiversity of the aquatic environment should be achieved through river basin management plans. River basin management and management schemes should constitute the planning tool for river basin waters. We consider this planning appropriate for the new climate-induced conditions. The guidelines lay down in a general and harmonious manner the objectives of water quality and quantity, aiming to ensure: good surface water quality or, for artificial or heavily modified water bodies, good ecological potential and good chemical status of surface waters.

References

1 Asfaw, A., Simane, B., et al.: Variability and time series trend analysis of rainfall and temperature in Northcentral Ethiopia: a case study in Woleka sub-basin. Weather Clim. Extrem. **19**, 29–41 (2018). https://doi.org/10.1016/j.wace.2017.12.002
2. Bojariu, R., Bîrsan, M., et al.: Schimbările climatice-de la bazele fizice la riscuri şi adaptare. Editura Printech, Bucureşti, Romania (2015)
3. Dascălu, S.I., Gothard, M., et al.: Drought-related variables over the Bârlad basin (Eastern Romania) under climate change scenarios. Catena **141**, 92–99 (2016). https://doi.org/10.1016/j.catena.2016.02.018
4. Donnelly, C., Greuell, W., et al.: Impacts of climate change on European hydrology at 1.5, 2 and 3 degrees mean global warming above preindustrial level. Clim. Chang. **143**(1–2), 13–26 (2017)
5. Duca, G., Xiao, H., Nedealcov, M.: Regional climatic changes and small rivers water quality in Republic of Moldovas South (Danube River basin). Present Environ. Sustain. Dev. **11**(1), 21–33 (2017)
6. Duca, G., Xiao, H., et al.: Dry periods impact on the surface water quality. Present Environ. Sustain. Dev. **11**(1), 5–20 (2017)
7. IPCC: The intergovernmental panel on climate change (2018). http://www.ipcc.ch
8. Nedealcov, M.: Resursele agroclimatice în contextul schimbărilor de climă. tipografia alina scorohodova, chişinău. Technical report (2012), 306 p. ISBN 978-9975-4284-8-4
9. Nedealcov, M., Railean, V., et al.: Climatic resources of the republic of Moldova, p. 76 (2013)
10. Ogallo, L.A., Omondi, P., et al.: Climate change projections and the associated potential impacts for Somalia. Am. J. Clim. Chang. **7**(02), 153–170 (2018). https://doi.org/10.4236/ajcc.2018.72011
11. Rodrigues, G.S., Putti, F.F., et al.: Climatological hydric balance and the trends analysis climatic in the region of Machado in Minas Gerais State, Brazil. Am. J. Clim. Chang. **7**(4), 558–574 (2018)
12. Stocker, T.F., Qin, D., et al.: Climate change 2013: the physical science basis. Intergovernmental Panel on Climate Change, Working Group I Contribution to the IPCC Fifth Assessment Report (AR5), vol. 25, p. 1535. Cambridge University Press, New York (2013)

The Impact of "Internet +" on the Business Models Transformation of Traditional Enterprises

Jinjiang Yan[1](✉), Yongyi Wang[1], Zhen Liu[1], Yong Huang[1], and Xinhui Wang[2]

[1] Business School, Sichuan University, Chengdu 610064, People's Republic of China
282095662@qq.com
[2] Southwest Minzu University, Chengdu 610041, People's Republic of China

Abstract. With the continuous penetration of the Internet into various industries, traditional enterprises have been hit by unprecedented shocks and began to gradually transform their business models. The aim of this study is to explore the extent of applications of Internet technology in traditional enterprises as well as the linkages between "Internet+" and the business model transformation by the case survey method and the regression analysis. The results revealed that the key to the transformation of the business model under the impact of the Internet is the optimization and adjustment of the internal structure of the company. On the other hand, increasing the external influence of the "Internet+" is also important.

Keywords: "Internet +" · Business model transformation ·
Traditional enterprises · Regression analysis

1 Introduction

The interest in the Internet has grown considerably since the development of industries is increasingly inseparable from network technologies. In China, the concept of "Internet +" was first proposed in 2012. It represents a new form of economic development and new business model, which integrates the Internet technology into all sectors of the economy and society, also enhances the innovation and productivity of the real economy [10]. For instance, "Internet+ agriculture" relates to a creative production method and agricultural supply chain management mode. A large number of studies have shown that "Internet +" has an impact on the transformation of business models, especially in the area of traditional enterprises. Manufacturing firms are migrating toward e-business technologies in order to lower their operating costs, raise productivity and quality, and respond rapidly to their customers' and other business partners' requirements [18].

In addition, "Internet +" also facilitates the supply-side reform in China. The structural reform on the supply side means that the focus of reform has

© Springer Nature Switzerland AG 2020
J. Xu et al. (Eds.): ICMSEM2019 2019, AISC 1002, pp. 335–347, 2020.
https://doi.org/10.1007/978-3-030-21255-1_26

shifted from the consumer side to the supply side, which is closely connected to the upgrading and business model transformation of enterprises. Moreover, the value proposition dimension in the business model is a reflection of the quality of supply. Although, recent works are starting to analyze the adoption and use of Internet within organizations and how these technologies support specific business processes, much of the existing research focuses on a view of Internet use [17], with very few studies examining the new concept "Internet +" along the business model. Thus, it is important to understand the key factors in "Internet+" that motivate the transformation of business model.

In this context, we investigate that how does "Internet +" affect the transformation of business models through an empirical analysis. The work is expected to serve as an assistant tool for managers of traditional enterprises in applying "Internet +" appropriately.

The remainder of this study is organized as follows: Section 2 briefly probes the research situation of business model and the adoption of Internet. Our hypothesis are also presented in this part. Section 3 details the methodology considered to establish the measurement of variables. Section 4 deals with the empirical analysis. Finally, conclusions are given in Section 5.

2   Literature Review and Hypothesis

The original concept of "Internet+" comes from the Chinese Internet industry. Entrepreneurs proposed that the Internet would become a new social infrastructure, simultaneously, an important way to promote the transformation and upgrading of traditional industries. The Chinese government [13] emphasized the advantages of the Internet and its infrastructure role. Alibaba Research highlighted the techniques contained in the "Internet+" [1], while the Tencent Research Institute underlined the new economic form generated by the integration of the Internet and traditional industries. Ouyang [16] elaborated on the detailed acting process of the Internet in traditional industries. But an unambiguous international definition for it does not exist. In view of this, we summarize the "Internet +" as a deep integration of Internet technologies including the mobile Internet, cloud computing, big data, and the Internet of Things, using the technology to optimize and integrate production factors to greatly enhance the efficiency of traditional industries such as capital, source allocation, production, organization, operations, etc., to achieve the new economic shape formed by the transformation and upgrading of traditional industries. The key point is innovation.

Most Foreign scholars have explored the impact of a single Internet technology on different industries. Bresciani et al. [5] highlight that knowledge management capabilities enhance alliance ambidexterity indirectly through firms' ICT (Information and communication technologies) capabilities. Wolfert et al. [19] has combed lots of literature related to the use of big data in smart farms, suggesting that The further development of data and application infrastructures (platforms and standards) and their institutional embedment will play a crucial role in the battle between these scenarios.

Business model refers to the path and method by which companies create value and gain revenue. The definition of business model can be roughly divided into four categories, mainly including: (a) Profit Model Theory [15], (b) System Theory [2], (c) Value Creation Theory [3], (d) Conceptual Logic Theory. The scholar of System Theory defines the business model on a system perspective. They think that the business model simply illustrates how the company conducts positioning and integration, arguing that the essence of the business model is that it is a concept. We believe that the definition of System Theory under the integration type can more fully explain the connotation of the business model. A competitive operating system that satisfies customers' needs and realizes customers' value through optimal realization of forms, and at the same time enables the system to achieve a total solution for sustainable profitability.

In general, the influencing factors of business model transformation of traditional enterprises are consistent with the driving forces and resistance of business model transformation. At present, many foreign scholars have studied the transformation of business models. Markides [12] focused on profitability, Eisenmann et al. [7] payed attention to bilateral markets, Clayton [6] thought that the transformation and innovation of business models is driven by internal factors. Meanwhile, Chinese scholars have also done a series of studies on this, Yu and others [20] have analyzed the impact of big data technology on business model transformation in the "Internet+" era. Liu and others [11] believe that "Internet +" has an impact on five aspects of the company's products and services, marketing strategies, information symmetry, and the value chain of the company. Although there are different directions and perspectives for the study of "Internet +" and business models, it is undoubted that "Internet +" has a certain influence on the transformation of traditional business models.

In the context of "Internet +", the biggest challenge for traditional enterprises comes from the disruptive Internet thinking. Hence, it will be a meritorious research to investigate how traditional enterprises find a way for sustainable development. In order to explore the connection between "Internet +" and the transformation of business models of traditional enterprises, we propose the following two hypotheses:

Hypothesis 1: The transformation of the business model of an enterprise must be started within the enterprise. That is, the construction of the "Internet+" within the enterprise has a positive effect on the transformation of the business model.

Hypothesis 2: The construction of "Internet +" related channels and their impact on business models have a negative effect.

3   Method

3.1   Measurement of "Internet +" Influence Degree

Based on the introduction and literature review, we choose six factors. The "Internet+" variable metrics are shown in Table 1. First-level indicators include Traditional factors and Innovative factors. The former one evaluates the extent to which traditional enterprises use Internet tools in terms of product sales, functional departments, and organizational structure. The secondary indicators consist of three indicators that are highly feasible in practice.

On the other hand, on the basis of the "Internet +" three-level evaluation index system, in order to unify the evaluation criteria, the more intuitive understanding was obtained and the basic qualitative evaluation criteria were combed. The results are shown in Table 2.

3.2   Measurement of Business Model Transition

According to the division of the business model canvas [14], and drawing on the research methods of Li [9] and others, in the constituent elements of the business model, this study selected 2 first-level indicators. metrics for business model transformation is shown in Table 3. In this sector, we will use the case survey method to measure each variable one by one and standardize the attribute information contained in the sample. This study refers to the measurement method of

Table 1. The measurement of "Internet +" variables

First-level indicators	Secondary indicators	Third-level indicators	No.
Traditional factors	Product sales	The level of online self-employment channel construction	1
		The level of "Third-party" online sales channel construction	2
	Functional department	The level of enterprise's official website construction	3
		The level of enterprise's APP construction	4
		The "WeChat Official Accounts" construction level of the enterprise	5
	Organization structure	Is it clear about the "Internet +" strategy	6
		Whether established a department or business unit related to "Internet +" or not	7
Innovative factors	The construction and cooperation of "Internet +" project	The level of "Internet +" project investment	8
		The level of cooperation with internet companies	9
	"Internet +" awareness	The strength of "Internet +" awareness	10
	"Internet +" strategy	The frequency of "Internet+" that enterprise seniors have mentioned in external publicity	11

the Likert 5-grade scale. Three investigators collectively score the variables mentioned above and also take into account the specificity of variables in the study. In the 1–5 sub-scale, added "0" score to improve the richness and reliability of data information [4,8].

After establishing the indicator system, we will start an empirical analysis through the data of expert scoring. The process of our analysis is shown in Fig. 1.

4 Empirical Analysis

4.1 Statistical Description

We selected the Cronbach alpha reliability test to test the reliability of the accuracy of the data obtained. The specific results are shown in Table 4. The coefficient value reached 0.784. The minimum value of the project deletion was also above 0.75, which was close to 0.8; the variance interpretation rate reached 74%, which was acceptable; The KMO value is 0.639, indicating that there is a certain relationship between the independent variables.

Table 2. The indicator evaluation system of "Internet +" influence level

No.	Code	Third-level indicators	Value
1	B1	The level of online self-employment channel construction	0–5
2	B2	The level of "Third-party" online sales channel construction	0–5
3	B3	The level of enterprise's official website construction	0–5
4	B4	The level of enterprise's APP construction	0–5
5	B5	The "WeChat Official Accounts" construction level of the enterprise	0–5
6	B6	Is it clear about the "Internet +" strategy	0/1
7	B7	Whether established a department or business unit related to "Internet +" or not	0/1
8	B8	The level of "Internet +" project investment	0–5
9	B9	The level of cooperation with internet companies	0–5
10	B10	The strength of "Internet +" awareness	0–5
11	B11	The frequency of "Internet+" that enterprise seniors have mentioned in external publicity	0–5

[a]The score item is 0–5, the higher the score, the greater the influence of Internet+ 0/1 is a binary variable

4.2 The Analysis of Business Model

The analysis of the correlation coefficient between the internal variables of the business model is shown in Table 5. The results show that within the business model, the maximum value of the correlation coefficient between variables is less

Table 3. The indicator evaluation system of business model transformation

	Secondary indicators	Code	Third-level indicators	Value
Change in value proposition	Change of target customers	A1	The degree of change in the target customer group compared with before	0–5
		A2	The degree of change in target market area compared with before	0–5
	Adjustments and changes in product/service portfolio	A3	The degree of change in product/service portfolio compared with before	0–5
Change in value support	Adjustments and changes in organizational structure	A4	The degree of change in organizational structure compared with before	0–5
	Adjustments and changes in R&D investment	A5	The degree of change in R&D direction compared with before	0–5
	Adjustments and changes in marketing	A6	The degree of change in sales costs compared with before	0–5
		A7	The degree of change in sales model compared with before	0–5
		A8	The degree of change in branding strategy compared with before	0–5
	Adjustments and changes in cooperation relationship	A9	The degree of change in relationship between enterprises and major suppliers compared with before	0–5
		A10	The degree of change in relationship between enterprises and major customer compared with before	0–5
		A11	The degree of change in relationship between enterprises and competitor compared with before	0–5
	Adjustments and changes in structure of income source	A12	The degree of change in income structure compared with before	0–5

[a]The score item is 0–5, the higher the score, the greater the degree of transformation

Table 4. Cronbach alpha reliability test

	AVE	VAR	CORR	Cronbach alpha
A3	18.730	59.814	0.594	0.751
A4	19.973	68.583	0.372	0.775
A5	19.162	64.362	0.442	0.767
A6	18.405	66.303	0.188	0.799
A7	17.811	65.658	0.267	0.786
A8	18.649	66.901	0.306	0.779
A9	18.216	60.341	0.487	0.762
A10	18.973	65.360	0.429	0.769
A11	18.892	65.544	0.493	0.765
A12	19.324	65.447	0.476	0.766
A13	18.649	64.401	0.484	0.764
A1	19.243	59.578	0.644	0.746
A2	18.514	63.646	0.367	0.775

Table 5. Correlation matrix

		A1	A2	A3	A4	A5	A6	A7	A8	A9	A10	A11	A12	A13
A1	A1	1	0.3685	0.5964	0.3758	0.2528	-0.0112	0.3103	0.4099	0.4972	0.2746	0.3476	0.2521	0.4373
A2	A2	0.3685	1	0.4873	0.0491	0.0504	0.0101	-0.0814	0.2750	0.2140	0.1598	0.2702	0.5474	0.1525
A3	A3	0.5964	0.4873	1	0.4682	0.1698	-0.0167	0.0053	0.3056	0.2977	0.3983	0.4963	0.4854	0.3260
A4	A4	0.3758	0.0491	0.4682	1	0.0812	0.0295	0.0830	-0.0876	0.3569	0.3481	0.3153	0.2158	0.2557
A5	A5	0.2528	0.0504	0.1698	0.0812	1	0.2583	0.3447	0.4251	0.3824	0.1455	0.3696	0.0840	0.2178
A6	A6	-0.0112	0.0101	-0.0167	0.0295	0.2583	1	0.2813	0.0104	0.1673	0.1750	-0.0090	0.2526	0.1028
A7	A7	0.3103	-0.0814	0.0053	0.0830	0.3447	0.2813	1	0.0596	0.2324	0.1302	0.0155	0.0164	0.3060
A8	A8	0.4099	0.2750	0.3056	-0.0876	0.4251	0.0104	0.0596	1	0.2754	0.0012	0.1696	-0.1536	0.1796
A9	A9	0.4972	0.2140	0.2977	0.3569	0.3824	0.1673	0.2324	0.2754	1	0.0933	0.1538	0.0857	0.3773
A10	A10	0.2746	0.1598	0.3983	0.3481	0.1455	0.1750	0.1302	0.0012	0.0933	1	0.5216	0.5654	0.1868
A11	A11	0.3476	0.2702	0.4963	0.3153	0.3696	-0.0090	0.0155	0.1696	0.1538	0.5216	1	0.4640	0.2864
A12	A12	0.2521	0.5474	0.4854	0.2158	0.0840	0.2526	0.0164	-0.1536	0.0857	0.5654	0.4640	1	0.3181
A13	A13	0.4373	0.1525	0.3260	0.2557	0.2178	0.1028	0.3060	0.1796	0.3773	0.1868	0.2864	0.3181	1

than 0.57, that is, there is no strong correlation between the internal variables of the business model.

The principal component analysis method is used to analyze the internal variables of the business model. The results are shown in Table 6, it shows the specific information of each linear combination of principal components. For example, the first principal component Eq. 1 can be expressed as:

$$\text{Prin} = 0.374966 * A1 + 0.262257 * A2 + \ldots + 0.291228 * A13 \qquad (1)$$

According to the correlation matrix in Table 6 and the steep slope map in Fig. 2, seven principal components were selected, and the proportion of variance explained was more than 85%. Based on this, we take the variance contribution rate of each factor as the weight to derive the comprehensive index of business model transformation (variable: business). The larger the index value, the higher the degree of business model transformation.

4.3   The Analysis of "Internet +"

For variable "Internet +", factor analysis is performed using the maximum variance orthogonal rotation method. The factors after rotation are shown in Table 7:

Table 6. Principal component analysis

	Prin1	Prin2	Prin3	Prin4	Prin5	Prin6	Prin7	Prin8	Prin9	Prin10	Prin11	Prin12	Prin13
A1	0.37497	0.14961	−0.20760	−0.19540	0.13960	−0.13730	0.33360	0.02789	−0.08500	−0.48500	−0.19600	−0.56430	−0.07500
A2	0.26226	−0.21270	−0.32220	0.36433	0.42520	0.04200	0.06200	−0.32870	0.05960	0.19480	0.42090	−0.05070	−0.36500
A3	0.38610	−0.17940	−0.22510	−0.07860	0.00820	0.08340	0.19112	0.17278	0.36960	0.02890	−0.49000	0.53371	−0.16300
A4	0.26188	−0.11030	0.12089	−0.59340	−0.09500	0.38754	0.06236	0.03065	0.31880	0.30320	0.35580	−0.22430	0.13400
A5	0.23315	0.39881	0.09794	0.28759	−0.41500	0.11791	−0.21470	−0.33290	0.21140	0.28610	−0.31500	−0.29090	−0.20300
A6	0.09990	0.16198	0.52213	0.37250	0.24180	0.05558	0.05041	0.40410	0.16780	−0.28700	0.10580	−0.00640	−0.12900
A7	0.14445	0.40056	0.38859	−0.05840	0.12650	−0.43710	0.44679	−0.30430	0.14430	0.08190	0.17910	0.29293	0.13990
A8	0.18643	0.35984	−0.45220	0.30067	−0.17900	0.02422	0.15817	0.43793	−0.08100	0.21600	0.23900	0.03312	0.42450
A9	0.28060	0.32809	−0.05020	−0.21330	0.21840	0.41287	−0.26150	−0.28560	−0.51600	−0.10500	−0.05500	0.32173	0.11780
A10	0.29096	−0.28380	0.30196	0.04468	−0.30200	−0.05720	0.29080	0.18286	−0.61200	0.28050	−0.00200	0.00050	−0.27600
A11	0.33271	−0.19250	0.01126	0.09273	−0.51000	−0.13060	−0.24090	−0.13570	0.09930	−0.53900	0.38420	0.19882	0.04730
A12	0.30678	−0.40690	0.22058	0.25964	0.22620	−0.07810	−0.13430	−0.13210	0.00480	0.09700	−0.25400	−0.16930	0.65490
A13	0.29123	0.14813	0.08770	−0.17870	0.24870	−0.47890	−0.58490	0.38865	0.02230	0.16300	0.06000	−0.02440	−0.19500

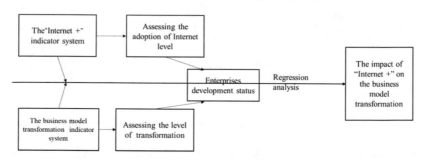

Fig. 1. A flow chart description of the research framework

Therefore, according to the rotated factor, the following variables are added: "Internet +" internal construction F1 (including B5, B6, B7, B8); "Internet +" channel construction and impact F2 (including B2, B3, B4, B9); interactive items Interaction between variable F5; The interaction item of F1 and F3 refers to variable F6; The interaction item of F1 and self-channel construction level F4 (B1) refers to variable F7; The interaction item of F2 and F3 refers to variable F8; The interaction item of F2and F4 refers to variable F9; The interaction item of F3 and F4 refers to variable F10.

The Pearson correlation coefficient shows that the correlation coefficient between the variables is 0.11028 maximum and the correlation is small. The VIF values for the variables are all less than 2.5. Therefore, there is less possibility of autocorrelation and multicollinearity problems.

Table 7. Factors after rotation

	F1	F2	F3	F4
B1	0.04432	0.10091	0.14749	0.56335
B2	0.29771	0.63983	0.25098	0.20076
B3	−0.0880	0.71772	0.38583	0.03065
B4	0.38187	0.58051	−0.05013	0.14141
B5	0.51474	0.32237	−0.13838	−0.2551
B6	0.69746	−0.08567	0.29412	0.31054
B7	0.74526	0.26065	0.10252	0.12324
B8	0.75389	0.37183	0.15704	−0.07106
B9	0.42308	0.72089	−0.14058	−0.0505
B10	0.37343	0.30368	0.65822	0.16069
B11	0.02927	0.00466	0.86406	0.17431

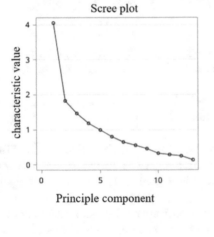

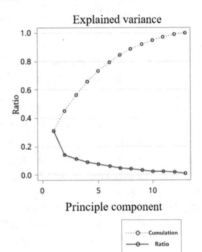

Fig. 2. Steep slope map

4.4 The Analysis of "Internet +" and Business Model

In order to verify the above assumptions, linear regression and hierarchical regression methods were used for regression analysis.

(1) Linear Regression

A stepwise linear regression method was chosen for further research. The regression results are shown in Table 8. The results show that F1, F2, F8, and F9 enter the model. The final results of the model are shown in Table 9.

From Table 9, we can conclude that the linear relationship between the business model and the "Internet +" variable is shown as follows:

Table 8. Stepwise regression

Step	Inter variable	Deleted variables	Introduced variables	Partial R2	R2	C(p)	F-value	Pr > F
1	F8		1	0.1752	0.1752	10.654	7.44	0.0099
2	F1		2	0.1823	0.3576	3.0036	9.65	0.0038
3	F2		3	0.0765	0.4340	0.9571	4.46	0.0424
4	F9		4	0.0508	0.4848	0.2706	3.15	0.0853

Table 9. Regression result

Variables	Parameter estimation	St.E	SSII	F	Pr > F
Intercept	0.0	0.08578	0.00349	0.01	0.9105
F1	0.4	0.09709	3.62871	13.35	0.0009
F2	0.0	0.10094	1.19256	4.39	0.0442
F8	0.0	0.12690	3.77221	13.88	0.0008
F9	0.0	0.13503	0.85658	3.15	0.0853

$$Business = 0.3548 * F1 - 0.21147 * F2 - 0.47283 * F8 - 0.23975 * F9$$

Only F1, F2, F8, and F9 entered the model, which leads to the following conclusions:

(a) "Internet +" external influence, self-channel construction level, interaction effects of F1 and F2, F1 and F3, F1 and F4 has no obvious effect on business model transformation;

(b) The "Internet +" internal construction has a positive effect on the "Internet +" transformation. In other words, if companies want to make business model changes faster and better, they should start with the internal "Internet +" impact, which also proves Hypothesis 1;

(c) "Internet +" channel construction and impact, interaction effects of F2 and F3, F2 and F4 has a certain negative effect on the impact of business model transformation, that is, enterprises constructs from outside is not a sensible choice. The reason is related to the rapid development of tripartite platforms (such as JingDong and TMall). There is a certain gap between the company's proprietary channel construction and the tripartite platform, whether it is traffic or sales. This also proves Hypothesis 2.

(2) Hierarchical Regression

The results of the regression analysis are shown in Table 10. Models from 1 to 10 are added to the F variables in order.

In Model 1, F1 is statistically significant, and F1 ("Internet +" external influence) has a certain role in promoting the "Internet +" business model.

Model 2 joins the F2 (self-operated channel construction level), and it can be found that R2 is reduced, and it has a certain inhibitory effect on business model transformation.

Model 3 joins the F3 ("Internet +" external influence), which is statistically significant and has a certain role in promoting the "Internet +" business model.

Model 4 joins the F4 (self-operated channel construction level), which has certain statistical significance and has a certain role in promoting the "Internet +" business model.

The interaction between the variables was added to Model 5 to Model 10. From the results, it can be seen that F7 (the interaction between F1 and F4) and F10 (the interaction between F3 and F4) has a certain role in promoting the "Internet+" business model, while the rest The interactive variables have a certain inhibitory effect on the "Internet+" business model.

Table 10. F-value and business model transformation

Variable	Model 1	Model 2	Model 3	Model 4	Model 5	Model 6	Model 7	Model 8	Model 9	Model 10
Constant	− 2.637*E-17	− 3.040*E-17	− 3.040*E-17	− 3.040*E-17	0.028	0.028	0.012	0.009	0.011	−0.013
F1	0.398	0.314	0.314	0.312	0.424	0.428	0.463	0.377	0.392	0.384
F2		−0.115	−0.116	−0.114	−0.087	−0.087	−0.107	−0.222	−0.210	−0.202
F3			0.005	−0.002	0.011	0.008	0.128	0.153	0.102	0.110
F4				0.052	0.053	0.053	−0.123	−0.107	−0.087	−0.120
F5					−0.320	−0.322	−0.452	−0.143	−0.118	−0.078
F6						−0.009	0.166	0.134	0.052	0.034
F7							0.709	0.396	0.218	0.097
F8								−0.471	−0.455	−0.453
F9									−0.220	−0.157
F10										0.217
R2	0.159	0.181	0.181	0.184	0.245	0.245	0.291	0.460	0.499	0.509
Adjust R^2	0.135	0.133	0.107	0.082	0.124	0.094	0.120	0.306	0.331	0.320
Change of R^2	0.159	0.023	0	0.003	0.061	0	0.046	0.169	0.039	0.100

In order to analyze the influence of the "Internet +" internal variables on the internal variables of business models, we also examine the relationship between "Internet +" and business models. The results show that the factor B5, B7, B8 have a positive influence on the business model variables. Among them, the factor B5 and B7 have the most positive effect on the degree of changes in the target customer group set by the company A1; the factor B6 and B8 have the most positive effect on the sales model of business model A9.

5   Conclusion and Discussion

5.1   Conclusion

Based on the two analysis methods above, we can see that the two hypotheses made in the previous have been verified. The key to the transformation of the "Internet+" business model is the optimization and adjustment of the internal structure of the company. On the other hand, it is to increase the "Internet+" external influence or to build a "Internet+" self-operated channel when the internal structure of the enterprise has changed to a certain extent.

5.2   Discussion

In order to provide valuable suggestions for the transformation and innovation of business models of traditional enterprises, this paper studied 37 representatives of traditional enterprises in Sichuan Province, collected data on the basis of measurement of relevant variables, and used SAS software to analyze the data. A series of conclusions have been drawn. These conclusions will play a guiding role for traditional enterprises in utilizing the development of the "Internet+" and can promote enterprises to further complete the transformation of business models, and at the same time, they also help the government to optimize the traditional industrial structure.

Due to the limitations of research time and technology, this paper only surveyed 37 representative traditional enterprises. In order to obtain more convincing conclusions, the number of samples can be further expanded; at the same time, the traditional business model transformation and "Internet+" should be considered more comprehensively. The relevant variables also contribute to the further deepening of the research. In future research, we will further discuss the differences and features before and after the transformation of traditional enterprises.

Acknowledgements. This research is funded by China National Social Science Fund Project(16BGL011).

References

1. "Internet +" research report. Technical report. Ali Institute (2015)
2. Amit, R., Zott, C.: Value creation in e-business. Strateg. Manag. J. **22**(6–7), 493–520 (2001)
3. Applegate, L.M.: E-business models: making sense of the internet business landscape. Information Technology and the Future Enterprise: New Models for Managers, pp. 49–94 (2001)
4. Bigliardi, B., Nosella, A., Verbano, C.: Business models in Italian biotechnology industry: a quantitative analysis. Technovation **25**(11), 1299–1306 (2005)
5. Bresciani, S., Ferraris, A., Del Giudice, M.: The management of organizational ambidexterity through alliances in a new context of analysis: internet of things (IoT) smart city projects. Technol. Forecast. Soc. Chang. **136**, 331–338 (2018)

6. Christensen, C.M.: The innovator's dilemma: when new technologies cause great firms to fail. Harvard Business Review Press, Boston (1997)
7. Eisenmann, T., Parker, G., Van Alstyne, M.W.: Strategies for two-sided markets. Harv. Bus. Rev. **84**(10), 92 (2006)
8. Hayes, J., Finnegan, P.: Assessing the of potential of e-business models: towards a framework for assisting decision-makers. Eur. J. Oper. Res. **160**(2), 365–379 (2005)
9. Li, L., Changwei, M., Hailin, L.: Nan Kai Business Review, vol. 18 (2015), (in Chinese)
10. Liu, B.: Development status of "internet plus" and its impact on China's economy. China's Collect. Econ. **24**, 47–54 (2015), (in Chinese)
11. Liu, J.G., Ma, D.Q., Chen, C.J.: Study on the "internet +" business model innovation path based on grounded theory-a case study of dropping behavior. Soft Sci. **7**, 30–34 (2016), (in Chinese)
12. Markides, C.: Disruptive innovation: in need of better theory. J. Prod. Innov. Manag. **23**(1), 19–25 (2006)
13. News, C.: Li Keqiang: formulating the "internet plus" plan to promote the healthy development of e-commerce (2015)
14. Osterwalder, A., Pigneur, Y.: Business model generation: a handbook for visionaries, game changers, and challengers. Wiley, New York (2010)
15. Osterwalder, A., Pigneur, Y., Tucci, C.L.: Clarifying business models: origins, present, and future of the concept. Commun. Assoc. Inf. Syst. **16**(1), 1 (2005)
16. Ouyang, R.H.: From "+internet" to "internet+" - how the technological revolution breeds a new type of economic society. Academic Front 05 (2015), (in Chinese)
17. Soto-Acosta, P., Colomo-Palacios, R., Popa, S.: Web knowledge sharing and its effect on innovation: an empirical investigation in SMEs. Knowl. Manag. Res. Pract. **12**(1), 103–113 (2014)
18. Soto-Acosta, P., Popa, S., Palacios-Marqués, D.: E-business, organizational innovation and firm performance in manufacturing SMEs: an empirical study in Spain. Technol. Econ. Dev. Econ. **22**(6), 885–904 (2016)
19. Wolfert, S., Ge, L., Verdouw, C., Bogaardt, M.J.: Big data in smart farming-a review. Agric. Syst. **153**, 69–80 (2017)
20. Yu, L.N.: Business model transformation in the era of big data. Inf. Syst. Eng. **12**, 109–110 (2014), (in Chinese)

Modeling the Barriers of Sustainable Supply Chain Practices: A Pakistani Perspective

Muhammad Nazam[1], Muhammad Hashim[2(✉)], Mahmood Ahmad Randhawa[1], and Asif Maqbool[1]

[1] Institute of Business Management Sciences, University of Agriculture, Faisalabad, Pakistan
[2] Department of Management Sciences, National Textile University, Faisalabad, Pakistan
hashimscu@gmail.com

Abstract. In recent years, manufacturing industries are adopting the concept of ecofriendly environment due to the emerging concept of sustainability in supply chain. In Pakistan, agriculture and textile sectors are contributing a major role for the development of economy. This study aims to identify sustainable outsourcing, sustainable production and distribution, sustainable competitiveness and innovation, sustainable buyer-supplier relationship, sustainable marketing and organizational culture, sustainable knowledge sharing, and sustainable technology practices as the seven most important supply chain management SCM barriers in different industries of Pakistan. Despite the importance of all these factors, industries need to identify the barriers hindering green supply chain management implementation. In this research, 42 sub-attributes, seven barrier categories relevant to sustainable supply chain management SSCM implementation were considered, with the help of literature and experts discussion. The core objective of this study is to develop the multi-attribute group decision making model to facilitate the strategic management of the industries in adoption of sustainable supply chain practices using fuzzy AHP approach. A practical case is used to check the robustness of the propose model. The key findings highlights that sustainable outsourcing factor is essential barrier during SSCM adoption. Finally, the results depict that proposed model is helpful for prioritizing the solutions of SSCM practices and suggest the roadmap to overcome on industrial barriers.

Keywords: Sustainable supply chain · Barriers · Fuzzy AHP

© Springer Nature Switzerland AG 2020
J. Xu et al. (Eds.): ICMSEM2019 2019, AISC 1002, pp. 348–364, 2020.
https://doi.org/10.1007/978-3-030-21255-1_27

1   Introduction

In recent years, sustainable supply chain management is getting the attention of researchers and practitioners throughout the globe. Adoption of sustainable supply chain management practices has been gaining popularity for the manufacturing sector because of its key objectives to reduce wastage and improve product life cycle. Pakistan as a developing country contrived to improve the infrastructure of industries and promoting the strategies for environmental development. As environmental pollution and polluted hazard chemicals are destroying the environmental resources of the Pakistan [8]. The concept of sustainable supply chain management SSCM is advanced concept in the business world. Now business is not solely depended on the profit but now the scenario has been changed. The pollution free environment is not considering important environment protection but also as an integral part of the business for the increasing profit and business horizons [9–11].

SSCM as advanced concept implemented in the developed countries and now these countries were adopted this SSCM practices. But the important question arises that these practices were developed to protect the green environment of the whole global. Now it is responsibility on the researches to identify the gaps and hurdles that hindrance the SSCM practices in the third or developing countries. This research work was conducted to highlight the SSCM practices in the Pakistan and to identify the potential of implementation and barriers [15,19]. Supply Chain Management SCM practices can be defined as a group of activities taken on by organizations for promoting its effective supply chain management. The practices of SCM have become very important prerequisite for taking competitive advantages in the current global environment and also enhancing profitably as well [1,2]. In current scenario, the field of logistics and SCM practices becomes the primal competitive component in worldwide manufacturing business. Nowadays, the mode of optimizing the SCM in organizations has been started among local and international companies. Latest technologies appliances are employed for this purpose. The supply chain is a coordinated set of different techniques to design and carry out all stairs in world-wide network practiced for acquiring raw materials, transform it into final goods, and deliver the final product to end customers. It admits chain-wide system for knowledge sharing, resource planning and synchronization, and performance measurements globally [3]. In todays globalization and intensive world-wide competition many apparel retailers are establishing strong supply chains to get the advantage from their competitors by providing the best value to their potential customers. Now the SCM practices have become very critical to manage risk, dynamism, and complexities of global sourcing. It is most important for industry to integrate the whole supply chain activities for getting maximum benefits [4,5]. Supply chain management SCM dramatizes every role of its members from supplier to end consumer and prior studies supported that competitive advantage can be achieved through the structured and well-designed supply chain system. But green supply chain management and its practices nowadays used as commonly stated terminology. The research gaps attracted the potential researches in this area or field. The potential research gaps included the green supply chain management GSCM practices and its implementation in developing countries and also highlighted the different barrier or hindrances [13]. SSCM practices offered various benefits to the large and globalized organizations and opened a way toward new environmental projects and initiatives for better goodwill. But question is here arising that how small and medium enterprises shell get benefit from it and what projects will be offered by SMEs regarding the environmental initiatives. Environmentally sustainable or GSCM has been emerged as a significant organizational viewpoint to achieve corporate profit and market share goals by eradicating the environmental risks and effects. It also been continued to improving ecological effectiveness of these organizations and their partners [14]. The synergistic linking of sustainability and supply chain management led toward the competitive and global dimensions in organizations. One example can elaborate it more precisely as multinational initiatives have established networks of suppliers globally to gain competitive advantage. The strictness in regulations and increased population along with consumer pressures as well, the industrial sector need to integrate environmental problems into regular and consistent practices and also developing their strategic planning program. It resulted as an integrating environmental problem into SCM and converted into an important agenda for manufacturers to achieve and sustain competitive advantage [17,18]. The field of SSCM is no doubt in its early development stages, both scholastically and practically. Scholastically it is treated as to persuade effectively and empirically advance in order for theory development and for this purpose some valuable and testable multi-item measurement scales will be needed [23]. Greater attention required only to focus on using multi-item latent constructs, evaluating their content validity and enhancing them through field-based testing [6]. Nazam et al. [12] investigated the risk assessment related to SSCM in textile industry of Pakistan. The purpose of this study was to examine and rank risks related to SSCM adoption applied to a practical case of textile manufacturing industries. The researchers used fuzzy AHP to compute the attribute weight of main and sub-category and Technique for Order Performance by Similarity to Ideal Solution TOPSIS approach to prioritize the barriers in uncertain environment. The findings depicted that procurement variable was most significant barrier followed by manufacturing, retailing, logistics and flexibility. As mentioned above, Pakistani enterprises should focus on the adoption of the SSCM practices to overcome on barriers but this process cannot be easily adopted without conducting the thorough research. The Pakistan as developing country has been focusing on the development of the SMEs through SMEDA project. The prior studies focused on the GSCM in the context of the SMEs and this will help to continue further researches on this topic to enhance the better implementation of GSCM practices in Pakistan. Every adoption process has different stages and the most critical stage is initial stage. Initial stage barriers have also been considered critical and important to be taken into consideration for proper implementation process [22]. This research fills the gap to structures the barriers in supply chain management by considering multi-attribute decision making problem under sustainable environment. Due to uncertainty and vagueness in the environment, it is an uphill task for the experts to provide their intellectual inputs regarding intensity of barriers in exact numerical values. In this perspective, fuzzy logic is important tool which is adopted for tackling the issues of subjectivity [10]. This research study develops fuzzy analytical hierarchy process AHP model to prioritize the barriers in supply chain. Furthermore, this study utilizes fuzzy AHP technique to compute weights intensity of the identified barriers. Finally, a practical case study is given to show the applicability of the proposed model. The present study was aimed at developing a roadmap for the sustainable policies affecting industrial productivity in Pakistan. The objectives of this were as follows: 1) To study the current practices of SSCM encountering supply chain barriers in different industries of Pakistan, 2) to study the gap between Pakistani and international practices of sustainable supply chain management and 3) to suggest recommendations for improving sustainable supply chain management practices in multiple industries. The remainder of this research work is divided as follows. Taking the introduction part which discusses the description of barriers, in Sect. 2, the problem statement is explained. The development of a conceptual framework for identifying and prioritizing the barriers can be described in Sect. 3. In Sect. 4, application of proposed model has been discussed. In Sect. 5, results and discussions are provided to demonstrate the feasibility of the proposed framework. Finally, the concluding remarks are provided in Sect. 6.

2   Key Problem Description

For managing sustainable supply chain, the manufacturing industries need to apply green procurement preferences to promote environmental initiatives in a systematic way. Nowadays, sustainability is highly relevant issue for exporters, as it significantly influences the Pakistani industries. In order to create sustainable business environment, organizations need to consider green issues in supply chain process of product procurement and its distribution to consumer in building a sustainable brand image [7]. The SSCM can be defined in simplest words as the major efforts to smoothly running the whole supply chain system along with fulfilling the environmental parameter or requirements in longer time. The innovations in the previous century only focused on the waste reduction for solely for the economic purposes but it lacks the focus on the environment. This was the research gap that now has attraction for the research opportunities. The recent century developed a new concept named as Green focusing on the protecting natural green environment. This new concept had changed the whole scenario of the research direction and gained excellent recognition [20]. Most Asian industries have been adopting the SSCM practices while replacing the conventional SCM practices. There were certain barriers that were highlighted during this conversion process. The whole supply chain systems were developed and developing according to SSCM practices. This implementation process is very complicated and it needs time and in-depth research to identify the weaknesses of the system while transferring from conventional to sustainable environment [21].

By developing the environment friendly product, industries can be sustainable socially and economically. Based on the academic and applied literature, it is quite evident that there has been a gap in the research to evaluate it more in the authentic results [16]. In the literature, few articles are available about the developing countries and particularly Pakistan [13]. Some of the researches were conducted in the India, Turkey, Iran and China about the green supply chain management practices adoption. In this research, well-structured questionnaires were designed to collect the data from different industrial sectors. The present problem deals in tackling the below proposed objectives;

1. To identify the barriers of supply chain management associated with the environmental sustainability at industrial context.
2. To evaluate the identi?ed essential barriers to priority by determining of their relative importance in adoption of sustainable practices.
3. To suggest roadmap regarding adoption of greenness in supply chain to ensure increased market share and sustained industrial environment.

Based on the above perspectives, most of the researches focused SSCM adoption, but the researchers failed to analyze that how to overcome on barriers against SSCM adoption. In recent years, mostly Pakistani organizations realize that modeling the barriers empirically in industrial supply chain play a vital role in the success of any business enterprise and overcoming of hurdles in a supply chain is the core process. Few Pakistani manufacturing organizations have implemented sustainable supply chain strategies through linkage with supply chain partners. But the ratio of success is very low due to certain barriers. Therefore, to overcome on these barriers a sustainability-based framework has been developed to rank the potential barriers along solutions in a stepwise manner [15].

Followings are the contributions of present research;

- Developed a fuzzy AHP model based on sustainability factors regarding adoption of sustainable supply chain practices.
- Provided empirical evidences on relationship between Sustainable Supply Chain Practices SSCP and eradication of barriers.
- Analysis of sustainable supply chain practices in the industrial sector of Pakistan.
- Provided guidelines to stakeholders and policy makers about their approach towards eco-environment.

This problem is categorized into three hierarchy decision process levels and the same is shown in Fig. 1. The three level hierarchical processes were as follows; Level-I) the overall goal of proposed problem, Level-II) this level shows category of main barriers, Level-III) this level represents the type of sub-barriers

Therefore, research is needed on the identification of essential barriers for SSCM adoption in Pakistani scenario. The present research conducted in the major industries of Pakistan and majority of the exporters were considered under study relevant to production and operational departments as well supply chain

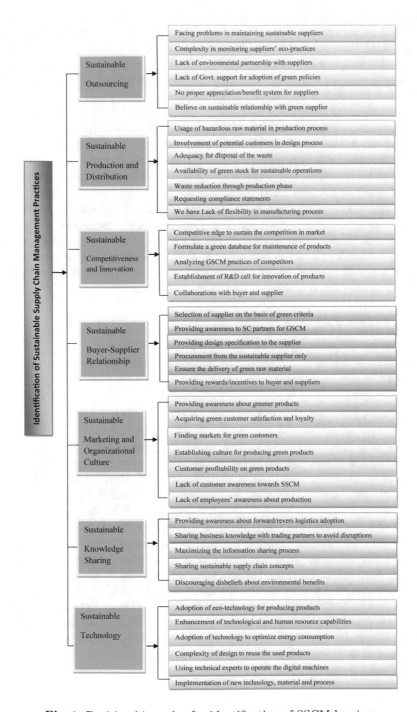

Fig. 1. Decision hierarchy for identification of SSCM barriers

managers. There are many reasons behind the failure of the adoption of SSCM practices in the organizations. These reasons were categorized into two classifications as internal and external barriers. The internal politics, lack of influence and institutional norms were included in the category of the internal barriers. The customer switching, lack of knowledge, etc. were included in the category of the internal barriers.

The hierarchical structure of these barriers is was very important step in the identification of process. This study considered sustainable outsourcing, sustainable production and distribution, sustainable competitiveness and innovation, sustainable buyer-supplier relationship, sustainable marketing and organizational culture, sustainable knowledge sharing, and sustainable technology as the seven most important SCM practices.

The major seven attributes comprising of 42 sub-attributes were identified after extensive review of literature review and discussion with supply chain experts. The experts were included from the Production and Operations, Research & Development, Merchandizing, Compliance and Audit Departments. Hence, this study offers a novel approach to understanding the barriers to SSCM implementation from a Pakistani industry perspective.

3   Solution Methodology

This section discusses about proposed framework and development of research instrument in the form of questionnaire.

3.1   Overview of Fuzzy AHP

The fuzzy AHP technique extends Saatys AHP by combining it with fuzzy logic and theory. In this method, fuzzy scales are adopted to check the relative intensity of the attributes in comparison with other attributes. Hence, a fuzzy judgment matrix can be structured and final output of alternatives is also depicted by using fuzzy variables. The feasible solution is obtained by providing the inputs in terms of linguistics expression using arithmetic operators. In this approach, employing fuzzy numbers to indicate the relative importance of one barrier type over the other, a fuzzy judgment vector is then obtained for each attribute. A fuzzy pair-wise comparison matrix is developed to compute the criterion weight. Table 1 represents the meaning of linguistic expressions in the form of fuzzy numbers and Table 2 shows the random consistency index RCI to compute the consistency ratio CR. Finally, experts are requested to provide their input in terms of linguistic expressions using Fig. 2. The linguistic expression further converted and analyzed for obtaining the weights vector.

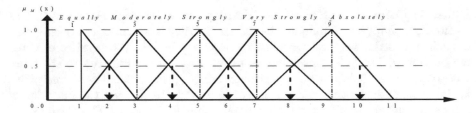

Fig. 2. Fuzzy membership function for linguistic expressions for attribute and sub-attribute

4 Application of Proposed Model

Research objectives were linked to effective policies of available resources through current supply chain practices; bridge the gap between domestic and international studies, suggest roadmap to increase productivity and profitability of the organizations in Pakistan. In addition, to highlight the challenges and prospects for the industrialist by making fuzzy logic analysis through field survey and experts of supply chain from different industries in Pakistan. This research has been conducted to evaluate the constructs of the sustainable supply chain management practices, supporting technology and implementation challenges.

4.1 Developing the Questionnaire

In this study, a cross-sectional research instrument was designed for survey. As the choice of suitable research design is the real essence of getting the solution of the research problem as well as the decision about the selection of the correct and right methods pertinent to the problem under study. The pre-testing questionnaire was circulated among industrial experts of various industries of Pakistan. After the inputs of industrial experts most essential barriers were identified using fuzzy AHP approach.

4.2 Collection of Data

This research selected and targeted the managers in the strategic and tactical levels from different industrial sectors in Pakistan. In order to initiate the research work, the objectives of the field survey, linked with the sustainable supply chain concepts regarding SSCM, were also introduced to target potential respondents by valid documents to assure that they clearly grasp multiple items in the field questionnaire, that includes the visions and concrete objectives of the proposed research, and also the usage of data for further evaluation.

Phase 1: Conducting Survey to Identify the Potential Barriers In this section, the proposed technique adopted for barriers assessment has been developed in industrial perspectives. The research methodology comprises of three major phases. The first phase requires problem-oriented industries to structure hierarchy of the whole attributes which are affecting the firms. The second phase is the systematic process which assists the experts to provide the intensity of inputs judgment regarding the attributes using fuzzy AHP. The third phase determines the scores vector of different attributes by analyzing them under seven different attributes namely; sustainable outsourcing, sustainable production and distribution, sustainable competitiveness and innovation, sustainable buyer-supplier relationship, sustainable marketing and organizational culture, sustainable knowledge sharing and sustainable technology. Finally, comparison of prioritization of results and policy implications has been discussed.

Phase 2: Identification of Essential Barriers for SSCM Implementations Research was conducted in the major industries of Pakistan and majority of the exporters were considered under study relevant to production and operational departments as well supply chain managers. The heterogeneous industrial sector is highly diverse and various complicated. In these days, the emerging industries have experienced a great deal of dynamic change with global sourcing and rising of price competition. Seven major attributes consisting of 42 sub-attributes were identified after detailed discussion with supply chain experts. The procedural ways of suggested methodology regarding selected study are given in Fig. 3.

Tables 1, 2 and 3 are used for assigning the inputs to the practices and further calculating the consistency ratio among the variables.

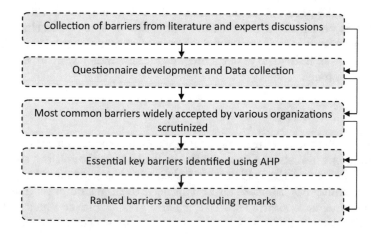

Fig. 3. Flowchart of the proposed research

Table 1. Scale for relative importance used in the pairwise comparison matrix

Intensity of importance	Fuzzy number	Linguistic variables	Triangular fuzzy numbers (TFNs)	Reciprocal of TFNs
1	$\tilde{1}$	Equally important	(1, 1, 3)	(0.33, 1.00, 1.00)
3	$\tilde{3}$	Weekly important	(1, 3, 5)	(0.20, 0.33, 1.00)
5	$\tilde{5}$	Strongly important	(3, 5, 7)	(0.14, 0.20, 0.33)
7	$\tilde{7}$	Very strongly important	(5, 7, 9)	(0.11, 0.14, 0.20)
9	$\tilde{9}$	Extremely more important	(7, 9, 11)	(0.09, 0.11, 0.14)

Table 2. The random consistency index

Size (n)	1	2	3	4	5	6	7	8
RI	0	0	0.52	0.89	1.11	1.25	1.35	1.4

Table 3. Fuzzy evaluation scores for alternative

Linguistic variables	Corresponding TFNs
Very poor (VP)	(1, 1, 3)
Poor (P)	(1, 3, 5)
Medium (M)	(3, 5, 7)
Good (G)	(5, 7, 9)
Very good (VG)	(7, 9, 11)

5   Results and Discussions

From Tables 4, 5, 6, 7, 8, 9, 10 and 11, the experts assign their input for construction of pair-wise comparison matrix. The input values were taken based on the fuzzy linguistic as well as numeral scale for making comparison with all major criterions by considering the scale of relative importance used in pairwise comparison matrix. The pair-wise comparison of main attributes was done by starting with the Sustainable Outsourcing C_1 criteria with Sustainable Production and Distribution C_2, Sustainable Competitiveness and Innovation C_3, Sustainable Buyer-Supplier Relationship C_4, Sustainable Marketing and Organizational Culture C_5, Sustainable Knowledge Sharing C_6 and Sustainable Technology C_7 at last. Similarly, same comparison of the other attributes done till the matrix is completely filled through experts opinions. Same procedure adopted to make comparison between sub-attributes.

Table 4. Pairwise comparison matrix of the major criterion used in SSCM practices

	C_1	C_2	C_3	C_4	C_5	C_6	C_7
C_1	1	2	3	2	3	2	3
C_2	0.5	1	3	2	3	2	2
C_3	0.33	0.33	1	2	2	2	2
C_4	0.5	0.5	0.5	1	3	2	3
C_5	0.33	0.33	0.5	0.33	1	2	2
C_6	0.5	0.5	0.5	0.5	0.5	1	2
C_7	0.33	0.5	0.5	0.33	0.5	0.5	1

After structuring a decision hierarchy, the weights of the criteria of SSCM initiative in supply chain were calculated by using fuzzy AHP and fuzzy set theory. The computation of arithmetic mean of the values found from the experts' evaluation and finally evaluation matrix was developed. The results obtained from the all sub-attribute pair-wise comparison matrices are shown in Table 12. Consistency ratios of all sub-criterion matrices with weight are computed to predict whether the inputs of experts are consistent or not. The acceptable value of CR should be less than 0.1 or 10% or equal.

Table 5. Pairwise comparison matrix of the sub-criteria with respect to the sustainable outsourcing attributes

	O_1	O_2	O_3	O_4	O_5	O_6
O_1	1	2	3	2	3	7
O_2	0.5	1	3	2	3	3
O_3	0.33	0.33	1	2	2	2
O_4	0.5	0.5	0.5	1	2	3
O_5	0.33	0.33	0.5	0.5	1	2
O_6	0.14	0.33	0.5	0.33	0.5	1

In Table 12, weights of assessment attributes and sub-attributes with respect to the case scenario for SSCM practices have been shown. Seven major attributes and 42 sub-attributes along weights and consistency values are shown with a final ranking. This table depicts the overall ranking of all sub-attributes in descending order.

Table 6. Pairwise comparison matrix of the sub-criteria with respect to the sustainable production and distribution attributes

	P_1	P_2	P_3	P_4	P_5	P_6	P_7
P_1	1	2	3	3	2	3	2
P_2	0.5	1	2	2	4	3	2
P_3	0.33	0.5	1	2	3	2	3
P_4	0.33	0.5	0.5	1	2	3	3
P_5	0.5	0.25	0.33	0.5	1	3	2
P_6	0.33	0.33	0.5	0.33	0.33	1	2
P_7	0.5	0.5	0.33	0.33	0.5	0.5	1

Table 7. Pairwise comparison matrix of the sub-criteria with respect to the sustainable competitiveness and innovation attributes

	C_1	C_2	C_3	C_4
C_1	1	3	4	2
C_2	0.33	1	3	2
C_3	0.25	0.33	1	2
C_4	0.5	0.5	0.5	1

Table 8. Pairwise comparison matrix of the sub-criteria with respect to the sustainable buyer-supplier relationship attributes

	B_1	B_2	B_3	B_4	B_5	B_6	B_7
B_1	1	2	2	3	2	3	2
B_2	0.5	1	3	5	3	4	2
B_3	0.5	0.33	1	2	3	5	2
B_4	0.33	0.2	0.5	1	2	3	3
B_5	0.5	0.33	0.33	0.5	1	4	2
B_6	0.33	0.33	0.2	0.33	0.25	1	2
B_7	0.2	0.5	0.5	0.33	0.5	0.5	1

After calculating the final weights, it is easy to infer that attribute sustainable outsourcing barriers have greater influence in decision-making of manufacturing sector. Similarly, the rest of attributes are also ranked in Table 12 and can be easily seen. In order to tackle the challenges in implementation of SSCM, the decision makers provide fruitful recommendation to evaluate the uncertainties of the case industries in holistic risky supply chain environment. Hence, the decision makers advised that SSCM should not limit to the eco-friendly environment aspects but also give emphasis on the non-eco criterion. In this way, the experts will be in a position to capture a transparent picture of the con-

Table 9. Pairwise comparison matrix of the sub-criteria with respect to the sustainable marketing and organizational culture attributes

	M₁	M₂	M₃	M₄	M₅	M₆	M₇
M₁	1	3	2	2	3	2	2
M₂	0.33	1	0.3	1	3	3	2
M₃	0.5	3	1	0.33	3	2	2
M₄	0.5	1	3	1	5	3	3
M₅	0.33	0.33	0.33	0.2	1	2	2
M₆	0.5	0.33	0.5	0.33	0.5	1	2
M₇	0.5	0.5	0.5	0.33	0.5	0.5	1

Table 10. Pairwise comparison matrix of the sub-criteria with respect to the sustainable knowledge sharing attributes

	K₁	K₂	K₃	K₄	K₅
K₁	1	0.33	2	3	2
K₂	3	1	3	2	5
K₃	0.2	0.33	1	2	2
K₄	0.33	0.5	0.5	1	2
K₅	0.5	0.2	0.5	0.5	1

Table 11. Pairwise comparison matrix of the sub-criteria with respect to the sustainable technology attributes

	T₁	T₂	T₃	T₄	T₅	T₆
T₁	1	3	9	1	5	3
T₂	0.33	1	7	0.2	5	4
T₃	0.11	0.14	1	0.11	0.2	3
T₄	1	5	9	1	7	5
T₅	0.2	0.2	5	0.14	1	2
T₆	0.33	0.25	0.33	0.2	0.5	1

text of SSCM implementation through the assessment process which can pave the way to improve the production level and sustain the competitive edge with customization.

6   Concluding Remarks and Policy Implications

6.1   Concluding Remarks

This study represents an important contribution to the development of sustainable supply chain practices by overcoming on barriers. A significant number of

Table 12. Weights of assessment criteria and sub-criteria with respect to the case scenario for SSCM practices

Major criterion	Major criterion weight	Sub-criteria	Consistency ratio (CR)	Ratio weight	Final weight	Ranking
Sustainable outsourcing	0.2735	O_1	0.064	0.3497	0.0957	1
		O_2		0.241	0.0659	2
		O_3		0.136	0.0372	6
		O_4		0.1326	0.0363	9
		O_5		0.0857	0.0234	16
		O_6		0.055	0.015	25
Sustainable production and distribution	0.2118	P_1	0.0988	0.2692	0.057	4
		P_2		0.2172	0.046	5
		P_3		0.1611	0.0341	12
		P_4		0.1322	0.028	13
		P_5		0.0927	0.0196	20
		P_6		0.0663	0.014	28
		P_7		0.0612	0.013	29
Sustainable competitiveness and innovation	0.1374	C_1	0.0996	0.4779	0.0657	3
		C_2		0.2561	0.0352	11
		C_3		0.1376	0.0189	22
		C_4		0.1284	0.0176	23
Sustainable buyer-supplier relationship	0.1425	B_1	0.0862	0.2493	0.0355	10
		B_2		0.2574	0.0367	8
		B_3		0.1701	0.0242	14
		B_4		0.1138	0.0162	24
		B_5		0.0986	0.0141	27
		B_6		0.0549	0.0078	35
		B_7		0.0559	0.008	34
Sustainable marketing and organizational culture	0.0871	M_1	0.0988	0.2557	0.0223	
		M_2		0.1365	0.0119	17
		M_3		0.1621	0.0141	30
		M_4		0.2294	0.02	26
		M_5		0.0757	0.0066	18
		M_6		0.0751	0.0065	37
		M_7		0.0654	0.0057	38
						39
Sustainable knowledge sharing	0.0854	K_1	0.0901	0.2303	0.0197	19
		K_2		0.4301	0.0367	7
		K_3		0.134	0.0114	31
		K_4		0.122	0.0104	32
		K_5		0.0836	0.0071	36

continued

Table 12. continued

Major criterion	Major criterion weight	Sub-criteria	Consistency ratio (CR)	Ratio weight	Final weight	Ranking
Sustainable technology	0.0622	T_1	0.0876	0.306	0.019	21
		T_2		0.163	0.0101	33
		T_3		0.0357	0.0022	42
		T_4		0.3837	0.0239	15
		T_5		0.0696	0.0043	40
		T_6		0.042	0.0026	41

organizations still consider SCM as being the same as integrated logistics management. Only few organizations have realized the importance of SCM, they lack an understanding of what means a comprehensive set of supply chain management practices. Measures to provide supply chain management practices in this research may be useful for managers of supply chain management in assessing their present supply chain practices. These can support the managers to discover the strengths as well as weaknesses of their supply chain management practices. The managers are able to conduct research and benchmarking of the level of adoption of SCM practices with suppliers and customers. Concerning the social implications, present study can contribute to a better understanding of supply chain and its management practices. This research will be very helpful for researchers in future studies of supply chain practices and investigate their relationships with other organization operations and outcomes for example competitive advantages and organization performance.

6.2   Policy Implications

This research is one the few studies which have identified common and essential practices along barriers faced by the different industrial sectors of Pakistan. The outcomes of this research can assist the CEOs of the firms who are willing to convert their traditional or conventional approach towards green one SSCM. Export figures of Pakistan's industries are dropping from many years, SSCM is one of the strategies that need to be adopted in order to sustain in the market. This study can be used by the Pakistani industry to enhance productivity, reduce waste and conserve energy. This research can help in prioritizing essential barriers and methods to remove them with the passage of time. Present study provides decision support tool to the industries while implementing SSCM and areas of improvements according to separate requirements of any firm. This study can assist multiple industries to eradicate essential barriers on preference basis for successful implementation of SSCM in Pakistan.

Acknowledgments. The authors wish to thank the anonymous referees for their helpful and constructive comments and suggestions. The work is supported by the Higher Education Commission of Pakistan, Islamabad (Grant No. No. 21-/780/SRGP/R&D/HEC/2016).

References

1. Carter, C.R., Dresner, M.: Purchasing's role in environmental management: Cross-functional development of grounded theory. J. Supply Chain. Manag. **37**(2), 12–27 (2001)
2. Carter, C.R., Rogers, D.S.: A framework of sustainable supply chain management: moving toward new theory. Int. J. Phys. Distrib. Logist. Manag. **38**(5), 360–387 (2008)
3. Chen, I.J., Paulraj, A.: Towards a theory of supply chain management: the constructs and measurements. J. Oper. Manag. **22**(2), 119–150 (2004)
4. Dan, B., Liu, F.: Study on green supply chain and its architecture. China Mech. Eng. **11**(4), 1233–1236 (2000)
5. Darnall, N., Jolley, G.J., Handfield, R.: Environmental management systems and green supply chain management: complements for sustainability? Bus. Strat. Environ. **17**(1), 30–45 (2008)
6. Diabat, A., Govindan, K.: An analysis of the drivers affecting the implementation of green supply chain management. Resour., Conserv. Recycl. **55**(6), 659–667 (2011)
7. Hashim, M., Nazam, M., Yao, L., Baig, S.A., Abrar, M., Zia-ur Rehman, M.: Application of multi-objective optimization based on genetic algorithm for sustainable strategic supplier selection under fuzzy environment. J. Ind. Eng. Manag. **10**(2), 188–212 (2017)
8. Kamran, A., Rizvi S.M.A.: Reason and trends for using packaged milk in pakistan: study of urban pakistani consumers. In: Proceedings of the Sixth International Conference on Management Science and Engineering Management, pp 909–924. Springer, Berlin (2013)
9. Kim, D., Cavusgil, S.T., Calantone, R.J.: Information system innovations and supply chain management: channel relationships and firm performance. J. Acad. Mark. Sci. **34**(1), 40–54 (2006)
10. Kumar, A., Zavadskas, E.K., Mangla, S.K., Agrawal, V., Sharma, K., Gupta, D.: When risks need attention: adoption of green supply chain initiatives in the pharmaceutical industry. Int. J. Prod. Res. 1–23 (2018)
11. Li, S., Ragu-Nathan, B., Ragu-Nathan, T., Rao, S.S.: The impact of supply chain management practices on competitive advantage and organizational performance. Omega **34**(2), 107–124 (2006)
12. Mau, M.: Supply chain management in agriculture–including economics aspects like responsibility and transparency/10, exploring diversity in the European agri-food system, pp. 28–31. Zaragoza, Spain (2002)
13. Mumtaz, U., Ali, Y., Petrillo, A., De Felice, F.: Identifying the critical factors of green supply chain management: Environmental benefits in pakistan. Sci. Total. Environ. **640**, 144–152 (2018)
14. Nazam, M., Xu, J., Tao, Z., Ahmad, J., Hashim, M.: A fuzzy ahp-topsis framework for the risk assessment of green supply chain implementation in the textile industry. Int. J. Supply Oper. Manag. **2**(1), 548 (2015)
15. Nourmohamadi Shalke, P., Paydar, M.M., Hajiaghaei-Keshteli, M.: Sustainable supplier selection and order allocation through quantity discounts. Int. J. Manag. Sci. Eng. Manag. **13**(1), 20–32 (2018)
16. Shekari, H., Shirazi, S., Afshari, M., Veyseh, S.: Analyzing the key factors affecting the green supply chain management: A case study of steel industry. Manag. Sci. Lett. **1**(4), 541–550 (2011)

17. Siddiqui, F., Haleem, A., Sharma, C.: The impact of supply chain management practices in total quality management practices and flexible system practices context: an empirical study in oil and gas industry. Glob. J. Flex. Syst. Manag. **13**(1), 11–23 (2012)
18. Singh, M., Kant, R.: Knowledge management barriers: An interpretive structural modeling approach. Int. J. Manag. Sci. Eng. Manag. **3**(2), 141–150 (2008)
19. Sloan, K., Klingenberg, B., Rider, C.: Towards sustainability: Examining the drivers and change process within smes. J. Manag. Sustain. **3**, 19 (2013)
20. Thuong, N.T.H., Zhang, R., Li, Z., Hong, P.T.D.: Multi-criteria evaluation of financial statement quality based on hesitant fuzzy judgments with assessing attitude. Int. J. Manag. Sci. Eng. Manag. **13**(4), 254–264 (2018)
21. Wu, K.J., Tseng, M.L., Vy, T.: Evaluation the drivers of green supply chain management practices in uncertainty. Procedia - Soc. Behav. Sci. **25**, 384–397 (2011)
22. Zhu, Q., Sarkis, J., Lai, Kh: Confirmation of a measurement model for green supply chain management practices implementation. Int. J. Prod. Econ. **111**(2), 261–273 (2008)
23. Zhu, Q., Geng, Y., Fujita, T., Hashimoto, S.: Green supply chain management in leading manufacturers: Case studies in japanese large companies. Manag. Res. Rev. **33**(4), 380–392 (2010)

Impact of Supplier's Lead Time on Strategic Assembler-Supplier Relationship in Pricing Sensitive Market

Jiayi Wei[1,2(✉)]

[1] The Shiyan School Attached to Shijiazhuang No. 2 Middle School,
Shijiazhuang 051430, People's Republic of China
[2] Shijiazhuang No. 17 Middle School, Shijiazhuang 050051,
People's Republic of China
jiayiwei2000@gmail.com

Abstract. Assemblers need to react to dynamic market in a timely manner in order to succeed in today's competitive environment. To achieve timely delivery, some assembler may decide to buy and stock (BS) components from its suppliers and some may choose to contract its suppliers under vendor-managed inventory (VMI). This paper examines the impact of supplier's lead-time on the assembler's strategic choice of its partnership with suppliers. Analysis indicates that assembler will generate highest profit under BS and lowest profit under VMI. In addition, assembler will get higher profit by contracting with the lower lead-time supplier under VMI.

Keywords: Inventory management · Vendor-managed inventory · Time-dependent pricing

1 Introduction

Due to the rapid development of global supply chains, it is crucial for assemblers to react to a dynamic market and deliver their products in the most cost competitive and timely manner in order to succeed in today's competitive environment. When the product price is time sensitive, timely delivery is essential since any delay of a shipment may lead to a price decrease for the assembled final product. Some assembler may decide to stock some components before they receive the actual order to achieve timely delivery. However, the severe obsolescence poses significant risk to the assembler when the client's order falls short of expectation. Consequently, effectively management of component inventory is quite important for the assembler in order to achieve superior cost and time performance. With the popularity of vendor-managed inventory (VMI) program, more and more suppliers starts to manage their own inventory at the customer's

© Springer Nature Switzerland AG 2020
J. Xu et al. (Eds.): ICMSEM2019 2019, AISC 1002, pp. 365–375, 2020.
https://doi.org/10.1007/978-3-030-21255-1_28

location where the suppliers are not paid for their delivery inventory until their goods are actually used. Companies such as Motorola, Intel, Lucent Technology are examples of VMI implementations.

Existing literature on component procurement strategy under time-dependent pricing either focus on the optimal component stocking quantities for components of an assembler-to-order product(see e.g., Hsu et al. [10]) or the pricing policy to coordinate the component suppliers under VMI scheme (see e.g., Fang et al. [8]). It is likely to happen in reality that the assembler may not keep the same relationship with all suppliers, buy and stock all components or contract under VMI scheme with all suppliers. The assembler may buy components directly from some and then contract with the others. This give rise to our main research question: given a group of suppliers differs in their lead time, which type the assembler will choose to buy and which to contract.

In this paper, we study the strategic assembler-supplier relationship management in an ATO environment with uncertain demand and time dependent product prices. Specifically, we consider a business setting where a contract assembler anticipating a future order for an ATO product composed of n different components with each of them provided by an independent supplier. The demand will be realized at time t^0, after which the products are delivered in multiple shipments whenever a batch of the product is assembled. However, the assembler will receive a different unit price for the products in each shipment with longer delivery time result in lower unit price. We assume that the assembler has abundant capacity and no shortage is allowed, thus the assembler must deliver the full order quantity. We also assume that the assembly time is negligible, so the initial shipment will be delivered to the client at time t^0. Our preliminary analysis indicates that assembler will generates the highest profit when he buy all the components from the suppliers and lowest profit when he contract with all the suppliers under VMI scheme. This is intuitive and make sense since the assembler will bear the risk of overstocking when he choose to stock the components by himself, which give rise to higher profit. When he transfers the risk to the suppliers, he need to offer a premium to the suppliers under the VMI contract in order to induce them to produce and stock. Another important result is that it would be better for the assembler to contract with the short lead time suppliers when all the suppliers are of equal cost. The reason is that the longer the supplier's lead time, the higher the premium the assembler need to offer.

2   Literature Review

Generally speaking, literature on ATO systems can be classified into two categories: centralized systems and decentralized systems. Song and Zipkin [14] present a general formulation of ATO systems [14]. Atan et al. [2] reviewed the recent literature on ATO systems and proposed some potential future directions. Some assume stochastic demand with known distribution; see e.g., Gurnani et al. [9], Song et al. [15]. Some consider deterministic demand with random component procurement lead time; see e.g., Chu et al. [5] and Proth et al. [13]. Research on decentralized assembly systems mainly focus on the contractual arrangements between assembler and suppliers. Ammar and Dolgui [1] examine an optimisation problem for component replenishment in two-level assembly systems under stochastic lead times. Deza et al. [7] studied component commonality for periodic review assemble-to-order systems [7]. Lee et al. [11] examine VMI systems with stockout-cost sharing under limited storage capacity and provides conditions on which VMI can coordinate the supply chain [11]. Mateen and Chatterjee [12] consider different ways of structuring the replenishment policy under VMI and develop models to exploit different cost trade-offs in the supply chain [12]. The VMI contract has been found to improve efficiency in the supply chain and increase total profits (Yu et al. [17]). Most literature assume a single order opportunity see e.g., Bernstein and DeCroiox [3] and Wang [16]. The main difference in our research is that we allow multiple deliveries of the final products which enable suppliers of different lead time to choose different production quantities.

Time dependent pricing is the key feature of our research that emphasis the importance of time-based competition. The competitive and time-sensitive market requires the assembler to carefully manage the product delivery time performance. See e.g., Cachon and Parker [4] and Boyaci and Ray [6]. A recent work by Hsu et al. [10] is one of the most relevant to this research. They develop and analyze an optimization model to determine the optimal stocking quantities for components of an ATO system with uncertain demand and the products price and components cost depend on their delivery time [10]. In our research we consider similar price-dependent setting; however we assume the component cost keeps the same as we focus on the suppliers' lead time difference. Another closely related research is Fang et al. [8]. Their model set up is also based on a price-dependent environment, where they explore how the assembler can use a VMI scheme to coordinate the component suppliers' production decisions. Efficient algorithm is developed to find the assembler's optimal pricing scheme. The main difference between our research and the above mentioned two literatures is that we assume assembler will choose some components to stock and deal with some other components suppliers under VMI contract instead of stock all, e.g., Hsu et al. [10] and contract with all the suppliers, e.g., Fang et al. [8,10]. We aim to find the effect of lead time in strategic assembler's choice of supplier's relationship.

3   Model Framework

For model tractability, consider a simple assemble to order system with one assembler and a single product consists of 2 components each produced by an independent supplier at a fixed unit cost c. We assume each final product require one unit of each component without loss of generality. The demand denoted by D is uncertain with $f(.)$ and $F(.)$ be the probability density function and cumulative distribution function respectively. Facing uncertain demand, the assembler will buy and stock some components from the corresponding suppliers, at the same time uses a VMI arrangement to manage the other suppliers. At some time before the actual order quantity is confirmed, the customer will first provides the BOM of the final product. The assembler need to decide the quantity to buy from the suppliers as well as the pricing scheme to the suppliers under VMI scheme, based on which the suppliers choose their production quantities. If the assembler want to contract with supplier denote as i, he will offer a pricing scheme P_i^t, which is decreasing in t, to encourage the supplier to produce the components Q_i before t^0 to achieve early product delivery for a higher product price, see Fang et al. [8]. If the assembler decides to by from supplier denote as j, he will order Q_j, see Hsu et al. [10].

We assume all suppliers have sufficient time and capacity to produce the initial quantity and the assembler's assembly time negligible. Suppose the suppliers can provide a reliable or constant supply lead time L_i. We number the suppliers such that $L_1 \leq L_2$, and we also denote $L_0 = 0$ for notation convenience. If the initial shipment is not sufficient to meet the demand due to shortage of one or more components, the corresponding suppliers will start a second batch of production at time t^0 in order to satisfy the remaining demand. When the assembler receives the additional components, he assemble and deliver it immediatel Denote P^0, P^1, P^2 be the corresponding unit prices received by the assembler. Recall that we have the assumption $L_1 \leq L_2$. We assume $P^0 \geq P^1 \geq P^2$ in order to model the assumption that longer delivery time lead to lower unit product price. Accordingly the contract supplier will receive a price of P_i^t $\{t = 0, 1, 2\}$ for each units of his component used at time $t^0 + L_t$, where P_i^t is a decreasing function in t. Each supplier has a reservation price, which is assumed equal to the cost c without loss of generality, to cover the minimum return for the product where $P_i^0 \geq P_i^1 \geq P_i^2 \geq c$. We assume all the suppliers' production cost keep the same during the whole process. We do not consider the holding costs as noted in Fang et al. [8] and there is no salvage value. We also assume the assembler incurs no cost for assembling and delivering the final product to the client [8]. All information include the demand distribution, product prices, production lead times, and costs is common knowledge to all parties.

The sequence of the decision making process is modeled as a Stackelberg game. The assembler, acting as the leader, moves first to specify the pricing scheme to the contract suppliers and also the quantity he will buy from the other suppliers. The contract supplier, acting as the follower, choose initial production quantity accordingly. For each given pricing scheme chosen by the assembler, the supplier's profit depends on his own production quantity Q_i and also the assembler's order quantity Q_j from other suppliers. y in order to get a higher price. Based on this model setup, there will be at most 3 possible delivery opportunities at $t^0, t^0 + L_1, t^0 + L_2$ respectively. A graphical illustration of the timeline of events is provided in Fig. 1.

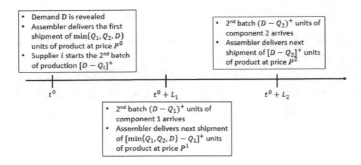

Fig. 1. Timeline of Events

Proposition 28.1. $Q_1 \leq Q_2$

Proposition 28.1 states that both the assembler and the contract supplier will never choose to stock more component than any others that has a longer production lead time. This results can be intuitively explained by the following statement. If demand $D \geq Q_1 \geq Q_2$, any component 1 in excess of Q_2 needs to wait for the second batch of component 2 to arrive before the final assembly. The revenue is of the same as the condition $Q_1 = Q_2$. In addition, if $Q_1 \geq D \geq Q_2$, it would lead to over stocking. Thus we should have $Q_1 \leq Q_2$.

Given the model set up, assembler can choose either to buy or contract under VMI with each supplier. This result in four strategies for the assembler as listed blow and analyzed in the following scenarios: buy from both suppliers (BB), contract with both suppliers (CC), contract with supplier 1 and buy from supplier 2 (CB), buy from supplier 1 and contract with supplier 1 (BC).

4   Model Analysis

4.1   BB: Buy from Both

Under this scenario, the assembler buy directly from the suppliers as in Hsu et al. [10]. He only need to decide Q_1^{BB} and Q_2^{BB} with the constraint that $Q_1^{BB} \leq Q_2^{BB}$.

$$
\begin{aligned}
Max\Pi_A^{BB} &= P^0 E[min\{Q_1^{BB}, D\}] + P^1 E[min\{Q_2^{BB}, D\} - Q_1^{BB}]^+ + P^2 E[D - Q_2^{BB}]^+ \quad (1) \\
&\quad - cE[max\{Q_1^{BB}, D\}] - cE[max\{Q_2^{BB}, D\}] \\
&= (P^2 - 2c)E(D) + (P^0 - P^1 + c)\int_0^{Q_1^{BB}} \bar{F}(x)dx - cQ_1^{BB} \\
&\quad + (P^1 - P^2 + c)\int_0^{Q_2^{BB}} \bar{F}(x)dx - cQ_2^{BB} \\
&\quad S.t. \ Q_1^{BB} \leq Q_2^{BB}
\end{aligned}
$$

Take the first and second order derivatives of Π_A^{BB} with respective to Q_1^{BB} and Q_2^{BB}, we can have $\frac{\partial \Pi_A^{BB}}{\partial Q_1^{BB}} = (P^0 - P^1 + c)\bar{F}(Q_1^{BB}) - c$, $\frac{\partial^2 \Pi_A^{BB}}{\partial^2 Q_1^{BB}} = -(P^0 -$

370   J. Wei

$P^1 + c)f(Q_1^{BB}) \leq 0$; $\frac{\partial \Pi_A^{BB}}{\partial Q_2^{BB}} = (P^1 - P^2 + c)\bar{F}(Q_2^{BB}) - c$, $\frac{\partial^2 \Pi_A^{BB}}{\partial^2 Q_2^{BB}} = -(P^1 - P^2 +$
$c)f(Q_2^{BB}) \leq 0$. Since Π_A^{BB} is jointly concave with respect to Q_1^{BB} and Q_2^{BB}, we
can set the first order derivative to zero to get the unconstraint optimal $Q_1^{BB'}$
and $Q_2^{BB'}$. If $Q_1^{BB'} \leq Q_2^{BB'}$, then $Q_1^{BB*} = Q_1^{BB'}$ with $\bar{F}(Q_1^{BB'}) = \frac{c}{P^0 - P^1 + c}$ and
$Q_2^{BB*} = Q_2^{BB'}$ with $\bar{F}(Q_2^{BB'}) = \frac{c}{P^1 - P^2 + c}$. Or else, set $Q_1^{BB} = Q_2^{BB} = Q^{BB}$ and
do the optimization again with the same routine, we can have $Q_1^{BB*} = Q_2^{BB*} = Q^{BB'}$ with $\bar{F}(Q^{BB'}) = \frac{2c}{P^0 - P^2 + 2c}$.

Denote $H(Q) = \frac{1}{\bar{F}(Q)}$, it increases with Q because $\bar{F}(x)$ is a decreasing
function, we can easily get the following result:

- If $P^0 - P^1 \leq P^1 - P^2$, $H(Q_1^{BB*}) = \frac{P^0 - P^1 + c}{c}$ and $H(Q_2^{BB*}) = \frac{P^1 - P^2 + c}{c}$
- If $P^0 - P^1 \geq P^1 - P^2$, $H(Q_1^{BB*}) = H(Q_2^{BB*}) = \frac{P^0 - P^2 + 2c}{2c}$

4.2 CC: Contract with Both

In this scenario, the assembler act as the leader who offers the pricing scheme
$P_i^0 \geq P_i^1 \geq P_i^2$ to supplier i. The suppliers act as follower and choose their own
production quantity simultaneously as in Fang et al. [8].

(1) Supplier problem

$$Max\Pi_{S_1}^{CC} = P_1^0 E[min\{Q_1^{CC}, D\}] + P_1^1 E[min\{Q_2^{CC}, D\} - Q_1^{CC}]^+ \tag{2}$$
$$+ P_1^2 E[D - Q_2^{CC}]^+ - cE[max\{Q_i^{CC}, D\}]$$
$$= (P_1^2 - c)E(D) + (P_1^0 - P_1^1 + c)\int_0^{Q_1^{CC}} \bar{F}(x)dx - cQ_1^{CC} + (P_1^1 - P_1^2)\int_0^{Q_2^{CC}} \bar{F}(x)dx$$

$$Max\Pi_{S_2}^{CC} = P_2^0 E[min\{Q_1^{CC}, D\}] + P_2^1 E[min\{Q_2^{CC}, D\} - Q_1^{CC}]^+ \tag{3}$$
$$+ P_2^2 E[D - Q_2^{CC}]^+ - cE[max\{Q_i^{CC}, D\}]$$
$$= (P_2^2 - c)E(D) + (P_2^0 - P_2^1)\int_0^{Q_1^{CC}} \bar{F}(x)dx + (P_2^1 - P_2^2 + c)\int_0^{Q_2^{CC}} \bar{F}(x)dx - cQ_2^{CC}$$

It is easy to check that Π_{Si}^C is concave in Q_i^{CC} and setting the first order deriva-
tive to zero we can easily get $\bar{F}(Q_1^{CC*}) = \frac{c}{P_1^0 - P_1^1 + c}$ and $\bar{F}(Q_2^{CC*}) = \frac{c}{P_2^1 - P_2^2 + c}$.

(2) Assembler problem

$$Max\ \Pi_A^{CC} = (P^0 - P_1^0 - P_2^0)E[min\{Q_1^{CC}, D\}] + (P^1 - P_1^1 - P_2^1)E[min\{Q_2^{CC}, D\} \tag{4}$$
$$-\ Q_1^{CC}]^+ + (P^2 - P_1^2 - P_2^2)E[D - Q_2^{CC}]^+$$
$$=\ (P^0 - P_1^0 - P_2^0)\int_0^{Q_1^{CC}} \bar{F}(x)dx + (P^1 - P_1^1 - P_2^1)\int_{Q_1^{CC}}^{Q_2^{CC}} \bar{F}(x)dx$$
$$+\ (P^2 - P_1^2 - P_2^2)(E(D) - \int_0^{Q_2^{CC}} \bar{F}(x)dx)$$
$$S.t.\ P_i^0 \geq P_i^1 \geq P_i^2 \geq c$$

One can easily get $P_1^2 = P_1^1 = P_2^2 = v$ and $P_2^0 = P_2^1$. Substitute back into
Π_A^{CC} we can have $\Pi_A^{CC} = (P^2 - 2c)E(D) + (P^0 - P^1 - (P_1^0 - P_1^1))\int_0^{Q_1^{CC}} \bar{F}(x)dx +$

$(P^1 - P^2 - (P_2^1 - P_2^2)) \int_0^{Q_2^{CC}} \bar{F}(x)dx$ Note that we have $\bar{F}(Q_i^{CC}) = \frac{c}{P_i^{i-1}-P_i^i+c}$ based on the supplier's optimal decision, which indicates a one-to-one correspondence between Q_i^{CC} and $P_i^{i-1} - P_i^i$. Thus Π_A^{CC} can be transformed to the following format to maximize Q_i^{CC}.

$$Max \ \Pi_A^{CC} = (P^2 - 2c)E(D) + (P^0 - P^1 + c(1 - \frac{1}{\bar{F}(Q_1^{CC})})) \int_0^{Q_1^{CC}} \bar{F}(x)dx \quad (5)$$

$$+ \ (P^1 - P^2 + c(1 - \frac{1}{\bar{F}(Q_2^{CC})})) \int_0^{Q_2^{CC}} \bar{F}(x)dx$$

$$S.t. \ Q_1^{CC} \le Q_2^{CC}$$

Denote $R(Q) = \frac{f(Q)}{F(Q)} \int_0^Q \bar{F}(x)dx$ and assume that $R(Q)$ is an increasing function of Q, which holds true for most distributions as listed in Fang et al. [8]. Take the first and second order derivatives with respect to Q_1^{CC} and Q_2^{CC}.

$\frac{\partial \Pi_A^{CC}}{\partial Q_1^{CC}} = (P^0 - P^1 + c(1 - \frac{1}{\bar{F}(Q_1^{CC})}))\bar{F}(Q_1^{CC}) - c\frac{f(Q_1^{CC})}{\bar{F}(Q_1^{CC})} \int_0^{Q_1^{CC}} \bar{F}(x)dx = (P^0 - P^1 + c)\bar{F}(Q_1^{CC}) - c(R(Q_1^{CC})+1); \frac{\partial^2 \Pi_A^{CC}}{\partial^2 Q_1^{CC}} = -(P^0 - P^1 + c)f(Q_1^{CC}) - c\frac{\partial R(Q_1^{CC})}{\partial Q_1^{CC}} \le 0$; $\frac{\partial \Pi_A^{CC}}{\partial Q_2^{CC}} = (P^1 - P^2 + c(1 - \frac{1}{\bar{F}(Q_2^{CC})}))\bar{F}(Q_2^{CC}) - c\frac{f(Q_2^{CC})}{\bar{F}(Q_2^{CC})} \int_0^{Q_2^{CC}} \bar{F}(x)dx = (P^1 - P^2 + c)\bar{F}(Q_2^{CC}) - c(R(Q_2^{CC})+1); \frac{\partial^2 \Pi_A^{CC}}{\partial^2 Q_2^{CC}} = -(P^1 - P^2 + c)f(Q_2^{CC}) - c\frac{\partial R(Q_2^{CC})}{\partial Q_2^{CC}} \le 0$

Since Π_A^{CC} is jointly concave with respect to Q_1^{CC} and Q_2^{CC}, which ensures the maximum of Π_A^{CC}. Set $\frac{\partial \Pi_A^{CC}}{\partial Q_1^{CC}} = 0$ and $\frac{\partial \Pi_A^{CC}}{\partial Q_2^{CC}} = 0$, we can get the unconstraint optimal $Q_1^{CC'}$ satisfying $\frac{P^0 - P^1 + c}{c} = \frac{R(Q_1^{CC'})+1}{\bar{F}(Q_1^{CC'})}$ and $Q_2^{CC'}$ satisfy $\frac{P^1 - P^2 + c}{c} = \frac{R(Q_2^{CC'})+1}{\bar{F}(Q_2^{CC'})}$. If $Q_1^{CC'} \le Q_2^{CC'}$, we can have $Q_1^* = Q_1^{CC'}$ and $Q_2^* = Q_2^{CC'}$. If not, we need to set $Q_1^{CC} = Q_2^{CC} = Q^{CC}$ and repeat the optimization process as listed above. we can have $Q_1^* = Q_2^* = Q^{CC'}$ with $\frac{P^0 - P^2 + 2c}{2c} = \frac{R(Q^{CC'})+1}{\bar{F}(Q^{CC'})}$.

Denote $K(Q) = \frac{R(Q)+1}{\bar{F}(Q)}$ and it is an increasing function with Q. Combing all the above results, we can get the final conclusion as below:

- If $P^0 - P^1 \le P^1 - P^2$, $K(Q_1^{CC*}) = \frac{P^0 - P^1 + c}{c}$ and $K(Q_2^{CC*}) = \frac{P^1 - P^2 + c}{c}$.
- If $P^0 - P^1 \ge P^1 - P^2$, $K(Q_1^{CC*}) = K(Q_2^{CC*}) = \frac{P^0 - P^2 + 2c}{2c}$

4.3   CB: Contract with Supplier 1

Under this scenario, the assembler contract with supplier 1 and buy directly from supplier 2. The assembler act as the leader who decides the quantity to buy from supplier 2 and offers a pricing scheme P_1^0, P_1^1, P_1^2 to supplier 1, who acts as the follower to decide the production quantity accordingly.

Given the pricing scheme, supplier 1's decision is of the same as the all contract condition. To be more specific, the $\bar{F}(Q_1^{CB*}) = \frac{c}{P_1^0 - P_1^1 + c}$.

For the assembler, he need to decide the pricing scheme as well as the quantity to buy from supplier 2 which is Q_2^{CB}.

$$Max \; \Pi_A^{CB} = (P^0 - P_1^0)E[min\{Q_1^{CB}, D\}] + (P^1 - P_1^1)E[min\{Q_2^{CB}, D\} - Q_1^{CB}]^+ \quad (6)$$
$$+ \; (P^2 - P_1^2)E[D - Q_2^{CB}]^+ - cE[max\{Q_2^{CB}, D\}]$$
$$S.t. \; P_1^0 \geq P_1^1 \geq P_1^2 \geq c$$

It is easy for us to verify that $P_1^2 = P_1^1 = c$, in this case the profit function can be written as below: $\Pi_A^{CB} = (P^2 - 2c)E(D) + (P^0 - P^1 - (P_1^0 - P_1^1))\int_0^{Q_1^{CB}} \bar{F}(x)dx + (P^1 - P^2 + c)\int_0^{Q_2^{CB}} \bar{F}(x)dx - cQ_2^{CB}$. From $\bar{F}(Q_1^{CB}) = \frac{c}{P_1^0 - P_1^1 + c}$, we can have $P_1^0 - P_1^1 = c(\frac{1}{\bar{F}(Q_1^{CB})} - 1)$. Since there is one to one correspondence between Q_1^{CB} and $P_1^0 - P_1^1$, the assembler's problem can be transformed as below:

$$Max \; \Pi_A^{CB} = (P^2 - 2c)E(D) + (P^0 - P^1 + c(1 - \frac{1}{\bar{F}(Q_1^{CB})}))\int_0^{Q_1^{CB}} \bar{F}(x)dx \quad (7)$$
$$+ \; (P^1 - P^2 + c)\int_0^{Q_2^{CB}} \bar{F}(x)dx - cQ_2^{CB}$$
$$S.t. \; Q_1^{CB} \leq Q_2^{CB}$$

With the same optimization methods as listed in the BB and CC case, we can easily ge the following final result:

- If $Q_1^{CB'} \leq Q_2^{CB'}$, $Q_1^{CB*} = Q_1^{CB'}$ with $K(Q_1^{CB'}) = \frac{P^0 - P^1 + c}{c}$ and $Q_2^{CB*} = Q_2^{CB'}$ with $H(Q_2^{CB'}) = \frac{P^1 - P^2 + c}{c}$.
- If $Q_1^{CB'} \geq Q_2^{CB'}$, $Q_1^{CB*} = Q_2^{CB*} = Q^{CB'}$ with $P^0 - P^2 + c + v = \frac{c(R(Q^{CB'}) + 2)}{\bar{F}(Q^{CB'})}$.
- $\Pi_A^{CB}(Q_1^{CB'}, Q_2^{CB'}) \geq \Pi_A^{CB}(Q^{CB'})$.

4.4 BC: Contract with Supplier 2

The assembler will decide the quantity to buy from supplier 1 and the pricing scheme to contract with supplier 2.

$$Max \; \Pi_A^{BC} = (P^0 - P_2^0)E[min\{Q_1^{BC}, D\}] + (P^1 - P_1^1)E[min\{Q_2^{BC}, D\} - Q_1^{BC}]^+ \quad (8)$$
$$+ \; (P^2 - P_2^2)E[D - Q_2^{BC}]^+ - cE[max\{Q_2^{BC}, D\}]$$
$$S.t. \; P_1^0 \geq P_2^1 \geq P_2^2 \geq c$$

Follow the same routine as contract with supplier 1 case, we can derive the final assembler's problem:

$$Max \; \Pi_A^{BC} = (P^2 - 2c)E(D) + (P^0 - P^1 + c)\int_0^{Q_1} \bar{F}(x)dx - cQ_1^{BC} \quad (9)$$
$$+ \; (P^1 - P^2 + c(1 - \frac{1}{\bar{F}(Q_2^{BC})}))\int_0^{Q_2^{BC}} \bar{F}(x)dx$$
$$S.t. \; Q_1^{BC} \leq Q_2^{BC}$$

It is easy to get the analogous final results as the contract with supplier 1 case. With the same optimization methods as listed in the BB and CC case, we can easily ge the following final result:

- If $Q_1^{BC'} \le Q_2^{BC'}$, $Q_1^{BC*} = Q_1^{BC'}$ with $H(Q_1^{BC'}) = \frac{P^0 - P^1 + c}{c}$ and $Q_2^{BC*} = Q_2^{BC'}$ with $K(Q_2^{BC'}) = \frac{P^1 - P^2 + c}{c}$.
- If $Q_1^{BC'} \ge Q_2^{BC'}$, $Q_1^{BC*} = Q_2^{BC*} = Q^{BC'}$ with $P^0 - P^2 + c + v = \frac{c(R(Q^{BC'}) + 2)}{F(Q^{BC'})}$.
- $\Pi_A^{BC}(Q_1^{BC'}, Q_2^{BC'}) \ge \Pi_A^{BC}(Q^{BC'})$

4.5   Comparison

Proposition 28.2. $\Pi_A^{BB*} \ge \Pi_A^{CB*} \ge \Pi_A^{BC*} \ge \Pi_A^{CC*}$

The assembler's profit under the BB structure is of the highest among the four possible structure and CC structure is of the lowest. The CB structure leads to a profit higher than BC but lower than BB. However the assembler's profit under BC structure is always higher than the CC structure.

Proof: Given assembler's profit under different structure listed above, it is intuitive that the total profit of the assembler and suppliers under the four different structure are equal with $\Pi_T = \Pi_A^{BB}$ and $Q_i^* = Q_i^{BB*}$, which explains the reason that Π_A^{BB} is of the highest among the four structure.

- $\Pi_A^{CB*} \ge \Pi_A^{BC*}$:

Based on the optimal decisions for the assembler under the BC and CB structure, note that we have $K(Q) = \frac{R(Q)+1}{F(Q)}$ and $H(Q) = \frac{1}{F(Q)}$. It is obvious that $K(Q)$ and $H(Q)$ are all increasing functions with $K(Q) \ge H(Q)$. Denote $x = P^0 - P^1$ and $y = P^1 - P^2$, we can easily have $y \le x$ for the graph $K^{-1}(1 + \frac{x}{c}) = H^{-1}(1 + \frac{y}{c})$. With Chain rule we can also get $\frac{dy}{dx} = \frac{K^{-1'}v}{H^{-1'}c} \ge 0$, which tells us that y always increases with x. The similar analysis applies to $H^{-1}(1 + \frac{x}{v}) = K^{-1}(1 + \frac{y}{c})$, which is essentially the inverse function of $K^{-1}(1 + \frac{x}{c}) = H^{-1}(1 + \frac{y}{v})$ thus they are symmetric with respect to the line $y = x$. The above analysis enables us to make the below conclusion:$K^{-1}(1 + \frac{x}{c}) \ge H^{-1}(1 + \frac{y}{c})$: $\Pi_A^{CB*}(Q^{CB'}) = \Pi_A^{BC*}(Q^{BC'})$; $K^{-1}(1 + \frac{x}{c}) \le H^{-1}(1 + \frac{y}{v})$ and $H^{-1}(1 + \frac{x}{c}) \ge K^{-1}(1 + \frac{y}{c})$: $\Pi_A^{CB*}(Q_1^{CB'}, Q_2^{CB'}) \ge \Pi_A^{BC*}(Q^{BC'})$ and $H^{-1}(1 + \frac{x}{c}) \le K^{-1}(1 + \frac{y}{c})$.

It is easy for us to have $Q_2^{CB*} = Q_2^* \ge Q_2^{BC*} \ge Q_1^{BC*} = Q_1^* \ge Q_1^{CB*}$. We can also rewrite the assembler's profit function under CB and BC structure as:
$$\Pi_A^{CB} = \Pi_T^{CB} + c(Q_1^{CB} - \frac{1}{F(Q_1^{CB})} \int_0^{Q_1^{CB}}) \; \Pi_A^{BC} = \Pi_T^{BC} + c(Q_2^{BC} - \frac{1}{F(Q_2^{BC})} \int_0^{Q_1^{BC}}).$$

Based on the optimal quantity relationship and also $p^2 - p^1 \ge p^1 - p^0$, we can easily have $\Pi_T^{CB*} \ge \Pi_T^{BC*}$. It is also easy for us to check that $cq - \frac{1}{Fq} \int_0^q$ decreases with q. Again because the quantity relationship $Q_1^{CB*} \le Q_2^{BC*}$, we can have $c(Q_1^{CB*} - \frac{1}{F(Q_1^{CB*})} \int_0^{Q_1^{CB*}}) \ge c(Q_2^{BC*} - \frac{1}{F(Q_2^{BC*})} \int_0^{Q_1^{BC*}})$. Adding the above two statements completes the proof that $\Pi_A^{CB*} \ge \Pi_A^{BC*}$ under this condition.

- $\bullet\, \Pi_A^{BC*} \geq \Pi_A^{CC*}$: The above analysis under CC structure indicates that $P^0 - P^1 \leq P^1 - P^2$, $K(Q_1^{CC*}) = \frac{P^0 - P^1 + c}{c}$ and $K(Q_2^{CC*}) = \frac{P^1 - P^2 + c}{c}$; and also $P^0 - P^1 \geq P^1 - P^2$, $Q_1^{CC*} = Q_2^{CC*} = Q_{CC'}$ with $K(Q_{CC'}) = \frac{P^0 - P^2 + 2c}{2c}$.

We can also rewrite the assembler's profit function as: $\Pi_A^{CC} = (P^2 - 2c)E(D) + (P^0 - P^1 + c)\int_0^{Q_1^{CC}} \bar{F}(x)dx - cQ_1^{CC} + (P^1 - P^2 + c(1 - \frac{1}{F(Q_2^{CC})}))\int_0^{Q_2^{CC}} \bar{F}(x)dx + c(Q_1^{CC} - \frac{1}{F(Q_1^{CC})}))$ It is easy for us to verify that $Q_1^{BC*} = Q_2^{BC*} = Q_{BC'} \geq Q_1^{CC*} = Q_2^{CC*} = Q_{CC'}$ when $P^0 - P^1 \geq P^1 - P^2$. Also $Q_1^{BC*} \geq Q_1^{CC*}$ and $Q_2^{BC*} = Q_2^{CC*}$ when $Q_1^{BC*} \leq Q_2^{BC*}$. When $Q_1^{BC*} \leq Q_2^{BC*}$ and $P^0 - P^1 \leq P^1 - P^2$, we will have $Q_1^{BC*} = Q_2^{BC*} \geq Q_2^{CC*} \geq Q_1^{CC*}$. One can easily check $q - \frac{1}{F(q)} \leq 0$, combining the quantity relationship we have above, we can easily conclude that $\Pi_A^{BC*} \geq \Pi_A^{CC*}$.

5   Conclusion

This paper examine the lead time influence on the assembler's strategic choice of its relationship with the component suppliers in the price sensitive market. The initial result we have indicates that the assembler will get the highest profit when he stock both the components and lowest profit when he choose to manage both suppliers under VMI contract. One important finding is that the assembler will prefer to choose the short lead time supplier under the VMI system compared to the long lead time supplier, which encourages the suppliers to decrease their lead time in order to be chosen as the contract supplier that can generate higher profit. This research assume that the retailer act as the leader in the whole game. We hope to find some other interesting insights when the component suppliers is more powerful. We also plan to extend the model to the condition that assembler choose a supplier to contract as sub-assembler that takes care of the component procurement issue. Combing our current model, we hope to provide some insights on the strategic choice of assembler-supplier relationship under price sensitive market.

References

1. Ammar, B., Doulgui, A.: Optimal order release dates for two-level assembly systems with stochastic lead times at each level. Int. J. Prod. Res. **56**(12), 4226–4242 (2018)
2. Atan, Z., Ahmadi, T., et al.: Assemble-to-order systems: A review. Eur. J. Oper. Res. **261**(3), 866–879 (2017)
3. Bernstein, F., DeCroix, G.A.: Decentralized pricing and capacity decisions in a multi-tier system with modular assembly. Manag. Sci. **50**, 1293–1308 (2004)
4. Boyaci, T., Ray, S.: Product differentiation and capacity cost interaction in time and price sensitive market. Manuf. Serv. Oper. Manag. **5**, 18–36 (2003)
5. Chu, C., Proth, J.M., Xie, X.: Supply management in assembly systems. Nav. Res. Logitst. **40**, 933–950 (1993)
6. Cachon, G., Parker, P.: Competition and outsourcing with scale economies. Manag. Sci. **48**, 1314–1333 (2002)

7. Antoine Deza, A., Huang, H., et al.: On component commonality for periodic review assemble-to-order systems. Ann. Oper. Res. **265**(1), 29–46 (2018)
8. Fang, X., So, K., Wang, Y.: Component procurement strategies in decentralized asseble-to-order systems with time-dependent pricing. Manag. Sci. **54**(12), 1997–2011 (2008)
9. Gurnani, H., Akella, R., Lehoczky, J.: Optimal order policies in assembly systems with random demand and random supplier delivery. IIE Trans. **28**, 865–878 (1996)
10. Hsu, V.N., Lee, C.Y., So, K.C.: Optimal component stocking policy for assemble-to order systems with leadtime-dependent component and product pricing. Manag. Sci. **52**(3), 337–351 (2006)
11. Lee, J.Y., Cho, R., Paik, S.K.: Supply chain coordination in vendor-managed inventory systems with stockout-cost sharing under limited storage capacity. Eur. J. Oper. Res. **248**(1), 95–106 (2016)
12. Mateen, A., Chatterjee, A.K.: Vendor managed inventory for single-vendor multi-retailer supply chains. Decison Support. Syst. **70**, 31–41 (2015)
13. Proth, J.M., Mauroy, G., et al.: Supply management for cost minimization in an assembly system with random component yield times. J. Intell. Manuf. **8**, 385–403 (1997)
14. Song, J.S., Zipkin, P.: Supply chain operations: Assembleto-order. In: Handbooks in Operations Research and Management Science, vol. 30, pp. 561–596. North-Holland, Amsterdam (2003)
15. Song, J.S., Xu, S., Liu, B.: Order fulfillment performance measures in an assembly-to-order system with stochastic leadtimes. Oper. Res. **47**, 131–149 (1999)
16. Wang, Y.: Joint pricing-production decisions in supply chains of complementary products with uncertain demand. Oper. Res. **56**, 1110–1127 (2006)
17. Yu, Y., Chu, F., Chen, H.: A stackelberg game and its improvement in a VMI system with a manufacturing vendor. Eur. J. Oper. Res. **192**(3), 929–948 (2009)

Effects of Customer Referral Programs on Mobile App Stickiness: Evidence from China Fresh E-Commerce

Qian Wang[1], Yufan Jiang[1], Chengcheng Liao[1], and Yang Yang[2(✉)]

[1] Business School, Sichuan University, Sichuan 610064, People's Republic of China
[2] Tourism School, Sichuan University, Sichuan 610064, People's Republic of China
371687937@qq.com

Abstract. Customer referral programs are an effective means of increasing customer acquisition and customer retention. Using a large-scale customer data set with 17, 116 observations from a Chinese fresh e-commerce app, we identify that participation in a referral program also increases customer stickiness. We apply propensity score matching to match each customer participating in the referral program with a similar customer (statistical twin) who did not participate. Then a regression model is constructed to compare their stickiness. Our empirical analysis shows that participation in a referral program increases stickiness of both the referrers and the referred customers by 54.13% and 10.88% respectively after matching. The results contribute to a growing literature on customer referral programs as a means for retaining and growing relationships with current customers. We also offer empirical evidence from China fresh E-commerce to contribute to the understanding of mobile app stickiness. The study is also important for marketers, who can differentiate customers with stickiness to realize more precise marketing.

Keywords: Customer referral program · Customer stickiness · Mobile app · Propensity score matching

1 Introduction

Customer referral programs in which the company rewards existing customers for bringing in new customers are increasingly forming part of a marketers' toolbox [31,32]. Without requiring up-front expenditure and connection data between customers, customer referral programs are easy to implement [5] with generating a referral code in mobile apps. For instance, Alipay app employs this strategy excellently by placing their "Invite a Friend, Get 20-99!" at the conspicuous place in their app. China Merchants Bank app also provides attractive coupons and gifts to customers who make successful referrals. Such programs exist in many industries, including online retailers, hotels, restaurants and clinics.

Prior studies on customer referral programs have confirmed them to be an effective and cost-efficient means for acquiring new customers with superior profitability for the company and high retention in both the short and the long run

© Springer Nature Switzerland AG 2020
J. Xu et al. (Eds.): ICMSEM2019 2019, AISC 1002, pp. 376–389, 2020.
https://doi.org/10.1007/978-3-030-21255-1_29

[1,32,33]. Other researches explore how referrals influence the recommender. They find that existing customers who participated in referral programs usually own a higher loyalty than those who do not [13]. Although it is increasingly well understood that customer referral programs acquire more valuable new customers and make existing customer more loyal, how referral programs affect their stickiness respectively haven't been studied yet. In 2017, 84% of consumers used their mobile phone to shop [26] and only sticky customers can make frequent in-app purchases. By increasing the frequency people visit an app and the duration of every visit, increased stickiness can be regarded as increasing the potential for in-app purchases [16]. Existing research already suggests that interaction is important to influence customer stickiness [24], but customer referral programs, especially operated in mobile apps, haven't been discussed as a specific form of interaction because of the absence of deeper insights.

Therefore, it is necessary to explore the effects of customer referral programs on stickiness in the context of mobile apps. More specifically, we answer two questions:

(1) Does participation in referral programs increase referrers' stickiness?

(2) Are customers acquired through a referral program more sticky than customers acquired from other methods?

We answer these questions by analyzing the data collected from an app of a fresh e-commerce company in China between June 1, 2015 and September 30, 2017. The data set includes 17, 116 customers' basic as well as referral information. Every customer participating in the referral program is match with a similar customer (statistical twin) who did not participate by propensity score matching. We obtain a smaller sample with 1, 178 observations then. Regression is conducted with both samples before and after matching and we compare stickiness of customers who participated in referral programs and those who did.

Our research offers three contributions. First, it theoretically and empirically strengthens the understanding of customer referral programs, as a means for retaining and growing relationships with current customers, by assessing referrers and referred customers simultaneously. Second, it extends the research on mobile app stickiness and it offers valuable information about how to improve customer stickiness by providing empirical evidence from China fresh E-commerce. Third, it provides practical guidelines for marketers to evaluate a customer referral program. More specifically, marketers can adopt more efficient referral programs based on the measurement of stickiness.

The rest of the article is organized as follows. We continue by the literature review describing stickiness and referral programs and develop hypotheses. Next, we analyze behavioral data obtained from a platform-based fresh e-commerce company, using a propensity score matching approach and multiple regression model. Then, we discuss the results. Finally, we conclude with implications for theory, practice and the limitations.

2   Theory and Hypotheses

2.1   Customer Stickiness

Customer stickiness is an intangible ability to keep customers coming back over a long period of time and helps create and maintain the competitiveness and sustainability of an organization [24]. Hallowell [14] argues that stickiness is longer and more frequent website visits. Li, Browne and Wetherbe [7] indicate that the definition of stickiness is the user's repetitive visits to their preferred websites based on their commitment to reuse. Rettie [29] points out that stickiness increases the frequency and duration of website visits. Koh and Kim [20] argue that stickiness increases both purchase intention and the number of ads watched by users. Bhatnagar and Ghose [4] and McCloskey [25] demonstrate that the longer users visit a website, the more likely they will make a purchase. Lin [23] tie both the dimensions of stickiness (duration of visits and repeat visits) to intention to transact via a website simultaneously. In the context of mobile apps, stickiness refers to the extent to which users frequently return to mobile apps for sustained use [28]. For the purpose of this study, we operationalize stickiness as the frequency of logging in.

Previous research has investigated the driving factors of stickiness. Perceived website's attributes [3,23], consumers' positive attitude to websites [23], consumer's satisfaction [3,21,24], commitment, trust and demographic characteristics [22,23] are all important to promote stickiness. Chiang and Hsiao [8] argue that continuance motivation and sharing behavior are critical antecedents of YouTube's stickiness. Zhang et al. [35] examine the effects of customer engagement with companies social networks on stickiness. Their empirical results show that customer engagement influences customer stickiness directly and positively as well as influences customer value creation indirectly.

In sum, it is widely accepted that interaction is a key factor affecting customer stickiness [24]. However, as a specific form of interaction between customers, the influence of customer referral programs on stickiness is still not clear. And existing literature has mainly focused on the stickiness of customers visiting websites not using apps.

2.2   Customer Referral Programs

Customer referral programs, a form of stimulated word of mouth that aimed to stimulate positive evaluation among existing customer bases [32], have become a popular way to acquire customers.

The benefits of referral programs are widely accepted and have been shown repeatedly in previous studies. For example, Garnefeld et al. [13] investigate the participation of referral programs increases existing customers' loyalty. For referrers, recommendation means a commitment. Davidow [10], Nyer and Gopinath [27] interpret word of mouth as a public commitment or public stance regarding an evaluation of the company or its products. Research in social psychology has shown that people who advocate a specific issue position are more likely to align their attitudes in the direction of that position. The commitment arises because people who recognize that they have endorsed a position will attribute favorability toward it. Commitments are strongest when made in front of large groups [11]. The public position customers taking is difficult to change when they make recommendations. Generally, referrer, referred customer and company all know that the referrer has made a recommendation and been rewarded after a successful referral. Thus, referral programs evoke public commitments with consistency effects [13]. A recommendation thus makes it more likely that the recommender remains loyal, which means that the recommender will repurchase a preferred product or service from the recommended company in the next time.

Thus, we hypothesize H1:

H1: Participation in referral programs increases referrers' stickiness.

For referred customers, Participation in referral programs enables them to get more information from an experienced source. The more knowledge or experience they possess, the less risk will be perceived by them [34]. When a referral relationship has been built, referred customers and people close to him or her are customers of a same company, which increases their trust in the company and strengthens the emotional bonds with it, as both balance theory and social closure theory predict [6]. And customers will mention friends who are customers in the same company when reflecting on their effect on a company [7]. Referred customers are likely to have a stronger sense of commitment and attachment to the firm. Hogan, Lemon and Libai [15] demonstrate that referral programs link acquisition and retention and make value assessments more accurate. Van den Bulte et al. [5] find that contribution margins of referring and referred customers vary systematically, relationship duration positively influences referral margins (especially after the referral), referred customers churn less as long as their referrer did not churn. Kumar, Petersen and Leone [33] make it possible to target the most profitable customers by computing customer referral value. And customers who are acquired through referrals are considered more valuable than other customers in both the short and the long run because they yield a higher contribution margin and higher retention rate than customers who are acquired through other channels [1,32]. In other words, referred customers are less likely to churn than nonreferred customers. Thus:

H2: Participation in referral programs increases referred customers' stickiness.

3   Method

3.1   Data and Measure

We obtained objective behavioral data on customers from a platform-based fresh e-commerce company in China. Referrers, who own a referral code, will earn 100 points for each successful referral and receivers will receive a coupon worth 5 in the mobile app. The data set on 1,7116 customers who participated at least once in the customer referral program includes basic information and behavior records of customers.

After registration, customers are assigned a user id. All of the customers come from Mianyang, Sichuan, China and registered from June 1, 2015 to September 30, 2017. The number of weeks that have elapsed since a customer registered is 7 130. For variables, definitions and measures, see Table 1.

3.2   Propensity Score Matching

Simply comparing the behavioral data from customers who either did or did not participate in a customer referral program would suffer from self-selection bias. We aim to determine differences in stickiness between customers who participated in the customer referral program and similar customers who didn't. However, we cannot know the outcome if customers had not participated. In addition to participation, customers who participated may differ substantially from those who did not, so simply comparing the outcome variables for both groups is not suitable.

To deal with it, we apply a matching technique using the propensity score [19], which is defined as the latent probability of receiving the treatment (i.e., participation in a customer referral program, in our case) given the covariates [30]. Based on the nearest-neighbor matching method, one type of propensity score, a set of closest controls for each treated case one at a time has been selected [9]. Then, we create differently ranked groups that are as similar to each other as possible. Imitating characteristics of treatment selection enables us to identify the treatment effect (i.e., the difference in stickiness) as if in a randomized experiment [2,12].

Customers participating in the customer referral program are either recommenders or receivers. To verify the hypotheses above, we have to match twice (The first treatment condition is attaining at least 1 successful referral for recommenders, and the second treatment condition is receiving a recommendation and registering for receivers). In previous studies, the factors influencing customers' willingness to recommend can be roughly divided into three categories. From the perspective of customer referral programs, it includes reward type, amount and allocation, etc. [18]; From the perspective of referrers' characteristics, it includes customer influence and professionalism, customer loyalty and satisfaction, as well as their own happiness threshold, etc. Points, the number of order and order amount can represent customer loyalty and satisfaction. Customers with higher points and level of activity are more likely to make recommendations [13]. From the perspective of referrals' characteristics, it includes the strength and paradigm of the recommended relationship [17]. Customers with close relationships (spouses or close friends) are more likely to be in the same urban area and are easier to participate in the customer referral program. Thus, we choose point, order, amount, and district as covariates.

3.3   Model

Based on the hypotheses and variable selection, we build a model to conduct multivariable linear regression analysis. It is expected to improve the fitting capability of the model and eliminate the heteroscedasticity by taking the natural logarithm of stickiness, tenure, point, order and amount in our model:

$$
\begin{aligned}
Ln\,(Customer\,stickiness) = \alpha_i &+ \beta_{i1}Referral + \beta_{i2}Referred + \beta_{i3}Container \\
&+ \beta_{i4}Ln\,(Tenure) + \beta_{i5}Client + \beta_{i6}Ln\,(Point) \\
&+ \beta_{i7}Ln\,(Order) + \beta_{i8}Ln\,(Amount) + \beta_{i9}District + \varepsilon_i
\end{aligned}
\tag{1}
$$

where α_i is a constant term in model i, $\beta_{i1} - \beta_{i9}$ are estimated coefficients of variables and ε_i is residual in model i. And the variables have been shown in Table 1.

Table 1. Variables, definitions and measure

Variables	Definitions	Measure
Customer stickiness	How often a customer logs into the app a week	The average frequency of logging into the app
Referral	Whether a customer attains a successful referral	Dummy, with 1 if a customer has at least 1 successful referral, 0 otherwise
Referred	Whether a customer is recommended to register	Dummy, with 1 if a customer receives recommendation and registers, 0 otherwise
Container	Whether there have been installed smart lockers near the customers' address	Dummy, with 1 if there have been installed smart lockers, 0 otherwise
Tenure	How many weeks that have elapsed since a customer registered	The number of weeks after registration
Client	Which clients used when customers register	Three dummies: Android, iOS, H5 and Others
Point	How many member points customers get when log in, attain a successful referral and purchase	Total number of points
Order	How many orders a customer placed	Total number of orders
Amount	How much a customer spent in purchasing	Amount of money spent in the app.
District	Where the customer comes from	Six dummies: Anxian, Fucheng, Gaoxin, Jingkai, Kechuang, Youxian

4   Results

4.1   Chi-Square Test

In Table 2, we have 589 observations in the first treatment and control group (i.e., whether customers have at least 1 successful referral) after matching. The chi-square test results show that the mean and standard deviation of each covariate in two groups have no significant difference (p > 0.1), which means we successfully matched.

After combining the two groups of samples into a new total sample, descriptive analysis was conducted on the main variables. See the results in Table 3.

Table 2. Comparison between groups after matching (Referrers)

	No referral (0)	At least 1 successful referral 1	P
N	589	589	
Point (mean (sd))	583.80 (939.06)	559.80 (864.41)	0.648
Order (mean (sd))	7.74 (13.96)	7.74 (12.93)	0.998
Amount (mean (sd))	302.08 (681.79)	271.98 (533.25)	0.399
District (%)			0.421
Fucheng	221 (37.5)	235 (39.9)	
Gaoxin	231 (39.2)	232 (39.4)	
Jingkai	18 (3.1)	25 (4.2)	
Kechuang	48 (8.1)	41 (7.0)	
Youxian	71 (12.1)	56 (9.5)	

Table 3. Summary statistics

Variables	Mean	Median	Max	Min
Stickiness	1.88	1.24	7.94	0.02
Tenure	64.95	54.54	129.16	9.28
Point	571.8	218	7922	20
Order	7.74	3	105	0
Amount	287.03	69.3	7000	0

In Table 3, we provide the summary statistics of main variables. The minimum stickiness is 0.02 and the maximum is 7.94, indicating that some customers hardly log in after registration, while some customers log in frequently with 8 times a week on average. In addition, if the number of orders or order amount is 0, it is necessary to add 1 to the original value in order to ensure that the antilog is greater than 0.

Table 4. Comparison between groups after matching (Referred customers)

	Customers receive a recommendation and register (0)	No recommendation (1)	P
N	4553	4553	
Point (mean (sd))	268.85 (436.33)	442.91 (581.01)	<0.001
Order (mean (sd))	3.18 (5.74)	2.59 (4.35)	<0.001
Amount (mean (sd))	93.32 (202.03)	63.27 (167.11)	<0.001
District (%)			<0.001
Anxian	2 (0.0)	2 (0.0)	
Fucheng	1638 (36.0)	2046 (44.9)	
Gaoxin	2182 (47.9)	1622 (35.6)	
Jingkai	324 (7.1)	487 (10.7)	
Kechuang	211 (4.6)	178 (3.9)	
Youxian	196 (4.3)	218 (4.8)	

Similarly, Table 4 shows the chi-square test results of the second treatment and control group (i.e., whether customers receive a recommendation and register). After matching, there are 4,553 samples in each group. All covariates have significant difference between two groups ($p < 0.001$), which means there is a match failure.

The reasons for the match failure can be summarized as following:

First, the method of propensity score matching is applicable to the conditions in which the number of observations that can be directly compared between the treatment group and the control group is very small. However, there are 4,553 observations, accounting for 26.6% of the total samples. Therefore, it is difficult to achieve the ideal non-difference matching.

Second, precise measurement and stable variables should be matched as much as possible. However, in some cases, there are too many variables measuring characteristics to make an ideal match. Some variables are controlled, but other variables are very different, which makes it impossible to compare the treatment group with the control group.

Therefore, when there are too many observations and too many variables used for matching, the matching fails.

4.2   Model Fitness

We use multi-variable linear regression model, which is widely used to analyze the correlation, significance, correlation direction and degree between dependent variables and multiple independent variables, to test our hypotheses.

Table 5. Main results

Variables	DV= Ln(stickiness)		
	Estimated coefficient		
	Model 1: Before matching (17116)	Model 2: After matching (1178)	
Referral	0.6055***	0.5413***	Baseline group: No referral
Referred	0.1384***	0.1088*	Baseline group: No recommendation
Container	−0.0025	−0.1203**	Baseline group: No container
Tenure	−0.5038***	−0.4600***	
H5(client)	−1.1559***	−1.0346***	Baseline group: Android
IoS(client)	−0.3883***	−0.4024***	
Others(client)	−0.6760***	−0.6204***	
Point	0.8329***	0.8013***	
Order	0.1063***	0.0863*	
Amount	−0.2087***	−0.2322***	
Fucheng(district)	−0.2707	–	Baseline group: Anxian
Gaoxin(district)	−0.1338	0.1036**	
Jingkai(district)	−0.1564	0.0703	
Kechuang(district)	−0.2834	−0.104	
Youxian(district)	−0.2691	−0.0655	
Intercept	−1.5223***	−1.4836***	
R^2	0.6544	0.7249	
F-statistic	2162 on 15 and 17100 DF	222.5 on 14 and 1163 DF	
P value	<0.001		
Signif. codes: $*p < 0.05. **p < 0.01. ***p < 0.001.$			

In Table 5, we present the results. Model 1 shows the result of analyzing the samples before matching(17116). The average stickiness of customers who attain least 1 successful referral is greater than those who have no referral by 60.55%

$(\beta_{11} = 0.6055, p < 0.01)$. Thus, H_1 is supported. The average stickiness of customers who receive a recommendation and register is also greater than those without recommendations by 13.84% $(\beta_{12} = 0.1384, p < 0.01)$. Thus, H_2 is supported. Model 2 shows the result of analyzing the samples after matching(1178). H_1 and H_2 are supported by the estimates in which both average stickiness of the referrer and the referrals are greater $(\beta_{21} = 0.5413, p < 0.001; \beta_{22} = 0.1088, p < 0.05)$.

Although the second treatment group failed to be matched for regression, it can be seen from the results of the above two models that acceptance of recommendation does have a significant positive impact on customer stickiness.

We also measure the degree of multicollinearity between independent variables by variance inflation factor (VIF). If one or more independent variables can be well linearly expressed by other independent variables, it will influence the estimated coefficient. In Tables 6 and 7, VIF values estimated by all

Table 6. Multicollinearity test of model 1

Variables	GVIF	Df	$GVIF^{(1/(2*Df))}$
Customer stickiness	1.22	1	1.1
Referral	1.29	1	1.13
Referred	1.01	1	1.01
Container	1.73	1	1.31
Tenure	1.6	3	1.08
Client	4.03	1	2.01
Point	5.44	1	2.33
Order	4.24	1	2.06
Amount	1.17	4	1.02

Table 7. Multicollinearity test of model 2

Variables	GVIF	Df	$GVIF^{(1/(2*Df))}$
Customer stickiness	1.03	1	1.02
Referral	1.19	1	1.09
Referred	1.01	1	1
Container	1.64	1	1.28
Tenure	1.62	3	1.08
Client	2.43	1	1.56
Point	2.71	1	1.65
Order	2.36	1	1.54
Amount	1.16	5	1.01

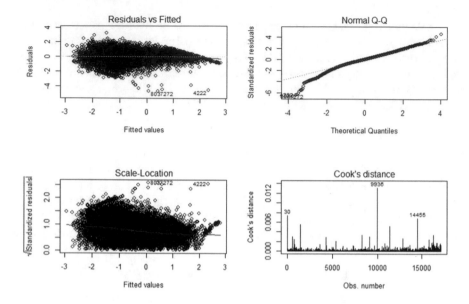

Fig. 1. Graphs of residual, QQ and Cook distance of model 1

parameters in both models are far less than 10, so the possibility of multi-collinearity is excluded.

4.3   Model Test

In order to select the optimal model, the main concern is whether the model meets three assumptions, including independence, normality and constant variance. Therefore, it is necessary to conduct model diagnosis on the two regression results and to find out whether there are abnormal values.

Figures 1 and 2 are graphs of residual, QQ and Cook distance obtained by testing the models respectively.

In Fig. 1, from the two residual graphs on the left, it can be seen that there is no observation with large residual, which means all observations can be fitted by the model, and the distribution of residual is disordered and has no obvious trend. The constant variance assumption is supported.

The QQ graph on the upper right obtained a scatter plot approximate to a straight line, which is considered to support the assumption of normality.

In the graph of Cook distance at the lower right, Cook distance Di reflects the difference between the final estimation caused by the inclusion and exclusion of the ith observation. As the graph shows, there is no outlier.

Therefore, it can be concluded that model 1 basically meets all the important assumptions of the linear model.

Similarly, Fig. 2 shows that model 2 basically meets all the important assumptions of the linear model. Therefore, both model 1 and model 2 pass the model test.

5   Discussion

By examining the impact of customer referral program participation on stickiness of referrers and referred customers, our research offers three contributions. First, focusing on referrers and referred customers in the same time, it promotes the understanding of customer referral programs empirically and theoretically, as a means for retaining and growing relationships with current customers. The stickiness of customers who attain at least 1 successful referral is greater than those who have no referral. Making Recommendation, equivalent to making a public commitment, results in that the recommender is more supportive of the products and services of an enterprise, which will be reflected in behavior. Second, the stickiness of customers who receive recommendation and register is also greater than those without recommendations. The relationship between referrers and referrals helps to reduce the risk of purchasing for new customers, it also reduces uncertainty and increases the frequency of new customers' logging in.

Second, it contributes to the research on mobile app stickiness and it offers valuable suggestion about how to increase customer stickiness with empirical evidence from China fresh E-commerce. Managers need to pay attention to the positive impact of customer referral programs on mobile app stickiness and formulate corresponding marketing plans to promote the recommendation intention of existing customers and the acceptance intention of new customers. Then the frequency of logging in and the rate of purchasing will be promoted.

Third, it provides actionable guidelines for marketers to evaluate a customer referral program. More specifically, marketers can adopt more efficient referral programs based on the measurement of stickiness. Customers who receive

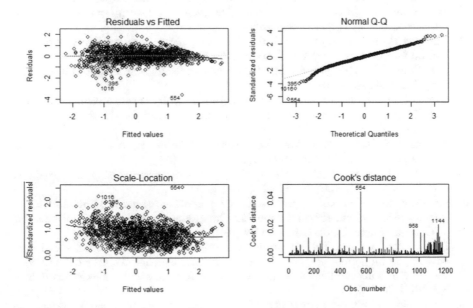

Fig. 2. Graphs of residual, QQ and Cook distance of model 2

recommendation and register are more valuable than those without recommendation, enterprises need to conduct more accurate marketing strategies for different customers, optimize the allocation of limited resources and save marketing costs.

At the same time, our research has two limitations that offer avenues for further research. First, it needs to consider more industry selections. Although the objective data comes from reality, it is only about fresh e-commerce industry. Conclusions applicable to other industries need to be further explored. Second, it needs flexible referral programs. The platform adopts the referral programs with the same type of reward and amount for all customers, and the form of referral programs that is most conducive to promote stickiness needs to be further studied.

Acknowledgements. This work was supported by the National Natural Science Foundation of China (Grant 71502019, 71472130), the Foundation for Humanities and Social Sciences of the Ministry of Education of China (Grant 17YJA630031), Sichuan University (Grant 20822041A4222).

References

1. Armelini, G., Barrot, C., Becker, J.U.: Referral programs, customer value, and the relevance of dyadic characteristics. Int. J. Res. Mark. **32**(4), 449–452 (2015)
2. Austin, P.C.: An introduction to propensity score methods for reducing the effects of confounding in observational studies. Multivar. Behav. Res. **46**, 399–424 (2011)
3. Bansal, H.S., Mcdougall, G.H.G., Dikolli, S.S., Sedatole, K.L.: Relating satisfaction to behavioral outcomes: an empirical study. J. Serv. Mark. **18**(4), 290–302 (2004)
4. Bhatnagar, A., Ghose, S.: An analysis of frequency and duration of search on the internet. J. Bus. **77**(2), 311–330 (2004)
5. Van den Bulte, C., Bayer, E., et al.: How customer referral programs turn social capital into economic capital. J. Mark. Res. **55**(1), 132–146 (2018)
6. den Bulte, C.V., Wuyts, S.H.K.: Social Networks in Marketing. MSI Relevant Knowledge Series (2007)
7. Tse, D.K., Chan, K.W.: Strengthening customer loyalty through intimacy and passion: Roles of customer-firm affection and customer-staff relationships in services. J. Mark. Res. **45**(6), 741–756 (2008)
8. Chiang, H.S., Hsiao, K.L.: Youtube stickiness: The needs, personal, and environmental perspective. Internet Res. **25**(1), 85–106 (2015)
9. Ho, D.E., Imai, K., et al.: Matchit: Nonparametric preprocessing for parametric causal inference. J. Stat. Softw. **42**, 1–28 (2011)
10. Davidow, M.: Have you heard the word? the effect of word of mouth on perceived justice, satisfaction and repurchase intentions following complaint handling. J. Consum. Satisf. Dissatisfaction Complain. Behav. **16**, 67–80 (2003)
11. Deutsch, M., Gerard, H.B.: A study of normative and informational social influences upon individual judgment. J. Abnorm. Soc. Psychol. **51**, 629–636 (1955)
12. Ea, S.: Matching methods for causal inference: A review and a look forward. Stat. Sci. **25**, 1–21 (2010)
13. Garnefeld, I., Tax, S.S.: Growing existing customers' revenue streams through customer referral programs. J. Mark. **77**(4), 17–32 (2013)

14. Hallowell, R.: The relationships of customer satisfaction, customer loyalty, and profitability: An empirical study. Int. J. Serv. Ind. Manag. **7**, 27–42 (1996)
15. Hogan, J.E., Lemon, K.N., Libai, B.: What is the true value of a lost customer? J. Serv. Res. **5**(3), 196–208 (2003)
16. Hsu, C.L., Lin, C.C.: Effect of perceived value and social influences on mobile app stickiness and in-app purchase intention. Technol. Forecast. Soc. Chang. **108**, 42–53 (2016)
17. Wirtz, J., Orsingher, C., et al.: The role of metaperception on the effectiveness of referral reward programs. J. Serv. Res. **16**, 82–98 (2012)
18. Jin, L., Huang, Y.: When giving money does not work: The differential effects of monetary versus in-kind rewards in referral reward programs. Int. J. Res. Mark. **31**(1), 107–116 (2014)
19. Joffe, M.M., Rosenbaum, P.R.: Invited commentary: Propensity scores. Am. J. Epidemiol. **150**(4), 327–333 (1999)
20. Koh, J., Kim, Y.G.: Knowledge sharing in virtual communities: an e-business perspective. Expert. Syst. Appl. **26**(2), 155–166 (2004)
21. Kurniawan, S.: Merged structural equation model of online retailer's customer preference and stickiness. WebNet World Conference on the WWW and Internet. 341–346 (2000)
22. Li, D., Browne, G.J., Wetherbe, J.C.: Why do internet users stick with a specific web site? A relationship perspective. Int. J. Electron. Commer. **10**(4), 105–141 (2006)
23. Lin, C.C.: Online stickiness: its antecedents and effect on purchasing intention. Behav. Inf. Technol. **26**(6), 507–516 (2007)
24. Khalifa, M., Limayem, M., Liu, V.: Online customer stickiness: A longitudinal study. J. Glob. Inf. Manag. **10**, 1–14 (2002)
25. Mccloskey, D.: Evaluating electronic commerce acceptance with the technology acceptance model. J. Comput. Inform. Syst. **44**(2), 49–57 (2004)
26. Nielsen: What's next for china's connected consumers (2018)
27. Nyer, P.U., Dellande, S.: Public commitment as a motivator for weight loss. Psychol. Mark. **27**(1), 1–12 (2010)
28. Racherla, P., Furner, C., Babb, J.: Conceptualizing the Implications of Mobile App Usage and Stickiness: A Research Agenda. Social Science Electronic Publishing (2012)
29. Rettie, R.: An exploration of flow during internet use. Internet Res. **11**(2), 103–113 (2001)
30. Rosenbaum, P.R., Rubin, D.B.: The central role of the propensity score in observational studies for causal effects. Biometrika **70**, 41–55 (1983)
31. Ryu, G., Feick, L.: A penny for your thoughts: Referral reward programs and referral likelihood. J. Mark. **71**(1), 84–94 (2007)
32. Schmitt, P., Skiera, B., Bulte, C.V.D.: Referral programs and customer value. J. Mark. **75**(1), 46–59 (2011)
33. Kumar, V., Petersen, J.A., Leone, R.P.: Driving profitability by encouraging customer referrals: Who, when, and how. J. Mark. **74**, 1–17 (2010)
34. Voyer, P.A.: Word-of-mouth processes within a services purchase decision context. J. Serv. Res. **3**, 166–177 (2000)
35. Zhang, M., Guo, L., Mu, H., Liu, W.: Influence of customer engagement with company social networks on stickiness: Mediating effect of customer value creation. Int. J. Inform. Manag. **37**(3), 229–240 (2017)

Impact of Supply Chain Management Practices on Organizational Performance and Moderating Role of Innovation Culture: A Case of Pakistan Textile Industry

Muhammad Hashim[1], Sajjad Ahmad Baig[1(✉)], Fiza Amjad[1], Muhammad Nazam[2], and Muhammad Umair Akram[3]

[1] Department of Management Science, National Textile University Faisalabad, Faisalabad, Pakistan
sajjad.baig@hotmail.com
[2] Institute of Business Management Sciences, University of Agriculture Faisalabad, Faisalabad, Pakistan
[3] Guanghua School of Management, Peking University, Beijing, People's Republic of China

Abstract. In today's competitive environment, the role of innovation is getting more popularity in organizations. Innovation and Supply chain management (SCM) play very important role for getting a competitive position in global environment. The study aims to investigate the impact of SCM practices on organizational performance through the moderating role of role of innovation culture. Data for investigation were collected through field survey from Pakistani textile industries and total response rate was 86% (236/275). Current study employed the Smart-PLS Analysis technique for data analysis. The results showed that SCM practices have significant impact on organization performance. However, the moderator makes the relationship stronger among SSP, CR, LIS, LIQ, ILP and organizational performance. While the innovation culture significantly moderates the relationship between SCM Practices and organization performance. In the era of China-Pakistan Economic Corridor (CPEC) the findings of current study provide insights into Mangers regarding how to improve the SCM projects for getting more benefits from ongoing projects. Further research can formulate additional measurements for practices of internal supply chain like quality management, supply integration and competitive advantage as a moderator.

Keywords: Supply chain management · Innovation culture · Organizational performance and Textile manufacturing Industry

© Springer Nature Switzerland AG 2020
J. Xu et al. (Eds.): ICMSEM2019 2019, AISC 1002, pp. 390–401, 2020.
https://doi.org/10.1007/978-3-030-21255-1_30

1   Introduction

In today's dynamic business environment, innovation and supply chain Management is considered as a vital source of organizational success. As many apparel
industries are establishing strong supply chains to get the advantage from their
competitors by providing the value-added products and potential customers.
Now-a-days, the SCM practices have become very critical for managing risk,
dynamism, and complexities of global sourcing. It is most important for industry to integrate the whole supply chain activities for getting maximum benefits.

Supply chain is a management of multiple functions performed by an organization includes sourcing and purchasing, manufacturing, logistics activities and
the strong coordination between supply chain partners. Substantial numbers of
organizations have started to realize that the supply chain is the primal for
establishing a profitable competitive edge in growingly crowded and competitive
markets [6,12,14,19,24]. Li et al [25] investigated five dimensions of supply chain
practice "strategic supplier partnership, customer relationship, level of information sharing, quality of information sharing, and postponement". The findings
conclude that higher levels of SCM practice can lead to increased competitive
advantage and enhance the organization performance.

The textile industries play a vital critical role for industrialist and under
developed economies for improving their living standards and employment. In
case of Apparel and textiles exports to USA, Pakistan is the fourth largest
supplier and recently exports approach to a figure of $1.9 billion. The textile
industry of Pakistan has a big contribution in the economy of the country (8.5%
of GDP). Currently there are number of textile units involved in producing
textile products for example about 1221 ginning units, 442 spinning units, 124
large spinning units, and 425 small units. The effective supply chain give the
number of advantages to textile industry for example lower inventories and costs,
higher productivity, shorter lead times, gain the higher profits and customer
loyalty [1].

In Pakistan, textile industry facing a number of issues, for example; outdated
manufacturing machines and there is no production and inventory planning. The
majority of the employees in industries are illiterate or less educated which holds
the position of manager. Even though our educational institutions related to textile are not emphasizing to SCM issues on practical grounds. The infrastructure
(Roads, Railways) is a major element of supply chain. Pakistani Railway infrastructure is not in a good condition that is becoming the reason of delays in
operation and transportation [2].

CPEC (China Pakistan Economic Corridor) project is good sign for the SCM
of textile value chain. It is a bilateral strategic project which will be a value
addition in the logistics and supply chain management of both countries. This
project is grabbing attention due to foreign direct investment in various sectors
comprising energy, infrastructure, rail and roads etc. China-Pak Economic Corridor will play a dynamic role for coordinating and integrating these activities
through buyer-supplier relations. It will support the goals of organizations which

are strategically aligned for working closely and aiming to lower the lead time as well as logistic cost.

2 Conceptual Background

2.1 Manufacturing Organization Performance

Organization performance is a measure that based on multiple dimensions like financial, customer satisfaction and market growth [45]. The organizations mainly have two types of goals short term and long term. Short term goals refer to increase the productivity while to reduce the inventory and lead time. Long term goals are refer to increase the market share and profit for all the partners of supply chain [39]. All the initiatives taken by the organizations including supply chain management ultimately lead to improve the organizational performance. Fraser (2006) [15] stated that a company can improve their performance by measuring and tracking financial and operational performance. Therefore, it is important to study the effect that supply chain management practices have on organizational financial performance.

A significant number of scholars have used financial and market criteria for measuring the organization performance, including return on investment, market share, profit margin on sales, the growth in sales, market share, return on investment and the competitive position in the market [37,42,47]. In line with this discussion, the present study is adopted the same instruments for measuring the organizational performance.

2.2 Strategic Supplier Partnership

Strategic supplier partnership (SSP) means a long term partnership between buyers and suppliers. It has a substantial impact on manufacturing operations like production costs, timely delivery, product quality and also strongly affects the organizational profitability or operational capabilities [17,20,32]. A strategic relationship emphasizes direct, long-term association and encourages mutual planning and problem solving efforts [46]. This strategic relationship is very helpful for promoting shared benefits among the business parties and ongoing participation in different key strategic areas like products, technology and markets [3]. Strategically adjusted organizations have ability to work closely together that helps them to eliminate the wasteful time and effort [25]. Thus, on the bases of above discussion we expect that:

Hypothesis 1. Strategic Supplier Partnership significantly link with organization performance.

2.3 Customer Relationship

Customer relationship (CR) is a one of key SCM practice among all practices [11]. The strong and closing CR permits an organization to differentiate its product or services from competitors, gain customer loyalty, and deliver valuable products to customers dramatically [31]. Vickery et al. [43] examined the importance of establishing a close customer relationship as an important practice of SCM integration to enable organizations to respond customer quickly.

After the above discussion our contention is that:

Hypothesis 2. Customer Relationship significant link with organizational performance.

2.4   Level of Information Sharing

There are two aspects of information exchange like quality and quantity. The two aspects are very important for supply chain management practices and in previous studies have been treated as an independent constructs [8,28]. Level (quantity aspect) information is critical and proprietary information about a supply chain partner to the extent that the information is communicated. It can vary from technical and strategic in nature, and from logistics activities to general market and customer information [41]. Significant numbers of researches have indicated that the key to the seamless SCM is making available undistorted and up-to-date marketing data at all SCM stages [32,36,40]. Stein and Sweat [31] conclude that SCM partners are able to work as a unit where the information is exchange with partners at regular bases. Together, they can respond with good understanding of end customer's needs and therefore faster to market changes. In light of this rationale, we expect that:

Hypothesis 3. Level of Information Sharing significant link with organizational performance.

2.5   Quality of Information Sharing

It means the accuracy, timeliness, adequacy, and credibility of data sharing in between SCM partners [9,23]. McCormack [26] quantified information by precision, rate, correctness, and easiness of forecasting data. Few researchers evaluated information quality by money, precision, and correctness. Vijayasarathy and Robey [44] evaluated level of information density and eminence. Information eminence is a vital component in getting advantageous from information sharing organism. Sum et al. [38] examined that information correctness is seriously influencing by efficiency of operations and their buyer services. McGowan [27] investigated that the system of information sharing is considered practicable when level of information sharing achieved a higher quality terms, promptly accessible, perfect and accurate. Therefore, It is very important practice of SCM that what, when and how the information are shared and with whom [22,33].

Hypothesis 4. Quality of Information Sharing significant link with organizational performance.

2.6   Internal Lean Practices

These practices mean to eliminate the waste like cost, time or any other useless activity in production process, characterized by reduced set-up times, small lot sizes, and pull-production [4, 7, 34]. The word "lean" means a system that uses less input to produce at a mass production speed, while offering more variety to the end customers. These studies emphasize on lean principles and practices for managing the supply chain in efficient and effective way. This is the motivation for using this dimension as one of the supply chain management practice for this study. The conclusion of this discussion is *Hypothesis 5*.

Hypothesis 5. Internal Lean Practices significant link with organizational performance.

2.7   Innovation Culture

Innovation culture can be defines as "shared common values, beliefs and assumptions of organizational employees that could facilitate the product innovation process". In such a situation, when an organizational culture or working environment encourages the employees' innovation capacity, tolerates risk, and supports personal growth and development [2], it may be called as an innovation culture. Todays, there is a keen interest of researchers and practitioner on the implementation of innovation and creativity in organization for gaining competitive advantage. More specially, the effects of an innovation culture on organizational performance [10, 16, 18, 30, 35].

In today's competitive environment, organizations are engaged in continues innovations and the development of knew knowledge and capabilities [6, 19, 25]. A firm's strategy for developing a new product is a main determinant of the organization's performance [29].

In line with above discussion, the present study is developed the following hypothesis:

Hypothesis 6. Innovation culture moderates the relationship between SCM Practices and organization performance.

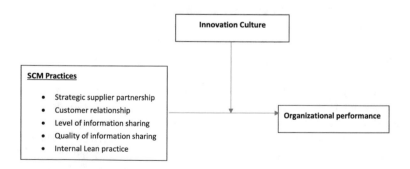

Fig. 1. Conceptual research model

3   Research Methodology

Population and Sampling

The textile industry has a major contribution in Pakistan economy. In manufacturing oriented organizations activities are managed by the cooperation of different departments. Therefore, different departments are identified for conducting the research study (procurement, stores, marketing and transport from Knitting, Weaving, Spinning, Dyeing and finishing, and Apparel manufacturing, and garment industries).

The target respondents consist of middle to top level managers in the relevant functional departments or in charge of supply chain practices. The sampling frame consists of 275 potential respondents from textile industries of Pakistan. The responses showed that a maximum number of respondents filled the questionnaire were a part of top and senior management. The total response rate is 53.5% (147/275). The study initially involves the review, a pre-test was given to professionals in the SCM field for finalizing the questionnaire. The measurement items of variables were used adopted by (G. Martin et al [5]) for innovation culture and S. Li et al. [25] for SCM practices and organizational performance.

4   Results and Discussion

The present study employed a SEM Analysis by using Smart-PLS 3.0 and run PLS-algorithm, PLS-blindfolding and PLS-bootstrapping to test all the Hypotheses. The reliability of the measurement model was checked through factor loadings and composite reliability. All the values of factor loading exceed the minimum acceptable values that are 0.7 [21]. As per the results of reliability, all the items have outer loading greater than 0.7. Current study have sample size of $n = 147$. The relationship between constructs and related items was specified by outer or measurement model although the structural model defined the relations between construct items as shown in Fig. 1. Therefore, the reflective measurement models' indicators achieve the satisfactory levels. Similarly, as shown in the Table 1 all the values of AVE-Average variance extracted and composite reliability are higher than the 0.50 threshold value, which provide support for the convergent validity of the measures.

The structural model analysis concentrates on the higher-order component task competence, which shows discriminant validity with all the other constructs [13]. The evaluation of measurement model confirms that all the construct measures are valid and reliable (Table 2). All the values for of AVE proves the discriminate of construct (Fig. 2).

Structural Model Estimation

According to Henseler et al., (2009) [21] the R^2 (coefficient of determination) is a measure of the predictive accuracy of the model. In presented study as shown

Table 1. Evaluation results of measurement model

Constructs	Items	Loading	Cronbach's alpha	Composite reliability	AVE
Strategic supplier partnership	SSP1		0.884	0.815	0.682
	SSP2	0.892			
	SSP3	0.818			
	SSP4	0.787			
	SSP5	0.804			
Customer relationship	CR1	0.915	0.854	0.765	0.545
	CR2	0.014			
	CR3	0.918			
	CR4	0.946			
	CR5	0.902			
Level of information sharing	LIS1	0.925	0.860	0.669	0.661
	LIS2	0.938			
	LIS3	0.941			
	LIS4	0.913			
Internal lean practices	ILP1	0.888	0.886	0.715	0.684
	ILP2	0.792			
	ILP3	0.815			
	ILP4	0.807			
	ILP5	0.831			
Innovation culture	IC1	0.755	0.874	0.797	0.593
	IC2	0.777			
	IC3	0.762			
	IC4	0.735			
	IC5	0.832			
	IC6	0.754			
Organizational performance	OP1	0.809	0.861	0.797	0.636
	OP2	0.830			
	OP3	0.768			
	OP4	0.802			
	OP5	0.778			
Level of information quality	LIQ1	0.931	0.847	0.761	0.562
	LIQ2	0.942			
	LIQ3	0.929			
	LIQ4	0.911			

in (Table 3 the value of Q^2 is greater than "0" which shows the stability of the model and the predictive relevance of inner model is satisfied.

As shown above in the Table 4, all the values of path coefficients indicate the strong positive significant relationship among the constructs. The SSP has significant and positive impact on Organizational Performance with (Path coefficient = 0.173, t-values = 7.520, P-value = 0.005). The CR has significant and positive impact on Organizational Performance with (Path coefficient = 0.060, t-values = 3.285, P-value = 0.004). The LIS has significant and positive impact on Organizational Performance with (Path coefficient = 0.251, t-values = 2.378, P-value =

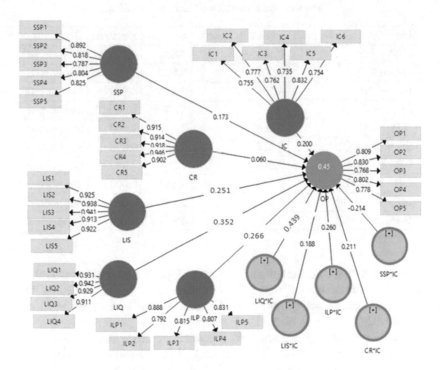

Fig. 2. The results of PLS algorithm

Table 2. Discriminant validity

	CR	IC	ILP	LIQ	LIS	OP	SSP
CR	0.682						
IC	0.056	0.770					
ILP	0.011	0.137	0.827				
LIQ	0.594	0.031	0.024	0.749			
LIS	0.650	0.040	0.033	0.649	0.813		
OP	0.091	0.207	0.174	0.110	0.086	0.797	
SSP	0.066	0.131	0.933	0.039	0.045	0.184	0.826

Table 3. Predictive accuracy and relevance of the model

Goodness of fit indices	R-Square (R^2)	(Q^2)
Organizational performance	0.45	0.42

Table 4. Path coefficient and t-Statistics

Research model's path	Path coefficients	t-values	P-values	Decisions
SSP→ OP	0.173	7.520	0.005	Supported
CR→OP	0.060	3.285	0.004	Supported
LIS→ OP	0.251	2.378	0.034	Supported
LIQ→ OP	0.352	11.566	0.000	Supported
ILP→ OP	0.266	3.119	0.000	Supported
SSP∗ IC→ OP	0.214	4.138	0.003	Supported
CR∗ IC→ OP	0.211	7.348	0.000	Supported
LIS∗ IC→ OP	0.188	4.553	0.000	Supported
LIQ∗ IC→ OP	0.439	2.987	0.000	Supported
ILP∗ IC→ OP	0.260	4.834	0.000	Supported

0.034). The LIQ has significant and positive impact on Organizational Performance with (Path coefficient = 0.325, t-values = 2., P-value = 0.005). The ILP has significant and positive impact on Organizational Performance with (Path coefficient= 0.266, t-values = 3.119, P-value = 0.000).

The moderation analysis show that Innovation culture significantly moderate the relationship between SSP, CR, LIS, LIQ, ILP. Innovation Cultures as make the relationship stronger among SSP, CR, LIS, LIQ, ILP and organizational performance. The results of SEM presented in Table 4 supporting all the proposed hypotheses. The *Hypothesis 1, 2 and 4* states that organizations with high levels of SCM practice through innovation culture have high levels of organizational performance.

5   Conclusions

This study presented a model for investigating the relationship between SCM practices innovation culture and organization performance. The comprehensive, reliable and valid instruments were developed for analyzing the proposed model. It was tested and verified by the statistical analysis including discriminant validity, composite reliability, path coefficient and t-Statistics. This study discuss the empirical grounds for supporting conceptual and prescriptive statements in the literature regarding the impact of SCM practices, innovation culture and measures of organizational performance.

In today's competitive environment, the role of innovation is getting more popularity in organizations. Organizations have started to realize that innovation play very important role for getting a competitive position and creating long-lasting advantages. The successful implementation of innovation leads an organization towards an industry leading performance. In short, the main objective of this study was to investigate the moderating role of innovation culture between SCM practices and organizational performance. The results of this study are encouraging and provide direction for further research. Concerning the social implications, the present study can contribute to a better understanding of supply chain and its management practices. The studied SCM practices are expected to encourage and facilitate further research.

Limitations and Future Directions

This study is about a textile manufacturing industries (an evaluation of SCM practices from the respective industries) in Pakistan. Such may not be suitable for distributors and retailers regarding constructs like internal lean practices. It is not easy to cover the all domain in one research study. Further research can formulate additional measurements for practices of internal supply chain like total quality management and competitive advantage as a moderator. The information quality level may be adversely affected by the chain length.

Acknowledgement. The authors would like to thank the Editor and the anonymous reviewers for their very valuable and helpful suggestions for improving the quality of present research.

References

1. Ahmed, Y.: Textile industry of Pakistan. Horizon securities SMC (2008)
2. Arsalan Supply chain management and textile industry of Pakistan. Arsalan's dawn article (2006)
3. Balsmeier, P., Voisin, W.J.: Supply chain management: a time-based strategy. Ind. Manag.-Chic. Then Atlanta **39**, 24–27 (1996)
4. Belfanti, N., Adoption of lean practices as management innovation. A review and conceptualisation. Int. J. Bus. Innov. Res. **18**(2):242–277 (2019)
5. Castro, M.D., Delgado-Verde, M.: The moderating role of innovation culture in the relationship between knowledge assets and product innovation. Technol. Forecast. Soc. Chang. **80**(2), 351–363 (2013)
6. Chan, A.T.L., Ngai, E.W.T., Moon, K.K.L.: The effects of strategic and manufacturing flexibilities and supply chain agility on firm performance in the fashion industry. Eur. J. Oper. Res. **259**(2) (2016)
7. Chavez, R., Gimenez, C.: Internal lean practices and operational performance: the contingency perspective of industry clock speed. Int. J. Oper. Prod. Manag. **33**(5), 562–5888 (2013)
8. Childerhouse, P., Towill, D.R.: Simplified material flow holds the key to supply chain integration. Omega **31**(1), 17–27 (2003)
9. Chizzo, S.A.: Supply chain strategies: solutions for the customer-driven enterprise. Supply Chain. Manag. Dir. **1**(1), 4–9 (1998)
10. Christensen, C., Raynor, M.: The innovator's solution: creating and sustaining successful growth. In: Harvard Business School Press (2003)
11. Dell, M.: The power of virtual integration: an interview with dell computer's Michael dell. interview by Joan Magretta. Harv. Bus. Rev. **76**(2):73–84 (1998)
12. Dubey, R., Altay, N.: Supply chain agility, adaptability and alignment: empirical evidence from the Indian auto components industry. Int. J. Oper. Prod. Manag. **38**(10), 129–148 (2018)

13. Hair, Jr., F.J., Sarstedt, M.: Partial least squares structural equation modeling (PLS-SEM) an emerging tool in business research. Eur. Bus. Rev. **26**(2), 106–121 (2014)

14. Flynn, B., Pagell, M., Fugate, B.: Survey research design in supply chain management: the need for evolution in our expectations. J. Supply Chain. Manag. **54** (2017)

15. Fraser, J.: Metrics that matter: uncovering KPIS that justify operational improvements. In: Research Project Carried out by Manufacturing Enterprise Solutions Association (MESA) and Presented at Plant2Enterprise Conference, pp. 24–25 (2006)

16. Govindarajan, V., Trimble, C.: Organizational dna for strategic innovation. Calif. Manag. Rev. **47**(3), 47–76 (2005)

17. Gunasekaran, A., Patel, C., Tirtiroglu, E.: Performance measures and metrics in a supply chain environment. Int. J. Oper. Prod. Manag. **21**(1/2), 71–87 (2001)

18. Hamel, G.: Leading the revolution. In: Plume (2002)

19. Hashim, M., Nazam, M.: Application of multi-objective optimization based on genetic algorithm for sustainable strategic supplier selection under fuzzy environment. J. Ind. Eng. Manag. **10**(2) (2017)

20. Hashim, M., Yao, L.: Multi-objective optimization model for supplier selection problem in fuzzy environment. In: Proceedings of the Eighth International Conference on Management Science and Engineering Management, pp. 1201–1213 (2014)

21. Henseler, J., Ringle, C.M., Sinkovics, R.R.: The use of partial least squares path modeling in international marketing. Soc. Sci. Electron. Publ. **20**(4), 277–319 (2009)

22. Hoek, R.I.V., Vos, B., Commandeur, H.R.: Restructuring European supply chains by implementing postponement strategies. Long Range Plan. **32**(5), 505–518 (1999)

23. Holmberg, S.: A systems perspective on supply chain measurements. Int. J. Phys. Distrib. Logist. Manag. **30**(10), 847–868 (2000)

24. Jones, C.: Moving beyond ERP: making the missing link. Logist. Focus **6**, 2–7 (1998)

25. Li, S., Ragu-Nathan, B.: The impact of supply chain management practices on competitive advantage and organizational performance. Omega **34**(2), 107–124 (2006)

26. McCormack, K.: What supply chain management practices relate to superior performance? In: DRK Research Team (1998)

27. McGowan, A.: Perceived benefits of abcm implementation. Account. Horiz. **12**(1), 31–50 (1998)

28. Mentzer, J.T., Min, S., Zacharia, Z.G.: The nature of interfirm partnering in supply chain management. J. Retail. **76**(4), 549–568 (2000)

29. Menzel, H.C., Aaltio, I., Ulijn, J.M.: On the way to creativity: Engineers as intrapreneurs in organizations. Technovation **27**(12), 732–743 (2007)

30. Michael, H.: Deep change. How operational innovation can transform your company. Harv. Bus. Rev. **82**(4), 84–93 (2004)

31. Moberg, C.R., Cutler, B.D.: Identifying antecedents of information exchange within supply chains. Int. J. Phys. Distrib. & Logist. Manag. **32**(9), 755–770 (2002)

32. Monczka, R.M., Petersen, K.J.: Success actors in strategic supplier alliances: the buying company perspective. Decis. Sci. **29**(3), 553–577 (2010)

33. Naylor, J.B., Naim, M.M., Berry, D.: Leagility: Integrating the lean and agile manufacturing paradigms in the total supply chain. Int. J. Prod. Econ. **62**(1C2), 107–118 (1999)

34. Womack, J.P., Jones, D.T.: Lean thinking. In: Simon and Schuster (1996)
35. Senge, P.M., Carstedt, G.: Innovating our way to the next industrial revolution. MIT Sloan Manag. Rev. **42**(2), 24 (2001)
36. Stein, T., Sweat, J.: Killer supply chains. Inf. Week **708**(9), 36–46 (1998)
37. Stock, G.N., Greis, N.P., Kasarda, J.D.: Enterprise logistics and supply chain structure: the role of fit. J. Oper. Manag. **18**(5), 531–547 (2000)
38. Sum, C.C, Yang, K.K.: An analysis of material requirements planning (MRP) benefits using alternating conditional expectation (ACE). J. Oper. Manag. **13**(1), 35–58 (1995)
39. Tan, K.C., Kannan, V.R., Handfield, R.B.: Supply chain management: supplier performance and firm performance. J. Supply Chain. Manag. **34**(3), 2 (1998)
40. Tompkins, J., Ang, D.: What are your greatest challenges related to supply chain performance measurement? IIE Solut. **31**(6), 66 (1999)
41. Towill, D.R.: The seamless supply chain - the predator's strategic advantage. Int. J. Technol. Manag. = J. Int. Gest.N Technol. **13**(1), 37–56 (1997)
42. Vickery, S., Calantone, R., Droge, C.: Supply chain flexibility: an empirical study. J. Supply Chain. Manag. **35**(3), 16–24 (2010)
43. Vickery, S.K., Jayaram, J.: The effects of an integrative supply chain strategy on customer service and financial performance: an analysis of direct versus indirect relationships. J. Oper. Manag. **21**(5), 523–539 (2004)
44. Vijayasarathy, L., Robey, D.: The effect of EDI on market channel relationships in retailing. Inf. Manag. **33**(2), 73–86 (1997)
45. Yamin, S., Gunasekaran, A., Mavondo, F.T.: Relationship between generic strategies, competitive advantage and organizational performance: an empirical analysis. Technovation **19**(8), 507–518 (1999)
46. Yoshino, M.Y., Rangan, U.S.: Strategic alliances: an entrepreneurial approach to globalization. Long Range Plan. **29**(6), 1241 (1995)
47. Zhang, Q.: Technology Infusion Enabled Value-chain Flexibility: A Learning and Capability-Based Perspective. Lambert Academic Publication (2010)

A Performance Assessment Framework for Baijiu Sustainable Supply Chain in China

Xianglan Jiang[1,2,3(✉)], Yinping Mu[2], and Jiarong Luo[4]

[1] Management School, Sichuan University of Science & Engineering,
Zigong 64300, People's Republic of China
xianglanjiang@163.com
[2] School of Management and Economics, University of Electronic Science
and Technology of China, Chengdu 611731, People's Republic of China
[3] Sichuan Technology & Business College, Dujiangyan 611830,
People's Republic of China
[4] School of Management and Economics, Southwest University of Science
and Technology, Mianyang 621010, People's Republic of China

Abstract. The production and consumption of Baijiu has great impact on environmental, social, and economic performance in supply chain management. In the face of industry issues like imbalance supply and demand, irrational structure, uneven regional development, resource consumption, and social wellbeing, there is an urgent need for companies to assess sustainability from supply chain perspective. The Chinese economy is developing at a good pace and there is an enormous demand for Baijiu in the country. However, research on the three sustainable dimensions is still sparsity in this field. This paper proposes a framework for identifying environmental, social and economic criteria for Baijiu supply chain, and applies fuzzy analytical hierarchy process (FAHP) for prioritizing criteria. The framework is applied in a real case study at a Baijiu company in China. Economic dimension has been reported as the most important dimension, social second and environmental third. Results from this framework could provide sustainable development and management insights for decision makers in Baijiu industry in China.

Keywords: Baijiu (Chinese liquor) · Performance assessment · Supply chain · Sustainability

1 Introduction

Sustainable supply chain management (SSCM) has been highly valued by academic and business circles due to rising consideration of social wellbeing and environment protection while supporting the economic goal of organizations in recent years [1–4]. Sustainable development is commonly defined as "development that meets the needs of present without compromising the ability of future

© Springer Nature Switzerland AG 2020
J. Xu et al. (Eds.): ICMSEM2019 2019, AISC 1002, pp. 402–414, 2020.
https://doi.org/10.1007/978-3-030-21255-1_31

generations to meet their own needs" [5], which has become a challenging and promising task in the new era [6]. The best recognized sustainability concept in business is the triple bottom line (TBL) principle, which defined as economic, environmental and social aspects [3,7]. Due to various pressures, firms increasingly endeavor to integrate the economic, ecological, and social dimensions of sustainability to achieve the triple bottom line following the introduction of the sustainability concept [8,9].

Baijiu (Chinese liquor) is one of the most famous distilled spirits in the world, which takes a significant part in Chinese people's life and Chinese culture [10]. Baijiu, brandy, whiskey, gin, vodka, and rum are the six most popular global distilled spirits [11]. Baijiu is regraded as a historical and cultural heritage, which can be used for Chinese people for a long time thanks to its good taste and healthy reason [12]. The Baijiu industry is a traditional industry that has played a very important role in the development of the Chinese economy.

- Most sustainable assessment researches has done with one or two dimensions, and very few have addressed the whole three dimensions [3,6]. Therefore, there is an urgent need to assess sustainable performance for Baijiu industry in China.
- Very few researches have appreciated the Baijiu performance assessment from the supply chain perspective [13]. Although there are many researches on the Baijiu product and techniques, few of them attach importance to supply chain management [10,12]. Therefore, it has become important to establish an evaluation framework for sustainable supply chain management in the Baijiu industry.
- Few sustainability criteria systems that concurrently consider the three dimensions have been established for the Baijiu supply chain [6,13]. In the Baijiu industry, the traditional criteria are deficiency for an effective evaluation of the sustainable performance of supply chains [11].

This paper makes three contributions. First, we develop a performance measurement framework with three dimensions of sustainability incorporated into the supply chain management. Second, we provide a model formulated as fuzzy analytical hierarchy process (FAHP) that could evaluate the performances. Third, we show how multiple sustainability objectives can be realized in Baijiu industry.

The remainder of the paper is organized into five sections. A literature review on sustainable supply chain and on sustainable performance assessment method is presented in Sect. 2. In Sect. 3, the assessment criteria of Baijiu SSCM are presented. In Sect. 4, the proposed method are presented. Section 5 presents the application example. And Sect. 6 concludes the study.

2   Literature Review

In this section, we review the sustainable supply chain management and performance assessment methods.

2.1   Literature on Sustainable Supply Chain Management

Sustainable supply chain management has many definitions [14,15]. Cooperet et al. defined Supply Chain Management (SCM) as the integration of business processes from end user through original suppliers that provides products, services and information that add value for customers [16]. Linton first linked sustainability and supply chains, and believed that sustainability must integrate issues and flows that extend beyond the core of supply chain management: product design, manufacturing by-products, by-products produced during product use, product life extension, product end-of-life and recovery processes at end-of-life [17]. Seuring and Müller defined sustainable supply chain management as the management of material, information and capital flows as well as cooperation among companies along the supply chain while taking goals from all three dimensions of sustainable development, i.e., economic, environmental and social, into account which are derived from customer and stakeholder requirements [1]. In sustainable supply chains, environmental and social criteria need to be fulfilled by the members to remain within the supply chain, while it is expected that competitiveness would be maintained through meeting customer needs and related economic criteria. This means that SSCM requires more collaboration among cooperative enterprises in order to achieve the supply chain goal and reach sustainable performance [18]. Economic, environmental and social factors must be taken into account and evaluate usual and important operations set of performance criteria [1,19]. While Carter and Rogers defined SSCM as the strategic, transparent integration and achievement of an organization's social, environmental, and economic goals in the systemic coordination of key interorganizational business processes for improving the long-term economic performance of the individual company and its supply chains [20]. This means that environmental performance and corporation social responsibility must be applied and encouraged by all the supply chain partners. In this paper, we defined sustainable supply chain management as the management of supply chain operations, resources, information, and funds in order to maximize the supply chain profitability while at the same time minimizing the environmental impacts and maximizing the social well-being.

2.2   Literature on Performance Assessment Methods

Performance measurement methods consider multiple criteria as sustainability performance encompasses multiple dimensions [8]. There are some contributions in this field: QR, ECR, BSC, SCOR, ABC, SaT and ROF.

Quick Response (QR) is a movement in the apparel industry to shorten lead time. Under QR, the retailer has the ability to adjust orders based on better demand information [21]. In 1992, efficient consumer response (ECR) was presented as a powerful tool for optimizing the supply chain performance within the US grocery industry [22]. ECR requires all parties in the supply chain to work together to fulfill consumer wishes 'better, faster and at less cost' [23]. The Balanced Scorecard (BSC) is a strategy performance management tool which tries to balance the four perspectives: financial, customers, internal business, and learning and growth [24]. Singh et al. used balanced scorecard to evaluate sustainability for manufacturing SMEs [25]. The Supply Chain Operations Reference Model (SCOR) aims at evaluating and improving SC performance based on the best practices and five generic metrics (reliability, responsiveness, agility, costs, and assets) [26]. ABC (Activity-Based Costing) is the emerging foundation of performance management. If you lift up the hood of a performance management system, you will find ABC powering performance measures for business activities [27,28]. Sink and Tuttle (SaT) model claims that the performance of an organisation is a complex interrelationship between seven performance criteria: profitability, productivity, quality, quality of work life, innovation, effectiveness, and efficiency [29,30]. Beamon thought a supply chain measurement system must place emphasis on three separate types of performance measures: resource measures (R), output measures (O), and flexibility measures (F) [31]. Each of these three types of performance measures has different goals. Resources goal is high level of efficiency, Output's is high level of customer service and flexibility's is ability to respond to a changing environment. These methods are mainly focusing on economical dimension, and rarely talked about environmental and social fields.

Although there are many tools to assess SCM practices, it is common to use the five metrics of SCOR Model to evaluate economic performance in academia field [8]. The SCOR model was created by the Supply Chain Council in 1996 as a standard framework for evaluating and improving supply chain performance which includes five generic metrics: reliability, responsiveness, flexibility, costs, and assets.

Most of SSCM models are focus on one dimension: the economic aspect or the environmental dimension [8]. Rare works integrate two sustainable aspects and few models integrate all the three dimensions [1]. Seuring divided the sustainable models into four categories: life-cycle assessment (LCA) based models, equilibrium models, multi-criteria decision making (MCDM), and applications of the analytical hierarchy process (AHP) [6]. Jiang et al. developed a network for Baijiu supply chain [13]. In this paper, we use fuzzy MCDM to evaluate the performance criteria of SSCM. Thus, we can conclude that there is insufficient integration and maturity in this field [1]. This paper seeks to rich this contribution by providing a framework for sustainable performance assessment of SCM practices. The framework develops and reinforces the concept of simultaneous assessment of the TBL sustainable dimensions.

3   Performance Assessment Criteria

Sustainable supply chain management is one of the key measures for modern management which emphases on sustainable development [32]. Sustainable assessment criteria of supply chain usually divided into three dimensions: economic performance, environmental protection and social responsibility based on TBL theory [33]. Inspired by the literatures and expert opinion, we suggest a model consider fifteen criteria to cover the three dimensions, five are related to the economic dimension and social dimension, and environmental dimension, respectively, as shown in Fig. 1

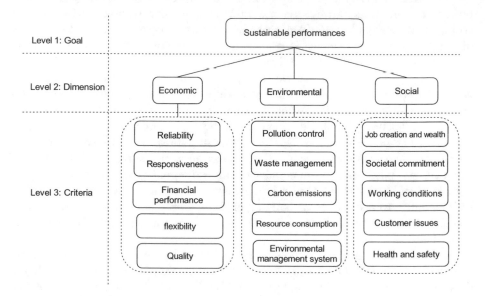

Fig. 1. Sustainable performance hierarchy structure

4 Methodology

In this section, the methodology is fuzzy AHP. AHP is a quantitative technique that structures a multi-criteria decision making solution introduced by Satty. Fuzzy AHP methods combine AHP with fuzzy set theory to solve hierarchical fuzzy problems. The fuzzy AHP can capture uncertain imprecise judgement of experts by handling linguistic variables. Recently the fuzzy AHP is widely used to solve multi-criteria decision problems in few other fields. Mardani et al. reviewed a total of 403 FMCDM papers published from 1994 to 2014 in more than 150 peer reviewed journals, according to the researches, papers were grouped into four main fields: engineering, management and business, science, and technology. Hybrid fuzzy MCDM in the integrated method and fuzzy in the individual section were ranked as the first and second methods in use [34].

Step1: Define scale of relative importance

Triangular fuzzy numbers $\tilde{1}$ to $\tilde{9}$ are used to present the scale program (Table 1). Take the imprecision of human qualitative assessments into consideration, the TFNs are defined with the corresponding membership function.

Step 2: Construct fuzzy pairwise comparison matrix

Pairwise comparison matrices are established among all the criteria in the dimensions of the hierarchy system based on experts' preferences, as showed in matrix $\tilde{A}$.

Table 1. Scale of relative importance used in the pairwise comparison matrix

Intensity of important	Linguistic variables	Membership function
$\tilde{1}$	Equally important/preferred	(1,1,1)
$\tilde{2}$	Between $\tilde{1}$ and $\tilde{3}$	(1,2,3)
$\tilde{3}$	Weakly important/preferred	(2,3,4)
$\tilde{4}$	Between $\tilde{3}$ and $\tilde{5}$	(3,4,5)
$\tilde{5}$	Strongly more important/preferred	(4,5,6)
$\tilde{6}$	Between $\tilde{5}$ and $\tilde{7}$	(5,6,7)
$\tilde{7}$	Very strongly important/preferred	(6,7,8)
$\tilde{8}$	Between $\tilde{7}$ and $\tilde{9}$	(7,8,9)
$\tilde{9}$	Extremely more important/preferred	(8,9,10)

$$\tilde{A} = \begin{bmatrix} \tilde{a}_{11} & \tilde{a}_{12} & ... & \tilde{a}_{1n} \\ \tilde{a}_{21} & \tilde{a}_{22} & ... & \tilde{a}_{2n} \\ \vdots & \vdots & \vdots & \vdots \\ \tilde{a}_{n1} & \tilde{a}_{n2} & ... & \tilde{a}_{nn} \end{bmatrix} \tag{1}$$

where $\tilde{a}_{ii} = 1, \tilde{a}_{ji} = 1/\tilde{a}_{ij}, \tilde{a}_{ij} \neq 0$.

Step 3: Evaluate fuzzy weights

To use geometric mean technique to define the fuzzy geometric mean and fuzzy weights of each criterion by Eq. 2.

$$\tilde{r}_i = (\prod_{j=1}^{n} \tilde{a}_{ij})^{\frac{1}{n}} \tag{2}$$

$$\tilde{w}_i = \tilde{r}_i / \sum_{i=1}^{n} \tilde{r}_i \tag{3}$$

where $\tilde{a}_{ij}$ is fuzzy comparison value of criterion i to criterion j, $\tilde{r}$ is the fuzzy geometric mean of the fuzzy comparison value of criterion i to each criterion and is the fuzzy weight of the ith criterion which can be indicated by a TFN, $\tilde{w}_i = (lw_i, mw_i, uw_i)$. The lw_i, lm_i and uw_i are the lower, middle and upper values of the fuzzy weight of the ith criterion.

Step 4: Check consistency

The consistency ratio (CR) for each of the matrix and overall inconsistency for the hierarchy are calculated in order to control the results of this method. When the crisp comparison matrix A is consistent, it means the fuzzy comparison matrix $\tilde{A}$ is also consistent. The consistency can be checked as follows:

(1) Calculate the largest Eigen value of the matrix by using Eq. (5)

$$Aw = \lambda_{\max} w \tag{4}$$

(2) The Consistency Ratio (CR) is used to estimate directly the consistency of pairwise comparisons. The CR is computed by using Eq. (6)

$$CR = \frac{CI}{RI} \tag{5}$$

$$CI = \frac{\lambda_{\max} - n}{n - 1} \tag{6}$$

where CI is consistency index. RI is random index, which is shown in Table 2, and n is matrix size.

Table 2. The random consistency index RI

Size (n)	1	2	3	4	5	6	7	8	9	10
RI	0	0	0.58	0.9	1.12	1.24	1.32	1.41	1.45	1.49

5   Case Example

Recently more and more Chinese organizations realize that sustainable development of SCM plays an important role in business success and that sustainable performances evaluation becoming a significant work for them. The paper is about a Baijiu company in China which takes sustainable performance assessment in the supply chain. The company wants to meet the economic, environmental and social demands in order to improve its corporate image and to serve the local economy. Management has to balance their profit motives with concerns for environment and social-economic dimensions.

5.1   Sustainable Criteria

To analyze the problem, a decision group of three experts which comprising senior manager, senior executive of SC members, and academician (professor) are formed for the evaluation and assessment of SSCM. The experts selected are highly skilled in their domains and are proficient in decision making. The selection of experts is decided on the basis of certain criteria like their individual academic, industrial and consultancy experiences, decision making skills, expertise in area etc. The sustainable performances of supply chain management are determined through literature review and expert opinion. In this study, fifteen sub-criteria of sustainable performances based on TBL are identified through literature review and intensive discussion with decision group members. There are three levels in decision hierarchy structure for this research. Sustainable performance assessment as the objective is in the first level of hierarchy. The triple bottom line: economic, environmental, and social dimensions in the second level, and fifteen criteria at third level, as shown in Fig. 1.

5.2   Weights Determination

In this step, the expert panel is asked to make pairwise comparisons of three dimensions of TBL and fifteen criteria by using linguistic variables in Table 1. Calculate the elements of synthetic pairwise comparison matrix by using the geometric mean method:

$$\tilde{a}_{ij} = (\tilde{a}_{ij}^1 \otimes \tilde{a}_{ij}^2 \otimes \cdots \otimes \tilde{a}_{ij}^3)$$

Take $\tilde{a}_{12}$ as an example,

$$\tilde{a}_{12} = (3.4760, 4.5789, 5.6462)$$

Through the same calculation process, we can get the synthetic pairwise comparison matrix of the dimensions in Table 3.

Fuzzy weights of dimension can be obtained by

$$\tilde{r}_1 = (\tilde{a}_{11} \otimes \tilde{a}_{12} \otimes \tilde{a}_{13})^{1/3} = (1.9086, 2.2823, 2.7377)$$

By using Eq. 3, we can get

$$\tilde{w}_1 = (0.6377, 0.6167, 0.6142)$$

with the same method, we can obtain the fuzzy comparison matrices of criteria economical, environmental and social are showed in Tables 4, 5 and 6.

The results obtained from the calculation based on pairwise comparison matrixes provided in Tables 3, 4, 5 and 6, are presented in Table 7.

we can obtain the fuzzy comparison matrices of criteria economical, environmental and social are showed in Tables 3, 4, 5 and 6. CR Values of all the matrixes are less than 0.1. Therefore, these matrixes are consistent. Quality was found to be the highest importance in order and followed by financial performance.

Table 3. Fuzzy pairwise comparison matrix of the dimensions

	Economic	Environmental	Social
Eco	(1, 1, 1)	(3.4760, 4.5789, 5.6462)	(2.0000, 2.5962, 3.6342)
Env	(0.1771, 0.2877, 0.2877)	(1, 1, 1)	(0.3969, 0.6057, 0.9086)
Soc	(0.2752, 0.3852, 0.5000)	(1.1006, 1.6510, 2.5198)	(1, 1, 1)
Weight	(0.6377, 0.6167, 0.6142)	(0.1369, 0.1509, 0.1434)	(0.2244, 0.2324, 0.2423)
MSw	0.6229	0.1441	0.2330

Table 4. Fuzzy pairwise comparison matrix for the economic criteria

	Eco1	Eco2	Eco3	Eco4	Eco5
Eco1	(1,1,1)	(1.8171,2.8845,3.9149)	(0.2027,0.2554,0.3467)	(2.2894,3.3019,4.3089)	(0.2184,0.2811,0.3969)
Eco2	(0.2554,0.3467,0.5503)	(1,1,1)	(0.1609,0.1926,0.2404)	(1.8171,2.8845,3.9149)	(0.1376,0.1598,0.1908)
Eco3	(2.8845,3.9149,4.9324)	(4.1602,5.1925,6.2145)	(1,1,1)	(4.8203,5.8480,6.8683)	(0.4055,0.5000,0.6934)
Eco4	(0.2321,0.3029,0.4368)	(0.2554,0.3467,0.5503)	(0.1456,0.1710,0.2075)	(1,1,1)	(0.1438,0.1682,0.2027)
Eco5	(2.5198,3.5569,4.5789)	(5.2415,6.2573,7.2685)	(1.4422,2.000,2.4662)	(4.9324,5.9439,6.9521)	(1,1,1)
Weight	(0.1243,0.1314,0.1396)	(0.0698,0.0707,0.0742)	(0.3277,0.3209,0.3195)	(0.0457,0.0444,0.0471)	(0.4325,0.4326,0.4197)
MSw	0.1318	0.0716	0.3227	0.0457	0.4283

Table 5. Fuzzy pairwise comparison matrix for the environmental criteria

	Env1	Env2	Env3	Env4	Env5
Env1	(1,1,1)	(1.0000,1.5874,2.0801)	(0.8736,1.4422,2.2894)	(0.4149,0.5503,0.7368)	(0.4642,0.6300,0.8736)
Env2	(0.4807,0.6300,1.0000)	(1,1,1)	(0.5848,0.7937,1.1856)	(0.3684,0.5503,0.7937)	(0.5000,0.6437,0.8550)
Env3	(0.4368,0.6934,1.1447)	(0.8434,1.2599,1.7100)	(1,1,1)	(0.2075,0.2646,0.3684)	(0.2154,0.2752,0.3816)
Env4	(1.3572,1.8171,2.4101)	(1.2599,1.8171,2.7144)	(2.7144,3.7798,4.8203)	(1,1,1)	(0.2154,0.2752,0.3816)
Env5	(1.1447,1.5874,2.1544)	(1.1696,1.5536,2.0000)	(2.6207,3.6342,4.6416)	(2.6207,3.6342,4.6416)	(1,1,1)
Weight	(0.1647,0.1728,0.1762)	(0.1261,0.1236,0.1310)	(0.1003,0.1007,0.1057)	(0.2280,0.2237,0.2250)	(0.3553,0.3507,0.3386)
MSw	0.1712	0.1269	0.1022	0.2256	0.3482

Table 6. Fuzzy pairwise comparison matrix for the social criteria

	Soc1	Soc2	Soc3	Soc4	Soc5
Soc1	(1,1,1)	(3.5569,4.5789,5.5934)	(3.9149,4.9324,5.9439)	(3.9149,4.9324,5.9439)	(0.9086,1.3867,1.9574)
Soc2	(0.1788,0.2184,0.2811)	(1,1,1)	(1.2164,1.5874,2.0274)	(0.1682,0.2027,0.2554)	(0.1788,0.2184,0.2811)
Soc3	(0.1682,0.2027,0.2554)	(0.4932,0.6300,0.8221)	(1,1,1)	(0.2554,0.3467,0.5503)	(0.1926,0.2404,0.3218)
Soc4	(0.1682,0.2027,0.2554)	(3.9149,4.9324,5.9439)	(1.8171,2.8845,3.9149)	(1,1,1)	(0.1788,0.2184,0.2811)
Soc5	(0.5109,0.7211,1.1006)	(3.5569,4.5789,5.5934)	(3.1072,4.1602,5.1925)	(3.5569,4.5789,5.5934)	(1,1,1)
Weight	(0.4014,0.4043,0.3982)	(0.0640,0.0617,0.0622)	(0.0583,0.0573,0.0610)	(0.1287,0.1297,0.1306)	(0.3191,0.3256,0.3324)
MSw	0.4013	0.0626	0.0589	0.1296	0.3257

Table 7. Final priority for the SCM sustainable performances

Dimension	Dimension	CR	Criteria weight	Relative weight	Relative weight	Global weight	Global weight
Economical	0.6229	0.0651	Eco1	0.1318	3	0.0439	8
			Eco2	0.0716	4	0.0239	12
			Eco3	0.3227	2	0.1076	5
			Eco4	0.0457	5	0.0152	15
			Eco5	0.4283	1	0.1428	1
Environmental	0.1441	0.0957	Env1	0.1712	3	0.0571	7
			Env2	0.1269	4	0.0423	10
			Env3	0.1022	5	0.0341	11
			Env4	0.2256	2	0.0752	6
			Env5	0.3482	1	0.1161	3
Social	0.2330	0.0603	Soc1	0.4013	1	0.1338	2
			Soc2	0.0626	4	0.0209	13
			Soc3	0.0589	5	0.0196	14
			Soc4	0.1296	3	0.0432	9
			Soc5	0.3257	2	0.1086	4

5.3   Results and Discussions

According to Table 3, we can see the proportion of the three dimensions. Economic dimension is 0.6229, environmental dimension is 0.1441, and social dimension is 0.2330. This means the experts think economic is the base for other dimensions. With the emphasis on the social wellbeing, social dimension takes the second rate. Although the environmental problem becomes more and more important in global economy, it still takes the last rate in the evaluation system. It may be related to the green production of Baijiu industry in recent years.

According to Table 4, we can see $Eco5 > Eco3 > Eco1 > Eco2 > Eco4$, it equals the weight of $Quality$ (0.4283) $> Financial performance$ (0.3227) $> Reliability$ (0.1318) $> Responsiveness$ (0.0716) $> Flexibility$ (0.0457). Quality is the most important criteria in the economic dimension, followed by financial performance. Reliability and responsiveness are the relative significance criteria not only for the focal company, but also for its supply chains. The Baijiu production has certain stability in Chinese economy, thus, flexibility is less important than other criteria.

According to Table 5, $Env5 > Env4 > Env1 > Env2 > Env3$ which means the weight of Environmental management system (0.3482) was ranked first, Resource consumption (0.2256) was ranked second, Pollution control was ranked third, Waste management (0.1269) was ranked fourth, and the final place of importance was Carbon emissions (0.1022). For the Baijiu company, in order to minimize the impact on the environment, the best way is implementing environmental management system. The focal company produces Baijiu, and its main raw materials are grains.

Due to Table 6, $Soc1 > Soc5 > Soc4 > Soc2 > Soc3$, Job creation and wealth (0.4013) was of the greatest importance, followed by Health and safety (0.3257). The third ranked criterion was Customers issues (0.1296), followed by Working conditions (0.0626), and Societal commitment (0.0589)was the last ranked criterion. With the development of the modern society, more and more Chinese companies pay attention to job creation and wealth. Along with a focus on safety, therefore, health and safety follows with job creation and wealth.

Considering the significance of global weights in sustainable performance evaluation, the ranking is identified as $Eco5 > Soc1 > Env5 > Soc5 > Eco3 > Env4 > Env1 > Eco1 > Soc4 > Env2 > Env3 > Eco2 > Soc2 > Soc3 > Eco4$ as presented in Table 7. According to the analysis of FAHP, Quality (Eco5), Job creation and wealth (Soc1), Environmental management system (Eco1), Health and safety (Soc5) are the top four most important criteria with the values of 0.1428, 0.1338, 0.1161, 0.1086, respectively. Flexibility, Societal commitment, Working conditions are the least important criteria with the values of 0.0457, 0.0589, 0.0626, respectively.

6   Conclusion

In this paper, we studied the performance assessment of the Baijiu supply chain from sustainable perspective. The empirical case study is presented to demonstrate the applicability of the proposed framework. Through literature review and expert opinion the three dimensions 15 criteria are identified. The results from the study indicate that economic is the most important dimension, and environmental is the least important dimension. The top three criteria are quality (Eco5), financial performance (Eco4), and reliability (Eco1), and the last three are responsiveness (Eco2), customers issues (Soc4), business practices (Soc5). As shown in the empirical case study, the proposed method is practical for evaluating sustainable performances in SCM. The framework could help Baijiu companies to identify criteria and realize sustainable supply chain management. And results from this framework could provide sustainable development and management insights for decision makers in Baijiu industry in China.

Acknowledgements. This work was supported by the National Natural Science Foundation of China under the Grant Nos.71772025, 71671118 and 71702156, the Humanities and Social Sciences of Ministry of Education of China (17YJC630098), the Sichuan Provincial Education Department of Key Research Base Key Program (No. CJZ17-02), Sichuan Social Science Key Research Base Project (Xq18C07), the Doctoral Fund Project of Sichuan University of Science and Engineering (Nos.2017RCSK21 and 2018RCSK01), the Sichuan Science and Technology Department Project (No.2019JDR0026), and the General Project of Sichuan Education Department (No.18SB0400).

References

1. Seuring, S., Müller, M.: From a literature review to a conceptual framework for sustainable supply chain management. J. Clean. Prod. **16**(15), 1699–1710 (2008)
2. Rajeev, A., Pati, R.K., et al.: Evolution of sustainability in supply chain management: A literature review. J. Clean. Prod. **162**, 299–314 (2017)
3. Kumar, G., Subramanian, N., Ramkumar, M.: Missing link between sustainability collaborative strategy and supply chain performance: Role of dynamic capability. Int. J. Prod. Econ. **203**, 96–109 (2018)
4. Reefke, H., Sundaram, D.: Sustainable supply chain management: Decision models for transformation and maturity. Decis. Support. Syst. **113**, 56–72 (2018)
5. Keeble, B.R.: The Brundtland report: Our common future. Med. War **4**(1), 17–25 (1988)
6. Seuring, S.: A review of modeling approaches for sustainable supply chain management. Decis. Support. Syst. **54**(4), 1513–1520 (2013)
7. Elkington, J.: Partnerships from cannibals with forks: The triple bottom line of 21st-century business. Environ. Qual. Manag. **8**(1), 37–51 (1998)
8. Boukherroub, T., Ruiz, A., et al.: An integrated approach for sustainable supply chain planning. Comput. Oper. Res. **54**, 180–194 (2015)
9. Brandenburg, M.: A hybrid approach to configure eco-efficient supply chains under consideration of performance and risk aspects. Omega **70**, 58–76 (2017)

10. Wang, H., Gao, Y., et al.: Characterization and comparison of microbial community of different typical Chinese liquor Daqus by PCRCDGGE. Lett. Appl. Microbiol. **53**(2), 134–140 (2011)
11. Shi, J., Xiao, Y., et al.: Analyses of microbial consortia in the starter of Fen liquor. Lett. Appl. Microbiol. **48**(4), 478–485 (2009)
12. Zheng, X., Han, B.: Baijiu, Chinese liquor: history, classification and manufacture. J. Ethn. Foods **3**(1), 19–25 (2016)
13. Jiang, X., Xu, J., et al.: Network design towards sustainability of chinese baijiu industry from a supply chain perspective. Discret. Dyn. Nat. Soc. (2018)
14. Chardine-Baumann, E., Botta-Genoulaz, V.: A framework for sustainable performance assessment of supply chain management practices. Comput. Ind. Eng. **76**, 138–147 (2014)
15. Mani, V., Gunasekaran, A., Delgado, C.: Enhancing supply chain performance through supplier social sustainability: An emerging economy perspective. Int. J. Prod. Econ. **195**, 259–272 (2018)
16. Cooper, M.C., Lambert, D.M., Pagh, J.D.: Supply chain management: more than a new name for logistics. Int. J. Logist. Manag. **8**(1), 1–14 (1997)
17. Linton, J.D., Klassen, R., Jayaraman, V.: Sustainable supply chains: An introduction. J. Oper. Manag. **25**(6), 1075–1082 (2007)
18. Govindan, K.: Sustainable consumption and production in the food supply chain: A conceptual framework. Int. J. Prod. Econ. **195**, 419–431 (2018)
19. Reefke, H., Sundaram, D.: Key themes and research opportunities in sustainable supply chain management-identification and evaluation. Omega **66**, 195–211 (2017)
20. Carter, C.R., Rogers, D.S.: A framework of sustainable supply chain management: moving toward new theory. Int. J. Phys. Distrib. Logist. Manag. **38**(5), 360–387 (2008)
21. Iyer, A.V., Bergen, M.E.: Quick response in manufacturer-retailer channels. Manag. Sci. **43**(4), 559–570 (1997)
22. Kotzab, H.: Improving supply chain performance by efficient consumer response? A critical comparison of existing ECR approaches. J. Bus. Ind. Mark. **14**(5/6), 364–377 (1999)
23. Ippolito, R.A.: Consumer reaction to measures of poor quality: Evidence from the mutual fund industry. J. Law Econ. **35**(1), 45–70 (1992)
24. Porter, M.: The Balanced Scorecard: Measures that Drive Performance. Harvard Business Review, USA (1992)
25. Singh, S., Olugu, E.U., et al.: Fuzzy-based sustainability evaluation method for manufacturing SMEs using balanced scorecard framework. J. Intell. Manuf. **29**(1), 1–18 (2018)
26. Garcia, D.J., You, F.: Supply chain design and optimization: Challenges and opportunities. Comput. Chem. Eng. **81**, 153–170 (2015)
27. Cooper, R., Kaplan, R.S.: Profit priorities from activity-based costing. Harv. Bus. Rev. **69**(3), 130–135 (1991)
28. Turny Peter, B.B.: Activity Based Costing The Performance Breakthrough. Kogan, UK (1997)
29. Sink, D.S., Tuttle, T.C.: Planning and Measurement in Your Organization of the Future. Industrial Engineering And Management (1989)
30. Tangen, S.: Performance measurement: from philosophy to practice. Int. J. Prod. Perform. Manag. **53**(8), 726–737 (2004)
31. Beamon, B.M.: Measuring supply chain performance. Int. J. Oper. Prod. Manag. **19**(3), 275–292 (1999)

32. Sarkis, J., Zhu, Q., Lai, K.H.: An organizational theoretic review of green supply chain management literature. Int. J. Prod. Econ. **130**(1), 1–15 (2011)
33. You, F., Tao, L., et al.: Optimal design of sustainable cellulosic biofuel supply chains: Multiobjective optimization coupled with life cycle assessment and input-Coutput analysis. AIChE J. **58**(4), 1157–1180 (2012)
34. Mardani, A., Jusoh, A., Zavadskas, E.K.: Fuzzy multiple criteria decision-making techniques and applications-Two decades review from 1994 to 2014. Expert. Syst. Appl. **42**(8), 4126–4148 (2015)

Spatial Evolution, Driving Factors and Comprehensive Development on Urban Agglomeration–A Case Study of Sichuan Province

Jialing Zhu, Quan Quan, Ming You, Sichen Xu, and Yunqiang Liu[✉]

College of Management, Sichuan Agricultural University, Chengdu 611130, People's Republic of China
liuyunqiang@sicau.edu.cn

Abstract. The study was in order to know the urban agglomerations' comprehensive capacity, driving forces and the spatial pattern of gravity; Taking cases of 4 urban agglomerations in Sichuan, an evaluation system was constructed in 3 dimensions: sociology, ecology and economics. The comprehensive carrying capacity level and driving forces were evaluated by entropy weight-TOPSIS, and revised gravity model was used to analyze the interaction forces and industry division; The results showed that the majority of the cities' comprehensive capacity increased and it still had a lot of room for improvement; industrial structure, population urbanization, urban infrastructure construction became the key drivers of the urban agglomerations development; the spatial pattern and industry division were not rational, diffusion effects of core cities were not marked; Some measures should be taken to achieve healthy urbanization of Sichuan, such as optimizing the structure of urban agglomeration, congregating development of the provincial capital city Chengdu, multi-point supporting of the secondary core cities and guiding of government.

Keywords: Urban agglomeration · Comprehensive carrying capacity · Driving forces · Gravitational pattern

1 Introduction

As a typical representative of cluster economies as well as a new carrier of promoting urbanization, urban agglomeration has become the most dynamic and potential core area in the pattern of socioeconomic development [14]. Consisting of 21 cities and autonomous prefectures, Sichuan, the largest province in Western China, covers four geomorphic types, namely mountains, hilly land, plains and plateaus. However, the previous coexistence of plain urban agglomeration and mountainous urban agglomeration has changed by means of industrial agglomeration, increase of urban density and construction of transportation network. At the same time, a series of problems have emerged, such as excessive resource

© Springer Nature Switzerland AG 2020
J. Xu et al. (Eds.): ICMSEM2019 2019, AISC 1002, pp. 415–429, 2020.
https://doi.org/10.1007/978-3-030-21255-1_32

consumption and environmental pollution. Urban agglomeration is a kind of advanced spatial organization of urbanization [8]. As an organic combination of social carrying capacity, ecological carrying capacity and economic carrying capacity, the promotion of comprehensive carrying capacity of urban agglomeration is influenced by various factors such as resource endowment, ecological environment, infrastructure and public service [17]. At the same time, transfer of resource elements and regional positive interaction also exert important influence on it [6]. Therefore, in order to understand the evolution process of urban agglomeration, especially urban spatial evolution, it is of great significance to analyze the characteristics of the carrying capacity of all subsystems and driving factors from the perspective of coordinated development.

For urban agglomeration, orderly development of space structure functions as an important symbol of coordinated development. Former researchers have conducted plenty of studies on urban space, especially on urban gravitation. Jimmy Q Stewart, who improved the gravity model with periodic significance, studied the migration mechanism of population by changing the mass in the model [12]. Taking GDP, population and space distance as parameters, H Matsumoto built a gravity model to analyze the network structure of the international airport [10]. Based on regional relationship from a multi-perspective angle, Genovese Bruno analyzed the mechanism of cross-regional selection of schools among Italian university students by using a modified gravity model [1]. Anthony G Wilson and Alexander D Singleton, who applied a spatial-interaction model, did relative research on the inequality in enrollment among higher education students [11,15].

Restricted by development stage, research on urban agglomeration in China lagged behind on its perspective and methods. The existing research on urban agglomeration is relatively weak, reflecting on that the interrelationship, features and mechanism among all subsystems are still immature. According to the present literature, the studies are mainly conducted on hardware factors, such as water, land and irreversible ecological environmental elements. In terms of method, the frequently-used methods are MSE, GIS, neural network model and fuzzy linear programming, all of which aim to evaluate the few mature urban agglomeration. Meanwhile, the analysis on comparison of temporal dynamics and spatial pattern is also rare related to the comprehensive carrying capacity of urban agglomeration in central and western regions of China.

2   Measure of Comprehensive Carrying Capacity in Sichuan Urban Agglomeration

2.1   Research Subject, Data Source and Evaluation Method

On the basis of "The 12th Five-Year Urbanization Development Plan of Sichuan", four urban agglomerations within Sichuan province are selected as research subjects. The four urban agglomerations are as follows: Chengdu plain urban agglomeration, southern Sichuan urban agglomeration, northeast Sichuan urban agglomeration and western Panzhihua urban agglomeration. Since 2005, a strategy of powerful industrial province was proposed, and the construction of four industrial zones was reinforced further within Sichuan province. On this basis, these urban agglomerations were planned. Considering the variability of indicators, annual data with intervals are applied. All the original data are from "Statistical Yearbook of China's Urban Construction" and "Statistical Yearbook of Chinese Cities". Some individual missing data is obtained by interpolation method from the given data.

In view of both natural and human factors, as well as the coordination of ecological, economic and social subsystems, the index systems are selected with higher frequency by learning from the evaluation index system of urban carrying capacity of Cheng Guangbin's and others' [4]. An evaluation index system of four urban agglomerations in Sichuan has been established (Table 1). The method of entropy TOPSIS, which is appropriate to make research on evaluation of regional carrying capacity by using both entropy method to measure the weight of indexes and TOPSIS method to determine the ranking of evaluation objects [3,9]. The main calculation steps for entropy TOPSIS are as follows:

Suppose the number of objects evaluated be m and the number of evaluation index of each evaluated object be n, a judgment matrix is constructed: $X = (x_{ij})_{m \times n}$, $(i = 1, 2, ..., m; j = 1, 2, ..., n)$; Standardized handling of judgment matrix: $y_{ij} = \frac{x_{ij} - \bar{x}_j}{s_j}$, in which $\bar{x}_j$ represents the mean value of index j; represents the standard deviation of index j. Translational coordinates: $X_{ij} = y_{ij} + 5$; To calculate entropy of information $H_j = -k \sum_{i=1}^{m} p_{ij} \ln p_{ij}$, in which $p_{ij} = \frac{X_{ij}}{\sum_{i=1}^{m} X_{ij}}$, $k = \frac{1}{\ln m}$. To define the weight of index j: $w_j = 1 - H_j \Big/ \sum_{j=1}^{n} (1 - H_j)$, in which $w_j \in [0, 1]$, and $\sum_{j=1}^{n} w_j = 1$; To calculate the weighted matrix: $R = (rij) \, m \times n$, $(i = 1, 2, ..., m; j = 1, 2, ..., n)$; To define the optimal index value r_j^+ and the worst index value r_j^- : $r_j^+ = \max_{1 \le j \le x} \{rij\} \, j = 1, 2, ..., n$; $r_j^- = \min_{1 \le j \le x} \{rij\} \, j = 1, 2, ..., n$. To calculate Euclidean distance from m cities to the optimal index value and to the worst index value r_j^-, $r_j^+ = \max_{1 \le j \le x} \{rij\} \, j = 1, 2, ..., n$, $r_j^- = \min_{1 \le j \le x} \{rij\} \, j = 1, 2, ..., n$. To calculate the relative close-degree from m cities to the optimal index value r_j^+: $Vi^+ = \sqrt{\sum_{j=1}^{n} (rij - r_j^+)^2} \, i = 1, 2, ..., m$.

When U_i is bigger, it shows that the close-degree between the comprehensive carrying capacity of the city and the optimal level is higher, on the contrary, it proves to be opposite.

Analysis of Evaluation Results In general, the development of the comprehensive carrying capacity in Sichuan urban agglomeration tends to be virtuous. According to the index system of comprehensive carrying capacity of urban agglomeration, the research subjects, namely the metric results of 19 cities are divided into four groups: higher carrying capacity, high carrying capacity, lower carrying capacity and low carrying capacity (Table 2) [2]. The critical point is obtained from plus-minus of half the standard deviation. Successively ranked first among the whole province, 6/7 of Chengdu plain urban agglomeration were at the level of higher or high carrying capacity in 2013 except for Meishan. However, the other 3 urban agglomerations, for its developmental stage, performed weaker in the comprehensive carrying capacity. Specifically speaking, nearly all the cities failed to achieve the level of higher or high in comprehensive carrying capacity other than 2 cities in northeast Sichuan urban agglomeration. Due to this, the holistic level of comprehensive carrying capacity was forced to decline area is

3 Analysis of Driving Factors about Comprehensive Development in Sichuan Urban Agglomeration

3.1 Analysis Model

As a systematic concept, the comprehensive development of an urban agglomeration can realize its own virtuous circle by means of competition and cooperation representing in the aspects of economics, society and ecological environment [7]. To further identify the main driving factors of the comprehensive carrying capacity of Sichuan urban agglomerations, the thesis adopts empirical research by using multiple linear regression analysis. In the research, the comprehensive carrying capacity is taken as dependent variable; 9 indexes, namely population growth rate, population density, proportion of employed population, land area of administrative area, green coverage rate of built-up area, comprehensive utilization of general industrial solid waste, sewage treatment rate, garbage disposal rate, the proportion of secondary industry in GDP, are selected as independent variables. The nine independent variables are sorted and selected according to the mean value of weight. By referring to Xiang Pengcheng's and Su Hongjian's index classification [5,13], the 9 independent variables are groups as three major driving forces: adjustment of industrial structure, urbanization of the population and construction of urban infrastructure. The formula of the model is as follows:

$$Y = a_0 + a_1 X_1 + a_2 X_2 + ... + a_i X_i + U$$

In the formula, a_i represents regression coefficient; Y represents dependent variable; X_i being independent variable and U being stochastic disturbance.

3.2 Results Analysis of Driving Factors

As the results of regression analysis shows (Table 3), the value of P, through test of regression equation F, is less than 0.05. It indicates that the regression

Table 1. Evaluation system of comprehensive carrying capacity

Target layer	Level-one index	Secondary index	Index attribute
Comprehensive carrying capacity of urban agglomeration	Social Index A1	B1: Growth rate of population%	–
		B2: density of population (p./sq.km.)	–
		B3: proportion of employed population	+
		B4: the number of hospitals, health-centers owned by every ten thousand people	+
		B5: the number of hospitals, health-centers owned by every thousand people	+
		B6: the number of doctors owned by every ten thousand people	+
		B7: the number of students equally divided among primary and secondary school teachers	–
		B8: the storage volume of public library every hundred people	+
		B9: the number of mobile phone users every hundred people	–
		B10: coverage of urban road per capita	+
		B11: the number of buses owned by every ten thousand people	+
	Ecological index A2	B12: land area of administrative region (sq.km.)	+
		B13: land area for city construction (sq.km.)	+
		B14: land area for residence (sq.km.)	+
		B15: total volume for urban water supply (10 000 tons)	+
		B16: total electricity consumption for urban residents (kwh)	+
		B17: electricity consumption per capita (kwh)	–
		B18: green area owned by every thousand people(hectare/every thousand people)	+
		B19: green coverage ratio of built-up area	+
		B20: emission of industrial SO2 every ten thousand people	+
		B21: comprehensive utilization of general industrial solid waste	–
		B22: sewage treatment rate	+
		B23: treatment rate of domestic garbage	+
	Economic index A3	B24: GDP per capita (yuan)	+
		B25: proportion of secondary industry in GDP	+
		B26: total retail sales of social consumer goods	+
		B27: sum of investments in fixed assets	+
		B28: average wage of staff and workers at post	+

Table 2. Comparison of carrying capacity among urban agglomeration

Urban agglomeration	City/ranking	2005	2008	2011	2012	2013
Chengdu plain urban agglomeration	Chengdu	0.580/1	0.594/1	0.584/2	0.586/2	0.582/3
	Deyang	0.502/9	0.467/12	0.461/11	0.454/8	0.510/9
	Mianyang	0.515/5	0.516/6	0.497/6	0.481/7	0.583/2
	Leshan	0.479/12	0.470/10	0.441/14	0.432/13	0.523/7
	Meishan	0.456/17	0.442/18	0.401/18	0.417/17	0.440/18
	Yaan	0.578/2	0.564/2	0.611/1	0.636/1	0.613/1
	Ziyang	0.502/8	0.515/7	0.513/4	0.489/5	0.523/8
Southern sichuan urban agglomeration	Zigong	0.493/10	0.482/9	0.472/9	0.449/10	0.487/10
	Luzhou	0.481/11	0.469/11	0.461/12	0.440/11	0.470/13
	Neijiang	0.463/16	0.453/15	0.479/8	0.451/9	0.475/12
	Yinbing	0.532/3	0.556/4	0.440/15	0.423/16	0.456/14
Northeast sichuan urban agglomeration	Guangyuan	0.505/7	0.531/5	0.512/5	0.491/4	0.532/6
	Suining	0.469/14	0.467/13	0.466/10	0.438/12	0.456/15
	Nanchong	0.454/18	0.449/17	0.427/17	0.414/18	0.451/16
	Guangan	0.473/13	0.460/14	0.429/16	0.431/14	0.450/17
	Dazhou	0.529/4	0.556/3	0.562/3	0.529/3	0.537/5
	Bazhong	0.464/15	0.453/16	0.479/7	0.482/6	0.538/4
Western panzhihua urban agglomeration	Panzhihua	0.512/6	0.503/8	0.443/13	0.428/15	0.479/11
	Xichang	0.423/19	0.412/19	0.389/19	0.397/19	0.405/19

Note: By verification, the influence of main urban area on the city accounts for a large proportion, so the scope of main urban

model has a significant linear relationship by verification. Among the independent variables, six of them, including population density, proportion of employed population, land area of administrative area, comprehensive utilization rate of general industrial solid waste, treatment rate of domestic waste and the ratio of secondary industry in GDP, prove to be significant at the level of 1%. However, the relationship shows no significance in terms of population growth rate. Standardized coefficient functions as a basis for determining the relative importance among variables. When the coefficient becomes greater, the degree of correlation between independent variables and dependent variables will turn out to be higher. In the model, the index of administrative land area (0.715) plays the most important part, mainly because of its essential land prominent role in urban survival and development; the next is population density (0.698),which shows the level of socio-economic development and the ability of infrastructure supply exert a significant impact on the urban comprehensive carrying capacity; the ratio of secondary industry in GDP, accounting for 0.388, also reflects that industry serves as the backbone of enhancing the overall economic strength and comprehensive competitiveness of Sichuan; proportion of employed population(0.357),being not only a reflection of the labor supply and demand but also a concentrated expression of the function of people's livelihood support, also plays a certain part in the development of the region. Other three ones indicate that ecological management and intensive utilization of resource should not be neglected in the process of urban expansion in future times.

By regression of the comprehensive carrying capacity to three urban agglomerations, the value of P by verification of regression equation P is less than 0.05, which shows that the regression result is significant (Table 3). Specifically speaking, population density, proportion of employed population and administrative land area in Chengdu urban agglomeration are proved to be significant at the level of 1%; population density and administrative land area in South Sichuan urban agglomeration are proved to be significant at the level of 1%, the other indexes except for proportion of employed population being significant at the level of 5%; for Northwest Sichuan urban agglomeration, it proves to be significant when treatment rate of household garbage being at the level of 1%, while population growth rate, administrative land area and green coverage rate of built-area being significant at the level of 5%. In terms of standardized coefficient, the population growth rate (1.085) and land area of administrative zone (0.865) of Sichuan urban agglomeration turns out to be the most significant, which implies that population urbanization plays a prominent role in improving the comprehensive carrying capacity. Various effects such as urban aggregation, scale and knowledge spillover foster the increase of economic efficiency and integrated development of more extensive area. Besides, the following three indexes such as proportion of employed population (0.305), ratio of secondary industry in GDP(0.249) and population growth rate (0.168) reflect that adjustment of industrial structure also have essential impact on the transformation of the comprehensive carrying capacity. The population density (−2.363) and land area of administrative zone (−1.999) in South Sichuan urban agglomeration are tested to be negative. The value indicates that unreasonable population urbanization has a reverse effect on the carrying capacity. In terms of treatment rate of domestic garbage (0.341) green coverage rate of developed areas(0.274)and sewage treatment rate (0.271), the driving effect comes second. As the main positive driving force for the improvement of comprehensive carrying capacity in the region, the demand for urban infrastructure construction surged along with the progress of urbanization. It is known that a powerful support of public service and a sufficient supply of infrastructure can help to reduce the cost of living and solve a series of social problems such as environmental pollution and ecological restoration, by means of which the overall structural driving capability of southern Sichuan can be improved. Within northeast Sichuan urban agglomeration, population urbanization, represented by administrative land area (0.819), functions as the important driving factor in improving the comprehensive carrying capacity of the local area; secondly, as symbols of urban infrastructure, treatment rate of domestic garbage (0.435) and green coverage rate of developed areas (0.306) are also essential boosters for improving urban carrying capacity. For the secondary industry accounting for the maximum proportion in GDP, industry is still the core support in West Pan urban agglomeration, and urban infrastructure is the second factor that drives its development.

Table 3. Regression analysis results of comprehensive carrying capacity of 3 urban agglomerations

Variables	Sichuan			Chengdu plain urban agglomeration			South sichuan urban agglomeration			Northeast sichuan urban agglomeration		
	Coefficient	t	P	Coefficient	t	P	Coefficient	t	P	Coefficient	t	P
C		152.249	0		82.026	0		136.439	0		67.734	0
Population growth rate	0.099	1.41	0.163	0.168*	1.876	0.072	0.207**	2.229	0.05	0.277**	2.213	0.039
Population density	0.698***	5.953	0	1.085***	5.92	0	−2.363***	−3.923	0.003	0.104	0.267	0.792
Proportion of employed population	0.357***	4.606	0	0.305***	3.577	0.001	0.068	0.438	0.67	0.069	0.469	0.644
Land area of administrative area	0.715***	5.876	0	0.865***	6.165	0	−1.999***	−3.18	0.01	0.819**	2.599	0.017
Green coverage rate of built-up area	0.141*	1.955	0.054	−0.053	−0.647	0.524	0.274**	3.139	0.011	0.306**	2.288	0.033
Comprehensive utilization of general industrial solid waste	−0.277***	−3.058	0.003	−1.152*	−2.054	0.051	0.187***	2.417	0.036	0.206	1.579	0.13
Sewage treatment rate	0.171**	2.15	0.035	0.034	0.462	0.648	0.271**	2.803	0.019	0.288	1.467	0.158
Garbage disposal rate	0.254***	3.259	0.002	0.141*	1.798	0.084	0.341**	3.018	0.013	0.435***	2.94	0.008
The proportion of secondary industry in GDP	0.388***	3.944	0	0.249*	1.865	0.074	0.260**	2.398	0.037	0.415	1.697	0.105
R^2	0.627			0.906			0.965			0.781		
Adjust R^2	0.585			0.872			0.933			0.683		
F	14.957			26.651			30.379			7.945		
P	0			0			0			0		

4 Analysis of the Spatial Pattern of Sichuan Urban Agglomerations

4.1 Modified Gravity Model

The development process of urban agglomeration is mainly manifested in the expansion of urban space [16]. In this model, the relative closeness between the comprehensive carrying capacity of each city and the optimal solution is defined as "quality", the geographically weighted mean distance as "distance", then, the function degree among cities within an urban agglomeration is determined. The calculation formula is as follows:

$$R_{ij} = GM_iM_j/d_{ij}^2 \tag{1}$$

In the formula, R_{ij} represents the gravitation of city i to city j, including social gravitation, ecological gravitation and economic gravitation; G as coefficient constant of gravity among cities and the value being 1, has no actual influence on measurement results; M_i and M_j refers to the social, ecological and economic quality between city i and city j; d_{ij} is defined as the distance from city i to city j.

Urban economic level determines the measure of economic gravitation among cities; Social ecological conditions determine the measure of social and ecological gravitation among cities. For ease of expression, the "quality" of a city refers to the degree of interaction among cities, namely M_i and M_j. In the previous study, the value of U_i has been obtained by TOPSIS method; then for ease of calculation, the final score of city quality is equal to 10 times of value U_i.

Distance indicators are used to reflect the spatial distance and convenience degree between two places as well as social cultural exchanges and identification. Generally speaking, geographical distance, social distance, psychological distance, political distance and cultural distance can be used as indicators of distance. As the research object is four urban agglomerations in Sichuan province, geographical distance is more appropriate to be chosen as indicator. Besides, the original linear distance is set as transportation distance. Considering the reachability and transportation mode of the geographic route, the mileage of transportation, by being given a weighting on average distance, replaces the linear distance, namely in the gravitation model. Among them, social gravitation is measured by passenger distance and economic gravitation by freight distance. The concrete equation is as follows:

$$d_{ij} = \sum_{k=1}^{2} \lambda_k D_{ijk} \tag{2}$$

In this equation, k represents the mode of transportation between two cities; when $k = 1$, it refers to road transport and when $k = 2$, it refers to railway transport. D_{ijk} represents the mileage from city i to city j when choosing the transportation mode k. λ refers to the possessing weight, which is determined by the proportion of transportation volume corresponding to the transportation mode of k in Sichuan Province. Due to its low frequency, air transportation, as well as the change of traffic costs is not taken into consideration in the study. In addition, as the comprehensive development of Panzhihua is obviously superior to the city of Xichang, the core judgment effect may be insignificant. Therefore, gravitation analysis will not be carried out to western Pan urban agglomeration.

4.2   An Analysis of Urban Gravity

In order to truly reflect the difference of the gravitation among these cities, the original value of urban gravitation index, ranging between [0, 1] (Fig. 5), is treated by adopting normalization processing. Then horizontal comparison is made to analyze the gravitational level of the cities within urban agglomerations. In addition, within each urban agglomeration, secondary core cities are established based on the relative sizes of their gravitational potential. By making full use of the comparative advantages of secondary core cities, it is expected that the sustained growth of entire cities and even urban agglomerations will be achieved.

An Analysis of Social Gravity   The overall level of social gravity index remains high in Chengdu plain urban agglomerations. It showed a steady upward trend before 2013 but a slight decrease in 2013. Among the indexes, Chengdu and Deyang appeared to be the most prominent, with its gravity fluctuating around 2. Since 2005, the gravity value of Chengdu has been far greater than that of other cities. Thanks to its obvious advantages on gravity, Chengdu plays a significant dominating role in the aspects of social life and economic organizations to its neighboring cities. The value of social gravity in south Sichuan fluctuated in a steady status, however, the difference of gravity index performed significant among these cities, with obvious polarization. Zigong's gravity value, ranking the highest among the annual data, proves its status of sub-core city in south Sichuan. Meanwhile, as an important transportation hub of Sichuan province, Zigong has taken advantage of its convenient traffic and transportation to diffuse some social elements to the outside city in an effective and balanced way. As a whole, social gravity values in northeastern Sichuan changed within a small range, maintaining a roughly stable state of "one-high and five-low". At present, most cities show the trend of low-speed growth with no significant spatial difference, besides, the degree of differentiation shares an identical relationship with the level of urban social development. It further illustrates that Nanchong, as a sub-core city in northeastern Sichuan, needs to make use of the relative advantages of its important traffic location thus guide the whole city group to develop healthily and efficiently.

An Analysis of Ecological Gravity   In view of the general tendency of the index changes of ecological gravity in Chengdu plain urban agglomeration, except for 2013, the gravity value among cities increased greatly. The growing rate of

Chengdu is about 0.3, that of Yaan and Ziyang being 0.2, both of which indicate there is great potential for ecological attraction among cities. Various types of new zones dominated by industrial space, such as Tianfu New Area, have firstly carried out space extension and made problems of traditional resources and environment regional. The change trend in the functions of regional ecological land is increasingly evident, which is in accordance with the increasing tendency of the ecological attraction among cities. In recent years, the index of South Sichuan urban agglomeration has maintained a stable state generally. Rivers function as important ecological linkages in this region. Along the Yangtze River and Tuo jiang River are several cities such as Neijiang, Luzhou and Yibin, which basically connect most cities of the agglomeration. As the ecological core of south Sichuan, Zigong has made great efforts to build a common ecological linkage space with other cities. Within northeast Sichuan urban agglomeration, the ecological gravity indexes showed a general fluctuating upward trend, among which Dazhou and Bazhong were the most prominent. As important ecological source region, Qinling-Bashan Mountains belong to non-constructive land with strict control in urban planning. Nanchong, the ecological core of northeast Sichuan, ought to play a positive exemplary role in protecting the ecological land rather than expand urban area in a disorderly way. At the same time, the development mode should be transformed to be an intensive ecotype from the previous extensive resource-oriented type.

An Analysis of Economic Gravity The value of Chengdu has always maintained above 2, closely followed by Deyang, above 1.8, Mianyang and Meishan range from 1.25 to 1.55, Leshan, Yaan and Ziyang are all less than 1. The values indicate the serious polarization on attractiveness among the cities. As prefecture-level cities, both Leshan and Ya'an exert great influence on the local economy. However, from the perspective of the overall gravity effect, the two cities perform a weaker economic role in Chengdu plain urban agglomeration. Dispersion force is more acted on spatial economic structures of Chengdu urban agglomerations. More prominent polarization on economic gravity value is shown in South Sichuan urban agglomeration. Due to the large gravity base of Zigong and Neijing, there is no evident inter-annual variation. Luzhou and Yibin, with a small gravity base, show a steady ascending tendency in recent years. As the economic core of south Sichuan, Zigong can make use of its absolute advantage of convenient transportation, such as traffic routes, to convey energy for urban development. With the increasing traffic accessibility, the gravity value of Nanchong, Dazhou and Bazhong have experienced obvious increase, among which the increasing rate of Bazhong being 0.2. Generally speaking, although Nanchong is the economic core of northeastern Sichuan, the urban economic attraction within the group is still relatively loose; the economic spatial structure of urban area is still at the stage of formation and development at the same time the network system of the center as well as the axis has not yet been completed.

Functioning as intrinsic factor affecting the nature of a city, the industrial structure, especially its layout and adjustment will speed up the evolution process of urban spatial form. The Krugman Index G_i is applied in weighing the industrial structural factors that affect inter-city economic linkage. A quantitative method is used to analyze the degree of specialization of industries, the possible matching relationships and strength of association (Table 4). When the value of is bigger, then it represents a higher specialization degree of industrial division among cities. However, a smaller value refers to a lower differentiation and specialization degree. The calculation formula of Krugman Index is as follows:

$$G_i = \sum_{p=1}^{n} |S_{ip} - S_{jp}| \tag{3}$$

The value of Krugerman index of the Chengdu plain urban agglomeration is generally highwhich implies that there is distinguished difference as well as a relative obvious complementarity in the industrial structure within the region. The high value of Krugman index between Chengdu and other cities coincides with its core status in the urban agglomeration. In addition, Mianyang owns a moderately high value. Compared with Chengdu, the value of Krugeman index of Deyang and Meishan is large, which illustrates their increasing awareness in strengthening the industrial division with core cities. By means of industrial division, the competitiveness is mitigated to a certain extent. On the contrary, the industrial division is relatively weak in Yaan and Leshan, which can be drawn from their Krugeman index comparing with Chengdu. The Krugman index between Ziyang and Chengdu is relatively high compared with other cities, reflecting a clear urban development plan and specific industrial division. In general, the Krugerman index is low in southern Sichuan urban agglomeration. Among the cities within the agglomeration, there is no obvious difference in industrial structure thus leads to poor complementarity. At present, the proportion of primary industry is less than 15% while the secondary industry accounts for 60%. The data reflects the obvious variation in the adjustment of industrial structure. Although the rapid improvement of secondary industry plays a vital role in driving the local economy, the insignificant industry specialization doesn't bring any virtuous interaction among the cities. The Krugman index of the cities in northeastern Sichuan urban agglomeration ranks at a medium level. On the whole, the industrial connection among cities appears to be weak with decentralized division.

5   Conclusion and Implication

(1) Generally speaking, the comprehensive carrying capacity of Sichuan urban agglomeration tends to be virtuous. However, the development within these agglomerations is not yet mature. Social carrying capacity, ecological carrying capacity and economic carrying capacity are important components of the comprehensive carrying capacity. Chengdu plain urban agglomeration, supported by the provincial capital, Chengdu, performs well in the carrying capacity.

Table 4. Krugman index of sichuan urban agglomerations

Chengdu plain urban agglomeration

	Chengdu	Deyang	Mianyang	Leshan	Meishan	Yaan	Ziyang
Chengdu							
Deyang	48.43						
Mianyang	36.03	17.36					
Leshan	46.67	4.8	20.4				
Meishan	48.29	5.88	12.26	10.54			
Yaan	45.65	5.16	12.2	8.2	3.36		
Ziyang	54.73	15.16	18.7	19.96	9.42	12.78	

South Sichuan Urban Agglomeration

2013	Zigong	Luzhou	Neijiang	Yinbing			
Zigong							
Luzhou	3.92						
Neijiang	13.28	9.36					
Yinbing	7.5	3.58	25.08				

Northeast Sichuan Urban Agglomeration

2013	Guangyuan	Suining	Nanchong	Guangan	Dazhou	Bazhong	
Guangyuan							
Suining	14.14						
Nanchong	15.84	8.76					
Guangan	8.54	5.66	8.82				
Dazhou	16.5	6.82	2.6	8.4			
Bazhong	4.16	18.3	18.36	12.7	19.02		

(2) The driving factors for the development of Sichuan urban agglomerations present a state of diversity. Altogether there are three major categories of driving factors, namely industrial structural adjustment, population urbanization and urban infrastructure construction. The first two factors are the key power in the development of Chengdu plain urban agglomeration. For south Sichuan urban agglomeration and northeast Sichuan urban agglomeration, the latter two factors play a more prominent role. Irrational population urbanization exerts a reverse reaction on urban development. The major driving forces for West Pan urban agglomeration are industrial structural adjustment and urban infrastructure construction.

(3) The existence of non-equilibrium of urban spatial layout causes its uncoordinated development and unreasonable spatial pattern within Sichuan urban agglomerations. Specifically, in recent years, the changing tendency of social gravity, ecological gravity and economic gravity are similar while the indexes vary greatly. Changes on gravity values are significant in Chengdu plain urban

agglomeration, reflected by Chengdu's neighboring cities' extreme dependency on Chengdu. Represented by polarization, the fluctuations in south Sichuan urban agglomeration are steady. Zigong and Neijing share close linkage with surrounding cities. For northeast Sichuan urban agglomeration, a relatively stable status with one-high and five-low occurs, in which Naichong performs the most actively.

Based on the above conclusions, it is suggested that a path of discrete agglomeration may be an effective choice for the development of Sichuan urban agglomerations under the circumstance of imbalanced condition. The suggestions are as follows:

(1) To optimize the system and structure of these urban agglomerations for stable and permanent development. In view of mutual dependency and coordination of social subsystem, economic subsystem and ecological subsystem, it is necessary for us to give consideration to resource conservation and environment protection in the process of developing urban agglomerations. Ecology should be taken as a restrictive factor in guiding the optimization of city structure. Only with a harmonious relationship among population, economy and ecology can a healthy urbanization be achieved.

(2) To establish the core status of provincial capital- Chengdu for concentrated development. With the help of spillover effect as well as the diffusion of the axis, the other secondary core cities can achieve positive interaction and advanced development. As a result, a radiation influence characterized by a spider web will take effect thus foster the overall development of Sichuan.

(3) To accelerate the development of secondary core cities for multipoint support. Specifically speaking, we should take advantage of the advanced service functions of these secondary core cities within Sichuan province, making them management as well as research and development center. Based on the difference on division of labor, the cities with the functions such as production, assembling and residence can be moved to the surrounding cities and towns. Hence a vertical functional system can be formed within a specific zone, leading to urban hierarchy for obvious comparative advantages of secondary cities.

(4) To strengthen government support and guidance for orderly coordination. By giving proper guidance and planning, the government can provide great help in the coordination of population urbanization and industrial structure adjustment. In this circumstance, each city can define its development direction, which will certainly shorten the time required for the formation of a complementary pattern at the same time increase the efficiency of the rational distribution of resources. In addition, with the assistance of government's macro-guidance, various constraints imposed by the complex and diverse geographical structure, such as regional infrastructure construction and transportation network construction, can be eliminated at an accelerated speed. Starting with intelligent transportation systems for green and efficient transportation, the coordinated development between regional economy and social ecology can be ultimately achieved.

Acknowledgements. This research work is supported by the Ministry of education "humanities and social sciences youth project" of China (15YJC630081), Sichuan social science research "twelfth five-year plan" project (SC14C027), the "Resource constraint and sustainable agricultural development" Sichuan postgraduate education reform and innovation team project (NCET-13-0921), and the Youth fund project of humanities and social science research of the ministry of education "study on economic cooperative development and spatial radiation effect of Chengdu-Chongqing urban agglomeration" (17YJC630136). Thanks for their support.

References

1. Bruno, G., Genovese, A.: A spatial interaction model for the representation of the mobility of university students on the italian territory. Netw. Spat. Econ. **12**(1), 41–57 (2012)
2. Chen, Y., Luo, J.: A review of the research on coordinated development of urban agglomerations. Urban Probl. **1**, 26–31 (2013)
3. Cheng, G., Zhang, P.: Study on evaluation of urban comprehensive bearing capacity of northern slope economic belt of tianshan mountain (in Chinses). Chin. Agric. Sci. Bull. **17**, 48–52 (2007)
4. Cheng, G., Shen, L., Long, W.: Comparative research on comprehensive carrying capacity of urban agglomerations along silk road economic belt in northwest China (in chinses). Econ. Geogr. **35**(8), 98–103 (2015)
5. Deng, R.: Carbon emission reduction performance evaluation of changzhu-zhuzhou-xiangtan two-oriented society pilot project—an empirical study based on the dual difference method (in chinses). Soft Sci. **30**(9), 51–55 (2016)
6. Fang, C., Yu, D.: Urban agglomeration: an evolving concept of an emerging phenomenon. Landsc. Urban Plan. **162**, 126–136 (2017)
7. Feng, T., Zhang, H., Hu, J., Xia, X.: Dynamics of green productivity growth for major chinese urban agglomerations. Appl. Energy **196**, 170–179 (2016)
8. Kang, Z., Li, K., Qu, J.: The path of technological progress for china's low-carbon development: evidence from three urban agglomerations. J. Clean. Prod. **178**, 644–654 (2018)
9. Li, G., Cheng, G.: The comprehensive evaluation of the graduate school's teaching quality based on entropy-weight and topsis methods **26**(3), 400–407 (2011)
10. Matsumoto, H.: International urban systems and air passenger and cargo flows: some calculations. J. Air Transp. Manag. **10**(4), 239–247 (2004)
11. Singleton, A., Wilson, A., O'Brien, O.: Geodemographics and spatial interaction: an integrated model for higher education. J. Geogr. Syst. **14**(2), 223–241 (2012)
12. Stewart, J.: An inverse distance variation for certain social influences. Science **93**(2404), 89–90 (1941)
13. Su, H., Wei, H.: Density effect, optimal urban population density and intensive urbanization (in chinses). China Ind. Econ. **10**, 5–17 (2013)
14. Sun, J., Li, Y., Gao, P.: A mamdani fuzzy inference approach for assessing ecological security in the pearl river delta urban agglomeration, Chinay. Ecological Indicators (2018)
15. Wilson, A.: The widening access debate: student flows to universities and associated performance indicators. Environ. Plan A **32**(11), 2019–2031 (2000)
16. Yan, S., Zhao, S.: Spatiotemporal dynamics of urban expansion in 13 cities across the jing-jin-ji urban agglomeration from 1978 to 2015. Ecol. Indic. **87**, 302–313 (2018)
17. Ye, Y.: Interpretation of the comprehensive bearing capacity of the city. Frontline **4**, 26–28 (2007)

Relationship Between Institutional Pressures, Green Supply Chain Management Practices and Business Performance: An Empirical Research on Automobile Industry

Jinsong Zhang[1]([✉]), Xiaoqian Zhang[1], Qinyun Wang[2], and Zixin Ma[3]

[1] School of Management, South-Central University for Nationalities, Wuhan 430074, People's Republic of China
zhangjinsong@scuec.edu.cn
[2] CIFI GROUP, Wuhan 430074, People's Republic of China
[3] College of Innovation and Entrepreneurship, South-Central University for Nationalities, Wuhan 430074, People's Republic of China

Abstract. With environmental pollution, resource waste and ecological imbalance gradually become the focus of global attention, green supply chain management (GSCM) has become an important strategic choice to reduce the environmental impact and improve operational performance so that enterprises can achieve sustainable development. Based on the institutional theory, this paper analyzes which institutional pressures motivate manufacturers to implement GSCM, and which business performance is positively affected by GSCM, and how institutional pressures influence on business performance. The moderating effect of enterprise scale and time of GSCM implementation on the relationship between GSCM practices and business performance is studied. Based on the questionnaire survey data in 224 manufacturers in automobile industry, the empirical research provides evidence that (a) institutional pressures have significant positive impact on GSCM practices and business performance, (b) GSCM practices also have significant positive impact on enterprise environment, economy and operational performance, and (c) enterprise scale and practice time both positively moderate the relationship between GSCM practices and business performance.

Keywords: Green supply chain management · Institutional theory · Business performance · Structural equation model · Supply chain management

1 Introduction

In recent decades, due to the pressure from regulations, customers and competitors, enterprises around the world have shown higher and higher concern about

© Springer Nature Switzerland AG 2020
J. Xu et al. (Eds.): ICMSEM2019 2019, AISC 1002, pp. 430–449, 2020.
https://doi.org/10.1007/978-3-030-21255-1_33

the impact of their own business on the environment under increasing globalization and fierce competition among enterprises [13]. Now, in order to reduce the impact of the business operations on the environment and improve business performance, manufacturers gradually regard Green Supply Chain Management (GSCM) as a feasible option to gain or maintain competitive advantages [7]. Existing researches have shown that many enterprises have realized that customers or other stakeholders don't always distinguish between a single enterprise and its partners in the supply chain. In a particular supply chain, core enterprises are usually responsible for the adverse environmental impacts of all organizations in the supply chain. As a result, they are often in charge of the environmental performance of the entire supply chain so that it is important to implement GSCM practices.

Institutional pressures are one of the most important factors to encourage enterprises to implement GSCM. According to many scholars, for example, Mathiyazhagan and Haq [21] believed that GSCM would be subject to the pressure from different stakeholders, including government regulations, customers at home and aboard, competitors, the neighboring community, non-governmental organizations, media, investors, and employees [21]. And enterprises hope to reduce the risk in the supply chain to environmental hazards or adverse publicity and government punishment, which are caused by non-compliance. So that enterprises can position themselves as environmentally responsible enterprises. As for the relationship between institutional pressures and environmental management practices, Mitra and Datta [22] have shown that institutional pressures are positively related to internal GSCM practices (ISO14001 certification) and external practices (investment recovery) [22]. And through subgroup analysis and meta-regression, Fang and Zhang [10] also have found that internal and external GSCM practices are positively related, and they are both positively related to business performance [10]. Most scholars believe that institutional pressures have made enterprises strengthen GSCM practices in China. And they emphasize the need for further research on institutional pressures, GSCM practices, and performance. This paper exactly seeks to make a contribution in this direction.

Since the reform and opening-up, the automobile industry has developed rapidly in China. With an annual sales volume of 29.942 million vehicles in 2017, China is the world's largest automobile producer. And the automobile industry has become one of the pillar industries of the national economy in China. However, the rapid development of the automobile industry has also led to serious and non-negligible environmental problems in China, which directly affect business performance. From the perspective of environment and resources, the GSCM implementation is necessary for the sustainable development of enterprises and society. Nevertheless, the research of Cao and Liu [34] has indicated that Chinese enterprises, including automobile manufacturers, are not active in GSCM implementation [34]. Institute of Public and Environmental Affairs (IPE) and Natural Resources Defense Council (NRDC) jointly released The CITI Index Evaluation Report on the Green Supply Chain in 2015, indicating that GSCM in China has not become a conscious behavior of enterprises. In this context, this paper studies the relationship between the implementation of GSCM practices and business performance of Chinese automobile manufacturers under institutional pressures, which can provide theoretical and practical significance for Chinese automobile manufacturers and policymakers.

The main innovations of this paper are as follows:

First, many empirical studies related to GSCM focus on effects of GSCM drivers or GSCM practices on business performance. Namely, scholars used to study only the front-end or the back-end. Even if someone simultaneously studies them, he is also separated the front-end and back-end, not mentioned system pressure influence on business performance. So it is yet to prove that the mediation of GSCM practices on business performance is full or partial. Therefore, on the basis of previous research, taking GSCM practices as the mediating variable of the relationship between institutional pressures and business performance, this paper studies the strength of mediating effect.

Second, this paper seeks to innovate on variables. Cheng et al. [5] summarized and classified GSCM practices into internal GSCM practices, external GSCM practices, and post-GSCM practices, among which post-GSCM practices mean reverse logistics [5]. Existing empirical studies on GSCM practices usually contain 3 to 4 types, and this study will consider 6 types of GSCM practices to investigate the relationship between GSCM practices and institutional pressures and business performance.

Third, many enterprises have a negative impact on business performance in their early efforts. With the extension of time of GSCM practices, recent studies show that there is a positive correlation between GSCM and performance. Therefore, while studying the influence of GSCM practices on performance, this paper added enterprise scale and the length of time of GSCM practices as moderating variables to analyze whether enterprise scale or practice time have a moderating effect on the relationship between GSCM practices and enterprise performance.

2   Theoretical Framing and Development of Hypotheses

2.1   Green Supply Chain Management and Institutional Pressures

In terms of the motivation for enterprises to implement GSCM, Tate et al. [30] thought that it would mainly come from pressures of various stakeholders and institutions [30]. Wu et al. [33] have pointed out that institutional pressures can affect the internal decision-making of an enterprise, and the enterprise strategy needs to follow the requirements of the institutional environment [33]. Exactly, owing to institutional pressures, stakeholders will force enterprises to adjust their business models, such as implementing environmental protection strategies. Namely, without institutional pressures, enterprises will not allocate resources for environmental management. Generally speaking, under intense institutional pressures, enterprises can improve or maintain their competitive advantages by allocating resources to implement environmental practices.

Institutional theory can be used to study how enterprises respond to external pressures brought by environmental protection problems [15]. DiMaggio and Powell [8] pointed out that enterprise strategy would be affected by the institutional environment, mainly including suppliers, resources, customers, governing authority and competitors [8]. Besides, Institutional theory holds that enterprises exist in a social network. And their behaviors and decision-making are often influenced by the various stakeholders in their social networks. Moreover, the theory claims the enterprises can meet the needs of its stakeholders to improve their competitiveness. And the institutional pressures mainly come from three aspects: coercive pressures, normative pressures, and imitative pressures [20]. Government agencies that are the major source of coercive pressures can compulsorily affect enterprise behaviors by formulating relevant laws, which include imposing fines and trade barriers and so on. For instance, The Environmental Protection Tax Law of the People's Republic of China implemented in 2018. Normative pressure prompts enterprises to carry out relevant practices in order to have people believe that their activities are lawful. Furthermore, Social normative pressure can explain environmental management practices of enterprises. Imitative pressures induce enterprises to try their best to learn the paths of successful competitors in the industry and consciously imitate their behavior.

According to relevant researches, this paper selected 15 institutional pressures: national environmental regulations (IP1), national regulations for resource conservation and protection (IP2), regional environmental regulations (IP3), regional regulations for resource conservation and protection (IP4), environmental regulations from exporting countries (IP5), export (IP6), sales to foreign customers (IP7), environmental requirements of domestic clients (IP8), environmental awareness of domestic consumers (IP9), construction of enterprise green image (IP10), media tracking the industry (IP11), public environmental protection awareness (IP12), competitors' environmental strategies (IP13), industrial professional associations (IP14) and environmental partnerships with suppliers (IP15). As for research on institutional environment and GSCM, the empirical results of some scholars have proved the impact of institutional pressures on enterprises' implementation of GSCM practices [33,36,37]. In China, manufacturers have been following different international trade standards, rules and requirements for environmentally friendly operations. And the impact of institutional pressures on enterprises' implementation of environmental management practices is very significant. Therefore, this paper proposes the following hypothesis.

H1. Institutional pressures positively influence the implementation of GSCM practices by Chinese automobile supply chain manufacturers.

2.2   Green Supply Chain Management and Business Performance

With reference to some previous studies [5,18,37], this paper divided GSCM practices into three main groups: 1 Internal GSCM practices, including internal environmental management and green design that can be implemented and managed by individual manufacturers; 2 External GSCM practices, including green procurement, green customer partnerships and investment recovery involving partners in the supply chain; 3 Post-GSCM practices, including reverse logistics involving supply chain partners and customers. Up to now, there are a lot of researches on these practices in the fields of production and operation management. In terms of the impact of implementing GSCM practices on business performance, Laari et al. [16], Ardian et al. [1] and other relevant scholars have conducted a large number of studies. And the results show that these practices enable enterprises to have higher performance in four aspects: economy, environment, operation, and society.

Regarding environmental performance, Lai and Wong [17] found that there was a positive correlation between GSCM and environmental performance in the production industry [17]. Zhu et al. [36] showed that internal GSCM positively correlated with environmental performance [36]. Tachizawa et al. [29] demonstrated that some external GSCM practices were also related to environmental performance through empirical research [29]. In addition, economic performance was the most important factor for enterprises to implement GSCM practices [35]. For economic performance, although more and more scholars are studying the relationship between environmental sustainability and business performance, the research results still don't clearly sure whether these practices will be beneficial to enterprises. For operational performance, previous studies have shown that environmental management within the organization can improve operational performance. For example, the adoption of internal GSCM can improve product quality and reduce delivery time to improve operational performance [17]. In addition, Azevedo et al. [2] evaluated five enterprises in the Portuguese automobile supply chain and found that there was a positive relationship between GSCM and operational performance while considering customer satisfaction and product quality [2]. Therefore, it is important to assess the impact of GSCM on business performance so as to know the priorities associated with business competition. For social performance, some studies believe that the implementation of GSCM can't bring the better corporate image and public relations. But Geng et al. [12] confirmed the significant correlation between internal GSCM and social performance through the meta-analysis [12].

In most studies of GSCM and performance, there are only one or two aspects. This research involves four aspects of performance research. The internal, external and post-GSCM practices of enterprises are considered in GSCM to study the possible positive relationship between these and economic, environmental, operational and social performance. Therefore, this paper proposes the following hypothesis.

H2. The implementation of GSCM has a positive impact on the performance of manufacturers in Chinese automobile supply chain.

2.3 Institutional Pressures and Business Performance

As for the relationship between institutional pressures and business performance, many studies have emphasized the direct relationship between institutional pressures and waste reduction, and shown that institutional pressures had a positive impact on enterprise environmental performance. As Zhu and Sarkis [25] once said, the existence of market (normative pressures) and regulation (coercive pressures) would have an impact on the environmental performance of organizations [25]. Simpson [28] believes that enterprises that fail to realize the waste recycling and the importance of recycling are at a disadvantage in the market competition [28]. Reducing wasted resources can better predict and effectively respond to institutional pressures, and ultimately improve the environmental performance of enterprises. Phan et al. [24] have found that institutional pressures have a positive impact on enterprise environmental performance [24]. Some scholars have also found a relationship between institutional pressures and enterprise social performance. For example, Hu et al. [14] studied the impact of institutional pressures and strategic response on business performance with 248 enterprises as effective samples and found that institutional pressures had a positive impact on business performance, and participatory strategy played a partial mediating role in institutional pressures and enterprise social performance [14]. From the perspective of interests of enterprise strategy, some scholars have found that when the top management's commitment to the natural environment is high, the relationship between institutional pressures and enterprise responses to pressures will strengthen. Therefore, this paper proposes the following hypothesis.

H3. Institutional pressures have a positive impact on the business performance of suppliers in Chinese automobile supply chain.

2.4   The Adjustment Effect of Implementation Time

GSCM is a kind of long-term behavior, which requires a certain amount of investment and time accumulation in order to obtain benefits and effectiveness. As a long-term practice, GSCM can only generate benefits and effectiveness by continuous investment and practices. The conclusion of earlier literature indicates that the improvement of environmental protection requirements will only make enterprises undertake more costs of prevention and management which increases its production and operation costs and product prices, and at the same time reduces the attention over its main advantages and cost input. Thus, it will hurt its interests and reduce its competitiveness in the industry and the overall social benefits [20]. That indicates a negative correlation between GSCM practices and performance [35]. However, with the passage of time, the increase of time for enterprises to implement GSCM practices, and the deepening of empirical research on GSCM, recent studies have shown a positive correlation between GSCM practices and business performance. As a result, the years of implementation of GSCM will have an impact on the relationship between the two. Therefore, this paper proposes the following hypothesis.

H4. The implementation time of GSCM practices moderates the relationship between GSCM practices and business performance.

2.5   The Moderating Effect of Enterprise Scale

Geng et al. [12] found that enterprise scale, industry type, ISO certification, export orientation, and other factors moderate the relationship between GSCM practice and business performance [12]. Among them, the survey results that enterprises of different sizes are quite different in the relationship between GSCM practices and business performance. However, some scholars believe that enterprise scale does not affect the relationship between GSCM and business performance [32]. Therefore, this paper seeks further research on the effect of enterprise scale on the relationship between GSCM and business performance. Here is the fifth hypothesis of this paper.

H5. Enterprise scale moderates the relationship between green supply chain management practice and business performance.

To sum up, based on the research of Vanalle et al. [31], this paper seeks further research and proposes the theoretical model as can be seen in Fig. 1.

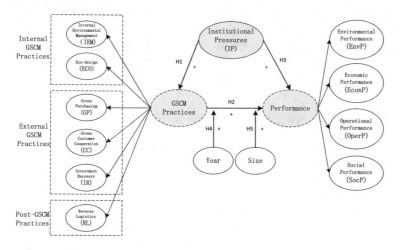

Fig. 1. A theoretical model of institutional pressures, GSCM practices and performance relationship based on SEM

3   Research Design

3.1   Construct Operationalization

In this paper, the scales with validity and reliability verified by kinds of literatures (mainly referring to literature [31]) are used as far as possible, which has adjusted and modified for conforming to the actual situation of domestic automobile industry (Table 1). The five-point scale (Likert 5) for evaluating these constructs ranges from strongly disagree (1) to strongly agree (5). There are two moderating variables. One is the practice time of GSCM that set the question "practice time of GSCM in your enterprise (Year)" including less than 2, 2–5, 5–10 years and more than 10 years. And the other one is enterprise scale that set the question "the current number of employees in your enterprise "Size" including below 100, 101–500, 501–1000, 1001–1500 and above 1500.

3.2 Data Collection

There are relatively high resource consumption, waste generation and the implementation of environmental management in the automobile industry. So it has gained highly wide attention to environmental protection and so on. Therefore, this study selected domestic automobile industry as the sample and took the middle and senior managers of manufacturers in this industry as the object. Thus, the research would study how this area is affected by institutional pressures, and how it affects business performance when enterprises practice GSCM. The research data collected by questionnaire are partly from the interviews conducted by mail, telephone and other forms. At the same time, this paper filtered the senior managers in the domestic automobile industry as respondents through wjx.cn, which is a professional questionnaire survey network in China. Ultimately, we issued 255 questionnaires in total and collected 249. Then we got rid of 25 invalid sheets and remained 224 valid ones.

4   Data Analysis and Hypothesis Testing

4.1   Reliability-Testing

This paper used statistical software (SPSS19.0) to test the reliability of the data obtained from the questionnaire survey, which considered Cronbach's alpha (α) as the index to measure the scale reliability. Generally, alpha ranging from 0.8 to 0.9 indicates that the questionnaire validity is very high. Alpha ranging from 0.7 to 0.8 indicates that validity is quite high. And if alpha is between 0.6 and 0.7, the validity is acceptable. There were 53 items in total. And the questionnaire data are very effective ($\alpha = 0.896$).

4.2   Results and Model Fitting Degree Analysis

(1) PLS path calculation and correction
    This paper used SmartPLS 3.0 to calculate the path for the data of valid sheets [31], then modified the model, and finally analyzed the degree of fitting of the results. To evaluate the impact of institutional pressures on the GSCM practices and business performance, a structural equation model with second-order latent variables was established according to the theoretical model proposed above. GSCM practices as second-order latent variable compose by first-order latent variable including internal environment management (IEM), product eco-design (ECO), green procurement (GP), green customer cooperation (CC), investment recovery (IR), and reverse logistics (RL). IEM and ECO

Table 1. Constructs and respective measures

Construct or latent variable (LV)	Measures or manifest variable (MV)	Measure code
Institutional pressures (IP)	National environmental regulations that include waste mission and cleaner production et al	IP1
	National regulations for resource conservation and protection	IP2
	Regional environmental regulations that include waste mission and cleaner production et al	IP3
	Regional regulations for resource conservation and protection	IP4
	Environmental regulations from exporting countries	IP5
	Export	IP6
	Sales to foreign customers	IP7
	Environmental requirements of domestic clients	IP8
	Environmental requirements of domestic customers	IP9
	Construction of enterprise green image	IP10
	Media tracking the industry	IP11
	Public environmental protection awareness that include community and non-governmental organizations et al.	IP12
	Competitors' environmental strategies	IP13
	Industrial professional associations	IP14
	Environmental partnerships with suppliers	IP15
Internal environmental management (IEM)	Top management commitment for GSCM	IEM1
	Top management support for GSCM	IEM2
	Cross-functional collaboration for environmental improvement	IEM3
	Existence of environmental management system	IEM4
	ISO 14001 certification	IEM5
Eco-design (ECO)	Products are designed to reducing material or energy consumption	ECO1
	Focus on reusing and recycling of products and component parts	ECO2
	Products are designed to avoid or reduce using hazardous materials and manufacturing processes with serious environmental impacts	ECO3
	Consideration for minimizing the waste of product manufacturing	ECO4

continued

Table 1. continued

Construct or latent variable (LV)	Measures or manifest variable (MV)	Measure code
Green Purchasing (GP)	Providing design specification to suppliers for product design documents with environmental requirements	GP1
	Requirements for suppliers to use environmentally friendly packaging that include biodegradable and non-hazardous	GP2
	Supplier environmental management audit	GP3
	Supplier's ISO 14001 certification	GP4
Green customer cooperation (CC)	Cooperation with clients for eco-design	CC1
	Cooperation with clients for cleaner production	CC2
	Cooperation with clients for green packages	CC3
	Cooperation with clients for using less energy during product transportation	CC4
	Cooperation with clients for product recovery	CC5
Investment recovery (IR)	Investment recovery (sale) of excess inventories/materials	IR1
	Sale of scrap and used materials	IR2
	Sale of excess capital equipment	IR3
	Perfect defective product circulation system	IR4
Reverse Logistics (RL)	Collection for recyclable products	RL1
	Separation of reusable products (component parts)	RL2
	Collection for reusable component parts and secondary use of them after repairing, refurbishing or remanufacturing	RL3
Environmental performance (EnvP)	Reduction of "Three waste"	EnvP1
	Reduction of consumption of dangerous, harmful, toxic substances	EnvP2
Economic Performance (EconP)	Decrease of energy cost	EconP1
	Decrease of waste discharge and disposal costs	EconP2
	Decrease of fee for environmental accidents	EconP3
Operational performance (OperP)	Increase amount of goods delivered on time	OperP1
	Decrease inventory levels	OperP2
	Decrease scrap rate	OperP3
	Improve product quality	OperP4
	Improve capacity utilization	OperP5

continued

Table 1. continued

Construct or latent variable (LV)	Measures or manifest variable (MV)	Measure code
Social performance (SocP)	Improve corporate image and social reputation	SocP1
	Improve employee satisfaction	SocP2
	Increase opportunities to sell products in the international market	SocP3

are internal GSCM practices. And GP, CC, and IR are external GSCM practices. And RL is post-GSCM practices. The first-order latent variable structure model can be used to obtain greater theoretical value and reduce the complexity of the model [3,9,19] There were four statistical quality parameters as follows that were used to reduce and verify the observed variables and latent variables in the model: 1 Loading ≥ 0.7; 2 CR (Composite Reliability) ≥ 0.7; 3 AVE (Average Variance Extracted) ≥ 0.5; 4 Cronbach's Alpha ≥ 0.5 [11,23]. Table 2 showed the relevant statistical quality indicators of the initial model.

The results in Table 2 show that the AVE values of latent variable including GSCM, IP and Performance were lower than 0.5, and the standard loading value of latent variable including IR and SocP were lower than 0.7. So the model could eliminate some first-order latent variables and some observed variables of IP. In the process of eliminating observation variables, starting from the one with low loading value, the variation of each statistic index value of the observation model was removed one by one. And finally, the modified model was determined. And finally, the modified model was determined. That modified model cut out two first-order latent variables (IR and SocP), and eight IP observed variables (IP2, IP4, IP5, IP8, IP9, IP11, IP12, and IP14), two IEM observed variables (IEM4 and IEM5), one GP observed variable (GP4), two CC observed variables (CC1 and CC4), and two OperP observed variables (OperP4 and OperP5).

(2) Model index fitting degree analysis

The results in Table 3 show the relevant statistical indicators of the modified model. The data suggested all observation variables: 1 Loading > 0.7; 2 Cronbach' s Alpha > 0.7. Those indicate that the measurement model has good reliability. Meanwhile, the AVE value and the CR value of all latent variables are greater than 0.7, indicating that the measurement model owns well convergent reliability. The total items of the scale are 31 after eliminating variables. And the alpha value of the total table was 0.857, indicating that the availability of the data obtained from the scale after removing variables was still very high. At the same time, in order to further verify the fitting degree of the model, this paper tested the discriminant validity of the measurement model. One method to evaluate the discriminant validity is to verify whether the variable has a higher factor loading coefficient than other variables. The results of cross factor loading analysis in Table 4 prove that the discriminant validity of the scale meets the requirements.

Table 2. Statistical quality indicators for the model

Constructs	Loading	AVE	Composite reliability	Cronbach's alpha
IEM	0.75	0.66	0.76	0.80
ECO	0.72	0.68	0.79	0.82
GP	0.85	0.63	0.74	0.73
CC	0.81	0.73	0.83	0.91
IR	0.58a		0.70	0.65a
RL	0.75	0.70	0.80	0.86
GSCM	–	0.42a	0.85	0.81
IP	–	0.40a	0.80	0.74
EnvP	0.79	0.58	0.73	0.71
EconP	0.86	0.54	0.72	0.71
OperP	0.74	0.59	0.75	0.73
SocP	0.43a	0.63	0.8	0.85
Performance	–	0.36a	0.84	0.8

a: The loading is below the required level.

4.3 Model Testing and Result Analysis

(1) Bootstrap significance testing Bootstrap resampling method was the common way to use in PLS path modeling. Through random resampling with a replacement of the initial samples, the same model estimation was carried out for each group of resamples. And this paper tested the important parameters in the PLS path model with the t-value of the estimated multi-group parameters. The null hypothesis of Bootstrap test assumes that the coefficient of a tested item is zero. If the results reject the null hypothesis, the coefficient is significantly non-zero.

This paper used SmartPLS 3.0 by Bootstrap to test the load coefficient and path coefficient of the model. When the Bootstrap value set to 1000, the results show in Table 5. If the absolute t-value is greater than 1.96, it is considered that significant under 5% confidence intervals. The results reject the null hypothesis, and it suggests that the coefficients of these models are not significantly equal to zero, and that relationship existing in the model is statistically significant. In the test of the mediating effect of GSCM practices on the relationship between IP and business performance, Bootstrap program was used to test the significance of the mediating effect [27]: 1 t-value = 2.558; 2 significance = 0.011.

(2) Analysis of model results

Through AMOS validation, we found that there was no investment recovery (IR) for implementing GSCM practices in the companies studied. IR is the ability of enterprises to obtain some economic advantages from environmental protection actions (increase revenue or reduce cost), while reverse logistics (RL) helps products to be reused to reduce enterprise cost [6]. Therefore, IR needs RL. According to the results of model validation, RL is relatively weak in the GSCM practices of the surveyed enterprises, which may affect the implementation of IR.

Table 3. Statistical quality indicators for the modified model

Constructs	Loading	AVE	Composite reliability	Cronbach's alpha
IEM	0.82	0.66	0.76	0.80
ECO	0.78	0.68	0.79	0.82
GP	0.71	0.63	0.74	0.73
CC	0.81	0.73	0.83	0.91
RL	0.74	0.70	0.80	0.86
GSCM	–	0.57	0.85	0.87
IP	–	0.69	0.80	0.79
EnvP	0.82	0.58	0.73	0.71
EconP	0.81	0.54	0.72	0.71
OperP	0.86	0.59	0.75	0.73
Performance	–	0.54	0.84	0.83

At the same time, although some studies have provided evidence for the positive impact of GSCM practices on social performance, no significant evidence has been observed in this study. The research results of Geng et al. [12] also showed that the correlation between GSCM practices and social performance was not significant, only internal environmental management had a significant impact on social performance, but supplier integration, product Eco-design, green customer cooperation, and reverse logistics had no significant correlation with social performance [12]. Through empirical research, Lai et al. [18] also found that the adoption of RL by Chinese manufacturers produced a lot of environmental and economic performance, but they didn't bring about social performance [18]. RL can improve the image and reputation of enterprises, and enhance the overall value of enterprises by improving their social performance [4]. Therefore, the conclusion in this paper may be that manufacturers fail to realize that green customer cooperation and eco-design are conducive to creating a better corporate image. Faced with the current situation that the friendly environment culture isn't deep enough in the domestic automobile industry, manufacturers should have in-depth communication with various stakeholders to identify some reverse logistics practices so that they can improve social performance [18].

First of all, IP1, IP3, IP6, IP7, IP10, IP13 and IP15 are proved to be the most important forms of institutional pressures leading to the implementation of GSCM practices in the surveyed enterprises. Among them, IP1 and IP3 are mandatory pressure, IP6, IP7, and IP10 are normative pressure, and IP13 and IP15 are imitative pressures. Seles et al. [26] believe that normative institutional pressures are more effective than coercive pressure in promoting enterprises to implement GSCM in more mature industries [26]. Therefore, the automobile supply chain studied in this paper can be considered to have a certain maturity, which also explained the positive effect of IP on operational performance. Secondly, for GSCM practices, internal environment management (IEM), product

Table 4. Cross factor loadings analysis.

Manifest variables	IP	IEM	ECO	GP	CC	RL	EnvP	EconP	OperP
IP1	0.76	0.39	0.09	0.21	0.19	0.28	0.39	0.27	0.27
IP3	0.85	0.37	0.4	0.17	0.22	0.08	0.24	0.07	0.14
IP6	0.8	–*	0.12	0.14	–*	*	0.15	0.29	0.22
IP7	0.9	0.34	0.31	0.27	0.26	0.21	–*	0.38	0.28
IP10	0.74	0.19	0.10	0.25	0.19	0.20	0.42	0.20	0.35
IP13	0.83	0.3	0.32	0.17	0.26	−0.05	0.22	0.27	0.20
IP15	0.96	0.43	0.2	0.06	0.42	0.01	0.15	0.36	0.30
IEM1	0.42	0.94	0.27	0.26	0.28	0.14	0.31	0.22	0.41
IEM2	0.27	0.79	0.27	0.19	0.47	0.22	0.25	0.15	0.27
IEM3	0.48	0.76	0.55	0.41	0.37	0.28	0.39	0.45	0.35
ECO1	0.12	0.45	0.97	0.41	0.29	0.05	0.25	0.38	0.29
ECO2	0.41	0.4	0.93	0.32	0.38	0.17	0.23	0.16	0.21
ECO3	0.21	0.3	0.73	0.42	0.34	0.23	0.27	0.26	0.28
ECO4	0.37	0.33	0.85	0.37	0.39	0.12	0.17	0.34	0.17
GP1	0.29	0.42	0.53	0.93	0.35	0.27	0.3	0.42	0.52
GP2	0.18	0.15	0.22	0.89	0.45	0.19	0.32	0.04	*
GP3	0.08	0.01	0.01	0.71	0.09	0.33	0.14	0.31	0.08
CC2	0.32	0.31	0.42	0.36	0.85	0.28	0.30	0.28	0.08
CC3	0.22	0.30	0.24	0.45	0.96	0.26	0.08	0.22	0.19
CC5	0.28	0.38	0.3	0.2	0.78	0.16	0.33	0.33	0.13
RL1	0.03	0.25	0.12	0.36	0.14	0.87	0.24	0.21	0.2
RL2	0.22	0.19	0.13	0.09	0.23	0.71	0.28	0.13	0.09
RL3	0.17	0.17	0.20	0.25	0.33	0.76	0.21	0.14	0.07
EnvP1	0.28	0.35	0.27	0.38	0.34	0.35	0.92	0.31	0.36
EnvP2	0.36	0.33	0.22	0.24	0.23	0.18	0.8	0.19	0.23
EconP1	–*	0.14	0.20	0.05	0.09	0.17	0.07	0.84	0.14
EconP2	0.50	0.32	0.35	0.31	0.39	0.09	0.23	0.87	0.22
EconP3	0.32	0.28	0.25	0.33	0.24	0.19	0.27	0.72	0.56
OperP1	0.32	0.43	0.34	0.29	0.12	0.07	0.3	0.56	0.94
OperP2	0.36	0.35	0.21	0.28	0.19	0.10	0.35	0.25	0.97
OperP3	0.24	0.11	0.07	0.35	0.07	0.24	0.07	0.29	0.83

* Indicates that the value is less than 0.01.

eco-design (ECO), green procurement (GP), green customer cooperation (CC) and reverse logistics (RL) have been proved to be GSCM practices of the surveyed enterprises. The influence of assemblers on suppliers is important for the surveyed enterprises to adopt GSCM, so the effective way to implement GSCM is

Table 5. Bootstrapping results for the model

Relation	t value	Sig	Relation	t value	Sig
GSCM->Performance (H2)	6.870	*	ECO4<-ECO	8.688	*
GSCM->EconP	3.169	*	GP1<-GP	7.825	*
GSCM->EnvP	3.185	*	GP2<-GP	2.015	0.044
GSCM->OperP	3.355	*	GP3<-GP	2.360	0.017
IP->GSCM (H1)	7.422	*	CC2<-CC	6.786	*
IP->CC	4.526	*	CC3<-CC	3.095	*
IP->ECO	5.700	*	CC5<-CC	4.510	*
IP->GP	5.038	*	RL1<-RL	2.861	*
IP->IEM	4.871	*	RL2<-RL	2.767	*
IP->RL	2.781	*	RL3<-RL	4.672	*
GSCM->CC	14.537	*	EnvP1<-EnvP	6.451	*
GSCM->ECO	13.117	*	EnvP2<-EnvP	2.022	0.043
GSCM->GP	11.853	*	EconP1<-EconP	2.275	0.023
GSCM->IEM	16.565	*	EconP2<-EconP	2.647	0.01
GSCM->RL	5.064	*	EconP3<-EconP	5.530	*
Performance->EconP	14.229	*	OperP1<-OperP	6.728	*
Performance->EnvP	9.972	*	OperP2<-OperP	3.009	*
Performance->OperP	18.432	*	OperP3<-OperP	2.688	*
IP->Performance (H3)	1.970	0.049	IP1<-IP	3.823	*
IEM1<-IEM	8.255	*	IP3<-IP	3.258	*
IEM2<-IEM	3.696	*	IP6<-IP	2.533	0.012
IEM3<-IEM	9.847	*	IP7<-IP	4.813	*
ECO1<-ECO	9.077	*	IP10<-IP	3.268	*
ECO2<-ECO	6.770	*	IP13<-IP	3.431	*
ECO3<-ECO	5.311	*	IP15<-IP	4.520	*

* = p-value < 0.01.

IEM and CC. The adoption of IEM is likely due to the fact that the vast majority of surveyed enterprises have mandatory ISO14001 certification. ISO14001 requires companies to constantly audit their suppliers, which may explain the green procurement practices of the companies under investigation. Product Eco-design refers to the action taken in the product development stage, aiming to minimize the environmental impact in the whole life cycle of the product. An important aspect of this approach is the use of intelligent design to promote product recycling, which is one of the fundamental characteristics is the closed-loop supply chain. National policies such as Technical Policy on Automobile Product Recycling and Utilization in 2006 and Standard Provisions on Compulsory Retirement of Motor Vehicles in 2012 are conducive to promoting the product eco-design and reverse logistics of domestic automobile manufacturers. Finally, EnvP, EconP, and OperP have been shown to have a positive impact on the surveyed enterprises through their GSCM practices.

As for the mediating effect of GSCM practices on the relationship between institutional pressures and business performance, the Bootstrap test results of the above model show that the direct effect of institutional pressures on business performance and the indirect effect of institutional pressures on business performance both pass the significance test. So GSCM practices play a partial mediating role in this relationship.

4.4 Moderating Effect Testing

To test whether the implementation time and enterprise scale of GSCM moderates on the relationship between GSCM and business performance, the moderating variable - Year and Size were respectively introduced into the adjusted structural model. A bootstrap program was used to test the significance of the moderating effect. The path coefficients of moderating effect show that value were more than 0 (Year value = 0.579, Size value = 0.536), and t-value were more than 1.96 (Year t-value = 3.239, Size t-value = 2.096), and the Year significance were 0.001 and the Size significance was 0.027. According to the results, the year of GSCM practices significantly moderates the relationship between GSCM practices and business performance, indicating that the implementation time of GSCM practices and enterprise scale both have a positive moderating effect on the relationship between GSCM practices and business performance of domestic automobile manufacturers. In other words, the longer the time of GSCM practices are, the more significant its impact on business performance will be. And the larger the enterprise scale is, the more significant the impact of GSCM practices on business performance will be. Therefore, hypothesis H4 and H5 establish.

5 Conclusions and Future Prospects

5.1 Main Conclusion

The above empirical results show that there is a structural relationship between institutional pressures and internal, external, post-GSCM practices and environmental performance, economic performance, operational performance, which supports the development of more active green environmental practices. Based on the PLS structural equation model, the following conclusions are drawn: (1) The state environmental protection laws and regulations, the regional environmental laws and regulations, the company product exports to overseas, sales of the enterprise products to foreign clients, establishment of the enterprise green image, the manufacturer's green production strategy on same product or substitute products and environmental partnerships with suppliers are the main institutional pressures in influencing enterprises to adopt GSCM practices. At the same time, the results show that these institutional pressures also have a positive effect on business performance. (2) The main GSCM practices adopted by the surveyed enterprises include internal environmental management, product eco-design, green procurement, green customer cooperation, and reverse logistics. (3) As for the relationship between these practices and performance, the results show that their implementations only have positive impacts on the economic, environmental and operational performance of the surveyed enterprises. And the relationship between GSCM and social performance cannot be proved. (4) Among the positive effects of GSCM practices on business performance, enterprise scale and the implementation time of enterprise green practice play a significant moderating role.

5.2   Management Implications

For the managers of Chinese automobile manufacturers, this paper proposes the following suggestions based on the results: (1) Enterprises' implementation of GSCM practices will significantly affect their environmental, economic and operational performance. Therefore, managers should form a correct view of GSCM practices and take them as a solution to enhance their competitiveness. Similarly, it is also important to recognize that pollution means poor utilization and they should treat it as a sign of problems in the production process. (2) Senior leaders of enterprises need to change their ideal. They should actively incorporate GSCM practices into enterprise strategy and practically integrate environmental awareness into various production and design processes. So they can achieve real product eco-innovation, and establish an inter-departmental collaboration mechanism to enhance internal environmental management practices, and improve resource utilization to reduce or even eliminate pollution. Through these, enterprises can improve their operational efficiency, economic returns, and social reputation, and then further enhance their competitiveness in the industry. (3) Leaders should always pay attention to the environmental protection policies issued by relevant departments, the trend of public opinion on environmental issues, and the environmental protection strategies of competitors. So they can formulate and adjust relevant GSCM practices and environmental protection strategies in time.

As for how the government makes appropriate policies to promote enterprises to implement GSCM practices, this paper will put forward the following suggestions based on the research results: (1) Faced with the situation that the front-end investment of GSCM practices is large and the payback period is long, it is difficult for enterprises to take the initiative to pay huge front-end investment and practical actions to implement GSCM without external forces. At the same time, it is often difficult for enterprises to find profits and competitive advantages after implementing GSCM. At this time, it is necessary for the government to formulate relevant environmental protection policies, including mandatory environmental and incentive environmental laws and regulations. Thus, it will force enterprises to take the initiative to implement GSCM practices and deepen the concept of green environmental protection into product design, procurement, reverse logistics, and other links. (2) It suggests that the government should intensify the implementation of relevant preferential policies, such as low-interest loans and fiscal subsidies. Meanwhile, agencies should not only support some small and medium enterprises to take the initiative to implement GSCM practices but also encourage them to substantially reduce pollution at the source instead of simply focusing on avoiding punishment. (3) The government can also improve the efficiency of law enforcement based on its own work, such as improving the speed of license approval and issuance, developing convenient channels for self-management. (4) The government should recognize and shoulder its due responsibilities, such as urging enterprises to assume social responsibilities and environmental protection, implementing existing policies in strict accordance with relevant regulations. In short, there are effective laws must be strictly enforced.

5.3   Research Limitations and Future Prospects

The limitation in this paper is that it only reflects the relationship among institutional pressures, GSCM practices and business performance in the supply chain of China's automobile industry. But it didn't introduce specific institutional pressures, specific GSCM practices and the interaction between various types of business performance. And nor did it analyze how various kinds of institutional pressures affect each other and how types of business performance affect each other. For example, at the present stage, whether does the product eco-design of Chinese automobile manufacturers positively correlate with their economic performance? And whether will enterprises indirectly affect economic performance by the positive way of improving environmental performance? In the meanwhile, the automobile industry adopts a hierarchical supplier management system in China. But this study doesn't focus on a certain level of automobile suppliers. Later studies can deepen the analysis from a specific level of suppliers to consider the relationship between suppliers and manufacturers.

In terms of the theoretical basis, the institutional theory explains how external drivers promote the practice of GSCM. Firstly, it is not clear how external drivers and internal factors interact to promote GSCM practices. Secondly, what kind of enterprises can be considered as the core enterprises in the supply chain? And what kind of mechanism should be established to stimulate these core enterprises still needs further research. Finally, the development of global supply chain provides the opportunity for enterprises from different countries to cooperate in the same supply chain. While enterprises are under the pressure of imitation, relevant green environmental protection practices are also diffused through this cooperation mechanism. However, this cooperation and diffusion mechanism needs to be further studied.

Acknowledgements. This work is supported by the technical innovation project (soft science research,No.2018ADC056) of Hubei science and technology department.

References

1. Ardian, Q., Zlatan, M.: Green supply chain management practices and company performance: A meta-analysis approach. Procedia Manuf. **17**, 317–325 (2018)
2. Azevedo, S.G., Carvalho, H., Machado, V.C.: The influence of green practices on supply chain performance: A case study approach. Transp. Res. Part E **47**(6), 850–871 (2011)
3. Becker, J.M., Klein, K., Wetzels, M.: Hierarchical latent variable models in pls-sem: Guidelines for using reflective-formative type models. Long Range Plan. **45**(5–6), 359–394 (2012)
4. Chan, R., He, H., et al.: Environmental orientation and corporate performance: The mediation mechanism of green supply chain management and moderating effect of competitive intensity. Ind. Mark. Manag. **41**(4), 621–630 (2012)
5. Cheng, Q., Zhou, Y.: Research on performance of management practice in green supply chain: A case study of an aluminum corporation (in chinese). China Soft Sci. **2017**, 10 (2017)
6. Corbett, C.: Extending the horizons: Environmental excellence as key to improving operations. Manuf. Serv. Oper. Manag. **8**(1), 5–22 (2006)
7. Daniel, K.F., Constantin, B.: Does sustainable supplier co-operation affect performance? Examining implications for the triple bottom line. Int. J. Prod. Res. **50**(11), 2968–2986 (2012)
8. Dimaggio, P.J., Powell, W.W.: The iron cage revisited: Institutional isomorphism and collective rationality in organizational fields. Am. Sociol. Rev. **48**(2), 147–160 (1983)
9. Edwards, J.R.: Multidimensional constructs in organizational behavior research: An integrative analytical framework. Organ. Res. Methods **4**(2), 144–192 (2001)
10. Fang, C., Zhang, J.: Performance of green supply chain management: A systematic review and meta analysis (in Chinese). J. Clean. Prod. **183**(S0959652618304), 906 (2018)
11. Fornell, C., Larcker, D.F.: Evaluating structural equation models with unobservable variables and measurement error. J. Mark. Res. **18**(1), 39–50 (1981)
12. Geng, R., Mansouri, S.A., Aktas, E.: The relationship between green supply chain management and performance: A meta-analysis of empirical evidences in asian emerging economies. Int. J. Prod. Econ. **183**(Part A), 245–258 (2016)
13. Govindan, K., Kaliyan, M., Kannan, D., Haq, A.N.: Barriers analysis for green supply chain management implementation in indian industries using analytic hierarchy process. Int. J. Prod. Econ. **147**(4), 555–568 (2014)
14. Hu, M., Ni, W.: Research on institutional pressure, strategic response and business performance (in Chinese). J. Ind. Technol. Econ. **35**(12), 60–67 (2016)
15. Jennings, P.: Ecologically sustainable organizations: An institutional approach. Acad. Manag. Rev. **20**(4), 1015–1052 (1995)
16. Laari, S., Toyli, J., et al.: Firm performance and customer-driven green supply chain management. J. Clean. Prod. **112**, 1960–1970 (2016)
17. Lai, K.H., Wong, C.: Green logistics management and performance: Some empirical evidence from chinese manufacturing exporters. Omega **40**(3), 267–282 (2012)
18. Lai, K.H., Wu, S.J., Wong, C.: Did reverse logistics practices hit the triple bottom line of chinese manufacturers? Int. J. Prod. Econ. **146**(1), 106–117 (2013)
19. Law, K.S., Wong, C.S., Mobley, W.H.: Toward a taxonomy of multidimensional constructs. Acad. Manag. Rev. **23**(4), 741–755 (1998)

20. Li, Y.: Institutional pressuresenvironmental innovation practices and firm performance - an institutional theory and ecological modernization theory perspective (in Chinese). Stud. Sci. Sci. **29**(12), 1884–1894 (2011)

21. Mathiyazhagan, N.A.: Analysis of the influential pressures for green supply chain management; adoption-an indian perspective using interpretive structural modeling. Int. J. Adv. Manuf. Technol. **68**(1–4), 817–833 (2013)

22. Mitra, S., Datta, P.P.: Adoption of green supply chain management practices and their impact on performance: An exploratory study of indian manufacturing firms. Int. J. Prod. Res. **52**(7), 2085–2107 (2014)

23. Peng, D.X., Lai, F.: Using partial least squares in operations management research: A practical guideline and summary of past research. J. Oper. Manag. **30**(6), 467–480 (2012)

24. Phan, T.N., Baird, K.: The comprehensiveness of environmental management systems: The influence of institutional pressures and the impact on environmental performance. J. Environ. Manag. **160**, 45–56 (2015)

25. Zhu, Q.: The moderating effects of institutional pressures on emergent green supply chain practices and performance. Int. J. Prod. Econ. **45**(18–19), 4333–4355 (2007)

26. Seles, B., Jabbour, C., Dangelico, R.: The green bullwhip effect, the diffusion of green supply chain practices, and institutional pressures: Evidence from the automotive sector. Int. J. Prod. Econ. **182**, 342–355 (2016)

27. Shrout, P.E., Niall, B.: Mediation in experimental and nonexperimental studies: new procedures and recommendations. Psychol. Methods **7**(4), 422 (2002)

28. Simpson, D.: Institutional pressure and waste reduction: The role of investments in waste reduction resources. Int. J. Prod. Econ. **139**(1), 330–339 (2012)

29. Tachizawa, E.M., Gimenez, C., Sierra, V.: Green supply chain management approaches: drivers and performance implications. Int. J. Oper. Prod. Manag. (2015). Forthcoming (11)

30. Tate, W.L., Ellram, L.M., Kirchoff, J.F.: Corporate social responsibility reports: A thematic analysis related to supply chain management. J. Supply Chain. Manag. **46**(1), 19–44 (2010)

31. Vanalle, R.M., Ganga, G., Filho, M.G., Lucato, W.C.: Green supply chain management: An investigation of pressures, practices, and performance within the brazilian automotive supply chain. J. Clean. Prod. **151**(Complete), 250–259 (2017)

32. Wong, C., Lai, K., Shang, K., Lu, C., Leung, T.: Green operations and the moderating role of environmental management capability of suppliers on manufacturing firm performance. Int. J. Prod. Econ. **140**(1), 283–294 (2012)

33. Wu, G., Ding, J., Chen, P.: The effects of gscm drivers and institutional pressures on gscm practices in taiwan's textile and apparel industry. Int. J. Prod. Econ. **135**(2), 618–636 (2012)

34. Yu, C., Liu, Z.: A feasibility analysis for practicing green supply chain management without government policy incentives (in Chinese). J. Ind. Eng. Eng. Manag. **31**(2), 119–127 (2017)

35. Zhu, Q., Sarkis, J.: Relationships between operational practices and performance among early adopters of green supply chain management practices in chinese manufacturing enterprises. J. Oper. Manag. **22**(3), 265–289 (2004)

36. Zhu, Q., Sarkis, J., Geng, Y.: Green supply chain management in china: pressures, practices and performance. Int. J. Oper. Prod. Manag. **25**(5), 449–468 (2005)

37. Zhu, Q., Sarkis, J., et al.: Institutional-based antecedents and performance outcomes of internal and external green supply chain management practices. J. Purch. Supply Manag. **19**(2), 106–117 (2013)

Non-cooperative Game Based Carbon Emission Reduction for Supply Chain Enterprises with a Cap and Trade Mechanism

Min Wang[1], Shuhua Hou[2,3], and Rui Qiu[2(✉)]

[1] Business School, Sichuan Agricultural University, Dujiangyan 611830, People's Republic of China
[2] Business School, Sichuan University, Chengdu 610064, People's Republic of China
qiuruicd@scu.edu.cn
[3] School for Environment and Sustainability, University of Michigan, Michigan, Ann Arbor 48109-1041, USA

Abstract. With the increasing attention of society to the environmental issues, carbon emission rights have become a new type of resource with certain commercial value, which has changed the production function and cost structure of the original enterprises. In the course of operation, enterprises have increased the consideration of carbon emission rights from the original focus on "raw materials and products (services)". Therefore, based on the carbon and trade mechanism, this paper deeply explores the decision-making of carbon emission reduction of the upstream and downstream enterprises in the supply chain and the choice of government carbon quota allocation scheme under the constraint of government carbon emissions, and uses game theory knowledge to analyze the effect of carbon emission reduction of upstream and downstream enterprises under the non-cooperative game. The results show that the emission reduction effect of the government's carbon quota per unit of product allocation is more obvious than that of the direct total amount restriction, and it is beneficial for sustainability of supply chain systems.

Keywords: Cap and trade mechanism · Supply chain enterprises · Supply chain · Non-cooperative · Sustainability

1 Introduction

In recent decades, climate change has become the focus all over the world. With the increasing frequency of extreme weather in the world, the environmental problems caused by it have seriously affected people's production and life [4–6]. The Intergovernmental Panel on Climate change (IPCC), established in 1988 by the United Nations Environment Program (UNEP) and the World Meteorological Organization (WMO), issued five series of climate change assessment reports

© Springer Nature Switzerland AG 2020
J. Xu et al. (Eds.): ICMSEM2019 2019, AISC 1002, pp. 450–459, 2020.
https://doi.org/10.1007/978-3-030-21255-1_34

between 1990 and 2014 [7,8]. These reports provide a detailed analysis of trends in global mean surface temperatures and sea levels as a result of climate change. In the latest assessment report, the Fifth Assessment report, the global mean surface temperature increased by $0.85\,°C$ over 130 years from 1880 to 2012. The 30 years from 1983 to 2012 were probably the hottest periods in more than one thousand years [9]. Since 1971, global glaciers have decreased by an average of 226 billion tons annually. The global sea level rose by $0.19\,m$ between 1901 and 2010. It can be seen from the contents of the report that the global climate change is becoming more and more violent, and the environmental problems caused by it are constantly threatening the survival of human beings.

For the implementation of the above-mentioned agreements, countries around the world have completed the corresponding carbon reduction policies based on their own actual situation. Current mainstream carbon reduction mechanisms include carbon taxes, cap-and-trade, and carbon subsidies [10–12]. Carbon tax is a prorated tax on a business based on its actual carbon emissions. Finland, Poland and other Nordic countries are the countries that implemented the carbon tax policy earlier, and the formulation and implementation of these policies are more effective [14–16]. The cap-and-trade system is that the government gives different initial carbon emission quotas to different enterprises and allows them to trade carbon quota in the carbon trading market [1–3]. If the initial carbon emission quota is excess, the excess amount can be sold in the market to improve its income; and if the initial quota is insufficient, it can be purchased in the market to meet the production demand. The European Union is the largest organization to implement a carbon and trade system in the world at present. Its successful experience has important reference significance for other countries to implement cap-and-trade policy and even the establishment of the world carbon trading market. By 2017, 46 carbon pricing policies had been implemented or planned. It includes 23 carbon trading markets dominated by local jurisdictions and 23 carbon tax policies at the national level. These policies cover 8 billion tons of carbon dioxide emissions, accounting for about 15% of the world's total emissions, of which carbon cap-and-trade systems account for nearly two-thirds [13].

Our government has been committed to energy-saving, emission-reduction and development of a low-carbon economy, and announced that carbon dioxide emissions will peak around 2030 and will strive to reach a peak as soon as possible. At the national level, China's National Program to Address Climate Change, the 12th Five-Year Plan for Energy Conservation and Emission Reduction and the corresponding quantified emission reduction targets have been formulated and implemented. At the same time, carbon trading, carbon tax (environmental tax) and other carbon regulatory measures have been introduced. In January 2012, the National Development and Reform Commission announced that it would carry out carbon emissions trading pilot projects in seven provinces and cities, including Beijing, Shanghai and Guangdong. After more than five years of pilot operation, by September 2017, there were more than 20 industries and more than 3000 enterprises and institutions have been incorporated into the carbon market. The cumulative turnover of emissions quotas was about 197 million tons of carbon dioxide equivalent and the cumulative turnover was about 4.516 billion yuan (Annual Report on China's Policies and Actions to Address Climate Change 2017). Meanwhile, the notice of "National Carbon Emission Trading Market Construction Plan (Power Generation Industry)" was issued, and it is planned to take the lead in running the national carbon trading market in power generation industry in 2018. The research framework of this paper is as follows: Sect. 2 establishes the supply chain non-cooperative game model under the carbon trading market; Sect. 3 analyzes the model and discusses the analysis results; Sect. 4 summarizes the research of this paper.

2   Modeling

In this section, the given carbon emission quota from the government, the upstream and downstream enterprises in the supply chain make optimal decisions based on several different emission reduction forms of carbon quota allocation schemes under the non-cooperative game is considered, aiming at maximizing the profit of the enterprise and the unit emission reduction rate is the decision variable.

2.1   Model Parameters

The parameters required in this paper are showed below.

Table 1. Please write your table caption here

e_i	Initial carbon emissions per unit product of enterprise i
c_i	Unit product production cost of enterprise i
p_ε	Carbon trading price
π_i	Profit of enterprise i
λ_i	Enterprise i emission reduction rate per unit product after product research and development (R&D)
σ_i	Unit carbon quota allocated by the Government to enterprises i free of charge
q	Output
a	Market capacity, Constant
w	Wholesale price of unit product
p	Retail price of unit product

2.2   Non-cooperative Game of Sustainable Supply Chain

In order to maintain environmental sustainability, the government has introduced some carbon emission policies. As a market-oriented environmental policy tool, cap and trade policies has been practiced and applied in many countries in the world, and its implementation effect has been widely praised by the governments of all countries. And in China, after four years of pilot projects in seven regions such as Beijing, Shenzhen and Hubei since 2013, and with the approval of the State Council in December 2017, the National Development and Reform Commission (NDRC) issued a national carbon emission trading market construction plan for the power generation industry. The introduction of the scheme marks the official start of the carbon emission trading system nationwide, and further promotes the development of China's economy towards green and low-carbon. It will play a positive role in promoting our economy to achieve a higher quality of development. Traffic green low-carbon is one of the important parts of economic green low-carbon in China. The basic information on current green low-carbon transportation is shown in Fig. 1.

The competition among enterprises in their respective industries is becoming increasingly fierce. The fierce competition has made enterprises realize that they must take the path of sustainable development. Therefore, these enterprises have to actively seek ways to make the supply chain system sustainable, and establish a sustainable supply chain system to meet the actual situation of enterprises for further implement sustainable supply chain management. At the enterprise level, under the restriction of government carbon emissions, enterprises actively carry out R&D investment to reduce their own carbon emissions to improve their own environmental conditions, win competitive advantages and achieve sustainable development of themselves and their supply chains. In recent years, the supply chain management of enterprises in developed countries has expanded from environmental concerns to corporate social responsibility concerns. There are two upstream and downstream enterprises in the supply chain, and there is a two-stage game process between enterprises. The first stage is for upstream and downstream enterprises to determine their own emission reduction investment level according to their respective carbon quotas, and the second stage is for upstream enterprises to determine their own product wholesale prices and downstream enterprises to determine their own product retail prices. The non-cooperative game of supply chain enterprises is as follows: the upstream and downstream enterprises respectively determine the level of emission reduction investment, and set the wholesale or retail price of products. Since non-cooperation means that the upstream and downstream enterprises determine their own investment level in emission reduction and set their own wholesale or retail prices, the second stage is a competitive game. The non-cooperative game between government and supply chain enterprises is shown in Fig. 2.

The profit function of downstream enterprises is as follows:

$$\pi_D = \{p - w - c_D - \mathrm{p}_e[(1 - \lambda_D)e_D - \sigma_D]\}\,(a - p) - \frac{1}{2}k\lambda_D^2.$$

Derivative of $\frac{\partial \pi_D}{\partial p} = a - 2p + w + c_D + p_e[(1 - \lambda_D)e_D - \sigma_D]$, and $\frac{\partial \pi_i{}^2}{\partial p^2} = -2 < 0$, then there is an optimum yield. The following results can be obtained by solving $\frac{\partial \pi_D}{\partial p} = 0$:

$$p^* = \frac{1}{2}\{a + w + c_D + p_e[(1 - \lambda_D)e_D - \sigma_D]\}.$$

The profit function of the upstream enterprise is:

$$\pi_U = \{w - c_U - p_e[(1 - \lambda_U)e_U - \sigma_U]\}(a - p^*) - \frac{1}{2}k\lambda_U^2$$
$$= \{w - c_U - p_e[(1 - \lambda_U)e_U - \sigma_U]\}(a - \frac{1}{2}\{a + w + c_D + p_e[(1 - \lambda_D)e_D - \sigma_D]\}) - \frac{1}{2}k\lambda_U^2.$$

> ➢ Air pollution is a critical public health threat in modern society.
> ➢ Road traffic emission has been a major contributor to air pollutions worldwide.
> ➢ Most of vehicles are charged by fuels.

Fig. 1. Basic information on current green low-carbon transportation

Derivative of w:

$$\frac{\partial \pi_U}{\partial w} = a - \frac{1}{2}\{a + w + c_D + p_e[(1 - \lambda_D)e_D - \sigma_D]\} - \frac{1}{2}\{w - c_U - p_e[(1 - \lambda_U)e_U - \sigma_U]\}$$
$$= \frac{1}{2}\{a + c_U - c_D + p_e[(1 - \lambda_U)e_U - (1 - \lambda_D)e_D - \sigma_U + \sigma_D]\} - w,$$

and $\frac{\partial \pi_i{}^2}{\partial p^2} = -1 < 0$, then there is an optimum yield. The following results can be obtained by solving $\frac{\partial \pi_U}{\partial w} = 0$:

$$w^* = \frac{1}{2}\{a + c_U - c_D + p_e[(1 - \lambda_U)e_U - (1 - \lambda_D)e_D - \sigma_U + \sigma_D]\}.$$

The following results can be obtained:

$$w^* = \frac{1}{2}\{a + c_U - c_D + p_e[(1 - \lambda_U)e_U - (1 - \lambda_D)e_D - \sigma_U + \sigma_D]\}. \tag{1}$$

$$p^* = \frac{1}{4}\{3a + c_U + c_D + p_e[(1 - \lambda_U)e_U + (1 - \lambda_D)e_D - (\sigma_U + \sigma_D)]\}. \tag{2}$$

$$q^* = \frac{1}{4}\{a - (c_U + c_D) - p_e[(1 - \lambda_U)e_U + (1 - \lambda_D)e_D - (\sigma_U + \sigma_D)]\}. \tag{3}$$

Formulas (1)–(3) are substituted into upstream and downstream enterprise profit functions respectively, the following results can be obtained:

$$\pi_U = \frac{1}{8}\{a - (c_U + c_D) - p_e[(1 - \lambda_U)e_U + (1 - \lambda_D)e_D - (\sigma_U + \sigma_D)]\}^2 - \frac{1}{2}k\lambda_U^2.$$

$$\pi_D = \frac{1}{16}\{a - (c_U + c_D) - p_e[(1 - \lambda_U)e_U + (1 - \lambda_D)e_D - (\sigma_U + \sigma_D)]\}^2 - \frac{1}{2}k\lambda_D^2.$$

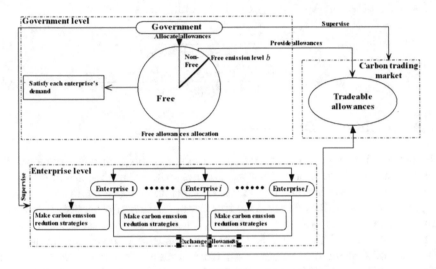

Fig. 2. Non-cooperative Game Framework between Government and Supply Chain Enterprises

Derivative for λ_U and λ_D, respectively, and the results are as follows:

$$\lambda_U = \frac{e_U p_e\{a - (c_U + c_D) + e_D p_e \lambda_D - p_e[e_U + e_D - (\sigma_U + \sigma_D)]\}}{4k - e_U^2 p_e^2}. \quad (4)$$

$$\lambda_D = \frac{e_D p_e\{a - (c_U + c_D) + e_U p_e \lambda_U - p_e[e_U + e_D - (\sigma_U + \sigma_D)]\}}{8k - e_D^2 p_e^2}. \quad (5)$$

The second derivatives of λ_U and λ_D are obtained respectively, $\frac{\partial \pi_U^2}{\partial \lambda_U^2} = \frac{e_U^2 p_e^2}{4} - k$, $\frac{\partial \pi_D^2}{\partial \lambda_D^2} = \frac{e_D^2 p_e^2}{8} - k$. The optimal solution exists only when $\frac{\partial \pi_U^2}{\partial \lambda_U^2}$ and $\frac{\partial \pi_D^2}{\partial \lambda_D^2}$ are less than zero.

The solution of (4), (5) equations can be obtained:

$$\lambda_U^* = \frac{2e_U p_e\{a - (c_U + c_D) - p_e[e_U + e_D - (\sigma_U + \sigma_D)]\}}{8k - p_e^2(2e_U^2 + e_D^2)}. \quad (6)$$

$$\lambda_D^* = \frac{e_D p_e\{a - (c_U + c_D) - p_e[e_U + e_D - (\sigma_U + \sigma_D)]\}}{8k - p_e^2(2e_U^2 + e_D^2)}. \quad (7)$$

Assume that the total amount of carbon emissions allocated by the government to enterprise i is E_i, then $p_e\sigma_iq \to E_ip_e$.

The profit functions of upstream and downstream enterprises are as follows:

$$\pi_U = [w - c_U - e_Up_e(1 - \lambda_U)](a - p) - \frac{1}{2}k\lambda_U^2 + E_Up_e. \tag{8}$$

$$\pi_D = [p - w - c_D - e_Dp_e(1 - \lambda_D)](a - p) - \frac{1}{2}k\lambda_D^2 + E_Dp_e. \tag{9}$$

The derivatives of λ_U and λ_D can be calculated as follows:

$$\lambda_U{}^{**} = \frac{2e_Up_e\{a - (c_U + c_D) - p_e(e_U + e_D)\}}{8k - p_e{}^2(2e_U{}^2 + e_D{}^2)}. \tag{10}$$

$$\lambda_D{}^{**} = \frac{e_Dp_e\{a - (c_U + c_D) - p_e(e_U + e_D)\}}{8k - p_e{}^2(2e_U{}^2 + e_D{}^2)}. \tag{11}$$

3   Analysis and Discussion

This section mainly analyzes the model of Sect. 2 to provide reference for better implementation of emission reduction. The majority of enterprises in china only rely on their own strength to implement sustainable supply chain management, and they are still in its infancy. But the implementation of sustainable supply chain management is not only affected by enterprises themselves, but also by the government, and these two factors comprehensively affect the implementation effect of sustainable supply chain management. In these two factors, the decision-making body of implementing sustainable supply chain management is the enterprise itself, while the government only acts as a guide and regulator to provide policy guarantee for enterprises to implement sustainable supply chain management. The government and enterprises work together and influence each other.

Generally speaking, the current way of free allocation of carbon quotas by the government is to allocate the total carbon emissions directly according to the historical output of enterprises.

Let $\Delta\lambda_i$ represent the difference between the optimal emission reduction rate per unit of output allocated by the government and the optimal emission reduction rate obtained by the government directly limiting the total amount of carbon emissions, which can be expressed as $\Delta\lambda_i{}^* = \lambda_i{}^* - \lambda$

Combining (4) and (5), the following results can be obtained:

$$V\lambda_U{}^* = \lambda_U{}^* - \lambda_U{}^{**} = \frac{2e_Up_e{}^2(\sigma_U + \sigma_D)}{8k - p_e{}^2(2e_U{}^2 + e_D{}^2)}.$$
$$V\lambda_D{}^* = \lambda_D{}^* - \lambda_D{}^{**} = \frac{e_Dp_e{}^2(\sigma_U + \sigma_D)}{8k - p_e{}^2(2e_U{}^2 + e_D{}^2)}.$$

Due to $e_i \geq 0, \sigma_i \geq 0, p_e \geq 0$ and $8k - p_e{}^2(2e_U{}^2 + e_D{}^2) \geq 0$, easily get $V\lambda_U{}^*_{(HC)} \geq 0$ and $V\lambda_D{}^*_{(HC)} \geq 0$. The government's carbon quota per unit of production allocation has a larger emission reduction rate than the direct total carbon emission limitation, and the difference is related to σ_I and P_ε.

The derivatives of σ_U and σ_D can be calculated as follows:

$$\frac{\partial V\lambda_U^*{}_{(HC)}}{\partial \sigma_U} = \frac{2e_U p_e^2}{8k - p_e^2(2e_U^2 + e_D^2)} > 0.$$
$$\frac{\partial V\lambda_D^*{}_{(HC)}}{\partial \sigma_D} = \frac{2e_D p_e^2}{8k - p_e^2(2e_U^2 + e_D^2)} > 0.$$

It is easy to get the monotone increasing relationship between the unit carbon quota and its difference. The bigger σ_i, the bigger $V\lambda_i*$, the higher the corresponding emission reduction rate. In other words, the more carbon quotas the government allocates to each enterprise in the supply chain, the more it can promote its emission reduction. Therefore, the government should choose the emission reduction scheme of allocating carbon quotas per unit to limit the carbon emission quotas of enterprises.

Even if the unit product has the most significant emission reduction effect, its optimal output will also increase, which will increase the total amount of products and further increase the total amount of carbon emissions. Therefore, if we want to keep the emission reduction rate constant and minimize the total carbon emissions, the best way is to reduce production, or even not to produce. At this time, the actual output has deviated from the optimal output under Nash Equilibrium, which will inevitably reduce the total profit of the supply chain.

For complex systems involving governments and supply chain enterprises, cap and trade policies should be applied. Without such a policy, supply chain firms will pursue the highest possible revenues and lack the incentive to continuously improve their emissions performance. A cap and trade policy can be developed by using the approach mentioned above. As discussed in discussion 1, such a policy could guide supply chain enterprises to curb carbon emissions. In addition, since the purpose of the cap and trade policy is to promote the development of energy-saving and emission-reduction technologies, carbon emissions should not be the only criteria emphasized by the government when formulating cap and trade policies, and it should pay attention to the satisfaction of supply chain enterprises rather than squeezing the profits of supply chain enterprises. The satisfaction of supply chain enterprises under a cap and trade policy is regarded as the goal of the government, and their profits are set as constraints to prevent any supply chain enterprises from losing because of the carbon trading price set by the government.

Secondly, governments can use the proposed model to choose their own cap and trade policies. In other words, governments can take full account of their actual situation and choose the cap and trade policies they need. Through the proposed method, the government can determine the appropriate carbon trading price to encourage supply chain enterprises to adopt improvement strategies. According to the government's decision, supply chain enterprises can choose appropriate emission reduction strategies to maximize profits. If enterprises choose to pursue the lowest carbon emissions, they will have to reduce production and sacrifice profits; however, if enterprises choose to maximize profits, it will inevitably produce more carbon emissions and sacrifice the environment.

Companies must balance their profits with their social responsibilities (i.e. carbon emissions) and governments should seize this feature to establish an effective incentive mechanism to promote their emissions reduction.

4   Conclusion

With the concept of environmental protection and sustainable development becoming more and more popular, greenhouse gas emissions and their economic effects have begun to receive widespread attention around the world. Taking total quantity control as the breakthrough point, it is an important part of the emission reduction policies of governments all over the world to gradually realize the significant transformation from carbon emission management to total emission control management. At the same time, the "win-win" policy of pursuing rapid economic development and environmental protection has gradually become the focus of social attention. In this form, carbon emissions trading emerged as a new trading mechanism, and just a few years it has become popular all over the world. China, as a big country with both industrialization and urbanization in accelerating development, how to seize this opportunity and develop green economy, promote energy saving and emission reduction, and fulfill its international obligations under the guidance of scientific development concept has become a major issue facing the government. Up to now, the research on carbon emissions trading is not perfect, and there is no theoretical guidance for the government to promote the formulation of policies related to the national carbon emissions trading system. Therefore, it is very urgent and necessary to carry out further research.

Based on the results of this study, we can see that: when considering carbon emission constraints, the government should first formulate an effective carbon quota allocation policy to encourage enterprises to be more proactive in reducing emissions. Secondly, for the carbon emission trading market, the government should focus on promoting emission reduction in the supply chain where the enterprises with more carbon emissions per unit product are located, and pay attention to the source of supply chain. Then the emission reduction costs of downstream enterprises and promote the implementation of emission reduction in an all-round way are reduced. Furthermore, enterprises should make the best decisions in terms of emission reduction level, the quantity of carbon quota sales, their own interests and social responsibilities. The enterprise should sacrifice the short-term benefit in exchange for the long-term sustainable development, and realize the corporate social responsibility. Only this way can the long-term benefit maximization of supply chain and the "win-win" situation of carbon emission minimization be realized. Last, the upstream and downstream enterprises in the supply chain should strengthen cooperation to achieve "win-win" of benefits and emission reduction. Then we can we achieve higher emission reduction efficiency, get more carbon emission rights and gain more benefits in the future economic society which pays more attention to environmental issues.

Acknowledgements. This research has been supported by the project of Research Center for System Sciences and Enterprise Development (Grant No. Xq18B02), Sichuan University (Grant No. 2018hhf-45), and China Scholarship Council (Grant No. 201806240126).

References

1. Kaarstad, O.: Norwegian Carbon Taxes and Their Implication for Fossil Fuels (1995)
2. Bryant, G.: Creating a level playing field? the concentration and centralisation of emissions in the european union emissions trading system. Energy Policy **99**, 308–318 (2016)
3. Brink, C., Vollebergh, H.R.J., Werf, E.V.D.: Carbon pricing in the eu: Evaluation of different eu ets reform options. Energy Policy **97**, 603–617 (2016)
4. Chen, X., Chen, J.: Supply chain carbon footprinting and responsibility allocation under emission regulations. J. Environ. Manag. **188**, 255–267 (2017)
5. Chevallier, R.: Climate change progress, the us withdrawal and what we can expect from cop23. J. Environ. Manag. (2017)
6. Chinn, L.N.: Can the market be fair and efficient? An environmental justice critique of emissions trading. Ecol. Law Q. **26**(1), 80–125 (1999)
7. Clò, S., Battles, S., Zoppoli, P.: Policy options to improve the effectiveness of the eu emissions trading system: A multi-criteria analysis. Energy Policy **57**(6), 477–490 (2013)
8. Coase, R.H.: The Problem of Social Cost (1960)
9. Convery, F.J., Redmond, L.: Market and price developments in the European union emissions trading scheme. Rev. Environ. Econ. Policy **1**(1), 88–111 (2007)
10. Criqui, P., Mima, S., Viguier, L.: Marginal abatement costs of co_2 emission reductions, geographical flexibility and concrete ceilings: An assessment using the poles model. Energy Policy **27**(10), 585–601 (1999)
11. Crossland, J., Li, B., Roca, E.: Is the European union emissions trading scheme (eu ets) informationally efficient? Evidence from momentum-based trading strategies. Appl. Energy **109**(2), 10–23 (2013)
12. Cruz, I.S., Katz-Gerro, T.: Urban public transport companies and strategies to promote sustainable consumption practices. J. Clean. Prod. **123**, 28–33 (2016)
13. Fang, G., Tian, L., et al.: How to optimize the development of carbon trading in china-enlightenment from evolution rules of the eu carbon price. Appl. Energy **211**, 1039–1049 (2018)
14. Hepburn, C.: Climate change economics: Make carbon pricing a priority. Nat. Clim. Chang. **7**(6), 389–390 (2017)
15. Herold, D.M., Lee, K.H.: Carbon disclosure strategies in the global logistics industry: similarities and differences in carbon measurement and reporting (2018)
16. Jiang, Z., Shao, S.: Distributional effects of a carbon tax on chinese households: A case of shanghai. Energy Policy **73**(10), 269–277 (2014)

A Comparative Study of Waste Classification Laws and Policies: Lessons from the United States, Japan and China

Lai Wei and Yi Lu[✉]

Business School, Sichuan University, Chengdu 610065, People's Republic of China
luyiscu@163.com

Abstract. With the rapid increase of municipal solid waste (MSW), the phenomenon of waste siege exists all over the world. To solve this problem, waste classification is wildly applied. Previous studies have shown that residents' support for laws and policies, as a critical factor, is often conducive to waste classification. At present, governments around the world have enacted waste classification laws and policies (WCLP) for MSW reduction. In order to construct a sound WCLP system, we analyzed and compared the typical WCLP systems of the United States, Japan and China. It shows that these three countries are focused on making laws and/or policies according to their own national conditions or specific situations. A more general legal system for the waste classification in various countries has been summarized. It would be beneficial to the thorny problem solving of waste siege faced by most of the countries and the new idea conceiving for global sustainable development.

Keywords: Waste classification and recycling · Laws and policies · Comparative analysis · Sustainable development

1 Introduction

With the rapid development of global social economy and technology, the living standard of residents is improving. However, rapid development also brings severe environmental problems, one of which is the increase of MSW worldwide [2, 19].

It was estimated that the world's 2.9 billion urban population produced about 3 million tons of domestic waste per day in 2000, and is expected to reach a peak of 11 million tons by 2025 with more than 6 million tons [5]. The cities all over the world produce 1.47 billion tons of domestic waste every year, of which only 15% are recycled, most of which are disposed of by landfills [14]. There are a thousand "waste hills" of various sizes covering an area of 5.33 km^2 at the outskirts of China's capital, Beijing [25]. "Waste siege" phenomenon has become an unavoidable problem resulting from the improper waste disposal [26]; not only in China, it is common in other developing countries [9].

© Springer Nature Switzerland AG 2020
J. Xu et al. (Eds.): ICMSEM2019 2019, AISC 1002, pp. 460–468, 2020.
https://doi.org/10.1007/978-3-030-21255-1_35

Even in some developed countries, this phenomenon still exists. For example, the annual MSW production in the United States increased from 88 million tons in 1960–254 million tons in 2007 [16].

MSW management is a common challenge the world faces in the 21st Century [24]. In the face of the severe situation of waste siege, waste classification is generally regarded as an important step of the solution. Scholars believe that in order to eliminate the dangers of MSW, the most essential step is to change public attitudes and to promote waste classification, which is considered to be the fundamental way to reduce the amount of waste disposal, improve the quality of living environment and solve the problem of waste siege [7].

Prior studies in the field of environmental psychology examined the key factors influencing recycling behavior [20,21]. Most of these studies focus on the direct behavior of recycling. Indirect behaviors, such as elected politicians and pro-government policies, have rarely been studied [4]. However, pro-environmental government policies, as an indirect factor, could significantly influence on people's direct environmental behavior [11]. Indirect behaviors, such as supporting, are a form of behavior. In this point, effective WCLP measures are considered to increase the attractiveness of pro-environmental behavior [18].

In this paper, we selected three representative countries, the United States, Japan and China, to analyze the WCLP systems, which are national indirect behaviors for waste recycling. National WCLP system of each country was briefly summarized and illustrated; these three cases were then comprehensively compared from three dimensions: mandates, inducements, and capacity-building. On this basis, a more general WCLP system was proposed. To a certain extent, this article summarizes the laws and policies system on waste classification in the United States, Japan and China then provides a new method for the future improvement of waste classification management system in other countries.

2 Method

Making laws and policies is a main means for a government to control the state macroscopically [1]. A range of programs and policy instruments are available to policy makers for waste managing and recycling [8]. Some scholars found that residents' perception of the effectiveness of laws and policies usually affects their recycling behavior [21]. Elmore [6] proposed four tools for policy: mandates, inducements, capacity-building, and system-changing. Based on Elmore's findings, some scholars have proposed three common policy tools used by decision makers in promoting environmentally friendly behavior: mandates, inducements and capacity-building [22].

On the basis of summing up the previous experience, in this paper, we divided the functions of government WCLP systems into the following dimensions: construction of sound recycling system (mandates), implementation of effective incentives or punishment measures (inducements), and popularization of waste classification education to residents (capacity-building); and take these three points as the indexes of comparative analysis. By using the analytical method [23], the WCLP systems of the United States, Japan and China were illustrated, and then compared from the dimensions of mandates, inducements and capacity-building. The aim of this paper is to find out the common characteristics and outstanding advantages of the construction of a sound WCLP system.

3 Three National WCLP Systems

3.1 WCLP System of the United States

Municipal trash collection programs emerged in the United States between the late 19th and early 20th century to address public sanitation [17]. There has been federal legislation on MSW since the 1960s [12]. Up to now, MSW legislation and management in the United States has formed a huge, complete and strict environmental regulation system composed of dozens of laws and thousands of regulations. From the point of the federal environmental laws and regulations system, the upper level is the National Environmental Policy Act, which is programmatic and operational. The lower levels of the system can be divided into two major laws and regulations: "pollution control", such as the Pollution Prevention Act, and "resources protection", such as the Resource Protection and Recycling Act.

In addition, under the premise of carrying out the federal legislation, the American States make their own legislations depend on local conditions. California has been fully implementing the General Hazardous waste Act since 2006, which forbids all residents and units of California from mixing hazardous waste into landfills.

The legal system of waste classification in the United States is shown in Fig. 1.

3.2 WCLP System of Japan

Japan has established a sound legal system of circular economy earlier in the world. According to the social situation and the characteristics of waste problem, Japanese government makes laws and policies with the times, and constantly revise and perfect them. In the 21st century, the laws on waste disposal in Japan become more and more perfect, and gradually form a three-tier legal system. The first level is the fundamental law, of which the most representative one is the Basic Law on Promoting the Establishment of a Circulating Society promulgated in 2000 as the basis of all kinds of environmental protection laws. By promoting it, Japan details the responsibilities and obligations of governments, businesses and citizens in waste disposal. The second level is comprehensive laws, such as the Law on Promoting Efficient Use of Resources. These laws put forward more detailed requirements for waste classification on the premise of basic laws. The third level is carefully classified as the specific product laws, such as the Appliances Recycling Act.

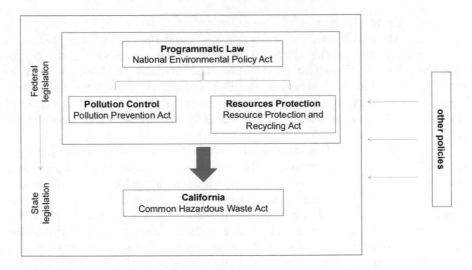

Fig. 1. The WCLP system of the United States

Japan's relevant laws on the violation of waste classification penalties are also very strict. With the establishment of evaluation and punishment mechanisms as the core, the state responsibility is strengthened in the law while the punishment measures for improper disposal of waste are improved. Under such strict legal supervision and punishment measures, Japan's waste management can be effectively dealt with.

The framework map of Japanese WCLP system is shown in Fig. 2.

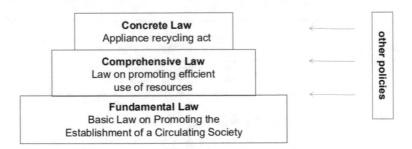

Fig. 2. The WCLP system of Japan

3.3 WCLP System of China

The legal system of waste classification in China is looser compared with those two mentioned before. The characteristics of China's is that there are more policies than laws. The main programmatic laws concerning waste classification are the basic law of environmental protection promulgated by the Central Committee (i.e. the Environmental Protection Law) and the single law on pollution prevention and control (i.e. the Solid Waste Pollution Prevention Law) [26]. In terms of relevant policies, there are "Municipal Solid Waste Management Measures" and "Municipal Appearance and Environmental Sanitation Management Regulations" promulgated by the Central Government as well. Moreover, local laws and regulations are formulated according to the actual situation and central order.

The MSW classification started late in China, and the corresponding legal system is not perfect compared with the developed countries. Although the waste classification is regulated in some legal provisions, the problems are still prominent. In March 2017, China's central government announced the Implementation Plan of Household Waste Classification System. In response to the call of the Central Government, many Provinces in China have issued policies related to waste classification. In this way, it makes up for the problem that Chinese waste classification system is too principled, lacking of operability and filled with unclear responsibility.

The map of Chinese WCLP system is as follows in Fig. 3.

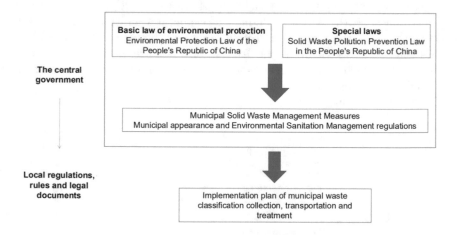

Fig. 3. The WCLP system of China

4   Comparison Analysis

In accordance with construction of sound recycling system (mandates), implementation of effective incentives or punishment measures (inducements), and popularization of waste classification education to residents (capacity-building), the WCLP systems of the United States, Japan and China are reviewed and compared.

From the point of the waste recycling system: The United States establishes the 4R principle, in the Resource Protection and Recycling Act, that is, Recovery, Recycle, Reuse and Reduction. It expands waste management from simple cleanup to integrated planning with separate recovery, reduction and reuse. The system of waste classification and recycling in the United States is relatively complete, and has gradually formed an industrial process of collection, recovery, treatment, processing and marketing. In Japan, the governmental rules on waste classification are very detailed, including not only the introduction of specific product recycling laws, but also strict provisions on waste collection days and specific delivery time. Similar to the US, in the Basic Law on Promoting the Establishment of a Circulating Society, Japan established the basic principles of a circular society by putting forward the 3R principal of Reduce, Reuse and Recycle of the waste. China's laws are too principled to compare with those of the United States and Japan. For instance, in the Municipal Solid Waste Management Measures, although reference is made to the MSW classification, there is no indication as to how it will be classified, which makes the operation of the law less feasible. But in the relevant policies, there are more specific regulations on the process of waste classification. In 2018, Chengdu Municipal Solid Waste Classification Implementation Scheme made detailed requirements for local waste classification standards, management responsibilities and recycling methods.

From the point of the incentives and penalties for residents: Taking the United States as an example, Philadelphia introduced the source classification of domestic waste incentive system in 2004, which has had a good effect. Meanwhile, Seattle's 2005 Waste Recycling Act imposed a 50 USD fine if the amount of recyclable waste in household waste exceeded 10%. In Japan, more emphasis is placed on punitive measures. For instance, for littering persons, a penalty of up to five years in prison and a fine of 10 million yen are imposed by the Waste Disposal Act. The law also requires citizens to actively report. At a mention of incentives, it through tax incentives, the implementation of compensation system or other ways. However, there is a lack of corresponding mechanisms in China. In the Implementation Plan of Household Waste Classification System, the incentive mechanism is implemented as an innovative system. Penalties are missing in laws or related policies.

Finally, as for the publicity and education for the residents: although the relevant laws of the United States do not attach direct emphasis to waste classification education for the public, the U.S. Environmental Protection Agency, the nation's top agency for waste and waste disposal, is in charge of it. Japan is more prominent in the propaganda and education of its residents with a whole system which the government takes the lead and all sectors of society respond positively to. For instance, Kawasaki Municipality popularizes waste classification and emission through the Kawasaki Municipal waste and Resource Classification and Disposal Method Manual. In China, the Central Government has made great strides in identifying, treating and educating the public about waste classification; thus, the residents' awareness is in growth. This work has been emphasized in the policies of both the central and local governments.

The comparison of the WCLP systems of the United States, Japan and China is shown in Table 1.

Table 1. Measurement model results and discriminant validity and correlations

Country	Recycling system (mandates) $(+, \pm, -)$	Inducement		Education (capacity-building) $(+, \pm, -)$
		Incentive $(+, \pm, -)$	Punishment $(+, \pm, -)$	
The United States	+	+	$\pm$	$\pm$
Japan	+	$\pm$	+	+
China	$\pm$	-	-	+

The table reflects observation by authors. (+), means strong; ($\pm$), means middle; (−), means weak

5   Conclusions and Suggestions

Clear recycling systems are the basis and prerequisite for effective MSW management [23]. The laws and policies of each country can ensure a complete recycling system to some extent. Nevertheless, as a developing country, the implementation of waste classification in China is still in infancy, and the regulations on recycling process and system are still lacking in seriousness.

The participation of residents and the effect of recycling have a positive relationship [13]. Both incentives and penalties can stimulate residents. The United States is more likely to use positive incentives to encourage residents to classify and recycle waste. Japan choose to use strict penalties to make residents aware of the necessity of circular economy. China has not done enough in this aspect, leading to a missing link in WCLP system.

Regarding the publicity and education of waste classification to the residents, Japan and China attach great importance to the popularization and promotion of waste classification knowledge. In the United States, this is not emphasized in law or policy, but as a duty of the government agency.

In the 21st century, along with the rapid growth of the economic, the MSW has become a global problem. Except to reducing waste generation from the source, recycling can also solve this problem [10]. Governments all over the world have enacted laws and policies for recycling waste in order to reduce waste of resources and create an environment for sustainable development [3]. A complete legal policy system with seriousness, practicability and maneuverability, needs to be continuously supplemented and perfected in practice. Through the comparative analysis of the three countries, we can see that for some developed countries such as the United States and Japan, where the waste classification started earlier, the relevant law and policy systems are relatively complete and have a focus on national conditions. While in some developing countries, such as China, waste classification is continuously being implemented, some aspects are still missing.

From the macro level, as the basis of waste classification, a clear recycling system is necessary. Countries should establish strict recycling systems in their laws and/or policies as a link to all parties involved in waste classification. As the United States has done, recycling waste into a complete industrial chain. Then, an effective incentive and/or punishment measure is the effective motive power for all people participating in waste classification. China's lack of this measure leads to a lack of incentive for residents taking the initiative to classify waste. Finally, in terms of publicity and education, Japan is a good example. The concept and consciousness of waste classification have become popular in Japan, and more importantly, Japan has not slackened its efforts in this regard. Good publicity and education can promote the subjective initiative of residents, greatly improve the effect of classification. Thus, in the process of waste classification management, countries should form a cooperative mechanism with citizen participation as the center and the government and society participating actively. This mechanism should be based on a variety of waste classification publicity and education, with a clear legal system of waste classification as the guarantee. And it should take strict punishment measures as external pressure, take effective incentive policy as the motive force.

Both developed and developing countries have the same attitude towards the current environmental problems [15]. Under the situation of the inevitable MSW increase in the world, how to carry out the waste classification more effectively will be the problem that all the countries in the world have to face. The method proposed in this paper may an effective approach in waste classification. And even further, in the process of global circular economy development in the future, every country needs to make continuous efforts.

Acknowledgements. This research is supported by the Humanity and Social Science Youth Foundation of Ministry of Education of China (Grant No. 17YJC630096).

References

1. Apip, A.: Study of law development policy in government act liability as state administrative tools. Public Policy Adm. Res. **4**(4), 76–84 (2014)
2. Chen, F., Luo, Z. et al.: Enhancing municipal solid waste recycling through reorganizing waste pickers: A case study in nanjing, china. Waste Manag. Res. J. Int. Solid Wastes Public Clean. Assoc. Iswa 734242X18766216 (2018)
3. Chen, M.F., Tung, P.J.: The moderating effect of perceived lack of facilities on consumers' recycling intentions. Environ. Behav. **42**(6), 824–844 (2010)
4. Courtenayhall, P., Rogers, L.: Gaps in mind: Problems in environmental knowledge-behaviour modelling research. Environ. Educ. Res. **8**(3), 283–297 (2002)
5. Daniel, H., Perinaz, B.T., Chris, K.: Environment: Waste production must peak this century. Nature **502**(7473), 615–7 (2013)
6. Elmore, R.F.: Instruments and strategy in public policy. Rev. Policy Res. **7**(1), 174–186 (2010)

7. Gamba, R.J., Oskamp, S.: Factors influencing community residents' participation in commingled curbside recycling programs. Environ. Behav. **26**(5), 587–612 (1994)
8. Grazhdani, D.: Assessing the variables affecting on the rate of solid waste generation and recycling: An empirical analysis in prespa park. Waste Manag. **48**, 3–13 (2016)
9. Han, H., Zhang, Z., Xia, S.: The crowding-out effects of garbage fees and voluntary source separation programs on waste reduction: Evidence from china. Sustainability **8**, 1–17 (2016)
10. Hopper, J.R., Yaws, C.L., et al.: Waste minimization by process modification. Waste Manag. **13**(3C4), 349–350 (1992)
11. Kollmuss, A.J.A.: Mind the gap: Why do people act environmentally and what are the barriers to pro-environmental behavior? Environ. Educ. Res. **8**(3), 239–260 (2002)
12. Peretz, J.H.: Waste management agenda setting: A case of incorrect problem definition? Waste Manag. Res. **16**(3), 202–209 (1998)
13. Peretz, J.H., Tonn, B.E., Folz, D.H.: Explaining the performance of mature municipal solid waste recycling programs. J. Environ. Plan. Manag. **48**(5), 627–650 (2005)
14. Pietzsch, N., Ribeiro, J.L.D., Medeiros, J.F.D.: Benefits, challenges and critical factors of success for zero waste: A systematic literature review. Waste Manag. **67**, 324 (2017)
15. Rauwal, M.C.F.K.S.: Environmental attitudes as predictors of policy support across three countries. Environ. Behav. **34**(6), 709–739 (2002)
16. Sidique, S.F., Joshi, S.V., Lupi, F.: Factors influencing the rate of recycling: An analysis of minnesota counties. Resour. Conserv. Recycl. **54**(4), 242–249 (2010)
17. Starr, J., Nicolson, C.: Patterns in trash: Factors driving municipal recycling in massachusetts. Resour. Conserv. Recycl. **99**, 7–18 (2015)
18. Steg, L., Vlek, C.: Encouraging pro-environmental behaviour: An integrative review and research agenda. J. Environ. Psychol. **29**(3), 309–317 (2009)
19. Storey, D., Santucci, L., Fraser, R., Aleluia, J., Chomchuen, L.: Designing effective partnerships for waste-to-resource initiatives: Lessons learned from developing countries. Waste Manag. Res. **33**(12), 1066–1075 (2015)
20. Tonglet, M., Phillips, P.S., Read, A.D.: Using the theory of planned behaviour to investigate the determinants of recycling behaviour: A case study from brixworth, uk. Resour. Conserv. Recycl. **41**(3), 191–214 (2004)
21. Wan, C., Shen, G.Q., Yu, A.: The role of perceived effectiveness of policy measures in predicting recycling behaviour in hong kong. Resour. Conserv. Recycl. **83**(83), 141–151 (2014)
22. Wan, C., Shen, G.Q., Yu, A.: Key determinants of willingness to support policy measures on recycling: A case study in hong kong. Environ. Sci. Policy **54**, 409–418 (2015)
23. Wen, X., Luo, Q., et al.: Comparison research on waste classification between china and the eu, Japan, and the USA. J. Mater. Cycles Waste Manag. **16**(2), 321–334 (2014)
24. Wilson, D.C., Velis, C.A.: Waste management - still a global challenge in the 21st century: An evidence-based call for action. Waste Manag. Res. J. Int. Solid Wastes Public Clean. Assoc. Iswa **33**(12), 1049 (2015)
25. Xin, T., Tao, D.: The rise and fall of a "waste city" in the construction of an "urban circular economic system": The changing landscape of waste in beijing. Resour. Conserv. Recycl. **107**, 10–17 (2016)
26. Yong, C.W., Lian, F.X.: Analysis of the barrier factors of municipal solid waste classification recycling. Adv. Mater. Res. **726–731**, 2618–2621 (2013)

Design and Implementation of Sustainable Supply Chain Model with Various Distribution Channels

YoungSu Yun[1(✉)], Anudari Chuluunsukh[1], and Mitsuo Gen[2]

[1] Department of Business Administration, Chosun University,
Gwangju, South Korea
ysyun@chosun.ac.kr
[2] Fuzzy Logic Systems Institute and Tokyo University of Science,
Fukuoka, Japan

Abstract. In this paper, we propose a sustainable supply chain (SSC) model with various distribution channels. For constructing the SSC model, (1) the minimization of total cost as economic issue, (2) the minimization of total amount of CO_2 emission as environmental issue, and (3) the maximization of total social influence as social issue are considered. Since the SSC model should have various distribution channels, (1) normal delivery, (2) direct delivery, and (3) direct shipment are also taken into consideration in it. A mathematical formulation is proposed to design the SSC model and it is implemented using hybrid genetic algorithm (pro-HGA) approach. In numerical experiments, several scales of the SSC model are presented and they are used to compare the performance of the pro-HGA approach with those of some conventional GA and HGA approaches. Experimental results prove that the pro-HGA approach is more efficient in solving the SSC model than the other competing approaches.

Keywords: Sustainable supply chain model · Economic · Environmental and social issues · Distribution channel · Hybrid genetic algorithm approach · Genetic algorithms

1 Introduction

One of the reasons for constructing supply chain (SC) model is to efficiently produce and distribute materials and products through the network of the SC model. Considering various distribution channels would become one of the options to reinforce the efficiency of production and distribution in the SC model. In general, three types of distribution channels (normal delivery: ND, direct delivery: DD, and direct shipment: DS) are considered in the SC model.

The ND, as a basic distribution channel, is generally used to send materials or products from a stage to the next one. For example, products are sent from manufacturer to distribution center (DC) in the SC model which consists of

© Springer Nature Switzerland AG 2020
J. Xu et al. (Eds.): ICMSEM2019 2019, AISC 1002, pp. 469–482, 2020.
https://doi.org/10.1007/978-3-030-21255-1_36

supplier, manufacturer, DC, retailer, and customer. However, some products can be directly sent from DC to customer not via retailer, and this kind of distribution channel is called as the DD. Another products can be also directly shipped from manufacturer to customer not via DC and retailer, which is called as the DS. The efficiency of the SC model with the three types of distribution channels mentioned above has been proved in many conventional literatures [4,5,9,9,12,21,21] suggested the SC model with various distribution channels. Each channel (i.e., the ND, DD and DS) was used to analyze the influence on product price and lead time. The experimental results proved that the duplicated use of the ND, DD and DS outperforms the single use of them in controlling product price and lead time. Chen et al. [1] proposed a SC model with various types of distribution channels. In first type, the ND is only considered in the SC model. All of the ND, DD, and DS are used in it as second types. In numerical experiments, the performance of each type was compared with each other, and experimental results showed that the SC model with the NN, DD, and DS is more efficient than that with the ND alone. Similar to Chen et al. [1], Yun and Chuluunsukh [21] also suggested the SC model which is composed of suppler, manufacturer, DC, retailer and customer. In the suggested SC model, the ND, DD, and DS are used for transporting products from a stage to the next one.

Among the main streams in the SC model, the principles of sustainable development have been mentioned in many literatures [2,3,14,19,20]. For the principles, three issues (i.e., economic, environmental and social issues) have been usually considered in constructing sustainable SC (SSC) model effectively. Economic issue is to consider either the minimization of the total cost or the maximization of the total profit resulting from the operation process of the SSC models [2,14,14,20] proposed the SSC model which consists of factory, warehouse, client and recovery center. In their proposed SSC model, the total cost (= total fixed cost + total transportation cost + total production cost + total human resource cost) as an objective is minimized. Arampantzi and Minis [2] suggested the SSC model with supplier, plant, DC, and customer for minimizing the total cost (= total investment cost + total operational cost). Similar to Mota et al. [14] and Arampantzi and Minis [2], Varsei and Polyakovskiy [20] also proposed the SSC model for wine industry in Australia. They considered supplier, winery, bottling plant, DC and demand center for constructing the SSC model. The minimization of the total cost (= total fixed cost + total transportation cost + total production cost + total purchasing cost + total storage cost) is taken into consideration. For various distribution channels, the DD which directly transports materials from supplier to bottling plant not via winery is considered, including the ND.

In environmental issue, the total amount or cost of CO_2 which is emitted when transporting and producing materials and products in the SSC model was usually considered to be minimized [2,15,19]. Varsei and Polyakovskiy [20] minimized the total amount of CO_2 generated by all involved facilities in the

SSC model. özceylan et al. [15] proposed the SSC model for automotive industry in Turkey. In the proposed SSC model, they considered the minimization of total amount of CO_2 emitted during transporting automotive and automotive-related components among each stage. In Arampantzi and Minis [2], total cost of CO_2 emission during production and transportation activities at each stage was minimized.

Social issue is to consider various social influences such as the number of created jobs at facilities, unemployment rate, the lost day caused by work's damages, and so on. [2,14,20]. Mota et al. [14] maximized the social issue which is composed of the number of jobs created at a facility, unemployment rate, population density and income distribution. Varsei and Polyakovskiy [20] maximized the social issue in terms of a set of social categories such as employment or impact on regions in the SSC model. Arampantzi and Minis [2] considered the minimization of the objective related to the generation of work opportunities which contribute to local community development, relate to less developed countries, improve employee satisfaction, and contribute to stable employment.

As mentioned above, conventional works have considered various distribution channels and three issues (economic, environmental and social issues) into their SC or SSC models. Unfortunately, however, most of them, except for Varsei and Polyakovskiy [20], did not consider the various distribution channels and three issues simultaneously in their works.

In this paper, we propose the SSC model with various distribution channels and three issues simultaneously. The proposed SSC model consists of suppliers, manufacturers, DCs, retailers, and customers at each stage. For various distribution channels, the ND, DD, and DS are used for transporting materials and products from a stage to the others. As economic and environmental issues, the total cost and the total amount of CO_2 emission when operating the SSC model are considered and minimized as objectives, respectively. Total social influence as a social issue is considered and maximized as another objective. Since three issues are considered as objectives simultaneously, the proposed SSC model is a multi-objective optimization problem. In Sect. 2, a conceptual structure of the proposed SSC model is showed. The mathematical formulation for the proposed SSC model is represented in Sect. 3. A hybrid genetic algorithm (pro-HGA) approach with genetic algorithm (GA), Cuckoo search (CS) and iterative hill climbing method (IHC) is proposed for implementing the mathematical formulation in Sect. 4. Numerical experiments using several scales of the proposed SSC model are presented and compared the performance of the pro-HGA approach with those of some conventional competing approaches in Sect. 5. Finally, some conclusions and future study direct are briefly summarized and suggested in Sect. 6, respectively.

2   Proposed SSC Model

The conceptual structure of the SSC model is shown in Fig. 1. The supplier sends materials to the manufacturer using the ND. At the manufacturer, products using materials are produced and some products are sent to the DC with $\alpha_1\%$ using the ND and the others are directly sent to the customer with $\alpha_2\%$ using the DS. At the DC, some products are sent to the retailer with $\beta_1\%$ using the ND and the others are directly sent to the customer with $\beta_2\%$ using the DD. The retailer sends products to the customer using the ND. Figure 1 also shows the use of three issues. As an economic issue, the total cost (= total fixed cost at the supplier, manufacturer, DC and retailer + total transportation costs at all distribution channels + total handling costs at the supplier, manufacturer, DC and retailer) are considered. The total amount of CO_2 emitted when producing products at the manufacturer and when transporting materials and products at all distribution channels are taken into account as an environmental issue. The numbers of created job opportunities, unemployment and the lost days caused by work's damages at the manufacturer are used as social issues.

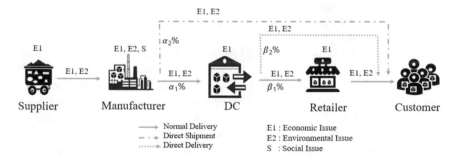

Fig. 1. A conceptual structure of the proposed SSC model

3   Mathematical Formulation

Some assumptions are considered for implementing the SSC model Yun and Chuluunsukh [21] as follows:

(1) Single type of product is only considered.

(2) The numbers of facilities considered at each stage of suppliers, manufacturers, DCs, retailers and customers are fixed and already known.

(3) Only one facility should be opened at each stage of suppliers, manufacturers, DCs, and retailers. However, all the customers are always opened.

(4) The fixed costs for operating the facilities considered at each stage of suppliers, manufacturers, DCs, and retailers are different and already known.

(5) The unit handling costs in the facilities considered at each stage of suppliers, manufacturers, DCs, and retailers are different and already known.

(6) The unit transportation costs considered between each stage are different and already known.

(7) The unit CO2 amounts emitted from transporting and producing materials and products at each stage are different and already known.

(8) The numbers of job opportunities, unemployment and the lost days at manufacturers are already known.

(9) The SSC model proposed in this paper is considered under steady-state situation.

Based on the above-mentioned assumptions, we define Index & Set, parameters, and decision variables as follows for designing the mathematical programming model for the SSC model:

Indices and Sets

s : index of supplier, $s \in S$.
m : index of manufacturer, $m \in M$.
d : index of DC, $d \in D$.
r : index of retailer, $r \in R$.
c : index of customer, $c \in C$.

Parameters

f_s : fixed cost at suppler s.
f_m : fixed cost at manufacturer m.
f_d : fixed cost at DC d.
f_r : fixed cost at retailer r.
h_s : unit handling cost at supplier s.
h_m : unit handling cost at manufacturer m.
h_d : unit handling cost at DC d.
h_r : unit handling cost at retailer r.
t_{sm} : unit transportation cost from suppler s to manufacturer m.
t_{md} : unit transportation cost from manufacturer m to DC d.
t_{mc} : unit transportation cost from manufacturer m to customer c.
t_{dr} : unit transportation cost from DC d to retailer r.
t_{dc} : unit transportation cost from DC d to customer c.
t_{rc} : unit transportation cost from retailer r to customer c.
q_{sm} : quantity transported from supplier s to manufacturer m.
q_{md} : quantity transported from manufacturer m to DC d.
q_{mc} : quantity transported from manufacturer m to customer c.
q_{dr} : quantity transported from DC d to retailer r.
q_{dc} : quantity transported from DC d to customer c.
q_{rc} : quantity transported from retailer r to customer c.
d_{sm} : distance between supplier s and manufacturer m.
d_{md} : distance between manufacturer m and DC d.

d_{mc} : distance between manufacturer m and customer c.
d_{dr} : distance between DC d and retailer r.
d_{dc} : distance between DC d and customer c.
d_{rc} : distance between retailer r and customer c.
c_s : capacity at supplier s.
c_m : capacity at manufacturer m.
c_d : capacity at DC d.
c_r : capacity at retailer r.
c_c : capacity at customer c.
$CO2_{et}$: CO_2 amount emitted in a vehicle per kilometer.
$CO2_{em}$: unit CO_2 amount emitted from manufacturing process at manufacturer m.
V_c : capacity that can be shipped in a vehicle.
w_{jo} : weight given to the created job opportunities.
w_{ue} : weight given to unemployment.
w_{ld} : weight given to the lost days caused by work's damages.
jo_m : number of the created job opportunities when manufacturer m uses technology t.
ld_m : lost days caused by work's damages during the use of technology t at manufacturer m.
ue_m : number of unemployment at manufacturer m.

Decision variables

x_s : takes the value of 1 if supplier s is opened or 0 otherwise.
x_m : takes the value of 1 if manufacturer m is opened or 0 otherwise.
x_d : takes the value of 1 if DC d is opened or 0 otherwise.
x_r : takes the value of 1 if retailer r is opened or 0 otherwise.
t_m : takes the value of 1 if technology t is used at manufacturer m or 0 otherwise.

First objective function $F_1(x)$ as an economic issue is to minimize total cost.

$$Min. \ F_1(x) = Total \ fixed \ cost \ (TFC) + Total \ handling \ cost \ (THC) \tag{1}$$
$$+ Total \ transportation \ cost \ (TTC),$$

$$TFC = \sum_s f_s \cdot x_s + \sum_m f_m \cdot x_m + \sum_d f_d \cdot x_d + \sum_r f_r \cdot x_r, \tag{2}$$

$$THC = \sum_s h_s \cdot c_s \cdot x_s + \sum_m h_m \cdot c_m \cdot x_m + \sum_d h_d \cdot c_d \cdot x_d + \sum_r h_r \cdot c_r \cdot x_r, \tag{3}$$

$$TTC = \sum_s \sum_m t_{sm} \cdot c_s \cdot x_s \cdot x_m + \sum_m \sum_d t_{md} \cdot c_m \cdot \alpha_1\% \cdot x_m \cdot x_d + \sum_m \sum_c t_{mc} \cdot c_m \cdot \alpha_2\% \cdot x_m$$
$$+ \sum_d \sum_r t_{dr} \cdot c_d \cdot \beta_1\% \cdot x_d \cdot x_r \sum_d \sum_c t_{dc} \cdot c_d \cdot \beta_2\% \cdot x_d + \sum_r \sum_c t_{rc} \cdot c_r \cdot x_r. \tag{4}$$

The TFC is calculated by the sum of the costs when the facilities at suppliers, manufacturers, DCs, and retailers are opened and established. The THC is the sum of the costs when materials or products at the facilities opened at suppliers, manufacturers, DCs, and retailers are handled. The TTC is the sum of the costs when materials and products are transported between each stage.

Second objective function $F_2(x)$ as an environmental issue is to minimize total amount of CO_2 emission.

$$Min.F_2(x) = Total\ amount\ of\ CO_2\ emission\ in\ transportation\ process\,(TCT) \tag{5}$$

$$+ Total\ amount\ of\ CO_2\ emission\ in\ production\ process\,(TCP)\,,$$

$$\begin{aligned}
TCT = &\left(\sum_s \sum_m d_{sm} \cdot x_s \cdot x_m\right) \cdot \left(\tfrac{c_s}{V_c}\right) \cdot CO_{2et} + \left(\sum_m \sum_d d_{md} \cdot x_m \cdot x_d\right) \cdot \left(\tfrac{c_m \cdot \alpha_1 \%}{V_c}\right) \cdot CO_{2et} \\
&+ \left(\sum_m \sum_c d_{mc} \cdot x_m\right) \cdot \left(\tfrac{c_m \cdot \alpha_2 \%}{V_c}\right) \cdot CO_{2et} + \left(\sum_d \sum_r d_{dr} \cdot x_d \cdot x_r\right) \cdot \left(\tfrac{c_d \cdot \beta_1 \%}{V_c}\right) \cdot CO_{2et} \\
&+ \left(\sum_d \sum_c d_{dc} \cdot x_d\right) \cdot \left(\tfrac{c_d \cdot \beta_2 \%}{V_c}\right) \cdot CO_{2et} + \left(\sum_r \sum_c d_{rc} \cdot x_r\right) \cdot \left(\tfrac{c_r}{V_c}\right) \cdot CO_{2et}\,,
\end{aligned} \tag{6}$$

$$TCP = \sum_m c_m \cdot x_m \cdot CO_{2em}. \tag{7}$$

The TCT stands for the total amount of CO_2 emitted during transportation process between each stage. The TCP means the total amount of CO_2 emitted during production process at manufacturers. In Eq. 6, the parameters values of V_c and $CO_2 et$ are randomly generated according to Paksoy et al. [15], özceylan et al. [16].

Third objective function $F_3(x)$ as a social issue is to maximize social influence.

$$\begin{aligned}
max.F_3(x) = &(weight\ given\ to\ the\ created\ job\ opportunity * number\ of\ job \\
&opportunities\ at\ manufacturer) - (weight\ given\ to\ unemployment \\
&* number\ of\ unemployment\ at\ manufacturer) \\
&- (weight\ given\ to\ the\ lost\ days\ caused\ by\ works\ damage \\
&* number\ of\ lost\ day\ caused\ by\ works\ damages\ at\ manufacturer) \\
= &(w_{jo} \cdot \sum_m jo_m \cdot x_m \cdot t_m) - (w_{ue} \cdot \sum_m ue_m \cdot x_m \cdot t_m) - (w_{ld} \cdot \sum_m ld_m \cdot x_m \cdot t_m)\,.
\end{aligned} \tag{8}$$

In Eq. 8, the parameter values for the created job opportunities, unemployment and the lost days caused by work's damages are randomly generated according to Devika et al. [6], Talaei et al. [18]. To optimize the three objective functions mentioned above, the following constraints should be taken into consideration.

$$\sum_s \sum_m q_{sm} \cdot x_s \cdot x_m - \sum_m c_m \cdot x_m \leq 0, \tag{9}$$

$$\sum_m \sum_d q_{md} \cdot \alpha_1 \% \cdot x_m \cdot x_d - \sum_d c_d \cdot x_d \leq 0, \tag{10}$$

$$\sum_m \sum_c q_{mc} \cdot \alpha_2 \% \cdot x_m - \sum_c c_c \leq 0, \tag{11}$$

$$\sum_d \sum_r q_{dr} \cdot \beta_1 \% \cdot x_d x_r - \sum_r c_r x_r \leq 0, \tag{12}$$

$$\sum_d \sum_c q_{mc} \cdot \beta_2 \% \cdot x_m - \sum_c c_c \leq 0, \tag{13}$$

$$\sum_r \sum_c q_{rc} \cdot x_r - \sum_c c_c \leq 0, \tag{14}$$

$$\sum_s x_s = 1, \tag{15}$$

$$\sum_{m} x_m = 1, \tag{16}$$

$$\sum_{d} x_d = 1, \tag{17}$$

$$\sum_{r} x_r = 1, \tag{18}$$

$$x_s = \{0,1\}, \quad \forall S, \tag{19}$$

$$x_m = \{0,1\}, \quad \forall M, \tag{20}$$

$$x_d = \{0,1\}, \quad \forall D, \tag{21}$$

$$x_r = \{0,1\}, \quad \forall R, \tag{22}$$

$$c_s, c_m, c_d, c_r, c_s \geq 0, \quad \forall s \in S, \forall m \in M, \forall d \in D, \forall r \in R, \forall c \in C. \tag{23}$$

Equations (9) to (14) stand for the quantity limitation for transportation between each stage. Equation (15) to (18) indicate that only one facility should be opened at each stage. Equations (19) to (22) show that each decision variable should take a value of 0 or 1. Equation (23) refers to non-negativity.

4 Pro-HGA Approach

As already known, most of complicated network problems including the SSC model have NP-complete nature [8,17], meta-heuristics approaches such as GA, CS and Tabu search have been adapted to find optimal solution in many literatures [7,13,22]. However, there exist many situations that most of conventional GA-based approaches do not well perform particularly [7]. Therefore, various hybrid approaches using GA and others have been developed [8,22]. In this paper, we also developed the pro-HGA approach to efficiently solve the SSC model in which it combines genetic algorithm (GA), Cuckoo search (CS) and iterative hill climbing method (IHC). The detailed implementation procedure of the pro-HGA approach in pseudo code is shown in Fig. 2.

5 Numerical Experiments

Three scales of the SSC network are presented for experimental comparison. The detailed information of the facilities considered at each stage is shown in Table 1. For example, in the scale 3 of Table 1, 60 suppliers, manufacturers, DCs, and retailers and 1 customer are considered. For comparing the performance of the pro-HGA approach, some conventional approaches are also used and their detailed information are summarized in Table 2.

All the approaches were programmed using MATLAB version 2014b and ran under a same computation environment (IBM compatible PC 1.3 Ghz processor-Intel core I5–1600 CPU, 4 GB RAM, and OS-X EI). The parameter settings for implementing them are as follows: total number of generations is 1,000, population size 20, crossover rate 0.5, and mutation rate 0.8. Number of host nest is 20, $\alpha = 1$, and $p_a = 0.25$ for the search of the CS used in the

procedure: pro-HGA approach
input: problem data, parameters
begin
　　$Gl_{best} = 0$
　　$t \leftarrow 0$　　// t: generation number
　　initialize parent population $P(t)$ by encoding routine;
　　evaluate $P(t)$ by decoding routine and keep the best solution GA_{best};
　　while (**not** termination condition)
　　　　produce offspring $O(t)$ from $P(t)$ by crossover and mutation routines;
　　　　evaluate $O(t)$ and locate the best solution GA_{best};
　　　　for each solution x_i in $O(t)$ **do**
　　　　　　generate a new solution x_{lev} from x_i by adapting Lévy flight routine;
　　　　　　randomly select a solution x_i in $O(t)$;
　　　　　　if $(F(x_{lev}) > F(x_i))$ **then** $C(t) \leftarrow x_{lev}$　　// $C(t)$: population of CS
　　　　end
　　　　worst solutions with fraction rate (fr_a) are abandoned;
　　　　randomly regenerate new solutions x_{new} as many as fr_a;
　　　　$C(t) \leftarrow x_{new}$
　　　　evaluate $C(t)$ and find the best solution CS_{best};
　　　　if $(F(GA_{best}) > F(CS_{best}))$ **then** $Gl_{best} \leftarrow GA_{best}$
　　　　　　else $Gl_{best} \leftarrow CS_{best}$
　　　　end
　　　　select the best solution x_v in $C(t)$;
　　　　randomly generate as many solutions as the population size in the neighborhood of x_v:
　　　　select the x_b with the best value among newly generated solutions;
　　　　if $(f(x_b) > f(x_v))$ **then** $C(t) \leftarrow x_b$
　　　　produce new $P(t+1)$ using $O(t)$ and $C(t)$ by adapt selection routine;
　　　　$t \leftarrow t+1$;
　　end
　　output: Gl_{best};
end;

Fig. 2. Detailed implementation procedure of the pro-HGA approach

Table 1. Three scales for experimental comparison

Scale	Supplier	Manufacturer	DC	Retailer	Customers
1	20	20	20	20	1
2	40	40	40	40	1
3	60	60	60	60	1

Table 2. Each approach for comparison

Approach	Description
GA	Conventional GA approach by Gen and Cheng [7]
HGA	Conventional HGA approach by Kanagaraj et al. [11]
pro-HGA	HGA approach with GA, CS and IHS proposed in this paper

HGA and pro-HGA approaches. The search range for the IHC is 0.5. These parameter values were obtained by fine tuning procedure of each approach. For mathematical formulation, some parameters are set as follows: $V_c = 10$, $CO2_{et} = 3$, $CO2_{em} = 0.5$, $w_{jo} = 0.1$, $w_{ld} = 0.1$, $w_{ue} = 0.1$, $jo_m = U[90, 100]$, $ld_m = U[20, 30]$, $ue_m = U[15, 20]$. Total 20 independent trials were carried out to eliminate the randomness of the search process of each approach. The performances of all the approaches are compared using various measures as shown in Table 3.

Table 3. Measure of performance

Measure	Description
$|S_j|$	Number of Pareto optimal solutions in reference solution set (S*) [10]
$R_{NDS}(S_j)$	Rates of Pareto optimal solutions in S* [10]
$DI_R(S_j)$	Average distance between Pareto optimal solutions and S* [10]
CPU time	Average CPU time after all trials

The values of $|S_j|$, $R_{NDS}(S_j)$ and $DI_R(S_j)$ shown in Table 3 are obtained after total 20 independent trials. For convenience of the analysis of experimental results, the mathematical formulation suggested in Sect. 3 is divided into three types (Problem 1, Problem 2 and Problem 3) as follows. The experimental results for each problem are summarized in Tables 4, 5 and 6.

Problem 1: min. $F_1(x)$ and min. $F_2(x)$
Problem 2: min. $F_1(x)$ and max. $F_3(x)$
Problem 3: min. $F_2(x)$ and max. $F_3(x)$

Table 4. Experimental results of each problem in Scale 1

Measure	Problem 1			Problem 2			Problem 3		
	GA	HGA	pro-HGA	GA	HGA	pro-HGA	GA	HGA	pro-HGA
$|S_j|$	2	0	3	1	0	1	2	0	2
$R_{NDS}(S_j)$	0.4	0	0.6	0.5	0	0.5	0.5	0	0.5
$DI_R(S_j)$	636	198	0	217	595	1,149	0	325	40
CPU time	6.24	6.28	6.57	6.3	6.28	6.57	6.24	6.28	6.57

In terms of the $|S_j|$ of Table 4, the number of the Pareto optimal solutions by the pro-HGA approach is greater than that by the GA approaches for the problem 1, which indicates that the search ability of the former is more efficient in terms of the S* than that of the latter. However, in the problems 2 and 3, the numbers by the pro-HGA approach are the same as those by the GA approach, and they all are greater than those by the HGA approach. The results of the $|S_j|$

Table 5. Experimental results of each problem in Scale 2

Measure	Problem 1			Problem 2			Problem 3		
	GA	HGA	pro-HGA	GA	HGA	pro-HGA	GA	HGA	pro-HGA
$\lvert S_j \rvert$	2	3	3	0	0	2	1	0	2
$R_{NDS}(S_j)$	0.26	0.37	0.37	0	0	1	0.33	0	0.67
$DI_R(S_j)$	311	0	390	786	808	0	41	836	1
CPU time	6.31	6.38	6.66	6.31	6.38	6.66	6.31	6.38	6.66

Table 6. Experimental results of each problem in Scale 3

Measure	Problem 1			Problem 2			Problem 3		
	GA	HGA	pro-HGA	GA	HGA	pro-HGA	GA	HGA	pro-HGA
$\lvert S_j \rvert$	2	1	3	1	2	2	0	1	4
$R_{NDS}(S_j)$	0.33	0.17	0.5	0.2	0.4	0.4	0	0.2	0.8
$DI_R(S_j)$	0	328	91	441	0	403	572	232	0
CPU time	6.5	6.51	6.82	6.5	6.51	6.82	6.5	6.51	6.82

have also influence on the rate of the Pareto optimal solutions in the S*, that is, in terms of the $R_{NDS}(S_j)$, the rate of the pro-HGA approach is higher than that of the GA approach in the problem 1, and those of the pro-HGA and GA approaches are higher than that of the HGA approach in the problems 2 and 3. In terms of the $DI_R(S_j)$, the pro-HGA approach shows significantly better result than the GA approach for the problem 1, since the shorter the average distance between the Pareto optimal solutions and the S* is, the better the approach is. However, the pro-HGA approach does not show any advantage against the GA and HGA approaches for the problems 2 and 3. In terms of the CPU time, all approaches have almost same search speed.

In terms of the $\lvert S_j \rvert$ of Table 5, the pro-HGA approach shows to be better performance than the GA and HGA approaches in the problems 2 and 3, though the result of the pro-HGA approach is equal to that of the GA approach in the problem 1. Similar results are also shown in terms of the $R_{NDS}(S_j)$ in the problems 1, 2, and 3. In terms of the $DI_R(S_j)$, the results of the pro-HGA approach are significantly better than those of the GA and HGA approaches in the problems 2 and 3, except for the problem 1.

In terms of the $\lvert S_j \rvert$ and $R_{NDS}(S_j)$ of Table 6, the results of the pro-HGA approach are superior to those of the GA and HGA approaches in the problems 1 and 3, though the pro-HGA approach has the same result with the HGA approach in the problem 2. In terms of the $DI_R(S_j)$, the GA approach is more efficient than the others in the problem 1, but the pro-HGA approach outperforms the GA and HGA approaches in the problems 2 and 3. Figures 3, 4 and 5 show the Pareto optimal solutions of the GA, HGA and pro-HGA approaches for the problems 1, 2 and 3.

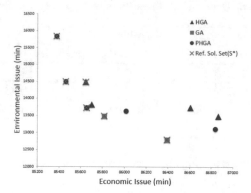

Fig. 3. Pareto optimal solutions of each approach when compared with the S* in Problem 1

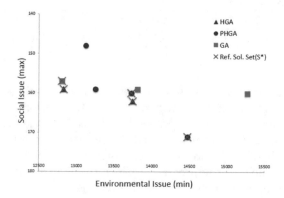

Fig. 4. Pareto optimal solutions of each approach when compared with the S* in Problem 2

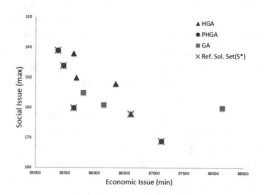

Fig. 5. Pareto optimal solutions of each approach when compared with the S* in Problem 3

In Fig. 3, we can see that two Pareto optimal solutions in the GA approach, one in the HGA approach, and three in the pro-HGA approach coincide with the S*, respectively. These results are the same as those of Table 6 in terms of the $|S_j|$. In Fig. 4, the number of the Pareto optimal solutions in the pro-HGA approach is the same as that in the HGA approach, and they outperform the GA approach. In Fig. 5, one Pareto optimal solution in the HGA approach and four in the pro-HGA approach are located in the S*, but in the GA approach, none is located in the S*.

Based on the experimental results of Tables 4, 5, 6 and Figs. 3, 4, 5, we can reach the following conclusion. In terms of the $|S_j|$, $R_{NDS}(S_j)$, and $DI_R(S_j)$, the pro-HGA approach has shown to be better performance than the GA and HGA approaches when the problem size increases from scales 1 to 3, though the former has not shown any merit in terms of the CPU time. In the Pareto optimal solutions of each approach when compared with the S*, the pro-HGA approach has located more Pareto optimal solutions than the GA and HGA approaches. This proves that the search scheme used in the pro-HGA approach is more efficient than those used in the GA and HGA approaches.

6 Conclusions

In this paper, we have shown the sustainable supply chain (SSC) model with three issues (economic, environmental and social issues) and various distribution channels simultaneously. The minimization of total cost, the minimization of total amount of CO_2 emission, and the maximization of total social influence have been considered for the economic, environmental and social issues, respectively. Three distribution channels (normal delivery, direct delivery, and direct shipment) have also taken into consideration in the SSC model. A mathematical formulation has been suggested to design the SSC model and it has been implemented using hybrid genetic algorithm (pro-HGA) approach. In numerical experiments, three scales of the SSC model have been presented to compare the performance of the pro-HGA approach with those of conventional GA and HGA approaches. Experimental results using various measures of performance have shown that the pro-HGA approach is more efficient than the other competing approaches.

However, since the scales used in the numerical experiments are relatively small sizes, larger scales will be used for comparing the performance of the pro-HGA approach. Also, more various hybrid approaches using GA with particle swarm optimization, Tabu search, Ant colony optimization, etc., will be considered. This will be left for future study.

References

1. Chen, X., Chuluunsukh, A., Yun, Y.S.: Design of closed-loop supply chain model with various transportation methods. In: Proceedings of APIEMS (Asia Pacific Industrial Engineering and Management Systems) Conference: pp. 18–23 (2017)

2. Arampantzi, C., Minis, I.: A new model for designing sustainable supply chain networks and its application to a global manufacturer. J. Clean. Prod. **156**, 276–292 (2017)
3. Barbosa-Póvoa, A.P., da Silva, C., Carvalho, A.: Opportunities and challenges in sustainable supply chain: An operations research perspective. Eur. J. Oper. Res. **268**(2), 399–431 (2018)
4. Chiang, W.K., Monahan, G.E.: Managing inventories in a two-echelon dual-channel supply chain. Eur. J. Oper. Res. **162**(2), 325–341 (2005)
5. Chiang, W.K., Chhajed, D., Hess, J.D.: Direct marketing, indirect profits: A strategic analysis of dual-channel supply-chain design. Manag. Sci. **49**(1), 1–20 (2003)
6. Devika, K., Jafarian, A., Nourbakhsh, V.: Designing a sustainable closed-loop supply chain network based on triple bottom line approach: A comparison of meta-heuristics hybridization techniques. Eur. J. Oper. Res. **235**(3), 594–615 (2014)
7. Gen, M., Cheng, R.: Genetic Algorithms and Engineering Optimization, vol. 7. Wiley, New York (2000)
8. Gen, M., Lin, L., et al.: Recent advances in hybrid priority-based genetic algorithms for logistics and scm network design. Comput. Ind. Eng. **125**, 394–412 (2018)
9. Hua, G., Wang, S., Cheng, T.E.: Price and lead time decisions in dual-channel supply chains. Eur. J. Oper. Res. **205**(1), 113–126 (2010)
10. Ishibuchi, H., Yoshida, T., Murata, T.: Balance between genetic search and local search in memetic algorithms for multiobjective permutation flowshop scheduling. IEEE Trans. Evol. Comput. **7**(2), 204–223 (2003)
11. Kanagaraj, G., Ponnambalam, S., Jawahar, N.: A hybrid cuckoo search and genetic algorithm for reliability-redundancy allocation problems. Comput. Ind. Eng. **66**(4), 1115–1124 (2013)
12. Lin, L., Gen, M., Wang, X.: Integrated multistage logistics network design by using hybrid evolutionary algorithm. Comput. Ind. Eng. **56**(3), 854–873 (2009)
13. Min, H., Ko, H.J., Ko, C.S.: A genetic algorithm approach to developing the multi-echelon reverse logistics network for product returns. Omega **34**(1), 56–69 (2006)
14. Mota, B., Gomes, M.I., et al.: Towards supply chain sustainability: Economic, environmental and social design and planning. J. Clean. Prod. **105**, 14–27 (2015)
15. Özceylan, E., Demirel, N., et al.: A closed-loop supply chain network design for automotive industry in turkey. Comput. Ind. Eng. **113**, 727–745 (2017)
16. Paksoy, T., Bektaş, T., Özceylan, E.: Operational and environmental performance measures in a multi-product closed-loop supply chain. Transp. Res. Part E: Logist. Transp. Rev. **47**(4), 532–546 (2011)
17. Savaskan, R.C., Bhattacharya, S., Van Wassenhove, L.N.: Closed-loop supply chain models with product remanufacturing. Manag. Sci. **50**(2), 239–252 (2004)
18. Talaei, M., Moghaddam, B.F., et al.: A robust fuzzy optimization model for carbon-efficient closed-loop supply chain network design problem: a numerical illustration in electronics industry. J. Clean. Prod. **113**, 662–673 (2016)
19. Taticchi, P., Garengo, P., et al.: A review of decision-support tools and performance measurement and sustainable supply chain management. Int. J. Prod. Res. **53**(21), 6473–6494 (2015)
20. Varsei, M., Polyakovskiy, S.: Sustainable supply chain network design: A case of the wine industry in australia. Omega **66**, 236–247 (2017)
21. Yun, Y.S., Chuluunsukh, A.: Environmentally-friendly supply chain network with various transportation types. J. Glob. Tour. Res. **3**(1), 17–24 (2018)
22. Yun, Y.S., Chuluunsukh, A., Chen, X.: Hybrid genetic algorithm for optimizing closed-loop supply chain model with direct shipment and delivery. New Phys.: Sae Mulli **68**, 683–692 (2018)

A Multiple Decision-Maker Model for Construction Material Supply Problem Based on Costs-Carbon Equilibrium

Rongwei Sun[1], Shuhua Hou[1], and Rui Qiu[1,2(✉)]

[1] Uncertainty Decision-Making Laboratory, Sichuan University, Chengdu 610064,
People's Republic of China
[2] School for Environment and Sustainability, University of Michigan, Ann Arbor,
Michigan 48109-1041, USA
qiuruicd@scu.edu.cn

Abstract. The rapid urbanization in China over the last decade has resulted in increased carbon emissions from the construction materials industry. Considering significant proportion of material purchase in total project costs and the massive amount of carbon emissions during construction material produce and transportation process, the material supplier selection has significant influence on the construction performance. This paper develops an multi-objective equilibrium strategy based bi-level programming model which fully reflects the conflicts between the construction company and the material suppliers and considers the trade-off between cost optimization and carbon emission reduction. An interactive solution approach with fuzzy random variables is designed. A practical case is given to demonstrates the efficiency of the formulated method.

Keywords: Construction material supply · Carbon emission reduction · Equilibrium strategy · Construction materials

1 Introduction

Construction materials are the basis of construction projects which account for 50 to 70% of a project's total costs. Therefore, the supply of construction materials is one of the most important in construction project, playing an increasingly important role on the efficient operation of the construction supply chain [12]. It was found that suppliers selection costs generally make up a significant percentage of the total industry supply chain costs, which shows that supplier selection plays such a critical role on the achievement of construction management objectives. Therefore, supplier selection has attracted significant research attention [2,5]. Raut, R D. et al. proposed a combined MCDM methodology in which Fuzzy analytic hierarchy process and linear programming are utilised for assigning weights of the criteria for supplier selection [9]. Safa, M. et al. develop an integrated construction materials management model to address the challenges

© Springer Nature Switzerland AG 2020
J. Xu et al. (Eds.): ICMSEM2019 2019, AISC 1002, pp. 483–491, 2020.
https://doi.org/10.1007/978-3-030-21255-1_37

in procurement and management of construction materials process by deploying principles of feasible materials management networks and a supplier selection process [10].

Meanwhile, an analysis of China's construction industry from 2005 to 2012 found that the construction sector carbon emissions had increased considerably since 2010 and that 73% of these being attributed to the construction materials produce and transportation stages. Therefore, carbon emissions mitigation methods for the produce and transportation of construction material are urgently needed [15].

In the construction material supply process, both of the construction company and material suppliers involve in decision-making process. However, the objectives of upper-level construction company are to achieve total costs minimization and carbon emission reduction under environmental protection policy while suppliers always pursuit for the total profits maximization, which will cause conflicts between different decision makers [3]. Equilibrium strategy, which has been used in many other fields, has been proved to be a effective tool to deal with such problems [1,7]. In this paper, those successful researches motivated to use equilibrium strategy based method to resolve the construction material supplier selection problem to guarantee the sustainable development of construction industry. Previous studies indicate that the bi-level programming is one of the most suitable tools for describing equilibrium strategy, which can fully consider multiple decision makers [11]. According to the above analysis, this paper proposes a equilibrium strategy based bi-level model to achieve trade-off between economy performance and carbon emission reduction under an uncertain environment.

Compared with previous researches, this paper fully considers the conflicts between the construction company and material suppliers and proposes an approach which integrated a bi-level model and uncertainty theory to effectively realize a equilibrium between stakeholders and deal with the uncertainties.

2   Modeling

Objective for construction company

As the upper-level decision maker, one of the most important objective for construction company is to minimize the total costs, which consist of procurement costs, inventory costs and transportation costs. Let $\widetilde{P}_{ij}$ be the unit material price of supplier i, W_j be the unit inventory price of project j, T_{ij} be the unit transportation price from supplier i to project j and x_{ij} be the decision variable that presents the purchase quantity decided by construction company. from the above, the total costs function of construction company is as follow:

$$\sum_{i=1}^{I}\sum_{j=1}^{J}P_i x_{ij} + \sum_{i=1}^{I}\sum_{j=1}^{J}W_j x_{ij} + \sum_{i=1}^{I}\sum_{j=1}^{J}T_{ij} x_{ij} \tag{1}$$

Under supervision of the department of environment protection, CC also needs to take carbon emission reduction into consideration. Let $\widetilde{CE}_i$ be the carbon emission coefficient of supplier i to produce a unit of product, $\widetilde{CT_{ij}}$ be transportation carbon emission coefficient from supplier i to project j. The total carbon emissions during the material supply process are:

$$\sum_{i=1}^{I}\sum_{j=1}^{J}\widetilde{CE_i}x_{ij} + \sum_{i=1}^{I}\sum_{j=1}^{J}\widetilde{CT_{ij}}x_{ij} \qquad (2)$$

Let D_j be the material demand of project j, so the purchase quantity constraint is

$$\sum_{i=1}^{I}x_{ij} \geq D_j. \qquad (3)$$

Let y_i be the decision variable that indicates if the supplier i is selected, K the number limitation of selected supplier, so there is

$$\sum_{i=1}^{I}y_i \leq K. \qquad (4)$$

Objective for supplier
As the lower-level decision maker, the objective of supplier i is to maximize the total profits, which is presented as sales revenue minus production costs and management costs. Let $\widetilde{H_i}$ be the unit production cost of supplier i and M_i be the management costs, the total profits of supplier i are as follow:

$$\sum_{i=1}^{I}\sum_{j=1}^{J}P_i x_{ij} - \sum_{i=1}^{I}\sum_{j=1}^{J}\widetilde{H_i}x_{ij} - \sum_{i=1}^{I}\sum_{j=1}^{J}M_i y_i \qquad (5)$$

Let C_i^L and C_i^U be the minimum and maximum production capacity of supplier i, so the sales quantity constraint is

$$C_i^L y_i \leq \sum_{j=1}^{J}x_{ij} \leq C_i^L y_i. \qquad (6)$$

Global model
There are many uncertain parameters in the proposed model including fuzzy randomness, which simultaneously describes both objective and subjective information as a fuzzy set of possible probabilistic models over a certain imprecision range, named fuzzy random variable (FRV). For instance, $\widetilde{CE_i}$ ranges from r_1 to r_3 with the most likely value r_2. Then, as parameter randomness can be expressed as a random distribution [6], the minimum value r_1 and the maximum value r_3 are selected as the left and right borders for the FRV $\widetilde{CE_i}$, with the most possible values for r_2 assumed to approximately follow Gaussian distribution, and then a double expected value operator (EVO) is employed to handle the FRV, which is presented as $E\left[\widetilde{CE_i}\right]$ [8, 13].

Let PE be the carbon emission data of history, α be the attitude of the authority towards carbon emissions reduction. To seek the trade-off between costs minimization and carbon emissions minimization, in this paper, the carbon emissions minimization objective is transformed into a constraint controlled within a range of acceptance [14]. The bi-level decision-making process is shown in Fig. 1.

$$\min F = \sum_{i=1}^{I}\sum_{j=1}^{J} \Gamma_i x_{ij} + \sum_{i=1}^{I}\sum_{j=1}^{J} W_j x_{ij} + \sum_{i=1}^{I}\sum_{j=1}^{J} T_{ij} x_{ij}$$

$$s.t. \begin{cases} \sum_{i=1}^{I}\sum_{j=1}^{J} E\left[\widetilde{CE_i}\right] x_{ij} + \sum_{i=1}^{I}\sum_{j=1}^{J} E\left[\widetilde{CT_{ij}}\right] x_{ij} \leq \alpha PE \\ \sum_{i=1}^{I} x_{ij} \geq D_j \\ \sum_{i=1}^{I} y_i \leq K \\ \max O_i = \sum_{i=1}^{I}\sum_{j=1}^{J} E\left[\widetilde{P_{ij}}\right] x_{ij} - \sum_{i=1}^{I}\sum_{j=1}^{J} E\left[\widetilde{H_i}\right] x_{ij} - \sum_{i=1}^{I}\sum_{j=1}^{J} M_i y_i \\ s.t. \left\{ C_i^L y_i \leq \sum_{j=1}^{J} x_{ij} \leq C_i^L y_i \right. \\ y_i = \{0,1\} \\ \forall i = 1,2,...,I. \\ \forall j = 1,2,...,J. \end{cases} \tag{7}$$

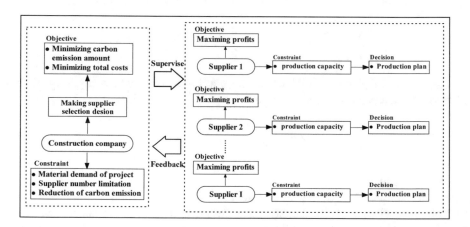

Fig. 1. The bi-level relationship between the construction company and the material suppliers

3   Case Study

To demonstrate the effectiveness and applicability of the proposed methodology, a practical construction materials supply problem in four typical material suppliers and two projects is given (Table 1).

3.1   Case Presentation and Data Collection

Before applying the modelling technique, practical parameter values were collected. The input data and parameters for the proposed model can be divided roughly into two categories: the crisp data and the uncertain data. The crisp data as shown in Tables 6, 2 and 3 are obtained from the information published by the companies and the Statistical Yearbook of the Chinese construction industry [4]. Uncertain parameters are estimated from expert experience which are considered as fuzzy parameters, as shown in Tables 4, 5. The storage cost and material demand of project 1 is 17 yuan/tonne and 424,780 thousand tonnes; the storage cost and material demand of project 2 is 19 yuan/tonne and 573,000 thousand tonnes. The number limitation of selected supplier K is 2. The upper limit of carbon emission PE is 2.65 million tonnes.

Table 1. Basic information for each supplier

Parameters	Supplier 1	Supplier 2	Supplier 3	Supplier 4
M_i (10^4 yuan)	12.890	10.570	11.230	12.154
C_i^L (10^4 tonnes)	28.941	32.571	23.768	29.696
C_i^U (10^4 tonnes)	68.952	71.384	65.135	70.826

Table 2. Distance between supplier i to project j

D_{ij} (km)	Supplier 1	Supplier 2	Supplier 3	Supplier 4
Project 1	42	35	36	37
Project 2	44	46	48	42

Table 3. Transportation cost between supplier i to project j

T_{ij} (km)	Supplier 1	Supplier 2	Supplier 3	Supplier 4
Project 1	37	48	41	39
Project 2	36	45	38	43

Table 4. Uncertainty parameters for each supplier

Parameters	Supplier 1	Supplier 2	Supplier 3	Supplier 4
$\widetilde{\widetilde{P_i}}$ (yuan/tonne)	(421,$\mathcal{N}$(468,21),483)	(387,$\mathcal{N}$(413,23),451)	(343,$\mathcal{N}$(382,21),397)	(411,$\mathcal{N}$(435,21),476)
$\widetilde{\widetilde{H_i}}$ (yuan/tonne)	(367,$\mathcal{N}$(382,24),403)	(299,$\mathcal{N}$(317,22),334)	(248,$\mathcal{N}$(267,23),288)	(305,$\mathcal{N}$(328,24),351)
$\widetilde{\widetilde{CE_i}}$ (kg/tonne)	(575,$\mathcal{N}$(630,25),684)	(652,$\mathcal{N}$(675,21),699)	(687,$\mathcal{N}$(712,23),751)	(635,$\mathcal{N}$(661,24),683)

Table 5. Unit transportation carbon emission from supplier i to project j

$\widetilde{CT_{ij}}$ (kg/km)	Supplier 1	Supplier 2	Supplier 3	Supplier 4
Project 1	(33,$\mathcal{N}$(37,8),41)	(43,$\mathcal{N}$(48,11),51)	(37,$\mathcal{N}$(41,7),45)	(37,$\mathcal{N}$(39,6),45)
Project 2	(31,$\mathcal{N}$(36,6),39)	(41,$\mathcal{N}$(45,5),49)	(32,$\mathcal{N}$(38,4),41)	(37,$\mathcal{N}$(43,7),47)

3.2   Results and Discussions

In this section, the results were calculated based on the different carbon emission control parameter α by running the proposed model on the MATLAB, as shown in Tabel 6, Figs. 2 and 3.

Table 6. Unit transportation carbon emission from supplier i to project j

	Total costs of CC (10^9 yuan)	Total carbon emission (10^6 tonnes)	Profits (10^6 yuan)			
			S1	S2	S3	S4
$\alpha = 1.00$	2.012	2.454		41.836	67.018	
$\alpha = 0.95$	2.012	2.454		41.836	67.018	
$\alpha = 0.90$	2.108	2.357	37.820			62.526
$\alpha = 0.85$	2.323	2.240	60.588			34.199

Note: CC and S denotes construction company and supplier.

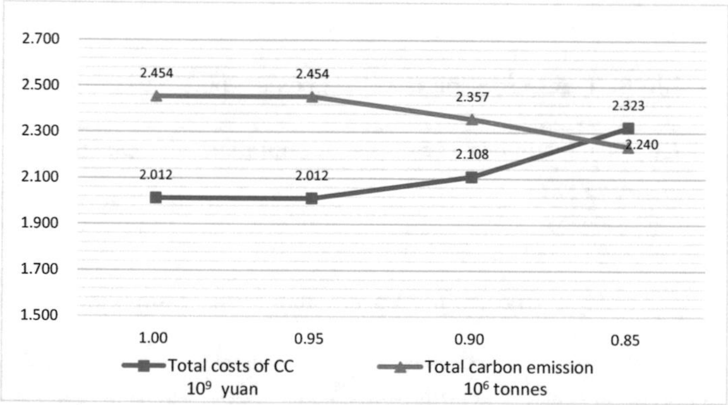

Fig. 2. The total costs and total carbon emission of construction company

When α=1.00, the total costs of construction company is 2.012×10^9 yuan, total carbon emissions are 2.454×10^6 tonnes. The selected suppliers are Supplier 2 which produce 4.248×10^5 tonnes materials for Project 1, and Supplier 3 which produce 5.730×10^5 tonnes materials for Project 2 respectively. When α=0.95,

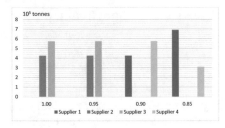
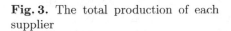

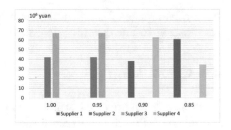

Fig. 3. The total production of each supplier

Fig. 4. The total profits of each supplier

the total costs of construction company is 2.012×10^9 yuan, total carbon emissions are 2.454×10^6 tonnes. The selected suppliers are Supplier 2 which produce 4.248×10^5 tonnes materials for Project 1, and Supplier 3 which produce 5.730 $\times 10^5$ tonnes materials for Project 2 respectively. When α=0.90, the total costs of construction company is 2.108×10^9 yuan, total carbon emissions are 2.357 $\times 10^6$ tonnes. The selected suppliers are Supplier 1 which produce 4.248×10^5 tonnes materials for Project 1, and Supplier 4 which produce 5.730×10^5 tonnes materials for Project 2 respectively. When α=0.85, the total costs of construction company is 2.323×10^9 yuan, total carbon emissions are 2.240×10^6 tonnes. The selected suppliers are Supplier 1 which produce 2.9008×10^5 tonnes materials for Project 1 and 3.9944×10^5 tonnes Project 1, and Supplier 4 which produce 1.347×10^5 tonnes materials for Project 1 and 1.7356×10^5 tonnes materials for Project 2 respectively.

From the results shown in Figs. 2, it shows that with the tightening of the carbon emission constraints, the total costs of construction company increases with the ratio of 0.00, 4.75 and 10.22%, while the total carbon emissions decreases with the ratio of 0.00, 3.92 and 5.00%. It can be seen that the slope rate for the changing total costs trend is bigger than that of the carbon emissions.

From the results shown in Figs. 3 and 4, it denotes that when construction company's attitude towards carbon emission control is relaxed ($\alpha = 1.00$ and $\alpha = 0.95$), Supplier 2 and Supplier 3 are selected to supply materials for projects, which can offer a lower price of materials but also have a poorer performance of carbon emission control during production process. When construction company's attitude towards carbon emission control becomes stricter ($\alpha = 0.90$), Supplier 1 and Supplier 4 are selected to supply materials for projects, which haver better performance in carbon emission control and more reasonable positions to projects, but also have a higher price of materials. When construction company's attitude towards carbon emission control becomes strictest ($\alpha = 0.85$), Supplier 1 and Supplier 4 are still selected. However, there is a obvious change that Supplier 1 supplies more materials for projects to meet needs. These results indicates that under the carbon emission policy environment, the competitive advantages of construction material suppliers should move from low-price orientation to low-carbon orientation.

4 Conclusion

In contrast to previous studies, this paper proposes a bi-level programming model to resolve the construction material supply problem focus on the conflicts between the construction company and material suppliers. Combined the cost minimization objective and carbon emission minimization objective, a bi-level model with fuzzy variables based on equilibrium strategy is developed to deal with the construction material supply problem. It takes the game between the construction company and suppliers into account by using a bi-level model. The results and related analysis of a case study with four suppliers and two projects were introduced to verify the applicability and effectiveness of the optimization model. This paper provided a construction material supplier mechanism with low-carbon orientation based on the game between the construction company and material suppliers.

Acknowledgement. This research has been supported by the project of Research Center for System Sciences and Enterprise Development (Grant No. Xq18B02) and Sichuan University (Grant No. 2018hhf-45).

References

1. Angulo, E., Castillo, E.: A continuous bi-level model for the expansion of highway networks. Comput. Oper. Res. **41**(1), 262–276 (2014)
2. Ayhan, M.B., Kilic, H.S.: A two stage approach for supplier selection problem in multi-item/multi-supplier environment with quantity discounts (2015)
3. Bemelmans, J., Voordijk, H., Vos, B.: Supplier–contractor collaboration in the construction industry. Eng. Constr. Arch. Manag. **19**(4), 342–368(27) (2013)
4. NBS: The statistical yearbook of the chinese constuction industry. National Bureau of Statistics of the Peoples Republic of China (NBS) (2016)
5. Hammami, R., Temponi, C., Frein, Y.: A scenario-based stochastic model for supplier selection in global context with multiple buyers, currency fluctuation uncertainties, and price discounts. Eur. J. Oper. Res. **233**(1), 159–170 (2014)
6. Kruse, R.: Statistics with Vague Data. Theory and Decision Library (1987)
7. Küçükaydin, H., Aras, N., Altınel, I.K.: Competitive facility location problem with attractiveness adjustment of the follower: A bilevel programming model and its solution. Eur. J. Oper. Res. **208**(3), 206–220 (2011)
8. Qiu, R., Xu, J., Zeng, Z.: Carbon emission allowance allocation with a mixed mechanism in air passenger transport. J. Environ. Manag. **200**, 204–216 (2017)
9. Raut, R.D., Bhasin, H.V., et al.: An integrated fuzzy-ahp-lp (fahlp) approach for supplier selection and purchasing decisions. Int. J. Serv. Oper. Manag. **10**(4), 400–425 (2011)
10. Safa, M., Shahi, A., et al.: Supplier selection process in an integrated construction materials management model. Autom. Constr. **48**(3), 64–73 (2014)

11. Sherali, H.D., Soyster, A.L., Murphy, F.H.: Stackelberg-nash-cournot equilibria: Characterizations and computations. Oper. Res. **31**(2), 253–276 (1983)
12. Udeaja, Emmanuel C.: A decision support framework for construction material supply chain management using multi-agent systems. J. Endod. **28**(3), 181–4 (2002)
13. Xu, J., Zhou, X.: Fuzzy-like multiple objective decision making. Stud. Fuzziness Soft Comput. **263**, 227–294 (2011)
14. Xu, J., Rui, Q., Lv, C.: Carbon emission allowance allocation with cap and trade mechanism in air passenger transport. J. Clean. Prod. **131**(Complete), 308–320 (2016)
15. Zhang, Z., Bo, W.: Research on the life-cycle CO_2 emission of China's construction sector. Energy Build. **112**, 244–255 (2016)

Do Firms Experience Enhanced Productivity After Cross-Border M&As?

Zihan Zhou[1], Lei Zhang[2], and Dongmei He[1(✉)]

[1] Business School of Sichuan University, Chengdu 610065,
People's Republic of China
hedongmei@stu.scu.edu.cn
[2] College of Computer Science of Sichuan University, Chengdu,
People's Republic of China

Abstract. This paper used a dataset of 86 cross-border M&A cases by Chinese listed firms from 2007 to 2012 and matched them with 81 domestic M&A cases to examine the post M&A performances. First, Probit and Logistic estimation methods were used to analyze whether the Chinese listed firm M&A decisions (cross-border or domestic) affected productivity levels, after which FGLS and OLS estimations were conducted to estimate the productivity changes five years before and after the merger years. It was found that firms involved in cross-border M&As have higher productivity than the domestic M&A group. However, in contrast with the domestic M&A, there was no significant productivity enhancements after the cross-border M&As. Further investigations found that state-owned enterprises (SOEs) performed better than privately-owned enterprises and that firms located in developed areas were more likely to experience significant productivity enhancements after the cross-border M&As.

Keywords: Cross-border M&A · Domestic locations · State ownership · Total factor productivity

1 Introduction

Cross-border mergers and acquisitions (M&As) have become more common due to the growth in the global economy and improvements in modern corporate governance and capital markets. The 2018 World Investment Report claimed that firms from emerging markets such as China were more likely to be involved in cross-border M&As when seeking to participate in the global market. The Chinese Outward Foreign Direct Investment Report reported that 765 Chinese cross-border M&As had been completed in 2016, with a total contract volume of 135 billion US dollars, an annual increase of 148% (Table 1). This significant growth in Chinese cross-border M&As was endorsed by the Chinese government's "Go Out" policy, which was aimed at encouraging Chinese firms to learn and compete in the global market, and especially to absorb advanced technologies

© Springer Nature Switzerland AG 2020
J. Xu et al. (Eds.): ICMSEM2019 2019, AISC 1002, pp. 492–509, 2020.
https://doi.org/10.1007/978-3-030-21255-1_38

and management skills. As a result, many listed firms, that had higher productivity and better corporate structures became more active in cross-border M&As, of which large SOEs (State-owned Enterprises) have been most dominant because of their closer government connections and available capital sources.

Table 1. Cross-border Chinese M&As from 2012–2016

Year	Contract volume (billion US dollar)	Annual increase (%)
2012	43.4	59.6
2013	52.9	21.9
2014	56.9	7.6
2015	54.4	−4.3
2016	135.3	148.6

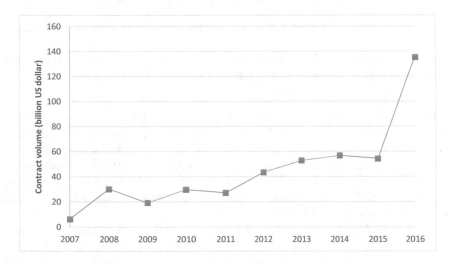

Fig. 1. Cross-border M&As by Chinese firms 2007–2016 *Note 1. Source* Chinese outward foreign direct investment report 2016

Behind these positive figures, however, are concerns about the firm's performances after the completion of the cross-border M&As as the primary M&A motivation is the strategic optimization of resources to enhance efficiency and improve overall performance [14]. Based on short term abnormal cumulative returns, there has been a general consensus by management researchers that cross-border M&As have a positive effect on a firm's performance [3,15]. However, as M&A is generally a longer term strategic move, performances should be assessed over both the short and long term. Therefore, the short-run analysis event method does not fully elucidate the M&A effect on overall firm performance over time. Most previous research has tended to examine post-M&A firm performance based on value creation or accounting-based profits, from which it was found that cross-border M&As were able to improve the financial performance of Chinese firms in terms of the return on assets or gross profit [4,12]. Although stock prices and accounting profits can be useful performance indicators, it could be argued that they are insufficient measures of organizational efficiency or productivity [26]. As one of the important motives behind emerging market M&As is to learn from foreign companies and enhance efficiency [6], measuring productivity would be a more appropriate approach when seeking to assess a firm's performance after a cross-border M&A.

Therefore, to examine the effect of cross-border M&As on firm performance, this paper selected 86 cross-border Chinese M&As from 2007 to 2012 to study performance changes five years before and five years after merger completion; that is, 11 years of performance data were gathered for each M&A case. Inspired by the studies of Bertrand and Betschinger [4] and Danbolt and Maciver [10], a control group was also introduced made up of 81 completed domestic M&As.

The following three questions drove the research in this paper:

(1) Are cross-border Chinese M&A firms more productive than domestic M&A firms?
(2) Does a cross-border M&A enhance a firm's productivity after the merger?
(3) Do state-ownership and the domestic location of the acquirer affect post-M&A productivity divergence?

This paper contributes to existing empirical analysis by comparing Chinese domestic and cross-border M&As using Total Factor Productivity (TFP) as the major dependent variable to evaluate the static and dynamic productivity changes before and after the M&As, and by linking the post-M&A performances with the firms' characteristics. This study can give guidance to future strategies for Chinese firms that may be planning a cross-border M&A, as well as providing a better understanding of Chinese M&A behaviors.

The remainder of this paper is organized as follows. A literature review is given in Sect. 2, after which Sect. 3 details the data resource and estimation procedures. Section 4 then presents the estimation results and a discussion as to what was observed, and Sect. 5 concludes the paper.

2   Literature Review and Hypotheses Development

Cross-border M&A firm performances have been widely discussed in previous research, most of which has made judgments based on the acquirer's post-M&A value creation or stock price. However, several authors have assessed pre- and post-M&A firm productivity to verify improvement's in firm efficiency, arguing that productivity assessments provide a much closer link between cross-border M&As and learning efficiency [5,21]. Solow [28] proposed that as TFP measured the pure impact of the technological progress after the tangible factors such as land, capital, and labor were removed, it was a better measurement for evaluating firm efficiency.

Productivity can be an important indicator for firms when deciding on cross-border M&As. Helpman et al. [17] ordered heterogeneous firms faced with foreign investment decisions by productivity and found that the least productive exited, third-tier productive firms served only domestic markets, second-tier productive firms served both domestic and export markets, and first-tier productive firms served the domestic market and undertook foreign direct investments. Nocke and Yeaple [23] focused on the relationship between cross-border M&As and firm productivity-levels, and instead of developing a linear productivity order, they predicted that both the most productive firms in R&D-intensive industries and the least productive firms in marketing-intensive industries would acquire foreign targets. Raff et al. [25] found that in Britain and Japan, the more productive a firm was, the more likely it was to choose greenfield investments rather than M&A. Contrary to previous conclusions however, Spearot and Alan [29] found that the more productive North American firms were, the more likely they were to adopt to cross-border mergers and acquisitions. Chinese firms that choose cross-border M&As need adequate finance and also good management skills and adaptability to be able to compete with other bidders and generate a synergy effect to bring value; therefore, firms with higher productivity are probably more suitable for these kinds of international take-overs. Thus our first hypothesis is:

Hypothesis 1. Firms that choose cross-border M&As have higher productivity than those that choose domestic M&As.

It is widely acknowledged that cross-border acquisitions enable emerging market firms to gain access to advanced technology and knowledge-intensive assets, which, in turn, may increase productivity. The enhancement in the emerging market acquirer's domestic productivity following a cross-border M&A implies that there is an obvious "reverse knowledge spillover" mechanism from the developed-nation targets to the emerging-market acquirer. Chen et al. [8] provided empirical evidence that the reverse spillover effect really existed, with emerging market MNEs with subsidiaries in host developed markets that are richer in technological resources exhibiting stronger technological capabilities domestically.

Bertrand and Capron [5] found that French firm domestic productivity was enhanced from cross-border M&A activities, and especially when there were learning opportunities in the target host country. Using panel data from 123 countries, Ashraf et al. [1] proved that M&As had a strong and positive effect on TFP in developed host countries, and Siegel and Simons [26] found that in Sweden, cross-border M&As led to improvements in firm performances and plant productivity.

However, Ataullah et al. [2] suggested that emerging market M&As reduced employee productivity due to their lack of absorptive capacity and experience in cross-border M&As. It was also found that cross-border acquisitions in less-developed countries and in culturally distinct countries reduced overall productivity. Specifically, based on data from Chinese acquirers, cross-border M&As

have not been found to significantly improve firm productivity [20]; however, domestic M&As that have no experience or integration problems have been found to have enhanced productivity [30].

Although some research has discussed the productivity effect of cross-border M&As, a comparison of firm productivity before and after the M&A has not yet been studied under a dynamic framework. Therefore, this paper adds to current research as it gives a dynamic performance analysis comparison five years before and after the completed mergers as well as a comparison study of cross-border and domestic M&A performance in China. Under this static and dynamic framework, we therefore test the following hypothesis:

Hypothesis 2. Cross-border M&As do not have a positive impact on a firms' productivity compared with domestic M&As.

Country-level, industry-level, and firm-level determinants are all important to a firm's post- M&A performance. Unfortunately, although there are many determinants that affect cross-border M&A performance, there has been little research that has discussed their effect on productivity. However, productivity effects can be inferred from the profitability determinant discussions because of the tight links between the two. We therefore made further discussion.

As dominant participants in Chinese cross-border M&As, large SOEs are more likely to undertake cross-border M&As for political purposes even though this may sacrifice their economic interests [9, 22]. It has been found that SOEs had worse performances compared with private firms and that SOE cross-border M&As can sacrifice minor shareholder benefits [9]. Therefore, to reduce domestic discrimination, private firms are more likely to produce abroad than SOEs [7].

However, it has also been argued that compared with private firms, as SOEs have strong political ties with the government, they face less financial constraints [24]. Du and Boateng [13] found that state ownership had a significant impact on shareholder value, with SOEs showing more abnormal returns for their Chinese acquirers, primarily because of a policy preference for SOEs rather than private firms. Accordingly, the following hypothesis addresses the impact of state-ownership on acquirer productivity:

Hypothesis 3. Cross-border Acquirers with dominant state-ownership experience post-M&A productivity enhancements.

The domestic location of the acquirers has also been found to be an important determinant. In China, there is a significant gap between the development levels of different regions, with the east coastal regions being more advanced due to the 1980s "Reform and Opening" policy. Using panel data from Chinese firms, it has been shown that the overseas foreign direct investment (OFDI) from the eastern regions had the largest positive reverse spillover effects on the TFP, while OFDI from the central and western regions was found to have no significant reverse spillover effects [19]. It was claimed that this distinction was a result of the imbalances and knowledge absorption capacity differences between the regional economies, which leads to our final hypothesis:

Hypothesis 4. Cross-border acquirers located in developed areas of China experience greater post M&A productivity enhancements.

3   Empirical Data and Model

3.1   Sample Selection

The cross-border M&A events data was extracted from the WIND database, which has information on; acquirers and target firm names, industry sectors, transaction dates, deal value, and payment methods. These events were cross-checked with the Thomson ONE M&A database to confirm each event and its current status. From 2007 to 2012, there were 206 completed cross-border Chinese listed firm M&As.

The following criteria were used to choose the M&A events for this study: (1) To ensure data availability, acquirers had to be listed on the Shanghai or Shenzhen Stock Exchanges; therefore, cases listed in Hong Kong or Taiwan were excluded as the accounting standards for formulating financial reports are different from Mainland China. (2) Only completed M&A events from January 2007 to December 2012 were considered, as the quantity and value of Chinese cross-border M&As rose significantly during this period. Further, to evaluate firm productivity 5 years after the merger, it was necessary to only include cross-border M&As that had been completed before 2012. (3) The sample was restricted to non-financial acquirers and targets as financial firms have different financial reporting systems and regulations that could bias the study results. (4) Targets from the British Virgin Islands and the Cayman Islands were excluded. (5) When firms had concluded more than one cross-border M&A cases during a year, only the acquisition with the highest transaction value was included. (6) Acquirers that were involved in multiple mergers within 3 months of each other were eliminated. Finally, 86 cross-border M&A cases were chosen for the research.

As with the studies of Bertrand and Betschinger [4] and Danbolt and Maciver [10], this paper also selected domestic M&As as the control group to compare productivity with the cross border M&As, the criteria for which was as follows: (1) The domestic M&A was completed in the same sample year. (2) All cases matched the cross-border M&A industries (industry classifications were in line with those in the WIND database). (3) No cross-border M&As were completed during the sample period. (4) Firm size (as represented by M&A volume and registered capital) were similar to the cross-border M&As. Based on Bertrand and Betschinger [4] and Danbolt and Maciver [10], this paper used M&A volume/registered capital of the acquiring firm as the criteria for the choice of the domestic M&As (error range within 10%).

After matching the data, an unbalanced panel was obtained from the final usable sample of 86 cross-border and 81 domestic M&As over an 11-year period, which included the five years before and the five years after the merger year. The sample industry distribution is detailed in Table 2.

3.2 Variables

Dependent variables

Total Factor Productivity (TFP_LP): TFP_LP was calculated using the LP method in Levinsohn and Petrin [18]. The natural logarithm form of the firms' net profit $(L\_netprofit)$, the natural logarithm form of the firms' number of employees (L_labour) and the natural logarithm form of firms' total assets $(L\_totalasset)$ were used for the computations.

Table 2. Sample number by industry

Industry	All	Cross-border	Domestic
Manufacturing	26	13	13
Materials	44	22	22
Consumption and services	41	23	18
Information technology	14	7	7
Energy	22	11	11
Medical	12	6	6
Real estate	8	4	4
Total	167	86	81

Independent variables

DT_{it}: This was a merger-dummy variable that analyzed whether the cross-border M&A had a static effect on firm productivity. As suggested by Dickerson et al. [12] and Diaz et al. [11], a value of 1 was taken for the merger year and the following 5 years, and a value of 0 was taken before the merger year.

DT_{jit}: This was a merger-dummy variable that analyzed whether the cross-border M&A had a dynamic impact on the firms' productivity, and considered the time lag between the cross-border M&A and any change in performance. DT_1 took a value of 1 one year after the acquisition and 0 in all other years, DT_2 took a value of 1 two years after the acquisition and 0 in all other years, and so on until DT_5, which took a value of 1 five years after the acquisition and 0 in all other years.

Control variables

The control variables and corresponding symbols are detailed in Tables 3, and 4 shows the descriptive statistics for the variables for the data estimation and data sources.

Table 3. Control variables and data sources

Variables	Definitions	Data Source
Total assets (ASS)	Measures all assets that a firm owns. Its ln value is LASS.	CSMAR
Tobin's Q (TOBINQ)	Computed as a ratio of the market capitalization against the total assets. This ratio was used to assess the changes in enterprise value.	CSMAR
Return on equity (ROE)	Computed as a ratio of the net income after taxes against the average equity value. This balance sheet ratio was used as a measure of profitability [11].	CSMAR
Leverage ratio (LEV)	Measures a firm's solvency, calculated as the ratio of total liabilities a gainst total assets.	CSMAR
Cross-border (CROSSBORDER)	Took the value 1 if the case was a cross-border M&A, and 0 if the case was a domestic M&A	WIND
State ownership (STOW)	Took the value 1 if it was a state-owned enterprise, and 0 otherwise.	CSMAR
East coastal area (EAST)	This was a dummy variable that indicated whether the firm was from a developed area (east coast area of China) or from a less developed area (inland area of China). Took the value 1 if the firm was located in the east coast region, and 0 otherwise.	Author's calculation
Developed province (DEVELOPED)	This was a dummy variable which served as a robustness check. Took the value 1 if the firm was located in a province with a higher than median GDP and 0 otherwise.	Author's calculation

3.3   Estimation Procedure

Single-Equation Probit Model:
To start with, single-equation Probit was employed to investigate the impact of productivity on the firms' M&A decisions. The equation was estimated using Probit and Logistic models. No significant differences were found between the two groups; $\beta_0, ..., \beta_6$ were the parameters being estimated.

To analyze the differences in the firm's productivity when they chose a cross-border or domestic M&A, the following equation was used (Hypothesis 1):

$$CROSSBORDER_{it} = \beta_0 + \beta_1 TFP\_LP_{it} + \beta_2 STOW_{it} + \beta_3 LASS_{it} + \beta_4 LEV_{it} + \beta_5 ROE_{it}$$
$$+ \beta_6 TOBINQ_{it} + \varepsilon_{it}(i = 1, 2, \cdots, 167; t = -5, -4, \cdots, 4, 5)$$

$$(1)$$

Table 4. Descriptive statistics for M&A cases and data sources (Cross-border versus Domestic)

Variables	Obs	Mean	Std. Dev	Min	Max	Data source
Cross-border						
TFP_LP	819	116.7	120.8	4.97	779.6	CSMAR
LASS	828	23.4	2.51	19.34	30.81	CSMAR
LEV	926	24.59	28.46	0.09	94.24	CSMAR
ROE	919	11.34	16.43	−137.9	142.4	CSMAR
TOBINQ	799	1.714	1.756	0.0651	18.5	CSMAR
Domestic						
TFP_LP	777	95.94	91.46	0.871	706.9	CSMAR
LASS	783	22.88	2.119	16.51	30.61	CSMAR
LEV	871	28.51	53.16	0.03	1,224	CSMAR
ROE	854	11	52.57	−559.2	852	CSMAR
TOBINQ	740	1.611	1.562	0.0514	12.11	CSMAR

Multivariate regression model (FGLS)

A panel data analysis was conducted to capture the static and dynamic nature of the firm's productivity after the cross-border M&As and to identify the factors that might have affected the productivity. By including DT_{jit} $(i = 1, 2, \cdots, 5)$ as an additional regressor, the static model was transitioned to a dynamic panel model. Therefore, the panel dataset consisted of a cross-section dimension (167 merger cases, 86 cross-border and 81 domestic, $i = 1, 2, \cdots, 167$), and a time dimension (11 periods: $t = -5, -4, \cdots, 4, 5$); therefore, there were 1700 observations, which was considered adequate to produce robust estimations for the scope of this analysis. Generally, autocorrelation, endogeneity, and heteroscedasticity problems are inherent in economic data sets. First, as some explanatory variables can be endogenous, the Ordinary Least Squares (OLS) estimators become biased and inconsistent. Second, unobserved panel-level effects (fixed effects) may be correlated with the explanatory variables, finally, the inclusion of lagged dependent variables can lead to autocorrelation. Following Diaz et al. [11], to deal with these issues and control the time effect and heteroskedasticity, a multivariate regression model was used and the regression coefficients estimated using feasible generalized least squares (FGLS). Further, to enhance the robustness of the result, IVOLS estimation with robust standard errors was employed to check the results.

To analyze the specific period in which a cross-border M&A exerted a static or dynamic impact on a firm's productivity, the following equation was considered: (solve Hypothesis 2)

$$TFP\_LP_{it} = \mu_0 + \beta_0 DT_{it} + \beta_1 STOW_{it} + \beta_2 LASS_{it} + \beta_3 LEV_{it} + \beta_4 ROE_{it} + \beta_5 TOBINQ_{it}$$

$$+ \sum_{j=1}^{5} C_j DT_{jit} + \varepsilon_{it}(i = 1, 2, \cdots, 167; t = -5, -4, \cdots, 4, 5)$$

$$(2)$$

which $\beta_0, ..., \beta_5$, C_j were the parameters being estimated; β_0 and C_j respectively measured the static and dynamic effects of the cross-border M&A.

To analyze the firm characteristics; size, efficiency, profitability, domestic location, and state ownership; that might affect firm productivity after the cross-border M&As, the following equation was used: (solve Hypothesis 3 and 4)

$$TFP\_LP_{it} = \mu_0 + \beta_0 DT_{it} + \beta_1 STOW_{it} + \beta_2 LASS_{it} + \beta_3 LEV_{it} + \beta_4 ROE_{it} + \beta_5 TOBINQ_{it}$$

$$+ \beta_6 EAST_{it} + \beta_7 DEVELOPED_{it} + \varepsilon_{it}(i = 1, 2, \cdots, 167; t = 0, \cdots, 4, 5)$$

$$(3)$$

4 Estimation Results and Discussions

First, the Probit estimation was run to identify the differences in firm productivity when the firms faced an M&A decision; domestic mergers (control group) or cross-border mergers (treatment group); in the same sample period. Then the static and dynamic impacts of the M&As on firm productivity were examined if the firms chose cross-border M&A. Finally, factors such as state-ownership and domestic location were selected to reveal the differences in the acquiring firm characteristics that contributed to post-M&A productivity divergence.

4.1 Do Acquiring Firms with Higher Productivity Tend to Seek Cross-Border M&As?

Table 5 gives the results of the single-equation Probit model, from which it was found that the TFP_LP was significantly positive, which indicated that firms involved in cross-border M&As often had higher productivity compared with the domestic group, which was in line with previous studies [17, 23, 29]. Compared with domestic M&As, cross-border acquisitions are generally more complex in nature, with the acquiring firms possibly faced with exchange rate risks, political risks and legal risks [15, 16]. Cultural conflicts can also increase the cost of post-cross-border M&A integration [27]. Therefore, firms with higher productivity would have a stronger ability to face the challenges associated with cross-border M&As. The Logistic estimation results showed the odd ratio for the two groups and the relationship to the explanatory variables, from which it was found that TFP_LP was significantly positive, indicating a greater odds ratio for the cross-border group.

Table 5. Single-equation probit and logistic model results

Model variables	Probit	Logistic
	Cross-border	Cross-border
TFP_LP	0.000840**	0.00137**
	−0.000358	−0.000585
STOW	−0.254***	−0.409***
	−0.0778	−0.125
LASS	0.0792***	0.127***
	−0.0192	−0.031
LEV	−0.000752	−0.00124
	−0.00128	−0.00206
ROE	−0.000382	−0.000626
	−0.000825	−0.00132
TOBINQ	0.0598**	0.0969**
	−0.0238	−0.0386
Constant	−1.754***	−2.813***
	−0.54	−0.875
Control year	Yes	Yes
Control industry	Yes	Yes
Correctly classified ratio	0.5899	0.59
Observations	1,532	1,532

Note 2. Robust standard errors in parentheses, ***$p < 0.01$, **$p < 0.05$, *$p < 0.1$

4.2 Static and Dynamic Productivity Impacts After Cross-Border M&As

Table 6 indicates that when the cross-border and domestic M&A cases were taken as subsamples, the coefficient for the merger-dummy variable was insignificant in the TFP_LP for the cross-border mergers, while the domestic M&As had significantly enhanced productivity. That means domestic M&As significantly enhanced productivity after merger when taking five years post-merger period as a whole.

When adding the merger-dummy variables $DT_{jit}, (DT_1, \cdots, DT_5)$ into the regression, for the domestic group, the coefficients for DT_2 to DT_5 were all significantly positive when related to TFP_LP, which indicated that a domestic M&A was able to enhance firm productivity starting from the second year after the merger (Table 7). However, the results from the cross-border group indicated no significant evidence for productivity enhancements in TFP_LP after the merger.

Therefore, it could be concluded that the cross-border M&A for the listed firms after the merger did not result in productivity improvements, which was in line with the results in Ataullah et al. [2] and Su and Xian [20], who found that due to the complexity and risks in global markets, acquirers from emerging markets such as China were either unable to absorb the acquired resources or market into their internal features within three to five years [30].

Table 6. Static FGLS estimations for TFP_LP

Model variables	FGLS		OLS	
	TFP_LP		TFP_LP	
	Cross-border	Domestic	Cross-border	Domestic
DT	−3.658	9.898*	−18.83**	0.978
	−5.253	−5.219	−8.764	−6.649
STOW	23.13***	14.26***	23.24***	5.832
	−2.814	−3.027	−6.227	−8.01
LASS	20.75***	14.60***	21.77***	19.18***
	−0.895	−1.13	−2.602	−3.319
LEV	−0.497***	−0.328***	−1.116***	−0.777***
	−0.0677	−0.0583	−0.159	−0.148
ROE	0.12	−0.00612	0.195	0.00953
	−0.0759	−0.0324	−0.236	−0.0713
TOBINQ	−0.426	2.639**	−2.81	6.654*
	−0.751	−1.039	−3.249	−3.595
Constant	−414.2***	−282.9***	−364.9***	−339.3***
	−26.42	−36.61	−59.34	−72.83
Control year	Yes	Yes	Yes	Yes
Control industry	Yes	Yes	Yes	Yes
Observations	798	734	702	634
R-squared			0.284	0.187
Number of id	86	81		

Note 3. Robust standard errors in parentheses, ***p<0.01, **p<0.05, *p<0.1

4.3 Further Discussion

Specific characteristics such as state ownership and domestic location were then taken into the regression with firm productivity. Tables 7 and 8 give the associated results for the post-merger period for the cross-border M&As, in which there were some interesting features, as discussed in the following.

Does state ownership affect firm productivity in the post-M&A period?

Table 7. Dynamic FGLS estimations for TFP_LP

| Model variables | FGLS | | OLS | |
| | TFP_LP | | TFP_LP | |
	Cross-border	Domestic	Cross-border	Domestic
DT1	1.424	6.297	−7.168	−0.0251
	−5.301	−4.986	−13.3	−10.93
DT2	−0.649	11.67**	−12.8	−1.499
	−5.983	−5.787	−12.41	−10.47
DT3	2.992	19.81***	−10.14	−1.091
	−6.434	−6.476	−13.11	−11.42
DT4	1.428	23.79***	−16.95	1.296
	−7.027	−7.099	−12.53	−12.56
DT5	6.149	28.67***	−13.26	−0.00582
	−7.656	−8.352	−12.82	−13.69
STOW	23.05***	14.04***	24.74***	5.613
	−2.674	−3.14	−6.115	−8.288
LASS	20.57***	15.05***	21.25***	19.25***
	−0.877	−1.143	−2.574	−3.329
LEV	−0.491***	−0.274***	−1.076***	−0.779***
	−0.0627	−0.0598	−0.159	−0.149
ROE	0.130*	−0.0168	0.256	0.01
	−0.0725	−0.0328	−0.24	−0.0718
TOBINQ	−0.346	3.377***	−3.009	6.657*
	−0.779	−1.023	−3.326	−3.575
Constant	−411.2***	−292.0***	−360.2***	−340.1***
	−25.75	−37.54	−59.74	−72.85
Control year	Yes	Yes	Yes	Yes
Control industry	Yes	Yes	Yes	Yes
Observations	798	734	702	634
R-squared			0.282	0.187
Number of id	86	81		

Note 4. Robust standard errors in parentheses, ***$p < 0.01$, **$p < 0.05$, *$p < 0.1$

Table 8 indicates that in the cross-border group, $STOW\_DT$ was positive and significant, which suggested that stated-owned enterprises (SOEs) performed better than privately-owned firms after the merger no matter whether the merger was cross-border or domestic, which was consistent with previous research [13]. This may have been because in general, state-owned firms in China are more likely to obtain preferential treatment and a favorable allocation of resources from the government, which enhances productivity.

Table 8. OLS and FGLS estimation results for state-owned acquirer performances

Model Variables	FGLS		OLS	
	TFP_LP		TFP_LP	
	Cross-border	Domestic	Cross-border	Domestic
DT	−14.18**	6.976	−41.75***	5.396
	−5.882	−5.904	−8.417	−8.482
STOW	11.78***	10.21**	−2.326	10.53
	−4.106	−4.455	−12.13	−10.18
$STOW\_DT$	19.71***	6.578	39.84**	−7.376
	−5.442	−5.47	−15.65	−12.96
LASS	20.75***	14.64***	21.56***	19.29***
	−0.87	−1.135	−2.603	−3.326
LEV	−0.497***	−0.327***	−1.103***	−0.778***
	−0.0689	−0.0589	−0.157	−0.148
ROE	0.112	−0.00585	0.219	0.00656
	−0.076	−0.0325	−0.236	−0.0708
TOBINQ	−0.287	2.938***	−2.51	6.666*
	−0.844	−1.062	−3.353	−3.601
Constant	−407.9***	−280.3***	−345.8***	−344.8***
	−27.35	−36.89	−59.78	−72.95
Control year	Yes	Yes	Yes	Yes
Control industry	Yes	Yes	Yes	Yes
Observations	798	734	702	634
R-squared			0.29	0.187
Number of id	86	81		

Note 5. Robust standard errors in parentheses, ***$p < 0.01$, **$p < 0.05$, *$p < 0.1$

Do firms in developed regions perform better in the post-M&A period?

Table 9 indicates that firms located in developed areas (East Coastal Area or Developed Province) tended to experience significant productivity enhancements after the cross-border M&A ($CROSSBORDER\_DT$ was significantly positive); however, domestic M&A firms located in the developed areas experienced significant productivity drops ($DOMESTIC\_DT$ was significantly negative), which revealed that firms from developed areas were more likely to experience enhanced productivity from a cross-border M&A [19]. Previous research has argued that firms from more developed areas have a better ability to internalize the advanced technologies and resources from target firms because of their greater emphasis on R&D and their previous experience. The robust check for the Developed Provinces showed similar results.

Table 9. FGLS estimation results for different domestic locations

Variables	East coastal area TFP_LP				Developed province TFP_LP			
CROSSBORDER_DT	8.241***	-4.198			7.133***	-14.04**		
	-2.024	-4.586			-2.421	-5.557		
DOMESTIC_DT			-11.28***	1.511			-7.532***	15.82**
			-2.033	-4.806			-2.422	-6.546
STOW	10.47***	18.71***	11.53***	19.26***	24.21***	-7.429	24.45***	-3.113
	-2.026	-4.205	-1.952	-4.165	-2.409	-5.403	-2.476	-5.551
LASS	16.58***	23.69***	16.91***	23.70***	16.40***	25.79***	16.64***	26.83***
	-0.789	-1.13	-0.757	-1.098	-0.716	-2.944	-0.72	-2.817
LEV	0.181***	-1.124***	0.139***	-1.102***	-0.367***	-0.490***	-0.395***	-0.372***
	-0.0394	-0.0819	-0.0381	-0.0808	-0.0409	-0.114	-0.0418	-0.11
ROE	0.156***	-0.00451	0.138***	-0.0057	0.0206	0.017	0.0182	0.0501
	-0.0423	-0.0502	-0.0409	-0.0492	-0.0317	-0.0501	-0.0305	-0.0564
TOBINQ	0.7	2.935**	0.426	2.995**	-0.213	6.968***	-0.14	7.953***
	-0.619	-1.361	-0.615	-1.361	-0.545	-1.947	-0.546	-1.968
Constant	-318.6***	-490.5***	-325.9***	-491.7***	-325.8***	-520.0***	-331.6***	-544.8***
	-22.86	-34.5	-21.83	-34.54	-46.72	-62.29	-45.8	-59.51
Control year	Yes	Yes	Yes	Yes	Yes	Yes	Yes	Yes
Control industry	Yes	Yes	Yes	Yes	Yes	Yes	Yes	Yes
Observations	743	783	743	783	1,270	256	1,270	256
Number of id	82	84	82	84	140	26	140	26

Note 6. Standard errors in parentheses, ***p<0.01, **p<0.05, *p<0.1

5   Conclusions and Implications

This paper used a dataset of 86 cross-border M&A cases from 2007 to 2012 to study the firms' M&A productivity performances, the main conclusions from which were as follows.

(1) There was an obvious productivity level effect for Chinese listed firm M&A decisions. The firms that had cross-border M&As were found to generally have higher productivity compared with the domestic M&A control group, which implied that firms with higher productivity would have stronger abilities to face the challenges associated with cross-border M&As.

(2) There were no obvious productivity improvements after the Chinese listed firms completed their cross-border M&As. The evaluation of the Chinese firms' static and dynamic productivity changes before and after the M&As found that in general, it was more difficult for firms to improve their productivity from cross-border M&As than from domestic M&As over the long-term. However, domestic M&As were found to enhance firm productivity five years after the merger. Emerging market acquirers such as Chinese firms often lack cross-border M&A experience, and therefore the complexity and risks in the global market become obstacles to the generation of synergy effects to improve productivity, which means that the acquired resources or markets are unable to be fully integrated within three to five years.

(3) Stated-owned enterprises (SOEs) performed better than privately-owned firms after both cross-border and domestic mergers.

(4) Firms located in the developed areas of China (East Coastal Area or Developed Provinces) tended to experience significant productivity enhancements after the cross-border M&As, while domestic M&A firms from the developed areas of China experienced significant productivity drops.

From this analysis of Chinese listed firm cross-border M&As, it was concluded that Total Factor Productivity (TFP) should be seen as an important index when designing cross-border M&A policies. In the future, cross-border M&As will continue to be the dominant method for Chinese firms looking to participate in the global market under the stimulation of the Chinese "Go out" policy. Although the analysis in this paper indicated that cross-border M&As might result in significant performance enhancements within five years, these results could be biased as we only used efficiency indicators for 11 years, which might not be long enough as firms often conclude cross-border M&As for strategic motivation.

From the results, some rational decision-making orientations can be given. First, to achieve better cross-border M&A performances, firms could first conclude a domestic M&A before investing abroad. Further, special policies are needed to assist private firms to actively participate in cross-border M&As as these firms are going to be the main international buyers in the future. Therefore, a more scientific decision-making support mechanism is needed to encourage rational, better-performing investments.

Table 10. List of provinces located in East Coastal Area

Liaoning	Jiangsu	Shanghai	Fujian	Zhejiang
Hebei	Shandong	Guangdong	Tianjin	Hainan

Table 11. List of Developed Provinces

Liaoning	Jiangsu	Shanghai	Fujian	Zhejiang
Hebei	Shandong	Guangdong	Anhui	Hubei
Neimenggu	Beijing	Sichuan	Henan	Hunan

Acknowledgements. This paper is partially supported by The National Natural Science Foundation of China (No.61563044, 61762074); National Natural Science Foundation of Qinghai Province (2017-ZJ-902).

References

1. Ashraf, A., Herzer, D., Nunnenkamp, P.: The effects of greenfield fdi and cross-border m&as on total factor productivity. World Econ. **39**(11), 1728–1755 (2016)
2. Ataullah, A., Hang, L., Sahota, A.S.: Employee productivity, employment growth, and the cross-border acquisitions by emerging market firms. Hum. Resour. Manag. **53**(6), 987–1004 (2014)
3. Aybar, B., Ficici, A.: Cross-border acquisitions and firm value: An analysis of emerging-market multinationals. J. Int. Bus. Stud. **40**(8), 1317–1338 (2009)
4. Bertrand, O., Betschinger, M.A.: Performance of domestic and cross-border acquisitions: Empirical evidence from russian acquirers. J. Comp. Econ. **40**(3), 413–437 (2012)
5. Bertrand, O., Capron, L.: Productivity enhancement at home via cross-border acquisitions: The roles of learning and contemporaneous domestic investments. Strat. Manag. J. **36**(5), 640–658 (2015)
6. Chang, S.C.: The determinants and motivations of china's outward foreign direct investment: A spatial gravity model approach. Glob. Econ. Rev. **43**(3), 244–268 (2014)
7. Chen, C., Yu, M., Tian, W.: Outward fdi and Domestic Input Distortions: Evidence from Chinese Firms (2016)
8. Chen, V.Z., Li, J., Shapiro, D.M.: International reverse spillover effects on parent firms: Evidences from emerging-market mnes in developed markets. Eur. Manag. J. **30**(3), 204–218 (2012)
9. Chen, Y., Young, M.: Cross-border mergers and acquisitions by chinese listed companies: A principal principal perspective. Asia Pac. J. Manag. **27**(3), 523–539 (2009)
10. Danbolt, J., Maciver, G.: Cross-border versus domestic acquisitions and the impact on shareholder wealth. J. Bus. Financ. Account. **39**(7–8), 1028–1067 (2012)
11. Diaz, B.D., Olalla, M.G.: The Effect of Acquisitions on the Performance of European Credit Institutions: Panel Data Analysis for the 90s. Social Science Electronic Publishing (2002)

12. Dickerson, A.P., Gibson, H.D., Tsakalotos, E.: The impact of acquisitions on company performance: Evidence from a large panel of uk firms. Oxf. Econ. Pap. **49**(3), 344–361 (1997)
13. Du, M., Boateng, A.: State ownership, institutional effects and value creation in cross-border mergers and acquisitions by chinese firms. Int. Bus. Rev. **24**(3), 430–442 (2015)
14. Dunning, J.H.: Trade, location of economic activity and the mne: A search for an eclectic approach. Int. Alloc. Econ. Act. **1023**, 203–205 (1977)
15. Gubbi, S.R., Aulakh P.S.: Do international acquisitions by emerging-economy firms create shareholder value? the case of indian firms. J. Int. Bus. Stud. **41**(3), 397–418 (2010)
16. Hai, Y.L., Deseatnicov, I.: Exchange rate and chinese outward fdi. Appl. Econ. **48**(51), 1–16 (2017)
17. Helpman, E., Yeaple, M.S.R.: Export versus fdi with heterogeneous firms. Am. Econ. Rev. **94**(1), 300–316 (2004)
18. Levinsohn, J., Petrin, A.: Estimating production functions using inputs to control for unobservables. Rev. Econ. Stud. **70**(2), 317–341 (2003)
19. Li, M., Liu, S.: Regional differences and threshold effects of fdi reverse technology spillover-a threshold regression analysis based on chinese provincial panel data. Manag. World **1**, 21–32 (2012). (in Chinese)
20. Li, S.U., Xian, G.: Does chinese enterprises' cross-border m&a promote productivity progress? China Economic Studies (2017)
21. Lichtenberg, F.R.: Siegel Dea (1987) Productivity and changes in ownership of manufacturing plants. Brook.S Pap. Econ. Act. **3**, 643–683 (1987)
22. Luo, Y., Tung, R.L.: International expansion of emerging market enterprises: A springboard perspective. J. Int. Bus. Stud. **38**(4), 481–498 (2007)
23. Nocke, V., Yeaple, S.: Cross-border mergers and acquisitions versus greenfield foreign direct investment: The role of firm heterogeneity. J. Int. Econ. **72**(2), 336–365 (2007)
24. Poncet, S., Steingress, W., Vandenbussche, H.: Financial constraints in china: Firm-level evidence. China Econ. Rev. **21**(3), 411–422 (2010)
25. Raff, H., Ryan, M., Stahler, F.: Firm productivity and the foreign-market entry decision. J. Econ. Manag. Strat. **21**(3), 849–871 (2012)
26. Siegel, D.S., Simons, K.L.: Assessing the effects of mergers and acquisitions on firm performance, plant productivity, and workers: newl evidence from matched employer employee data. Strat. Manag. J. **31**(8), 903–916 (2010)
27. Slangen, A.H.L., Hennart, J.F.: Do multinationals really prefer to enter culturally distant countries through greenfields rather than through acquisitions? the role of parent experience and subsidiary autonomy. J. Int. Bus. Stud. **39**(3), 472–490 (2008)
28. Solow, R.M.: Technical change and the aggregate production function. Rev. Econ. Stat. **39**(3), 554–562 (1957)
29. Spearot, A.C.: Firm heterogeneity, new investment and acquisitions. J. Ind. Econ. **60**(1), 1–45 (2012)
30. Yang, Y.: Research on corporate m&a and total factor productivity change. Hum. Resour. Manag. **11**, 23–29 (2016). (in Chinese)

Future of India-China Relations: A Key Role to the Global Economy

Nitin Kumar[(⊠)] and Sita Shah

Business School, Sichuan University, Chengdu 610064, People's Republic of China
nitinkumar1811@outlook.com

Abstract. This study is aimed at exploring the influence of India-China relationship on the global economy and its extent of control on global economy dynamics. Due to the mass population, the two countries have the potential to increase and influence the global economy. In this study, an extensive literature review, export and import and competitiveness index secondary data were used to explore the current position of India and China in the global economy. A SWOT analysis was used to analyze the future relation. The findings show that the growth of business between India and China is positive but there is still an imbalance of trade between the two countries. It also indicates that Indias economy is getting despondency and more competitive, that directly effect on global economy. According to the Global Competitiveness Index (GCI) score report, China has a better rank compared to India, so the strong relationship can help India improve its GCI Index. Finally, the economy relations can be improved by introducing better public services and good policymaking.

Keywords: India · China · Global economy · Trade · Future relation · Competitiveness index

1 Introduction

In the global economy, India and China are the rapidly growing power to lead the economy. Decline the economic progress of the India and China has a vast influence on the global economy from the last three decades the influence both are positive in terms of economic contribution in the global economy [13]. It is possible that in the outcome of the recent economic crisis and current gradual recovery would have a major role in global economic growth [22]. Although both countries have several features in common and their economic takeoff differs in timing, strength and development process. But the political system is the main difference between India and China. At this age of globalization, the relationship between economics has an important in the global economy [15].

Both countries went through similar developments, but performances differ with intensity and timing. This article shows their trade, which moves to India and China relations and sentiments regarding future development. This study

© Springer Nature Switzerland AG 2020
J. Xu et al. (Eds.): ICMSEM2019 2019, AISC 1002, pp. 510–522, 2020.
https://doi.org/10.1007/978-3-030-21255-1_39

focuses on the future relationship between these two countries and the key role to the global economy [18]. In a long-run perspective, India and China have an advantage from beginning to make a good business environment and international relation, though they undertake the trade liberalization strategies during sufficiently stable economies and strong to face global competition [19]. India-China relations passed melodramatic changes over the past 3 decades, for long last interest of stability and peace in India-China and Asia, both countries are helping each other and exchange cooperation in diverse areas. In trade and economic, and for economic development as well [25]. Improvements in diplomatic and trade relations have led to strong economic growth in India and China. The relationship of them is a little complex in past mistrust, but the relation in trade is improving continuously. India and China are playing out competing strategies, to reach the level of a giant position, in terms of their peaceful relations and development in the Asia-Pacific region. They should have to keep good relations with each other for their economic development [13].

First, more than half of the absolute global GDP growth now falls on both countries, which affects significantly the structure of the global consumption of services and goods [24]. Until 2015, China was the worlds fastest-growing major economy, with growth rates averaging 10% over 30 years. Though few studies have been done on similar issues still there is lack of sufficient research works to address the issue "how the relationship between India and China can influence on the global economy and what extent" [25]. Therefore, this study intends to fill the research gap by exploring the future of India-China relations and the role in the global economy and its extent of control on global economy dynamics. This paper has been organized into several sections. Section 2, describes the theoretical perspective and dimensions of the previous research in this field. Sections 3 and 4 describe the methodology and the ranking of India and China in the global economy in terms of competitiveness index. Section 5 uses SWOT analysis for exploring the trade relationship between India-China. Section 6 presents the findings of the topic. In the final section, the article summarized the finding and recommends some policy measure for the future of India-China relationship [17].

2   Literature Review

In 2017 China and India are the two giants populous nations in the world with 1.409 billion and 1.338 billion populations, respectively, so together they account for over one-third of the worlds population, which have long histories, have been lagging behind in many areas compared to more developed countries [25]. As an Asian giant both of the countries have the strategic vision to grip the potential to make an important role globally based on the resemblance of strategic interests [28]. Amrita Narlikar (2017) observed that Prime Minister (PM) of India Mr. Narendra Modi (Mr. Modi) involves in dealing with corruption and reforming Indias perverse laws of labor, and land acquisition laws, providing a single-window approach to investors, undertaking agricultural reform, and so forth. And in contrast to China, all these will have to be confronted with a human face, not only for moral reasons but for all the political reasons that go with being a democracy. But if the BJP is able to make even some headway on the programmed of what its manifesto referred to as the Economic Revival, the gains of growth will be very considerable for Indian citizens, also for the international community. Indian development will matter internationally, and it will matter in more interesting ways than the simplistic Goldman Sachs story of the benefits of growth markets providing drivers for the world economy [18].

Gancheng Zhao (2016) investigated that as a rising power, India is embracing ever more ambitious goals. Since won the general election in a landslide victory of the Bharatiya Janata Party (BJP) and PM of Mr. Modi came in to power in May 2014, India has been pursuing its goal of an Indian century with its democratic institutions dividends of young population and market potentials, and his commitments to leading India to meet the peoples expectations, also reveal some of Indias external goals. Mr. Modi commitment constituents and the world at large conveyed a clear message that India was determined to regenerate itself as a nation. Although Indias status as a global power remains debatable, the international community seems to have accepted such a prospect for India in the coming decade [22].

Salitskii et al. (2013) opined that India and China became constant greater when the two giant countries had relatively successfully overcome the global financial crises of 2008–2009, while their global GDP share continued to grow. Between the two, India has been slower to develop. The international influence of both states is also growing, and China is often referred to as a new economic superpower [24]. Similarly, Salahuddinin Ayyub (2012) states that with China's joining of World Trade Organization (WTO), bilateral trade between India and China has speed up due to lower tariffs. Although in a few fields these two countries are competitors there are many areas, where they can do achieve their goal by cooperation. Their nature of production shows that for bilateral trade and development there is a huge potential [3].

While India adopted the Westminster style of democracy with a multi-party system, and China became a Communist society with a single party rule and limited emphasis on civil rights and liberties. India is the seventh largest and second most populous country in the world, with nearly 70% of the population of India directly or indirectly dependent on farming and agriculture. Poverty level of India is higher than Chinas and its workforce is less trained and disciplined [19].

At present, India is ranked 3rd largest economy in Asia, and under the 25th largest economies ranking, India is holding 4th position in the world, with a 2016 GDP of $8.7 trillion [12]. Through the hard work is not only put Asia on the track to sustained growth, but it also reinforces the historical role in the global economy, as a leader for the 21st Century. Statistics predict that India became the third-largest consumer economy, due to consumption became triples to $4 trillion by 2025. India demand increased for oil, while China will increase its energy consumption. In terms of the demand for energy, the competition between them is increasing gradually which may invite any conflict in the future [23]. They are already taking the initiative to collaborate on focusing energy demand through a joint agreement and tries to shape an "Organization of Petroleum Importing Countries" [29]. As well as, to meet the rising demand for fuel in India, due to passenger vehicles, demand is growing year on year, and its economy grows, India needs to double its refining capacity by 2040 [7]. A number of elements of the economic infrastructure of India require administrative reform especially bureaucracy. Foreign direct investment (FDI) is a key target of India for getting attraction from international investor which can influence the economic growth of the country. At the beginning of 2017, the first few months, India has suffered a big trade deficit with China and it was about \$10.60 billion [12]. Since the price of Chinese product is lower than India, due to China manufacturing supremacy, which directly influences the domestic product India, and as well as some Indian companies were bound to leave the market, where China is far more keep than India in terms of trade disparity and the possibility of a free trade agreement (FTA) [5]. India has the potential to make striving aims for rapidly changing economics which can gain a new dimension with making a friendly relationship with China. In the 18th century, there was almost the same participation found in a global economy where the place was India, Europe, and China. But in the 21st century unfolds both India and China stand self-confident to recover their weight in the global trade. Thus, Indias economic attract with China remain a crucial dimension of its economic and foreign policy [18]. The fear of Chinese expansionism was a cause to make strategic ties with India, and accordingly, western countries actions follow different diplomacy in case trade but it increases Indias diplomatic space [1].

3   Methodology

The study assesses the history and current moves in India and China relations using SWOT (i.e. strength, weakness, opportunities and threats) analysis, under the strength review the Cooperation, efforts, growth, trade, agreement, investment, aspects and ranking, and for weakness, corruption, trade deficit, and internal problems, for opportunities with the relations, potential, investment and as well as for threats also, and simple ranking method using the Global competitiveness index (GCI) data and assessments of growth rate based on percentage analysis of Indias and Chinas export trade have been determined from 2013 to 2017. Trade details and sentiments regarding the future relations of India-China assessed and analyzed [26].

4   The Position of India and China in the Global Economy

4.1   India in Global Economy

As a large country, India is continuously done its policy reform for improving export and import in the international arena. Though the share of business in global level of India is lower than China, it was still a broad prospect to improve the position due to a large population, technology, and business-friendly policy [28]. The tariff on the agricultural product is about 40.8% and a non-agricultural product is about 12.1% which shows its trade position in the business sector. Due to trade liberalization, the global economic leadership is moving to Asia and India seems to be a leader in business as like China [27]. China has already invested a large amount of belt and road initiative, infrastructure and communication sector, comparing these issues India is not faster like China. Friendly trade environment can help to make a better relationship between two countries as well as the control of trader in global trade liberalization [2].

According to the GCI score report, China has a better rank than India. The study shows GCI data from the year 2013 to 2017. In these five years, the rank of India is lower than China. As per GCI in 2013 and 2014, China rank was 29th, whereas India was in 59 and 60. Similarly, in 2015, 2016 and 2017, the rank of China was 28th, respectively whereas India was 71st and 55th. GCI indicates that the trend of the business position in India is improving gradually while China has a better position in the last five years [26]. GCI score for five financial years as a rank of India and China are presented in Fig. 2.

4.2   China in Global Economy

China has a good reputation in the global market due to their product features, prices, and business policy. According to GCI, China sells their products all around the world. The contribution and share of China in the global market is better than other countries. Though China is a giant population country they motivate peoples capability for work [30]. So, China already made human capital largely by engaging their huge population in trade. China already initiated to build Belt and Road initiative (BRI) for making a communication network from Asia, to Europe and Africa which will enable them to make openness of the market and control the market [11]. This BRI is not only connected the people but also connects their business to Chinese business market. It will bring a tremendous positive change in the business sectors which ultimately help China to control over the global economy [16]. In the last decade due to tremendous growth rates, China has gained a good status in the global economy. Actually,

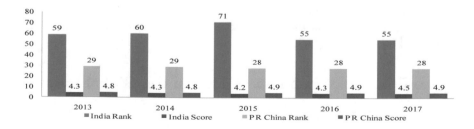

Fig. 1. Global competitiveness index data

China achieved the first position whereas the US, Japan, and India place second, third and fourth position in the global economy [29].

4.3   Some India-China Trade Statistics

Among the most encouraging recent developments in India and China ties is the rapid growth in bilateral trade [21]. As observed in Fig. 2, there has been tremendous growth in trade between India and China between 2013 and 2017. These export and import as the bilateral trade paint very impressive picture year on a year percentage rate of growth as depicted in Fig. 2. It is also observed that India export to China much higher in 2017 as compared to previous years [6]. Economic growth of India and China is better than previous years. Which indicate that the future relationship of India-China will play a key role in the global economy.

5   SWOT Analysis of India and China Relationship

India and China aim to become major powers in the Asia-Pacific but of course, not the sole power. It would require many decades for either India or China to replace the current sole power, The United States [27]. They have theyre own economic and strategic force limitations, that they would have to overcome to reach the level of The United States, which remains the major player in the Asia-Pacific region. To reach the level of superpower status and given that both are playing out competing strategies, India and China have yet to cross many stages in terms of their peaceful development and relations with Asia and African states in the Asia-Pacific region [2]. This becomes an extremely difficult and competitive process for the two regional rivals. The strategies that are played out in the Asia-Pacific region are based on competition, cooperation, and containment between the regional powers (i.e. India and China) [18]. As India and China emerge in international politics, they are strengthening their economies and modernizing their deterrent forces. While playing out competing strategies, India and China are developing economic and strategic relationships

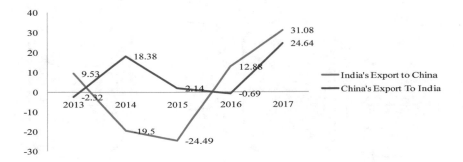

Fig. 2. The percentage growth rate of exports in India-China from 2013 to 2017

with strategically important countries. India and China relations are built on the basis of strengths and opportunities, and represent the threats and weakness in the current environment in Fig. 3 [17].

5.1   Strengths

They are the new drivers of the global economy, and have a strong and positive business growth regular agreement signing for doing the business of various products, the complementary aspects and the overall ranking of India-China are in a better position in the global economy [24]. India the largest business hub in the world after China, It is a better opportunity for China to invest in India and make a positive and strong relationship in the future [10]. It is possible through the co-operation of both countries. Both are giants populous countries and the fastest growing economies in the world. Growth in diplomatic and economic both countries influence has increased the significance of their bilateral relationship [21]. The main strength of the trade relationship between the two countries is co-operation between them, investment decision and treaty, the joint level business relationship by organizing various levels of meeting and memorable of understanding (MOU) signing. In the view of trade growth, China is the largest import trading partner of India as per report issued by GOI in 2017 [31].

Their economic strengths are widely considered complementary. In services and Information technology, China perceived to be strong. India is strong in goods like hardwares while China is stronger in software. A physical market of China is much stronger than India, where the Indian financial market is much stronger than China. Both the country has certain historical interactions as well, for example, the spread of religion (i.e. Buddhism) in China from India and Silk Road trade [19]. It is showing that both countries have good future relations and will play a key role in the global economy.

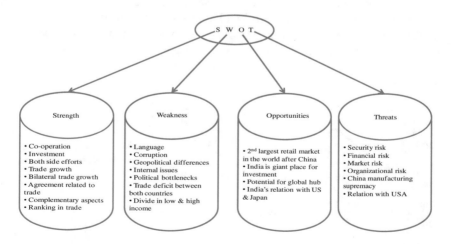

Fig. 3. A SWOT analysis of India-China relationship. Source: adapted after Muthiah

5.2   Weakness

Though the relationship between India and China is satisfactory in case of trade there is still some weaknesses are remaining. The study found some weakness of the trade relationship between them such as language and culture berries between the country's people, various levels of corruption in both countries especially for trade, geopolitical differences, internal and political issues, political bottlenecks, trade deficit between two countries, and divide in low and high income [20]. Language is a big weakness because both countries have limited speaker, who can speak both nation language but all the things depend on the fast, immediate and superior action by the government. Corruption in India is creating internal issues, and trade deficit. The political situation in India is very worse and it is the major weakness of India compare to China [16]. The cultural and political systems of each country are totally different from China. The political system in India is ruled by the worlds largest democracy and China ruled by a single party. These differences are also a major weakness for both countries [19]. Internal problems within China and the not have freedom due to the senior personnel nature of the government are likely to affect its trade strategy. Chinas exchange-rate policy, financial infrastructure and the inefficiency of its state-owned enterprises these problems are also related. Other boundaries include its China government politics, short of democracy, a lack of public opposition, environmental destruction, indiscriminate, little free press and government control on the Internet. Both countries, corruption and human rights abuses are also a big issue [19]. The study found that previously trade balance between India-China fluctuated yearly, but in 2017 Indias export to China increases [6].

5.3   Opportunities

In global trade, there are still opportunities for both countries to make the relationship better and control over the global market. This study identified some opportunities regarding trade relationship as India is the giant and largest business hub globally after China. India and China will become the largest and fastest growing markets in the worlds due to mass consumption of Industrial raw materials, capital goods, energy, and other goods etc.

Over the last few years, China has the most remarkable "go-global" world vision strategy to move upmarket equally technologically and industrially and in a context of increasing Chinese investments in different countries [29]. It shows that China has a good opportunity to invest and create a job opportunity in India and make a positive and strong relationship [17]. Like China is the top biggest market in the world whereas India has the second largest market hub for trading due to its population and both the countries provide a giant offer for the business, both of them are potential for the global hub, and both the countries have a good relationship with other potential countries [14]. Both countries bilateral trade is about to achieve magnificent growth. China will develop its service industry to move some China-based manufacturing may move to India, mainly at low-end export manufacturing. India and China will present an Asian manufacturing and services powerhouse to service the global community [19].

5.4   Threats

Business always bears some risk. Some of the prominent business risks such as security risk, financial risk, market risk, and organizational risk. In the case of international trade, the risk is increasing day by day in various dimensions. Though all the international trade is done by following The Uniform Customs and Practice for Documentary Credits (UCPDC), and other international business rules but still some risks are embedded in every step. Security risk can happen in terms of information communication technology, physical risk especially during transportation of the product. This kind of risks is available in international trade. Besides, the trade rules and regulations are varying from country to country which also creates some problems during the trade. The traders of both countries should be alert and careful during trade [21]. China is the largest trading partner of India, not only India almost globally has China lead the world in export due to manufacturing supremacy. In the domestic context, Chinas relations with its neighbors are really important since these relationships affect the domestic view, both strategically and economically. The fact of that, in recent years China has sought to strengthen its relations with almost all neighbors [29].

6   Trading Relationship Between India and China

China is the largest trading partner of India, not only India almost globally has China lead the world in export due to manufacturing supremacy. China exports different items that consist export of to India are organic chemicals, electronics, machinery, tools, nuclear reactors, equipments, boilers, mineral fuels, silk, oils, and machinery etc. India also imports some value-added products from China [21]. India comes under the China top 10 largest trading partners globally. Where India hold a 1st position in the highest major products import from China as shown in Fig. 4 [8]. While China continues to enjoy a huge satisfactory balance of trade, however, both countries are growing rapidly and can drive the world economy in future with a pool of the worlds largest skilled workforce [18]. To increase the overall exports more efforts are being made for expanding the trade with an emphasis on products and services, enhancing market access, and other barriers. This is done through the bilateral meetings between both countries. China has a good opportunity to invest and create a job opportunity for youngsters in India and make a strong future relationship [14]. Both countries exporters are encouraged to show their products by attending to trade fairs in both countries. The trade deficit of India as increased during the last three years with top 10 countries [8]. The two Asian leaders are serving to move out of the world recession through their, good relations, trade, and import [4]. Mentionable that India imports a large part of their necessary products from China and it was 16.86% as a single country as shown in Fig. 4. This data indicates that India has great business confidence in China in terms of their regular trade.

Since 1980 economic development of both countries, especially focusing on the growth and openness of the relationship with reference to international trade and FDI [6]. Also, GDP Indias has expanded every year averagely over the

Fig. 4. Indias major sources of imports

decade, as compared to China. The key factors that India is expected to do for the developments aim to sustain its economic growth in order to exceed Chinas economy and come to the top [2]. Globally the economic activities in both countries have become increasingly significant, owing to which both the countries are trying to accelerate the development by expanding their economic base.

7 Discussion

The earlier relationship between India and China was falling behind due to many aspects like language, culture, corruption, internal issues, trade deficit, political issues, and border vision etc. Nowadays both countries are focusing and try to fill all gaps between them, now both countries main vision is to focus on development and maintain the relationship in the future [24]. As shown in Fig. 4, Indias import top major products from China [8]. Through the trade vision, China is the largest trading partner of India, not only India almost globally has China lead the world in export due to supremacy in manufacturing. The study observed that China top 10 global trading partners imported the most Chinese shipment by dollar value during 2017, India hold 7th position out of 10 countries [31].

There has been great growth in trade between both countries between 2013–2017. This export-import as the bilateral trade shows a very impressive picture on a yearly percentage rate of growth as depicted in Fig. 2. It is also observed that compared to previous years India export to China much higher in 2017 [6]. The economic growth of both countries is better than previous years, which indicate that the future relationship of India-China will play a key role in the global economy. China and India are the fastest growing countries, have emphatically made their presence felt in the rapidly globalizing world economy. Both the countries are among the major countries in the global population as both the nations contribute 37% of the worlds population and 6.4% of the value of world output and income at current prices and exchange rates [8]. Economic growth going rapidly, and large foreign direct investments (FDI) and growing trade along with the whole world are some of the unusual features of both the countries of development. Both countries in relation to their political background, nature of development especially economic and its approaches like exchange rate arrangements opted by both the countries [5]. In this trade relations between both the countries and with the rest of the world becomes attention-grabbing. India and China trade relationship is comparatively stronger in terms of imports and exports among the other countries. Major items that involve Chinese exports to India are organic chemicals, electrical machinery, equipments, boilers, cement, mineral fuels, silk, and oils [4]. The determinants of fastest growth of the two countries are identified in this study which has been presented in Fig. 5.

The study found that the major determinants of the trade relationship between India and China are policy and relations, trade statistics, free trade agreement, and border diplomacy. India had an unfavorable trade balance with China except 2017. While China continues to enjoy a huge satisfactory balance of trade [5]. However, both countries are growing rapidly and can drive the world economy in the future with a pool of the worlds largest skilled workforce [9]. To increase trade, need efforts to improve the relationship between both countries. Both the countries have decided to work in a group with open hand that will resolve their old border dispute and builds a new framework. According to different time periods and comparative perspectives, the economic development of both countries has been mostly examined. Since 1980, GDP of India and China have increased to 10 and 6% annually respectively. While the US, Europe, and Japan have been significantly lower economic growth [8].

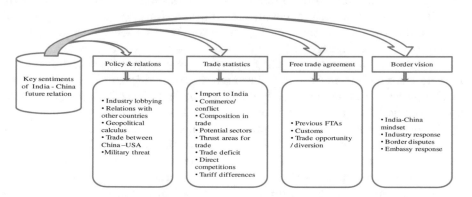

Fig. 5. Determinants of the future relations of India and China. Source: adapted after Muthiah

8   Conclusion

The growth of business between India and China have a positive impact but there is still an imbalance of trade between two countries and as per GCI ranking report India rank is lower than China. Their economic growths are impressive economic growth as they are increasing their global trade by openness. Based on trade liberalization, they not only manufactured goods but also increased their trades and relation by adopting well-designed industrial and international trade policies. In some extents, China is more advanced than India especially electronic sectors. The GCI shows that China secured 28th position by achieving 4.9 but India secured 55th position by achieving 4.3 in 2015–2016. The article further point out that both countries are integrating into the global economy through their positive growth effects. However, the growth of business between India and China is positive there is still an imbalance of trade between two countries and as per GCI ranking report India rank is lower than China. The findings also reveal that Indias economy is getting despondency and more competitive, that directly impacts positively on the global economy. With the good relationship, in future both China and India can push and transform global economy to a greater height. Finally, the inequalities of trade can be reduced by introducing better public services and good willing of policymakers.

References

1. Scott, D.: The rise of india: UK perspectives. Int. Aff., 165–188 (2017)
2. Arif, S.M.: A history of Sino-Indian relations: from conflict to cooperation. Int. J. Polit. Sci. Dev. 1(4), 129–137 (2013)
3. Ayyub, S.: Indo-China trade relations: present trends and future prospects. Iqra Int. Manag. J., 34–51 (2012)
4. Bhandari, D., Shrimali, G.: The perform, achieve and trade scheme in India: an effectiveness analysis. Renew. Sustain. Energy Rev., 1286–1295 (2017)
5. Bussière, M., Mehl, A.: China's and India's roles in global trade and finance: twin titans for the new millennium? EBSCO Publishing, 14–29 (2008)
6. Government of India Ministry of Commerce. Indias foreign trade (2017)
7. Coxhead, I., Jayasuriya, S.: China, India and the commodity boom: economic and environmental implications for low-income countries. World Econ., 525–551 (2010)
8. Factbook, C.: The world factbook. Central intelligence agency (2011)
9. Hölscher, J., Marelli, E., Signorelli, M.: China and India in the global economy. Econ. Syst. 34(3), 212–217 (2010)
10. Kapur, A.: Can the two Asian giants reach a political settlement? Asian Educ. Dev. Stud. 5(1), 94–108 (2016)
11. Khan, Z.: China-India growing strides for competing strategies and possibility of conflict in the Asia-Pacific region: China-India competition for Asia-Pacific. Pac. Focus 31(2), 232–253 (2016)
12. Kong, H.: Special section on India-China relations in the early 20th century. 173–175 (2014)
13. Marelli, E., Signorelli, M.: China and India: openness, trade and effects on economic growth. Eur. J. Comp. Econ. 8(1), 129–155 (2011)
14. Mazumdar, S.: India's economy: some reflections on its shaky future. Futures 56, 22–29 (2014)
15. Menon, S.: China, the world and India. China Rep. 52(2), 129–137 (2016)
16. Menon, S.: Some thoughts on India, China and asia-pacific regional security. China Rep. 53(2), 188–213 (2017)

17. Muthiah, K.: CHINDIA—the changing times of China and India bilateral relations. Manag. Res. Rev. **33**(1), 23–40 (2009)
18. Narlikar, A.: India's role in global governance: a modification? Int. Aff. **93**(1), 93–111 (2017a)
19. Overholt, W.H.: China and the evolution of the world economy. China Econ. Rev. **40**, 267–271 (2016)
20. Pant, H.V.: Rising China in India's vicinity: a rivalry takes shape in Asia. Camb. Rev. Int. Aff., 1–18 (2016)
21. Perkins, J.O.N.: Macro-Economic Policy. Routledge (2018)
22. Roy, A., Mathur, S.K.: Bilateral trade costs and growth of trade in services: a comparative study of India and China. In: Theorizing International Trade (2017)
23. Rui, W., Yong, G., Liu, W.: Trends of natural resource footprints in the BRIC countries. J. Clean. Prod. **142**, 775–782 (2016)
24. Salitsky, A., Shakhmatov, A.: China and India: new drivers of the global economy? Her. Russ. Acad. Sci. **3**, 453–458 (2013)
25. Saran, S.: Changing dynamics in India-China relations. China Rep. **53**(2) (2017)
26. Schwab, K.: The global competitiveness report 2017–2018. World Economic Forum (2017)
27. Schwartz, N.D., Abrams, R.: Advisers work to calm fearful investors. N. Y. Times **25**, 5–6 (2015)
28. Singh, S.: India-China relations: perception, problems, potential. South Asian Surv. **15**(1), 83–98 (2008)
29. Wang, H., Miao, L.: China Goes Global. Palgrave Macmillan, London (2016)
30. Weber, E.: Economic reform, social development and conflict in India. Reg. Sci. Policy Pract. **4**(3), 207–230 (2012)
31. Workman, D.: Chinas top trading partners (2019)

Effects of Autocracy and Democracy on FDI's Inflows

Tahir Yousaf[1], Qurat ul Ain[2(✉)], and Yasmeen Akhtar[3]

[1] Business School, Sichuan University, Chengdu 610065, People's Republic of China
[2] School of Public Finance and Taxation, Southwestern University of Finance and Economics, Chengdu 610065, People's Republic of China
quratulain_ad36@yahoo.com
[3] University of Sargodha, Sargodha, Pakistan

Abstract. This study aims to examine the effect of Autocracy and Democracy on Foreign Direct Investment (FDI) inflows in 15 Asian developing countries. For empirical examination, this study used panel data over the period 1995 to 2015 by employing dynamic panel data. The study shows a significant and positive relationship with autocracy and significant and negative relation with democracy. The empirical finding suggests that the country can attract more FDI in the presence of more autocratic institutions and less democratic institutions. Among the controlling variables trade, market capitalisation, population and literacy rate are significant with democracy while trade becomes insignificant with the main variable democracy.

Keywords: FDI · Autocracy · Democracy · Asian developing countries · Dynamic panel data · Developing countries

1 Introduction

The past decades have observed a considerable increase in the stock of Foreign Direct Investment (FDI) throughout the world. Only in developing countries, the stock of FDI expanded from about \$10 billion in 1986 to over \$99 billion in 1995. Although some parts belong to developing democratic countries and some to developing autocratic ones. This case creates a question among researchers and policymakers "Does democracy ease foreign direct investment?" To answer this question, two points of views found about how democracy influence FDI. At the one side, democratic institutions might have a positive effect on FDI because democracy gives checks and balances on elected officials, and this in result decreases random government intervention, decreases the risk of policy turnaround and strengthening property right protection (North and Weingast 1989; Li 2009) [15]. Olson [17] mentioned that traditional democracies, through administrative restraint and judicial sovereignty, guarantee property rights, which makes a safe, steady and attractive environment for foreign

© Springer Nature Switzerland AG 2020
J. Xu et al. (Eds.): ICMSEM2019 2019, AISC 1002, pp. 523–533, 2020.
https://doi.org/10.1007/978-3-030-21255-1_40

investors to invest. According to Olson, democracy is more attractive to FDI than autocracy [17].

Conversely, multinational corporations like better to invest in autocratic countries for the following three reasons. First, democratic restraints over elected politicians inclined to weaken the monopolistic positions. Second, these restraints further avert host governments from offering magnanimous financial and fiscal stimulus to foreign investors. Third, wide access to elected officials and broad political participation provide an institutionalized path through which local business can look for conservation. In every case, the increased pluralism ensured by democratic institutes creates policy consequences that decrease the multinational organizations level of freedom in the host developing country (Li and Resnick 2003) [11]. Although both autocrats and democratic leaders may take economic incentives from FDI, autocrats face less public pressure than democratic leaders if they select to secure foreign investors and investments from pressures such as higher labour incentives, labour conservation and non-friendly taxation schemes.

While the economic determinants of FDI for growing countries have been examined to a substantial degree, it is somewhat surprising that the significance of variations in a democratic structure in host countries has received comparatively less attention. But there is much less published literature on FDI-democracy nexus and what is comprehensible in these papers is that no unanimity has been observed about the reaction of democracy on FDI. For whole means, there come into sight three different findings in the existing literature; those that profess an inverse effect, those that profess a positive effect, and those who say there is no effect. The objectives of this study are as follows to determine the type of relationship between autocracy and democracy with FDI in Asian developing countries and to analyze the role of FDI inflows in country economic growth.

The scheme of the study is as follows: Sect. 1 provides the background and objectives of the study. Section 2 provides the literature review on the FDI and its determinants and role of the political regime in FDI. This chapter also discusses the findings of recent studies on FDI determinants. Section 3 explains the methodology of the study. It includes a discussion on the data collection method and of the variables used in the study. The discussion also includes the econometric methods used for the study and econometric problems related to the data are also discussed. Section 4 consists of the discussion the results derived from the model. Finally, Sect. 5 presents the conclusion, limitations and possible extensions to the study.

2 Literature Review

The two main characteristics of the political economy include the flow of political democracy and the rise of economic globalisation. Harms and Ursprung [8] in their study on 62 developing and emerging market economies found that MNEs are more likely to be attracted by countries where democracy is encouraged [8].

Oneal [18] observe no relationship between the political regime and US FDI outflows [18]. Li and Q [10] and Resnick [19] shows that democracy has a negative impact on FDI [10]/[19]. Li and Resnick [11] explain how democratic institutions impede FDI inflows [11]. These include democratic constraints over elected politicians weakening the oligopolistic or monopolistic positions of MNEs, preventing host governments from offering generous financial and fiscal incentives to foreign and finally the broad access to elected officials and full political participation offer institutionalized avenues through which indigenous businesses can seek protection. Furthermore, the authors propose that democracy also prevents the state from predatory rent-seeking.

The determinants of FDI flows are locational or pull factors such as the size of the market, and push factors, related to the conditions in the source country. Conventionalists emphasized that market size, taxes, trade policies, exchange rate and interest rate policies, production costs, infrastructure adequacy etc. influence the location decision of multinational enterprises in the global market place. Recently, foreign direct investment by multinationals of developed countries is considered important for transition as well as developing economies as they bring capital and advancement in technology and management know-how. Chen et. al (2018) finds that FDI drops significantly in election years, when policy uncertainty increases. The negative effects of uncertain policy on FDI also rely on the ratio of democratization and the political system. In democracies and countries with the Assembly elected president, the decline of FDI in election years is far more pronounced [4].

Du, Lu and Tao [6] determine the role of regional economic institutions and traditional factors in the location choice of foreign direct investment (FDI) [6]. In their study, they compare the sensitivities of FDI from six major areas (Hong Kong, Taiwan, US, EU, Japan and Korea across China's region. They find that FIEs (Foreign-invested enterprises) that are institutionally or culturally remote from Chinaexhibit a stronger aversion to regions with weaker economic conditions. Mathur and Singh [13] find that foreign investors consider economic freedoms in making decisions about where to locate capital [13]. Their study declares that democratic countries receive less FDI flows if economic freedoms are not guaranteed.

Jensen [9] mentioned the two reasons for which 'democratic accountability' may attract FDI [9]. First, democratic governments produce more credible and consistent policies, which reduce the risk of fundamental policy reversals such as nationalization, expropriation and renegotiation of tax rates. Second, in democratic regimes, political leaders are held accountable for their actions by the firms. If governments go back on the contracts after the investment has been made, MNCs can react by refusing to invest in the future. Therefore, the fear of the loss of reputation can lead to more consistent policies towards MNCs. Bano et al. [2] argued that the post-financial crisis period specifies that terrorism and energy scarcity are the main drivers of decrease in FDI inflows to Pakistan. Inflation market size, inflation, and exchange rates affect FDI inflows positively [2].

Mengistu and Adhikary [14] in their study examined the effects of six components of good governance on FDI inflows in 15 Asian countries for the period 1996–2007. Six components of good governance include voice and accountability, regulatory quality, human capital, infrastructure, lending rate and GDP growth rate [14]. They revealed six components of good governance, political stability and absence of violence, government effectiveness, and the rule of law and control of corruption as key determinants of FDI inflows. Bellos and Subasat [3] investigated the link between governance and foreign direct investment for 14 transition countries [3]. By applying a panel gravity model, they suggest that the lack of good governance does not deter the foreign direct investment. On the other hand, it encourages foreign direct investment. FDI has a positive and significant effect only in the case of non-European countries. Moreover, the contribution of R&D is higher than that of human capital and FDI in all cases (Tsamadias et. al 2018) [20]. Guerin and Manzocchi [7] investigated the effect of political regime on bilateral FDI flows from advanced to emerging countries [7]. Their results suggest that democracy has a positive effect on FDI inflows from developed to emerging countries. They also state that effect of democracy on FDI works through the total factor productivity channel in emerging countries. Mathur and Singh [13] mentioned the association between foreign direct investment, democracy and corruption [13]. They suggested that foreign investors concerned about economic liberalism, rather than political liberty, in making decisions about where to put capital. Recently, Agbloyor, Gyeke-Dako, Kuipo and Abor [1] finds that the relationship between FDI, economic growth and institutional quality may diverge based on country-specific characteristics such as natural resource endowment and financial development [1].

Okfar et al. [16] investigated the influence of democracy on FDI inflow in Sub-Saharan Africa [16]. Their findings suggest that as Sub-Saharan countries have shown advancement towards a stronger and more efficient democracy FDI inflow declines. Furthermore, the harmonized interest of the host country has now taken the place of profit maximization interest of the MNCs, exploiting the profit margin shrinks of MNCs. Libman [12] considers the impact of sub-national political systems on economic growth by applying the case of Russian regions [12]. He studies the influence of democracy on economic performance and size of the bureaucracy on economic outcomes. He observed a nonlinear relationship between democracy and economic growth. Furthermore, he states that Regions with high levels of democracy, as well as strong autocracies, perform better than hybrid regimes.

3   Methodology

3.1   Data Collection Methods

The section discusses the data collection methods and the sample size used to conduct this study. The current study about the impact of autocracy or democracy on FDI is based on the Asian Developing countries. The data is collected for economic and non-economic factors of these countries. Fifteen stateswere selected for analysis about the impact of autocracy or democracy on FDI.

3.2 Sample

The sample was selected from Asian countries for analyzing and generalizing the impact of autocracy and democracy on FDI in developing countries. The data set comprises of all the developing countries of Asia. A secondary method of data collection was used where the information required was obtained from the websites of the World Bank, International Transparency index, the respective country's official databases. Panel data sets analysis uses sequential blocks or cross-sections of data where within each resides a time series. Panel data in this study has two dimensions spatial and temporal. The spatial dimension in panel data is a composite of the cross-section dimension and in the current study consists of 15 Asian developing countries. On the contrary, the temporal dimension in the research uses some observations of each variable for each year. Data for was collected for 1995–2015 current study, so it covers 20 years. The econometric results are based on dynamic panel data Heteroscedasticity or specification errors were removed from panel data by the white diagonal method (White 1980) [21]. Furthermore, multi-collinearity from the CGVF models was removed by replacing problem variables with new variables in the model. Finally, autocorrelation in this type of data was removed by taking the remedial measure of autoregressive treatment.

3.3 Regression Analysis

Regression will be used to reveal the relationship between autocracy, democracy and FDI in developing countries. The regression will specify the relationship among the dependent variable, independent variables and control variables used in this study. The ordinary least square (OLS) estimation will be used to diminish the residuals of the models for the current study. OLS estimation minimises the residual of the model and enables the sample regression function to explain the maximum portion of the population regression function (Cuthbertson 1996) [5]. The estimated equations for the model are

$$
\begin{aligned}
Log\,(FDI) = C &+ \beta_1\,(autocracy) + \beta_2 log\,(MKTcap) + \beta_3 \log\,(trade) \\
&+ \beta_4 \log\,(gdp) + \beta_5 \log\,(exports) + \beta_6 log\,(imports) \\
&+ \beta_7 \log\,(SeaAccess) + \beta_8\,(LITrate) + \beta_9\,(Population) + \mu
\end{aligned}
\tag{1}
$$

$$
\begin{aligned}
Log\,(FDI) = C &+ \beta_1\,(democracy) + \beta_2 log\,(MKTcap) + \beta_3 \log\,(trade) \\
&+ \beta_4 \log\,(gdp) + \beta_5 \log\,(exports) + \beta_6 log\,(imports) \\
&+ \beta_7 \log\,(SeaAccess) + \beta_8\,(LITrate) + \beta_9\,(Population) + \mu
\end{aligned}
\tag{2}
$$

3.4   Variables of the Study

The study is composed of variables that may affect these flows, such as the democracy, autocracy, market capitalisation, trade, imports, exports, population, and literacy rate and sea access in these developing countries. Among these variables, FDI (foreign direct investment) is a dependent variable for the study. Democracy and autocracy are the two main variables in the above mentioned two models. The set of control variables include market capitalisation, trade, imports, exports, population, and literacy rate and sea access. The discussion on the selection and measurement of these variables is given as follows

Foreign Direct Investment (FDI)

An FDI inflow includes net FDI inflows, showing the amount of inward investment by foreigners less investment taken out of the country by foreigners. FDI is taken as a dependent variable in the study.

Democracy and Autocracy

For assessing the impact of political regime on FDI, the main independent variable of the study determines the level of democracy and autocracy in a country. For that purpose, a widely used democracy or autocracy composite indicator is used. According to index, a country is defined as a democracy (or autocracy) if the widely-used composite indicator of regime type from POLITY IV (Marshall and Jaggers 2000), computed as the difference between the 10-point democracy index (DEMOC) and the 10-point autocracy index (AUTOC), is greater than or equal to 6 (or smaller than or equal to -6). The index unit score is between the range -10 (strongly autocratic) and $+10$ (strongly Democratic).

Imports

The term import measures the amount of good or service brought in from one country to another country in a legal way. Imported goods or services are basically provided by foreign producers to domestic consumers.

Exports

Exports denote the number of goods and services produced in the home country sold in other markets. Export of commercial quantities of goods is also subjected to the involvement of the customs authorities.

Sea Access

The quality of infrastructure in recipient countries such as roads, highways, ports, communication networks, electricity, etc., is a critical determinant of FDI. This study uses a dummy variable of sea access by following the literature and is expected to have a positive association with FDI. The variable is coded as 1if countries have seaports otherwise zero.

Trade

The trade variable in this study includes the sum of exports and imports. The log of tradies used as an independent variable in the study as a measure of openness in a country. In trade, there is the transfer of ownership of goods and services from one person to another.

Population

Population refers to a measure of country annual population's growth percentage.

Literacy Rate

The quantity and quality of basic education received by the population, which is increasingly important in today's economy.

Gross Domestic Product (GDP)

GDP measures of the size of the markets. Domestic demand in the host country can play a crucial role in attracting 'market seeking' FDI, where the primary objective of MNCs is to serve the domestic market. Following the literature, this study uses per capita real GDP as a proxy for the domestic market size.

Market Capitalization

Market capitalization is the total value of the issued shares of a publicly traded company; it is equal to the share price times the number of shares outstanding.

4 Results and Discussion

The chapter deals with the information and statistical methods applied to the sample data. This chapter explains the results of the research about the relationship between autocracy and democracy with FDI in detail. The chapter reports the descriptive statistics and correlation analysis results. This chapter presents the multiple regression analyses and finally, the results are concluded.

4.1 Descriptive Statistics

The descriptive statistics are calculated for analyzing the basic features of variables selected in the data set. The values of mean and standard deviation for each variable are presented in Table 1.

Table 1. Descriptive statistics

	Minimum	Maximum	Mean		Std. Deviation	Variance
	Statistic	Statistic	Statistic	Std. Error	Statistic	Statistic
Fdi	−4.55	1.851	6.82384	1.2632	2.20608	4.867
Gdp	240.409	5.93	2.2289	3.56503	6.28701	3.953
Imports	0	1.52	6.40175	9.22337	1.636987	2.68
Exports	0	1.75	6.99157	1.09927	1.951026	3.807
Trade	0	3.27E+12	1.33993	2.02085	3.58666	1.286
Market capitalization	2.00E+07	6.00E+12	1.73E+11	4.02E+10	6.44E+11	4.146
Dem	0	10	4.65	0.204	3.607	13.009
Auto	−88	10	1.77	0.529	9.366	87.722
Sea access	0	1	0.87	0.019	0.34	0.116
Population	492891	1.00E+09	2.07E+08	2.14E+07	3.79	1.437
Literacy rate	34	99.4	7.8589	1.053042	18.68964	349.303

Table 1 shows that countries in the sample have a mean FDI of \$6.823842E9. The minimum value of FDI for sampled 15 countries is −4.5500E9 and maximum

value is 1.8510E11. The mean value of selected countries GDP is terms of US $ is
$2.228984E11. The mean value for imports is $6.401757E10 while the mean value
of exports amounts to $6.991571E10. The values of exports and imports amount
the total trade of these countries. The overall mean of trade in the selected Asian
countries is $1.339930E11 and market capitalization accounts to the mean value
of $1.73E11.

4.2   Regression Analysis

The regression analysis in the model is based on the dependent variable, independent variables and set of control variables. As discussed in Chap. 3, the FDI
is the dependent variable used in this study. The independent variables in the
current study include autocracy and democracy. Multiple regression analyses are
performed to test all the hypotheses for the models of the study.

4.3   Regression Analysis with Autocracy

Regression analysis is performed firstly by taking autocracy as a main independent variable with FDI. Table 2 presents the regression results of the model with
the autocracy performed on SPSS.

Table 2. Regression analysis with autocracy

Variables	Coefficients	t-statistic	Sign
(Constant)	4.296	2.379	0.018
Ln exports	−1.178	−0.95	0.343
Ln imports	1.24	0.998	0.319
Sea access	3.728	4.531	0.000
Ln market capitalization	0.099	3.004	0.003
Population	3.052E−09	3.994	0.000
Literacy rate	0.097	5.917	0.000
Auto	0.132	4.674	0.000
R-squared	0.338		
Adjusted R-squared	0.323		
F-Statistic	22.345		

The model results presented in Table 2 shows that the value of R squared
is 33.8% for the selected model. The value of R squared .338 depicts that only
33.8% of the variation in the dependent variable is explained by the independent variables of the model. The 66.2% variations remain unexplained by these
independent variables (Autocracy, exports, imports, population, market capitalisation and literacy rate and sea access). Finally, the value for F-statistic is

22.345 and is significant at 5% level of significance. The significance of f statistics shows that the model is stable and reliable.

The coefficient of autocracy shows that autocracy has a significant and positive relationship with the FDI in developing Asian countries. The value of coefficient is 0.132. This value is significant at 1% significance level.

The values of the Log of imports and Log of exports reveal an insignificant relationship between them with FDI. However, the values of the log of market capitalization, sea access, population and literacy rate show the positive and significant association of these variables with FDI.

Sea access is a dummy variable in the study coded as 1 if countries have seaports otherwise zero. The value of the coefficient is 3.728 which are significant at 1% level of significance.

4.4   Regression Analysis with Democracy

Regression analysis is performed secondly by taking democracy as a main independent variable with FDI. Table 3 presents the regression results of the model with the democracy performed on SPSS.

Table 3. Regression analysis with democracy

Variables	Coefficients	t-statistic	Sign
Constant	−3.324	−1.263	0.207
Ln trade	12.993	4.112	0.000
Ln exports	−7.361	−4.111	0.000
Ln imports	−5.57	−2.402	0.017
Sea access	3.853	4.587	0.000
Ln market capitalization	0.117	3.427	0.001
Population	2.871E−09	3.663	0.000
Literacyrate	0.092	5.531	0.000
Democracy	−0.234	−2.767	0.006
R-squared	0.352		
Adjusted R-squared	0.335		
F-Statistic	20.689		

The model results presented in Table 3 shows that the value of R squared is 35.2% for the selected model. The value of R squared 0.352 depicts that only 35.2% of the variation in the dependent variable is explained by the independent variables of the model. The 64.8% variations remain unexplained by these independent variables (Democracy, exports, imports, population, market capitalisation and literacy rate and sea access). Finally, the value for F-statistic is

20.689 and is significant at 5% level of significance. The significance of f statistics shows that the model is stable and reliable.

The coefficient of democracy shows that democracy has a significant and negative relationship with the FDI in developing Asian countries. The value of coefficient is −0.234. This value is significant at 10% significance level.

Sea access is a dummy variable in the study coded as 1 if countries have seaports otherwise zero. The value of the coefficient is 3.858 which are significant at 1% level of significance. The value of sea access variable is almost close to the one calculated in the model of FDI with autocracy.

5 Conclusion

This study aims to examine the effect of Autocracy and Democracy on Foreign Direct Investment (FDI) inflows in 15 Asian developing countries. For empirical examination, this study used panel data over the period 1995 to 2015 by employing dynamic panel data. The study shows a significant and positive relationship with autocracy and significant and negative relation with democracy. The empirical finding suggests that the country can attract more FDI in the presence of more autocratic institutions and less democratic institutions. Among the controlling variables trade, market capitalization, population and literacy rate are significant with democracy while trade becomes insignificant with the main variable democracy. Moreover, democracy is not a favourable situation for foreign direct investment while autocracy is a good situation to attract foreign direct investment.

References

1. Agbloyor, E.K., Gyeke-Dako, A., et al.: Foreign direct investment and economic growth in SSA: the role of institutions. Thunderbird Int. Bus. Rev. **58**(5), 479–497 (2016)
2. Bano, S., Zhao, Y., et al.: Why did fdi inflows of Pakistan decline? from the perspective of terrorism, energy shortage, financial instability, and political instability. Emerg. Mark. Financ. Trade **55**(1), 90–104 (2019)
3. Bellos, S., Subasat, T.: Governance and foreign direct investment: a panel gravity model approach. Int. Rev. Appl. Econ. **26**(3), 303–328 (2012)
4. Chen, K., Nie, H., Ge, Z.: Policy uncertainty and FDI: evidence from national elections. J. Int. Trade Econ. Dev., 1–10 (2018)
5. Cuthbertson, K.: The expectations hypothesis of the term structure: The UK interbank market. Econ. J. **106**(436), 578–592 (1996)
6. Du, J., Lu, Y., Tao, Z.: Regional institutional strength and FDI location choice in China: implications for East Asian FDI source countries/areas. Tech. rep., Working Paper Series 9 (2008)
7. Guerin, S.S., Manzocchi, S.: Political regime and fdi from advanced to emerging countries. Rev. World Econ. **145**(1), 75–91 (2009)
8. Harms, P., Ursprung, H.W.: Do civil and political repression really boost foreign direct investments? Econ. Inq. **40**(4), 651–663 (2002)

9. Jensen, N.M.: Democratic governance and multinational corporations: political regimes and inflows of foreign direct investment. Int. Organ. **57**(3), 587–616 (2003)
10. Li, Q.: Democracy, autocracy, and expropriation of foreign direct investment. Comp. Polit. Stud. **42**(8), 1098–1127 (2009)
11. Li, Q., Resnick, A.: Reversal of fortunes: democratic institutions and foreign direct investment inflows to developing countries. Int. Organ. **57**(1), 175–211 (2003)
12. Libman, A., Vinokurov, E.: Is it really different? patterns of regionalisation in post-soviet central Asia. Post-Communist Econ. **23**(4), 469–492 (2011)
13. Mathur, A., Singh, K.: Foreign direct investment, corruption and democracy. Appl. Econ. **45**(8), 991–1002 (2013)
14. Mengistu, A.A., Adhikary, B.K.: Does good governance matter for FDI inflows? evidence from Asian economies. Asia Pac. Bus. Rev. **17**(3), 281–299 (2011)
15. North, D.C., Weingast, B.R.: Constitutions and commitment: the evolution of institutions governing public choice in seventeenth-century England. J. Econ. Hist. **49**(4), 803–832 (1989)
16. Okafor, E.E.: Youth unemployment and implications for stability of democracy in Nigeria. J. Sustain. Dev. Afr. **13**(1), 358–373 (2011)
17. Olson, M.: Dictatorship, democracy, and development. Am. Polit. Sci. Rev. **87**(3), 567–576 (1993)
18. Oneal, J.R.: The affinity of foreign investors for authoritarian regimes. Polit. Res. Q. **47**(3), 565–588 (1994)
19. Resnick, A.L.: Investors, turbulence, and transition: democratic transition and foreign direct investment in nineteen developing countries. Int. Interact. **27**(4), 381–398 (2001)
20. Tsamadias, C., Pegkas, P., Mamatzakis, E., Staikouras, C.: Does R&D, human capital and FDI matter for TFP in OECD countries? Econ. Innov. New Technol. **28**(4), 386–406 (2019)
21. White, H.: Using least squares to approximate unknown regression functions. Int. Econ. Rev. **21**, 149–170 (1980)

Part III
Oganizational Strategy

Strategic Management Model for Academic Libraries – The Case Study of Ilma University, Karachi

Asif Kamran[✉], Farhana Shoukat, Nadeem A. Syed, and Sheeraz Ali

Department of Management Sciences, ILMA University, Karachi, Pakistan
asifkamrankhan@gmail.com

Abstract. This study explored the acts of Strategic Management in academic libraries. The paper depends on a study that was led in 2015. Using ILMA University formerly Institute of Business and Technology (IBT) library as case study, it expected to figure out the way of moving ideal models in data administration, set up the key difficulties and discovered how Strategic Management was drilled.More imperative information was concurred through meetings with pioneers of academic libraries in the ILMA University), a business School in Karachi. A Strategic Management model was designed for academic libraries of the University to conquer the hindrances that they may confront in their future systems. Models give a framework to envisioning reasonable movement. Educational libraries have had long-standing associations with the investigation environment. In the blink of an eye, as they tune in setting up future purposes, chairmen may be all around served by the orchestrating perspectives being grasped by school research executives and imperative authorities. The Model, which is displayed in this paper, is an eventual outcome of an extensive audit of Strategic Management compositions. It gives a tenet to insightful libraries' pioneers to grasp Strategic Management gauges and practices in their libraries and to beat the obstacles that they may stand up to in their future strategies.

Keywords: Model · Strategic management · Library · University · Students

1 Introduction

Academic libraries oblige the assorted needs of scholars, researchers, technocrats, specialists, students, and others actually and professionally put resources into advanced education.

Strategic Management is a progressing procedure worried with the ID of key objectives, vision, mission and targets of an association alongside an investigation of its present circumstance, create fitting systems, put these methodologies without hesitation, and assess, adjust or change these techniques when required [6].

© Springer Nature Switzerland AG 2020
J. Xu et al. (Eds.): ICMSEM2019 2019, AISC 1002, pp. 537–545, 2020.
https://doi.org/10.1007/978-3-030-21255-1_41

Strategic Management is the critical instrument in the administration of university libraries. The investigative and successful key management concerns the acknowledgment of strategic management and future improvement of university libraries [8].

There has been minimal deliberate observational exploration concentrating on how scholastic libraries ought to strategically react with a specific end goal to stay significant. All around, academic libraries have been trying to maintain focused capabilities in the changing environment [5].

Libraries must keep on demonstrating their quality to the college and illustrate that the extremely huge venture made in the library is all around coordinated and well spent, a speculation in our structures, staff, and accumulations, as well as in the academic accomplishment of students and personnel [4].

ILMA University has a business school in Karachi was set up in 2001 in the private division by the Global Education Consultants Society (GEC) to advance higher education key for Pakistan's financial advancement. There are three campuses of University in different parts of Karachi and all campuses have separate libraries.

This paper intends to introduce a Model for the Strategic Management process in the Business School libraries, as an endeavor to assist academic librarians: with adopting Strategic Management standards and practices in Business School libraries; to conquer the hindrances that they may confront in their future systems; and to give a rule to figuring and actualizing procedures and also for the execution estimation process.

This paper means to present a Model for the Strategic Management process in the ILMA University libraries, as an attempt to help academic librarians: with receiving Strategic Management measures and practices in ILMA University libraries; to overcome the deterrents that they may stand up to in their future frameworks; and to give a guideline to figuring and realizing strategies furthermore for the execution estimation process.

2 Literature Review

Models give a structure to envisioning viable activity [1]. It is clear that library administrators must utilize administration instruments to run academic libraries' services. For measuring the quality of academic libraries and data administrations; or services of libraries, Mixed-model CAF-BSC-AHP and PAQ-SIBI-USP was produced by Melo and Sampaio [9].

Scholarly libraries and their establishments arrangement data administrations concentrated on the present and developing needs of their clients, and abstain from putting resources into what is not, or is no more, vital to them [14].

Scholarly libraries can't get ready for the future or position themselves on grounds until they comprehend their changing parts in the flow learning and research environment. Understanding and assessing library utilization designs and formative ways are essentials to figuring a basic and fitting reaction to across the board, fast changes in advanced education [15].

Every scholarly library exists in a domain in which advancement and examination are the standard. It is along these lines intelligent to construe that scholarly libraries are very much situated to take an interest and go for broke with a specific end goal to contend in conveying administrations to scientists and educators alike [12].

For libraries to keep on accomplishing their part and reason they must react to the advanced, money related and societal changes by comparatively difficult their own particular manners of working [16].

Quick advancements in innovation, and additionally changes in regions, for example, insightful correspondence, information administration, and advanced education instructional method are influencing client desires and driving scholarly libraries to grow new assets and administration regions [13].

Strategic planning served as administration apparatus to give libraries in the study with course and concentrate; then again, there remains a solid requirement for pioneers to impart the significance of consolidating assessment into the arranging procedure [10].

Libraries need to play a strategic part in supporting scholastics. The strategic drivers for change shift from organization to foundation: from the need to decrease expenses and the need to adjust administrations to the open doors new building improvements bring, to the need to make joined administrations, and the need to modernize and guarantee adaptability for what's to come [3].

Strategic objectives converge with Bold Aspirations by centering the Libraries on dynamic and deliberate mix into showing and learning exercises, grant and research endeavors, and financing and backing of workforce exploration as comprehensively as could be expected under the circumstances [2].

Libraries were chosen in view of accessible web archives that gave confirmation of (1) eminent models of examination administrations in the libraries' web locales, arranging archives, vital arrangements, association graphs; (2) customary or more up to date structures reference administration that are situated as one piece of a by and large vital cluster of examination administrations; and (3) probability that the talked with library would add to a more up to date library calling wide model of exploration administrations [11]. They provide a panorama of how academic libraries can be thinking about their role, relative to the mission of their institutions and their contribution to society [7].

3   Methodology

Content investigation of writing gave profitable data about past models and ideas of key administration. More imperative information was concurred through meetings with pioneers of academic libraries in the ILMA University, has a business School in Karachi. SM model was designed for academic libraries of conquer the hindrances that they may confront in their future systems.

4 Strategic Management Model

Demonstrating the key administration procedure permits all who are included to offer a typical casing of reference for their management exercises. It additionally allows organizers to screen progress toward the advancement of a ton of results.

This paper introduced a Strategic Management Model for the Business School libraries, as an endeavor to assist academic librarians: with adopting Strategic Management standards and practices in ILMA University libraries; to conquer the hindrances that they may confront in their future systems; and to give a rule to figuring and actualizing procedures and also for the execution estimation process.

Figure 1 show the Strategic management model explains the following hierarchy:

▶ Arranging the originator
o Choosing development leader
• Provost for academic affairs
• Provost council
• Academe tic department
• University faculty library committee
• University office of research services
• University office of department

▶ Scanning the environment
o Scanning the target areas
• University and library mission
• Library organizational structures and staff
• University priorities
• Library funding levels and sources
• Services needs of faculty
• Services needs of students
• Library technology
• Library facilities and equipment

▶ Examine the strategic options
o Library improvement inventiveness

▶ Planning unit arrangement
o Library development targets
o Financing, formation & recruitment
o Ongoing planning and evaluation

▶ Acknowledgment of agenda
o Library acceptance
o University acceptance

▶ Implementing the strategic plans
o Stakeholders
o Implementation stages

4.1   Arranging the Originator

In Fig. 1, arranging the originator is from time to time talked about by scholars on vital arranging, be that as it may, the idea given to picking the best individuals to do arranging is basic for the achievement of the whole process. The library organization imagines the imperative authoritative arranging structure and after that chooses the best organizers accessible. Power is given to organizers in a particular charge. Power for the arranging configuration can take after customary various level lines or useful divisions, depending on the boss manager's affinity for overseeing arranging exercises.

Vital arranging works on the suspicion that individuals with comparative inspirations can concede to what their common reason ought to be and can shape gainful organizations that will propel a mutual hobby. On the off chance

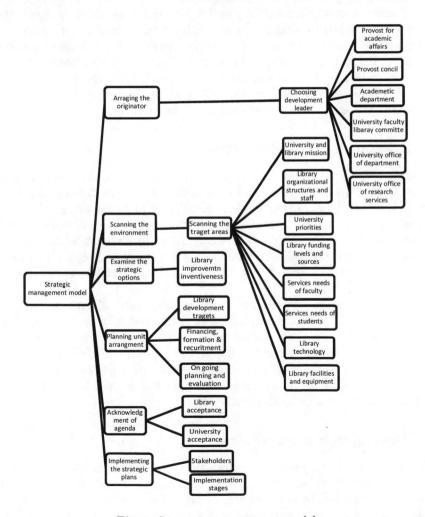

Fig. 1. Strategic management model

that long haul achievement is to be acknowledged, it is important that awareness, advocacy, and acknowledgment of required change include the library's
boss accomplices and significant electorates. Research libraries have moved from
being supply to request driven operations, and this new arrangement requires
extensively based affiliations. This incorporates the situating of key library overseers on arranging and choice making groups of on-and off campus associations.
Library overseers search for chances to pass on their message and responsibilities.

4.2   Scanning the Environment

Scanning the environment is a standard component of Strategic Management.
They distinguish sway regions, and additionally current conditions and future
elements that will probably influence the organization or unit. In Fig. 1, scanning
of environment ought to concentrate on those ranges applicable to propelling the
mission of the university and the library. The data assembled can build up an
identifying so as to arrange structure the key players and bringing into center
their needs, issues, suppositions, and opportunities. Data is acquired from various
sources, including librarians, campus chairmen, staff, and understudies. They are
asked where the library is today, where it ought to be in five a long time, and
the progressions expected to move in the sought heading.

4.3   Examine the Strategic Options

The focal points and disservices of different choice ways are recognized and
looked at. In asset improvement, situations can be cast in different ways. One
strategy is to direct regard for the results of different planning levels, for example,
decrease objectives, insignificant objectives, and most extreme objectives. Different methodologies incorporate a business as usual or incremental approach, an
"entrance over proprietorship" introduction, a scaremonger choice (e.g., purchase
no books and decrease all building hours), and an investigation and advancement
point of view.

Targets outer to the library, for example, corporate affiliations, individual
giving and the college enrichment, should be addressed. However they are introduced, an examination of vital alternatives ought to unite the best situations
and permit organizers to prescribe a few strategies at distinctive levels. The situations permit all to see what has been suggested from a specific arrangement
of circumstances. Library managers can then set reasonable objectives that will
direct the library for an amplified time frame.

4.4   Planning Unit Arrangements

In Fig. 1, at this stage, organizers adjust the objectives of the library with the
objectives of the library's supporters and the mission of the university. Library
executives coordinate the best thoughts of arranging members with the library's
restricted assets so as to accomplish a predefined result. They must settle on

how advancement endeavors will be financed, engaged, sorted out, and staffed. A project for consistent arranging and assessment is made.

The vital arranging record will give the premise to the case explanation utilized as a part of raising money exercises. Of specific significance is the enunciation of financing objectives. Library needs must be effortlessly comprehended and attractive to contributors. In the meantime, these projects and undertakings need to speak to precisely the needs of the library. On the off chance that financing is accomplished for territories not perfect with arranging objectives, the library gets to be receptive and wanders from the reasons that have been carefully built with the assistance of bodies electorate and accomplices.

4.5   Acknowledgement of the Agenda

The acknowledgment period of key arranging is frequently disregarded by organizers. The contrasts between key arranging and different sorts of arranging should be caught on. Operational arrangements are the method for accomplishing institutional objectives as per the financial backing introduced to the unit. Strategic arrangements decide the particular destinations for accomplishing those objectives. Unless key arranging is directed to begin with, both operational and strategic planning is restricted to the heading set for the unit by an outside office.

Acknowledgment relies on upon illuminating partners about what is being arranged and how their own objectives are progressed by it. Planning a library arrangement, notwithstanding when it includes agents from effect territories, concerns the explanation of working exercises at one authoritative level. Since university force is described by the sharing of power and assets, the acknowledgment stage includes acceptance from others that the arrangement is beneficial and can be advanced as a major aspect of the college motivation. Supporting accomplices, both on and off grounds, will have distinctive arranging and planning instruments from those of the requesting unit, and modification will be made so as to adjust the arrangements, calendars, and spending plans of all concerned.

4.6   Implementing the Strategic Plan

Selection of a key arrangement can be thought to have happened when principals confer their assets to the progression of basic objectives. More vital than the measure of upgraded income is the foundation of a joint arranging process which will serve the common needs of members far into what's to come.

Implementation happens in the accompanying stages:

- Precisely picked organizers imagine the thoughts for change and venture that vision to the unit's numerous constituents.
- Arrangements are championed that will change the unit's future.
- All stakeholders know the advantages and outcomes of the arrangement and feel that the objectives are feasible.

- Stakeholders give endorsement of a calendar for usage of the arrangement.
- Principals bolster and embrace both the arrangement and the vital arranging procedure. They execute targets of the arrangement and contribute their assets to the accomplishment of common objectives.

Strategic management involves understanding the strategic position of an organization, strategic choices for the future and managing strategy in action. Strategic management involves exploring and management of an organizational corporate strategy. It also involves modeling and analyzing the overall corporate strategy of the system to include the strategic position of the organization, strategic choices by the organization and strategy in action within and around the organization

5 Conclusion

Models give a system to picturing viable activity. Scholastic libraries have had long-standing connections to the exploration environment. Presently, as they take part in setting up future purposes, administrators may be all around served by the arranging points of view being embraced by college research directors and important specialists.

The Model, which is exhibited in this paper, is an aftereffect of a far reaching review of Strategic Management writings. It gives a rule to scholarly libraries' pioneers to embrace Strategic Management standards and practices in their libraries and to beat the snags that they may confront in their future procedures.

Future Research

A Strategic Management model was designed for academic libraries of the University to conquer the hindrances that they may confront in their future systems. Models give a framework to envisioning reasonable movement. Educational libraries have had long-standing associations with the investigation environment. In the blink of an eye, as they tune in setting up future purposes, chairmen may be all around served by the orchestrating perspectives being grasped by school research executives and imperative authorities. The Model, which is displayed in this paper, is an eventual outcome of an extensive audit of Strategic Management compositions. It gives a tenet to insightful libraries' pioneers to grasp Strategic Management gauges and practices in their libraries and to beat the obstacles that they may stand up to in their future strategies.

References

1. Birdsall, D.G., Hensley, O.D.: A new strategic planning model for academic libraries. Coll. Res. Libr. **55**(2), 149–59 (1994)
2. Ellis, E.L., Rosenblum, B.: Positioning academic libraries for the future: a process and strategy for organizational transformation. In: Proceedings of the IATUL Conferences United States: Purdue University, p 13 (2014)

3. Feret, B.: Library as a hub. Changing roles and functions of academic library. In: 2011 IATUL Proceedings New Zealand: Purdue, e-Pubs:35 (2011)
4. Franklin, B.: Aligning library strategy and structure with the campus academic plan: a case study. J. Libr. Adm. **49**(5), 495–505 (2009)
5. Gichohi, P.M.: The strategic management practices in academic libraries in kenya: the case of USIU library. Libr. Philos. Pract. 1–27 (2015)
6. Hijji, K.Z.A.: Strategic management model for academic libraries. Procedia - Soc. Behav. Sci. **147**(147), 9–15 (2014)
7. Kaspar, W.A.: Strategic Management and Situational Awareness in Academic Libraries (2018)
8. Li, H., Wang, N.: Study on strategic management of universities based on competitive intelligence. In: International Conference on Management and Service Science (2011)
9. Melo L.B., Sampaio, M.I.: Quality measures for academic libraries and information services: two implementation initiatives-mixed-model CAF-BSC-AHP and PAQ-SIBI-USP (2006)
10. Piorun, M.: Evaluation of strategic plans in academic medical libraries. Libr. Inf. Sci. Res. **33**(1), 54–62 (2011)
11. Piorun, M.: Reference Service at an Inflection Point: Transformations in Academic Libraries, pp. 491–499. Indianapolis, IN (2013)
12. Pongrácz Sennyey, K.K.: Libraries and the network: some considerations on how libraries are affected by the network: 1–18 (2011)
13. Saunders, L.: Academic libraries' strategic plans: top trends and under-recognized areas. J. Acad. Libr. **41**(3), 285–291 (2015)
14. Troll, D.A.: Changes in usage, usability, and user support. Fourth Northumbria 349–354 (2001)
15. Troll, D.A.: How and why libraries are changing: what we know and what we need to know. Portal Libr. Acad. **2**(1), 99–123 (2002)
16. Wade, M.: Re-inventing the Library-the Role of Strategic Planning, Marketing and External Relations, and Shared Services at the National Library of Scotland. Creating a Culture for Innovation and Change-Management and Marketing with Academic and Research Libraries Scotland: Management and Marketing with Academic and Research Libraries 94 (2012)

Research on the Factors Affecting the Delisting of Chinese Listed Companies

Yanyan Zhang[✉]

School of Economics, Sichuan University, Chengdu 610065,
People's Republic of China
10288348@qq.com

Abstract. This paper focus on the main factors affecting the uncap-listed companies, thus determining the influencing factors of the delisting companies and the overall implementation effect of the delisting system. Based on the data of 391 companies in China's A-share listed companies from 1989 to 2016, the Logistic model was used to study the factors affecting the delisting of listed companies. The results show that the proportion of state-owned shares, the proportion of tradable shares, the degree of loss and the age of listing are the main influencing factors of the delisting of listed companies, but have no significant relationship with the profitability, growth ability and solvency indicators on behalf of the company's comprehensive operational capabilities, which means the current delisting system of Chinese stock market was ineffective.

Keywords: Listed companies · Delisting · Factors · Factors analysis

1 Introduction

China's securities market reform is in the midst of a tough period. After the new Nine-Point guideline" proposed a perfect delisting system, the reform of the delisting system of listed companies was put on the agenda. At present, the primary task of the reform is to examine the delisting links and the overall implementation effect, and to identify the defects of the current delisting system. Perfect delisting system is conducive to achieving the survival of the fittest, improving market effectiveness and protecting the legitimate rights and interests of investors. When an abnormal situation occurs in a listed company, causing it to be delisted, the stock will be treated specially and the ST Mark will be placed before the stock. In general, special treatment is an essential part of the delisting of listed companies. The company must turn losses into profits within the specified time before it can apply for cancellation of special treatment, otherwise it will face the dilemma of being delisted. Therefore, ST (special treatment) companies usually actively promote the cancellation of special legal conditions to achieve the successful uncap of the company's stock. Cancellation Special treatment, which is, uncap is one of the common means of avoiding delisting. The influencing factors and implementation effects of the cap removal are of vital

© Springer Nature Switzerland AG 2020
J. Xu et al. (Eds.): ICMSEM2019 2019, AISC 1002, pp. 546–556, 2020.
https://doi.org/10.1007/978-3-030-21255-1_42

importance to the implementation of the delisting system. In order to avoid delisting, ST's motive for remove caps is very strong.

The main innovation of this paper is that the research method is different from the existing results. It is not based on whether the delisting is the standard for dividing the sample company, but the endogenous character of the delisting company is analyzed based on whether or not the cap is removed. To determine the influencing factors of the delisting of listed companies in China, Specifically, this paper takes the data of 391 companies in China's A-share listed companies from 1989 to 2016 as a sample, and constructs a logistic model for empirical analysis. The general structure of the thesis is: firstly, based on the basic theory, the variables and samples are selected, and then the theoretical hypothesis of the influencing factors is made from the existing literature and theory. The empirical analysis part selects the significant variables through descriptive statistics and regression analysis. Finally, test the results and draw conclusions and recommendations.

2 Literature Review

In recent years, there are three main researches on the factors affecting delisting in foreign literatures. (1) the first is to divide the reasons for delisting into passive delisting and active delisting, and to study their causes separately. For example, Martinez [1] summarizes and demonstrates the reasons for delisting by literature research. Konno [2] studies the impact of equity concentration and company size on listed companies' active delisting. (2) the second is to discuss different countries from a national perspective to set a delisting standard. For example, Charitou [3] demonstrated the equity structure standard in the delisting rules of the New York Stock Exchange. Zhou [4] compared the delisting standards between China and the United States, and proposed suggestions for China. (3) the third is to use the empirical method to build a delisting early warning model based on the delisting factors. For example, Garcia [5] and Hwang [6] explored this issue from a financial and non-financial perspective. It can be seen that the special treatment links unique to China's delisting system have not been reflected in foreign literature.

The domestic literature has a lot of research on the factors affecting the capping and delisting of listed companies. The main conclusions are as follows: (1) the government subsidies obtained by the company can significantly help the company get out of trouble [7]. (2) the company background, company size, financial leverage, and the shareholding ratio of the largest shareholder significantly affected the company's cap [8]. (3) the factors that really affect the realization of ST uncaps are not the size of the company and the level of profitability, but the macroeconomic situation, the nature of controlling shareholders, non-recurring gains and losses, and other listed companies themselves cannot directly control factor [9]. (4) when the company is ST, whether it can get help and take off the cap of ST depends on the background of the controlling shareholder of the company, the proportion of controlling shareholders, the size of the listed company, etc., and the quality of the company itself is not too more relationships [10]. It can be seen that the domestic literature has more theoretical analysis and less empirical analysis on the factors affecting delisting. There are more studies on delisting as the classification criteria in empirical analysis, but less research on the extraction criteria as the classification criteria; the system is also very imperfect.

This paper pays more attention to which intrinsic features of St company significantly affect the possibility of being saved, rather than focusing on which external means can be used to remove the cap, as in the existing literature, so when constructing the econometric model, the variables are all the state before the company is St, not the state after the company is St.

3   Sample Selection and Variable Settings

According to the current delisting system, the delisting procedure of listed companies in China is shown in Fig. 1. When a listed company terminates its listing; it usually starts with special treatment. If the listed company is specially treated, it may apply for cancellation of special treatment (i.e., uncap), but if the situation deteriorates further, it will face the suspension of listing. After that, companies that are suspended from listing may face two outlets: resumption of listing or termination of listing.

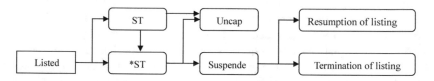

Fig. 1. Delisting procedures for listed companies

This paper selects 442 listed companies that have been specially treated from 1998 to 2016 as sample companies, and divides the sample companies into uncapped companies and non-uncapped companies according to whether they have experienced cancellation of special treatment. There are two major categories of ST companies. Specifically, 442 companies that were once listed by ST were divided into two categories of sample companies, namely 266 uncapped companies that have experienced the cancellation of special treatment and 176 non-uncapped cap companies that have not experienced the cancellation of special treatment. And they are used as the initial sample of the two major categories of research. In order to make the selection of samples more reasonable, companies with more serious data missing in the initial sample were excluded. In the end, 391 companies were selected as research samples, including 260 uncapped sample companies and 131 non-uncapped sample companies. All data are from the GuoTaiAn database and the Wind information database.

This paper mainly examines the main factors affecting ST's cap removal through empirical methods, which means that what kind of ST companies are easier to be uncapped. This paper mainly examines the characteristics of the sample company before the ST, and the time value of the sample data is also defined in the year before the listed company was first ST or the year of ST. The specific research variables and value methods are shown in Table 1:

Table 1. Variables and definitions

Variable name	Symbol	Definition and value method
Uncap or not	Recovery	Whether ST company successfully uncap, it takes 1 or no.
Company Size	Size	The natural logarithm of the company's total assets by the end of ST
Degree of loss	Loss	Net assets per share of the company at the end of the year
Age of listing	Age	The year the company was ST - the year the company was listed
Major shareholding ratio	First	The shareholding ratio of the company's largest shareholder by the end of ST
State-owned shares	State	Number of state shares/total shares of the company by the end of ST
Proportion of tradable shares	Float	Number of tradable shares/total number of shares of the company by the end of ST
Profitability	Roe	The company's return on equity before the end of ST
Growth ability	Growth	The growth rate of the company's operating income by the end of ST
Solvency	Debt	The company's asset-liability ratio at the end of the previous ST

4 Theoretical Analysis and Research Hypothesis

Hypothesis 1. The size of the company is positively related to the results of the cap removal. Ramanujam [11] argues that the larger the company, the more resources it can call, so the size of the company is positively related to the probability that the company's performance will turn to success. Witmer [12] and Stein [13] argue that the possibility of a small company delisting is greater from the perspective of corporate liquidity. Hongyi [14] and Fei [15] also believe that the company's operating performance is significantly positively related to the company's size. The research by Fama [16] and He [17] concluded that companies with smaller scales are more likely to turn to success. Based on the mainstream view of the theoretical community, we get hypothesis 1.

Hypothesis 2. The smaller the company's loss, the easier it is to remove the cap. The original intention of the cap removal system is to restore the stock status of ST companies with good performance to normal. Then, the lower the degree of loss in ST companies and the higher the level of business performance, the more likely it is to remove the cap.

Hypothesis 3. The smaller the company's listing age, the greater the possibility of obtaining assistance to achieve uncap. Foreign scholars have conducted a lot of research on the relationship between company age and performance. Guo [18] and Audretsch and Elston [19] through empirical analysis, it is believed that the company's performance growth rate increases with its age. There are few differences in the existing research conclusions. Yasuda [20] believe that the longer the time to market, the higher the proportion of the company's losses; the company's growth is negatively related to the company's listing age. According to the mainstream view of the theoretical circle, we get hypothesis 3.

Hypothesis 4. The higher the shareholding ratio of the company's major shareholders, the easier it is to obtain assistance to achieve uncap. Berle, Means [21] and Wu [22] believe that major shareholders have a positive effect on the performance growth of listed companies. Thomsen [23] and Hui [24] argue that there is a non-linear relationship between equity concentration and firm performance, and that company performance will only increase when equity concentration is moderate. Combined with the actual situation of China's securities market, make the hypothesis 4.

Hypothesis 5. The higher the proportion of the company's state-owned shares, the greater the possibilities of obtaining aid to achieve uncap. China's securities market has long been dominated by state-owned shares and state-owned legal person shares, and state-owned listed companies in the market are in a relatively strong position. Once the company is facing difficulties, when the actual controller of the listed company is the subject of the state, the rescue channel will be wider and the rescue will be relatively stronger. Therefore, we have the hypothesis 5.

Hypothesis 6. The lower the proportion of the company's tradable shares, the greater the possibility of obtaining assistance to achieve the cap removal. China's securities market has experienced a long period of split shareholding, and more than two-thirds of the shares cannot be listed for circulation. Since the tradable shares can be listed and traded, when the company faces a financial crisis, the higher the proportion of the company's tradable shares, the faster its stock price will fall through the transaction. In addition, since tradable shares are more dispersed than non-tradable shares, the higher the proportion of tradable shares, the lower the efficiency of external rescue measures.

Hypothesis 7. The higher the return on net assets and the growth rate of operating income. The lower the asset-liability ratio, the more likely it is that the company will receive assistance to achieve uncap. The company's profitability, growth ability and solvency respectively affect the company's operating performance to a certain extent, and they are usually positively related to the company's performance. According to financial practice, we represent the company's profitability, growth ability and solvency by ROE, operating income growth rate and asset-liability ratio. Therefore, we assume the hypothesis 7.

5 Empirical Analysis Results

5.1 Descriptive Statistics

In order to analyze the correlation between various internal factors of ST and its cap removal results, the two groups of sample companies were first tested for mean difference by independent sample T test, which was used for the uncapped company (Recovery=1) and the non-uncapped company (Recovery = 0). Their some important features will be compared below. Table 2 is a descriptive statistical list of the indicators of the two groups of sample companies. From the Table, we can see that there are significant differences in the mean values of other variables except for the two variables of company size and profitability.

5.2 Logistic Model Regression Analysis and Results

This article will use the dichotomous logistic regression analysis to determine the main internal factors affecting ST's cap removal, that is, which ST companies are more likely to receive assistance to avoid delisting. A large number of analytical practices show that the logistic regression model can well meet the modeling requirements for categorical data, so it has now become the standard modeling method for categorical dependent variables. Taking $logit(P)$ as dependent variable, a Logistic regression model containing P independent variables is set up as follows:

$$\log it\,(P) = \beta_0 + \beta_1 x_1 + \cdots + \beta_p x_p. \tag{1}$$

This is the Logistic regression model, from which we can get:

$$P = \frac{exp\,(\beta_0 + \beta_1 x_1 + \cdots + \beta_p x_p)}{1 + exp\,(\beta_0 + \beta_1 x_1 + \cdots + \beta_p x_p)} \quad 1 - P = \frac{1}{1 + exp\,(\beta_0 + \beta_1 x_1 + \cdots + \beta_p x_p)}. \tag{2}$$

β_0 is a constant term, and β_1 (i $= 1 \cdots P$) is the regression coefficient for each variable. Whether the ST Company takes the recovery as the explanatory variable, the Model 1 introduces all the explanatory variables into the model, and the Model 2 is obtained by introducing the variables one by one into the model.

From the regression analysis results in Table 3, it is known that among the selected explanatory variables, there are four variables that have significant influence on predicting whether ST companies can remove caps. They are the degree of loss, company age, state-owned shares, and tradable shares proportion. The characteristics of the regression coefficients of these four explanatory variables indicate two points: First, the direction of the correlation, that is, whether the four variables are positive or negative for the ST Company's cap removal. Specifically, since the regression coefficients of the variable Loss and the variable State are positive, these two variables have a positive effect on the uncap; the regression coefficients of the variable Age and the variable Trade are negative, and these two variables are negative effects for the cap removal. The second is the size of the correlation, that is, according to the size of the regression coefficient of each variable, it can be determined that the influence of the four variables on the cap removal of the ST company is from the largest to the smallest. the proportion of state-owned shares, the degree of loss, the age of the company and the proportion of tradable shares. The specific analysis is as follows:

(1) The higher the proportion of state-owned shares is, the greater the possibility that ST companies will remove their caps. The regression coefficient of the variable State is 1.873, which means that for each unit of the company's state-owned shares, the probability of taking off the cap increases by $0.25*1.873 = 0.468$ units, and 0.25 is the marginal rate of change of the conditional probability. The coefficient of the variable State is larger than the coefficient of other variables, so the influence of the proportion of state-owned shares on the removal cap of ST

Table 2. Descriptive statistics

Explanatory variables	Whether to remove the cap	Number of samples	Mean	Standard deviation	T value	Sig.
Size	0	131	20.7029	0.84798	0.619	0.536
	1	260	20.6435	0.91927		
Loss	0	128	0.2578	0.43915	−7.268	0
	1	260	0.6154	0.48744		
Age	0	131	7.9400	3.07000	4.373	0
	1	260	6.5800	2.55300		
First	0	131	31.1208	13.29194	−5.761	0
	1	260	40.0636	16.61133		
State	0	131	0.1800	0.20800	−6.089	0
	1	236	0.3300	0.25300		
Trade	0	129	46.2023	14.66659	4.161	0
	1	257	40.0785	11.31480		
Roe	0	107	−1.5649	3.23924	−1.572	0.118
	1	248	−1.0174	2.40332		
Growth	0	130	−0.2946	0.49729	−3.826	0
	1	259	−0.0951	0.46009		
Debt	0	131	0.9326	0.97723	3.088	0.002
	1	260	0.6647	0.24803		

Note: The homogeneity of the variance of each variable sample has passed the Levene test and is directly included in the results of the T test

Table 3. Logistic regression results

Variable	Model 1			Model 2		
	β	Sig	$\exp(\beta)$	β	Sig	$\exp(\beta)$
Constant	4.735	0.176	113.885	2.195	0	8.983
Size	−0.134	0.462	0.875			
Loss	0.426	0.001	1.531	0.486	0	1.627
Age	−0.143	0.008	0.867	−0.152	0.002	0.859
First	0	0.937	0.999			
State	1.873	0.008	6.508	1.736	0.003	5.673
Trade	−0.023	0.055	0.977	−0.026	0.014	0.974
Roe	−0.032	0.588	0.968			
Growth	0.438	0.169	1.55			
Debt	0.182	0.824	1.2			
Predictive accuracy	74.90%			75.10%		

companies is the largest among the selected variables. The reason may be that the higher the proportion of state-owned shares is, the association between listed companies and the government will be more. The company with the government background will have some potential huge advantages in the special background of the Chinese stock market. Hypothesis 5 is confirmed.

(2) The lower the loss of listed companies, the easier it is to remove the cap. Here, the listed company is represented by the company's net assets per share at the end of the year, indicating the degree of loss of the ST Company. The lower the net assets value per share, the higher the company's loss. The regression results show that the regression coefficient of the variable Loss is positive, that is to say, the higher the company's net asset per share, the lower the degree of loss, the greater the possibility that the company will remove the cap. There may be some reasons: on the one hand, companies with lower losses are more likely to turn losses. Such companies can improve their business capabilities and improve their performance by adjusting their internal corporate governance structure, changing management mechanisms, and broadening financing channels. On the other hand, companies with lower losses are more likely to attract assistance from the main body of the rescue. On the other hand, companies with huge losses are hard to return, and it is difficult to become an object of external rescue. As a result, companies with lower losses will be more likely to achieve cap removal through self-help or external assistance. Hypothesis 2 is confirmed here.

(3) Companies with smaller listings are more likely to remove their caps. The results of the regression analysis showed that the company's age of listing was negatively correlated with ST's cap removal results. Combined with the existing research conclusions, the younger the listed company, the more likely it is to remove the cap, mainly because of the following aspects:

Judging from the company's own operating conditions, older companies are more likely to enter a recession than younger companies, and it is more difficult that companies in a recession are trying to remove their caps in the short term. Second, this may be a result of the weakening or reversal of the earnings management capabilities of listed companies. Chinese listed companies generally have a phenomenon of earnings management in the process of IPO or share allotment. The company's earnings management ability is limited after all. If the degree of earnings management is too large, it will face a larger turn in the next year's reversal. The pressure has even caused losses in the company's performance. The longer the time on market is, the greater the pressure on the earnings management that the company may face, and the greater the resistance of ST tries to remove the cap. Third, the cause of this phenomenon may also be related to the subjective preference of the external rescue object. For example, the government is more willing to subsidize the representative enterprises of high-tech industries, and has set up special financial subsidy funds for such listed companies in terms of policies, and the age of such companies is generally small. Hypothesis 3 is confirmed.

(4) ST companies with a low percentage of tradable shares are more likely to remove their caps. Under the system that state-owned shares and legal person shares cannot be circulated, the stock price of the secondary market is generally high. Moreover, the higher the proportion of non-tradable shares in the company's shares, the higher the share price allocated to the tradable shares. The higher the stock price is, the higher the value of the shell resources is, and the greater the incentives for listed companies are. That is to say, when a company with a high proportion of non-tradable shares faces the risk of delisting, it is more likely to get out of the crisis. Hypothesis 6 is confirmed.

(5) The size of the company, the shareholding ratio of major shareholders, the company's profitability, growth ability and solvency are not significant in predicting whether or not to remove the cap.

6   Research Conclusions and Policy Recommendations

The main conclusions of this paper are: ST Company's successful cap removal depends mainly on the company's state-owned shares, the proportion of tradable shares, the degree of loss and the age of listing, and there is no significant relationship with indicator of profitability, growth ability and solvency of the company's comprehensive operational capabilities. It can be seen that the delisting system did not achieve the original intention of the system, which design for the survival of the fittest in the implementation of the special treatment segment, that is, the delisting system was "ineffective".

The policy recommendations in this paper are: Since the delisting system of listed companies in China has been deeply affected by institutional problems, China's securities market has always had two major institutional problems, namely, "taking state-owned shares and state-owned legal person shares as the main body" and "taking non-tradable shares as the main body". The listed company's shareholding structure, such structural characteristics are contrary to the most basic principles of the market economy, leading to the irrational structure of China's securities market and the government's excessive involvement in the securities market. At the same time, it is also the institutional roots of the market mechanism of the securities market that cannot be effectively implemented, including the nuisance of the implementation process of the delisting system. It distorts the original intention of the delisting system and weakens the implementation effect of all aspects of the delisting system.

According to the analysis conclusions of this paper, the proportion of state-owned shares and the proportion of tradable shares have a significant impact on whether ST can achieve the uncapping, but many factors related to the quality of the company itself have little impact on the uncap. It can be seen that the existing delisting system in China clearly reflects the shortcomings of these two major institutional problems in the special handling of revocation. Under the profound influence of the institutional problems, China's delisting system is still difficult to achieve the design effect of inferior storage and purification of the market. Therefore, in order to improve the implementation effect of the delisting system in China's securities market, we should first solve the institutional problems of the market, gradually increase the degree of mercerization of the securities market, and lay a good institutional foundation for the implementation of the delisting system.

Funding. The periodical results of national social science foundation: "The development and optimization research of the delisting system of listed companies under the background of registration system reform (17BJL093)". This paper is also funded by fundamental research funds for the central universities of Sichuan University, "Research on the fluctuation of market prices of agricultural products (selfresearch in 2019 - Economic 016)".

References

1. Martinez, I., Serve, S.: Reasons for delisting and consequences: a literature review and research agenda: reasons for delisting and consequences. J. Econ. Surv. **31**(3), 733–770 (2017)
2. Konno, Y., Itoh, Y.: Why do listed companies delist themselves voluntarily? J. Financ. Manag. Prop. Constr. **23**(2), 152–169 (2018)
3. Charitou, A., Louca, C., Vafeas, N.: Boards, ownership structure, and involuntary delisting from the new york stock exchange. J. Account. Public Policy **26**(2), 249–262 (2007)
4. Zhou, Y.: A comparative study of china and the us delisting system. Am. J. Ind. Bus. Manag. **10**, 855–863 (2017)
5. García, H.: Financial performance and stock prices into delisting companies in MSE. Innovaciones de negocios **19**, 85–105 (2017)
6. Hwang, I.T., Sun, M.K., Jin, S.J.: A delisting prediction model based on nonfinancial information. Asia-Pac. J. Account. Econ. **21**(3), 328–347 (2014)
7. Shanshan, Y.: Government subsidies and corporate financial distress recovery - an empirical analysis based on "uncapped caps" of ST listed companies. Econ. Res. Ref. **28**, 69–76 (2015). (in Chinese)

8. Cheng, W., He, W.: An empirical study on the performance of st companies' cap removal under delisting pressure. Friends Account. **02**, 53–59 (2014). (in Chinese)
9. Ke, F., Li, W.: Empirical research on the implementation effect of china's delisting system. J. Beijing Technol. Bus. Univ. (Social Science Edition), 78–88 (in Chinese) (2014)
10. Yun, F., Liu, Y.: Empirical analysis of the implementation effect of the delisting system of listed companies. J. Financ. Econ. **02**, 133–143 (2009). (in Chinese)
11. Ramanujam, V.: Environment context, organizational context, strategy and corpomte turn around. University of Pittsburgh, Pittsburgh (1984)
12. Witmer, J.: Why do firms cross-delist? an examination of the determinants and effects of cross-delisting. Financial Markets Department, Bank of Canada, Working Paper (2006)
13. Stein, V., Wiedemann, A.: Risk governance: conceptualization, tasks, and research agenda. J. Bus. Econ. **86**(8), 813–836 (2016)
14. Xia, H.: Internal audit quality, enterprise scale and corporate performance: empirical study of listed companies based on panel data. J. Cent. Univ. Financ. Econ. **6**, 71–78 (2016). (in Chinese)
15. Fei, X., Wang, H.: Firm size, market competition and implementation performance of r&d subsidy. Sci. Res. Manag. **39**(7), 43–49 (2018). (in Chinese)
16. Fama, E.F., French, K.R.: The cross-section expected stock returns. Finance **47**(2), 427–465 (1992)
17. He, Y.: Corporate liabilities, social responsibility and corporate performance. Account. Mon. **06**, 3–10 (2018)
18. Guo, M.: Industrial agglomeration, enterprise age and government subsidies. Finance **36**, 52–56 (2017)
19. Audretsch, D.B., Elston, J.A.: Can institutional change impact high-technology firm growth?: evidence from germanys neuer markt. J. Prod. Anal. **25**(1), 9–23 (2006)
20. Yasuda, T.: Firm growth, size, age and behavior in Japanese manufacturing. Small Bus. Econ. **24**(1), 1–15 (2005)
21. Berle, A.A., Means, G.C.: The Modern Corporation and Private Property: A Reappraisal. Macmillan, New York (1932)
22. Wu, W.: Ownership structure, diversification and corporate performance. Discuss. Mod. Econ. **7**, 99–109 (2018). (in Chinese)
23. Thomsen, S.: Ownership structure and economic performance in the largest european companies. Strateg. Manag. J. **21**(6), 689–705 (2000)
24. Hui, C.: Optimized allocation method of service elements based on fuzzy evaluation information. Stat. Decis.-Mak. **2**, 176–180 (2018). (in Chinese)

Customer Relationship and Efficiency Analysis of the Listed Air Companies in China

Ying Li[1], Zelin Jin[1], Hongyi Cen[1], Yung-ho Chiu[2(✉)], and Jian Jiao[3]

[1] Wangjiang Road No. 29, Chengdu 610064, People's Republic of China
[2] Department of Economics, Soochow University, 56, Kueiyang St., Sec. 1,
Taipei 100, Taiwan
echiu@scu.edu.tw
[3] Chinese Academy of Sciences, Zhongguancun Nansijie No. 4,
Beijing 100190, People's Republic of China

Abstract. With the rapid development of globalization and the improvement of residents' income levels, the scale of air transport business has grown steadily, resulting in a substantial increase in customer traffic. The degree of marketization of airlines has been continuously improved, and operational efficiency has been significantly improved. Chinese air passenger transport market has gradually reached the level of development in North America and Europe. However, major domestic airlines face fierce competition from domestic and international counterparts. Building good customer relationship efficiently matters. This study uses a two-stage dynamic DEA method to evaluate the impact of customer complaints on airline operational efficiency by comparing the financial data of seven listed airlines in China from 2014 to 2017. The network data dynamic analysis method was used to analyze and evaluate the operating efficiency and productivity index of the seven companies. The results of the study indicate that customer complaints have a positive and significant impact on the airline's operational efficiency. In addition, The smaller Shandong Airlines, Hainan Airlines and Jixiang Airlines have higher total efficiency scores than the three largest airlines in ChinaAir China, Southern Airlines and Eastern Airlines. The operating efficiency of the latter three companies is less than 0.4, which is a lot of room for improvement. The airline with high operating efficiency considering customer satisfaction is Spring Airlines, and the company has a trend of continuous improvement in efficiency.

Keywords: Chinese listed airline company · Customer complaints · Two-stage dynamic DEA

© Springer Nature Switzerland AG 2020
J. Xu et al. (Eds.): ICMSEM2019 2019, AISC 1002, pp. 557–573, 2020.
https://doi.org/10.1007/978-3-030-21255-1_43

1   Introduction

With fast growing economy and globalization, the scale of customer traffic has increased significantly. It has led to a steady increase in the scale of air transport business. The airline's operational efficiency has been significantly improved, and the marketization of the industry has continued to increase, making the air transport industry a great development and becoming an important industry in social development and economic growth.

Air service industry in China maintains a steady and rapid development in the future. While, the quality of service of Chinese airlines is declining. The customer complaint rate of major airlines continued to increase and arrive a highest level in 2017 [1]. Airline customer's satisfaction will affect on airline performance and the competitiveness of airlines [2,16].

This study applied a network dynamic research model to analyze the efficiency of Chinese listed air company to study the service quality of the companies and its impact on the efficiency of the companies.

2   Literature Review

There are some previous studies which are focus on business performance assessment for aviation or aviation-related industries with traditional DEA models or regression analysis methods. Domestic and foreign scholars use DEA to study the operation and financial performance of airlines mainly including CCR-DEA method, SUP-CCR-DEA method, network RAM model, Malmquist index method and other models. They identified areas for improvement in technical efficiency and scale efficiency for each sample company in different years [9,12]; more relative studies could be the following: Jiang Youhui and Wen Jun used the financial data of the five major airlines in China in 2006, using the CCR-DEA method and the SUP-CCR-DEA method to analyze [7]. Tan Yushun and Chen Senfa used the operational indicators of 16 airlines in 2008 to evaluate the operational performance of air transport companies using a simple network DEA, and ranked the overall performance of different airlines [18]. Yu Jian analyzed the production and operation data of China's five major airlines during the period 2002–2006 with Malmquist index-based method to estimate the changes in the total factor productivity of China's airline industry [19]. The results show that the average growth rate of China's total factor productivity in 2002–2006 was 1.2%. Productivity has generally improved, and this growth has been affected by both overall efficiency improvements and technological advances. Zhang Yanna and Xu Yuefang adopted the DEA congestion model and the output-oriented BCC model to study the 2014 financial data and production and operation data of 11 Chinese airlines [22]. The study found that Air China's scale returns remain unchanged, and Okay Airways is in a state of increasing scale returns. Other airlines except the two airlines have varying degrees of congestion. It is further found that in addition to Air China and Okay Airways, nine other airlines have invested in redundancy and insufficient output.

Some studies focus on the measurement of the Total Factor Productivity (TFP) of the aviation industry in China [20]; The study found that although the total factor productivity of the US and China aviation industry increased during 1998–2006, the drivers of growth were not the same. The main driving force for the development of the US airline industry is technological change and innovation. Li Ye collected the actual operational data of 22 global airlines from 2008 to 2014 in order to study the reasons for the differences in efficiency among airlines using the weakly processed network RAM model [11]. Chen Hui collected the data of China's five major airlines from 2004 to 2008 by studying the efficiency of China's civil aviation transportation industry, and calculated the total factor productivity index, comprehensive technical efficiency value, technical efficiency value and scale efficiency by DEA method [4]. Value, etc. In addition, the study analyzes the status quo and trends through charts and concludes that: 1. The overall industrial efficiency of the civil aviation transportation industry is showing a growing trend, which mainly depends on the growth of the national economy, the gradual relaxation of government regulation, and the establishment of a modern management system for airlines. However, the efficiency of the industry is highly susceptible to external emergencies. 2. The reorganization of civil aviation transportation industry will bring certain negative impacts to the civil aviation transportation industry in the short term, resulting in a large number of personnel and asset redundancy and technological regression, which will eventually lead to a decline in the productivity of the entire industry. 3. If the airline management concept is not changed in time, it will lead to low level of business competition, blind competition between each other, and lack of motivation and pressure for innovation. However, with the relaxation of government regulation, enterprises have greater operational autonomy and enthusiasm. The improvement in efficiency is just around the corner. 4. The impact of government regulation on the efficiency of China's civil aviation transportation industry is very large, so it is necessary to relax government regulation. Liu Jingbu used the data from the total passenger traffic and passenger traffic, civil aviation freight volume and freight volume from 1987 to 2008 for statistical analysis. The research uses the input-oriented CCR model to analyze the technical efficiency, pure technical efficiency, scale efficiency and scale return of civil aviation from 1987 to 2008 [13]. Zhang Xu conducted a study on the performance evaluation of air transport production operations in China, collected actual data on the operation of the air transport industry from 1990 to 2005, and used the SUP-DEA model and the CCR-DEA model to calculate the relative efficiency, scale efficiency, assessing the operational performance of China's air transport over the years [21]. The study found that operations in the 1990s were ineffective. Operational performance has been relatively effective since the 21st century, and operational conditions have improved relatively. However, there is a serious excess capacity, low fuel consumption efficiency, and the relative lack of drivers restricts the development of civil aviation. In short, air cargo is extremely inefficient.

Except for the above researches, more application research of DEA model on the service industry are mainly focus on performance assessment of the shipping company [5,6,8,14,15]. A few researches on customer satisfaction has begun to appear. Most scholars use the questionnaire survey method to collect data and conduct customer satisfaction with airline service through questionnaire surveys of customers [3,10,17].

The previous studies rarely consider the impact of the overall environmental inter-period linkage on the performance of the aviation industry. Such input factors as customer satisfaction and education and training have great performance in aviation-related industries, but few literature studies in the past. Further previous studies support less enough for the area of air companies service efficiency improvement based on customer complain.

This study collected customer complain data in airline listed companies in China as customer satisfaction index and investment in education and training in the same companies, with application of a two-stage dynamic data envelopment analysis, in order to analyze the operational efficiency of aviation industry. The data including the financial data, customer complaints, human capital investment and other data of the seven listed airlines in China from 2014 to 2017 are collected from public sources. The impact of customer complaints on airline operational efficiency was evaluated. The network data dynamic analysis method was used to analyze and evaluate the operating efficiency and productivity index of the seven companies.

3   Research Methods

The most widely used models for DEA are the CCR model, the BCC model, and the SBM model. The CCR model is based on the assumption of fixed-scale compensation. The output is positively correlated with the input. The efficiency of each decision-making unit is the ratio of the linear combination of output and input. However, in actual situations, the scale returns are not the same under different production scales. The BCC model has been improved to calculate pure technical efficiency (PTE) and scale efficiency (SE) by measuring technical efficiency (TE). In 2007, Fare, Grosskopf and Whittaker proposed a network data envelopment analysis model. They used sub-production technology as a sub-decision unit and considered that they formed the production process, and then used the traditional CCR and BCC modes to find the optimal solution. In 2009, Tone and Tsutsui further proposed a weighted SBM network data envelopment analysis model. "The linkage between the various departments of the decision-making unit is used as the analysis basis of the network DEA model, and each department is regarded as the sub-decision unit, and the SBM model is used to find the optimal solution." In 2013, Tone and Tsutsui proposed a weighted SBM dynamic network data envelopment analysis model, with inter-period activities as linkability. Inter-period activities can be divided into four types, desirable, undesirable, and discretionary. Change), non-discretionary (non-changeable) as the basis for the analysis of the dynamic DEA model, "The DEA model is divided into three forms, namely, input-oriented, output-oriented, and non-oriented, and then use SBM mode to find the optimal solution."

The following is the dynamic network DEA basic mode and solution description.

3.1   Basic Model Definition of SBM

Suppose there are n $DMU_j(j = 1,\ldots,n)$, each DMU has k divisions ($k = 1,\ldots,K$), and there are T time periods ($t = 1,\ldots,T$). Each DMU has inputs and outputs in the t period, through Carry over (link) to the next period $t + 1$.

Let m_k and r_k represent the input and output of each department K, and use $(k, h)\, i$ to represent the departments k to h, L_{hk} to represent the k and h department collection, input and output, and the connection and existence period are defined as follows:

(1) Inputs and outputs

$X_{ijk}^t \in R_+ (i = 1,\ldots,m_k; j = 1,\ldots,n; K = 1\ldots,K; t = 1,\ldots,T)$ representing the input item i of the DMU_j division in the t period.

$y_{rjk}^t \in R_+ (r = 1,\ldots,r_k; j = 1,\ldots,n; K = 1\ldots,K; t = 1,\ldots,T)$: representing the output item r of DMU_j in the t period.

If part of the output is not ideal, it is considered as the input of division k .

Links

$Z_{j(kh)t}^t \in R_+ (j = 1; \ldots; n; l = 1; ..; L_{hk}; t = 1; \ldots; T)$: representing the link connecting k division to h division in DMU_j in t period, where L_{hk} is the number of items linked from k to h.

$Z_{j(kh)t}^t \in R_+ (j = 1; \ldots; n; l = 1; ..; L_{hk}; t = 1; \ldots; T)$ Carry-overs

$Z_{jkl}^{(t,t+1)} \in R_+ \ (j = 1,\ldots,n; l = 1,..,L_k; k = 1,\ldots k, t = 1,\ldots,T - 1)$: representing the span of DMU_j k division to h division in t to $t + 1$, where L_k is the number of items during the k division span.

The following are the mathematical formulas of the basic model:

Production possible

$$x_k^t \geq \sum_{j=1}^n x_{jk}^t \lambda_{jk}^t (\forall k, \forall t) \tag{1}$$

$$y_k^t \leq \sum_{j=1}^n y_{jk}^t \lambda_{jk}^t (\forall k, \forall t)$$

$$z_{(kh)l}^t \geq, =, \leq \sum_{j=1}^n z_{j(kh)l}^t \lambda_{jk}^t (\forall l, \forall (kh)_l, \forall t)(Output of t - phase k division)$$

$$z_{(kh)l}^t \geq, =, \leq \sum_{j=1}^n z_{j(kh)l}^t \lambda_{jh}^t (\forall l, \forall (kh)_l, \forall t)(Input of t - phase k division)$$

$$z_{kl}^{(t,t+1)} \geq, =, \leq \sum_{j=1}^n z_{jkl}^{(t,t+1)} \lambda_{jk}^t (\forall k_l, \forall k, t = 1,\ldots,T - 1)(Inter - period of t period)$$

$$z_{kl}^{(t,t+1)} \geq, =, \leq \sum_{j=1}^n z_{jkl}^{(t,t+1)} \lambda_{jk}^{t+1} (\forall k_l, \forall k, t = 1,\ldots,T - 1)(Inter - period of t + 1 period)$$

$$\lambda_{jk}^t \geq 0(\forall j, \forall k, \forall t) \sum_{j=1}^n \lambda_{jk}^t = 1(\forall k, \forall t),$$

Equation (1) represents the economies of scale (return-to-scale)

(2) DMU_o Defined as follows:

$DMU_o(o = 1 \ldots n) \in p$, Expressed as follows, the input and output limits Eq. (2) are as follows.

$$x_{ok}^t = X_k^t \lambda_k^t + s_{ko}^{t-}(\forall k, \forall t) \tag{2}$$
$$y_{ok}^t = Y_k^t \lambda_k^t - s_{ko}^{t+}(\forall k, \forall t)$$
$$e\lambda_k^t = 1(\forall k, \forall t)$$
$$\lambda_k^t \geq 0, s_{ko}^{t-1} \geq 0, s_{ko}^{t+} \geq 0, (\forall k, \forall t)$$

3.2   Links Restrictions

The four restrictions of Links are as follows:
   (1) Variable situationfree link value case, LF
   If the input or output is continuous

$$Z_{(kh)free}^t \lambda_h^t = Z_{(kh)free}^t \lambda_k^t (\forall (k, h) free, \forall t) \tag{3}$$
$$Z_{(kh)free}^t = (Z_{1(kh)free}^t, \ldots, Z_{n(kh)free}^t) \in R^{L_{(h)free} \times n}$$

And the existing link value and the value of the changeable link, the relationship is as in Eq. (4).

$$Z_{o(kh)free}^t = Z_{(kh)free}^t \lambda_k^t + S_{o(kh)free}^t \tag{4}$$

(2) Unchangeable situation (Non-discretionary fixed link value case, LN)

$$Z_{o(kh)fix}^t = Z_{(kh)fix}^t \lambda_h^t (\forall (k, h) fix, \forall t) \tag{5}$$
$$Z_{o(kh)fix}^t = Z_{(kh)fix}^t \lambda_k^t (\forall (k, h) fix, \forall t)$$

(3) Input situation ("as-input" link value case, LB)

$$Z_{o(kh)in}^t = Z_{(kh)in}^t \lambda_k^t + S_{o(kh)in}^t ((kh)in = 1, \ldots, link in_k) \tag{6}$$

(4) Output situation ("as-output" link value case, LB)

$$Z_{o(kh)out}^t = Z_{(kh)out}^t \lambda_k^t - S_{o(kh)out}^t ((kh)out = 1, \ldots, link out_k) \tag{7}$$

3.3   Inter-period

The variability across the period is that the output in the t period becomes the input of the $t + 1$ period, which can be divided into four activities as follows:

1. Desirable (Good) carry-over case (CG)
2. Undesirable (bad) carry-over case (CB)
3. Discretionary (Free) Carry-over case (CF)
4. Non-discretionary (Fixed) carry-over case (CN)

Its mathematical equation is as in Eq. 8.

$$\sum_{j=1}^{n} z_{jk_1\alpha}^{(t,(t+1))} \lambda_{jk}^{t} = \sum_{j=1}^{n} z_{jk_1\alpha}^{(t,(t+1))} \lambda_{jk}^{t+1} (\forall k; \forall k_l; t = 1,\ldots,T-1) \quad (8)$$

α stands for good, bad, changeable or unchangeable.

And each different type of restriction is as in Eq. (9):

$$Z_{ok_lgood}^{(t,(t+1))} = \sum_{j=1}^{n} z_{jk_lgood}^{(t,(t+1))} \lambda_{jk}^{t} - s_{ok_lgood}^{(l,(l+1))} k_l = 1,\ldots,ngood_k; \forall k; \forall t) \quad (9)$$

$$Z_{ok_lbad}^{(t,(t+1))} = \sum_{j=1}^{n} z_{jk_lbad}^{(t,(t+1))} \lambda_{jk}^{t} - s_{ok_lbad}^{(t,(t+1))} k_l = 1,\ldots,nbad_k; \forall k; \forall t)$$

$$Z_{ok_lfree}^{(t,(t+1))} = \sum_{j=1}^{n} z_{jk_lfree}^{(t,(t+1))} \lambda_{jk}^{t} - s_{ok_lfree}^{(t,(t+1))} k_l = 1,\ldots,nfree_k; \forall k; \forall t)$$

$$Z_{ok_lfix}^{(t,(t+1))} = \sum_{j=1}^{n} z_{jk_lfix}^{(t,(t+1))} \lambda_{jk}^{t} - s_{ok_lfix}^{(t,(t+1))} k_l = 1,\ldots,nfix_k; \forall k; \forall t)$$

$$s_{ok_lgood}^{(t,(t+1))} \geq 0, s_{ok_lbad}^{(t,(t+1))} \geq 0, s_{ok_lfree}^{(t,(t+1))} : free(\forall k_l; \forall t)$$

(a) The objective Function

The overall efficiency:

$$\theta_0^* = \min \frac{\sum_{t=1}^{T} W^t \left[\sum_{k=1}^{K} W^k \left[1 - \frac{1}{m_k + linkin_k + nbad_k} (\sum_{i=1}^{m_k} \frac{s_{iok}^{t-}}{x_{iok}^t} + \sum_{(kh)_l=1}^{linkin_k} \frac{s_{o(kh)_lin}^t}{z_{o(kh)_lin}^t} + \sum_{k_l=1}^{nbad_k} \frac{s_{ok_lbad}^{(t,(t+1))}}{z_{ok_lbad}^{(t,(t+1))}}) \right] \right]}{\sum_{t=1}^{T} W^t \left[\sum_{k=1}^{K} W^k \left[1 + \frac{1}{r_k + linkout_k + ngood_k} (\sum_{r=1}^{r_k} \frac{s_{rok}^{t+}}{y_{rok}^t} + \sum_{(kh)_l=1}^{linkout_k} \frac{s_{o(kh)_lout}^t}{z_{o(kh)_lout}^t} + \sum_{k_l=1}^{ngood_k} \frac{s_{ok_lgood}^{(t,(t+1))}}{z_{ok_lgood}^{(t,(t+1))}}) \right] \right]} \quad (10)$$

(b) period and division efficiencies

Period and division efficiencies are as follows:

(b1) Period efficiency:

$$\tau_0^{t*} = \frac{\sum_{k=1}^{K} W^k \left[1 - \frac{1}{m_k + linkin_k + nbad_k} (\sum_{i=1}^{m_k} \frac{s_{iok}^{t-}}{x_{iok}^t} + \sum_{(kh)_l=1}^{linkin_k} \frac{s_{o(kh)_lin}^t}{z_{o(kh)_lin}^t} + \sum_{k_l=1}^{nbad_k} \frac{s_{ok_lbad}^{(t,(t+1))}}{z_{ok_lbad}^{(t,(t+1))}}) \right]}{\sum_{k=1}^{K} W^k \left[1 + \frac{1}{r_k + linkout_k + ngood_k} (\sum_{r=1}^{r_k} \frac{s_{rok}^{t+}}{y_{rok}^t} + \sum_{(kh)_l=1}^{linkout_k} \frac{s_{o(kh)_lout}^t}{z_{o(kh)_lout}^t} + \sum_{k_l=1}^{ngood_k} \frac{s_{ok_lgood}^{(t,(t+1))}}{z_{ok_lgood}^{(t,(t+1))}}) \right]}$$

(b2) division efficiency:

$$\delta_{0k}^* = \frac{\sum_{t=1}^{T} W^t \left[1 - \frac{1}{m_k + linkin_k + nbad_k} (\sum_{i=1}^{m_k} \frac{s_{iok}^{t-}}{x_{iok}^t} + \sum_{(kh)_l=1}^{linkin_k} \frac{s_{o(kh)_lin}^t}{z_{o(kh)_lin}^t} + \sum_{k_l=1}^{nbad_k} \frac{s_{ok_lbad}^{(t,(t+1))}}{z_{ok_lbad}^{(t,(t+1))}}) \right]}{\sum_{t=1}^{T} W^t \left[1 + \frac{1}{r_k + linkout_k + ngood_k} (\sum_{r=1}^{r_k} \frac{s_{rok}^{t+}}{y_{rok}^t} + \sum_{(kh)_l=1}^{linkout_k} \frac{s_{o(kh)_lout}^t}{z_{o(kh)_lout}^t} + \sum_{k_l=1}^{ngood_k} \frac{s_{ok_lgood}^{(t,(t+1))}}{z_{ok_lgood}^{(t,(t+1))}}) \right]} (\forall k)$$

(b3) division period efficiency:

$$p_{Ok}^{t*} = \frac{1 - \frac{1}{m_k + linkin_k + nbad_k}(\sum_{i=1}^{m_k} \frac{s_{iok}^{t-}}{x_{iok}^t} + \sum_{(kh)_l=1}^{linkin_k} \frac{s_{o(kh)_l in}^t}{z_{o(kh)_l in}^t} + \sum_{k_l=1}^{nbad_k} \frac{s_{ok_l bad}^{(t,(t+1))}}{z_{ok_l bad}^{(t,(t+1))}})}{1 + \frac{1}{r_k + linkout_k + ngood_k}(\sum_{r=1}^{r_k} \frac{s_{rok}^{t+}}{y_{rok}^t} + \sum_{(kh)_l=1}^{linkout_k} \frac{s_{o(kh)_l out}^t}{z_{o(kh)_l out}^t} + \sum_{k_l=1}^{ngood_k} \frac{s_{ok_l good}^{(t,(t+1))}}{z_{ok_l good}^{(t,(t+1))}})} (\forall k; \forall t)$$

$$Z_{ol_k}^{(0,1)} = \sum_{j=1}^{n} Z_{jlk}^{(0,1)} \lambda_{jk}^l (\forall l_k) \tag{11}$$

From the above results can be obtained: the overall efficiency, the period efficiency, the division efficiency and division period efficiency.

Productivity Index Model Theory

The MPI value indicates the growth and variation of the Total Product Productivity (TFP) of the DMU. It mainly discusses the progress or backwardness of the marginal technical efficiency from period 1 to period 2 and is described as follows:

In the period between t and $t+1$, the ratio of sector efficiency, where if DCU > 1 indicates a progressive trend, DMU = 1 maintains the status quo, and DCU < 1 indicates a declining trend .

(a) Divisional catch-up index (DCU): In the period between t and $t+1$, the ratio of sector efficiency, where if DCU > 1 indicates a progressive trend; DMU = 1 maintains the status; DCU < 1 indicates a declining trend.

$$DCU = \gamma_{ok}^{t \to t=1} = \frac{\rho_{ok}^{t+1*}}{\rho_{ok}^{t*}} \quad (t = 1, \cdots, T-1; k = 1, \cdots, K; o = 1, \cdots, n) \tag{12}$$

(b) Divisional frontier-shift index (DFS):
Same as the non-radio MPI model described earlier, as shown in Eq. 13.

$$DFS = \sigma_{ok}^{t \to t+1} = \sqrt{\sigma_{ok}^t \sigma_{ok}^{t+1}} \tag{13}$$

(c) Divisional Malmquist index (DMI):
The product of the department's DCU and DFS, so the equation represents Eq. (14).

$$DMI = DCU \times DFS = \mu_{ok}^{t \to t+1} = \gamma_{ok}^{t \to t+1} \sigma_{ok}^{t \to t+1}$$
$$(t = 1, \cdots, T-1; k = 1, \cdots, K; o = 1, \cdots, n) \tag{14}$$

(d) Overall Malmquist index (OMI): The weighted geometric mean of the sector MPI, expressed as Eq. (15).

$$OMI = \mu_o = \prod_{k=1}^{K} (\mu_{ok})^{w_k} \quad (o = 1, \cdots, n) \tag{15}$$

μ_{ok} is the weighted geometric mean of $\mu_{ok}^{t \to t+1}$ $(t = 1, \cdots, T-1)$, $w_k \geq 0, \sum w_k = 1$.

4   Empirical Results and Analysis

4.1   Data

After comprehensively collecting all listed airline data, the study selected seven airlines based on the integrity of the data, including: Southern Airlines, Air China, Eastern Airlines, Spring Airlines, and auspicious Airlines, Shandong Airlines, Hainan Airlines. The seven airlines, as research samples, conducted research on data from 2014–2017. The company's operating financial data comes from the annual reports and financial statements of the companies published on the official website. The customer complaint indicators are from the Civil Aviation Administration of China, and the market value-earnings ratio and other data are from GuoTaiAn Database.

4.2   Indicator Definition

The main indicators used in this study include:

(1) Training investment
   Training is the most direct human capital investment activity of enterprises. Human capital creates more enterprise value through training. Therefore, training investment is used to measure the investment cost of human capital. This paper intends to replace the training investment with the employee education funds in the annual report.
(2) Customer complaints
   This study uses the customer complaint rate of each airline published by the Civil Aviation Administration of China as an indicator to measure customer complaints.
(3) Number of employees
   The number of employees can be seen in the size of the company, representing the amount of human capital in the real estate company.
(4) Fixed assets The aviation industry is a high investment category. Fixed assets such as aircraft equipment purchase and land investment are the key elements for the expansion of the production and operation scale of aviation companies.
(5) Operating costs Operating costs are the basis for airline companies to earn sales by completing sales. In order to achieve higher returns, companies must consider how to reduce operating costs and thus increase competitiveness.
(6) Operating income Good operating income can bring about an increase in the company's profitability and reflect the ability of aviation companies to earn income from their business activities.
(7) Profit before tax Profit before tax is the total profit and loss of the company in the production and operation process, and is an important indicator to measure the efficiency and management level of the company.
(8) Market value and price-earnings ratio The market value of a listed company refers to the total value of the shares in which the issued shares are calculated at market prices. The stock market price is closely related to the company's operating performance, and the two influence each other (Fig. 1).

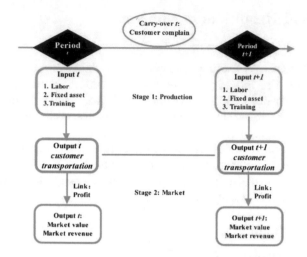

Fig. 1. Airline two-stage dynamic research process

4.3   Narrative Statistics of Input and Output Indicators

The following table shows the statistical analysis of airline input and output indicators. It can be seen that Table 1 Statistical analysis of input and output indicators for each year. As Figs. 2–3 shows the data analysis of the number of fixed assets employees and education and training inputs of the airlines. The results show that the situation of the sample airlines is quite different. It can be seen that in addition to Hainan Airlines, most companies' fixed asset investment continues to rise. Among them, Eastern Airlines has seen a large increase in fixed assets investment, and the total amount also surpassed other airlines in 2017. China International Airlines' fixed asset investment has surpassed other airlines in 2014, and has since increased, but it has not changed much in 2015 and 2016. In 2017, fixed assets investment has increased significantly, slightly lower than that of China Eastern Airlines. the company. The relative size of Spring Airlines, Hainan Airlines and Jixiang Airlines is relatively small, but it is still possible to see the increase in fixed assets investment of Spring Airlines and Juneyao Airlines.

In terms of the number of employees, the number of employees in all companies has increased over the years, the largest increase is in Southern Airlines, and Spring Airlines has seen a significant increase in 2017. From the perspective of education investment, Eastern Airlines' education and training investment has continued to rise, and it has increased rapidly in 2017, far exceeding other airlines. Southern Airlines education and training investment was the highest among all airlines in 2014, although it appeared in 2015. It fell, but it was both rising in 2016 and 2017 and there was a high investment. The investment in education and training of Spring Airlines, Air China and Shandong Airlines is much smaller than that of other airlines.

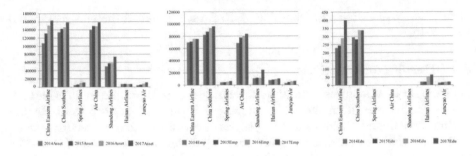

Fig. 2. Statistical analysis of airline input indicators

The graph shows the company's passenger traffic, sales revenue, pre-tax profit, market capitalization, price-to-earnings ratio and customer complaints. It can be seen that the number of customer complaints from various companies has continued to increase, and in 2017, the increase was the fastest and the increase was large. Among them, Eastern Airlines customers complained the most, far more than other airlines, but other airlines also reached the most complaints from customers in 2017. Southern Airlines customer complaints were the highest among all sample companies in 2016, although the indicator is also increasing in 2017, but it is smaller than Eastern Airlines.

It can also be seen that the passenger capacity of China Eastern Airlines, China Southern Airlines and Air China continued to rise from 2013 to 2016, but it has seen a significant decline since 2017. The passenger load of Spring Airlines, Shandong Airlines and Hainan Airlines continued to rise, and the Junyao Airlines experienced a small decline in the final year.

From the sales revenue and pre-tax profit, we can see that most of the two indicators of the company are on the rise. In 2017, the sales revenue of China Southern Airlines increased rapidly and exceeded the international airlines. The sales revenue of the parent company, Eastern Airlines, was weak, and the increase was smaller in 2017 than in the previous year. From the perspective of pre-tax profits, Air China has the largest growth in 2017 and is much larger than other airlines. Spring and Autumn Airlines' pre-tax profit fluctuated significantly and reached its highest level in 2015, but it declined in 2016. Although it increased in 2017, it was still lower than 2015.

In terms of market value, except for the market value of Jixiang Airlines and Spring Airlines continued to decline in the next three years, other companies showed a decline in market value in 2016. However, the market value rebounded again in 2017, and the increase was very large. Among them, the international airlines had the largest market value growth and far exceeded other airlines.

But most companies' P/E ratios are not performing well. It can be seen that most companies' P/E ratios are fluctuating. Only Shandong Airlines' price-earnings ratio rebounded in 2017 and the rate of increase was very high. The price-earnings ratio of other companies in 2017 was significantly lower than that

568   Y. Li et al.

of previous years. Jixiang Airlines' 2017 P/E ratio dropped the most compared with 2016.

4.4   Results and Analysis

(1) Comparative analysis of total performance scores of airlines in 2014–2017

As can be seen from the table below, the total efficiency scores of Hainan Airlines, Shandong Airlines and Juneyao Airlines are 1 for four consecutive years, indicating that the efficiency improvement space of these companies is 0. The other four airlines have lower efficiency scores and have large difference.

The company with the lowest total efficiency score is Spring Airlines, which ranks the last in all airlines, only 0.0076. The overall efficiency score of international airlines is slightly better, but it is only about 0.33. The improvement space is still very large. Eastern Airlines and China Southern Airlines have similar total efficiency scores, and they all have an efficiency score close to 0.2.

From the perspective of time development, the total efficiency scores of companies from 2014 to 2017 also have different trends. It can be seen that Spring Airlines' lowest efficiency score in 2014 is only 0.0019, but it has reached 1 from 2015 to 2017, and the efficiency has improved rapidly. In the next three years, the improvement space is 0.

International airlines' annual efficiency scores fluctuate significantly. In 2014, the total efficiency score was 1, there is no room for improvement, but the total efficiency score in 2015 was only 0.25, and in 2016 it fell to 0.21, but in 2017 it reached 1.

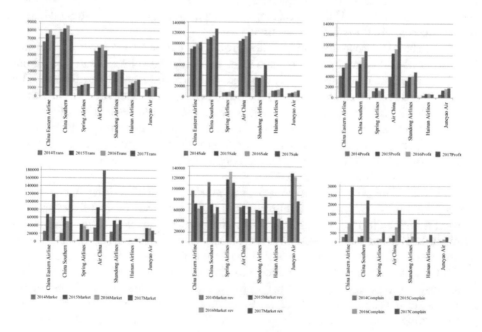

Fig. 3. Statistical analysis of airline output indicators

The efficiency score of Eastern Airlines continued to decline from 2014 to 2017, from 0.32 in 2014 to 0.1 in 2015, and only about 0.14 in 2016. It will rebound slightly in 2017, but only about 0.18.

The situation of Southern Airlines is similar. In 2014, its efficiency score was around 0.33, but it dropped to 0.177 in 2015 and fell further to 0.1 in 2016. It rebounded slightly in 2017, but it was only around 0.22.

Table 1. Total airline efficiency score for 2014–2017

DMU	Overall score	Rank	2014(1)	2015(1)	2016(1)	2017(1)
Eastern Airlines	0.1734	6	0.3197	0.1485	0.1413	0.1785
Southern Airline	0.1808	5	0.3287	0.1773	0.1125	0.2249
Spring Airlines	0.0076	7	0.0019	1	1	1
Air China	0.33	4	0.4242	0.2482	0.2121	1
Hainan Airlines	1	1	1	1	1	1
Shandong Airlines	1	1	1	1	1	1
Juneyao Airlines	1	1	1	1	1	1

(2) Comparative analysis of the improvement of production indicators of various companies

Judging from the improvement space of airline customers' complaints, the need for improvement of Eastern Airlines is lower than that of 2014, indicating that its customer complaints have declined, but by 2017 it has risen to 33%, and the improvement space is expanding. The space for improvement of airlines continues to decline. International aviation has declined since 2014. After 2015, the space for improvement is 0. The improvement space for other airlines is 0 (Table 2).

Table 2. Airline customer complaints improvement

DMU	Overall score	Rank	2014(%)	2015(%)	2016(%)	2017(%)
Eastern Airlines	0.1734	6	−45.14	−12.76	0	−33.68
Southern Airline	0.1808	5	−48.99	0	−29.93	−20.43
Spring Airlines	0.0076	7	0	0	0	0
Air China	0.33	4	−32.04	−17.85	0	0
Hainan Airlines	1	1	0	0	0	0
Shandong Airlines	1	1	0	0	0	0
Juneyao Airlines	1	1	0	0	0	0

(3) Comparative Analysis of Productivity Indexes of Companies

From the two-stage productivity index score, the score of less than 1 in the first stage indicates that the efficiency improvement is generally decreasing. If it is greater than 1, it indicates that the efficiency improvement is generally rising; the second stage is less than the score, indicating that the efficiency of the production and operation phase is constant. Declining, greater than 1, indicates that its production and operation efficiency is increasing.

The above table is the score of the two-stage productivity index of each airline. From the total score of the productivity index, it can be seen that among all airlines, Spring Airlines' performance continues to improve and progress the most. In one stage, its Malmquist index is 1.183, to the second stage, its Malmquist index is 4.869.

Judging from the total score of the Productivity Index, Spring Airlines ranked first and above 1, indicating that the company's performance improvement is very good. Southern Airlines and International Airlines are in second and third respectively, but below 1, which indicates that these companies are performing better, but progress is slow. Shangdong Airlines and Hainan Airlines have relatively low productivity index scores, ranking seventh and sixth.

From the two-stage productivity index, the best stage MPI is also Spring Airlines, which exceeds 1, indicating that the progress is very large, but the MPI of the first stage is lower than the second stage, indicating that the company's performance management in other fields is also very effective. The first-stage productivity index of Jixiang Airlines ranked second, but the productivity index in the second stage dropped significantly, to the fifth.

The first-stage scores of Jixiang Airlines and Shandong Airlines were higher than the second-stage, indicating that the performance improvement of the two companies in customer service was better, although the progress was not great, but the regression was slow.

Hainan Airlines' productivity index is not high in two stages, only about 0.45 in one stage and only 0.5 in the second stage. Although the MPI of the second stage is higher than the first stage, the MPI score of the second stage is also very low, indicating that the customer is satisfied. Performance management in terms of degree and other aspects needs to be improved and improved.

The first phase of MIA of China International Airlines is much lower than the second phase, indicating that there is still room for improvement in customer service, and performance improvement in other aspects is much better, slightly better than the first stage (Table 3).

Table 3. Airline Productivity Index score for 2014–2017

DMU	Overall score	Rank	Div1 MPI	Div2 MPI	Overall MPI	Rank Malmquist
Eastern Airlines	0.1734	6	0.9091	0.9488	0.9287	4
Southern Airline	0.1808	5	0.9401	1.0316	0.9848	2
Spring Airlines	0.0076	7	1.1828	4.8705	2.4002	1
Air China	0.33	4	0.5384	1.7664	0.9752	3
Hainan Airlines	1	1	0.4488	0.5456	0.4948	6
Shandong Airlines	1	1	0.6429	0.3806	0.4947	7
Juneyao Airlines	1	1	0.9433	0.5963	0.75	5

(4) Analysis of factor correlation

From the correlation analysis, it can be seen that under the P-Value, 5% significant level, the impact of customer complaints on the airline's business performance efficiency is positively and significantly affected, indicating that the reduction of customer complaints is conducive to the improvement of airlines' business performance.

However, customer complaints have a positive or negative impact on the market value of the company. Under the P-Value, 5% significant level, the market value of the firm has a positive and significant impact on the airline's operating efficiency. It shows that the increase in the market value of the company has a positive impact on the operating efficiency of the airline (Table 4).

Table 4. Correlation analysis

	Complain	Market value	efficiency
Complain	1.00000***	0.28834	0.47875**
	(p-value < 0.001)	(p-value = 0.2050)	(p-value = 0.0281)
Market value	0.28834	1.00000***	0.72286***
	(p-value = 0.2050)	(p-value < 0.001)	(p-value = 0.0002)

***means $p < 0.01$; **means $p < 0.05$

5   Conclusions and Management Significance

This study applied network dynamic DEA model to analyze the overall operating performance and productivity index of the seven listed airlines from the year of 2014 to 2017. The data including the financial data, customer complaints, human capital investment and other data of the seven listed airlines in China from 2014 to 2017 are collected from public sources. The impact of customer complaints on airline operational efficiency was evaluated. The network data dynamic analysis method was used to analyze and evaluate the operating efficiency and productivity index of the seven companies. In general, although the seven airlines occupy a major share of the Chinese private aviation market, there are still various problems in the performance and performance improvement of these companies.

(1) There are no airlines which has a overall efficiency score of 1. It shows that almost all the listed airlines should keep improve its management lever with strong sustainable development capabilities in China's aviation enterprises. (2) Only Spring Airlines has performed better compared to other companies. The company has made great progress in the continuous improvement of performance. From the year to year, the company has been making continuous progress. (3) Although Hainan Airlines, Shandong Airlines and Jixiang Airlines

have a total performance score of 1, indicating that there is no room for improvement, but their productivity index is below 1, indicating that there are certain problems in the sustainability of these three companies.

In summary, it can be seen that in addition to the three companies with a total performance of 1 in China's listed airlines, several other companies still have a lot of room for performance improvement.

The managerial suggestion could be as follows:

(1) Further investment on human capital development is necessary for most of the companies. Eastern Airlines and China Southern Airlines have significant room for improvement in education investment and passenger traffic, indicating that it is necessary to increase investment in human resources training and development to enhance human capital and improve employee productivity. Even better, improving the quality of service, which can attract more customers, ultimately improve the business performance.

(2) Culture construction for service quality improvement and customer satisfaction are one of the most important aspects of airline performance management goals. Through the construction of corporate culture, the improvement of employees' job satisfaction and work involvement will help improve their work ability and attitude, thus improving service quality and customer satisfaction.

(3) Social responsibility should be taken as one of the important managerial activities in the companies. Due to the size and market share of China International Airlines, China Southern Airlines and Eastern Airlines in the domestic civil aviation market, the social impact of these airline services is also relatively large. Increasing investment in human resources training and development, and improving human capital and other measures will benefit the productivity and job satisfaction of these companies, thereby enhancing the company's customer satisfaction. Reducing complaints and increasing investment in improvement are also more conducive to the improvement of the company's business reputation. Ultimately, these methods will improve the performance of the company.

Acknowledgements. This study is supported by the National Natural Science Foundation of China (NO.71773082); Sichuan Province Philosophy and Social Sciences Project (No. 2016SC16A006).

References

1. Administration CNA (2017). http://www.caac.gov.cn/index.html
2. Chen, F., Zhao, Y.: Study on airline customer satisfaction. Value Eng. **36**(2), 6–8 (2017a)
3. Chen, F., Zhao, Y.: Study on airline customer satisfaction. Value Eng. **36**(2), 6–8 (2017b)
4. Chen, H.: Research on the efficiency of China's civil aviation transportation industry based on DEA method. Hefei University of Technology (2010)
5. Felício, J.A., Caldeirinha, V., Dionísio, A.: The effect of port and container terminal characteristics on terminal performance. Marit. Econ. Logist. **17**(4), 493–514 (2015)

6. Gutiérrez, E., Lozano, S., Furió, S.: Evaluating efficiency of international container shipping lines: A bootstrap dea approach. Marit. Econ. Logist. **16**(1), 55–71 (2014)
7. Jiang, Y., Wen, J.: Evaluation of operational performance of major airlines in china based on dea model. Sci. Technol. Eng. **1005**, 1175–1178 (2010)
8. Kang, H.W., Wang, G.W.Y., Bang, H.S., Woo, S.H.: Economic performance and corporate financial management of shipping firms. Marit. Econ. Logist. **18**(3), 317–330 (2016)
9. Li, J., Yang, C.: Research on operational efficiency evaluation of china's civil aviation industry based on ahp and dea. Contemp. Econ. **6**, 110–111 (2014)
10. Li, Y.: Empirical study on factors affecting customer satisfaction in civil aviation transportation market based on structural equation model. Nanjing University of Aeronautics and Astronautics (2010)
11. Li, Y.: Improved DEA model and application of airline efficiency evaluation. Dalian University of Technology (2016)
12. Li, Y., Yu, J., Wu, Y.: Comparative study on total factor productivity of Chinese and American airlines industry. J. Xidian Univ. (Social Science Edition) **5**, 19–26 (2008)
13. Liu, J.: Analysis and evaluation of China civil aviation operation efficiency based on DEA. J. Wuhan Univ. Technol. (Information and Management Engineering Edition) **33**(3), 483–487 (2011)
14. Low, J.M.W.: Capacity investment and efficiency cost estimations in major east asian ports. Marit. Econ. Logist. **12**(4), 370–391 (2010)
15. Rajasekar, T., Deo, M.: Does Size Influence the Operational Efficiency of the Major Ports of India? - A Study. Social Science Electronic Publishing (2014)
16. Song, S.: Research on customer satisfaction of China's low cost airlines based on SEM. Civil Aviation University of China (2017)
17. Song, S.: SEM-based customer satisfaction research for low-cost airlines in China. Civil Aviation University of China (2017)
18. Tan, Y., Chen, S.: Performance analysis of air transportation company based on network dea. J. Southeast Univ. (Natural Science Edition) **4105**, 1114–1118 (2011)
19. Yu, J.: Analysis of total factor productivity of china's airlines industry based on malmquist index. J. Beijing Inst. Technol. (Social Science Edition) **6**, 43–46 (2007)
20. Zhang, W.: Research on the evaluation of operational efficiency of low-cost airlines in China based on DEA. Nanjing University of Aeronautics and Astronautics (2012)
21. Zhang, X.: Performance evaluation of china's air transportation production operation based on dea model. Silicon Val. **6**(11), 118–119 (2013)
22. Zhang, Y., Xu, Y.: Analysis of china airlines efficiency based on dea congestion model. J. Civ. Aviat. Flight Univ. China **25**(6), 9–14 (2014)

How Social Factors Drive Electronic Word-of-Mouth on Social Networking Sites?

Muhammad Sohaib[1], Peng Hui[1], Umair Akram[2(✉)], Abdul Majeed[3], and Anum Tariq[1]

[1] School of Economics and Management, Beijing University of Posts and Telecommunications, Beijing, People's Republic of China
[2] Guanghua School of Management, Peking University, Beijing 1008713, People's Republic of China
akram.umair88@pku.edu.cn
[3] School of International Trade and Economics, University of International Business and Economics, Beijing, China

Abstract. Evolution of social media provides a great opportunity for marketers to investigate social factors that influence customer's engagement in electronic Word-of-Mouth (eWOM) via social networks. Given the collaborative nature of Social Networking Sites (SNSs) such as WeChat, Weibo, and QQ. This study investigates how social factors relate to eWOM in SNSs. The homophily and tie strength are found as important antecedents of eWOM in SNSs environment. The results confirm that the homophily have positive direct influence on eWOM. In addition to that, the homophily have positive indirect impact on eWOM, through mediation of the tie strength. This study represents a unique effort to focus on the combined direct and indirect effects of the homophily on eWOM. The implications for practitioners and marketers are discussed.

Keywords: Social factors · Social networking sites · SNSs · Social media · Homophily · Tie strength · Electronic word-of-mouth · EWOM

1 Introduction

The advent of web 2.0 has given rise to social media that become most popular channels for dissemination of information among marketers and customers across the globe [9]. According to Kaplan and Haenlein [13], social media are "a group of Internet-based applications that build on the ideological and technological foundations of Web 2.0, and that allow the creation and exchange of User Generated Content". Social media involves a range of online information-sharing platforms including social networking sites (SNSs), online review sites,

© Springer Nature Switzerland AG 2020
J. Xu et al. (Eds.): ICMSEM2019 2019, AISC 1002, pp. 574–585, 2020.
https://doi.org/10.1007/978-3-030-21255-1_44

online discussion forms, and blogs. Specifically, much attention has been paid to
SNSs due to its social and collaborative nature.

In China, impressive growth of internet has given rise to 616.5 million SNSs
users [2,22]. Millions of netizens use SNSs to obtain or forward the useful information. The most famous social media includes; WeChat, QQ, Weibo, RenRen,
Youku, Dianping, and Doupan. Specifically, Tencent and Sina Weibo are leading groups in social media. Tencent's total revenues were USD11,135 million
in second quarter (Q2) of 2018, an increase of 30% year-on-year. Sina Weibo's
total revenues were USD426.6 million, an increase of 68% year-on-year. In second quarter of 2018, WeChat and Weibo have gained 1057.7 million and 431
million Monthly Active Users (MAU) respectively. On the other hand, QQ has
803.2 million MAU less than the first quarter of 2018 [5,20]. Figure 1 explains
the detailed information of MAU. Given the importance of SNSs, this research
focuses on SNSs as emerging platforms for customer-to-customer communication, namely known as Word-of-Mouth (WOM).

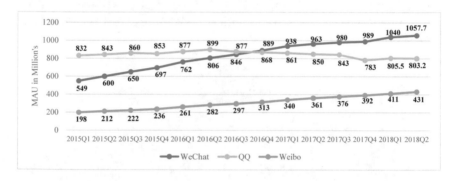

Fig. 1. MAU of WeChat, QQ, and Weibo

The significance of WOM has been well recognized in the marketing literature
[6,20]. WOM is explained as the act of trading marketing associated information
among customers, and plays a vital role in shaping customer attitude and behavior towards products and services [4]. Customers usually trust on consumer's
generated information than company-generated messages. Because of this, they
often depend on consumer's opinions to make purchase decisions. The dawn of
internet-based media has given the rise to electronic Word-of-Mouth (eWOM).
Abubakar et al. [1] defined eWOM as "any positive or negative statement made
by present or previous customers about a product, service or company, which
is made available to large audiences via the Internet". eWOM can take place
through a wide range of online media such as emails, blogs, review sites, forums,
virtual communities, chat rooms, instant messaging platforms, and SNSs [7,25].
From previous empirical investigations found that eWOM can be categorized
into two forms which are review generating factors and impacts of eWOM [8,21].
The main two directions for eWOM research found in the literature are review

generating factors and eWOM impacts. Figure 2 presents the detail of these two directions.

SNSs provides an ideal platform for customers to share their opinions about products and services with their circle of friends, colleagues and other intimate connections. This also enables customers to involve in certain behaviors such as opinion seeking, opinion giving, and opinion passing. Although, SNSs have the capacity to shape eWOM in the virtual-market, why and how eWOM take place in social domain need to be examined.

Accordingly, an investigation of SNSs as eWOM platform is needed. SNSs enable its users to involve in building and maintaining relationships by exchanging eWOM. Therefore, this research attempts to investigate social factors that influence on customers engagement in eWOM in SNSs. The empirical model of this study has integrated Social Exchange Theory (SET) and conceptualized the homophily and the tie strength as determinants of eWOM. This empirical work provides the three contributions. First, it attempts to investigate the direct effects of the homophily and the tie strength on eWOM, with SET. Second, it examines indirect effect of the homophily through the tie strength on eWOM. Lastly, it provides theoretical and practical insights of social relationships and eWOM via SNSs.

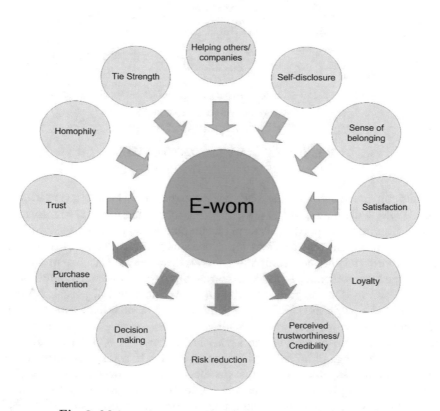

Fig. 2. Main review generating factors and eWOM impacts

2   Conceptual Framework

Figure 3 explains the empirical model. Conceptually, eWOM in SNSs can be divided into three forms: opinion seeking, opinion giving, and opinion passing [15]. In SNSs, interactive environment enables vibrant and interactive eWOM, where an individual customer can take on multiple roles such as the opinion seeker, the provider, and the transmitter. Opinion seekers looking for information and advice from others before a purchase. On the other hand, opinion providers influence on others attitudes and behaviors. Opinion passing is a behavior that more likely to occur in social media environment that facilitates multidirectional communication. It involves forwarding of an original opinion from one source to another [12]. According to the SET, the behavior depends on the assessment of the benefit. The individual chooses the most beneficial one [16].

In SNSs, individuals share and forward the useful information either non- psychological benefits such as gifts, greeting cards, or psychological benefits such as self-esteem, love, trust, friendship, reputation, helping others, and social interdependence [17]. Hence, identifying the motives to share information in SNSs is of great importance, as it can encourage information sharing behavior. From previous studies, we found that the homophily and the tie strength are key social factors. These motivational factors significantly influence on customers intentions to seek, provide, and forward eWOM in the context of Chinese SNSs.

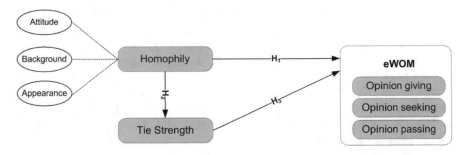

Goodness of fit statistics: χ^2 / d.f. = 2.06 (χ^2 = 684.713, d.f. = 332), GFI =.936, AGFI =.911, CFI = .965, IFI = .943, NFI = .920, NNFI = 0.938, RMSEA = 0.045

Fig. 3. Empirical model of eWOM

$$eWOM = \beta0 + \beta1(AT) + \beta2(BG) + \beta3(AP) + \beta4(HP) + \beta5(TS) + \varepsilon \quad (1)$$

Notes: eWOM = electronic Word-of-Mouth; AT = Attitude; BG = Background; AP = Appearance; HP = Homophily; TS = Tie Strength

2.1   Homophily

Homophily refers to the degree of similarity or congruency between interactions of two individuals [4]. Since individuals willing to socialize and communicate with those who share similar attributes, usually called as social homophily. It is more likelihood to occur the interpersonal communication between two homo-characteristic individuals C that is, homophilous [23]. Consequently, exchange of information occurs between individuals who share some common characteristics such as family and friends.

The communication between the sender and receiver increase as the similarity between them increases. This increased flow of information is facilitated by the homophily [8]. Specifically, SNSs friends are tend to be similar in demographics such as age, as well as in perceptual attributes such as values and beliefs [14]. Thus, higher level of the similarity in attributes can increase likelihood to engage in eWOM behavior. It also can indirectly influence on eWOM behaviors via the tie strength in SNSs.

Hypothesis H1: Homophily with their contacts has a positive effect on eWOM in SNSs (a) opinion giving (b) opinion seeking (c) opinion passing.

Hypothesis H2: Homophily exerts a positive effect on the tie strength in SNSs.

2.2   Tie Strength

Tie strength refers to "the potency of the bond between members of a network" [19]. Social ties can be categorized into weak or strong. Strong ties, such as friends or family members, establish closer relationships that are within social circle of an individual, and are enthusiastic to provide emotional support. Weak Ties, are commonly among less personal affiliations, such as colleagues or acquaintances with strangers, and are able to facilitate information-seeking [24]. Previous studies found that the tie strength plays an important role in peer communication on social media [8,19]. The strong or weak ties can merely influence on customer's behavior. Strong ties significantly effect on the individuals than weak ties. However, SNSs connective features permit to expand personal circle to external groups that increases influence of weak ties. This stimulates the flow of communication among customers and spread product or service-related information, thereby encouraging eWOM behavior [4]. Accordingly, the hypothesis is proposed to investigate the effects of the tie strength on eWOM in SNSs.

Hypothesis H3: Tie strength with their contacts has a positive effect on eWOM in SNSs (a) opinion giving (b) opinion seeking (c) opinion passing.

3   Method

An online-based survey was conducted to collect the data. The content validity of adopted measures was confirmed in related studies. Since, target audience of the study were in China, original measures were translated from English to

Chinese version. In order to avoid translation ambiguities, the help of two linguistic experts were taken. A pilot study of 45 participants was conducted to understand the instruments that were well understood. This only suggests minor improvements. The changes were applied to the modified items.

An online survey link was posted on famous social media such as WeChat and QQ. This survey form was composed of three sections. The first section contained survey guidelines and articulated gratitude to participants from authors. And then participants were asked to answer all questions by recalling their experiences related to most frequently used SNSs. The second part of the survey contained demographic questions. The final section constitutes the measurement items. This study was not subject to common method variance (CMV) because of technology-related topics [18]. However, some efforts were still made to minimize the effects of CMV. The respondents were instructed that there were no wrong or right answers. The constructs name and relationships of measurement items were not revealed for accuracy of results.

3.1 Sample Procedure

A sample of 418 responses were collected over the four months from March–June 2018. The 21 responses were disregarded due to less time (less than two minutes) was taken to answer the online survey questionnaire. A final sample of 397 was considered valid for measurement of model, see Table 1. The sample was considered appropriate as Hair et al. [11] suggested ratio of 20:1. This study empirical model contains nine constructs; hence, a sample size of more than 180 was considered as appropriate.

3.2 Measures

The construct of eWOM was operationalized with three first-order dimensions: opinion giving, opinion seeking, and opinion passing. Each dimension was examined with three items adopted from Chu and Kim [4]. Tie strength was measured with three items adopted from Chu and Kim [4]. In context of this study, the homophily construct was operationalized with three first-order dimensions: attitude, background, and appearance. For each dimension was measured with four items adopted from Chu and Kim [4]. All measurement items of eWOM were assessed on seven-point Likert scale ranging from 'strongly disagree' (1) to 'strongly agree' (7). However, the tie strength and the homophily measurement items were measured on seven-point semantic-differential scale.

Table 1. Demographics of survey respondents (N = 397)

	Frequency (n)	Percentage (%)
Gender		
Male	204	51.39
Female	193	48.61
Age		
18–25	181	45.60
26–35	135	34.01
36–45	52	13.09
46 and above	29	07.30
SNSs*		
WeChat	395	99.49
QQ	234	58.94
Weibo	178	44.83
Users daily average time consumption in SNSs		
< 30 min	83	20.91
30 min–1 h	168	42.32
1–2 h	120	30.23
> 2 h	26	06.54
Top 3 Topic in SNSs*		
Events/News	341	85.89
Gossips/Rumors	338	85.13
Entertainment	294	74.81
Top 3 activities in SNSs	Mean**	Std. Deviation
Reading news, and commenting on the wall	5.33	1.71
Posting comments on the wall	4.80	1.65
Chatting (e.g. WeChat chat)	4.56	1.43

Notes: *Respondents select multiple SNSs/topics ** 7-point scale

4   Results

4.1   CFA, Reliability and Validity

The empirical model was measured with confirmatory factor analysis (CFA), via using AMOS 22. All values of CFI, IFI, and NFI were above the 0.9; RMSEA values were below than 0.08; AGFI values were higher than 0.8; and 2 / d.f. values were less than 3. The model fit indices (2 / d.f. = 1.93 (2 = 764.533, d.f. = 395), GFI = 0.939, AGFI = 0.911, CFI = 0.954, IFI = 0.948, NFI = 0.955, NNFI = 0.936, RMSEA = 0.043) met the threshold value and indicated goodness of fit.

The internal consistency of measures was assessed by using composite reliability and Cronbach's alpha. All values of CR and Cronbach's alpha were above the 0.7 threshold level, suggested satisfactory reliability [11]. The validity of constructs was measured with convergent and discriminant validity. Convergent validity was measured with standardized loadings and average variance extracted (AVE), see Table 2. All standardized loadings were above 0.7 and AVE values were more than 0.5, explained adequate convergent validity [3]. Moreover, correlation matrix diagonal values were above the off-diagonal values (correlation between any two variables), indicated satisfactory discriminant validity, see Table 3.

Table 2. Means, standard deviations, and correlations among study variables

Construct	Items	Mean	Loading	CR	AVE
Attitude*	atd01	3.62	0.81	0.83	0.66
	atd02	3.80	0.82		
	atd03	3.36	0.75		
	atd04	3.48	0.73		
Background*	bgd01	3.72	0.86	0.84	0.72
	bgd02	3.23	0.79		
	bgd03	3.68	0.83		
	bgd04	3.96	0.85		
Appearance*	apr01	3.11	0.72	0.81	0.68
	apr02	3.54	0.75		
	apr03	3.48	0.84		
	apr04	3.30	0.81		
Tie Strength	tsr01	4.18	0.82	0.85	0.74
	tsr02	4.43	0.84		
	tsr03	4.64	0.88		
Opinion giving	opg01	5.28	0.90	0.90	0.81
	opg02	5.11	0.93		
	opg03	5.34	0.95		
Opinion seeking	ops01	5.41	0.94	0.92	0.83
	ops02	5.06	0.89		
	ops03	5.22	0.92		
Opinion passing	opa01	4.67	0.86	0.87	0.78
	opa02	4.71	0.85		
	opa03	4.55	0.89		

Note: *Homophily dimensions

Table 3. Discriminant validity

	HP	TS	OG	OS	OP
HP	0.83				
TS	0.60**	0.86			
OG	0.08	0.68**	0.90		
OS	0.14	0.71**	0.87**	0.91	
OP	0.11	0.56*	0.79**	0.82**	0.88

Notes: * P < 0.05; ** P < 0.01; diagonal values are the square root of AVE for each construct. HP = Homophily; TS = Tie Strength; OG = Opinion Giving; OS = Opinion Seeking; OP = Opinion Providing

4.2   Structural Model

The fit indices of model $\chi2/d.f. = 2.06$ ($\chi2 = 684.713$, d.f. = 332), GFI = 0.936, AGFI = 0.911, CFI = 0.965, IFI = 0.943, NFI = 0.920, NNFI = 0.938, RMSEA = 0.045) shows goodness of model fit. Hereafter, constructs direct effects were measured. The hypothesized variables associations were statistically explained by standardized path coefficients. Figure 3 explains the fit indices and paths.

Homophily direct effects on eWOM in SNSs (a) opinion giving, ($\beta = 0.12$) (b) opinion seeking ($\beta = 0.16$) (c) opinion passing, ($\beta = 0.06$) were found insignificant. Therefore, H1 was not supported. On the other hand, the homophily ($\beta = 0.35$) revealed positive significant effect on the tie strength. This outcome supported the H2. The effect of tie strength on eWOM in SNSs (a) opinion giving, ($\beta = 0.42$) (b) opinion seeking ($\beta = 0.40$) (c) opinion passing, ($\beta = 0.38$) was found positively significant. Hence, H3 was accepted. These outcomes explained the homophily and the tie strength are important predictors of eWOM in SNSs context.

Following H2-H3, the hypotheses was proposed homophily indirect effect on eWOM mediate by tie strength. To measure the mediating role of the tie strength, relative to Sobel statistical test, scores of Z were 2.10362, 3.07328, and 2.39173 respectively, which were higher than 1.96. These results explain that the tie strength positively and significantly mediates between the homophily and eWOM (a) opinion giving, (b) opinion seeking, (c) opinion passing. Table 4 describes the corresponding hypotheses results.

Table 4. Path coefficients and corresponding hypotheses results

Path		
H1a: Homophily → opinion giving	0.12	Insignificant
H1b: Homophily → opinion seeking	0.16	Insignificant
H1c: Homophily → opinion passing	0.06	Insignificant
H2: Homophily → tie strength	0.35**	Sig. Positive
H3a: Tie strength → opinion giving	0.42***	Sig. Positive
H3b: Tie strength → opinion seeking	0.40**	Sig. Positive
H3c: Tie strength → opinion passing	0.38*	Sig. Positive

Notes: * $P < 0.05$; ** $P < 0.01$; diagonal values are the square root of AVE for each construct. HP = Homophily; TS = Tie Strength; OG = Opinion Giving; OS = Opinion Seeking; OP = Opinion Providing

5   Discussion and Conclusion

In China, businesses are increasingly using SNSs for marketing. This phenomenon fosters to investigate the social factors that influences on eWOM in SNSs in China. The study considers the key social factors that effects on eWOM occurred in Chinese SNSs. The purpose of this study is to investigate direct effects of the homophily and the tie strength on eWOM in Chinese SNSs environment. Furthermore, the mediating role of the tie strength between the homophily and eWOM is identified. The results explain that the homophily has insignificant direct impact on eWOM. However, the homophily exerts a positive indirect influence on eWOM through the tie strength. This study outcomes are consistent with previous researches [4,8]. In one study found that the homophily has negative influence on eWOM behaviors, in the USA [4]. Hence, businesses in USA should pay attention to create groups for diverse SNSs users. In contrast, Chinese strongly believe in relationships which is known as the tie strength. The strong ties positively mediates between the homophily and eWOM. These findings provide an idea for strong marketing campaigns in SNSs. Firms in China should take advantages from groups of similar SNSs users (e.g. using background, interests, perceptions, demographics, attitudes and behaviors).

Tie strength was found to be a positive predictor of eWOM in SNSs environment. The results support the previous studies [8,10,23]. Moreover, the outcomes are not very unexpected as in China more importance is given to relationships. Also, SNSs user's associations (e.g. list of friends) reflects significant role in collectivistic societies. Therefore, relationship closeness of an individual would influence other's eWOM. The results suggests that successful marketing campaigns design should include ways for customers to engage in interpersonal talks. For example, enhancing and using the tie strength. This could be effective to engage SNSs users in eWOM, in China.

The study theoretically contributes into the eWOM literature. The empirical investigation of social factors as antecedents of eWOM in SNSs are well understood. Furthermore, the SET is extended by incorporating empirical model of this study. It enhances our understanding about social factors, they are the homophily and the tie strength in SNSs environment. In China, competitive nature of businesses needs to know in depth about collectivistic culture. The strength of social bonding engages customers in eWOM behaviors in SNSs. This would be helpful to build strong relationship ties between customers and businesses. The findings also provide useful insights about eWOM behaviors in SNSs for marketing policy makers, social media (e.g. WeChat, Weibo, and QQ), and online businesses (e.g. www.taobao.com and www.jd.com).

The present empirical investigation has few limitations. The results generalizability maybe limited because data was collected through convenience sampling technique. Future studies should use random, quota sampling methods, which could provide interesting findings. Furthermore, only two cities of China (Beijing and Shanghai) were focused for data collection. Future research may include other big cities of China. This would provide better precision of results. This study focused on social relationships due to collaborative nature of SNSs. The studies in future can investigate others motivational factors such as trust, interpersonal influence, voluntary self-disclosure, and individual differences on customers engagement in eWOM. Future studies could also compare the customers' behavior across the country and other social media sites. For example, micro-blogging sites (Xiohongshu, Mogujie), video-sharing sites (Youku, Douyin, Bilibili), and review-generating sites (Dianping, Meituan).

References

1. Abubakar, A.M., Ilkan, M., Sahin, P.: eWOM, eReferral and gender in the virtual community. Mark. Intell. Plan. **34**(5), 692–710 (2016)
2. Akram, U., Hui, P., et al.: Impulsive buying: a qualitative investigation of the phenomenon. In: Proceedings of the Tenth International Conference on Management Science and Engineering Management, pp. 1383–1399. Springer (2017)
3. Akram, U., Hui, P., et al.: Shopping online without thinking: Myth or reality. In: International Conference on Management Science and Engineering Management, pp. 15–28. Springer (2018)
4. Chu, S.C., Kim, Y.: Determinants of consumer engagement in electronic word-of-mouth (eWOM) in social networking sites. Int. J. Advert. **30**(1), 47–75 (2011)
5. CIW (25 Aug 2018) Top social networking mobile apps in China in Q2 2018, led by WeChat, QQ, Weibo (2018)
6. Erkan, I., Evans, C.: The influence of eWOM in social media on consumers' purchase intentions: an extended approach to information adoption. Comput. Hum. Behav. **61**, 47–55 (2016)
7. Erkan, I., Evans, C.: Social media or shopping websites? the influence of ewom on consumers' online purchase intentions. J. Mark. Commun. **24**(6), 617–632 (2018)
8. Farías, P.: Identifying the factors that influence eWOM in SNSS: the case of Chile. Int. J. Advert. **36**(6), 852–869 (2017)

9. Felix, R., Rauschnabel, P.A., Hinsch, C.: Elements of strategic social media marketing: a holistic framework. J. Bus. Res. **70**, 118–126 (2017)
10. Gvili, Y., Levy, S.: Consumer engagement with eWOM on social media: the role of social capital. Online Inf. Rev. **42**(4), 482–505 (2018)
11. Hair, J.F., Black, W.C., et al.: Multivariate Data Analysis, vol. 6 (2006)
12. Hayes, J.L., King, K.W.: The social exchange of viral Ads: referral and coreferral of Ads among college students. J. Interact. Advert. **14**(2), 98–109 (2014)
13. Kaplan, A.M., Haenlein, M.: Users of the world, unite! the challenges and opportunities of social media. Bus. Horiz. **53**(1), 59–68 (2010)
14. Kim, S., Kandampully, J., Bilgihan, A.: The influence of eWOM communications: an application of online social network framework. Comput. Hum. Behav. **80**, 243–254 (2018)
15. King, R.A., Racherla, P., Bush, V.D.: What we know and don't know about online word-of-mouth: a review and synthesis of the literature. J. Interact. Mark. **28**(3), 167–183 (2014)
16. Li, J.: Knowledge sharing in virtual communities: a social exchange theory perspective. J. Ind. Eng. Manag. (JIEM) **8**(1), 170–183 (2015)
17. Mahapatra, S., Mishra, A.: Acceptance and forwarding of electronic word of mouth. Mark. Intell. Plan. **35**(5), 594–610 (2017)
18. Mazzarol, T., Sweeney, J.C., Soutar, G.N.: Conceptualizing word-of-mouth activity, triggers and conditions: an exploratory study. Eur. J. Mark. **41**(11/12), 1475–1494 (2007)
19. Mittal, V., Huppertz, J.W., Khare, A.: Customer complaining: the role of tie strength and information control. J. Retail. **84**(2), 195–204 (2008)
20. Sohaib, M., Hui, P., Akram, U.: Impact of eWOM and risk-taking in gender on purchase intentions: evidence from Chinese social media. Int. J. Inf. Syst. Change Manag. **10**(2), 101–122 (2018)
21. Sohaib, M., Hui, P., et al.: Understanding the justice fairness effects on ewom communication in social media environment. Int. J. Enterp. Inf. Syst. (IJEIS) **15**(1), 69–84 (2019)
22. STATISTA (16 Aug 2018) Number of social network users in China from 2015 to 2022 (in millions)
23. Van Esch, P., Northey, G., et al.: The moderating influence of country of origin information seeking on homophily and product satisfaction. J. Promot. Manag. **24**(3), 332–348 (2018)
24. Yan, Q., Wu, S., Zhou, Y., Zhang, L.: How differences in ewom platforms impact consumers' perceptions and decision-making. J. Organ. Comput. Electron. Commer. **28**(4), 315–333 (2018)
25. Yen, C.L.A., Tang, C.H.H.: Hotel attribute performance, eWOM motivations, and media choice. Int. J. Hosp. Manag. **46**, 79–88 (2015)

Online Impulse Buying of Organic Food: Moderating Role of Social Appeal and Media Richness

Anum Tariq[1], Changfeng Wang[1], Umair Akram[2(✉)], Yasir Tanveer[1], and Muhammad Sohaib[1]

[1] School of Economics and Management, Beijing University of Posts and Telecommunications, Beijing 100876, People's Republic of China
[2] Guanghua School of Management, Peking University, Beijing 100871, People's Republic of China
akram.umair88@pku.edu.cn

Abstract. While antecedents of impulsive buying behavior have been addressed by existing literature, the current study investigates the impulse buying of organic food through indirect effect of social appeal and media richness of website on relationship between consumer attitude and online impulse buying behavior. In total, 270 individuals from Beijing, China were approached and in response 197 valid responses were collected. Data were collected in the span of two months (September and October, 2018) through online questioannaire placed on Wechat, a famous social media platform. Results revealed a significant moderation of social appeal as social communities and forums influence consumer attitude in both cognitive and affective extents. A websites with highly media richness can increase impulsive decisions and can convert intentions into buying. Results instigate professionals of virtual marketing to review schemes to deal modern consumers by developing websites with social learning mechanism, user-friendlier and visually appealing to push organic intake.

Keywords: Social appeal · Media richness of website · Online impulse buying behaviour · Organic food · Consumer awareness

1 Section Heading

Social commerce has been emerged for over one decade; it is a novel form of e-commerce that is mediated by social media centered on offline and online environments [20]. Social commerce incorporates virtual communities that vitalize user interaction and user generated content provided by plentiful applications involving product recommendation to a friend, review, discussion panels and writing/rating a review [11], which assists customers to apprehend their online buying purpose and prepare them to make more informed and precise buying decision based on their social knowledge and experience. According to CNNIC

© Springer Nature Switzerland AG 2020
J. Xu et al. (Eds.): ICMSEM2019 2019, AISC 1002, pp. 586–599, 2020.
https://doi.org/10.1007/978-3-030-21255-1_45

[3], the penetration rate of Internet in China had increased to 51.7% up 1.3% from December 2015 with 710 million netizens. The popularity of electronic device usage is changing the buying patterns as they are letting consumers to be more informed about their buying decision through online information screenings, over 60% of Internet users considers online access to know about products and brands.

According to Global Mobile Consumer Survey, the cellphone ownership rate in China is 96%, which constitutes it as in integral part of their life. 93%-awakened users check their phones within one hour on less. Generation of late 80s and onward is the main consumption force, they spend more time on social media and therefore probably more effected by WeChat moments of friends and different "opinion leaders" on social media platforms [6]. About 40% of online shopping is impulsive [17]. China represents the fastest growing organic market but consumer attitudes of organic food products through social commerce is less acknowledged. Nowadays, consumers are moderately involved in social groups by which they interact with others to have their opinions and comments about products and evaluate products accordingly that help them to realize the product performance. Modernization and industrialization of agricultural food production has supplemented a rapid socio-economic development of China. Chinese agriculture is more relied on modern agriculture chemicals that result in augmented health hazards and environmental concerns. Perilous amounts of chemicals and food safety instances have been spotted in food production [18]. These concerns have lead Chinese consumers to buy more environmental certified products amid organic products. Consumer's attitude towards organic buying has been grounded on ecological, ethical and quality concerns for production of food. Additionally, information disposal of organic food upsurges consumer knowledge, a key factor in organic food buying [16].

Globally, many consumers adopt "You are what you eat" that triggered demand for green/organic products. Organic consumption has been endorsed not only in Europe and North America, but also in China and India [14]. Consumers are extra observant in intangible values of organic food due to its distinctive qualities, they are unable to find product related information at their own: they have to be propagated by vendors. E-commerce can be advantageous in enhancing the product's intangible value proposition that results in increased profitability of purchases. So, websites are key factors in online buying behaviors [2]. Highly media richen websites influences consumer browsing behaviors leads them to impulsive shopping, product appearance is a critical factor to browse and buy. Chinese consumers are frequently using Internet to acquire information about food safety that stirred their willingness to buy organic food. Website features symbolize building blocks that help to express positive signals about consumers' perception and attitude and online product display is a critical motivating factor to browse and buy more products [2]. The information gap amongst consumers' perceived and actual quality is considered to be a key obstacle in online organic buying. In this perspective, scholars have studied the social appeal and media richness engaged in online buying of organic food China.

Inappropriate product presentation and insufficient information intensifies perceived risk of product performance that ceases buying decision [13]. However, previous studies have focused on product browsing attributes, while credibility characteristics have been branded for organic food. Therefore virtual presentation and social appeal towards organic food impulse buying has been overlooked. Therefore, the purpose of this research study is to examine moderating relationship of media richness and social appeal on the relationship of consumer attitude towards organic food and online impulse buying behavior

2   Literature Review

2.1   The Theory of Planned Behavior (TPB)

The Theory of planned behavior [1] proposed that behavioral control and subjective norm control intentions. This theory involves; (a) attitude formation, (b) percieved behavioral and (c) subjective norms. It is based on individuals' intention to execute a certain behavior that is subjective to individuals' attitude, perceived behavioral control and subjective norms. Social norms considered to be the influence of social factors, especially social pressure of others felt on selection a definite behavior. Buying decisions of organic personal care products were influenced by others' experiences, while applying TPB.

2.2   Media Richness Theory

Media richness is branded as "Ability of information to change understanding within a time interval" [5] that spurs communication effect. Media richness classifies the ambiguity of transmitted messages. Conventionally, face-to-face communications has been a richest medium while text interaction stood leanest. It is exclusively used for confusing or indistinct messages as it includes abundant information and assorted cues. Recently, the rationale of online product presentation has been explained by this theory. Online product presentation is key factor in forming consumer perception that dominate attitudes and buying intentions because of internet spatial limitations [19]. Online product projection is vibrant, which not only edifies value and quality but also notably enriches buying preference that effects perception, attitude and buying intentions.

2.3   Organic Consumption Through Social Commerce

Social commerce is a product of social networking commercial drive based on social interaction. This social interaction grounds an information exchange platform like product feedback and preferences. It has been a most learned platform that unfolds health topics, social information sharing, recommendations and referrals that ultimately influences social commerce. After Europe, now Asian consumers have also shown great interest in organic consumption [11]. A huge numbers of mobiles have been created by advanced Internet technology of China.

Chinese has preferred online shopping because of its distinctive pros, as they feel safer. Online shopping provides extensive information about products available online, thus saves their time and they felt safe as well [12]. Organic food includes products free of artificial chemicals like antibiotics, herbicides, pesticides, and inherently revised organisms [18]. Literature has several terms to recognize organic food, such as "local", "fresh", "natural" and "pure". Wechat (a Chinese social networking application) has been very helpful for information dissemination like sharing experiences, exchanging ideas, technology and current market information [2]. Following text has highlighted key factors of consumer attitude towards organic consumption.

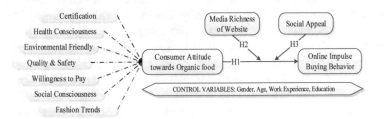

Fig. 1. Hypothetical model

2.4   Contributing Factors Towards Organic Consumption

Health consciousness, In recent times, consumers have been quite aware of health related topics and so conscious about their eating habits that necessitate them to opt nutritional food in their diet [14]. *Quality and safety,* Organic food is considered to be of superior quality due to its known production practices as No chemicals and crop-preserving residues were used, which makes it safer [8]. *Certification,* It is also key factor that incites organic consumption. Authentication and validation of products by certified institution/government is key to gain consumer trust [18]. *Willingness to pay,* Perception of price is imperative in purchase of organic food, but how much this factor effects Chinese consumers is still mysterious. Nearly 60% price premium is higher in china for organic vegetables [18]. *Environmental and ethical consumerism* has also shaped todays' individual behavior [14]. These factors have developed a moral obligation and positive attitude towards environment that is ultimately developed into "eco-friendly products" and "green products" food. Social consciousness, buying behaviors are related with individual's personality, each consumer behaves differently and consumes according to his/her personality [20]. *Fashion trends,* In Italy and USA, consumption of organic food has considered being a status symbol, stemming in exclusive and expensive social trends in elite societies of some courtiers [14].

2.5   Consumers' Attitude and Online Impulse Buying Behavior of Organic Food

Organic food includes items that are free from synthetic compounds and chemicals such as fertilizers, antibiotics, herbicides, pesticides, and genetically modified organisms. In household matters, parents are highly concerned about what should be bought for children [18]. Buying behaviors are related to individuals' personality; each consumer behaves differently and consumes according to his/her personality. Intentions to buy organic food are best predicted by attitude and found to be a positive and significant link between attitude and intentions. Attitude towards a brand is based on consumers' preferences and brand evaluation that depicts ones' likes and dislikes [14]. Chinese consumers are found to be less informed about naturalness', usually found this information on labels and food descriptions, hence carrying suspicious attitude towards organic food certification due to less organic food advertisement traditional channel [18]. Internet is found to be a most frequent channel used by Chinese to browse information about food safety. Media providing quality cues like nutrition content, production process, natural and animal welfare, shopping experiences and online reviews helps in development of consumer attitude that influences its purchase [16]. Proliferation of e-commerce activities have widespread the online impulse buying in consumers, an immediate and unplanned decision based on sudden strong feelings about a particular product and organic buying decisions have also been influenced by subjective and personal norms. Ease of access, shopping anytime of the day and self-service are considered to be drivers of online impulse buying that has freed consumers from experiencing physical store constraints. Furthermore, online shoppers are more spontaneous and impulsive [2]. H1: Consumer attitude towards organic food leads to higher online impulsive buying.

2.6   Media Richness of Website as Moderator

Many research studies have verified environmental impact on impulsive buying of consumers, but mostly on traditional shopping. On the other hand, online shopping is short of smell, texture and atmospheric features of product. In online shopping context, webmorpherics' (e.g. graphics, color, text, audios and videos) including website design attributes may trigger online impulse buying. Website cues have been empirically tested on online impulse purchase in past, such as website quality, media format, visual appeal and representational delight, most of them are alike and interrelated [11,20]. Media richness of website impacts online shopping as similar to good services and low price [2]. Online buyers evaluate quality of service by site's interface design and prefer to visit and buy more from better designed websites. Online shopping is merely based on text description provided by websites, as online shopping does not allow consumers to try or touch products that create hurdles in crating customer perception. Media richness offers broad and deeper sensory breadth permitting consumers to recognize products' descriptions through more diverse channels [8]. The ability to fulfill expectations of consumers by physical design and esthetics of traditional store is analogous to ability of website to fascinate and retain e-consumers by vigilant development of website displays and interfaces.

Online product presentation through depictions, videos, and 3D images can thoroughly present adequate features of organic food and its production process, which can ultimately enhance consumers' belief and develop their interest in organic food [19]. A resourceful website information probably has a positive effect on online impulse buying. Website design has been studied as a determinant of trust, satisfaction, product/vendor quality, attitude enjoyment, arousal and emotions, online shopping excellence and online impulse buying. Contrarily, limited research, if any, exists on the effect of website personality on consumers' impulse buying which encourage individuals to purchase organic product, a novel character of this study. Hence, the following hypothesis is formulated: H2: Media richness moderates the effect of consumer attitude on online impulse buying behavior.

2.7   Social Appeal as Moderator

Buying behaviors are subject to individuals, each consumers behaves differently and consumes according to his/her personality. Some are curious about societal welfare that shaoes their behavior towards food and health choices [20]. Societal trends have forced eating habits towards healthier and ones' well-being responsibilities that effect organic food buying decisions. Subjective norm is a perceived social pressure that engages in or avoids ones' behavior, a total accessible normative beliefs based on others' expectations (family, friends, and significant others etc.). "important others" organic preference have impact on one's organic buying intention as they feel confident doing so. Impulse buying is an unplanned purchase; a result of comparing alternate purchase intentions with actual outcome also triggered by product encounters [17]. Consumers' attitude is vital predictor in buying sustainable seafood besides recommendations and influences of friends and family [18]. Environmental concerns, feelings and positive image also have positive impact of purchase intentions in green products [13] as subjective norms builds organic consumption. Today, consumers are more sensible to have healthy lifestyle by eating sustaining food.

Social appeal is found to be an influential uniqueness that develops individual product preferences. Buying patterns are influenced by perceptions of society and their social identities [2] that reflects functional significance for customers like pro-social status of being truthful, valued cohort and status. So such consumptions explicit eco-friendliness and concerns for environmental preservation by observing social pressure [12]. Social media offers many health topics followed by awareness about a better quality life. Information sharing, recommendations, and referrals allows consumers to exchange product feedback and supplementary information [17]. Social appeal is remarkably affects ones' emotions, beliefs and behaviors. Organic food consumption is considered to be a status symbol with exclusive and expensive social trends in elite societies. It symbols a purchasing power and luxurious life styles of high disposable income consumers [2]. Based on above literature we hypothesis.

H1: social appeal moderates the relationship of consumer attitude and online impulse buying behavior.

3 Research Methodology

3.1 Sample and Procedure

Study sample includes only those responses who have online buying experience of organic food and rest were eliminated. The respondents mostly bought organic dairy products like milk, cheese and butter along with organic rice, organic vegetables and organic eggs. In the next section, demographics were obtained and especially asked about their social media usage and its frequency. In the last section, latent variables of this study have been asked to test the latent hypotheses. First, it was ensured that respondents have experienced organic food buying through online channels and it has been done through a screening question of nominal scale (yes/no).

3.2 Data Collection

Data were collected in the span of two months (September and October, 2018) through online questioannaire placed on Wechat, a famous social media plateform in China [12]. The online survey was conducted in Beijing, China. The online questionnaire was translated into Chinese. So, to have pure and understood responses, researchers got it translated by a Native Chinese speaker. Then, it was translated back into English as well and both versions were compared to eliminate inconsistencies to guarantee the actual meanings of the questions. An online questionnaire is considered to be the most appropriate in online context because of its geographic impartiality, low budgeted and quick responsiveness. Cross-sectional data collection technique was followed to realize the structural relationship among all constructs [12]. A quantitative research method was applied to test proposed hypothetical model of this study (Fig. 1). Three sections structured a complete online questionnaire which is used to collect responses.

3.3 Descriptive Statistics

In total, 270 respondents were approached, nevertheless 197 valid responses received (response rate = 72.9%) with 71 males (36%) and 126 females (64%) respondents. Female respondent were found to be on higher side. They are frequently using social media platform for organic food as compared to males. Around half of the respondents were above 36 year of age (Table 1). Whole sample size is well educated as minimum qualification asked was graduation. Amos 22.0 was utilized to test adequacy of measurement model.

Table 1. Sample demographics ($n = 197$)

		n			$n(\%)$
Gender	Male	71	Log-on frequency	Daily	173 (0.88)
	Female	126		2–4 times a week	24 (0.12)
Age	25 or less	28		2–4 times a month	-
	26–35	72		Once a month	-
	36–45	43		≤ 1 h	2 (0.10)
	Above 46	54		2 h	23 (0.11)
Education	Graduation	77	Log-on duration	3 h	113 (0.57)
	Masters	103		4 h	45 (0.23)
	Doctoral studies	17		≥ 5 h	14 (0.07)

3.4 Measures

Consumer attitude, was measured on seven items developed by Gil et al. [9] and Locke et al. [15]. Media richness of Website was measured by thirteen items scale adapted by Cyr & Bonanni [4] and Ganguly et al. [8] which were measured in three strands; (1) visual design, (2) information design and (3) Navigation design. Social appeal is measured on four items scale, developed and modified to fit in the context of online shopping by Lockie et al. [15]. Online impulse buying behavior is measured on five items scale developed by Verhagen and van Dolen [8]. All responses were measured on seven point Likert scale from 1 (strongly disagree) to 7 (strongly agree).

4 Results

4.1 Confirmatory Factor Analyses (CFA)

Independent and dependent factors were classified, evaluation of hypothetical research model and validity analysis were done through CFA. CFI and RMR levels were tested to study acceptable fit levels as CFI = 0.90 and RMR = 0.05–0.08 were considered to be an indicator of good fit. CFA analysis was applied to confirm factor structure extracted by EFA to improve overall model fitness. Amos 22.0 were used and all factors as independent variables were involved in proposed hypthetical framework. Four factors model was examined followed by others factor models. With CFA calculating measurement reliability and validity of latent variables, the four factors model determined the satisfactory threshold CMIN/df; 2.315, p = 0.01), CFI = 0.942, RMSEA = 0.056, RMR = 0.076, GFI = 0.977, AGFI = 0.993, NFI = 0.937. All values represented satisfactory model fit indices. Additionally, all the factor loadings were statistically significant.

4.2 Reliability, Convergent Validity and Discriminant Validity

Convergent validity and reliability were measured by following metrics; Cronbach's alpha; Composite reliability and average variance extracted. Results indicated adequate reliability and convergent validity evaluated on following conditions [10]; (a & b) All measurement item loadings and Cronbach's alpha >0.70 and CR of each variable >0.80. For internal consistency among measurement items, Cronbach's alpha was ranging from 0.79 to 0.91, CR for all construct were >0.7 as recommended. Fronell and Larcker [7] method was applied to assess discriminant validity.

The square root of (AVE) between a construct and its measures must be greater than the correlations between the construct and any other construct in the research model and it was found greater than inter-correlations with another construct. All values of AVE were >0.5 demonstrating that measurement items were converged on same construct (Table 2). Construct reliability and validity significantly braced hypothesized model.

Table 2. Measurement model results and discriminant validity and correlations

Constructs mean	SD	Item loading	AVE [a]	Composite reliability [b]	α	CA	MR	SA	OIBB
CA (7) 5.01	1.173	0.76–0.87	0.810	0.840	0.850	−0.84			
MR (13) 4.91	1.368	0.79–0.90	0.720	0.880	0.790	0.429	−0.85		
SA (4) 4.61	1.110	0.84–0.93	0.800	0.870	0.830	0.551	0.403	−0.890	
OIBB (5) 5.39	1.289	0.81–0.92	0.880	0.800	0.910	0.473	0.378	0.349	−0.940

Goodness of Fit statistics: CMIN/df; 2.315, p = 0.01), CFI = 0.942, RMSEA = 0.056, RMR = 0.076,GFI = 0.977, AGFI = 0.993, NFI = 0.937

Notes: a Average variance extracted (AVE) = (Summation of the square of the factor loadings)/(summation of the square of the factor loadings) + (summation of the error variances); b Composite reliability = (square of the summation of the factor loadings)/(square of the summation of the factor loadings) + (square of the summation of the error variances); Acronyms: Consumer attitude towards organic food (CA), Media richness of website (MA), Social Appeal (SA), Online impulse buying behavior (OIBB); The brackets () scores diagonal are the square root of AVEs of the individual constructs

4.3   Hypotheses Testing

An associative relationship between all constructs of research model were assessed and built by a structural relationship model. Structure equation modeling (SEM) and CFA were applied for model evaluation. The estimated coefficient of determination assessed by SEM analysis was similar to that found in the regression analysis. The results shows outstanding structural model fit indices: CMIN/df: 2.473, p = 0.001, CFI = 0.98, GFI = 0.987, RMSEA = 0.074, RMR = 0.041, AGFI = 0.987, NFI = 0.958.

H1 represents direct hypotheses between consumer attitude and OIBB. Consumer attitude towards organic food based on based seven antecedents has a significant positive impact on OIBB ($R^2 = 0.470$, $\beta = 0.729$, $p < 0.001$; t-value = 14.191). This means 47% variance in OIBB is explained by consumer attitude towards organic food. Thus, H1 is supported. Direct effects of moderators were also assessed on OIBB before further moderation analysis. Media richness of website has found to have significant direct effect on OIBB ($\Delta R^2 = 0.503$, $\beta = 0.337$, $p < 0.05$; t-value $= 4.349$). Social appeal also found significant on OIBB ($\Delta R^2 = 0.511$, $\beta = 0.214$, $p < 0.001$; t-value $= 8.310$).

For moderation, three step hierarchical moderation analysis were utilized. First, independent variable is regressed on dependent variable, followed by regression of independent variable with moderator and in the last step, interaction obtained by product of both independent and moderator variables is entered. For H2, R^2 of CA is 0.470 with significant and positive value of media richness of website. In third step, ΔR^2 characterized the integration term (CA X MR) was 0.033 which means consumer attitude and media richness of website jointly explained 3.3% change in OIBB. The total interaction effect was $\Delta R^2 = 0.507$ represented 50.7 change in OIBB ($\beta = 0.441$, $p < 0.001$; t-value $= 2.672$). Increase in R^2 from 0.503 to 0.507 depicts significant moderating effect of media richness. Thus, H2 is supported (Table 3).

Table 3. Moderation analysis of media richness of website

	β	t-value	F	R^2	Adjusted R^2	Δ^2
CA	0.729***	14.191	1256.49**	0.470	0.470	0.470***
MR	0.337**	4.349	610.91***	0.503	0.510	0.033***
CA×MR	0.441***	2.672	413.82***	0.507	0.519	0.004***

Notes: F- statistics are for overall models. CA = Consumer attitude towards organic food, MR = Media Richness of website, Dependent variable = Online impulse buying behavior. Control variables are age, gender, education and income. *p 0.05 **$p < 0.01$ ***$p < 0.000$.

For H3, social appeal is significantly effects OIBB ($\beta = 0.214$). A significant interaction of CA and SA on OIBB was found ($\beta = 0.571$, $p < 0.001$; t-value $= 3.647$). The interaction term (CA X SA) was $\Delta R^2 = 0.006$ which means consumer attitude and social appeal jointly explained 6% change in OIBB with overall 51.7% of variance in OIBB (R2 = 0.517 and $\beta = 0.571$). Thus, H3 is supported (Table 4).

Table 4. Moderation analysis of social appeal

	β	t-value	F	R^2	Adjusted R^2	Δ^2
CA	0.729***	14.191	1256.49**	0.470	0.470	0.470***
MR	0.214***	8.310	692.67**	0.511	0.520	0.041***
CA×MR	0.571***	3.647	454.82***	0.517	0.529	0.006***

Notes: F- statistics are for overall models. CA = Consumer attitude towards organic food, SA = Social Appeal, Dependent variable = Online impulse buying behavior. Control variables are age, gender, education and income. *$p < 0.05$ **$p < 0.01$ ***$p < 0.000$.

5   Discussion

Strong demand of consumers is grounded on grave food safety has offered e-commerce a novel outlet for organic deals. Chinese Internet usage has become the foremost source of information to gain food protection knowledge, an effective and delightful way to have more abundant product related information. This study has focused online impulse buying of organic food based on social appeal and media richness of website. The hypothesized model was fully supported by results. Specifically acceptance of H1: where it was found that using social commerce, consumer attitude towards organic food has a positive and significant impact on online impulse buying behavior. In other words, consumers' attitude towards organic food becomes more positive, its online impulsive buying would also increase. Consumers usually join social communities only for sake of their own specific needs, by observing others' knowledge and experience and through interaction with them ultimately affect their attitude towards products and websites. Chinese consumers are skeptical while buying online as they distrust food producers, so they search more information for alternative evaluation. Online ratings, reviews and others' shopping experiences play a significant role in organic food sale. A thorough review may have concrete information, which may build confidence to make effective decision. Online reviewer can also reflect their involvement that ultimately assists others in browsing more quality information and shaping their buying decisions. It has been primarily due to widespread use of social media that significantly influences consumers' buying behavior.

Accordingly, online product presentation would increase impulsiveness by using striking layouts, compact fonts, perky colors and appropriate graphics. Contrary to this, if visual aids such as pop-up messages, animated banners and twinkling texts may amuse consumers that leads them to negative perception which ends at switching buying decision. Since this study examined online buying of organic food, it is important for website to use portraits, videos and 3D descriptions that can intensely present products' look and quality and influencing consumers' buying decisions by dipping risk and uplifting their trust in the website information. Interestingly navigation of website has been studied as a highest irritating factor in online shopping [2], because online consumers preferred to have a intuitive and uncomplicated navigation design which helps them to complete transaction with least exertion.

6   Practical Implications

The leading contribution of this study is to endorse the influence of social appeal on consumers' attitude while online impulsive buying of organic food. Social appeal can significantly increase impulsive buying of organic food. Social communities and forums influence consumer attitude in both cognitive and affective extents. Consumers do participate in online social groups, by observing others' knowledge and experience and through interaction ultimately can affect their attitude and beliefs. Social recommendations, Rating and reviews have also help to build attitude towards organic consumption by elucidating product/seller quality, utilitarian shopping experience dipping customers' indecision about buying decision.

The second contribution is to approve the impact of media richness of website on consumer attitude while purchasing organic food online. The study realizes that organic food knowledge presented by a high media richness websites can increase impulsive decisions and can convert intentions into buying. This denotes that website with pleasant visual design may more attract customers. A website with site maps, direct and shortcut links and one click operations can minimize visitors' effort to complete buying process. Website visitor has to perform minimal task to achieve their browsing goal. In case of organic food, information asymmetry has caused market failure, so informative website can influence pragmatic consumers' perception towards organic consumption. Websites with high media richness can reduce perceived risk in organic buying; organic consumer expects "naturalness" and Chinese consumers generally obtain such information through food labels and product descriptions. This information fertility stimulates consumers' senses in realizing expedient characteristics of organic food. Moreover, information cues like nutrition, production methods, eco-friendly and animal welfare also influences consumer attitude and buying intentions. If the contents delivered by websites are coherent with what socially a consumer is observing, it will further compel one to make impulse buying decisions.

7   Research Implications & Conclusion

This study contributes knowledge to organic consumption research in several ways. First, it offers an enhanced model for online impulse buying based on consumer attitude with indirect effect of social appeal and highly media richer website personality. Second, the role of social appeal as the theoretical foundation of this research reaffirms the significant effect on impulse buying of organic food. Social commerce has transformed consumer behavior; consumers' decisions are more relied on others' posted contents. The research model sheds light on how social appeal influences customers' attitude towards impulsive buying. Use of referrals and reviews, product experience and, learning from forums and communities are found to be more important in consumers' decision quality of organic food. From a managerial viewpoint: results advocate that managers can safely institute a link between product and consumer's identities to achieve preliminary market acceptance. Marketing managers must initiate tactics that use influential contents like "important others", whose opinions may trigger consumption choices. Using social media, health communities can serve as mode of communication to modify value proposition of prospect organic consumers. Developing knowledge communities could be a creative way to realize the customer knowledge value. Such long-term communities would be advantageous in development of trust, mutual benefit, and belongingness. Providing comprehensive information through social media can easily and quickly develop an attitude towards organic consumption. Because social media is considered to be an extraordinary medium, paying more attention on social media marketing communications that positively affects their cognitive and purchase intentions. Lastly, for online marketers, intense competition in cybernetic market and sustaining an appealing and proficient website is critical for attraction and retention of customers. Tedious and infective website interface can be quickly rejected with a single mouse click, a big challenge for them. Additionally, The trick for effective trigger of impulse buying is to give consumers adequate exposure to the relevant sustainable stimuli with least physical, mental and time effort by crafting a cordial shopping experience. The study stipulates empirical evidence of the significance and applicability of the social dimensions in development of organic attitude.

References

1. Ajzen, I.: From Intentions to Actions: A Theory of Planned Behavior (1985)
2. Akram, U., Hui, P., et al.: Factors affecting online impulse buying: evidence from Chinese social commerce environment. Sustainability **10**(2), 352 (2018)
3. (CNNIC) CINIC Internet statistical reports (2016)
4. Cyr, D., Bonanni, C.: Gender and website design in e-business. Int. J. Electron. Bus. **3**(6), 565–582 (2005)
5. Daft, R., Lengel, R.: Organizational information requirements, media richness and structural design. Manag. Sci. **32**(5), 554–571 (1986)
6. Duffett, R.G.: Influence of social media marketing communications on young consumers' attitudes. Young Consum. **18**(1), 19–39 (2017)
7. Fornell, C., Larcker, D.: Evaluating structural equation models with unobservable variables and measurement error. J. Mark. Res. **18**(1), 39–50 (1981)
8. Ganguly, B., Dash, S., et al.: The effects of website design on purchase intention in online shopping: the mediating role of trust and the moderating role of culture. Int. J. Electron. Bus. **8**(4), 302–330 (2010)
9. Gil, J.M., Gracia, A., Sánchez, M.: Market segmentation and willingness to pay for organic products in Spain. Int. Food Agribus. Manag. Rev. **3**(2), 207–226 (2000)
10. Hair, J., Black, W., et al.: Multivariate Data Analysis; A Global Perspective. Pearson Education Inc (2010)
11. Hajli, N.: Social commerce constructs and consumer's intention to buy. Int. J. Inf. Manag. **35**(2), 183–191 (2015)
12. Jin Jslhlhly, S., Li, Y.: Preferences of Chinese consumers for the attributes of fresh produce portfolios in an e-commerce environment. Br. Food J. **119**(4), 817–829 (2017)
13. Mccarthy, B., Liu, H., Chen, T.: Innovations in the agro-food system: adoption of certified organic food and green food by Chinese consumers. Br. Food J. **118**(6), 1334–1349 (2016)
14. Rana, J., Paul, J.: Consumer behavior and purchase intention for organic food: a review and research agenda. J. Retail. Consum. Serv. **38**, 157–165 (2017)
15. Stewart, L., Kristen, L., et al.: Choosing organics: a path analysis of factors underlying the selection of organic food among Australian consumers. Appetite **43**(2), 135–146 (2004)
16. Thogersen, J., Barcellos, M., et al.: Consumer buying motives and attitudes towards organic food in two emerging markets. Int. Mark. Rev. **32**(3/4), 389–413 (2015)

17. Verhagen, T., Dolen, W.: The influence of online store beliefs on consumer online impulse buying: a model and empirical application. Inf. Manag. **48**(8), 320–327 (2011)
18. Xie, B., Wang, L., et al.: Consumer perceptions and attitudes of organic food products in Eastern China. Br. Food J. **117**(3), 1105–1121 (2015)
19. Yue, L., Liu, Y., Wei, X., et al.: Influence of online product presentation on consumers' trust in organic food a mediated moderation model. Br. Food J. **119**(12), 2724–2739 (2017)
20. Zhang, K., Benyoucef, M.: Consumer behavior in social commerce: a literature review. Decis. Support. Syst. **86**, 95–108 (2016)

Research on the Advertising Diffusion Effectiveness on Microblog and the Influence of Opinion Leaders

Dan Zhang[1], Chuanpeng Xu[1], Malian Shuai[2], Wenyu Xiong[1], Wen Jiang[1], Dong Xu[3], Yue He[1(✉)], and Weiping Yu[1]

[1] Business School, Sichuan University, Chengdu 610064,
People's Republic of China
807706482@qq.com
[2] NetEase Computer System Limited Company, Hangzhou 310006,
People's Republic of China
[3] Sichuan Huaxin Modern Vocational College, Chengdu 610107,
People's Republic of China

Abstract. Advertising Microblog promotes and publicizes products or services by texting the information about then for a marketing promotion or expanding. Users' negative attitude toward advertising makes choosing the right advertising node particularly important. This paper takes Roseonly as a case. Firstly, an estimation of advertising diffusion effectiveness based on the emotion and wideness-hotness is set up. Secondly, *KPrank* algorithm based on the *PageRank* is proposed to recognize opinion leaders considering the node characteristics and the importance in the network. Finally, it studies the function of leader of opinion in the advertising diffusion. The results show that users of Roseonly have low emotion value and wideness-hotness value. The opinion leaders authenticated by impersonal is more than the personal. At the same time, the paper finds those leaders of opinion have more influence on the wideness-hotness. Also, the leaders of opinion can take positive influence to other carriers or audience.

Keywords: Advertising microblog · Diffusion effectiveness · Leader of opinion · Wideness-hotness · Sentiment · Marketing

1   Introduction

The concept of "microblogging" began in 2006 with the creation of Twitter. In China, Sina microblogging has opened the market through the celebrity effect quickly, as of September 2017, microblogging monthly active users totaled 376 million [6]. The review and forwarding process in microblogging platform not only allows the public to participate in the dissemination of information, but also shorts the distance of relationship between the public and the business or organization. It has become the network of enterprises to promote their products

J. Xu et al. (Eds.): ICMSEM2019 2019, AISC 1002, pp. 600–615, 2020.
https://doi.org/10.1007/978-3-030-21255-1_46

and services because of its unique wide range of social user groups and virus spread of information dissemination speed, and at the same time, it is gradually transformed from a typical social type into a media type.

Advertising microblogging is a marketing form that the company presents products, services or brand information through text on microblogging, which is in order to promote the company's reputation and also improve the marketing efficiency. The existing advertising microblogging form mainly includes product advertising, promotion advertising, brand advertising and activity advertising [14]. Compared to web advertising, advertising microblogging is more authentic and vivid, which giving users a sense of the product and brand when interacting with the advertising microblogging. And through the spread between the users, the product and brand will become more popular and persuasive. The emergence of diversified advertising microblogging, as well as the user's negative attitude towards the advertisements and instinctive rejection may make some users feel tired, or may cause the unnecessary cost of waste from advertisers in the future. The paper mainly studies the mechanism of advertising microblogging communication, finds out the key nodes in advertisement transmission, and evaluates the propagation effect to improve the precision marketing efficiency.

The main innovation of the article is to propose a two-dimensional evaluation system based on the degree of emotion and wideness-hotness of the advertisement, so that the communication effect evaluation is more accurate and more objective. Meanwhile the paper tried to find the impact of opinion leaders on the effect of communication, which for enterprises or organizations in the microblogging network to promote the brand, products and services with a more comprehensive theoretical support. The opinion leader recognized the improved *KPrank* algorithm based on *PageRank* principle, which is considered with both the node network centrality and the node's own attributes, so that the calculation of the node's influence will be more comprehensive.

The main contributions of this paper include:

(1) Proposing a two-dimensional evaluation system based on the degree of emotion and wideness-hotness of the advertisement, so that the communication effect evaluation is more accurate and more objective.

(2) Trying to find the impact of opinion leaders on the effect of communication, which for enterprises or organizations in the microblogging network to promote the brand, products and services with a more comprehensive theoretical support.

(3) The opinion leader recognized the improved *KPrank* algorithm based on *PageRank* principle, which is considered with both the node network centrality and the node's own attributes, so that the calculation of the node's influence will be more comprehensive.

2   Literature Review

Researchers are mostly based on qualitative methods and questionnaire to study advertising on twitter. Fogel [11] researched on trust for twitter direct-to-consumer prescription medication advertisements by questionnaire. Komodromos [18] employed a qualitative study with 20 in-depth interviews to explore the factors/strategies that influence students' acceptance of e-shopping. To enhance the retail outlet operational efficiency on twitter. Pandey [2] researched on social media marketing impact on the purchase intention of millennials. Adetunji [20] examined the relationships between the social media marketing communications by questionnaire. Voorveld [24] examined how consumers' engagement with social media platforms drives engagement with advertising embedded in these platforms through survey. Chu [8] qualitatively analyzed the current state of knowledge on electronic word-of-mouth in advertising research. Anagnostopoulos [4] proposed a methodology for performing targeted outdoor advertising by leveraging the use of social media. Other researchers adopt quantify method. To explore whether early propagators of trending topics respond to advertising messages, Lambrecht [1] collaborated with a charity and a fashion firm to target ads at consumers who embraced a Twitter trend early in its life cycle by posting about it, and compared their behavior to that of consumers who posted about the same topic later on. Dai [9] collected e-cigarette related tweets that found the higher prevalence of e-cigarette advertising was associated with states with better tobacco control impact and lower youth smoking prevalence to study the negative effect of social media advertisements. In general, advertising on twitter are almost adopt qualitative method, so scholars will have a broader research and development space.

Microblogging spread effects mean the impacts of articles to users and society on microblogging. Some scholar adopt qualitative methods to find factors what influence twitter spread (Zheng [26] and Swasy [22]) and investigated the reasons for spreading information in forms of retweets (Firdaus [21]). Other scholar adopt quantitative and machine learning methods. Bakal [5] conducted to quantitatively assess how health related tweets diffuse in the directed follower friend Twitter graph through the retweeting activity. Varshney [23] presented a Bayesian network based approach for solving the problem that predicting the probabilities of diffusion of a message through the links of a social network. But all over these, there is still a lack of emotional factors on the role of the study on the effect of communication.

The nodes of microblogging platform refer to microblogging users, and the researches of the microblogging node are focused on opinion leaders. At present, the studies of opinion leaders are mainly divided into two directions. The first one is the identification indicators of opinion leaders. Lopez [17] found the number of followers and people or brands they follow is key to identify loyal opinion leaders in brand communities developed in Twitter. Cui [15] put forward a method that identifies the opinion leaders according to the change of user features and outbreak nodes. The second one is based on the node centricity of social network structure to measure the nodes' influence.

Some scholars made a few researches on the nodes' influence. Riddell [12] performed a network analysis of emergency physicians and ranked them on three measures of influence: in-degree centrality, eigenvector centrality, and betweenness centrality. Litterio [3] proposed a matrix for the classification of the individuals based on the combination of eigenvector centrality and betweenness centrality. Mauricio [19] built a retweet network where the centrality metrics degree, betweenness and pagerank were calculated to identify which users most influenced the social movement.

So according to the three aspects of the relevant literature above, following deficiencies could be found:

(1) Most of the researches about advertising microblogging communication are based on qualitative analyse;

(2) Most of the information communication networks are based on the user's attentions and been concerned in the relationship, but the researches of forwarding relationship are weak;

(3) The existing research lacks the influence of the nodes itself and the centrality of the network structure on the influence of the nodes;

(4) The current researches on the effect of information dissemination ignore the emotional factors.

This paper studies the microblogging advertising effects assessment and opinion leaders on the role of communication at *Sina platform* with Roseonly product advertising microblogging. The addition of emotional factors to the assessment of the dissemination of the system, makes the evaluation results more comprehensive. And this paper also presents a method based on *PageRank* principle, which considers the influencing factors of nodes and the centrality of the network structure to identify the opinion leaders. Finally, the roles of opinion leaders in communication are measured based on the effect of communication and the position of opinion leaders. The results of the research can provide a basis for accurate marketing and help companies make decisions.

3   Methods and Materials

Based on the existing research status and problems, this chapter introduces the research ideas of the paper, including data acquisition and processing, advertising microblogging communication effect evaluation and opinion leader recognition algorithm.

3.1   Data Acquisition and Processing

First, *Sinaopenplatform*, *JAVASDK*, is used to crawl each forwarding node properties and relationships in the forwarding process about the advertising microblogging of Roseonly through *SinaAPI*. And the initial data is cleaned to help analysis.

Then, Chinese word segmentation system, *ICTCLAS*2015, is used to do the word segmentation of the microblogging text, which developed by the China Institute of scientific computing technology. Therefore, the article uses the "*NTUSD* - Simplified Chinese Emotional Polarity Dictionary" proposed by LiZhao [16] to mark the acquired emoticons for labeling.

After all these, some filtering work for those words which can't express the emotions, single words or high frequency words are done. Comprehensively the filter vocabulary from the Harbin Institute of Technology, Sichuan University and other universities, as well as Baidu stop word list, are selected to match the experimental text, then the words which can be found in these filter word lists are filtered out.

3.2   Evaluation on the Effect of Advertising Microblogging

Foreign research on the effect of information dissemination focuses on the psychological and behavioral impact of the audience. Chen Yuan [25] refers to the four ladder patterns of cognition, sentiment, attitude and behavior to assess the spread of advertising, according to social psychologists Lavidge and Steiner's L & S model (Lavidge & Steiner Mode) in his study, as shown in Fig. 1.

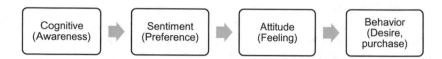

Fig. 1. The advertising effectiveness of ladder model

This paper combines the mode of communication effect, and evaluates the effect of advertising microblogging in two dimensions, the degrees of sentiment, wideness and hotness. The measurement of the degree of sentiment is mainly to identify the emotions in the content of the forwarded text of each advertisement microblogging. The degree of wideness-hotness mainly evaluates the effect from the quantitative data index in the microblogging forwarding network to measure the cognitive and behavioral aspects of the information dissemination.

(1) Sentiment Analysis

From the researches of existing information dissemination effect assessment, it can be seen that the studies of the sentiment are not enough [5, 21–23, 26]. This paper uses the content analysis method [7] to analyze the user's forwarded text content based on the sentiment dictionary to dig the emotion of the forwarding user. This paper sets the emotional value of the positive adjective and verb to +1, the negative adjective and the verb's emotional value to −1, and sets the adjective and adverb with the degree adverb ±1 on the basis of the original. Since the format of the expression symbol is [expression + text], when there are N expressions in the analysis unit, the corresponding value will ±N in the category. Let two compilers encode 50% of the forwarding content and test the consistency, and then the results are as evaluation criteria to the sentiment classification results of the content analysis machine software (ROST Content Mining6).

The sentiment degree of each microblogging is defined as E_i, the cumulative sum of the emotional values of each analysis unit. And the formula is as follows:

$$E_i = \sum_{n=1}^{n} e_n \qquad (1)$$

e_n is the emotional value of $n - th$ analysis unit for the $i - th$ forwarding microblogging.

(2) Dissemination Analysis of Wideness and Hotness

The wideness is horizontal to describe the range of propagation, while the hotness is vertical to describe the extent of the impact of the spread. It indicates a higher popularity of microblogging text while the range is wider and the impact is deeper. The article selected three indicators to assess microblogging of the wideness and hotness, the expression is as follows:

$$G_i = \alpha C_i + \beta R_i + \delta U_i \tag{2}$$

The paper firstly summarizes the data of each original microblogging, and uses the deviation standardization to carry on the dimension processing to the data. Through the first transformation of the original data, the index value is mapped to [0, 1]. C is the sum of the number of comments of the original microblogging and its forwarding microblogging, and R is the number of forwarding of each the original microblogging. Since a forwarding user may forward microblogs for multiple times, there may be multiple forwarding nodes, so the number of forwarding users (U) is added to the microblogging. α, β and δ represent the weight coefficients of G_i, and the weight coefficients are determined by $CRITIC$ method. This method not only considers the impact of the size of the index on the weight, but also avoids the conflict between the indicators, which is more scientific relative compared with Entropy Method and Deviation Method [13].

After determining the weight, G_i is calculated according to the formula for each microblogging. And the emotional values are made to be the vertical axis, while wideness and hotness to be the horizontal axis. Then the two-dimensional evaluation system of advertising microblogging dissemination is established.

3.3 Opinion Leader Identification

With the rapid development of microblogging, the concept of opinion leader is accepted by most people and the studies about opinion leader are increasingly diversified. The paper identifies the opinion leader by measuring the influence of the nodes, and proposes the *KPrank* algorithm.

KPrank is similar with the *PageRank* algorithm. Combined with the convergence of *PageRank* calculation and the principle of random walk model, the forwarding node itself is considered as the control factor. A node i is in the propagation network, the forwarding probability of the number of non-forwarding fans is λ_i. This forwarding probability is determined by the node i itself in the whole information dissemination, which is more biased towards information sharing or bias towards information acquisition. And ϕ is a node parameter that is calculated according to the following formula.

$$\phi = N_{followers}/N_{friends} \tag{3}$$

$N_{followers}$ represents the number of fans for each fan of the node, and $N_{friends}$ represents the number of concerns for each fan of the node.

When ϕ is at the following conditions, the value of forwarding probability λ_i is as follows:

$$\begin{cases} \phi = 1, \lambda_i = 0.5 \\ \phi < 1, 0 \le \lambda_i < 0.5 \\ \phi > 1, \lambda_i > 0.5 \end{cases} \quad (4)$$

In this paper, when $\phi > 1$, λ_i is 0.7, and when $\phi < 1$, λ_i is 0.3.

The time difference which the user posts information to the fans to forward the information is set as the node propagation time, and the shorter the propagation time of the node is, the higher the responsiveness of the user's fans concerned the users, which also says that the user's appeal is relatively higher. The microblogging comments are more, then the microblogging released the user's attention is higher, the effective dissemination of information will be stronger [10].

In summary, the probability of forwarding, the propagation time and the number of comments are directly related to the node's own attributes of the forwarding in the information dissemination network, and directly show the influence of the node on the dissemination of information. Therefore, it can be set that the node attribute influence degree is $\omega_i(C_i, \frac{t_i}{T}, \lambda_i)$, C denotes the number of comments, denotes the propagation time, and λ denotes the forwarding probability. T represents the length of time the information source is released to the last forwarding time. In the calculation, the number of comments after the dimension is selected.

In the microblogging forwarding network, the user can be forwarded more than one person concerned about the blog, which not only reflects the user's activity, but also indirectly reflects the attention of the user's influence and appeal. Thus, the adjacency matrix of node i to j user forwarding times is:

$$L_{ij} = \begin{cases} n, \text{ when user i forwards the microblog of user j for n time} \\ 0, \text{ when user i doesn't forward the microblog of user j} \end{cases} \quad (5)$$

Combined with the theory of *PageRank* principle, and considered the contribution of forwarding behavior to information dissemination and the influence of node's own attributes, the paper constructs the following *KPr* to measure the influence of nodes on the whole information dissemination.

$$K\Pr(P_i) = \sum_{p_j \in M(p_i)} \frac{L_{ji} * K\Pr(Pj)}{L_T} + \omega_i \quad (6)$$

$M(p_i)$ represents the set of forwarding i nodes, and L_T represents the total number of times the article is forwarded.

$$\omega_i(C_i, \frac{\Delta t_i}{T}, \lambda_i) = \sqrt{C_i * \frac{\Delta t_i}{T} * \lambda_i} \quad (7)$$

The forwarding node of this paper takes micro-blog ID as the unique value, but a user may forward multiple microblogging, so multiple forwarding nodes may be just one user to play. The formula for setting the user $KPr(P)$ is as follows:

$$K\Pr(P) = \frac{\sum_{i=1} K\Pr(P_i)}{n} \tag{8}$$

Where n represents the number of microblogs forwarded by user P in the forwarding network, and $K\Pr(P_i)$ represents the KPr value of the $i-th$ forwarding node that the user P forwards.

4   Results

Based on the theory and technology described earlier, according to the research scheme designed as above, the paper first collects data from *Sinaplatform* through *API* interface and preprocesses it, then establish the two-dimensional assessment system of emotional-wideness-hotness propagation effect to assess the spread of advertising microblogging effect. Finally, Java is used to improve the *PageRank* algorithm-the *KPrank* algorithm to identify the opinion leader and analyze the impact of the opinion leader position on the propagation effect.

4.1   Data Collection and Preprocessing

By the virtue of "only once for one person" rules and "believers love, love is the only" brand concept, Roseonly makes it well-known by microblogging marketing. It sells the ordinary flowers rose to dozen in a box at the price of 999 yuan, an eternal rose to reach the high price of 699 yuan, and at a short period time it becomes "instant microblogging celebrity". Because it's advertising microblogging has a certain representation, it is selected as a research object.

Using *SinaAPI*, the data of Roseonly from February 1, 2013 to December 24, 2014 are accessed, which totally related to 1536 microblogging data, including 121 advertising microblogging, 59905 forwarding node information, which including microblogging release time, original microblogging id, original microblogging content and other 24 attribute values.

After data cleaning, the effective forwarding nodes collected from original 121 ads microblogging turn 58033, and the effective use rate turns 98.16%. 121 ads microblogging are published by 39 users, including 15 non-certified users, 24 authenticated users, and including Roseonly official microblogging released 83 advertising microblogging, accounting for 68.60%. And then pretreatment of the collected comments on the content of the word segmentation and stop words, etc., to remove the emotional analysis is useless or which even has interference information on.

4.2   Analysis and Evaluation of the Effect of Advertising Microblogging

(1) Sentiment Evaluation

Based on the research design in Chap. 2, the article uses the emotional dictionary to identify the user's micro-text emotion. At the same time, the content analysis method is used to do the second classification of the emotional polarity, which gets emotional recognition results of 121 microblogging. Figure 2 is the comprehensive emotional values of the frequency distribution map of 121 advertising microblogging, As can be seen from the figure, the minimum emotional value is 0, which tends to be neutral. Others are positive, which mean positive emotions. According to the emotional value distribution from the frequency, it could find that emotional values are mainly concentrated in the 0.3–0.7, accounting for about 85% of the total, and the emotional value of 0.4 advertising microblogging accounted for 5% of the total. From the way of calculating the emotional value of the study design, when the emotional value is closer to ± 1, the more obvious the emotion is, and the closer to 0, the more neutral emotion is. From the current pattern of distribution, Roseonly advertising microblogging in most of the 121 microblogging are with positive emotions, but the overall emotional value of 0.49 on average, shows that its positive emotional intensity is not high. The selection of advertising microblogging standards directs to the real social advertising standards, so the performance of the emotional may be the reality of people's advertising, and is directly related to the exclusion. But from the whole emotional point of view, Roseonly advertising microblogging in the transmission to the various forwarding users has brought the positive impact, but each forwarding user's preference for Roseonly products is more than exclusion.

(2) Calculation of wideness-hotness

According to the research design in Chap. 3, the paper first quantifies the data of the three indexes: the sum of the number of reviews (C), the number of forwarding numbers (R), and the forwarding users Number (U). Then, the correlation coefficient and standard deviation between the four indexes are calculated, and then the weights of each index in $G_i = \alpha C_i + \beta R_i + \delta U_i$ are calculated according to CRITIC method, which shows $\omega_i = (C_i, R_i, U_i) = (0.18, 0.27, 0.55)$. Finally, the formula for the degree of wideness-hotness is as follows:

$$G_i = 0.18\,C_i + 0.27R_i + 0.55\,U_i \tag{9}$$

According to the *CRITIC* method, the number of forwarding users is the most important in the information communication and forwarding network, which weight is 0.55, followed by the number of forwarding (0.27) and the number of comments (0.18).

Forwarding user refers to the user who forwards the microblogging, and the forwarding node refers to the new microblogging information with the original microblogging information generated by forwarding in the forwarding network. From the definition point of view, the forwarding behavior of the forwarding user constitutes a forwarding node, which is unique. A forwarding node can only be generated by one forwarding user, but a forwarding user can generate many forwarding nodes. The more users are forwarded, the wider the audience is, but the more forwarding node does not represent the broader audience of information, since the forwarding node is determined by the number of forwards, rather than the number of forwarding users, and the more forwarding nodes, the reason may be due to the more forward in the premise of the number of fixed forwarding users. This kind of forwarding is weaker than the increase in the number of forwarding users, and the diffusion of information is weak. At the same time, the comments in the microblogging forward is a kind of user's attitude and emotional performance, only the information and the current viewer is ideological resonance or shock, the viewer's inner feelings will change, and thus affect the behavior. From another point of view, when the number of comments becomes more, the information in the forward user network in the attention and popularity will go higher.

(3) Advertising Microblogging Spread Effect

Combined with the degree of emotion and wideness-hotness, the emotional degreeCwideness-hotness propagation effect evaluation of two-dimensional map is built, in which the $Y - axis$ for the emotional degree, the $X - axis$ for the wideness-hotness, and the overall emotional average emotional value $Y = 0.49$ and the overall wide average $X = 0.16$, two straight lines for adjusting X and Y axes are drawn. The two-dimensional evaluation chart is shown in Fig. 3.

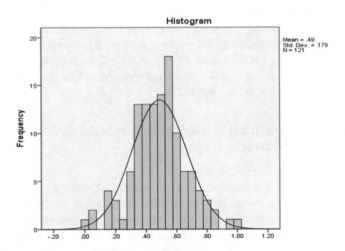

Fig. 2. Comprehensive emotional value of the frequency distribution chart on the 121 advertising microblogging

From 121 Roseonly product advertising microblogging statistical distribution, it can be seen that most concentrate in the third quadrant, that is, emotional and wideness-hotness are below the average, accounting for 38.84%, this area of advertising microblogging in the entire region, is a low degree of wideness-hotness and positive emotions are weak, in the reality of society, people are often given a certain degree of exclusion of advertising, the paper selection of advertising microblogging are direct advertising type, and thus in the emotional strength and wideness hotness, most still slightly biased in the weak feelings of low light. Followed by the emotional value is greater than the average and wideness-hotness less than the average of the second quadrant, accounting for 31.41%, the wideness-hotness of this region is low but the positive emotional expression is very strong, which indicates that the users prefer to this product, and low wideness-hotness, may be due to the limitations of these ads microblogging audience. While the first quadrant degree of emotion and wideness-hotness are higher than the average number of heat micro-blog accounted for only 19.84% of the total, users' emotions of this area and have strong performance and wideness-hotness is also high, which means this area has the best micro-blog advertising communication effect in the transmission network. And the mainstream of the core consumer group is likely to exist in the region. The last one is the fourth quadrant, accounting for 9.91%, wideness-hotness is greater than the average and the emotional degree is less than the average, although the region has a high degree of wideness-hotness, but the expression of positive emotions is slightly weak, although the area of the audience is large, but the dissemination of information on the user's preference for this product is not obvious, which possible because that the products in the promotion and publicity don't stimulate the users and thus have a lack of strong ideological resonance or impact, regional distribution is shown in Fig. 4. From the whole point of view, when the wideness-hotness is greater, the emotional value will not become greater always, wideness-hotness is a breadth of microblogging spread, and emotional degree is a microblogging in the proliferation of the users' thoughts or attitudes depth, microblogging spread effect should be integrated from the spread breadth and depth to measure, both are indispensable.

4.3 Opinion Leader Role in Roseonly Advertising Microblogging Forwarding Network

(1) Opinion Leader Identification
According to the *KPrank* algorithm proposed in Sect. 2.3, the paper further studies the nodes with large information dissemination in the microblogging network to extract the core nodes in the information dissemination, and provide effective theoretical support for the late product or brand promotion strategy.

According to the formula (8), the *KPr* value of each node is obtained, and the *KPr* value of each node is clustered by means of K value. Finally, the values are clustered into 4 categories, as shown in Table 1.

From the clustering center value, cluster 1 is the node group with the highest influence of nodes, and the clustering center value is 1.0627. Therefore, the paper defines the opinion leader of the forwarding network in the cluster 1 where the *KPr* value is at the top 20, are shown in Table 2, and the highest of it is 1.5183 and the lowest is 1.0513.

From the *KPr* value descending in the top 20 list of view, five are personal identities, and four individuals belonging to the real name certification (Rose

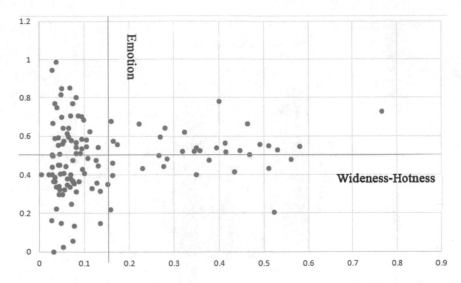

Fig. 3. Two-Dimensional assessment of sensitivity – wideness-hotness propagation effect

Table 1. KPr clustering results

Clustering category	Number of cases	Clustering center value
1	40	1.0627
2	313	0.8069
3	35404	0.0015
4	2794	0.4623
Effective number	38551	
Missing number	0	

MM, Wang Xiaochuan, Bao Fan, etc.), while the other 15 are non-personal status, but Belong to the community class identity, there are regional type (Shenzhen size, eat and drink IN Tianjin Tianjin, etc.), interest type (film assembly number, Sina fashion, etc.), professional type (entrepreneur magazine) and so on. It also shows that the KPr value calculated in the forwarding node user's influence as a whole, social identity of the user is more likely to be the opinion leader in the forwarding network.

(2) An Analysis of the role of Opinion Leaders in Communication

Through a brief analysis of the position of the opinion leaders in the network, the related microblogging spread and the corresponding opinion leader and the KPr value are shown in Table 3.

From the position of the top 20 nodes descending from the KPr value, 45% of the opinion leaders are located in the first quadrant, that is, the positive sensibility is more obvious, and the wideness-hotness is also higher on the microblogging

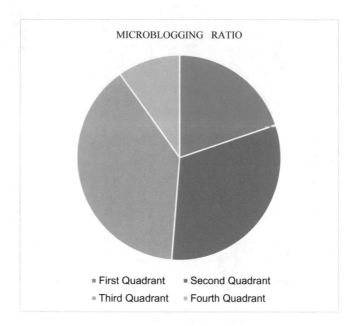

Fig. 4. Each quadrant area ratio distribution of microblogging number

Table 2. Node KPr value of the top 20

Node ID (the last five digits)	Username	KPr	Node ID (the last five digits)	Username	KPr
41140	Rose MM	1.5183	43110	999 letters for myself	1.1402
12480	Things of Shenzhen	1.3623	34630	Shanghai headlines	1.1402
62880	Hot-tops of Chengdu	1.2580	99020	Sina fashion	1.1402
69840	Global hot fashion style	1.1731	30330	Xiaochuan Wang	1.1402
32800	Entertainment in Shenzhen	1.1407	26430	Fashion ladies	1.1402
69100	Entrepreneur magazine	1.1402	27530	Tiny wisdom in life	1.1382
95790	Entertainment in Tianjin	1.1402	92930	Ship above movies	1.0532
68750	People fashion life	1.1402	90070	lisa_jiang tingting	1.0520
69780	All love to comment-group	1.1402	58430	Baofan	1.0520
29230	sdssds5666	1.1402	21240	Gemini whisper	1.0513

Table 3. Advertising micro-blog communication effect

Quadrant distribution	Emotional value	Wide-heat value	Number of opinion leaders	KPr value
1	0.56580	0.44210	1.8	1.2410
2	0.74350	0.05802	1.2	1.1307
3	0.39220	0.09860	1.0	1.0521
4	0.44705	0.49835	1.0	1.1402

information propagation path, the average KPr reaches the highest (1.2410); 10% is located in the fourth quadrant, that is, positive emotional is weak, and the wideness-hotness is higher on microblogging information dissemination path, and the average KPr value is at the second (1.1402); In the second quadrant with a strong emotional performance and a relatively wideness-hotness, where has 30% of the opinion leaders, the average KPr value is the lowest of all quadrants as 1.0521; and the third quadrant with less emotional and low wideness-hotness has 15% of the opinion leaders with an average KPr of 1.1307.

Through KPr value and the wideness-hotness value in quadrants, the correlation coefficient is 0.669 (P = 0.06, confidence 10%) by the correlation analysis, which indicates that the influence of opinion leaders, KPr, has impact on the wideness-hotness and shows a significant positive effect. The article then analyzes the emotion of comments sent by the opinion leaders, and finds that the emotions of comments are consistent with the emotional value of the whole blog, and they are all positive, which further shows that the emotion is contagious and interlinked. The opinion leaders have some guiding effects in the emotional dimension of information dissemination.

To sum up, from the distribution of opinion leaders, the influence of opinion leaders has a clear positive impact on the information dissemination of the wideness-hotness, and opinion leaders of the emotional tendencies have a sense of the same emotional leadership on their fans.

5   Conclusions

Based on the advertisement microblogging network in Sina microblogging platform, this paper studied the effect of information dissemination in the forwarding network and the effect of opinion leaders on the communication effect.

The study found that advertising microblogging information dissemination is mainly affected by both the audience and the forwarding carrier as an information carrier, and it is a small number of core users in the kind of communication object. Opinion leaders in the non-personal status are significantly more than the number of personal identification. The paper evaluated the effect of information dissemination by establishing the two-dimensional evaluation system of emotion degree and wideness-hotness, found that Roseonly product advertising microblogging forwarding users of advertising microblogging have positive emotional performances, but most of the advertising microblogging positive emotional value is low, that is, positive emotional performance is not obvious, while the overall level of wideness-hotness is low, which may be related to the daily user on the exclusion of advertising. Finally, according to the opinion leader in the distribution of microblogging advertising, it is found that the influence of opinion leaders has a strong positive effect on the breadth of information dissemination, and the emotional tendencies of opinion leaders have a certain emotional effect to their fans.

The main innovation of the article is to propose a two-dimensional evaluation system based on the degree of emotion and wideness-hotness of the advertisement, so that the communication effect evaluation is more accurate and more objective. Meanwhile the paper tried to find the impact of opinion leaders on the effect of communication, which for enterprises or organizations in the microblogging network to promote the brand, products and services with a more comprehensive theoretical support. The opinion leader recognized the improved *KPrank* algorithm based on *PageRank* principle, which is considered with both the node network centrality and the node's own attributes, so that the calculation of the node's influence will be more comprehensive. Because the existing Sina microblogging open platform *API* has a strict access to data permissions restrictions, the data collected by the paper is only the latest 2000 data, but for those who forward more than 2000 microblogging, there is a certain lack of data. So, the future research should be further improved in data collection and data processing. And the article only studied a microblogging platform in a product, Roseonly, late research can be incorporated into more types of product advertising microblogging to conduct a comparative study of whether there are differences.

Conflict of Interest: The authors declare that they have no conflict of interest.

References

1. Lambrecht, A., Tucker, C., Wiertz, C.: Advertising to early trend propagators: evidence from twitter. Mark. Sci. **37**(2), 177–199 (2018)
2. Pandey, A.: Social media marketing impact on the purchase intention of millennials. Int. J. Bus. Inf. Syst. **28**, 62–147 (2018)
3. Litterio, A.M., Nantes, E.A., et al.: Marketing and social networks: a criterion for detecting opinion leaders. Eur. J. Manag. Bus. Econ. **26**(3), 347–366 (2017)
4. Anagnostopoulos, A., Petroni, F., Sorella, M.: Targeted interest-driven advertising in cities using twitter. Data Min. Knowl. Discov. **32**(3), 1–27 (2016)
5. Bakal, G., Kavuluru, R.: On quantifying diffusion of health information on twitter (2017)
6. Beijing SWDC 2017 Weibo user development report (2017)
7. Li, B.Q.: Describe the content of communication features for dissemination of research hypothesis testing-content analysis introduction (part2). Contemp. Commun. **51**(1) (2000)
8. Chu, S.C., Kim, J.: The current state of knowledge on electronic word-of-mouth in advertising research. Int. J. Advert. **37**(1), 1–13 (2018)

9. Dai, H., Deem, M.J., Hao, J.: Geographic variations in electronic cigarette advertisements on twitter in the United States. Int. J. Public Health **62**(4), 1–9 (2017)
10. Zhu, G.F.: A method of calculating the influence of micro-blog users based on domain. Ph.D. thesis, Chongqing, The Computer Science school of Southwest University, Chongqing (2014)
11. Fogel, J., Adnan, M.: Trust for pharmaceutical company direct-to-consumer prescription medication advertisements. Health Policy Technol. **7**, 26–34 (2018)
12. Riddell, J., Brown, A., et al.: Who are the most influential emergency physicians on twitter? West. J. Emerg. Med. **18**(2), 281–287 (2017)
13. Wang, K., Song, H.Z.: Comparative analysis of three objective weight empowerment. Technoeconomics Manag. Res. (6) (2003)
14. Gang, L.: Advertising communication strategy analysis under we-media environment. Guide Bus. **15**, 116–233 (2012). (in Chinese)
15. Cui, L., Pi, D.C.: Identification of microblog opinion leader based on user feature and interaction network. In: Web Information System & Application Conference (2015)
16. Zhao, L.: Microblog sentiment analysis model based heat topic detection on public opinion. Ph.D. thesis, Lnazhou, The Computer Science school of Lanzhou University, Lanzhou (2013)
17. Lopez, M., Sicilia, M.: Sicilia m.identification of loyal opinion leaders on twitter. Cuadernos De Gestion **17**(1) (2017)
18. Marcos, K., Tap, P., Alhaji, A.M.: Influence of online retailers' social media marketing strategies on students' perceptions towards e-shopping: a qualitative study. Int. J. Technol. Enhanc. Learn. **10**, 218–234 (2018)
19. Mauricio Yagui, M.M., Monsores Passos Maia, L.F., et al.: Data mining of social manifestations in twitter: analysis and aspects of the social movement "bela, recatada e do lar" (beautiful, demure and housewife). Infocomp J. Comput. Sci. **17**(1), 30–44 (2018)
20. Adetunji, R.R., Rashid, S.M., Ishak, M.S.: Social media marketing communication and consumer-based brand equity: an account of automotive brands in Malaysia. J. Komun. Malays. J. Commun. **34**, 1–19 (2018)
21. Firdaus, S.N., Chen, D., Sadeghian, A.: Retweet: a popular information diffusion mechanism - a survey paper. Online Soc. Netw. Media **6**, 26–40 (2018)
22. Swasy, A.: A little birdie told me: factors that influence the diffusion of twitter in newsrooms. J. Broadcast. Electron. Media **60**(4), 643–656 (2016)
23. Varshney, D., Kumar, S., Gupta, V.: Predicting information diffusion probabilities in social networks: a Bayesian networks based approach. Knowl. Based Syst. **133**(S0950705117303), 180 (2017)
24. Voorveld, H.A.M., Van Noort, G., et al.: Engagement with social media and social media advertising: the differentiating role of platform type. J. Advert. 1–17 (2018)
25. Chen, Y., Yuan, Y.H.: The research on information communication effect of microblog. J. Inf. Resour. Manag. **37**, 74–77 (2012)
26. Zheng, Z.Y., Yang, H.C.: Factors influencing message dissemination through social media. Phys. Rev. **97**(6), 1–6 (2018)

What Affects the Innovative Behavior of Civil Servants? Survey Evidence from China

Rongrong He[1] and Shuaifeng Li[2(✉)]

[1] School of Public Administration, Sichuan University, Chengdu,
People's Republic of China
[2] Hainan University, Haikou, People's Republic of China
lishuaifeng415@hainu.edu.cn

Abstract. To explore the factors influencing the innovative behavior of
civil servants, a hypothetical model was established in this study on the
basis of literature research and in-depth interview. To verify the research
hypotheses, a questionnaire survey was conducted on 322 Chinese civil
servants and statistical analysis was made on the sample data using
SPSS22.0 and MPLUS7. The results show that: (1) Prosocial leadership
silence and bureaucratic culture can only indirectly influence the innovative behavior through the expected innovation gains; (2) Job autonomy
can not only directly predict the innovative behavior, but also indirectly
influence the innovative behavior through the expected innovation gains.
The research results suggest that government agencies should advocate
the silent leadership style, and take measures to restrain the bureaucratic
culture and improve the job autonomy of civil servants so as to stimulate
their innovative behavior.

Keywords: Innovative behavior · Prosocial leadership silence ·
Bureaucratic culture · Job autonomy · Expected innovation gains ·
Government agencies

1 Introduction

Innovation is not only reflected in the strategic layout of national development, but also in organizational management and individual behavior. Innovative behavior is the key resource for maintaining organizational competitive
advantage and sustainable development [5]. For public organizations, innovation
becomes increasingly urgent due to both external and internal pressures [10].
However, the phenomenon of lazy politics such as "no merit but no fault" is not
uncommon in Chinese government organizations. Civil servants initiative innovative behavior is easy to be regarded as a challenge to the present system. The
lack of innovative behavior exerts a negative impact on government management
and society development. Therefore, it is of great practical significance to explore

© Springer Nature Switzerland AG 2020
J. Xu et al. (Eds.): ICMSEM2019 2019, AISC 1002, pp. 616–627, 2020.
https://doi.org/10.1007/978-3-030-21255-1_47

the factors that elicit innovative behavior of civil servants. Internally, it is conductive to improving the operational efficiency of government organizations and creating an innovative organizational culture. Externally, it helps to respond to citizens rising expectations of responsive and accountable government, and improve the credibility of the government.

To further explore the influencing factors of civil servants innovative behavior, the present study respectively introduces three variables including prosocial leadership silence, bureaucratic culture and job autonomy based on the literature research and depth interview. Using first-hand survey data from more than 20 Chinese local government agencies, we examine whether the three variables have significant impacts on civil servants innovative behavior. Furthermore, based on the social expectation theory which suggests that an individuals motivation depends on the product of valence and expectancy, we introduce the variable of "expected innovation gains". We argue that through affecting the risks and benefits expectation of innovative behavior, prosocial leadership silence and job autonomy positively influence innovative behavior, while bureaucratic culture has negative effects on innovative behavior.

The present paper mainly has the following innovations and contributions: First, in China, the previous research on innovative behavior mainly focused on knowledge employees of enterprises [8] and universities [18]. The innovative behavior of civil servants has not been paid its due attention. We conduct an empirical research in the public sector and attempt to discover the special factors that influence the innovation behavior of civil servants. Second, although leadership style has been proved to be an important factor in subordinates innovative behavior [19], the research on silent leadership style which is very traditional in Eastern culture is still quite scarce. We establish the relationship between silent leadership style and subordinates innovative behavior based on the empirical research. Third, previous studies of bureaucratic culture have typically employed a qualitative approach [15], but we conduct a quantitative empirical study.

This paper is organized in three parts. First, we establish the research hypotheses based on literature review and depth interviews. Second, after a brief introduction of the sample and measurement, we test our hypotheses through carrying out regression analysis and mediation effect test. Finally, we discuss the theoretical value and practical implications of our results. Through this research, we hope to provide a new interpretation path for civil servants innovative behavior. Moreover, it is also expected to provide guidance for formulating scientific policies and adopting reasonable leadership style.

2   Literature Review and Hypotheses

Innovative behavior is the process by which employees generate, introduce, and apply useful novel ideas or things in the organizational activities [9]. It includes many stages, such as problem identification, creativity generation, sponsorship seeking, alliance building, innovation model building and large-scale implementation [12]. Though it is not recognized by the organizations reward system, it still has a positive effect on organizations, teams and individuals.

2.1   Prosocial Leadership Silence and Innovative Behavior

Silence is one of the traditional Chinese leadership styles. Chinese scholars Huang Gui and her collaborators divided leadership silence into five dimensions, respectively, defensive silence, prosocial silence, test silence, trickery silence and prestige silence. Prosocial silence is a silent behavior that leaders reserve information and opinions in order to maintain a harmonious relationship and better listen to their subordinates. It is reflected in the following aspects: leaders do not express their opinions beforehand, letting employees solve problems by themselves and not pointing fingers at subordinates [6]. Based on social exchange theory, in order to maintain this high-quality communication model, subordinates will increase organizational citizenship behavior such as innovation in return. Additionally, previous researches have also demonstrated the influence of leadership style on innovation. For example, Xie et al. (2018) found that transformational leadership can positively influence the atmosphere of innovation by building trust [17]. Based on the above analysis, the following research hypothesis can be proposed:

H1 Prosocial leadership silence positively affects the innovative behavior of civil servants.

2.2   Bureaucratic Culture and Innovative Behavior

Organizational culture is the values, beliefs and norms of behavior shared by members of the organization. Wallach (1983) divided organizational culture into three dimensions, respectively, innovative culture, supportive culture and bureaucratic culture [14]. Bureaucratic culture often means impersonalization in the process of organizational operation. In the strong bureaucratic culture, civil servants must abandon their personal feelings, and operate like a screw in the whole bureaucratic machine, with discretion not being allowed. According to the theory of planned behavior [1], attitude, subjective norm and perceived behavioral control can affect individuals behavior by influencing the behavior intention. Therefore, bureaucratic culture which including implicit values and behavioral norms can inhibit civil servants innovative behavior to a certain extent. Additionally, the influence of organizational culture on innovative behavior has also been proved by previous studies. For example, Naranjo-Valencias research in Spain found that adhocracy culture can affect the radical product innovation by influencing innovative behavior [11]. Based on the above analysis, we propose the following research hypothesis:

H2 Bureaucratic culture negatively affects the innovative behavior of civil servants.

2.3   Job Autonomy and Innovative Behavior

Job autonomy refers to the degree to which employees can control and decide on their own work methods, work schedules and work standards [3]. Generally, civil servants job autonomy refers to the right of decision-making in accomplishing their tasks. Based on the theory of self-determination [4], the intrinsic motivation and self-determination degree of individuals are important factors affecting behavior occurrences and outcomes. Therefore, the autonomy of civil servants in their work can effectively stimulate their intrinsic motivation and self-efficacy, and then they are willing to shoulder responsibility of promoting organizational innovation and improving organizational effectiveness. Previous researches have provided preliminary evidence for the relationship between job autonomy and innovative behavior. For example, the research conducted by Wu et al. (2014) found that when the openness to experience and proactive personality are controlled, job autonomy still has a significant positive impact on innovative behavior [16]. Based on the above analysis, we propose the following research hypothesis:

H3 Job autonomy positively affects the innovative behavior of civil servants.

2.4 The Mediating Role of Expected Innovation Gains

Expected innovation gains refer to the individuals prediction of the benefits that can be brought about by innovative behavior. It includes the performance improvement expectation and image improvement expectation. The present study mainly measures the expected innovation gains through image improvement expectation.

First, this study believes that prosocial leadership silence can positively affect the expected innovation gains. Prosocial leadership silence can help to create a relaxing and harmonious personnel relationship, work atmosphere and organizational culture, making civil servants feel less nervous, anxious and frightened. In such environment, civil servants can feel more encouragement and support for innovation. Therefore, prosocial leadership silence positively affects the expected innovation gains.

Second, bureaucratic culture is characterized by authority compliance and rule dependence. The present study believes that bureaucratic culture can negatively affect the expected innovation gains. In the bureaucratic culture, to avoid the arbitrariness of management, all members must act in accordance with the programmed organizational rules. Only by following the organizational rules can individuals survive and develop in the organization. In such kind of culture, initiative innovation is easy to suffer from negative consequences. As a civil servant has said in an interview, "unless the leaders want you to innovate, just do your own job honestly, or even just drink tea or read newspapers at least without being excluded".

Third, job autonomy is the degree of self-determination in work and this study believes that job autonomy can positively affect the expected innovation gains. Based on the theory of self-determination, job autonomy will positively influence the expected innovation gains through creating positive emotions and enhancing the self-efficacy.

According to the above analysis, prosocial leadership silence and job autonomy can positively affect the expected innovation gains while bureaucratic culture has negative effects. Furthermore, this study considers that the expected innovation gains can positively affect civil servants innovative behavior. Based on the public choice theory, civil servants are also rational economic people, who decide their behavior according to the judgment of cost and benefit. Therefore, the positive expectation of innovation results can mobilize their innovative behavior to a certain extent. Besides, according to Fromms expectancy theory, an individuals motivation depends on the product of valence and expectancy. Valence refers to the value of achieving goals to satisfy ones personal needs and the expected innovation gains are the individuals estimate of the innovation valence. In the same case of expectancy, the innovation valence is directly proportional to the innovative behavior of civil servants.

Therefore, this study believes that the prosocial leadership silence, bureaucratic culture and job autonomy can not only directly affect the civil servants innovative behavior, but also exert indirect effects through the expected innovation gains. Preliminary evidence has been provided by previous studies. For example, the empirical research from Kao et al. (2015) proved that the expected image gains play a mediating role between transformational leadership and service innovative behavior [7]. Based on the above analyses, the following hypotheses are proposed:

H4a The expected innovation gains play a mediating role between prosocial leadership silence and innovative behavior.

H4b The expected innovation gains play a mediating role between bureaucratic culture and innovative behavior.

H4c The expected innovation gains play a mediating role between job autonomy and innovative behavior.

Based on the above hypotheses, the following theoretical framework was formed (shown in Fig. 1).

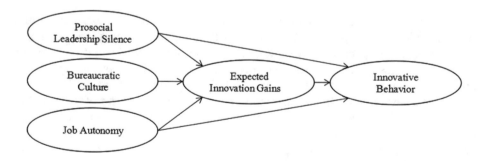

Fig. 1. The theoretical framework of the study

3   Research Methods

In order to test the above research hypotheses, we obtained the empirical data through questionnaire surveys on the local government civil servants. On this basis, we used SPSS22.0 and MPLUS7 for data cleansing and statistical analysis.

3.1 Sample and Data Collection

The respondents were from more than 20 Chinese local government agencies in Chengdu, Ganzi, Bazhong, Zhengzhou, Kunming, Hangzhou, and Guiyang. The investigation was conducted from June to August in 2018. A total of 428 questionnaires were distributed and 386 questionnaires were returned. The recovery rate was 9.2%, after the questionnaire screening, and 322 valid questionnaires were finally obtained.

Among the 322 respondents, 63.7% are men, and 36.3% are women. In terms of age, 25.5% are under 30 years old, 43.8% are between 30 and 40 years old, 23.0% are between 40 and 50 years old, and 7.7% are over 50 years old. In terms of education, 10.5% have a junior college degree or below, 73.0% have a bachelor degree and 16.5% have a master degree or above. Regarding rank, staff members account for 5.9%, assisting roles of sections or equivalents account for 23.3%, and leading roles of sections or equivalents occupy 25.8%. In terms of work time, 12.4% work in the current unit for less than one year, 36.3% for 1–5 years, 22.7% for 5–10 years, 17.7% for 10–20 years, and 1.9% for 20 years and above.

3.2 Measurement

In this study, all latent variables were measured by the maturity scales developed by domestic and foreign scholars, which have been proved to have high reliability and validity in previous studies. Besides, we modified and deleted some items according to actual needs in this study. Referring to the previous studies, gender, age, education, rank, and work time were set as control variables. All items were measured by a five-point Likert scale from strongly disagree to strongly agree.

Innovative behavior was measured by 6 items from Scott and Bruce's scale [12], which measured innovative behavior from the aspects of problem establishment, conception generation, innovation support seeking, and implementation of the innovation plan. Sample items include "Generating creative ideas", "Promoting and championing ideas of others", and "Developing adequate plans and schedules for the implementation of new ideas". In addition, two items were deleted after confirmatory factor analysis, and Cronbachs alpha is 0.837 in this study.

Prosocial leadership silence was measured by 3 items from Huang et al. [6], including "Leaders do not express their opinions in order to listen to their subordinates", "Leaders do not order or interfere with subordinates in order to show their trust". The Cronbachs alpha is .806 in this study.

Bureaucratic culture was measured by 8 items from Wallach [14]. We revised some of the items through translation and back-translation and pre-investigation. Sample items include "My organization structure is hierarchical", "In my organization, formal procedures generally govern what people do" and "My organization has a power-oriented structure". One item was deleted after the CFA. The Cronbachs alpha is .880 in this study.

Job autonomy was measured by 3 items from Spreitzer (1995) [13]. Sample items include "If I were to do something innovative, my image in the organization would be enhanced" and "Participating in the implementation of new ideas will improve my image in the organization". The Cronbachs alpha is .873 in this study.

The expected innovation gains were measured by 4 items from Ashford et al. (1998) [2]. Sample items include "If I were to do something innovative, my image in the organization would be enhanced" and "Participating in the implementation of new ideas will improve my image in the organization". The Cronbachs alpha is .860 in this study.

4   Results

4.1   Test of Common Method Bias

Harmans single-factor test method was applied to test the common method bias in this study. We put all measurement items into the model for exploratory factor analysis. The results showed that all the measurement items can be categorized into five constructs with eigenvalues being greater than 1.0, and accounting for 68.918% of the total variance. The first construct only accounts for 28.674% of the total variance (below the recommended level 40%), indicating that the common method bias in this study is acceptable.

4.2   Reliability and Validity

We conducted confirmatory factor analysis to assess the reliability and validity of the measurement scales. As shown in Table 1, all items factor loadings are higher than the criterion of 0.50. The values of composite reliability ranges from 0.815 to 0.884 (above the recommended level 0.70). The average variance extracted scores of the constructs ranges from 0.529 to 0.713, which are higher than the criterion of 0.50. These results indicated that convergent validity of our measurement is very good. We further compared the square roots of the AVE scores (bold numbers on the diagonal in Table 1) and correlations of constructs to assess the discriminant validity. The results showed that the square roots of the AVE scores are greater than the correlations among the constructs, thereby confirming the five constructs in this study have good discriminatory validity.

4.3   Descriptive Statistics and Correlations

The descriptive statistical results and correlation coefficients of the main variables in this study were shown in Table 2. The correlation matrix between the variables is above the table, and the mean and standard deviation of the five variables are at the bottom of the table.

As shown in Table 2, prosocial leadership silence is significantly positively correlated with the expected innovation gains ($r = 0.235, p < 0.001$), but not significantly correlated with innovative behavior ($r = 0.067, p > 0.05$). There exists a significant negative correlation between bureaucratic culture and the expected

Table 1. Loadings, reliability and validity.

Variables	Factor loading	Composite reliability	AVE	Discriminant validity				
				1	2	3	4	5
1. Prosocial leadership silence	0.612 0.870	0.815	0.6	0.775				
2. Bureaucratic culture	0.508 0.838	0.884	0.529	−0.263	0.727			
3. Job autonomy	0.708 0.909	0.88	0.713	0.15	−0.239	0.844		
4. Expected innovation gains	0.710 0.846	0.86	0.607	0.293	−0.327	0.315	0.779	
5. Innovative behavior	0.682 0.885	0.84	0.571	0.109	−0.218	0.452	0.499	0.756

innovation gains ($r = -0.292, p < 0.001$), and also between bureaucratic culture and innovative behavior ($r = -0.202, p < 0.001$). Job autonomy is significantly positively correlated with the expected innovation gains ($r = 0.291, p < 0.001$), and also with innovative behavior ($r = 0.424, p < 0.001$). The expected innovation gains are significantly positively correlated with innovative behavior ($r = 0.440, p < 0.001$). Moreover, these results provide preliminary evidence for the subsequent hypotheses testing.

Table 2. Descriptive statistics and correlations.

Variables	1	2	3	4	5
1. Prosocial leadership silence	1				
2. Bureaucratic culture	−0.228***	1			
3. Job autonomy	0.142*	−0.188***	1		
4. Expected innovation gains	0.235***	−0.292***	0.291***	1	
5. Innovative behavior	0.067	−0.202***	0.424***	0.440***	1
Mean	3.002	2.173	3.128	3.155	3.179
Std. deviation	0.997	0.815	1.031	0.868	0.899

Note 1.: $*p < 0.05, **p < 0.01, ***p < 0.001$

4.4 Hypotheses Testing

Direct Effect Test Since the main research variables in this study are latent variables, we used structural equation model (SEM) to fit the data, then set bootstrap value in Mplus, and tested whether the direct effects are significant through point estimation and interval estimation. We set prosocial leadership silence, bureaucratic culture and job autonomy as independent variables, and innovative behavior as the dependent variable, to construct a structural equation model. The results showed that the model fit is very good ($\chi^2 = 487.731, df = 280, \frac{\chi^2}{df} = 1.742 < 3, CFI = 0.939 > 0.9, TLI = 0.931 > 0.9, RMSEA = 0.048 < 0.08, SRMR = 0.066 < 0.08$). The influence of prosocial leadership silence on innovative behavior is positive, yet not significant $\beta = 0.035, p > 0.05$, 95% confidence interval includes zero) and thus H1 is not supported. The effect of bureaucratic culture on innovative behavior is negative, but also not significant ($\beta = -0.134, p > 0.05$, 95% confidence interval includes zero), indicating that H2 is not supported either. Job autonomy has a positive and significant effect on the innovative behavior ($\beta = 0.334, p < 0.001$, 95% confidence interval doesnt include zero) and thus H3 is supported.

Mediation Effect Test In this study, we used the bias-corrected nonparametric percentile bootstrap method to test the three mediating effects (H4a, H4b, H4c). Referring to the test procedure proposed by Zhao et al. (2010) [20], we set the bootstrap sampling times as 5000 in Mplus, and directly tested whether the interaction terms $a * b$ are significant.

In Table 3, PS stands for prosocial leadership silence, BC stands for bureaucratic culture, JA stands for job autonomy, EIG stands for the expected innovation gains and IB stands for innovative behavior. Moreover, a is the effect of the independent variable on the mediator variable, b is the effect of the mediator variable on the dependent variable, c is the total effect of the independent variable on the dependent variable, c' is the direct effect, and $a * b$ is the indirect effect.

Table 3. Descriptive statistics and correlations.

Mediation path	Total effect (c)	Direct effect (a)	Direct effect (b)	Direct effect (c')	Indirect effect (a*b)	Indirect effect 95% confidence interval
PSEIGIB	0.026	0.162**		-0.04	0.066*	[0.023,0.131]
BCEIGIB	-0.128	-0.272**	0.405***	-0.018	-0.110**	[-0.217, -0.046]
JAEIGIB	0.332***	0.190**		0.255***	0.077**	[0.032, 0.145]

Model fit $\chi^2 = 445.618, df = 274, \chi^2/df = 1.626, CFI = 0.950, TLI = 0.942, RMSEA = 0.044, SRMR = 0.051$

Note 2.: $*p < .05, **p < .01, ***p < .001$

As shown in Table 3, prosocial leadership silence has a significant indirect effect on innovative behavior through the expected innovation gains ($a * b = 0.066, p < .05$, 95% confidence interval doesnt include zero), so H4a is supported. Bureaucratic culture also has a significant indirect effect on innovative behavior through the expected innovation gains ($a * b = -0.110, p < 0.01$, 95% confidence interval doesnt include zero), so H4b is supported. Additionally, the indirect effect of job autonomy on innovative behavior through the expected innovation gains is also significant ($a * b = 0.077, p < 0.01$, 95% confidence interval doesnt include zero), so H4c is supported too.

5 Conclusions and Discussion

5.1 Research Conclusions

In this study, we took 322 civil servants as the participants and discussed the influencing factors of civil servants innovative behavior. The results showed that: (1) Prosocial leadership silence can promote the innovative behavior of civil servants by positively affecting the expected innovation gains. (2) Bureaucratic culture can inhibit civil servants innovative behavior through negatively affecting the expected innovation gains. (3) Job autonomy can not only directly and positively affect the innovative behavior of civil servants, but also produce indirect effects by promoting the expected innovation gains.

5.2 Theoretical Value

The main theoretical value of this study can be described as the following aspects: first, it enriches and expands the theory of leadership style. Second, it provides empirical support for the social exchange theory, planned behavior theory and self-determination theory. Third, it innovatively links prosocial leadership silence and bureaucratic culture with innovative behavior, which can contribute to expanding the interpretation path of civil servants innovative behavior.

5.3 Practical Implications

The present research has some practical implications for government organization. First, the innovative behavior of civil servants is conducive to stimulating the vitality of government organizations, and leaders should attach the great importance and fully encourage the innovative behavior of subordinates. Second, leaders should pay attention to the positive influence of prosocial silent behavior, trust and authorize subordinates in the institution design and daily work. Third, the governments should be alert to the negative impact of bureaucratic culture, and try to introduce market competition mechanism for creating innovative organizational culture. Finally, the governments should pay attention to enhancing civil servants expected innovation gains through multiple ways such as institutional incentives, cultural cultivation and leadership style transformation.

6 Limitations and Future Research

In addition, the present study also has some limitations. For example, prosocial leadership silence and bureaucratic culture have only indirect effects on innovative behavior, with the main effects not being significant, for which the reason should be further explored in the future. Besides, only the mediator variable of expected innovation gains is added to the model. we should try to add more mediator variables based on different theoretical foundations, such as public service motivation, which can provide more explanation paths for the civil servants innovative behavior. Finally, all the variables are measured at the individual level in this research, and we should conduct multi-level research in the future, measuring leadership style and organizational culture at the organizational or team level.

Acknowledgements. We sincerely thank Ph.D. candidate Fu Yaping from the School of Public Administration of Sichuan University for the help in data collection, and be grateful for the language modification from Ph.D. candidate Liao Qi in the School of Foreign Languages of Sun Yat-Sen University.

References

1. Ajzen, I.: The theory of planned behavior. Organ. Behav. Hum. Decis. Process. **50**(2), 179–211 (1991)
2. Ashford, S.J., Rothbard, N.P., et al.: Out on a limb: the role of context and impression management in selling gender-equity issues. Adm. Sci. Q. **43**(1), 23–57 (1998)
3. Breaugh, J.A.: The measurement of work autonomy. Hum. Relat. **38**(6), 551–570 (1985)
4. Deci, E.L., Koestner, R., Ryan, R.M.: A meta-analytic review of experiments examining the effects of extrinsic rewards on intrinsic motivation. Psychol. Bull. **125**(6), 627–668 (1999)
5. Han, L., Zeng, S., et al.: Bridging the gaps or fecklessness? a moderated mediating examination of intermediaries effects on corporate innovation. Technovation p S0166497218301226 (2018)
6. Huang, G., Fu, C.G., Guan, X.H.: Construction and measurement of leadership silence dimensions in organizations. Manag. World **7**, 122–129 (2015)
7. Kao, P.J., Pai, P., et al.: How transformational leadership fuels employees service innovation behavior. Serv. Ind. J. **35**(7–8), 448–466 (2015)
8. Li, H.: A study on influence of work resources to employee innovative behavior: based on conservation of resources theory. J. Nanjing Tech Univ. (Soc. Sci. Ed.) **17**(06), 69–80 (2018)
9. Liu, Y., Shi, J.T.: A study on the relationship between the effects of the organizational innovative climate and those of motivational preference, on employees innovative behavior. Manag. World **10**, 88–101+114+188 (2009)
10. Miao, Q., Newman, A., et al.: How leadership and public service motivation enhance innovative behavior. Public Adm. Rev. **78**(1), 71–78 (2018)
11. Naranjo-Valencia, J.C., Jimenez-Jimenez, D., Sanz-Valle, R.: Organizational culture and radical innovation: does innovative behavior mediate this relationship? Creat. Innov. Manag. **26**(4), 407–417 (2017)
12. Scott, S.G., Bruce, R.A.: Determinants of innovative behavior: a path model of individual innovation in the workplace. Acad. Manag. J. **37**(3), 580–607 (1994)
13. Spreitzer, G.M.: Psychological empowerment in the workplace: dimensions, measurement, and validation. Serv. Ind. J. **35**(8), 1442–1465 (1995)
14. Wallach, E.J.: Individual and organizations: the culture match. Train. Dev. J. **37**(2), 29–36 (1983)

15. Wang, Y.P.: The potential limitation of Confucian political ethics and its impact on Chinese bureaucracy culture. J. Party Sch. Tianjin Comm. CPC **04**, 80–85 (2014)
16. Wu, C.H., Parker, S.K., et al.: Need for cognition as an antecedent of individual innovation behavior. J. Manag. **40**(6), 1511–1534 (2014)
17. Xie, Y., Wei, X., et al.: Leadership style and innovation atmosphere in enterprises: an empirical study. Technol. Forecast. Soc. Change **31**, 257–265 (2018)
18. Xing, N.N.: The influence of organizational learning on innovative performance of university scientific researchers from the perspective of sor. Rev. Econ. Manag. **34**(06), 86–94 (2018)
19. Yahia, N.A., Montani, F., Courcy, F.: The role of stressors on innovation behavior: when superior empowering leadership protects the innovation potential of workers. Psychol. Trav. Organ. **24**(1), 51–67 (2018)
20. Zhao, X., Lynch, J.G., Chen, Q.: Reconsidering Baron and Kenny: myths and truths about mediation analysis. J. Consum. Res. **37**(2), 197–206 (2010)

The Study on the Influence of Online Interactivity on Purchase Intention on B2C Websites: The Interference Moderating Role of Website Reputation

Rongjia Su[1(✉)] and Dianjie Liang[2]

[1] College of International Studies, Sichuan University, Chengdu, Sichuan, People's Republic of China
suzansrj@163.com
[2] Business School, Nottingham Trent University, Nottingham, UK

Abstract. As online shopping becomes increasingly popular, a growing number of B2C websites attempt to increase consumer purchase intention through interacting with consumers. This research aims to investigate the influence of online interactivity on purchase intention on B2C websites with focus on the interference moderating role of website reputation. Data are collected through questionnaires and regression analysis is applied to examine the moderating and mediating effects. Findings reveal that online interactivity has a positive impact on trust when website reputation is high. On the contrary, online interactivity has a negative impact on trust when website reputation is low. Meanwhile, trust mediates the influence of online interactivity on purchase intention. Therefore, B2C websites should not blindly pursue interaction with consumers without regard to website reputation, because it can be counterproductive.

Keywords: Online interactivity · Website reputation · Trust · Purchase intention · Interference moderating effect

1 Introduction

Online shopping has gained momentum recently. The report released by China Internet Network Information Center (CNNIC) in Beijing on 20 August 2018 shows online shoppers using online payment account for 71% of internet users, indicating that online shopping and payment enjoy wide application among internet users. E-commerce boasts fast growth, and performs a pivotal role in coordinating structural reform, creating employment opportunities, and facilitating rural revitalization [4].

A variety of products, convenience, lower costs are perceived as the merits of online shopping [42]. Consumers enjoy wider options afforded by more services and products online [24]. However, Online shopping is considered to be unsafe by

© Springer Nature Switzerland AG 2020
J. Xu et al. (Eds.): ICMSEM2019 2019, AISC 1002, pp. 628–640, 2020.
https://doi.org/10.1007/978-3-030-21255-1_48

plenty of consumers [35]. The uncertainty in the transaction and payment lead to the higher perceived risk. Uncertainty is a major hindrance for consumers to shop online [42]. A number of studies have proved that online interactivity can significantly alleviate perceived risk, improve trust and raise purchase intention [30,34,36]. Thomas et al. [35] pointed out that interactivity can significantly improve trust and trust can significantly affect purchase intention. Online interactivity exerts a positive effect on purchase intention through trust [34]. Van Noort et al. [36] found online interactivity could generate flow experience for online shopper, which could affect purchase intention indirectly. Online interactivity increase the patronage intention significantly [7]. Nevertheless, Liu and Shrum [17] suggested, a high degree of online interactivity could undermine the positive attitude towards website for inexperienced users. Other researchers pointed out that online interactivity does not exert a significant effect on purchase intention. Therefore, previous studies have concluded inconsistent results on whether online interactivity affects purchase intention.

Regarding trust in online shopping, some studies have confirmed the effect of online interactivity on trust [17,34,40]. Hence, an increasing number of websites attempt to increase online interactivity with consumers. However, the results vary. Some websites have registered satisfactory results while others found counterproductive. Most of existing studies have added other variables such as the website design, website security and order fulfillment to jointly enhance trust. Few studies have systematically investigated factors affecting the relationship between online interactivity and trust. For example, low interactivity may explain low level of trust towards websites. While interactivity is high, trust may still be low, indicating that the influence of interactivity on trust is more complex than we have expected, a new approach should be adopted to investigate the relationship between the two variables.

To address the gap, this study highlights the role of website reputation, and proposes that the effect of online interactivity on trust is neither positive nor negative; it depends on the level of website reputation. This research aims to investigate the interference moderating role of website reputation on the relationship between online interactivity and trust. Interference moderating effect can be defined as when the level of moderator is high, the effect of the independent variable on dependent variable is positive. Conversely, when the level of moderator is low, the effect is negative [19]. This research also examines the mediating role of trust in the relationship between online interactivity and purchase intention.

The contribution of this research is threefold. First, this research can enrich the literature of online interactivity, providing better understanding on relationship between online interactivity, trust and purchase intention. Another theoretical contribution is the inclusion of website reputation as the moderator. The effect of online interactivity on trust depends on the level of website reputation. Third, the findings of this research offer practical implications for practitioners to run B2C websites better. It can help decision makers to identify the appropriate level of interactivity with consumers in line with website reputation, thus preventing unnecessary online interactivity that would undermine consumer trust towards the website.

2   Literature Review and Hypotheses

2.1   The Interference Moderating Role of Website Reputation

According to Wu [41], online interactivity can be divided into actual interactivity and perceived interactivity. The former one refers to the physical features of interactive technologies on the websites, while the latter one is a mental concept, referring to the interactivity experienced and felt by users. This research focus on perceived interactivity, because a great number of researchers agreed that perceived interactivity is the determinant of consumer response [21,22,32]. Voorveld and van Noort [37] confirmed that a higher degree of online interactivity leads to more positive attitude towards website brand. Wu et al. [40] revealed the importance of online interactivity on consumer trust. Suntornpithug and Khamalah [34] pointed out that online interactivity between sellers and buyers can facilitate communication, and enhance trust. As a result, consumers are apt to believe that the website will fulfill orders, protect consumer privacy and personal information. However, other researchers found the effect of online interactivity could be affected by user experience and involvement. Under the circumstance of high involvement, a high degree of online interactivity will undermine the positive attitude of inexperienced users towards the website [17].

Furthermore, Pappas [24] explained that various factors could affect consumer purchase decision in the complicated online environment, a fresh perspective is needed to evaluate the combinations of factors. Consumers tend to predict the result before shopping online, positive prediction could increase the willingness to purchase [14]. When internal characteristics of websites are not enough for consumers to make accurate prediction, external characteristics of websites become a crucial factor. Website reputation, as a primary external characteristic, could render assurance to consumers when shopping online, which could reduce perceived risks and ease the concern about the privacy violation [15], thus enhancing consumer trust towards the website. Likewise, Jin et al. [13] confirmed the positive effect of website reputation on trust.

Consumers could gain better understanding of product information, refund and exchange policy through interacting with online service of the website with high reputation. In the process of interactivity, consumers can learn more about the capability of the website to fulfill orders and the willingness to prioritize consumer needs, thereby increasing the trust towards the website. In contrast, constant interactivity between the website with low reputation and consumers would be counterproductive, undermining trust towards the website. This is because frequent interactivity initiated by a low-reputation website may make consumers feel that the goal of the website is to promote and sell products rather than care about consumer needs and benefits.

Therefore, we posited that

H1: website reputation has interference moderating effect on the relationship between online interactivity and trust

H1a: when website reputation is high, online interactivity has a positive impact on trust

H1b: when website reputation is low, online interactivity has a negative impact on trust.

2.2   The Mediating Effect of Trust

Security is the main concern of consumers over online shopping. Owing to the greater uncertainty of online shopping, trust has been regarded as a primary factor affecting purchase intention when shopping online [25,33,42]. According to Suntornpithug and Khamalah [34], trust towards the website can be referred as the willingness to transact online with potential risks, for consumers believe that the reliability of the website can ensure a good result. Ling et al. [16] proved that trust has a positive effect on the customer online purchase intention. Wu et al. [42] pointed out that trust plays a crucial role in customer behavioral decision. The study of Chiu et al. [2] illustrated that trust is closely related to purchase intention. Similarly, Djahantighi and Fakar [5] stated that trust significantly affects purchase intention. When consumers trust the website, they are confident of the capability to fulfil orders, which contributes to the increase in purchase confidence and purchase intention [38].

Purchase intention in this study can be defined as the degree to which consumers would like to buy goods or services online [3,43]. Some studies support the view that trust is a vital mediating variable influencing purchase intention [1,10]. Wu et al. [40] provided empirical evidence that online interactivity exerts an indirect impact on purchase intention through affecting consumer trust and attitude towards website.

Therefore, we posited that

H2: trust towards the website has a positive effect on purchase intention

H3: trust towards the website mediates the relationship between online interactivity and purchase intention.

Research model is displayed in Fig. 1. This paper proposes that website reputation has interference moderating effect on the relationship between online interactivity and trust; trust mediates the relationship between online interactivity and purchase intention.

3   Methodology

3.1   Questionnaire Design and Pretest

The questionnaire consists of two parts. The first part is to measure variables. All items were measured on a 7-point Likert scales (1 = strongly disagree, 7 = strongly agree). There are four items measuring online interactivity, adopted from Rayport and Jaworski [29] and Lu [18] five items measure website

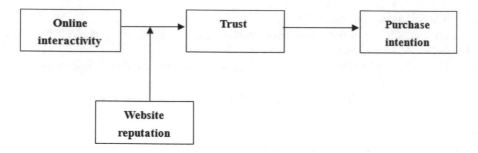

Fig. 1. Research model

reputation, derived from the work of Newburry [23] Petrick [27] and Fombrun et al. [8] four items measuring trust were taken from the work of McKnight et al. [20], Gefen et al. [9] and Pavlou [26]. Extrapolating from the work of Putrevu and Lord [28] and McKnight et al. [11], four items were used to assess purchase intention. The second part collects personal information on gender, age, and education.

This research is interested in investigating online purchase intention so the respondents need to have rich experience in online shopping. According to the report issued by CNNIC [4] 20–29 year old age group is the major force in online shopping. Most of them are college students or young employees. On the one hand, college students have not earned income, and rely on parents for stipend. Hence, they spend money carefully and often choose the economical online shopping. On the other hand, those who have just started work are preoccupied by working tasks and do not have enough time to shop in traditional stores. They spend more time in front of computers and often choose the convenient online shopping. This study mainly targets these two groups of people.

This research studies B2C websites. According to the survey of iResearch [11], trading volume on B2C websites reached 3.6 trillion RMB in 2017, accounting for 60.0% of online shopping volume; the growth rate was 40.9% exceeding that of C2C market (15.7%). This figure shows the importance of B2C websites in online shopping. In addition, website reputation is imperative for a B2C website as it usually represents a company, unlike a C2C website consisting of numerous small sellers. According to the report on Chinese online shopping market issued by iResearch [12], B2C websites with large market share including Tmall.com, JD.com, Suning.com, and Vip.com were listed as options in the questionnaire, respondents first choose a B2C website in which they do the shopping most and then answer questions related to that website. To ensure the validity of the scale, a pretest was carried out. 80 questionnaires were issued, 36 copies were collected and showed good reliability with Cronbach's over 0.7.

3.2   Data Collection

Online and offline questionnaires were handed out in three cities including Chengdu, Hangzhou, and Shenzhen. Respondents were mainly undergraduates, postgraduates and MBA students in universities. The period lasts from August, 2017 to May, 2018. 800 questionnaires were distributed, 423 were collected and 276 were valid, representing 34.5% of valid response rate. Samples were described in Table 1.

Table 1. Descriptive analysis of samples

		N.	Proportion
Gender	Males	151	54.7%
	Females	125	45.3%
Age	<20 years old	82	29.7%
	20–25 years old	179	64.9%
	26–30 years old	13	4.7%
	31–35 years old	1	0.4%
	>35 years old	1	0.4%
Education	Undergraduates	114	41.3%
	Postgraduates	162	58.7%
B2C website visited most	Tmall.com	129	46.7%
	JD.com	78	28.3%
	Suning.com	28	10.1%
	Vip.com	15	5.4%
	Others	26	9.4%

4 Results

This study utilizes SPSS 20.0 to analyze and process the data derived from questionnaires. Reliability analysis, factor analysis and regression were used to test the hypotheses.

4.1 Reliability and Factor Analysis

Cronbach's of online interactivity, website reputation, trust and purchase intention exceed 0.80, suggesting satisfactory reliability. Validity analysis comprises content-related validity and structural validity. All items in the survey derive from empirically tested scale in literature, which ensure good content-related validity. Results from factor analysis show that all items loaded significantly between 0.73 and 0.89 on the variables they intended to represent (P<0.001), suggesting satisfactory structural validity (Table 2).

Table 2. Reliability and validity analysis

Variables	Number of items	Cronbach's	Factor loading
Online interactivity	4	0.87	0.79–0.86
Website reputation	5	0.86	0.76–0.82
Trust	4	0.82	0.73–0.86
Purchase intention	4	0.89	0.84–0.89

4.2   Hypotheses Testing

As suggested by Wen et al. [39] when independent variables and dependent variables are continuous variable, hierarchical regression can be used to test moderation effect. The first step is the regression of the independent variable on the dependent variable and the moderator, generating R_1^2; the second step is the regression of the independent variable, the moderator and the interaction term on the dependent variables, obtaining R_2^2. If R_2^2 is significantly larger than R_1^2, or the interaction term is significant, the moderation effect is significant. Mediation effect is tested by the method recommended by Wen et al. [39]: (1) regression of the independent variable on the dependent variable, regression coefficient is significant; (2) regression of the mediator on the dependent variables, regression coefficient is significant; (3) regression of the independent variable on the mediator, regression coefficient is significant; (4) regression of the independent variable and the mediator on the dependent variable, the coefficient of the mediator is significant. Meanwhile, if the coefficient of the independent variable becomes insignificant with the inclusion of the mediator, then there is full mediating effect; if the coefficient of the independent variable becomes smaller, then there is partial mediating effect.

The Interference Moderating Effect of Website Reputation This study uses hierarchical regression analysis to explore the interference moderating effect of website reputation on the relationship between online interactivity and trust. Results are shown in Table 3. Model 1 tests the effect of online interactivity on trust. Results suggest that online interactivity has a significantly positive effect on trust ($\beta = 0.207, P < 0.001$). Model 2 tests the influence of online interactivity and website reputation on trust, demonstrating website reputation has a significantly positive effect on trust ($\beta = 0.396, P < 0.001$). Model 3 tests the moderating effect of website reputation on the relationship between online interactivity and trust, manifesting website reputation has moderated the relationship between online interactivity and ($\beta = 0.099, P < 0.05$).

What is worth noting is that the effect of online interactivity on trust turns from positive to negative with the inclusion of interaction term in model 3. This phenomenon can be explained from two aspects. First, according to the study of Dong et al. [6], the coefficient of the independent variable is different when interaction term is added, suggesting the moderation effect is significant. After

Table 3. The moderation effect

	Model 1	Model 2	Model 3
Constant	4.487	2.855	4.820
Online interactivity	0.207***	0.101*	−0.427*
Website reputation		0.396***	0.022
Online interactivity*website reputation			0.099*
R^2	0.065	0.196	0.215
ΔR^2	0.062	0.19	0.206
F value	19.202***	33.303***	24.836***

Note: $*p < .05, **p < .01, ***p < .001$

interaction term is added, the coefficient of online interactivity becomes from 0.207 ($P < 0.001$) to −0.427 ($P < 0.05$), demonstrating the moderation effect is significant and further support H1. On the other hand, the change from positive to negative effect shows that website reputation changes the direction of relationship between online interactivity and trust, which is called pure moderator in the research of Sharma and Gur-Arie [31]. To clarify how website reputation changes the direction of relationship between online interactivity and trust. This research divides website reputation into a group of high reputation (mean + a standard deviation) and the other group of low reputation (mean − a standard deviation). Moderation effect is show in Fig. 2.

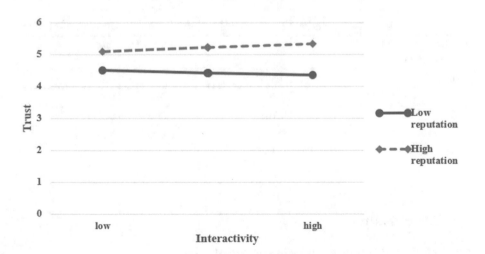

Fig. 2. Interference moderation effect

It is shown in Fig. 2 that when website reputation is low, online interactivity has a negative influence on trust, suggesting a high level of online interactivity reduces customer trust towards the websites. Therefore, H1b is supported. When website reputation is high, online interactivity has a positive impact on trust, indicating a high level of online interactivity increases customer trust towards the websites. Therefore, H1a is proved. As can be seen, website reputation changes the direction of relationship between online interactivity and trust. When the moderator has high value,the independent variable has a positive impact on the dependent variable; When the moderator has low value,the independent variable has a negative impact on the dependent variable, this moderator is considered to have interference moderating effect [19].

The Mediating Effect of Trust As suggested by Wen et al. [39], regression was conducted to test whether trust mediates the relationship between online interactivity and purchase intention. First step is to build the relationship between purchase intention and online interactivity (IV→DV), the coefficient is significant ($\beta = 0.255, P < 0.001$). In the second step(M→DV), trust has a significantly positive influence on purchase intention ($\beta = 0.864, P < 0.001$), H2 is then supported. In the third step(IV→M), trust is significantly affected by online interactivity ($\beta = 0.207, P < 0.001$). In the fourth step, with the inclusion of trust, the influence of online interactivity on purchase intention drops from 0.255 to 0.082, but still significant($P < 0.05$), suggesting partial mediation and supporting H3 (Table 4).

Table 4. The mediate effect

	Step 1	Step 2	Step 3	Step 4
	Purchase intention	Purchase intention	Trust	Purchase intention
Online interactivity	0.255***		0.207***	0.082*
Trust		0.864***		0.838***
R^2	0.094	0.702	0.065	0.711
ΔR^2	0.09	0.701	0.062	0.709
F value	28.323***	644.324***	19.202***	335.218***

Note: $*p < .05, **p < .01, ***p < .001$

5 Conclusions

5.1 Conclusions and Managerial Implications

Although numerous studies investigate the role of trust in online shopping and prove the positive effect of online interactivity on trust [17,34,40], these studies do not fully explain why the effect of online interactivity varies from websites to websites. Some websites incur consumer skepticism through interactivity, while others enhance consumer trust. Plenty of existing literature are interested in adding variables which can improve trust, including web design, order fulfillment and web security, few studies focus on exploring factors that could change the direction of relationship between online interactivity and trust. This research sheds new light on the relationship between online interactivity and trust. This research introduces website reputation as a moderator and confirms that the influence of online activity on trust is neither positive nor negative. It relies on the degree of website reputation. Website reputation has interference moderating effect on the relationship between online interactivity and trust. Increasing online interactivity improves customer trust when website reputation is high. By contrast, increasing online interactivity decreases customer trust when website reputation is low. This research reveals the relationship between online interactivity, trust and website reputation from a new perspective, and enrich the research on online shopping.

Moreover, this research has practical implications on how to manage and operate B2C shopping websites. Considering the interference moderating effect of website reputation, website should adopt appropriate level of online interactivity in accordance with website reputation instead of blindly pursuing frequent interaction with customers. When website reputation is high, a website should establish customer service to interact with consumers before and after the sale, instantly answer questions and solve problems. Meanwhile, great importance should be attached to customer reviews and suggestions. Proper settlement of disputes and complaints can improve relationship with customers and promote their trust towards the website. In this way, customers are convinced that the website is fully capable of meeting their needs, safeguarding their interests, thus increasing purchase intention on this website.

On the other hand, when a website is at the starting-up stage without reputation being built, or the reputation of website has already been destroyed. The priority of the website should be put on the improvement of website reputation rather than desperately interact with customers, because this may lead customers to doubt the purpose of the interaction. Privacy protection on the website may also be suspected, thus trust towards the website will drop. Great efforts should be made to honestly describe information about products and services, ensure the quality, deliver orders on time and fulfill responsibilities on exchange and refund. Interactivity can be raised after website reputation is improved.

5.2 Limitations and Future Direction

Limitations in this research are shown as follows. Firstly, this study collected data from the perspective of consumers, and focus on the perceived interactivity, and perceived website reputation. Future research can gather data from B2C websites, calculate the frequency of interactivity between websites and consumers, and examine the qualifications of websites granted by a third party. Secondly, the sample mainly covers college students and young people just starting work, so the conclusions drawn from this research may not be applied to other groups of consumers. Future research can study the online shopping habits of other age groups and occupations. Finally, this study concerns the impact of website-related variables on purchase intention without considering other variables such as income and economic development in different cities. Future research can incorporate these variables to study the impact on purchase intention.

References

1. Chen, L.: The Study on Website Features and Consumer Purchase Intention: Mediating Role of Trust. National Taipei University (2007)
2. Chiu, C.M., Chang, C.C., et al.: Determinants of customer repurchase intention in online shopping. Online Inf. Rev. **33**(4), 761–784 (2009)
3. Chu, C.W., Lu, H.P.: Factors influencing online music purchase intention in Taiwan: an empirical study based on the value-intention framework. Internet Res. **17**(2), 139–155 (2007)
4. CNNIC: The 42th report on china internet development (2018)
5. Djahantighi, F., Fakar, E.: Factors affecting customer's trends for reservation foreign hotels via internet in Iran. Int. Bull. Bus. Adm. **7**(2), 6–14 (2010)
6. Dong, W.: The application of moderator to management research in China. Chin. J. Manag. **12**, 1735–1743 (2012)
7. Etemad-Sajadi, R.: The impact of online real-time interactivity on patronage intention: the use of avatars. Comput. Hum. Behav. **61**, 227–232 (2016)
8. Fombrun, C.J., Gardberg, N.A., Sever, J.M.: The reputation quotient SM: a multistakeholder measure of corporate reputation. J. Brand Manag. **7**(4), 241–255 (2000)
9. Gefen, D., Karahanna, E., Straub, D.W.: Trust and TAM in online shopping: an integrated model. MIS Q. **27**(1), 51–90 (2003)
10. Guan, H., Dong, D.: The empirical study on forming mechanism of consumer-based e-tail brand equity. China Bus. Mark. 37–39 (2008)
11. iResearch: 2017 annual data on e-business and logistics (2017)
12. iResearch: Report on 2017 Chinese online shopping market (2017)
13. Jin, B., Yong Park, J., Kim, J.: Cross-cultural examination of the relationships among firm reputation, e-satisfaction, and e-trust, and e-loyalty. Int. Mark. Rev. **25**(3), 324–337 (2008)
14. Kim, J., Lennon, S.J.: Effects of reputation and website quality on online consumers' emotion, perceived risk and purchase intention: based on the stimulus-organism-response model. J. Res. Interact. Mark. **7**(1), 33–56 (2013)
15. Li, Y.: The impact of disposition to privacy, website reputation and website familiarity on information privacy concerns. Decis. Support Syst. **57**, 343–354 (2014)
16. Ling, K.C., Chai, L.T., Piew, T.H.: The effects of shopping orientations, online trust and prior online purchase experience toward customers' online purchase intention. Int. Bus. Res. **3**(3), 63 (2010)
17. Liu, Y., Shrum, L.: A dual-process model of interactivity effects. J. Advert. **38**(2), 53–68 (2009)
18. Lu, F.: The Study on Determinants of Initial Trust Towards Websites. Zhejiang University (2005)
19. Luo, S., Jiang, Y.: Management Survey Research Methodology. Chongqing University Press (2014)
20. McKnight, D.H., Choudhury, V., Kacmar, C.: The impact of initial consumer trust on intentions to transact with a web site: a trust building model. J. Strateg. Inf. Syst. **11**(3–4), 297–323 (2002)

21. McMillan, S.J., Hwang, J.S.: Measures of perceived interactivity: an exploration of the role of direction of communication, user control, and time in shaping perceptions of interactivity. J. Advert. **31**(3), 29–42 (2002)
22. Mollen, A., Wilson, H.: Engagement, telepresence and interactivity in online consumer experience: reconciling scholastic and managerial perspectives. J. Bus. Res. **63**(9–10), 919–925 (2010)
23. Newburry, W.: Reputation and supportive behavior: moderating impacts of foreignness, industry and local exposure. Corp. Reput. Rev. **12**(4), 388–405 (2010)
24. Pappas, I.O.: User experience in personalized online shopping: a fuzzy-set analysis. Eur. J. Mark. **52**(7/8), 1679–1703 (2018)
25. Pappas, I.O., Kourouthanassis, P.E., et al.: Sense and sensibility in personalized e-commerce: how emotions rebalance the purchase intentions of persuaded customers. Psychol. Mark. **34**(10), 972–986 (2017)
26. Pavlou, P.A.: Consumer acceptance of electronic commerce: integrating trust and risk with the technology acceptance model. Int. J. Electron. Commer. **7**(3), 101–134 (2003)
27. Petrick, J.F.: Development of a multi-dimensional scale for measuring the perceived value of a service. J. Leis. Res. **34**(2), 119–134 (2002)
28. Putrevu, S., Lord, K.R.: Comparative and noncomparative advertising: attitudinal effects under cognitive and affective involvement conditions. J. Advert. **23**(2), 77–91 (1994)
29. Rayport, J.F., Jaworski, B.J.: E-Commerce. McGraw-Hill Higher Education (2001)
30. Roy Dholakia, R., Zhao, M.: Retail web site interactivity: how does it influence customer satisfaction and behavioral intentions? Int. J. Retail Distrib. Manag. **37**(10), 821–838 (2009)
31. Sharma, S., Durand, R.M., Gur-Arie, O.: Identification and analysis of moderator variables. J. Mark. Res. **18**(3), 291–300 (1981)
32. Song, J.H., Zinkhan, G.M.: Determinants of perceived web site interactivity. J. Mark. **72**(2), 99–113 (2008)
33. Stouthuysen, K., Teunis, I., et al.: Initial trust and intentions to buy: the effect of vendor-specific guarantees, customer reviews and the role of online shopping experience. Electron. Commer. Res. Appl. **27**, 23–38 (2018)
34. Suntornpithug, N., Khamalah, J.: Machine and person interactivity: the driving forces behind influences on consumers' willingness to purchase online. J. Electron. Commer. Res. **11**(4), 299–325 (2010)
35. Thomas, M.R., Kavya, V., Monica, M.: Online website cues influencing the purchase intention of generation Z mediated by trust. Indian J. Commer. Manag. Stud. **9**(1), 13–23 (2018)
36. Van Noort, G., Voorveld, H.A., Van Reijmersdal, E.A.: Interactivity in brand web sites: cognitive, affective, and behavioral responses explained by consumers' online flow experience. J. Interact. Mark. **26**(4), 223–234 (2012)
37. Voorveld, H., van Noort, G.: Moderating Influences on Interactivity Effects. Advances in Advertising Research, vol. III, pp. 163–175. Springer (2012)
38. Yang, Q., Liu, Y., Han, C.: The study on mechanism of customer trust in online shopping. J. Intell. **30**(2011), 197–201 (2011)
39. Wen, Z., Chang, L., Hau, K.T., Liu, H.: Testing and application of the mediating effects. Acta Psychol. Sin. **36**(5), 614–620 (2004)
40. Wu, G., Hu, X., Wu, Y.: Effects of perceived interactivity, perceived web assurance and disposition to trust on initial online trust. J. Comput. Mediat. Commun. **16**(1), 1–26 (2010)

41. Wu, G.M.: The role of perceived interactivity in interactive ad processing. The University of Texas at Austin (2000)
42. Wu, W.Y., Ke, C.C., Nguyen, P.T.: Online shopping behavior in electronic commerce: an integrative model from utilitarian and hedonic perspectives. Int. J. Entrep. **22**(3), 1–16 (2018)
43. Zhang, H., Bai, C., Li, C.: The analysis on consumer purchase intention-comparision on theory of reasoned action and theory of planned behavior. Soft Sci. **9**, 130–135 (2011)

Turnover: Organizational Politics or Alternate Job Offer?

Aimon Iqbal[(⊠)], Abdullah Khan, and Shariq Ahmed

Khadim Ali Shah Institute of Technology Karachi, Karachi, Pakistan
aimon.iqbal@hotmail.com

Abstract. Reducing turnover has always been a key issue in any industry. Reducing turnover will help organizations to retain their key employees, thus contributing to the higher productivity and possibly profitability as well. In this study, an insight has been contributed into the relationship between perceived organizational politics and alternate job offering higher pay with turnover amongst the employees of Bank Alfalah (Karachi). A cross-sectional study with a deductive approach was carried out. The responses collected through survey questionnaires from the employees of Bank Alfalah revealed that turnover has a significantly positive relationship with perceived organizational politics and alternate job with higher pay. It was also found out that office politics is a stronger factor to convert turnover intention into decision. Findings in this research offer important implications for the banking industry, as well as for theory in research driven by organizational politics and higher pay offered by alternate job.

Keywords: Alternate job with higher pay · Organizational politics · Turnover · Pakistan

1 Introduction

Employee retention is an effort by the organizations to provide a conducive environment which supports the existing staff to stay with the organization. Nowadays the issue of retaining employees is becoming complicated. Employee turnover has been in limelight from the beginning, it is an important and sensitive issue in the banking sector of Pakistan. Banks have been facing the risk of losing their key and high performers. Apart from many other operating factors, which causes turnover, the two main factors which have been studied in this research are; perceived organizational politics and alternate job offer. Eleven categories have been figured out that defines the negative impact of turnover on the organizations. From these eleven few are; High Financial cost, Success and/or failure of organization, Productivity lost and workflow interruptions etc. [37]. To cut down the negative impact, organizations need to know why employees leave an organization. Many researchers are interested to know these reasons and are still interested in this topic as it is a complicated one. In the current study we try

© Springer Nature Switzerland AG 2020
J. Xu et al. (Eds.): ICMSEM2019 2019, AISC 1002, pp. 641–653, 2020.
https://doi.org/10.1007/978-3-030-21255-1_49

to get an insight about the relationship of turnover with organizational politics and alternate job offer with higher pay.

Alternate job with higher pay is a factor that might tiger the need to switch the job. Money has always been a source of attraction to human. An Indian study contributed that an alternate job opportunity with higher pay can increase the rate of turnover [8]. This paper gets an insight into this regard and also attempts to check the more influencing factor among the said factors. Along with other correlates alternate job offer was considered to have a positive relationship with turnover [12]. Dissatisfied employees search for alternatives, in which they compare the cost and benefits of future job with that of current job and then decide to quit the job [45]. It is also contributed to the field of HRM that compensation influences retention [40]. In this research we would dig deeper inside to know the relationship between the alternate job offered by another organization with higher pay and the turnover.

There is another aspect which is to be revealed through this research and that is of the perceived organizational politics which has some impact over turnover intension. Mintz Berg defines Organizational Politics as "Individual or group behavior that is informal, ostensibly parochial, typically divisive, and above all in a technical sense, illegitimate" [32]. Organizational politics creates feeling of discomfort among employees. Perceived organizational politics and turnover intention has a positive relationship [27]. Perceived organizational politics and turnover intension are directly proportional [28]. Organizational politics can influence the employee to make the decision of turnover. Even skilled and key employees could feel compelled to leave the organization in which they are working. The politics in any organization could have adverse effects on the human resource of the company because as a result employees feel uncomfortable and are always in seek of alternate jobs. Perceived organizational politics is one of the concerns which give some impact over the intension of employee to leave an organization. The perceived organizational politics also indirectly effect the job performance of employees, it becomes higher if despotic behavior of supervisor is evident [36].

In both the cases it is quite possible that separation may happen. In this study we will investigate that which of the two factors (alternate job offer with higher pay or perceived organizational politics) play more influential role as compared to other in increasing turnover in Bank Alfalah Limited Karachi. The purpose of this research is to cut down the rate of turnover in banking sector of Karachi with special reference to Bank Alfalah Limited, so as to reduce the turnover in Bank Alfalah. Banking sector in Pakistan constitutes of nearly 31 banks, which includes four foreign banks, five belongs to public sector and remaining 22 are local private ones. Top six banks have the major stake of the banking asset, including Bank Alfalah. These banks successfully make up to 57 percent of the currency deposits yearly. The banks in Pakistan are governed by State Bank of Pakistan.

2   Literature Review

2.1   Turnover

Turnover can be defined as the amount of employees who leaves an organization [30]. William Mobley defines the turnover as an act in specific time which is a physical separation from an organization [33]. If turnover persists in an organization it increases its impact on productivity which becomes negative in an attenuated manner [14]. The reasons for turnover was studied in detail and reported that there are 26 variables which are standing behind turnover, they segregated these variables in 3 categories as External Correlates, Work-related Correlates and Personal Correlates. In their detailed conclusion they added that perception of alternate employment is positively related to turnover [12]. Alternate employment was then studied with the other attractions, so Mobley and others reported a conceptual model of turnover, along with other variables, they specially emphasized on attraction expected utility of alternate and concluded that it plays an important role in turnover process and it need further research [34]. The turnover of employees is not good for organizations as the fear of productivity decline comes up. It has been added recently in literature review that turnover significantly negatively affect productivity causing millions of dollars materialistic cost [35]. Evidence is available in past study that says that turnover is one of the most strong causes of decline in productivity as well as sagging the morale of employees in public and private sectors [2]. Apart from alternate job the second factor which is under consideration is office politics. A study found out that there is a positive relationship between office politics and turnover [10]. This study would help organizations in policy making about Retention of employees as well as making the environment conducive for employees to continue working. Retaining high performers or the talented staff in the organizations is a critical and sensitive issue which can lead to higher productivity. Higher productivity means higher profitability [31].

2.2   Perceived Organizational Politics (POP)

Politics in organizations is not a new concept; it has its roots in the history. The concept started to emerge in 1960s when Burn coined the concept of micropolitics he has defined politics when people are being used as resources in any competing situation [9]. A scholar believes contrary to Burns that politics perception is related to ones job, employing organization and demographic characteristics [19]. Researchers found that office politics can be one of the reasons for turnover [24], they found that there is a significant positive relationship between Perceived Organizational Politics (POP) and turnover intention [39]. Another study shows that organizational justice and Perceived organizational politics play significant role over employee turnover intentions [41]. A study conducted in Pakistan reveals that even high psychological capital does not significantly affect the strong and positive relationship of politics-turnover intentions [1]. When employees feel frustrated due to politics and there is nothing which can be done then they chose to keep quite. A recent addition to literature says that there is a positive relationship between silence of the employees and perceived organizational politics [25]. A small group of researchers claim in their study that if the organizational politics is reduced then important organizational and individual contributions can be brought up [43], so by taking care of politics, productivity can be increased through proper leadership style [23]. The reason for productivity decline due to office politics may be turnover and absenteeism [18].The drawbacks of office politics includes The sabotaging of self-expression and supportive management [26]. A survey in U.S retail industry industry shows that internal integration is directly influenced by perceived organizational politics [43].

Construct 1: There is a relationship between Perceived organizational politics and Turnover

Measuring Perceived Organizational Politics The perceived organizational politics was measured by using the scale presented by Kacmar and Carlson [30]. It three sub dimensions which are the General political Behavior, Go-along-to-get-along and pay and promotion policies. For measuring General Political Behavior we used Colleagues around you build themselves up by pulling others and There is an influential group/individual that no one can cross. For Go-Along-to-get-along Agreeing with influential group/individual is best alternative and It is safer to think what you are told, than to make up your own mind were used for measurement. For Pay and promotion policies the 2 items were used as follows; When it comes to pay raise and promotion decisions, policies are irrelevant and Promotions around here are not valued much because how they are determined is so political.

Hypothesis

H_0: There is no relationship between organizational politics and Turnover (Or μ of organizational politics = μ of Turnover)

H1: General political Behavior has a negative relationship with turnover

H2: Go-along-to-get-along has a negative relationship with turnover

H3: Pay and promotion policies has a negative relationship with turnover

2.3   Alternate Job Offer with Higher Pay

A recent research contributed that along with other factors alternate job opportunity with higher pay increases the attrition in call center industry [8]. The great deal of work on employee turnover has been done by Cotton and Tuttle. They identified 26 variables which the research proved to be leading to the employee turnover [12]. From these 26 variables pay satisfaction has a strong and negative relationship with turnover intention, a study revealed in Pakistani Banks [4]. Unlike with turnover Pay satisfaction has a positive relationship with organizational commitment, if employees are not satisfied with pay they develop an intention to leave the organization [44] and [11]. Similar to pay satisfaction, pay rise satisfaction is negatively associated with turnover this means the higher the pay the lower the turnover will be [42]. Pay dispersion also leads to the likelihood for managers to leave the organizations in which they are employed [6]. Pay satisfaction is so important for employee retention that if an executive is receiving lower pay than that of the relative market, the chance of turnover increases [16]. A group of researchers who has worked on pay satisfaction, job satisfaction and organizational commitment concludes that pay satisfaction is directly as well as indirectly related to the turnover intention [29]. An alternate job attracts an employee psychologically and pulls him away from the current job out of self-interest [7]. The organizations should keep in mind that Compensation has a significant impact over turnover [17].

Construct 2: There is a relationship between Alternate job offer with higher pay and Turnover.

Measuring Alternate Job Offer Mobley was one of the many scholars who stressed the importance of including the factor of alternate job offer when studying turnover [33]. In the study done by Arnold and colleagues 26 variables were found out to be the contributing factors to measure the turnover [12]. From these 26 variables one was availability of alternatives, which belonged to the External Correlate category identified by Cotton and Tuttle. This variable was measured by following item At my current age, education, occupation, and the general economic condition, chance of finding a suitable position in some other organization is very high. Responses were recorded on a 5 point scale from Strongly Agree to Strongly Disagree [3]. Now Alternate job offer in our model becomes variable instead of construct. The reason behind is that it is measurable directly.

Hypothesis

H_0: Alternate job Offer has no relationship with turnover (Or μ of alternate job offer = μ of turnover).

H4: Alternate Job offer has a negative relationship with overall Job Satisfaction.

H5: Alternate Job offer has a negative relationship with Satisfaction with Pay.

The model describes that Perceived organizational politics and alternate job offer with higher pay are the two independent variables considered in this study. The turnover is studied as the dependent variable in this research. It is to check the relationship between perceived organizational politics and Alternate job offer having higher pay with turnover. The study also checks which of the independent variable have more influence over turnover. Figures 1 and 2 show the theoretical framework and research model of this paper, respectively.

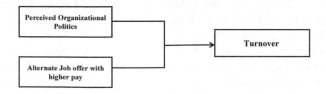

Fig. 1. Theoretical framework

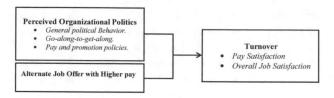

Fig. 2. Research model

3   Methodology

This is a cross-sectional study with considerations of understanding the causal effect of variables. The method used to define this paper is descriptive and causal as well the paper is design in quantitative pattern. Problem was explored from the secondary data which is literature from the past published papers. Furthermore to get facts and figures to justify the paper empirically we have used the primary data [5] and the tool used to collect the primary data i.e. Questionnaires which were structured and close ended. The target population is from the staff working in the branches of Bank Alfalah presented in the premises of Karachi. The given sample size is 20 staff members who became the part of my study. Moreover the non-probability convenience sampling is used to collect the data through survey. The questions regarding perceived organizational politics and alternate job offer with higher pay were included in the questionnaire. The question design on the nominal and ordinal scaling included options from 1-strongly Agree to 5-Strongly Disagree. The reliability was checked respectively of all the constructs using Cornbach Alpha method through SPSS. The reliability was found quite satisfactory. Correlation Coefficient, Regression analysis and other tests were used to find the core relationship between been the said variables.

4   Analysis and Interpretation

4.1   Statistical Approach

The purpose of this study is to investigate the relationship between organizational politics and alternate job offer with higher pay with turnover. To establish this relationship a survey was conducted.

The data was collected from males as well as females. The genders of respondents are presented in Table 1.

The Table 1 shows that 85.5% were male who became a part of sample and the remaining 14.4% were females. The designations which belonged to the respondents have been summarized in the Table 2.

This table shows that most of the respondents were officers. 74 of the respondents were line managers and similarly 45 belonged to the post of senior manager.

A reliability analysis was carried out for assessing internal consistency of scale item. To assess it we utilized Cronbachs Alpha whose estimate was, $\alpha = 0.80$ indicating good stability in the questionnaire.

Table 1. Frequency distribution of gender of respondents

Gender		
	Frequency	Percentage
Male	124	85.51%
Female	21	14.48%
Total	145	100%

Table 2. Frequency distribution of the gender of respondents

Designations	
Positions	Frequency
Officer	93
Line Manager	74
Senior Manager	45
Total	145

The measure of degree of linear association between the two variables in the sample is depicted in Table 3 below:

Table 3. Correlations

Correlations				
		Turnover	Politics	Alternate_job
Turnover	Pearson Correlation	1	0.635**	0.523**
	Sig. (2-tailed)		0	0
	N	145	145	145
Politics	Pearson Correlation	0.635**	1	0.361**
	Sig. (2-tailed)	0		0
	N	145	145	145
Alternate_job	Pearson Correlation	0.523**	0.361**	1
	Sig. (2-tailed)	0	0	
	N	145	145	145

**Correlation is significant at the 0.01 level (2-tailed)

The Table 3 shows a significant moderate relationship between organizational politics and turnover ($r = 0.635$, $p < 0.001$). It also shows that the turnover is also moderately related to alternate job with higher pay, but this relationship is relatively weaker than the former one ($r = 0.523$, $p < 0.001$). There was no negative or zero linear dependence observed between variables. Thus it can be said that the two variables are positively correlated with turnover. Another thing worth noticing is that organizational politics has high computed value compared to that of alternate job offer i.e. r = 0.635 which indicates a positive and moderate correlation. So office politics was found to be more influential comparatively.

The Technique of paired sample T test was also utilized with the help of SPSS to test the constructed hypothesizes (Table 4).

Table 4. Correlations

Paired samples test

		Paired differences					t	df	Sig.(2-tailed)
		Mean	Std. deviation	Std. error mean	95% confidence interval of the difference				
					Lower	Upper			
Pair 1	Turnover-Alternate_job	0.34943	0.7909	0.06568	0.2196	0.47925	5.32	144	0
Pair 2	Turnover-Politics	0.01055	0.7951	0.06603	−0.11996	0.14106	0.16	144	0.873

The Paired samples T test shows that H0 of office politics has been rejected, $t(144) = 0.160$, $p \leq 0.873$. This means that there is a difference between the mean of politics and turnover. Or in other words office politics have been effecting on turnover. H_0 made for alternate job offer with higher pay has not been rejected $t(144) = 5.320$, $p \leq 0.001$. This means that alternate job offer has not been influencing the turnover.

The technique of multivariate regression analysis is utilized to estimate combined effect of alternate job offer with higher pay and organizational politics on the turnover.

$$\text{Turnover}_t = 0.828 + 0.331\text{Alternate\_job}_t + 0.422\text{Politics}_t + \mu_t$$

Where μ_t represents the error term. All these estimation for construction of model equation were undertaken at 5% significance level. Besides necessary statistics for estimated models are represent in the Table 5.

Table 5. Coefficients[a]

Model	Unstandardized coefficients		Standardized coefficients	t	Sig.	Collinearity statistics	
	B	Std. Error	Beta			Tolerance	VIF
(Turnover)	0.828	0.249		3.321	0.001		
Alternate job	0.331	0.091	0.264	3.631	0	0.87	1.15
Politics	0.422	0.07	0.44	6.053	0	0.87	1.15

The probability value (*p* value) and standardized beta of the estimated coefficients are found within the significant ranges indicating or showing the stability of the estimated model. The summary of the model constructed is illustrated in the Table 6.

Table 6. Model Summary

Model	R	R square	Adjusted R square	Std. error of the estimate	Change statistics					Durbin-Watson
					R square change	F Change	df1	df2	Sig. F change	
1	0.609a	0.657	0.701	0.54624	0.347	37.757	2	142	0	1.836

The coefficients of multiple determinations determine the percent of the variance in the dependent variable explained jointly by the independent variables whose value for this model is 65.7% which indicates that variations in turnover is jointly explained by attributes of alternate job offer with higher pay and organizational politics. The value of DW statistics = 1.836 shows that there is no serial correlation in the error terms.

The determined statistics for estimated model had a significance level of 0.01% , this indicates that the data is perfect for drawing a conclusion on the population parameter as the p-value is less than the significance level. Furthermore, the computed F-value (F stats = 150.757, p < 0.001) is far higher than the critical value (i.e. 150.531 > 2.122) is an indication that the independent variables have significant effects on the Turnover, as shown in Table 7. Graphs under Figs. 3 and 4 also authenticate the stability of the estimated model.

Table 7. ANOVA

Model		Sum of squares	df	Mean square	F	Sig.
1	Regression	32.52	2	16.26	150.757	0.000a
	Residual	61.153	142	0.431		
	Total	93.673	144			

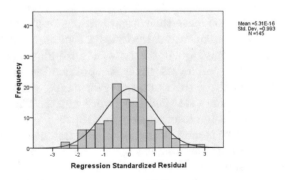

Fig. 3. Histogram plot of residual confirming normality

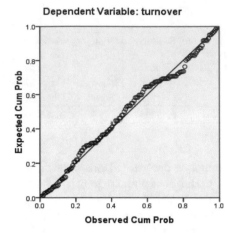

Fig. 4. Normal P-P Plot of regression residual normality in error term

5   Conclusion & Recommendation

5.1   Conclusion

This study is aimed at checking the relationship of perceived organizational politics and higher pay offered by alternate job with the turnover. The study also aims to know that which of these factors is more influential to convert the turnover intention into the decision. So to conclude it can be said that firstly the perceived organizational politics and alternate job offered have significant and positive relationship with turnover. This means that when organizational politics increases the chance of employee turnover will also increases. Moreover organizational politics has more influence over turnover as compared to higher pay offered by an alternate job. The results provided the evidence that people leave organizations more because of the organizational politics as compared to that of the Higher Pay offered by an alternate job, in the context of Bank Alfalah. So it can be said that people leave their organizations more when they become frustrated because of the organizational politics rather than only preferring the higher pay offered by alternate job. Job satisfaction is indirectly proportional to the office politics [22]. The employees prefer the politics free environment. Such an environment gives credit of their better performance, rather than taking it away. When the employees are satisfied they work better. The outcomes of the employees are also effected due to office politics [13].

Employees believe that money or higher pay is not everything. Employees prefer a trustworthy working environment in which no one tears others to make themselves up. Supervisors with abusive supervision become factors to increase the chance of turnover [4]. Our study is in line with a study conducted in Lahore, Pakistan concluded that office politics is found to be the main cause of the turnover intention [24]. Another important aspect is perception of politics, if employees feel that they are surrounded by office politics than it is quite possible that high performers indulge in the politics game or may think to quit. Supervisors perceptions of politics in their organizations provoked the feelings of turnover [15].

5.2   Recommendations

Retaining employees is such a sensitive issue that if not resolved on time then it may cost key employees to leave organizations. In a case where turnover is high due to office politics, then introducing leadership can solve the matter to some extent. Leadership is a behavior through which the team members are guided to achieve set objectives. It is not about giving orders, in fact it is about showing the followers how to do a certain assigned work. Leadership curtails the negative behaviors and lock in the positive and expected behaviors. It is also evident from literature that introducing leadership style will increase the outcomes of employees. The responsible leadership style increases outcomes by employees through reducing the office politics [23]. Employee turnover can be decreased by introducing Responsible Leadership which increases the organizational commitment [20,38].

Another thing is the organizational commitment can also decrease the turnover. It can be said on the basis of research that when organizational commitment is increased then organizational politics and job tension are decreased [21]. Organizational commitment is the loyalty to work with one organization, instead of switching jobs. The employees believe in the promise to give their best output to achieve the organizational objectives. There may be two ways to increase organizational commitment through intrinsic and/or extrinsic rewards. The example of intrinsic reward can be; motivating employees through appreciation and recognition. The extrinsic reward may include monetary benefits. For example; Realizing the good performance and giving rewards (salary increments, bonuses and alike) to high performer can reduce turnover rates and increase the organizational commitment. These rewards are the return of good performance and compatible rewards can satisfy the employees. And satisfied employees show high organizational commitment.

References

1. Abbas, M., Raja, U., et al.: Combined effects of perceived politics and psychological capital on job satisfaction, turnover intentions, and performance. J. Manag. **40**(7), 1813–1830 (2014)
2. Abbasi, S.M., Hollman, K.W.: Turnover: the real bottom line. Public Pers. Manag. **29**(3), 333–342 (2000)
3. Arnold, H.J., Feldman, D.C.: A multivariate analysis of the determinants of job turnover. J. Appl. Psychol. **67**(3), 350 (1982)
4. Azeem, M., Humayon, A.A.: The impact of pay satisfaction, job stress, and abusive supervision on turnover intention among banking employees (2017)
5. Bell, E., Bryman, A., Harley, B.: Business Research Methods. Oxford University Press (2018)

6. Bloom, M., Michel, J.G.: The relationships among organizational context, pay dispersion, and among managerial turnover. Acad. Manag. J. **45**(1), 33–42 (2002)
7. Bretz Jr., R.D., Boudreau, J.W., Judge, T.A.: Job search behavior of employed managers. Pers. Psychol. **47**(2), 275–301 (1994)
8. Budhwar, P.S., Varma, A., et al.: Insights into the Indian call centre industry: can internal marketing help tackle high employee turnover? J. Serv. Mark. **23**(5), 351–362 (2009)
9. Burns, T.: Micropolitics: mechanisms of institutional change. Adm. Sci. Q. 257–281 (1961)
10. Chhetri, P., Afshan, N., Chatterjee, S.: The impact of perceived organizational politics on work attitudes: the moderating role of leader-member-exchange quality. In: Politics and Social Activism: Concepts, Methodologies, Tools, and Applications, pp. 1229–1242. IGI Global (2016)
11. Choudhury, R.R., Gupta, V.: Impact of age on pay satisfaction and job satisfaction leading to turnover intention: a study of young working professionals in India. Manag. Labour Stud. **36**(4), 353–363 (2011)
12. Cotton, J.L., Tuttle, J.M.: Employee turnover: a meta-analysis and review with implications for research. Acad. Manag. Rev. **11**(1), 55–70 (1986)
13. Cropanzano, R., Howes, J.C., et al.: The relationship of organizational politics and support to work behaviors, attitudes, and stress. J. Organ. Behav. Int. J. Ind. Occup. Organ. Psychol. Behav. **18**(2), 159–180 (1997)
14. Sophie, D.W., Slvbivs, M.E.: Employee turnover (volatility) and labor productivity. WSE Report 7 (2015)
15. Daskin, M., Tezer, M.: Organizational politics and turnover: an empirical research from hospitality industry. Tourism **60**(3), 273–291 (2012)
16. DeConinck, J.B., Stilwell, C.D.: Incorporating organizational justice, role states, pay satisfaction and supervisor satisfaction in a model of turnover intentions. J. Bus. Res. **57**(3), 225–231 (2004)
17. Falk, J., Karamcheva, N., et al.: Comparing the effects of current pay and defined benefit pensions on employee retention. Working paper 2018-06. Technical Report (2018)
18. Ferris, G.R., Kacmar, K.M.: Perceptions of organizational politics. J. Manag. **18**(1), 93–116 (1992)
19. Gandz, J., Murray, V.V.: The experience of workplace politics. Acad. Manag. J. **23**(2), 237–251 (1980)
20. Haque, A., Fernando, M., Caputi, P.: The relationship between responsible leadership and organisational commitment and the mediating effect of employee turnover intentions: an empirical study with Australian employees. J. Bus. Ethics 1–16 (2017)
21. Hochwarter, W.A., Perrewé, P.L., et al.: Commitment as an antidote to the tension and turnover consequences of organizational politics. J. Vocat. Behav. **55**(3), 277–297 (1999)
22. Hochwarter, W.A., Kiewitz, C., et al.: Positive affectivity and collective efficacy as moderators of the relationship between perceived politics and job satisfaction. J. Appl. Soc. Psychol. **33**(5), 1009–1035 (2003)
23. Kacmar, K.M., Andrews, M.C., et al.: Ethical leadership and subordinate outcomes: the mediating role of organizational politics and the moderating role of political skill. J. Bus. Ethics **115**(1), 33–44 (2013)
24. Kafeel, N., Alvi, A.K.: What play significant role in increasing turnover intention: job hopping or perceived organizational politics? (2015)

25. Khalid, J., Ahmed, J.: Perceived organizational politics and employee silence: supervisor trust as a moderator. J. Asia Pac. Econ. **21**(2), 174–195 (2016)
26. Kiewitz, C., Hochwarter, W.A., et al.: The role of psychological climate in neutralizing the effects of organizational politics on work outcomes. J. Appl. Soc. Psychol. **32**(6), 1189–1207 (2002)
27. Kim, J.K., Fu, Y.: The effect of perceptions of organizational politics on turnover intention and organizational citizenship behavior in Chinese convergency companies. J. Digit. Converg. **13**(8), 177–189 (2015)
28. Labrague, L., McEnroe-Petitte, D., Gloe, D., Tsaras, K., Arteche, D., Maldia, F.: Organizational politics, nurses' stress, burnout levels, turnover intention and job satisfaction. Int. Nurs. Rev. **64**(1), 109–116 (2017)
29. Lum, L., Kervin, J., et al.: Explaining nursing turnover intent: job satisfaction, pay satisfaction, or organizational commitment? J. Organ. Behav. Int. J. Ind. Occup. Organ. Psychol. Behav. **19**(3), 305–320 (1998)
30. Mayhew, R.: Employee turnover, definitions and calculations. The Houston (2013)
31. Mellat Parast, M., Fini, E.E.H.: The effect of productivity and quality on profitability in US airline industry: an empirical investigation. Manag. Serv. Q. Int. J. **20**(5), 458–474 (2010)
32. Mintzberg, H., Mintzberg, H.: Power In and Around Organizations, vol. 142. Prentice-Hall, Englewood Cliffs, NJ (1983)
33. Mobley, W.H.: Some unanswered questions in turnover and withdrawal research. Acad. Manag. Rev. **7**(1), 111–116 (1982)
34. Mobley, W.H., Griffeth, R.W., et al.: Review and conceptual analysis of the employee turnover process. Psychol. Bull. **86**(3), 493 (1979)
35. Moon, K., Bergemann, P, et al.: Manufacturing Productivity with Worker Turnover. Social Science Electronic Publishing (2018)
36. Naseer, S., Raja, U., et al.: Perils of being close to a bad leader in a bad environment: exploring the combined effects of despotic leadership, leader member exchange, and perceived organizational politics on behaviors. Leadersh. Q. **27**(1), 14–33 (2016)
37. Phillips, J.J., Connell, A.O.: Managing Employee Retention (2003)
38. Pless, N.M., Maak, T.: Responsible leadership: pathways to the future. In: Responsible Leadership. Springer, pp. 3–13 (2011)
39. Poon, J.M.: Situational antecedents and outcomes of organizational politics perceptions. J. Manag. Psychol. **18**(2), 138–155 (2003)
40. Sarkar, J.: Linking compensation and turnover: retrospection and future directions. IUP J. Organ. Behav. **17**(1) (2018)
41. Shafiq, M., Khan, N.U., et al.: Organizational justice mitigates adverse effects of perceived organizational politics on employees turnover intentions at workplace. J. Manag. Info **4**(1), 6–11 (2017)
42. Tekleab, A.G., Bartol, K.M., Liu, W.: Is it pay levels or pay raises that matter to fairness and turnover? J. Organ. Behav. Int. J. Ind. Occup. Organ. Psychol. Behav. **26**(8), 899–921 (2005)
43. Thornton, L.M., Esper, T.L., Autry, C.W.: Leader or lobbyist? How organizational politics and top supply chain manager political skill impacts supply chain orientation and internal integration. J. Supply Chain Manag. **52**(4), 42–62 (2016)
44. Vandenberghe, C., Tremblay, M.: The role of pay satisfaction and organizational commitment in turnover intentions: a two-sample study. J. Bus. Psychol. **22**(3), 275–286 (2008)
45. Walton, J.: Exploring reasons for employee turnover: a case study of the retail industry in Atlanta, Georgia. Ph.D. thesis, Northcentral University (2018)

Research on the Influence Mechanism of eWOM on Selection of Tourist Destinations—The Intermediary Role of Psychological Contract

Mo Chen[✉], Jingdong Chen, and Weixian Xue

Faculty of Economy and Management, Xi'an University of Technology, Xi'an 710054,
People's Republic of China
chenmo19921201@hotmail.com

Abstract. In the context of a growing amount of e-travel bookings, eWOM (electronic word-of-mouth), as a source of information, mainly affects the destination selection of tourists by influencing their psychologies. There is a "non-written" psychological contract for the customers and the service providers of the tourist destinations, which might influence the choice of tourists' destinations by eWOM. This paper aims to explore psychological contract's influence on the mechanism of eWOM on the selection of tourist destinations, also to improve the service quality of travel companies and other eWOM-related industries, and furthermore to establish an effective and incentive response mechanism for management of eWOMs in tourism destinations. With a sample of over 284 tourists from a wide range of research sites, this study positively supports the results and probes into the influence of the test which linked the tourist destination selection with eWOM. The results suggest that a psychological contract is playing an intermediary role in such relationships.

Keywords: eWOM · Tourist destination selection ·
Psychological contract · Service quality

1 Introduction

In the global economic downturn, the tourism market still maintains a strong growth rate. International tourist arrivals grew 7.0% in 2017, which is the highest increase since the 2009 global economic crisis and well above UNWTO's long-term forecast of 3.8% per year for the period of 2010–2020. Meanwhile, with the use of the internet, the development of tourism is closely related to the Internet. Online booking websites like Booking and Agoda give us platforms to learn about destination-related information and to purchase services. Moreover, there are also many tourists willing to share to travel experiences and photos on social networking sites (SNSs). The travel-related eWOM on SNSs could overcome spatiotemporal limitations and spread to all corners in the social network [14].

© Springer Nature Switzerland AG 2020
J. Xu et al. (Eds.): ICMSEM2019 2019, AISC 1002, pp. 654–667, 2020.
https://doi.org/10.1007/978-3-030-21255-1_50

Electronic word-of-mouth (eWOM) ranges from the post-purchase evaluation on travel reservation platforms to the reviews shared by users SNSs. eWOM offers a valuable perspective to analyze information dissemination and its influence on users and followers.

eWOM acts as carriers of travel information and is an important source of information for tourists. However, there are various kinds of word-of-mouth on the online platform, which means the authenticity and quality of information that cannot be guaranteed. Therefore, source trustworthiness could play important roles in consumer' decision making [20], especially after the emergence of Internet Water Army (a large amount of paid posters who write inauthentic reviews) which dramatically undermines the authenticity and value of user reviews [1]. Although some tourists intuitively comment on destinations and real tourism experiences both while traveling and in retrospect, the effectiveness of the reviews is not always trustworthy. There is more than a single expression of emotion in their eWOM. Some word-of-mouth has been forwarded and widely disseminated, which has a huge impact on tourists, while others have little influence on tourists' decision-making. Therefore, eWOM no longer affects the tourist destination choice intention only in positive or negative ways. More information needs to be filtered through the subjective judgment of the word-of-mouth listener. This paper aims to explore the psychological judgment process mechanism of tourist groups' decision making by reading eWOM information.

2   Literature Review and Hypotheses

eWOM is defined as any words or discussions regarding certain goods, services, or enterprises, either positive or negative, and that is accessible by anyone online [9]. eWOM has a unique feature compared with traditional face-to-face communication taking place in a virtual, eWOM occurs in an Internet-enabled environment. However, they share the same characteristic: recording information about products and services. There are two main research topics of eWOM and word-of-mouth. One is the research on the comment motivation of the originator of word-of-mouth, whose topics surrounded effects, altruism, self-interest, and reciprocation. The other is the research on the influence of the listener of word-of-mouth, whose theme focuses on customer loyalty, product evaluation, purchasing decision, consumer empowerment and product acceptance [13]. Thus, first and foremost, the subject of the study should be identified. Tourists were studied as word-of-mouth listeners.

In the research of eWOM listeners, it is found that psychological factors are critical. The research on customer satisfaction revealed that perceived value and risk has never stopped. There is research proved that the interaction effect of product involvement and perceived risk has a positive correlation on positive word-of-mouth intention and behavior [8].

Psychological contracts are broader than legal contracts, and they include perceptual, unwritten, and implicit terms that cannot be explicitly incorporated in a legal contract. There are several ways to categorize psychological contracts

[23]. A widely accepted typology views psychological contracts as either transactional psychological contracts or relational psychological contracts [22]. From the perspective of the formation of a psychological contract, social networks play vital roles in developing congruent psychological contracts among new employees [5]. The Internet made the social network easier than before, and as a result, the psychological contract of the network consumer groups are more closely. It is the reason that the study of a psychological contract should be introduced into this study. When consumers buy a product or service, it is no longer a simple individual decision. They form a consensus which is a "non-written" psychological contract in their online discussions.

However, where there is a psychological contract, there is a violation of the psychological contract. Based on the literature that has validated the fundamental role of PCV (psychological contract violation). With the deepening of marketing research, the study of psychological contract violation has been applied to help our understanding of buyer-seller relationships of online marketplaces [21].

2.1   Theoretical Framework

In the study of the eWOM information dissemination, the model of the adoption of the eWOM information is put forward by using the technology acceptance model for reference [24]. As shown in Fig. 1, eWOM as an informing effect on tourists' decision. When they decide on the destination, it means that they have adopted some of the WOM information online which they read. The process of tourism destination selection is a continuous screening process. At the same time, the tourists' judgment on the usefulness of eWOM is to analyze whether the information is related to the psychological contract with their mind. Finally, base on which they think related to the psychological contract, they judge whether their psychological contract reached or not, and make the decision on tourism destination.

Fig. 1. The conceptual framework of the research.

2.2   Hypothesis Development

From our literature review of the change in eWOMs' influence mechanism, the subjective perception from customers as the variable was shown to be an essential facilitator of change. The psychological contract is the judgment of consumer psychology. So, the psychological contract transforms the influence mechanism of the positive and negative eWOMs through the attitudes affecting consumer decision-making into the influence mechanism of consumers' subjective judgment on the degree of acceptance of eWOM information. eWOMs listener's level of any message acceptance depends on not only the characteristics of eWOMs' source and the content of eWOMs' quality, but also their psychological contract. Therefore, the following hypotheses are proposed.

H1: The quality of eWOM has a positive impact on the psychological contract for tourists.

The promises and obligations of service agents' psychological contracts are succinctly expressed as one side of the services triangle. It can be found that service agents should keep promises to their customers [25]. eWOM is the customer's feedback on service performance. The quality of eWOM is the quality evaluation of the content of eWOM. Hence, hypothesis 1 is proposed.

H2: The credibility of eWOM sources has a positive impact on the psychological contract of tourists.

Source credibility of consumer reviews is a commonly examined concept in the eWOM literature. The credibility of the eWOM source has been described as the degree of believed authenticity of the information, or as the willingness of the speaker to communicate authentically and genuinely. In the B2C market, user trust will affect the social contract structure [16]. The psychological contract is also a kind of contract. Distorted information has an impact on contracts, so the authenticity of information has an impact on psychological contracts.

H3: Psychological contract has a positive effect on tourists' willingness to choose a tourist destination.

Psychological contract violation with the community of sellers in a marketplace will decrease the buyer's perceived effectiveness of the marketplace's institutional structures [21]. According to this study, it can be proposed that a psychological contract can promote tourists' willingness to purchase tourist destination services so that they will choose to travel there in the future.

H4: The quality of eWOM information has a positive impact on the willingness to choose a tourist destination.

The intention of tourism destination choice is necessarily the willingness of tourists to buy the products and services provided by the tourism destination. Therefore, the promotion of purchase intention to the choice of tourist destination can explain that the quality of online word-of-mouth may play a decisive role in the choice of a tourist destination. The quality of eWOM is mainly measured from the perspective of information characteristics, such as their relevance, adequacy, objectivity and so on. Higher quality reviews will increase consumers' willingness to buy [11].

H5: The credibility of eWOM sources has a positive impact on the willingness to choose a tourist destination.

The reliability of the network community as the eWOM concentration place will affect the effectiveness of the eWOM in the community [19]. The credibility of eWOM sources is affected by the strength of the relationship between the source and the recipient, the eWOMs communication platform, the perceived usefulness of the website and the perceived risk and the consumers' trust. The consumer trust in an online shopping website positively predicts their perceived credibility of the consumer reviews posted on the site, suggesting a transfer of perceived trustworthiness from the shopping site to the consumer reviews [12].

H6: The effect of the quality of eWOM sources on the selection of tourist destination is partially mediated by psychological contract.

H7: The effect of credibility of eWOM sources on the selection of tourist destination is partially mediated by psychological contract.

Several researchers have explored the mediating effect that attitude has in the relationship of eWOM and purchase intention. However, there is almost no test for the mediating effect of the psychological contract. The mediating effect that effective response and cognitive response have in the relationship of eWOM and purchase intention has been demonstrated [6]. The two dimensions of the psychological contract (transactional psychological contract and relational psychological contract) can also be understood as two dimensions of the cognitive and affective division. The transactional psychological contract is a kind of contract that emphasizes the exchange of tangible wealth, the specific transaction content and the relative short-term contract. Compared to transactional psychological contracts, relational psychological contracts emphasize not only the material return but also the satisfaction of the customer emotional needs.

3   Methodology

For structural equation modeling (SEM), a minimum sample size of 200 is recommended for accurate analysis. Therefore, we decided a sample size of 300 for this study. These study participants were internet users in China who had been involved in word-of-mouth communication regarding electronic products. We used Wechat and Weibo to disseminate our online questionnaire. Finally, there were 284 complete responses collected. The questionnaire is improved from the existing scale to make it more suitable for tourism scenarios. Descriptive statistical analysis of recovered sample data is as follow Table 1.

Due to the nature of the questionnaire distribution platform, a large number of college students and graduate students responded to the questionnaire. That is the reason why there are 226 respondents in the 19–30 age group. However, it does not make much difference to the results in our research. A majority of they reported that they favored word-of-mouth as a source (50.5%) to learn about product information over other marketing communication sources [30].

This questionnaire is divided into three parts. The first part is the respondent demographics which include respondents' gender, age, family monthly income, educational background. The second part investigates the internet habits of respondents (e.g., Which platform do you commonly use for tourist destinations eWOMs?). It is in order to help them recall the scenes of reading eWOMs (e.g., Which tourist destination have you been searching lately on your most commonly used online platform?). In this part, the subjects will focus on a tourist destination that they have planned to visit in the future. According to the existing research, the distance between tourists and tourist destinations [18], the image of tourist destinations [2] and other factors will affect the choice of

Table 1. Sample descriptive statistical analysis.

Demographic characteristic	n	Percent
Gender		
Male	125	44.01
Female	159	55.99
Age		
<18 years	6	2.11
19–30 years	226	79.6
31–40 years	9	3.17
41–50 years	18	6.34
>50 years	25	8.80
The highest level of education achieved		
High school or technical secondary school	29	15.76
University or college	144	50.70
Postgraduate	111	39.08
Family monthly income (RMB)		
1000–4999	114	40.14
5000–9000	90	31.69
>9000	80	28.17
Average number of hours spent online per day		
<2 h	43	15.14
2–5 h	138	48.59
5–8 h	66	23.24
>8 h	37	13.03
Commonly used website (multiple selection∗)		
SNS (social networking sites)	227	79.93
Booking websites	143	50.35
Tourism forum	60	21.13

Profile of respondents (n = 284)

tourist destinations. However, this study focuses on the impact of online word-of-mouth on the choice of a tourist destination and only studies the changes in tourism willingness of the tourist destination that customers pay attention to after reading eWOM. That is the reason for designing this part. The third part, as Table 2. Shows, the abbreviations of the variables in brackets, is consisted of five-point Likert scales, ranging from 1 (strongly disagree) to 5 (strongly agree), and was used to measure the unobserved variables. At present, the psychological contract scales selected in this research are widely used in the Chinese literature review of the research service and product purchase, which avoids Chinese respondents misunderstanding in the translation of the English version of the psychological contract scale.

Table 2. Questionnaire items.

Variable	Item	Loadings	CR	AVE	Source
Quality of eWOM(Q)	High content relevance	0.717	0.953	0.508	Lee (2007) [21]
	The date of publication is very timely	0.741	14.899		
	Meaning accuracy	0.738	12.039		
	The content is complete	0.747	11.227		
Source credibility of eWOM(S)	The source of eWOMs is reliable	0.751	10.994	0.873	Cheung (2008) [3]
	The source of eWOMs is trustworthy	0.778	12.238		
	The source of eWOMs is professional	0.744	13.232		
Psychological contracts(P)	Transactional psychological contracts (T)			0.646	Haicheng Luo (2005) [27]
	I believe the prices in this tourist destination are reasonable and reasonable	0.768	10.901		
	I believe the tourist destination can provide a convenient and quick service	0.847	14.975		
	I believe I can experience a wealth of entertainment in the resort	0.828	16.464		
	Relational psychological contracts (R)			0.734	
	I believe the tourist destination regulatory authorities will correct the mistakes in time	0.805	12.373		
	I believe the services provided by the tourist destination are assured and reliable	0.865	15.416		
	I believe I feel respect and friendship when I travel to this tourist destination	0.808	14.738		
	I believe the tourist destination related services not only care about my travel experience but also attach great importance to my feedback	0.755	11.064		
Destination choice intention(D)	I am going to travel to a destination in the future	0.596	12.029	0.649	Lee (2005) [28]
	In the future, I will recommend the destination to my relatives and friends	0.651	12.391		
	I will revisit the destination in the future	0.668	12.825		

AVE average variance extracted, *CR* composite reliability

CFA was conducted to assess internal consistency (reliability), item loadings (convergent validity), and discriminant validity. Each item was modeled as a reflective indicator of its hypothesized latent construct. The constructs were allowed to co-vary in the CFA model. The measurement models were evaluated using maximum likelihood estimation. Hair (2010) [29] suggested that the loadings, average variance extracted and composite reliability should be used to assess convergent validity. They suggested that loadings should be >0.5, AVE > 0.5, and CR > 0.7 to achieve convergent validity. The results of this analysis, shown in Table 2, indicated that all loadings for the model were more than 0.5, the AVE were all more than 0.5, and the composite reliabilities were all above 0.7. Thus convergent validity was confirmed.

4   Results

4.1   Structural Model

The Cronbach's value for this measure which has 17 items is 0.938. Descriptive statistics, including zero-order correlations between the key variables, are reported in Table 3. The correlation results indicated construct validity, as all variables are positively correlated as anticipated. A structural equation modeling technique was utilized to test research hypotheses and explore research questions.

Table 3. Correlation of latent variables (P < 0.01**)

	Q	S	P	D
Q	1	0.710**	0.660**	0.547**
S	0.710**	1	0.694**	0.510**
P	0.660**	0.694**	1	0.572**
D	0.547**	0.510**	0.572**	1

The initial path model demonstrated a relatively poor model fit ($\chi^2/df = 2.742, \text{GFI} = 0.883, \text{AGFI} = 0.841, \text{CFI} = 0.933, \text{RMSEA} = 0.080$), with two paths failing to achieve statistical significance. Because these two indexes (RMSEA > 0.05, GFI < 0.90) are not suitable, a new modified model was obtained by adding a new significant correlation path between the residual and the initial model. These paths were found between the following variable pairs: (1) the quality of eWOM and psychological contracts ($\beta = 0.274, p = 0.142$), and (2) the credibility of eWOM sources and the choice of tourism destination ($\beta = 0.272, p = 0.200$). A revised model corrects the correlated error (see Fig. 2), which resulted in a good model fit.

A close look shows that the quality of eWOM was positively related ($\beta = 0.262, p < 0.001$) to psychological contracts, and so was the credibility of eWOM sources ($\beta = 0.416, p < 0.001$). Thus, H1 and H2 were validated. The relationship between the credibility of eWOM sources and the selection

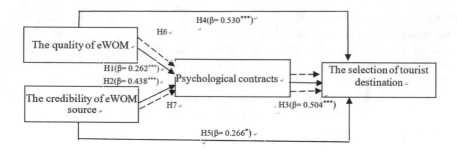

Fig. 2. Revised model $(p < 0.05*, p < 0.01 * *, p < 0.001 * **)$

of tourism destination was positive $(\beta = 0.530, p < 0.001)$, as was the relationship between psychological contracts and the choice of tourism destination was positive $(\beta = 0.504, p < 0.001)$. Therefore, H5 and H3 were supported. The credibility of the eWOM source was positively related $(\beta = 0.262, p < 0.05)$ and the selection of tourism destination were positive.

To test the sixth and seventh hypotheses we used the Sobel test. The multicollinearity of variables is tested before regression. The result of multicollinearity VIF (Variance Inflation Factor) mean = 1.95 > 1, but VIFmax = 2.93 < 10 (see Table 4). Thus, it proved that the data could be regressed without a serious problem of approximating multicollinearity. The Sobel test and the Aroian test seemed to perform best in a Monte Carlo study [15] and converge closely with sample sizes greater than 50 or so. In the sixth hypothesis, the hypothesis of H0 can be rejected, because its Z = 9.603 > 1.96, (a = 1.508, Sa = 0.107, b = 0.328, Sb = 0.025). Moreover, for the same reason, H7 was supported (Z = 9.975 > 1.96, a = 1.474, Sa = 0.096, b = 0.328, Sb = 0.025).

Thus, It can be concluded that the psychological contract has a mediation effect on the quality of eWOM sources and the selection of tourism destination. It still plays an intermediary role to affect the quality of eWOM sources and the selection of tourism destination.

Table 4. Result table of multicollinear

Variables	VIF	1/VIF
Q	2.33	0.429295
S	2.75	0.364278
P	2.93	0.340937
D	2.43	0.411159
Gender	1.06	0.942351
Education	1.02	0.976079
Income	1.10	0.907421
VIF mean	1.95	

4.2   Control Variable Analysis

In the study, if the control variable is not controlled, some changes will occur to the dependent variable. Therefore, only by controlling all variables other than independent, can we find out the causal relationship in the experiment. Finally, the control variables selected in this paper should be analyzed.

According to existing studies, demographic characteristics will have an impact on customers sharing their traveling experiences and the influence of other users on the internet [10]. Therefore this paper selected demographic characteristics as a control variable (education background, gender, and income) to set up a model. This section will explore whether these control variables affect eWOMs information (the quality of eWOM and the source credibility of eWOM), the psychological contract and the selection of tourist destination, by using Student's t-test and ANOVA (Analysis of Variance).

The results of the three control variables are obtained. (1) After the analysis of variance, there is no significant difference in the impact of education background on eWOM information ($F = 0.211, p = 0.810 > 0.05$), the psychological contract ($F = 323, p = 0.724 > 0.05$) and the selection of tourist destination($F = 0.180, p = 0.836 > 0.05$). (2) There was a difference in gender in the choice of tourist destinations in the t-test ($T = -2.424, P = 0.016 < 0.05$), which also indicated that the control variables were valid. (3) There is a difference in the perception of whether a psychological contract is reached by tourists of different income levels ($1 = 1000 - 4999, 2 = 5000 - 9000, 3 => 9000, F = 2.027, P = 0.044 < 0.05$).

5   Discussion and Conclusion

This study has attempted to explore eWOM and its impacts on tourists' choices of destination. The intermediary role of the psychological contract has been found. The result provides a deeper and richer portrait of the relationship between eWOM and destination selections. The proof hypothesized (see Table 5) in the paper can enrich the impact mechanism of tourism consumer behavior and decision-making, and better illustrate that the research of eWOM is complementary to the traditional marketing and promotion strategies.

Firstly, the influence of the quality of eWOM on the intention to choose a tourist destination is more significant than the influence of the source credibility of eWOM on the intention to choose a tourist destination. This research conclusion is also reflected in the studies of Cheung (2008) [3], Eagly (1993) [4] all of which have proved that the information quality of eWOM directly affects consumers' purchase intention.

Secondly, compared with the source credibility of eWOM, the eWOM quality has a higher impact on the selection of tourism destination ($\beta = 0.405$). It shows that the credibility of the eWOM source is more critical than eWOMs quality for tourists' psychological contracts. With the increasingly complex online communication environment, the credibility of eWOM sources is not as convincing as the perception of the quality of word-of-mouth itself.

Thirdly, in this paper, the psychological contract has a positive effect on the tourists' willingness to choose tourist destinations. A reverse proof of the conclusion of another study of Jinhong Gong (2014) [7] proves the positive effect of psychological contract in tourism situation. However, the psychological contract in a tourist's mind is changeable. From the psychological contract shows what visitors pay attention to at this stage. For different periods, tourists pay attention to the content of psychological contract and focus on the dimensions of a tourism destination is not necessarily the same.

The fourth, the psychological contract has a mediation effect on the quality of eWOM sources and the selection of tourism destination. Meanwhile, the psychological contract has a mediation effect on the credibility of the eWOM source and the selection of tourism destination. Also, previous studies showed that eWOMs affected consumer behavior by consumer attitudes. The results show that the influence path of online word-of-mouth to consumer behavior can also be considered from the perspective of the psychological contract.

Table 5. Hypotheses testing. ($p < 0.05*, p < 0.01**, p < 0.001***$)

Relationships		Unstd. estimate	Std. estimate	S.E.	C.R.	P	Supported
H1	P<—Q	0.259	0.262	0.192	1.352	***	Supported
H2	P<—S	0.405	0.438	0.179	2.262	***	Supported
H3	D<—P	0.621	0.504	0.262	2.368	***	Supported
H4	D<—Q	0.660	0.530	0.118	5.584	***	Supported
H5	D<—S	0.306	0.266	0.243	1.262	0.017	Supported

Model fit: $\chi^2/df = 2.351$, GFI $= 0.903$, AGFI $= 0.867$, TLI $= 0.919$, CFI $= 0.949$, RMSEA $= 0.050$, RMR $= 0.49$

From the perspective of marketing, relevant tourism management departments of tourist destinations, enterprises, and businesses of tourist destinations should take the following measures to improve the image of tourist destinations better and achieve successful marketing of tourist destinations: (1) Encourage visitors to write high-quality eWOMs. (2) Extract useful information and lesson out of eWOMs and gain insights to improve their services [17]. (3) Timely understand and update the content of tourists' psychological contract.

At the same time, as tourists, we should leave a high-quality eWOM after traveling. Only in this way can our travel experience help other tourists to understand the specific situation of the tourist destination, and to build a good Internet interactive environment between tourists.

6   Limitations and Implications

The study has some limitations that suggest opportunities for future research.

(1) Sample selection is influenced by the distribution platform and audience, resulting in a smaller number of people aged 31–40. In future research, we can expand the sample size, to achieve universal empirical research conclusions.
(2) Although the psychological contract scale is mainly used to improve the existing scale produced, the scale of the context of tourism between the two dimensions of discrimination is not high. The psychological contract scale of tourism destination can be better tested and improved in the future, and the characteristics of the scale can add.
(3) People's psychological contract may be different because of different laws and regulations, social and cultural background. Future research can expand sample size, increasing the survey object of different countries, and comparing the similarities and differences of the research results.
(4) In the future, we can conduct in-depth research based on the effects of control variables. Due to the sample limitations of the questionnaire, some demographic variables under the questionnaire sample size is small, which will affect the study of control variables. In the future, we can study the moderating effect of control variables (e.g. sex) on the whole model.

Acknowledgements. This paper is supported by the Shaanxi Province Social Science Fund (No. 2016R015) and Research on Promoting to Construct Cultural Powerhouse by Cultural-oriented Tourism-Based on Customer Marketing No. 2016R015).

References

1. Guo, B., Wang, H., et al.: Detecting the internet water army via comprehensive behavioral features using large-scale e-commerce reviews. In: 2017 International Conference on Computer, Information and Telecommunication Systems (CITS), pp. 88–92 (2017)
2. Chen, N., Funk, D.C.: Exploring destination image, experience and revisit intention: a comparison of sport and non-sport tourist perceptions. J. Sport Tour. **15**(3), 239–259 (2010)
3. Cheung, C.M.K., Lee, M.K.O., Rabjohn, N.: The impact of electronic word-of-mouth. Internet Res. **18**(3), 229–247 (2008a)
4. Eagly, A.H., Chaiken, S.: The Psychology of Attitudes. Harcourt, San Diego (1993)
5. Engel, K.L., Cortina, J.M.: The influence of one's social network on psychological contract formation. Diss. Theses Gradworks **14**(3), 115–124 (2009)
6. Gao, L., Li, W., Ke, Y.: The influence of online word-of-mouth on consumers' purchase intention in social commerce: the mediating role of emotional response and the moderating role of curiosity. J. Manag. Eng. **4** (in Chinese) (2017)
7. Gong, J., Xie, L., Peng, J.: How travel service dishonesty affects customer trust: the role of psychological contract violation and corporate reputation. Tour. Trib. **29**(4), 55–68 (2014). (in Chinese)

8. Ha, H.Y.: The effects of consumer risk perception on pre-purchase information in online auctions: brand, word-of-mouth, and customized information. J. Comput. Mediat. Commun. **8**(1) (2010)
9. Hennig-Thurau, T., Gwinner, K.P., Walsh, G., Gremler, D.D.: Electronic word-of-mouth via consumer-opinion platforms: what motivates consumers to articulate themselves on the internet? J. Interact. Mark. **18**(1), 38–52 (2010)
10. Hernndezmndez, J., Muñozleiva, F., Snchezfernndez, J.: The influence of e-word-of-mouth on travel decision-making: consumer profiles. Curr. Issues Tour. **18**(11), 1001–1021 (2015)
11. Jimnez, F.R., Mendoza, N.A.: Too popular to ignore: the influence of online reviews on purchase intentions of search and experience products. J. Interact. Mark. **27**(3), 226–235 (2013)
12. Lee, J., Park, D.H., Han, I.: The different effects of online consumer reviews on consumers' purchase intentions depending on trust in online shopping malls. Internet Res. **21**(2), 187–206 (2011)
13. Litvin, S.W., Goldsmith, R.E., Pan, B.: Electronic word-of-mouth in hospitality and tourism management. Tour. Manag. **29**(3), 458–468 (2008)
14. Luo, Q., Zhong, D.: Using social network analysis to explain communication characteristics of travel-related electronic word-of-mouth on social networking sites. Tour. Manag. **46**(46), 274–282 (2015)
15. Mackinnon, D.P., Warsi, G., Dwyer, J.H.: A simulation study of mediated effect measures. Multivariate Behav Res **30**(1), 41 (1995)
16. Magrane, J.R.: Personal information sharing with major user concerns in the online B2C market. In: ACM SIGMIS Conference on Computers & People Research (2015)
17. Mellinas, J.P., Reino, S.: eWOM: The Importance of Reviews and Ratings in Tourism Marketing (2018)
18. Nicolau, J.L., Ms, F.J.: The influence of distance and prices on the choice of tourist destinations: the moderating role of motivations. Tour. Manag. **27**(5), 982–996 (2006)
19. Park, D.H., Lee, J., Han, I.: The effect of on-line consumer reviews on consumer purchasing intention: the moderating role of involvement. Int. J. Electron. Commerc. **11**(4), 125–148 (2007)
20. Park, H.L., Zheng, X., Josiam, B., Kim, H.M.: Personal Profile Information as Cues of Credibility in Online Travel Reviews (2013)
21. Pavlou, P.A., Gefen, D.: Psychological contract violation in online marketplaces: antecedents, consequences, and moderating role. Inf. Syst. Res. **16**(4), 372–399 (2005)
22. Rousseau, D.M., Tijoriwala, S.A.: Assessing psychological contracts: issues, alternatives and measures. J. Organ. Behav. **19**(S1), 679–695 (2010)
23. Sels, L., Janssens, M., Brande, I.V.D.: Assessing the nature of psychological contracts: a validation of six dimensions. J. Organ. Behav. **25**(4), 461C488 (2004)
24. Sussman, S.W., Siegal, W.S.: Informational influence in organizations: an integrated approach to knowledge adoption. Inf. Syst. Res. **14**(1), 47–65 (2003)
25. Zeithaml, V., Bitner, M.J.: Services marketing. McGraw-Hill, New York (2000)
26. Lee, E.J.: Deindividuation effects on group polarization in computer-mediated communication: the role of group identification, public-self-awareness, and perceived argument quality. J. Commun. **57**(2), 385–403 (2010)
27. Luo, H.: The psychological contract in the marketing context and its measurement. Bus. Econ. Manag. **25**(6), 37–41 (2010). (in Chinese)

28. Lee, C., Lee, Y., Lee, B.: Korea's destination image formed by the 2002 World Cup. Ann. Tour. Res. **32**(4), 839–858 (2005)
29. Hair, J.F., Black, W.C., Babin, B.J.: Multivariate Data Analysis: A Global Perspective. Pearson Education (2010)
30. On Campus Research: Behavior and Trends of Student Consumers. National Association of Campus Stores (2012)

A Study on the Impact of Customer Engagement on Continued Purchase Intention for Online Video Websites VIP Service

Jingdong Chen[✉] and Wenxin Xu

Faculty of Economy and Management, Xi'an University of Technology, Xi'an 710054, People's Republic of China
912321926@qq.com

Abstract. Along with the development of network video technology and the gradually improvement of users' demand, the online video website user-paid market has shown great potential and vitality as an emerging blue ocean. In order to explore users' intention to purchase online video websites VIP service continuously, the SEM is utilized to carry out an empirical research on the drive and influence of customer engagement on continued purchase intention based on the multi-dimensional customer engagement perspective of cognition, emotion and behavior with the introduction of perceived value as a mediator variable. Empirical results show that customer's emotion and behavior engagement have positive impacts on perceived value; utilitarian value and hedonic value of perceived value have significant positive impacts on continued purchase intention. While cognition and emotion as well as behavior engagement also have directly promoting effects on continued purchase intention. Focusing on multi-dimension customer engagement, the research has not only enriched the relevant theories of continuous purchase intention, but also provided effective theoretical guidance for improvement of online video websites VIP service.

Keywords: Customer engagement · Continued purchase intention · Perceived value · VIP service · SEM

1 Introduction

With the maturity of Internet technologies and the rapid spread of mobile devices, various Internet applications are constantly emerging. Among many Internet products, network video has become an important way for people to entertain and acquire information with its intuitive display and rich content. In recent years, along with the enhancement of copyright protection awareness of film and television works in the online video market, Chinese video websites have begun to develop at a high speed. After nearly a decade of market testing,

© Springer Nature Switzerland AG 2020
J. Xu et al. (Eds.): ICMSEM2019 2019, AISC 1002, pp. 668–682, 2020.
https://doi.org/10.1007/978-3-030-21255-1_51

online video website VIP services have gradually become a trend. Therefore, it is necessary to enrich the academic research in this context. By browsing the relevant research, most scholars regard video VIP service as one of the profit methods of video websites, taking video website as the research perspective to explain the production, development status and problems. However, few scholars explore the VIP purchase intention from the perspective of users. It is imminent for video website operators to study which factors affect VIP services purchase intention, especially affecting continued purchase intention.

With the development of SNS, customer's demand has begun to be diverse and versatile. When customers purchase the product or service, they pay more attention to their perceived value, which is realized by the interaction with sellers (i.e. enterprises). Thus, a closer relationship between customers and businesses is required. In fact, the relationship between customers and enterprises is no longer limited to simple trading relationships, service-dominant (S-D) logic and relationship marketing paradigms are being strengthened, which raises the concept of "customer engagement". The existing literature has in-depth researches on the concepts and dimensions of customer engagement, which are mainly derived from the theory. For example, Forrester Consulting defined customer engagement in 2008 as "establishing in-depth relationships with customers to drive purchase decisions, interaction and participation". At present, the empirical research of customer engagement is mainly to explore and verify the antecedent and consequences. Marbach et al. [15] analyzed the positive impact of customer engagement on customer perceived value; most scholars believe satisfaction, loyalty, trust, attachment, self, brand connection, etc. may be the consequences of customer engagement. Multiple sources indicate that these factors have a positive impact on users' continued use behavior.

Based on this, combining the characteristics of online video websites, this study explores the impact of customer engagement on continued purchase intention for online video websites VIP service with the introduction of perceived value as a mediator variable, through literature analysis, questionnaire survey and data analysis. From the perspective of customers, the research extends the empirical research of customer engagement to the emerging online video website market, enriching the academic research results of customer engagement. In addition, it provides theoretical guidance for further development of online video websites VIP service.

2   Literature Review

2.1   Customer Engagement (CE)

The use of the term "engagement" has been traced back to the 17th century, when it was used to describe a number of notions, including a moral or legal obligation, tie of duty, betrothal, employment, and/or military conflict [5]. In the last two decades, the term "engagement" has been used extensively in fields including psychology, sociology, political science, and organizational behavior. For example, while "civic engagement" has been studied in sociology, "social engagement" in the field of psychology, "student engagement" in educational psychology, "employee engagement" and "stakeholder engagement" in the organizational behavior/management literature [7]. The initial use of the term "engagement" in the business practice discourse was traced back to Appelbaum [1]. He pointed out that CE consists of rational loyalty and emotional attachment. Since then, "engagement" has quickly become a hotspot in the field of marketing, which has formed a series of concepts, such as customer engagement, consumer engagement, customer engagement behaviors, and customer branding engagement.

The concept of CE is defined differently in business practitioners and academics. From the perspective of organization, the business practitioners define CE as "enhanced interaction and promotes the emotional, psychological or material investment activities of customers in the brand". However, the academics interpret CE as a systematic approach to the tendency of customers with organizational representatives and other customers to participate in collaborative knowledge exchange processes. The conceptualizations identified in a literature review are summarized in Table 1.

Table 1. Engagement conceptualizations and dimensionality in the marketing literature

Authors	Concept	Definition
Patterson et al. [17]	Customer engagement	The level of a customer's physical, cognitive, and emotional presence in their relationship
Pham and Avnet [18]	Engagement behavior	Finds that engagement "seems to be inferred from a pattern of action or withdrawal with respect to a target object"
Higgins and scholer [8]	Engagement	A state of being involved, occupied, fully absorbed or engrossed in something (i.e. sustained attention), generating the consequences of a particular attraction or repulsion force. The more engaged individuals are to approach or repel a target, the more value is added to or subtracted from it
Vivek et al. [20]	Consumer engagement	The intensity of an individual's participation & connection with the organization's offerings & activities initiated by either the customer or the organization
Holllebeek [10]	Customer brand engagement	The level of a customer's motivational, brand-related and context-dependent state of mind characterized by specific levels of cognitive, emotional, and behavioral activity in brand interactions

Note: Source: According to relevant literature

When it comes to CE measurement, we need to define its dimensions. The existing academic research of the CE dimension is divided into two major categories: single/multiple dimensions. 40% of scholars identify CE as a single dimensional concept (Brodie 2011) [5]. The definition of single dimension is divided into two types: the emphasis on the psychological or behavior level. For example, Pham and Avnet [18] mainly defines CE from the perspective of behavior, and believes that measuring with behavior dimension is beneficial to observation analysis. Beckers et al. [3] believe that CE reflects the changes in the customer's psychological activities, which drive the behavior. More scholars believe that multi-dimensionality can better reflect the meaning of CE. Patterson et al. [17] proposed four specific CE components, including absorption, dedication, vigor and interaction; Vivek [20] proposed that CE includes three dimensions: enthusiasm, conscious participation and interaction. Hollebeek [11] proposes three dimensions of cognition, emotion and behavior of CE. The level of customer-enterprise or customer-brand engagement is different in different environments. Among them, the three-dimensional definition of cognition, emotion and behavior is more recognized, gradually becoming the standard to measure CE.

2.2   Perceived Value (PV)

The concept of "perceived value" has been favored by scholars in various fields since its appearance, especially in marketing field. PV stems from consumer behavior science, and it is "the overall assessment of the effectiveness of a product by consumers through perceived revenue and perceived sacrifice" [24]. Regarding the definition of PV, different scholars have different definition with different perspectives. In summary, there are two main viewpoints: rational and perceptual. The rational view believes customers pursue the maximization of product/service utility (i.e. focus on functionality). However, the perceptual propose customer consumption is a subjective consciousness, accompanied by sundry symbolic and aesthetic principles (i.e. focus on hedonicity [22]).

Researches about PV focus on the measurement and driving factors, the definition of perceived value dimension is the key procedure. At present, most scholars agree with the multi-dimensional theory of perceived value, and build a multi-model of perceived value. For example, Sheth et al. [19] proposed five specific PV components: functional value, social value, emotional value, cognitive value, and conditional value; Holbrook [9] believes that the dimension of PV includes eight types: convenience, quality, success, reputation, taste, appearance, morality and faith. Babin et al. [2] argue that there are two motives to pursue rewards, one is conscious pursuit of practical results, the other is related to unconscious, immediate sensory responses. Therefore, utilitarian value (UV) and hedonic value (HV) can be used to indicate the reward for shopping.

2.3   Continued Purchase Intention

Purchase intention or purchase behavior has always been a popular research content in marketing field. According to Fishbein's definition, purchase intent is the subjective probability or possibility of a consumer purchasing a particular product [14]. With the development of Internet economy and the popularity of B2C/B2B platforms, just focusing on purchasing is not enough to cope with the volatile market environment [23]. At present, the problem that enterprises need to pay attention to is to attract and retain customers, which is also the focus of academic research. Therefore, the concept of "continued purchase intention" has become the key in marketing research field.

Concepts similar to continued purchase intention include users' continuance intention (UCI) and consumer's repurchase intention (CRI). UCI originated from Information System Continuance, which is an extension of Information System Acceptance Theory and a new development in the study of the adoption and use of information systems. When it comes to consumer satisfaction and post-purchase behavior, Expectation Confirmation Theory believes that repurchase intention is the key to whether the customers/consumers are willing to purchase the product or use the service again. Conceptually, there is no essential difference between UCI and CRI, which are used to study different objects in different situations. Compared to UCI and CRI, continued purchase intention not only pays attention to number of purchases, but also is on a stable relationship between the customer and shopping system. Only when customers resonate with the company providing the product or service cognitively or emotionally and the product or service meets their functional or emotional needs will they have a continuous willingness to purchase. Jones [13] believes that customers will form an intention to purchase again based on the actual perceived value after purchasing or using a certain product and service. In 1998, the scholar also proposed a theoretical model of the UCI.

3   Research Hypothesis and Model Construction

3.1   Customer Engagement and Continued Purchase Intention

According to Vivek's definition about CE dimension, CE is a combination of cognition engagement, emotion engagement and behavior engagement. According to the theoretical construction of CE from majority of scholars, loyalty, trust, attachment, etc. may be CE consequences, which have a positive impact on customer purchase intention from many documents, so it can be speculated that CE has a positive impact on continued purchase intention. In addition, some scholars found that consumers' active participation and promotion have formed an online community commitment through researching on brand communities of social networking sites, which affected purchase and word-of-mouth marketing [21]. The reason why a customer purchases or continuously purchases VIP service on video websites is that VIP service is required, and the essential need is a kind of feeling or perception (i.e. the so-called cognition engagement); on the other hand, customers believe that they will get emotional satisfaction and comfort, so that they are willing to continue to buy, and even recommend to friends or families (i.e. the emotion and behavior engagement). Based on the theoretical derivation, the authors hypothesize the following:

H1a: Cognition engagement has a significant positive impact on continued purchase intention;

H1b: Emotion engagement has a significant positive impact on continued purchase intention;

H1c: Behavior engagement has a significant positive impact on continued purchase intention.

3.2   Customer Engagement and Perceived Value

Utilitarian value refers to a reasonable and efficient way to judge the effectiveness of a product or service in a rational and efficient manner in the consumer experience. Chan proposes that in the co-creation activities of a service sector, the quality of service, the customized service and the high degree of customer control can be perceived through the economic value of customer experience. The study found that the customer's motivation is dependent on its expected value, the highly interactive and participatory nature of the engagement affects the customer perceived value. When both parties believe that each other is reliable, it will generate trust, then form the engagement. The higher degree of engagement, the better customers can be satisfied in the interaction with enterprise. This satisfaction may be material or spiritual, that is, gaining perceived value (utilitarian/hedonic value).

H2a Cognition engagement has a significant positive impact on UV

H2b Emotion engagement has a significant positive impact on UV

H2c Behavior engagement has a significant positive impact on UV

$H2a'$ Cognition engagement has a significant positive impact on HV

$H2b'$ Emotion engagement has a significant positive impact on HV

$H2c'$ Behavior engagement has a significant positive impact on HV.

3.3   Perceived Value and Continued Purchase Intention

Dodds and Monroe[6] pointed out that the willingness of an individual to perform a certain behavior is directly affected by the perceived value of the outcome of the behavior in "value-intention" model. Hsiao et al.[12] prove that customer perceived value has a direct impact on customers' repeated purchase intention, using the continuous use of mobile advertising as background. Overby and Lee[16] divide perceived value into two dimensions of utilitarian value and hedonic value, and apply it to the study of customer online shopping behavior. The results show that utilitarian value has more influence on purchase intention than hedonic value. Customers will be measured by previous purchase experiences before making a continuous purchase decision. Only when the customer feels practical and emotional satisfaction will there be a willingness to continue to purchase. This satisfaction is the so-called perceived value.

H3a: UV has a significant positive impact on continued purchase intention;

H3b: HV has a significant positive impact on continued purchase intention.

Based on the hypotheses, the theoretical model is shown in Fig. 1.

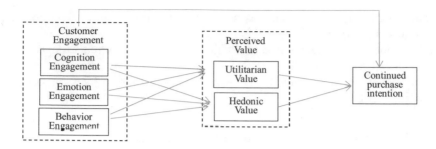

Fig. 1. Theoretical model

4   Research Design and Empirical Analysis

4.1   Variable Design and Measurement

A structured questionnaire was used to measure the constructs (after refinement and instructions adapted to the characteristics of research background). The questionnaire consists of two parts: demographic characteristics and measurement items. This paper takes the online video website VIP service as the object, and designs variable measurement items, involving CE, PV and continued purchase intention. Measures of CE was created for the current study based upon the scale of Hollebeek[11]. Items for PV was based upon the scale in Overby and Lee[16]. The scale of continued purchase intention is based on Bhattacherjee[4] and Lin et al. (2012). After the pre-test and adjustment, the questionnaire determines the final items (see Table 2).

4.2   Data Collection and Statistical Analysis

This article aims to study the VIP purchase on various online video sites, including Youku, Tencent Video, iQIYI, Mango TV et al. Through the distribution of questionnaires in the interactive areas of these video websites and social platforms such as QQ, WeChat and Weibo, a total of 350 copies were published, and 331 responses were collected, of which 303 were valid (60% female; 62% undergraduate; 63% purchased with 50% purchased monthly). The effective recovery rate was 87%. The results of the descriptive statistical analysis are shown in Table 3.

4.3   Reliability and Validity Analysis

The three indicators of the reliability test are Cronbach's α, Composite Reliability (CR) and Average Variance Extracted (AVE). The criterion is: Cronbach's $\alpha > 0.7$; $CR > 0.7$; $AVE > 0.5$. The reliability and validity test results of the measurement model in this paper are shown in Table 4. It can be seen from the above that the reliability test analysis results meet the above criteria, indicating that the scale has good internal consistency.

Table 2. Variable measurement scale

Variable	Items	Authors
Cognition	COG1: Using VIP gets me to think about that video website	Hollebeek [11]
	COG2: I think about online video website a lot when I'm using it	
	COG3: Using VIP service stimulates my interest to learn more about this online video website	
Emotion	EMO1: I feel very positive when I use VIP service	
	EMO2: Using VIP service makes me happy	
	EMO3: I feel good when I use VIP service	
	EMO4: I'm proud to use VIP service	
Behavior	BEH1: I spend a lot of time using VIP service from the online video website, compared to other other alternative services	
	BEH2: Whenever I'm using VIP, I usually use that video website	
	BEH3: Online video website is one of the entertainments I usually use when I use VIP service	
Utilitarian Value	UV1: The price of the VIP I purchased from this online video website are at the right level, given the quality	Overby and Lee [16]
	UV2: When I make a purchase from this online video site, I save time	
	UV3: The VIP I purchased from the online video site were a good buy	
	UV4: This online video website offers a good economic value	
Hedonic value	HV1: Making a purchase totally absorbs me	
	HV2: This online video website doesn't just sell VIP-it entertains me	
	HV3: Making a purchase from this online video site "gets me away from it all"	
	HV4: Making a purchase from this video website truly feels like "an escape"	
Continued purchase intention	CPI1: I'm willing to buy VIP from this online video site continuously	Bhattacherjee [4], Lin et al. (2012)
	CPI2: I prefer to buy the VIP from video site continuously rather than buy other alternative services (such as traditional paid TV)	
	CPI3: I am willing to recommend this VIP service	

Table 3. Descriptive analysis of sample features

Item	Type	Number of samples	Percentage (%)
Gender	Female	181	59.70
	Male	122	40.30
Education	High school and below	79	26.07
	Undergraduate	189	62.38
	Master degree and above	35	11.55
Buy VIP situation	Purchased	198	63.35
	Not purchased	105	36.65
Purchase service type	Monthly	98	49.49
	Season	62	31.31
	Annually	38	19.20

The aggregate validity is judged by the factor load, CR and AVE. It is also required that CR and AVE are greater than 0.7 and 0.5 respectively, and the criteria for determining the factor load, Hair et al. suggest that the normalization factor load is above 0.5 and reaching a significant level is acceptable. According to the results shown in Table 4 the paper satisfies these judgment criteria, indicating that the measurement model has good polymerization validity.

From the specific fitting results in Table 4, standard factor loadings of each index or dimension on the corresponding latent variable are between 0.711 and 0.938, and all pass the significance test at $P < 0.001$ level, so the validity of the scale can meet the requirements.

4.4 Path Analysis and Hypothesis Verification

The next step in the analysis involved testing of the structural model and corresponding theoretical relationships. In this study, AMOS 17.0 is used to analyze the path of the research model, and the path coefficient between each variable in the model and its significant performance are calculated. The path coefficient value reflects the influence degree of one variable on the other variables and can also be used. To test the search hypothesis proposed in chapter "Research on the Container Optimization of Communication Products". The path coefficient analysis between the variables in this study and the corresponding T and P values are shown in Table 5.

In the hypothesis test, when the absolute value of the T value is greater than its critical value (1.96), it means statistically significant, and there is a significant influence relationship between the two variables. And the P value indicates the level of significance that rejects the null hypothesis.

Table 4. Reliability and validity test results

Variable	Index	Cronbach's alpha	Standardized factor load	CR	AVE
Cognition	COG1	0.819	0.762	0.808	0.584
	COG2		0.711***		
	COG3		0.841***		
Emotion	EMO1	0.902	0.84	0.913	0.636
	EMO2		0.857***		
	EMO3		0.791***		
	EMO4		0.862***		
Behavior	BEH1	0.905	0.824	0.959	0.885
	BEH2		0.938***		
	BEH3		0.873***		
Utilitarian value	UV1	0.828	0.802	0.894	0.586
	UV2		0.810***		
	UV3		0.816***		
	UV4		0.825***		
Hedonic value	HV1	0.785	0.822	0.804	0.589
	HV2		0.713***		
	HV3		0.733***		
	HV4		0.824***		
Continued purchase intention	CPI1	0.886	0.909	0.884	0.579
	CPI2		0.832***		
	CPI3		0.813***		

Note: *** means P<0.001

As the results can be seen from Table 5, cognition, emotion, and behavior of customer engagement have a significant positive impact on continued purchase intention $(\beta = 0.825, p < 0.001; \beta = 0.726, p < 0.001; \beta = 0.808, p < 0.001)$, so the hypotheses H1a-c are supported. While cognition engagement has statistically non-significant impacts on utilitarian value $(\beta = 0.070, p > 0.001)$ and hedonic value $(\beta = 0.045, p > 0.001)$, the effect of emotion and behavior engagement on utilitarian value $(\beta = 0.389, p < 0.001; \beta = 0.426, p < 0.001)$ and hedonic value $(\beta = 0.402, p < 0.001; \beta = 0.318, p < 0.001)$ are significant. So, the hypotheses H2a, $H2a'$ are not supported, but H2b, $H2b'$, H2c, $H2c'$ are supported. Both utilitarian and hedonic value $(\beta = 0.780, p < 0.001; \beta = 0.786, p < 0.001)$ are positively related to continued purchase intention, so H3a-b are supported.

Table 5. Path coefficient analysis result

Paths	Standardized coefficients (β)	t value	p value
Cognition engagement continued purchase intention	0.825	19.16	***
Emotion engagement continued purchase intention	0.726	13.8	***
Behavior engagement continued purchase intention	0.808	19.332	***
Cognition engagement utilitarian value	0.07	1.12	0.262
Emotion engagement utilitarian value	0.389	4.495	***
Behavior engagement utilitarian value	0.426	6.987	***
Cognition engagement hedonic value	0.045	0.844	0.094
Emotion engagement hedonic value	0.402	7.386	***
Behavior engagement hedonic value	0.318	4.363	***
Utilitarian value continued purchase intention	0.78	12.5	***
Hedonic value continued purchase intention	0.786	11.838	***

Note: *** means P<0.001

4.5 Mediating Effect Test

According to the causal stepwise regression method proposed by Baron& Kenny, Chinese scholar Zhonglin [25] also proposed a method to test mediator based on this: $Y = cX + \beta_0; Z = aX + \beta_1; Y = bZ + \beta_2$ (Y: dependent variable; X: independent variable; Z: mediate variable). If "c" , "a" and "b" are significant, indicating that the influence of the independent variable on the dependent variable is partially realized by the mediate variable. Then path coefficient c' of the dependent variable is obtained after the introduction of the intermediate variable, if c' is significant and c is significantly reduced, then the effect of the independent variable on the dependent variable only partially through the intermediary variable; if c' is not significant, then the effect of the independent variable on the dependent variable is completely influenced by the intermediate variable.

In order to test the mediating effect of perceived value between CE and continued purchase intention, this study uses the above-mentioned causal stepwise regression method to verify the mediating effect of perceived value, and constructs the following multiple regression model:

$CPI = \beta_1 + c'E + bPV;$

$CPI = \beta_2 + c'B + bPV.$

According to Table 5, the positive influence of cognition on utilitarian value and hedonic value is not significant, and does not satisfy the premise of the above-mentioned method to test mediator (i.e. "a" is no significant), so no longer discuss. After adding the perceived value, the impact of emotional and behavioral engagement on continued purchase intention is still significant, and the standardized coefficient of the independent variable is significantly reduced. Therefore, it can be considered that the perceived value plays a partial mediating role in emotion and behavior engagement for continued purchase intention (Table 6).

Table 6. Mediating role of perceived value

Label	Path	CE's β	Significance	Mediating role
1	E→CPI	0.726	0.000	Partial mediation
	E→PV→CPI	0.145	0.000	
2	B→CPI	0.808	0.000	Partial mediation
	B→PV→CPI	0.240	0.000	

Note: CE: customer engagement; E: emotion engagement; B: behavior engagement; PV: perceived value; CPI: continued purchase intention

5 Results and Discussions

5.1 Model Results

This paper uses AMOS and SPSS data analysis software to analyze and test the collected data through empirical research methods. The test results are shown in Table 7.

Based on the above empirical results combining with the research background and objects, the following conclusions are made:

(1) CE can positively affect continued purchase intention of online video website VIP. The concept of customer engagement is a dynamic process of interaction and value creation with the company. Because of the cognitive, emotional, and behavioral connections to the online video sites, users are willing to continue to purchase VIPs. Highly engaging customers are more willing to continue to purchase video site VIPs, because they can trust the site and get a thrill from the VIP service, which allows them to continuously purchase the service. If customers have an emotional connection to the online video site, that is, non-transactional behavior from the cognitive, emotional and behavioral aspects, this will increase the likelihood that the customer will purchase VIP product or service of the video site on the website. The customer's purchase intention is enhanced. Not only that, when they have relevant product or service needs, they will firstly think of the online video website which has been engaged with them and to purchase its VIP service continuously;

(2) Perceived value exerted a significant impact on user's VIP purchase intention. Utilitarian value and hedonic value can clearly express the perception of video website users. Only when they have a functional and entertaining experience from VIP services can they promote the willingness to continue purchase.

Table 7. Hypothesis test result

Label	Hypothesis	Test results
H1a	Cognition has a significant positive impact on the continued purchase intention	√
H1b	Emotion has a significant positive impact on continued purchase intention	√
H1c	Behavior has a significant positive impact on continued purchase intention	√
H2a	Cognition has a significant positive impact on utilitarian value	×
H2b	Emotion has a significant positive impact on utilitarian value	√
H2c	Behavior has a significant positive impact on utilitarian value	√
$H2a'$	Cognition has a significant positive impact on hedonic value	×
$H2b'$	Emotion has a significant positive impact on hedonic value	√
$H2c'$	Behavior has a significant positive impact on hedonic value	√
H3a	Utilitarian value has a significant positive impact on continued purchase intention	√
H3b	Hedonic value has a significant positive impact on continued purchase intention	√

From the conclusion that the functional value has the most significant impact on continued purchase intention, the practical value of the service, that is, the usefulness of the service, will be an important criterion for users to evaluate the service for users who choose VIP instead of free service. Utilitarian value has a significant positive impact on VIP purchase intention, and this conclusion is in line with the research findings of scholars in other information systems. Video quality, loading speed, resource richness, and related value-added services are all important factors that influence users' willingness to pay;

(3) In view of the impact of CE on perceived value, the impact of cognition engagement on perceived value is not significant. Exploring the reasons can be considered from the following two aspects. On the one hand, compared with the two dimensions of emotion and behavior, the cognitive dimension reflects the more superficial psychological state in theory, rather than the content obtained after careful consideration. However, the acquisition of perceived value is obtained after experience and discrimination, so the impact of cognition on perceived value is minimal. On the other hand, online video websites provide a large number of VIP services, and users need to measure the perceived value through actual use, so they prefer behavior rather than psychological.

5.2   Implications

According to the conclusions of this paper, CE has a positive impact on continued VIP service's purchase from online video websites, with a key role of perceived value. Therefore, managers of online video websites should fully understand this point. How to attract and retain users on online video sites has two aspects: one is to improve the quality of website content and optimize the use experience, which can satisfy the user's practical value and enjoyment value; the second is to enhance the interaction and deep contact between users and websites. The engagement between users and websites can be formed by closer contact, thus retaining the user to increase the company's revenue.

From the perspective of CE, this paper empirically studies the influencing factors of the intention of purchasing VIP services continuously from online video websites, introducing the perceived value as a mediator. Although the article has refined the previous related research and achieved certain results, there are still some problems, such as the lack of optimization of the model fitting index, due to the constraints of subjective and objective conditions, which need to be further studied.

Acknowledgments. This paper is supported by the Shaanxi Province Social Science Fund (No. 2016 R015) and Research on Promoting to Construct Cultural Powerhouse by Cultural-oriented Tourism-Based on Customer Marketing (No. 2016R015).

References

1. Appelbaum, A.: The constant customer. J. Dalian Univ. Technol. (2011) (in Chinese)
2. Babin, B.J., Darden, W.R., Griffin, M.: Work and/or fun: measuring hedonic and utilitarian shopping value. J. Consum. Res. **20**, 644–656 (1994)
3. Beckers, S.F.M., Risselada, H., Verhoef, P.C.: Customer engagement: a new frontier in customer value management. Handb. Serv. Mark. Res. **2**(6), 97–120 (2014)
4. Bhattacherjee, A.: Understanding information systems continuance: an expectation-confirmation model. Mis Q. **25**(3), 351–370 (2001)
5. Brodie, R.J., Hollebeek, L.D., et al.: Customer engagement: conceptual domain, fundamental propositions & implications for research. J. Serv. Res. **14**(3), 252–271 (2011)
6. Dodds, W.B.: Effects of price, brand, and store information on buyers' product evaluations. J. Mark. Res. **28**(3), 307–319 (1991)
7. Gaoshan, W., Xin, Z., et al.: The impact of electronic service quality on customer conformity: the mediating effect of customer perceived value. J. Dalian Univ. Technol. (2) (2019)
8. Higgins, E.T., Scholer, A.A.: Engaging the consumer: the science and art of the value creation process. J. Consum. Psychol. **19**(2), 100–114 (2009)
9. Holbrook, M.B.: Consumer value: a framework for analysis and research. Advances in Consumer Research (1996)
10. Hollebeek, L.D.: Demystifying customer engagement: exploring the loyalty nexus. J. Mark. Manag. Forthcoming. (2011) DOIurl10.1080/0267257X2010500132
11. Hollebeek, L.D., Glynn, M.S., Brodie, R.J.: Consumer brand engagement in social media: conceptualization, scale development and validation. J. Interact. Mark. **28**(2), 149–165 (2014)
12. Hsiao, W.H., Chang, T.S.: Understanding consumers' continuance intention towards mobile advertising: a theoretical framework and empirical study. Behav. Inf. Technol. **33**(7), 730–742 (2014)

13. Jones, T.O., Sasser, W.E.: Why satisfied customer defects. Harv. Bus. Rev. **71**, 88–99 (1995)
14. Lapierre, J.: Customer-perceived value in industrial contexts. J. Bus. Ind. Mark. **15**(2), 122–140 (2000)
15. Marbach, J., Lages, C.R., Nunan, D.: Who are you and what do you value? Investigating the role of personality traits and customer-perceived value in online customer engagement. J. Mark. Manag. **32** (2016)
16. Overby, J.W., Lee, E.J.: The effects of utilitarian and hedonic online shopping value on consumer preference and intentions. J. Bus. Res. **59**(10), 1160–1166 (2006)
17. Patterson, P., Yu, T., De, R.K.: Understanding customer engagement in services. In: Advancing Theory, Maintaining Relevance, Proceedings of ANZMAC 2006 Conference, Brisbane, pp. 4–6 (2006)
18. Pham, M.T., Avnet, T.: Rethinking regulatory engagement theory. J. Consum. Psychol. **19**(2), 115–123 (2009)
19. Sheth, J.N., Newman, B.I., Gross, B.L.: Why we buy what we buy: a theory of consumption values. J. Bus. Res. **22**(2), 159–170 (1991)
20. Vivek, S.D.: A scale of consumer engagement. Doctor of Philosophy Dissertation, Department of Management & Marketing, Graduate School, The University of Alabama, UMI (2009)
21. Wei, L., Xiaoting, F.: Research on the causes and consequences of online brand conformity. Soft Sci. **32**(06), 125–128+144 (2018)
22. Wen-Jun, Z., Ming, Y.I., Xue-Dong, W.: Research on continued participation intention of social Q&A platform users-based on the view of perceived value. Inf. Sci. **35**(2), 69–74 (2017)
23. Wu, L., Yiaoxin, X., Ting, C.: The impact of user perceived value on the willingness to pay on the online paid question and answer platform analysis of the regulating effect based on past behavior. Journalism Mass Commun. Mon. **10**, 92–100 (2018)
24. Zeithaml, V.A.: Consumer perceptions of price, quality, and value: a means-end model and synthesis of evidence. J. Mark. **52**(3), 2–22 (1988)
25. Zhonglin, W., Lei, Z., et al.: Testing and application of the mediating effects. Acta Psychol. Sin. **05**, 614–620 (2004)

A Study of Museum Experience Evaluation from the Perspective of Visitors' Behavior

Shuangji Liu, Yongzhong Yang$^{(\boxtimes)}$, and Mohsin Shafi

Business School, Sichuan University, Chengdu 610065, People's Republic of China
yangyongzhong116@163.com

Abstract. Museums are attractive learning, tourism, leisure and entertainment destinations, and visiting a museum has become a way of life for many peoples. Both visitors and museum officials are increasingly concerned with the experience. This paper constructs and empirically tests a museum experience evaluation model from the perspective of visitors' behavior. Based on the first-hand information obtained from network questionnaire, the research findings are as follows: (1) The visiting behaviors of "ants" and "grasshoppers" have significant influence on the learning experience and aesthetic experience, "Butterfly" and "fish" style visiting behavior have significant influence on entertainment experience; (2) All four kinds of visiting behaviors have significant influence on content generation, Ant-like visiting behavior has significant influence on content enhancement. This paper discusses the conclusions of the research, the main theoretical contributions and the implications for museum management, the next step of the research is prospected.

Keywords: Visitor behavior · Museum experience ·
Experience evaluation · Content enhancement

1 Introduction

The word "experience" is a hot topic in various fields nowadays. From the purchase experience of customers, mobile phone users' experience, car driving experience, hotel consumption experience, to dazzling courses experience, more and more attention has been paid to people's experience in various fields. The field of museum is without exception. As an important space carrier carrying the excellent traditional Chinese culture, today museums conform to a spiritual value appeal for peoples. Museum experience and evaluation of experience are increasingly valued by visitors and museum managers [18].

In recent years, the research on museum experience has gradually increased. Antón et al. [1] explored the experience from the perspective of co-creation and studied the experience value co-creation process based on the visitor's plan and knowledge perspective. While, Su et al. [11] explored the experience from the perspective of thinking about the failure of museum services, and extracted 12

© Springer Nature Switzerland AG 2020
J. Xu et al. (Eds.): ICMSEM2019 2019, AISC 1002, pp. 683–693, 2020.
https://doi.org/10.1007/978-3-030-21255-1_52

experience dimensions of visitors. Furthermore, Antón et al. [2] examined the museum visit factors that prevent or further visitor satiation. After a thorough review of literature, it has been observed that there is a lack of discussion on museum experience from the perspective of visitor's behavior. Visiting behavior is an external manifestation of visitor psychology [13]. Research on visitor's behavior helps us to understand psychological factors such as visitor's needs, hobbies, attitudes, experiences and so on. It also helps us to launch personalized special exhibitions to meet visitors' experience needs to the greatest extent [16]. In this context, it is necessary to deeply study the museum experience and experience evaluation from the perspective of visitors' behavior and reveal the influence mechanism of the museum experience evaluation from the perspective of visitors' behavior.

2   Literature Review and Research Hypothesis

2.1   Research on Visiting Behavior

The research on the behavior of museum visitors started in 1897, when a German scholar Eph He Naier tried to obtain visitors' opinion on visiting exhibits through question-and-answer method [12]. Since then, Yale university professor Edward Robinson focused his research on "fatigue" museum phenomenon, particularly, the visitors in the exhibition for museum design problems arising in the process of the phenomenon of physical and mental fatigue, which opened up a new research territory of the museum visitors' behavior [16]. Zancanaro et al. [17] conducted a qualitative study designed to identify the different behavior of visitors in the museum. Based on the movement of visitors in the museum space, the authors summarized four basic visitors' behavior: Ant, Butterfly, Fish and the Grasshopper. The "Ant visitors" follow a long way in the museums; and are willing to walk and spend a lot of time to visit the museum. They are generally interested in visiting all the exhibits in the museum, and they are also interested in the details of the exhibits. They are willing to listen to the facilitator, and a large number of people does not affect their visit. The "Butterfly" visitors are also interested in visiting all the exhibitions, and their paths can be redirected at any time. If there are many people in an exhibition, they will choose to visit the part later. They are also willing to listen to the facilitator, if the exhibits are close to each other, they will be very depressed. The "Fish" visitors do not want to walk a lot in the museum, and they prefer to stand in the center of the exhibition hall. The "Grasshopper" visitors are familiar with the museum's exhibits and have clear plans and preferences before their visit. They do not want to listen to the facilitators, and they are willing to go a long way to visit the exhibits that they planned [4].

Most of these studies on the behavior of visitors are qualitative descriptions and how visitors' behavior affect the experience of visitors has not been explored yet in the literature, which the present study aims to cover. Based on the four visiting behaviors proposed by Zancanaro et al. [17], namely, "Ant", "Fish", "Butterfly", "Grasshopper", this study investigates the influence of different visiting behaviors on museum experience.

2.2   Research on Museum Experience

The word "experience" originated from ancient Greek philosophy and emerged from the classical aesthetics of medieval Germany. It is referred to the feeling that participants get and leave a deep impression on their brains [15]. Bowsijk et al. [3] defined experience as something that transcends daily life, or something that make it unforgettable, which helped individuals enrich their experiences. From the perspective of consumers, experience was the overall evaluation of the purchase and used of goods or services [10]. Lijuan [7] described that experience is an action to try something in order to discover the unknown, acquire knowledge and confirm the hypothesis. It is a kind of personal experience that people get in contact with the environment.

Gilmore and Pine [9] in their study on *ExperienceEconomy*, classified the experience value in four dimensions i.e escape, aesthetics, entertainment and learning, such as visitors left ordinary residence (escape), feeling different visual culture and the aesthetic feeling of cross-regional customs (aesthetics), spent and enjoyed it (entertainment), these processes also had sensory impact and psychological feelings, and spiritual progress (learning).The book also pointed out that the most successful travel experiences should be synchronized to include the four aspects mentioned above, that was, the "sweet spot" where the four intersect. Falk and Dierking [5] first described what was the museum experience in his book natural history experience. The definition of a museum experience was as follows: "It is referred to the motivation of visitors began to produce to visit the museum visitors all ideas in the process of the end of the activity, the actual action and memory of the whole. Since then, western studies on museum experience had basically followed this concept. Levent and Pascual-Leone [6] described in their works that the multi-sensory experience of touching, smelling and listening to artworks could enrich the experience of museum visitors, and crowded, queuing, bad sound, tired feet, lacked of seat, difficulties of positioning and navigation and the uncertainty in understanding art will affect the museum experience of idcal effect.

Based on the review of the above relevant literature, this paper mainly adopts the description of Falk and Dierking [5] in their book natural history experience to define the museum experience, which referred to the whole idea, practical action and memory of visitors from the beginning to the end of the museum activity. For the dimensionality division of museum experience, Pine and Gilmore's [9] classification of experience value is mainly adopted, namely learning experience, entertainment experience, aesthetic experience and escapist experience.

2.3   Research Hypothesis

The knowledge, plan, motivation and behavior of visitors before the visit will affect the experience of visitors. Antón et al. [1] explored the co-creation process of the value of museum experience based on visitors' plans and knowledge. Su et al. [11] extracted 12 experience dimensions of visitors from the perspective of thinking about the failure of museum services. Antón et al. [2] explored the importance of experience from the perspective of the reasons for preventing or promoting visitors to the museum. Based on the perspective of visitor behavior, this paper investigates the influence of visitors' behavior on museum experience. Thus following research hypotheses are proposed:

(1) Influence of Visitors' Behavior on Visitor Experience

Different visiting behaviors may have different influences on visitors' experience. Antón et al. [1] pointed out that the motivations and behaviors of visitors before visiting may affect the experience of visitors differently. In addition, most researchers believe that there was a positive correlation between visitors' behavior and visitors' experience [8]. Therefore, following four hypotheses are proposed:

Hypothesis 1. "Ant" visiting behavior has a positive impact on visitors' experience.

Hypothesis 2. "Fish" visiting behavior has a positive impact on visitors' experience.

Hypothesis 3. "Butterfly" visiting behavior has a positive impact on visitors' experience.

Hypothesis 4. "Grasshopper" visiting behavior has a positive impact on visitors' experience.

(2) The Influence of Visitors' Behavior on the Generation and Reinforcement of Content after the Visit

When visitors are about to leave the museum, each visitors visiting experience may be different, how to evaluate the quality of experience that visitors activities, some scholars on social platforms through visitors to upload photos to evaluate the experience of visitors (Huy Quan Vu et al., 2017) [14], some through content generation and content two indicators to evaluate the experience of visitors, including content generation refers to the visitors after the visit, will in time in further through the network to find information about the museum, more looking forward to participate in the activities organized by the museum next time, when appropriate, the museum management should also be given some suggestions to improve the quality of work; Content intensification means that when visiting again, the experience is further deepened [1]. Therefore, this paper proposes the following last two hypothesis:

Hypothesis 5. The four visiting behaviors have a positive impact on content generation.

Hypothesis 6. The four visiting behaviors have positive effect on content reinforcement.

Based on the above theories and relevant assumptions, we propose a museum experience evaluation model based on the perspective of visitors' behavior, as shown in Fig. 1.

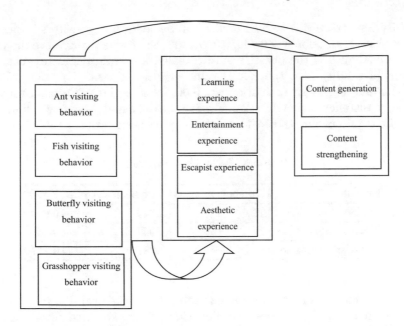

Fig. 1. Museum experience evaluation model based on the perspective of visitors' behavior Source: Authors' drawing

3 Empirical Research and Results

3.1 Descriptive Statistics

The data was collected through questionnaire. The questionnaire comprised of visitors' behavior scale, visitors' experience scale and after the visit of evaluation scale. All items were measured using *FivePiontLikert* scales with value ranging from 1 (Strongly Disagree) to 5 (Strongly Agree). In total, 182 questionnaires were collected, out of which 172 were valid with a response rate of 94.5%.

The proportion of respondents under the age of 25 years accounts for 40.81%, 26 to 35 years accounts for 29.53%, 36 to 45 years accounts for 14.07%, 46 to 55 years accounts for 10%, and 56 to 65 years accounts for 5.59%. Among them 27.91% of respondents are male and 72.09% are female. The educational level of respondents' accounts for 12.33% of students in high school degree or below, 61.27% in junior college degree, and 26.4% have a graduate degree. From the perspective of career distribution, students account for 46.16%, business leaders or managers account for 8.49%, professional and technical personnel accounts for 19.65%, teachers and doctors accounts for 5.58%, civil servants account for 5.12%, and others account for 20.12%.

3.2 Data Test and Factor Analysis

It was assumed that there was a causal relationship between visitors' behavior and experience, as well as post-visit content reinforcement and content generation. Therefore, SPSS 18.0 software was used for statistical analysis. In terms of reliability test, $Cronbach's Alpha$ of visitors' behavior scale, visitors' experience and post-visit evaluation was 0.749, 0.903 and 0.887 respectively, since the values are greater than 0.7, the questionnaire revealed a certain degree of reliability. In terms of validity test, the KMO value of visitors' behavior scale, visitors' experience and post-visit evaluation was 0.90, 0.869 and 0.867 respectively. The internal consistency of the questionnaire is relatively high, and the validity also meets the requirements, thus it is suitable for further analysis (Table 1).

Table 1. Reliability statistics

Variable	Number of items	Cronbach's Alpha	KMO
Visitors' behavior	16	0.749	0.690
Visitors' experience	12	0.903	0.869
Post-visit evaluation	6	0.887	0.867

From the Factor analysis four main factors with eigenvalues greater than 1 were extracted from the visitors' behavior scale, which explained 66.3186% of the total variance of the original variables. The four factors, namely, "Ant" visiting behavior, "Fish" visiting behavior, "Butterfly" visiting behavior, "Grasshopper" visiting behavior were based on the theoretical assumptions. Similarly, four factors with eigenvalues greater than 1 were extracted from the visitors' experience scale explaining 77.302% of the total variance, the extracted factors are learning experience, entertainment experience, aesthetic experience and escapist experience. While, from the post-visit evaluation scale only two factors with eigenvalues greater than 1 were extracted, explaining 71.154% of the total variance, namely, content generation and content strengthening.

3.3 Empirical Results

The results indicates that all four types of visitor's behavior, namely, Ant, Fish, Butterfly and Grasshopper have a significant positive impact on content generation, $\beta = 0.229 \; p < 0.01 \; \beta = 0.105 \; p < 0.01 \; \beta = 0.188 \; p < 0.01 \; and$ $\beta = 0.135, \; p \; < \; 0.05$. Among them, the "Ant" visit behavior has the greatest influence on content generation, followed by the "Butterfly" and "Fish" visit behavior. Furthermore, the results indicates significant positive impact of "Ant" visiting behavior on content strengthen, $\beta = 0.118, p \; < \; 0.01$. In addition, "Ant" and "Grasshopper" visiting behaviors positively and significantly affected the learning and aesthetic experience, $\beta = 0.389 \; p \; < \; 0.01 \; \beta = 1.503 \; p < 0.01 \; \beta = 0.125 \; p < 0.01 \; and \; \beta = 0.091,$ $p \; < \; 0.01$ respectively. Moreover, the "Butterfly" and "Fish" visiting behavior positively and significantly affected the entertainment experience, $\beta = 0.059, p \; < \; 0.01$ and $\beta = 0.118, p < 0.05$ respectively. The regression analysis results of visitor's behavior and visitor's experience, content strengthening and content generated have been presented in Table 2 and Table 3.

Table 2. Regression analysis results of visitors' behavior, content generation and content enhancement

The dependent variable	The independent variables	B	T	Sig	Collinear statistics	
					Tolerance	VIF
Content strengthening	(constant)					
	"Ant"	0.118	1.861	0.001	1.000	1.000
	"Fish"	−0.103	−1.457	0.119	1.000	1.000
	"Butterfly"	0.069	1.022	0.304	1.000	1.000
	"Grasshopper"	0.032	0.471	0.629	1.000	1.000
Content generation	(constant)					
	"Ant"	0.229	5.198	0.000	1.000	1.000
	"Fish"	0.105	1.727	0.005	1.000	1.000
	"Butterfly"	0.188	2.901	0.004	1.000	1.000
	"Grasshopper"	0.135	2.182	0.029	1.000	1.000

Table 3. Regression analysis results of visitors' behavior and visitors' experience

The dependent variable	The independent variables	B	T	Sig	Collinear statistics	
					Tolerance	VIF
Learning experience	(constant)					
	"Ant"	0.389	6.315	0.000	1.000	1.000
	"Fish"	−0.026	−0.424	0.672	1.000	1.000
	"Butterfly"	−0.025	−0.413	0.680	1.000	1.000
	"Grasshopper"	1.053	1.867	0.007	1.000	1.000
Entertainment experience	(constant)					
	"Ant"	0.207	3.280	0.061	1.000	1.000
	"Fish"	0.059	0.959	0.008	1.000	1.000
	"Butterfly"	0.118	1.708	0.022	1.000	1.000
	"Grasshopper"	0.065	1.067	0.245	1.000	1.000
Escapist experience	(constant)					
	"Ant"	0.219	3.466	0.091	1.000	1.000
	"Fish"	0.050	0.539	0.525	1.000	1.000
	"Butterfly"	0.068	1.349	0.174	1.000	1.000
	"Grasshopper"	0.204	3.119	0.101	1.000	1.000
Aesthetic experience	(constant)					
	"Ant"	0.125	1.728	0.005	1.000	1.000
	"Fish"	−0.021	−0.168	0.866	1.000	1.000
	"Butterfly"	0.078	1.484	0.136	1.000	1.000
	"Grasshopper"	0.091	1.236	0.002	1.000	1.000

4   Conclusions and Implications

4.1   Conclusions

This study focuses on the question of "evaluation of museum experience under different visiting behaviors". In recent years, few studies have been conducted on the museum experience. For instance, Antón et al. [1] explored the experience from the perspective of co-creation, and studied the experience value co-creation process based on the visitor's plan and knowledge perspective. Su et al. [11] explored the experience from the perspective of thinking about the failure of museum services. While, Antón et al. [2] investigated the importance of experience from the perspective of the reasons for preventing or promoting visitors to the museum. Based on the above studies, this paper further discusses the evaluation of museum experience under different visiting behaviors from the perspective of visitor's behavior. The results supported some of the hypotheses which lead to important conclusions.

The findings of this study indicates that different behaviors of visitors have a significant effect on different experiences, for instance, "Ant" and "grasshopper" visiting behaviors have a significant positive relationship with learning and aesthetic experiences, while, "Butterfly" and "Fish" visiting behaviors have a significant positive correlation with entertainment experience. The findings further indicates that people with "Ant" visiting behavior are more careful about their visit, while with "grasshopper" visiting behavior has a clear plan before the visit, and these actions lead them to more learning experience and aesthetic experience. People with "Butterfly" and "Fish" visiting behavior are relatively casual, and pay more attention to entertainment experience.

Furthermore, people with "Ant", "Grasshopper", "Butterfly" and "Fish" visiting behaviors have significant positive correlation with content generation, moreover, "Ant" visiting behavior has a significant positive relationship with content strengthening. The empirical results indicate that the museum management personnel should arrange interactive activities in a timely manner in order to discuss the feelings of the visitors. People with "Ant" visiting behavior are more careful about their visit, carefully, thus their experience and feelings must be stronger.

4.2   Main Contributions

(1) Expand the New Perspective of Museum Experience Evaluation

Previous studies explored the co-creation of experience value based on visitors' knowledge, plans etc. and the experience from the perspective of thinking about the failure of museum services including visit factors that prevent or further visitor satiation [1, 2, 11], however, this study contributed to the museum experience literature by examining the evaluation of museum experience from the perspective of visitors' behavior.

(2) Enrich the Research on the Evaluation of Museum Experience

Several studies have been conducted on the evaluation of experience in the field of tourism and marketing, but there is no specific research in the field of museums. This paper studies the evaluation of experience in museums from the perspective of visitors' behavior, which further enrich the relevant research on the evaluation of experience.

(3) A Museum Experience Evaluation Model Based on Visitor Behavior Perspective is Proposed

Zancanaro et al. [17] had summarized four basic visiting behaviors based on the movement of visitors in the museum space, but how these four specific visiting behaviors quantitatively affect the experience has not been studied. Based on the previous available literature, this paper proposes a museum experience evaluation model based on the perspective of visitors' behavior, that is, the identification of visitors' four visiting behaviors before the visit, various experiences during the visit, and experience evaluation after the visit.

4.3   Management Enlightenment

(1) Focus on the Quality of Visitors' Experience

The experience quality of visitors is the focus of the management of the museum. By constantly identifying various experiences of visitors and improving, the experience quality of visitors, many people can be attracted to visit the museum frequently, which will increase the comprehensive benefits of the museum.

(2) Optimize the Display of Exhibits

Different visiting behaviors have different influences on different experiences, different visiting behaviors having different paths in the museum. The display of exhibits should be optimized to allow people to have more and higher quality experiences within a certain visiting time.

(3) Feedback of Visitors' Experience in Time

During and after the visit, visitors may leave messages on the website of the museum or *WeChat* official account to express their feelings. Museum staff should pay attention to reply in time, continuously improve the service and maximize the quality of visitors' experience.

4.4   Research Limitations and Prospects

This paper is an exploratory study on the evaluation of museum experience from the perspective of visitors' behavior. It mainly studies the relationship between visitors' behavior and visiting experience, and between visiting behavior and content generation and content strengthening after visiting based on the four different visiting behaviors. Since there were few quantitative studies on the evaluation of museum experience from the perspective of visitors' behavior, this paper conducts exploratory research on the basis of existing literature. Despite of the contribution to theory and practice, there remain some limitations mainly due to research conditions that provide room for future research. First, the categories of visiting behaviors were based on the limited available literature, still there is a need of in-depth thinking and further research to investigate whether these four behaviors cover all aspects. Second, the data was collected only from art museums which limits the generalizability of the findings, future studies can consider other types of museums to generalize the findings.

In addition, the research on the evaluation of museum experience from the perspective of visitors' behavior needs to be further expanded. Moreover, it might be interesting to see the relationship between visitors' behavior and experience under different demographic characteristics such as gender and income.

Acknowledgments. We gratefully acknowledge the research support received from the Key Project of the National Social Science Fund (18AGL024).

References

1. Antón, C., Camarero, C., Garrido, M.J.: Exploring the experience value of museum visitors as a co-creation process. Curr. Issues Tour. **21**(12), 1406–1425 (2018a)
2. Antón, C., Camarero, C., Garrido, M.J.: A journey through the museum: visit factors that prevent or further visitor satiation. Ann. Tour. Res. **73**, 48–61 (2018b)
3. Boswijk, A., Thijssen, T., Peelen, E.: The Experience Economy: A New Perspective. Pearson Education (2007)
4. Brida, J.G., Disegna, M., Scuderi, R.: The behaviour of repeat visitors to museums: review and empirical findings. Qual. Quant. **48**(5), 2817–2840 (2014)
5. Falk, J.H., Dierking, L.D.: The Museum Experience. Routledge, London (2016)
6. Levent, N., Pascual-Leone, A.: The Multisensory Museum: Cross-disciplinary Perspectives on Touch, Sound, Smell, Memory, and Space. Rowman & Littlefield, Lanham (2014)
7. Lijuan, L.: Study on intention analysis of visitors' participationin co-creating value in xiangshan park. Ecol. Econ. **12**, 32 (2013)
8. Otto, J.E., Ritchie, J.B.: The service experience in tourism. Tour. Manag. **17**(3), 165–174 (1996)
9. Pine, B.J., Gilmore, J.H.: The experience economy. Hard. Bus. Rev. **76**(6) (1998)
10. Sheng, C.W., Chen, M.C.: A study of experience expectations of museum visitors. Tour. Manag. **33**(1), 53–60 (2012)
11. Su, Y., Teng, W.: Contemplating museums service failure: extracting the service quality dimensions of museums from negative on-line reviews. Tour. Manag. **69**, 214–222 (2018)
12. Tsiropoulou, E.E., Thanou, A., Papavassiliou, S.: Quality of experience-based museum touring: a human in the loop approach. Soc. Netw. Anal. Min. **7**(1), 33 (2017)
13. Vincent, F.Y., Lin, S.W., Chou, S.Y.: The museum visitor routing problem. Appl. Math. Comput. **216**(3), 719–729 (2010)
14. Vu, H., Luo, J., Ye, B., et al.: Evaluating museum visitor experiences based on user-generated travel photos. J. Travel Tour. Mark. **35**(4), 493–506 (2018)
15. Wright, P.: The quality of visitors experiences in art museums. New Museol, pp. 119–148 (1989)

16. Yoshimura, Y., Sobolevsky, S., Ratti, C., Girardin, F., Carrascal, J.P., Blat, J., Sinatra, R.: An analysis of visitors' behavior in the louvre museum: a study using bluetooth data. Environ. Plan. B Plan. Design **41**(6), 1113–1131 (2014)
17. Zancanaro, M., Kuflik, T., Boger, Z., Goren-Bar, D., Goldwasser, D.: Analyzing museum visitors behavior patterns. In: International Conference on User Modeling, pp 238–246. Springer, Berlin (2007)
18. Yang, Y.: An introduction to creative management. Econ. Manag. Publ. House (2018)

Changing Preference Aspects from Traditional Stores to Modern Stores

Abdullah Khan[1(✉)], Shariq Ahmed[1], and Farhan Arshad[2]

[1] Management Science, KASB Institute of Technology, Karachi, Pakistan
abdullah@kasbit.edu.pk
[2] Management Science, Shaheed Zulfikar Ali Bhutto Institute of Science and
Technology, Karachi, Pakistan

Abstract. The retail market of Pakistan is transforming from traditional to modern retail outlets. Thus, the modern stores in Pakistan capturing the large market share by giving the varieties of product to the consumer. As the research is conducted to analyze those factors which effect on consumer store choice from traditional to modern stores. The method utilized in research is quantitative with a deductive approach. The study analyzes the impact of price, store location, product convenience and visual merchandising on consumer store choice. A close-ended questionnaire was used as a data collection tool. The sample included over 329 respondents, including each individual visiting the stores for their grocery shopping. Further, the SEM is also used to identify the variation among the variables. The study identified that the independent variables (factors) are having an impact on the dependent variable (store preference). Further the statistical model identifies that in the current environment, people prefer modern stores.

Keywords: Environment · Location · Consumer store preference · Traditional market · Modern store

1 Introduction

In the modern era, the traditional family stores are now replacing with the modern retail store. This development of retail transformation is becoming the significant topic for the researches to analyze the behavior of the consumer that why they are not preferring their traditional retail store to purchase their consumable items [8]. Research studies explain the weak spot of the traditional retailing system and examine the boundaries of the modern format's preference. Thus, it has long relied on that traditional techniques of marketing for capturing shopper's attention were in the drift as the shoppers are purchasing their consumable items from modern stores. Also, consumers' motivations were recognized by their values to purchase their consumable items. Although few buyers place their prominence on scant cash and reside from disasters, others focus on obtaining products and accepting values' [18]. The modern stores are now the

© Springer Nature Switzerland AG 2020
J. Xu et al. (Eds.): ICMSEM2019 2019, AISC 1002, pp. 694–704, 2020.
https://doi.org/10.1007/978-3-030-21255-1_53

attractive leading outlets for confined fresh manufacture to compare with local stores. Furthermore, modern stores in developing countries connect in association marketing and set-up enthusiastic supply chains with minor farmers, as contrasting to relying on traditional retail associations in markets. The modern store technique is to focus on value mainly perceptible to customers where the official quality principles are low and not imposed or mistrust [12]. Moreover, the self-service shopping, store environments, displaying price, and violent marketing discounts are flattering the strategic competitive edge [2]. The growing speed of globalization has rationalized the retail setting globally. Modern stores are becoming the large organizational structure of the retail market with a massive range of product variety. Stores are gradually substituting the local stores by providing the variety of products to the buyer. The grouping of low price, product convenience, and attractive merchandising technique are the key characteristics of these modern stores to attract the consumer [4]. In the present setting, the US is devoid that the retail trade pioneer with a large role of approximately 66% to GDP of US. Wal-Mart is not only the huge modern replace yet it is moreover incorporated one of the major restructured associations on the globe. This phenomenon isn't just restricted to the boundaries of united states, as the crafts state like Pakistan is considering a batch of pledge and propose of GDP, In Pakistan is currently at a stage where consumers need a display of stuff and retail organization, buyers' boost in spending their obsessed the retail partition growth rate to 7.2% among 1999–2002 [13]. This end has encouraged the theme of a move in store choice amongst the consumer who before the beginning of modern shops from the traditional retailers. It is proposed that a query about the future viewpoint and trade of local retailers in Pakistan chiefly in the cities like Lahore, Karachi, and Islamabad. Pakistan is belonging to the developing nation which materialization the retailing sector around 42 billion dollars and still growing.

1.1   Problem Statement

This research identified the current state of the retail division that has been developed in the past few years and becomes a focal economic commotion. Several types of retailing outlets contain a chain of exacting store, supermarket, local shops, and factory outlets. These modern stores coat a large range of goods, products, and services with huge organizations contributing thousands of products in one place as compared to the modern stores traditional retailing in Pakistan is facing difficulties to survive in modern world. The modern shop's section in the growth of a financial system which can't be equalized, yet one article is furthermore exceptionally convention retail sets or conventional retailing mock-up submission such a huge amount of straight employment to a big bit of the population [8]. Thus these homogeneous or modern retail associations try to build buyers that they approach the store again and again by offering concessions. This research study then aimed towards the effect made contemporary retailing shops over buyer store preference and moreover its philosophical consequence at the trade of conventional outlets.

1.2 Objective of Study

Following research, objectives were developed to find out the Pakistani customer preferences over the modern and traditional stores.

1. To analyze the effect of store location over the consumer store preference.

2. To establish the relation of product price and store ambiance on buyers store selection.

3. To analyze the association among conveniences of product accessibility impact the customer store selection.

4. To obtain the effect of visual merchandising on buyers interest and concern of buyers for store selection.

2 Literature Review

Store selection has been identified that buyers lean to choose those shops for their purchasing which present discounts. Moreover, several buyers are also liable towards that shop which presents the minimal price and effort, a better range of products, easy access and quick depart service [6]. Past researches suggested that the buyers attitude for local stores is altering and deflecting towards the modern store or not, the beginning of such huge and contemporary retail shops has given buyers more option in stipulations of grocery purchasing, this occurs when functional in Pakistani circumstance can obtain more composite, both merchant and consumer at present state are appraised the fact and it isn't apparent that which motivators are pouring the fact frequently to the modern shops, consumer are additionally tending to the contemporary trade frequently owing to the atmosphere of the shops [16]. However, there isn't any fact from purchaser side for any sort of reliability to some retail stores which establishes that the purchaser chosen the modern style stores but still they purchase from conventional shops while there isn't any of the time restraint. The modern style retail shops take benefit from the purchaser confusion to transfer consumer store preference to modern stores by sculpting and develop diverse price discounts and price composition for key grocery products [21]. Additionally, in the modern stores, where most of the sections are located in city center region and are connected with a wide variety of goods, including a vast deal of meandering, and obtaining more goods at lesser cost. Some consumer feels that purchasing in these stores is beneficial, as they regulate their additional efforts by walking in these modern shops space to their one-stop shopping plan. Individuals' obtain report that can help to develop the judgment precision of mutual first- preference prospects and the replacement chances of the Exogenous exchange model the foundation catalog level is low [11].

2.1 Factor Impacting the Store Preferences

This research study by [14] put a lot of importance on the fundamentals that can believe their component in individuals' preference towards the modern stress and the exceptionality of shops that can manage the purchasers' essential business process. The basics integrate are product price, visual merchandising (product position, Promotional products materials), product convenience and district of store and position business offers. An additional research study illustrates the manipulate of shop ambiance on an impulsive acquiring choice among consumers and the outcome exposed that amongst the consumer with less purchasing power, finances factors such as low prices, voucher, and supportive shop supporters were additional likely to control precipitate purchasing behavior [7]. Moreover, factors with an impressive consequence which contain fresh perfume and drying strength only be obliging in maintenance consumer for a huge store but it was suspect that it influences the consumer's inclination purchasing behavior straight [5]. An excellent shopping knowledge, shaped by the store variables, construction trust during the reliable performance and attractive financial and emotional switching expenses yield extended reliability and high confidence. Thus the traditional stores are opposing with modern shops with high connection standards and societal standard [7].

Price – A research study of purchaser attitude on the modern trade and expediency stores in Pakistan outline that purchaser like to purchase low participation FMCG goods expediently commencing the modern retail as they suggest more revisiting about expenditure and support [21]. Recently a sale, advertising has established to be a significant tool for merchants and investigators. The attractive free time commotions stem from the clutch of significant sales, advertising in constructing a modern marketing strategy. Price and promotion were classified into several sorts of tender such as deals, promotions, purchaser promotions, intended at the influence the final purchaser to create the purchase [9]. The pricing plan is the key advertising technique when contending with your opponent; it has also been the feasible result on closing stages purchaser's choice in trade of an exacting good. The diminutive purchaser price reduction is an entrenched retailing method with which mainly retailers, whether global or local influence purchaser to buy more of an exacting product. It originated that 65% purchaser locate the best shopping transaction in malls with a range of value cut tenders, gift parcels [19]. Several retailers have executed software that assists them in cost associated decision making. Traders can give cost cuts at the end purchaser without any cooperation on the excellence that consumer perceives by increasing their delivery and purchasing system [21].

Visual Merchandising – Along with the various characteristics of a store that assist to exchange window buyers into real clientele while visual merchandising achieve the greatest position, the other aspect was also not far following. Traders frequently see shelf position to influence unit sales; this is why shelf position is used deliberately to improve the buyer's visual observation which can eventually enlarge the sales [22]. The research study carries out on the buyers in Pakistan designated that 90% of consumers sturdily oblique that by applying effective visual merchandising observe the association can execute better advertising strategies for their goods. Data showed that only 6% of consumers said that effective visual merchandising has no outcome on the marketing approach of

goods [15]. Visual merchandising is vital for essentially receiving buyer's deliberation. 83% of data collected by a human instigate from propinquity. Visual
merchandising also act as a significant part in producing the brand attentiveness
among consumers, the formulation of the product value as well as the general
advertising strategy of the goods or services [22].

Store Location – The placement of an outlet is very serious in a consumer buying method, composite decision making procedure, amplified prices
and restricted flexibility accessible is a chosen factor in a traders' plan, and a
good propinquity plan can guide to a thriving project for a trader yet if its
general plan is average [10]. Retail gravitation" associates that consumer choice
between retail stores (Groups of stores) is lined by the outlets, center attractiveness, which enhance with a center's corporeal amount, but reduces with its
space from the consumer home [17]. The closest accumulate to consumers are
likely to acquire since from there; consumer favor to purchase from that shop
which approaches within their proximity. The penchant of a traditional shop in
the last was suitable for this easy fact that its position was in the neighborhood
of consumers. Consumer seek store positioned near their residence or closer to
their neighborhood, a consumer favorite for this one characteristic is supplementary than other feature because it expenses them fewer, saves their point, and
is more suitable, it is also has been established from different investigators that
a modern retail organization.

Convenience – Suitable goods are buying with very less consideration
method, which creates consumer attention for the easiest selection of the shopping experience. Consumers also worth the easy entree to the ledge for product
obtain along the site of the modern outlet. The apparent and real-time obses-
sive to shop opposing product is huge deliberation when purchasing is stands
on convenience exclusively [1]. According to a study; Shopping tendency of US
consumer are more complex, they expect a high stage of product and services
quality [3]. The purchaser is no longer appear and shopping for fabric and other
family goods from the tiny local marketplace, relatively their shopping favorite
has a change to the big supplies in malls [3]. Consequently, outlet offers reasonably less range of products but center on exacting product class. They do not
suggest any price endorsement rather they use expediency and affordability as
their main attribute and devices [20]. Hence, it is obvious due to city lifestyle
altering trends; the consumer will forever appear for expediency where all can
be complete and buy at one exacting position with the slightest bother. Thus,
on the basis of literature following research hypothesis was proposed:

Hypothesis a1 (Ha1): The Store location has created an impact on shopper's
store preferences.

Hypothesis a2 (Ha2): The Product convenience determines the shopper's
change their store preferences.

Hypothesis a3 (Ha3): The price has an effect on the decision of shopper's for
store preferences.

Hypothesis a4 (Ha4): The visual merchandising has impacted over the decision of shopper's store preference.

3   Methodology

This research study is presented in cross sectional study design to define the current competition among the new modern style store and traditional stores. The quantitative method is used to identify the result of design hypothesis to the defined problem. The explanatory research method is used to find out the effect among the defining factors used as an independent variable (Store Location, Convenience, Price and Visual Merchandising) over the dependent variable (Store Preference). The quantitative and empirical evidence are used to identify the cause and effect relationship among the dependent and independent variables by the help of primary data which is being collected with the structured close-ended questionnaire. Further, the respondent selected for this research study are the buyers who choose to purchase from local or international modern design retail shops. The target population of this study includes the individuals having the purchasing power. Moreover, the target population is divided into a demographic division of gender and age; as the respondent belongs to the vicinity of Naheed, imtiaz, hyperstar, metro and some other small stores presented in Karachi, Pakistan. The Sample population of this research is 329 respondents selected with the help of non-probability convenience sampling techniques. Thus, the analyses used to compute on the statistical software are SPSS and AMOS. The analysis techniques used in this study to identify the solution of the defined objectives are simulation, correlation, ANOVA and coefficient analysis to further clarified the relationship of the variables.

4   Analysis and Discussion

4.1   Structural Equation Modeling

The study identified the relationship among the factors Store Location, Convenience, Price and Visual Merchandising with customer store preference. Figure 1 Structural Equation Model of Customer Store Preference based on the connection of buyers, store selection depends on different decision factors, in this study the 4 driven factors used as the part of this study to test the significant relationship on buyers store preference, Convenience (Product Availability). Furthermore, the location of the store is also the significant factor which assists the consumer to select the store for purchasing as important as the location the Price is also the key point in changing the behavior to select the store. Thus, with the bulk amount of product these retailers have an edge to offer better discount and low pricing to the consumer as compared to the traditional store's retailers. Consumer store preference is also driven by the visual merchandising (shelf placement of products) which helps them to visualize the product before purchasing.

Moreover, the model identifies the regression values of the driven factors with store preference which is computed as 0.49 and identifies that the driven factors having their variation on customer decision to select the store for purchasing.

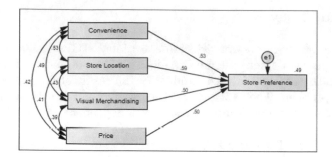

Fig. 1. Structural equation model of customer store preference

Table 1. Correlation among independent variable with store preference

Independent variable	Correlation with store preference
Convenience	0.535**
Store location	0.589**
Visual merchandising	0.505**
Price	0.506**

The correlation identification of the structural equation model and computed correlation table values identify the association of an independent variable with store preferences as store location is having the highest correlation with store preference as well as the continence is also having the positive relationship among with driven the customer decision. Thus price and visual merchandising are playing its part to create the positive, moderate relationship with the store preference as the correlation values also computed in the correlation Table 1.

4.2   Customer Store Preference in Karachi

The data are computed to analyze the store preference of the target population to have their interest in purchasing their monthly grocery from the stores. The data computed from the respondent indicated that 12% of the respondent preferred the NAHEED store to purchase their monthly grocery (Table 2).

However, 28% of the respondent prefer the purchasing from the IMTIAZ store. The results further indicated that 23% of the population are still in preference to purchasing from conventional stores as well as the 77% population of Karachi, Pakistan is in favor of modernized superstores where they can purchase all kinds of products from the single shop.

Table 2. Correlation among Independent variable with Store preference

Store	Frequency	Percent (%)
NAHEED	41	12
IMTIAZ	92	28
HYPERSTAR	61	19
METRO	58	18
OTHERS	77	23
TOTAL	**329**	**100**

4.3   Multi-regression Among Independent Variable with Store Preference

Multiple regression is computed to analyses the impact and the variation of independent variable Store Location, Convenience, Price and Visual Merchandising the driven the behavior of the individual to change their preference for purchasing products to fulfill their needs. However, the regression equation is a drive from the research model. Where value represents the gradient and the x value represents independent variables.

However, the tested values in Table 3 identify the correlation of the independent variable with the dependent variable is positive and strong and positive correlation R = 0.702. Thus, the regression value indicated the variation of independency on the dependent variable of store choice where the R square value is 0.493 which is around 49.3% of the variation. Although the adjusted value of regression is 48.7%.

Table 3. Regression table

R	0.702a
R square	0.493
Adjusted R square	0.487
F - value	78.743
F significance	0.000b

Furthermore, the F – value is tested to identify the model fitness of this research study where the f significance value is less than 0.05 and show the significance of the model in this research study.

In Table 4 the computed values highlight the coefficient values of the research model and individually identify the variation of a single driven variable with the dependent variable. The result computes that the location of the stores is having a high variation which is around 32% and having the most driven factor that can change the behavior of the individual to choose their store. Furthermore, the

Table 4. Regression Table

Model	Unstandardized coefficients		Standardized coefficients	t	Sig.
	B	Std. error	Beta		
(Store preference)	0.531	0.235		2.257	0.025
Convenience	0.134	0.039	0.174	3.429	0.001
Store location	0.289	0.045	0.319	6.487	0
Visual merchandising	0.151	0.037	0.194	4.088	0
Price	0.175	0.036	0.225	4.922	0

price is also having 22.5% of variation on the dependent variable. The further significance value identifies that the computed result is less than the tabulated value which is 0.05 which interpret that all the null hypothesis of Ha1, Ha2, Ha3, and Ha4 are rejected and the result indicated that the selected driven variable is having their impact to change the behavior of the consumer to change their store preference for purchasing decision making.

5   Conclusion

This research study identifies that the modern stores clarify diverse motivational factors for the consumer to store preference: some driven factors were tested in this study, which is reasonable pricing, easy location, visual merchandising and product convenience. This research study identified the strong connection among location and store preference as the location for any modern store is an impotent factor for the consumer to select any store for purchasing. Thus, the identified results also indicated the positive correlation and variation of factors with consumer's store preference on modern style retail store over traditional retail. Further, the study shows that if the store location is easily accessible to the consumer more individual comes into the store for purchasing; pricing is the second driven variable that shows positive correlation and impacts over store preference. Thus, there is also the relationship exist with product convenience and visual merchandising with store preference. This indicated that in the current scenario, individual in Pakistan more prefer the modernized style supermarket where they can shop any kind of product in a single shop. Thus the result achieves that if retailers need more highlighted profit they might focus some of the tested factors in this study which are the location, price, visual merchandising and product convenience. Thus, modern style stores in Pakistan is also having an edge of huge walk-ins nowadays, so they have the opportunity of bulk selling product, that's why they offer different discount to the consumer which is difficult for the traditional store to offer them to their customers. Moreover, we can define the shoppers' preference for traditional to modern stores and it stays for a prolonged period of time because almost 77% of the target population was shopping from the new retail outlets. Moreover, the retail business has displayed significant development in past times and also faces many challenges and risk.

References

1. Comyns, B.: Impact of visual merchandising on university of new Hampshire students. Honors Theses Capstones (2012)
2. Chang, E., Luan, B.: Chinese consumers' perception of hypermarket store image. Asia Pac. J. Mark. Logist. **22**(4), 512–527 (2010)
3. Jhamb, D., Kiran, R.: Emerging retail formats and it's attributes: an insight to convenient shopping. Glob. J. Manag. Bus. Res. **12**(2) (2012)
4. DH, B.: Drivers of superstore shopping: a case study of Faisalabad city (2013)
5. Urvashi, G., Tandon, V.K.: Changing consumer preferences from unorganized retailing towards organized retailing: a study in Jammu. Res. Gate (2013)
6. Geistfeld, L.V., Paulins, V.A.: The effect of consumer perceptions of store attributes on apparel store preference. J. Fash. Mark. Manag. **7**(4), 371–385 (2003)
7. Mishra, H.G., Sinha, P.K., Koul, S.: Customer dependence and customer loyalty in traditional and modern format stores. Indian Bus. Res. **9**(1), 59C78 (2018)
8. Hollands, S., Campbell, M.K., et al.: Association between neighbourhood fast-food and full-service restaurant density and body mass index: a cross-sectional study of Canadian adults. Can. J. Public Health **105**(3), 172–178 (2014)
9. Peter, J.P., Olson, J.C.: Consumer behavior & marketing strategy (2010)
10. Karadeniz, M.: The importance of retail site selection in marketing management and hypothetical approaches used in site selection. Nav. Eng. J. **5**(3) (2009)
11. Wan, M., Huang, Y., et al.: Demand estimation under multi-store multi-product substitution in high density traditional retail. Eur. J. Oper. Res. **266**(1), 99–111 (2018)
12. Martinez, M.G., Poole, N.: The development of private fresh produce safety standards: implications for developing mediterranean exporting countries. Food Policy **29**(3), 229–255 (2004)
13. Hefer, Y.: Channel switching behavior from traditional grocery stores to branded grocery chains in Karachi - leadership & management - the best way to share & discover documents. DocGoNet **12**(2) (2014)
14. Wan, M., Huang, Y., et al.: The seven ps of marketing and choice of main grocery store in a hyperinflationary economy. Contemp. Mark. Rev. **5**(2) (2013)
15. Singh, N.: The impact of visual merchandising on consumer behaviour in comparison with luxury and retail brands. Int. J. Bus. Adm. Res. Rev. **3**(1) (2016)
16. Sinha, P.K., Banerjee, A.: Store choice behaviour in an evolving market. Int. J. Retail. Distrib. Manag. **32**(10), 482–494 (2004)
17. Dunne, P.M., Lusch, R.F., Carver, J.R.: Retailing. Cengage Learning, Mason, OH (2013)
18. Ramanathan, S., Dhar, S.K.: The effect of sales promotions on the size and composition of the shopping basket: regulatory compatibility from framing and temporal restrictions. J. Mark. Res. **47**(3), 542–552 (2010)
19. Thakur, S., Sharma, S.K., Kumar, D.: A study of consumer behavior towards organized food retail storess in Bhopal city. Int. J. Manag. Res. Rev. **2**(6) (2016)

20. Sinha, P.K., Kar, S.K.: Insights into the Growth of New Retail Formats in India (2010)
21. Sunanto, S.: Modern Retail Impact on Store Preference and Traditional Retailers in West Java. Social Science Electronic Publishing (2013)
22. Hefer, Y.: Visual merchandising displays - practical or ineffective? Int. Retail. Mark. Rev. **9**(1), 73–78 (2013)

The Link Between Heterogeneity in Employment Arrangements, Team Cohesion and Team Organizational Citizenship Behavior: A Moderated Mediation Model

Xuan Wang, Yanglinfeng Zheng, and Xiaoye Qian[(⊠)]

Business School, Sichuan University, Chengdu 610064, People's Republic of China
xyqian@scu.edu.cn

Abstract. With the worldwide popularity of flexible employment arrangements in the last two decades, how to manage the employment blending team has aroused more research interests. This study investigates a moderated mediation model linking heterogeneity in employment arrangements with team organizational citizenship behavior based on social identity theory. Collecting data from four Chinese medical organizations, we provide evidence that heterogeneity in employment arrangements influences team organizational citizenship behavior through team cohesion, and the relationship between heterogeneity in employment arrangements and team cohesion is moderated by leader humility.

Keywords: Heterogeneity in employment arrangements · Team cohesion · Team organizational citizenship behavior · Leader humility

1 Introduction

Nowadays, firms are more inclined to use nonstandard employment arrangements, such as dispatched and temporary work, to reduce labor cost and enhance employment flexibility. However, the use of nonstandard employment arrangements has brought new challenges to team management in these organizations. On the one hand, the attitudes and performance of nonstandard workers differ from those of standard workers [8], making it hard to manage and motivate these two different kinds of workers at the same time. On the other hand, introducing nonstandard workers into the work team may activate employment-status-based group faultlines [10] and influence team interaction [5,6], further affecting team effectiveness. Hence, a better understanding of the influence of heterogeneity in employment arrangements (standard and nonstandard employment) in a work group will help managers better deal with the aforementioned issues of nonstandard employment, and consequently encourage the employers to adopt nonstandard employment arrangements.

© Springer Nature Switzerland AG 2020
J. Xu et al. (Eds.): ICMSEM2019 2019, AISC 1002, pp. 705–716, 2020.
https://doi.org/10.1007/978-3-030-21255-1_54

Existing literature mainly focuses on the individual level (e.g. nonstandard workers) [8,21], while research regarding the employment blending team (i.e., the work team is composed of both standard and nonstandard workers) is far from sufficient. Meanwhile, present research of the employment blending team suggests that heterogeneity in employment arrangements is associated with team members' turnover intentions, trust in peers, relations toward peers and supervisors, and helping behaviors [5]. Yet it is not clear how heterogeneity in employment arrangements impacts team outcomes.

Our study aims to respond to these limitations in extant literature. First, we investigate the influence of heterogeneity in employment arrangements at the team level. In specific, we choose team organizational citizenship behavior (OCB) as the outcome variable because team OCB is susceptible to conflicts and dysfunctional intrateam relationships [7]. Second, drawing from social identity theory, we try to offer a reasonable explanation of the impact of heterogeneity in employment arrangements on team OCB. We argue that heterogeneity in employment arrangements can cause negative interactions among sub-groups inside the team, which influences team cohesion and then team OCB. Third, we highlight the role of leader humility as a moderator of the model. Leader humility has been evidenced to be related to team effectiveness [16], yet not investigated in the context of employment blending team. It's important to explore how leader humility affects the relationship between heterogeneity in employment arrangements, team cohesion and team OCB.

2   Model Development and Hypotheses

2.1   The Impact of Heterogeneity in Employment Arrangements on Team OCB

Social identity theory suggests that people tend to categorize themselves and others into different groups to maintain positive social identities [4]. In firms, nonstandard workers often differ from standard workers in skills, position levels and tasks. Hence, employees' work arrangement difference causes social categorization inside the team easily.

In the employment blending team, greater proportions of nonstandard workers make the boundary of the two sub-groups more prominent [19]. According to in-group favoritism and out-group discrimination [18], team members are less likely to help those categorized as out-group, thus inhibiting team OCB directed toward individuals (OCBI). Meanwhile, team members, especially standard workers, might experience a decrease of job satisfaction and organizational loyalty [1], resulting in team OCB directed toward organization (OCBO) decreasing. We propose the following hypotheses:

Hypothesis 1a: Heterogeneity in employment arrangements has a negative effect on team OCBO.

Hypothesis 1b: Heterogeneity in employment arrangements has a negative effect on team OCBI.

2.2   Team Cohesion as a Mediator in the Heterogeneity in Employment Arrangements-Team OCB Relationship

Team cohesion involves a sense of belonging to the team and feelings of morale associated with membership in the team [3]. With more nonstandard workers being introduced to the team, conflicts and competitions are intensified, harming intrateam relationships [5]. Standard workers might feel their social status being threatened and opportunities being taken while nonstandard workers still have negative social identities and feel excluded from the team. Salient group boundary caused by employment arrangements inhibits the understanding between standard and nonstandard workers and leads to low commitment of team members [9], thus compromising team cohesion. Team members with low team cohesion will withdraw from the team psychologically [20], further diminishing team OCB. Drawing on the preceding arguments, we hypothesize:

Hypothesis 2: Heterogeneity in employment arrangements has a negative effect on team cohesion.

Hypothesis 3a: Team cohesion mediates the relationship between heterogeneity in employment arrangements and team OCBO.

Hypothesis 3b: Team cohesion mediates the relationship between heterogeneity in employment arrangements and team OCBI.

2.3   Leader Humility as a Moderator

Humility is defined as "an interpersonal characteristic that emerges in social contexts that connotes (a) a manifested willingness to view oneself accurately, (b) a displayed appreciation of others' strengths and contributions, and (c) teachability" [14]. High levels of leader humility will promote team integration by empowering leadership behaviors, improve followers' commitment via cultivating supportive organizational contexts [12]. In addition, team members will see the leader with high levels of humility as their role model and emulate the leader's humble behaviors [13]. When team members learn to acknowledge their own mistakes and spotlight others' strengths, salience of the group boundary will decrease, further decreasing bias [11]. This suggests that leader humility functions as a moderating variable of the relationship between heterogeneity in employment arrangements and team cohesion. Hence, we propose that:

Hypothesis 4a: The negative association between heterogeneity in employment arrangements and team cohesion is moderated by leader humility, such that it is stronger for lower than for higher levels of leader humility.

We propose that leader humility further moderates the "heterogeneity in employment arrangements-team cohesion-team OCB" relationship. As discussed above, leaders with high levels of humility help promote team integration and influence team members though social contagion [12,13]. In this case, the effect of heterogeneity in employment arrangements on team cohesion is weaker, resulting in a weaker indirect effect on team OCB. This leads to the following hypotheses:

Hypothesis 4b: The indirect effect of heterogeneity in employment arrangements on team OCBO via team cohesion is moderated by leader humility, such that the indirect effect will be mitigated when leader humility is high.

Hypothesis 4c: The indirect effect of heterogeneity in employment arrangements on team OCBI via team cohesion is moderated by leader humility, such that the indirect effect will be mitigated when leader humility is high.

The conceptual model is shown in Fig 1.

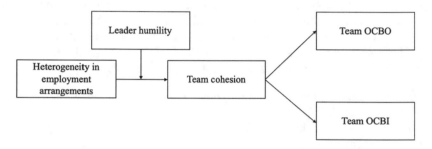

Fig. 1. The conceptual model of the study

3 Method

3.1 Sample and Procedure

We collected survey data from 355 workers in 66 teams across four Chinese medical organizations. To ensure leaders' ratings for all team members, we obtained the name list from the HR department before the survey. We then wrote the name of each team member on each questionnaire and numbered the questionnaires in a special way (Table 1).

We distributed questionnaires to team members and team leaders on site. The team member survey asked participants to provide self-ascriptive ratings on their employment status and demographic variables. The team leader survey requested leaders to rate their employment status and demographic variables. It also asked them to assess team cohesion and each team member's OCBO and OCBI. Three team leaders failed to assess team cohesion, rendering the teams' surveys unusable. Therefore, the final sample consisted of 63 teams.

The average actual team size was 5.70, ranging from two to 15 team members. A representative team member was male (25.98%), 37 years of age, and with an average organizational tenure of 11.47 years. A representative team leader was male (33.33%), 41 years of age, and with an average organizational tenure of 14.81 years.

Table 1. Descriptive statistics of the sample

Variables	Sample size	M	SD
Team leader			
Age	63	40.59	6.05
Tenure	61	14.81	8.83
Team member			
Age	249	36.92	8.59
Tenure	250	11.47	8.93
Variables	**Type**	**Sample size**	**Proportion**
Team leader			
Gender	Total	63	100.00%
	Male	21	33.33%
	Female	42	66.67%
Education level	Total	63	100.00%
	University graduate and above	47	74.60%
	College graduate	11	17.46%
	Junior college graduate	3	4.76%
	High school graduate and below	2	3.17%
Employment status	Total	63	100.00%
	Standard	61	96.83%
	Nonstandard	2	3.17%
Team member			
Gender	Total	254	100.00%
	Male	66	25.98%
	Female	188	74.02%
Education level	Total	253	100.00%
	University graduate and above	148	58.50%
	College graduate	70	27.67%
	Junior college graduate	27	10.67%
	High school graduate and below	8	3.16%
Employment status	Total	254	100.00%
	Standard	222	87.40%
	Nonstandard	32	12.60%

3.2   Measures

All measures used a Likert response scale ranging from 1 (strongly disagree) to 5 (strongly agree). We translated all measures into Chinese following a standard back-translation procedure.

Heterogeneity in Employment Arrangements. Both team leaders and team members reported their employment statuses. We then categorized them into two: One is the standard worker, including permanent worker; The other is the nonstandard worker, including contract and dispatched worker. We assessed heterogeneity in employment arrangements using Blau's [2] index:

$$H = 1 - \sum_{i=1}^{2} p_i^2 \tag{1}$$

Where p_i denotes the proportion of standard or nonstandard workers in a team, $H \in (0,1)$.

Leader Humility. We assessed leader humility with a scale originally developed by Owens et al. [14] and later modified by Owens et al. [15]. Team members rated leader humility. Sample items include "My leader admits it when he or she makes mistakes" and "My leader shows a willingness to learn from others". Cronbach's alpha was 0.96. The responses were aggregated at the team level (ICC1 = 0.14; ICC2 = 0.40; r_{wg} = 0.96).

Team Cohesion. We used the referent-shift model to measure team cohesion with a six-item scale developed by Bollen and Hoyle [3]. Team leaders rated team cohesion. Sample items include "Team members feel a sense of belonging to the team" and "Team members feel that they are members of the team". Cronbach's alpha was 0.93.

Team OCBO. Team leaders assessed each team member's OCBO using a seven-item scale developed by William and Anderson [22]. Sample items include "Attendance at work is above norm" and "Give advance notice when unable to come to work". Cronbach's alpha was 0.85. ICC1 and ICC2 were 0.60 and 0.86, respectively. r_{wg} was 0.97. We aggregated OCBO at the team level.

Team OCBI. Team leaders assessed each team member's OCBI using a seven-item scale developed by William and Anderson [22]. Sample items include "Help others who have been absent" and "Help others who have heavy work loads". Cronbach's alpha was 0.90. ICC1 and ICC2 were 0.55 and 0.83, respectively. r_{wg} was 0.98. We also aggregated OCBI at the team level.

Control Variables. We control for (1) Team size; (2) Team mean tenure; (3) Team type: 1 = team of specialties and sub-specialties, 0 = team of support staff.

4   Result

4.1   Descriptive Results

Table 2 presents the descriptive statistics and the bivariate correlations of the study variables. As shown in the table, heterogeneity in employment arrangements was significantly related to team cohesion ($\beta = -0.31$, p < 0.05), team

Table 2. Means, standard deviations, and correlations among study variables

Variable	M	SD	1	2	3	4	5	6	7	8
1. Heterogeneity in employment arrangements	0.11	0.18	1.00							
2. Team cohesion	4.20	0.67	−0.31**	1.00						
3. Team OCBO	4.18	0.62	−0.28**	0.71***	1.00					
4. Team OCBI	3.88	0.63	−0.30**	0.76***	0.86***	1.00				
5. Leader humility	3.44	0.53	−0.15	0.28**	0.32***	0.37***	1.00			
6. Team size	5.70	3.34	0.07	−0.09	−0.12	−0.27**	−0.25**	1.00		
7. Team mean tenure	12.04	5.07	0.02	0.06	0.08	−0.05	−0.04	0.11	1.00	
8. Team type	0.43	0.50	0.27**	−0.16	−0.17	−0.07	0.31**	−0.36***	−0.31**	1.00

Note: *p < 0.1, **p < 0.05, ***p < 0.01, n = 63

OCBO ($\beta = -0.28$, p < 0.05) and team OCBI ($\beta = -0.30$, p < 0.05). Team cohesion was significantly related to team OCBO ($\beta = 0.71$, p < 0.01) and team OCBI ($\beta = 0.76$, p < 0.01). These results provided preliminary evidence for our hypotheses.

4.2 Confirmatory Factor Analyses

We conducted CFA to confirm fitness of the measurement model. The results revealed that the four-factor model ($\chi^2_{(428)}$ = 794.55, RMSEA = 0.12, CFI = 0.84, TLI = 0.83, SRMR = 0.07) provided a better fit than the other three models.

4.3 Model Testing

We proposed that heterogeneity in employment arrangements would be negatively related to team OCBO (H1a) and team OCBI (H1b). To test Hypotheses 1a and 1b, we used ordinary least squares (OLS) regressions. As shown in Table 3, Model 4 included all the control variables. Then we entered heterogeneity in employment arrangements into the model (Model 5) and found that its effect on team OCBO was marginally significant ($\beta = -0.80$, p < 0.1), supporting our Hypothesis 1a. Similarly, the result showed that the relationship between heterogeneity in employment arrangements and team OCBI was marginally significant ($\beta = -0.88$ p < 0.1; see Model 8), supporting our Hypothesis 1b (Table 4).

Hypotheses 2 and 3 proposed that heterogeneity in employment arrangements would be negatively related to team cohesion (H2) and team cohesion would mediate the relationship between heterogeneity in employment arrangements and team OCBO (H3a) and team OCBI (H3b). As shown in Table 3, heterogeneity in employment arrangements was significantly associated with team

Table 3. Regression results

Variables	Team cohesion			Team OCBO			Team OCBI		
	Model 1	Model 2	Model 3	Model 4	Model 5	Model 6	Model 7	Model 8	Model 9
Constant	4.49***	4.44***	4.42***	4.47***	4.43***	1.65***	4.46***	4.42***	1.40***
Team size	−0.03	−0.02	−0.02	−0.04	−0.03	−0.02	−0.06**	−0.05**	−0.04**
Team mean tenure	0.00	0.01	0.00	0.00	0.01	0.00	−0.01	−0.01	−0.01
Team type	−0.29	−0.15	−0.28	−0.29	−0.18	−0.09	−0.26	−0.14	−0.04
Team cohesion						0.62***			0.68***
Heterogeneity in employment arrangements		−1.03**	−0.62		−0.80*	−0.16		−0.88*	−0.18
Leader humility			0.43**						
Heterogeneity in employment arrangements ×Leader humility			1.65*						
R^2	0.05	0.12	0.24	0.07	0.11	0.52	0.11	0.16	0.62

Note: *$p < 0.1$, **$p < 0.05$, ***$p < 0.01$, n = 63

Table 4. Results of confirmatory factor analyses

Model	χ^2	df	χ^2/df	RMSEA	CFI	TLI	SRMR
One-Factor model	1622.22	434	3.74	0.21	0.48	0.45	0.20
Two-Factor model (Team OCBO and Team OCBI are combined)	1141.07	433	2.64	0.16	0.69	0.67	0.22
Three-Factor model (Team OCBO and Team OCBI are combined)	837.01	431	1.94	0.12	0.82	0.81	0.07
Four-Factor model	794.55	428	1.86	0.12	0.84	0.83	0.07

cohesion ($\beta = -1.03$, $p < 0.05$; see Model 2); therefore, H2 is supported. When both heterogeneity in employment arrangements and team cohesion were entered into the model simultaneously (Model 6, Model 9), heterogeneity in employment arrangements dropped from significance indicating that team cohesion mediated the relationship between heterogeneity in employment arrangements and team OCB. We further conducted Sobel's test [17] to assess the mediation. The results showed that the intervening effects of team cohesion on the relationship between heterogeneity in employment arrangements and team OCBO (Sobel z = -1.98, $p < 0.05$) and team OCBI (Sobel z = -2.01, $p < 0.05$) were both significant. Therefore, H3a and H3b are supported.

We also proposed that the relationship between heterogeneity in employment arrangements and team cohesion, as well as the relationship between heterogeneity in employment arrangements, team cohesion and team OCB would be moderated by leader humility (H4). Firstly, as shown in Table 3, we found that the interaction effect between heterogeneity in employment arrangements and leader humility relative to team cohesion was marginally significant ($\beta = 1.65$, p < 0.1; see Model 3), supporting Hypothesis 4a. We plotted the interactive effect on team cohesion, as shown in Fig 2.

Secondly, we adopted a bootstrapping analysis (n = 1,000). As shown in Table 5, We found that the indirect effect of heterogeneity in employment arrangements

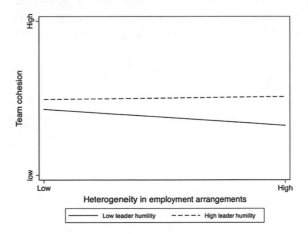

Fig. 2. The interactive effect of heterogeneity in employment arrangements and leader humility on team cohesion

Table 5. Conditional indirect effects between heterogeneity in employment arrangements and team OCBO via team cohesion

Predicor	B	SE	z	p
Team cohesion				
Constant	3.66	0.64	5.74	0.00
Heterogeneity in employment arrangements	−6.29	2.88	−2.19	0.03
Leader humility	0.24	0.17	1.42	0.16
Heterogeneity in employment arrangements × Leader humility	1.65	0.85	1.95	0.05
Team OCBO				
Constant	1.67	0.55	3.04	0.00
Team cohesion	0.54	0.09	6.14	0.00
Heterogeneity in employment arrangements	−4.09	2.08	−1.97	0.05
Leader humility	0.12	0.12	0.96	0.34
Heterogeneity in employment arrangements × Leader humility	1.19	0.61	1.97	0.05
Leader humility	Boot indirect effect	Boot SE	Boot z	Boot p
Conditional indirect effect at leader humility = M ± 1 SD				
−1 SD (−0.53)	−1.01	0.37	−2.75	0.01
M (0.00)	−0.50	0.30	−1.69	0.09
+1 SD (0.53)	0.01	0.42	0.02	0.99

Note: n = 63 teams. The data of heterogeneity in employment arrangements and leader humility are grand mean centered. For the limitation of the space, control variables are not reported in this table. Bootstrap sample size = 1,000

on team OCBO was significantly negative only when leader humility was at a low level (-1 SD; bootstrapped indirect effect $= -1.01$, $p < 0.05$). When leader humility was high, the indirect effect became insignificant ($+1$ SD; bootstrapped indirect effect $= 0.01$, ns). Similarly, as shown in Table 6, the indirect effect of heterogeneity in employment arrangements on team OCBI was significantly negative only when leader humility was at a low level (-1 SD; bootstrapped indirect effect $= -1.16$, $p < 0.01$). When leader humility was high, the indirect effect became insignificant ($+1$ SD; bootstrapped indirect effect $= 0.01$, ns).

Table 6. Conditional indirect effects between heterogeneity in employment arrangements and team OCBI via team cohesion

Predicor	B	SE	z	p
Team cohesion				
Constant	3.66	0.64	5.74	0.00
Heterogeneity in employment arrangements	−6.29	2.88	−2.19	0.03
Leader humility	0.24	0.17	1.42	0.16
Heterogeneity in employment arrangements × Leader humility	1.65	0.85	1.95	0.05
Team OCBI				
Constant	0.99	0.51	1.93	0.05
Team cohesion	0.64	0.08	7.78	0.00
Heterogeneity in employment arrangements	−0.35	1.93	−0.18	0.86
Leader humility	0.18	0.11	1.61	0.11
Heterogeneity in employment arrangements × Leader humility	0.08	0.56	0.14	0.89
Leader humility	Boot indirect effect	Boot SE	Boot z	Boot p
Conditional indirect effect at leader humility = M ± 1 SD				
−1 SD (−0.53)	−1.16	0.39	−2.96	0.00
M (0.00)	−0.58	0.33	−1.77	0.08
+1 SD (0.53)	0.01	0.47	0.02	0.99

Note: n = 63 teams. The data of heterogeneity in employment arrangements and leader humility are grand mean centered. For the limitation of the space, control variables are not reported in this table. Bootstrap sample size = 1,000

5 Conclusion

In this study, we examined the moderated mediation model linking heterogeneity in employment arrangements, team cohesion and team OCB. We found that heterogeneity in employment arrangements impacted team OCB via team cohesion. We also found that leader humility acted as a moderator of the mediation relationship.

Our findings suggest that despite the benefits of adopting nonstandard employment arrangements, organizations should be aware of the negative influence of heterogeneity in employment arrangements in a work group. We recommend that organizations take actions to promote team cohesion and team OCB for employment blending teams, such as team building activities. Additionally, appointing a humble leader to the employment blending team might mitigate the negative effect of heterogeneity in employment arrangements.

Some limitations of the study should be noted. First, the cross-sectional data cannot prove a causal link between the variables. Future research should use longitudinal data to avoid the problem. Second, this study investigated the mediating role of team cohesion. However, there could also be alternative mechanisms drawing from other theoretical lens. We advise scholars to explore these mediators to better understand the influence of heterogeneity in employment arrangements on team effectiveness.

Acknowledgments. The authors thank for the support by the National Natural Science Foundation of China (Grant No.71402108 and No.71872117), the Philosophy and Social Science Fund of Sichuan Province (Grant No. SC17B053) and the Sichuan University Special Research Project under "Double First-Class" Initiative (Grant No. SKSYL201703).

References

1. Banerjee, M., Tolbert, P.S., DiCiccio, T.: Friend or foe? The effects of contingent employees on standard employees' work attitudes. Int. J. Hum. Resour. Manag. **23**(11), 2180–2204 (2012)
2. Blau, P.M.: Inequality and Heterogeneity: A Primitive Theory of Social Structure, vol. 7. New York (1977)
3. Bollen, K.A., Hoyle, R.H.: Perceived cohesion: a conceptual and empirical examination. Soc. Forces **69**(2), 479–504 (1990)
4. Brewer, M.B.: Intergroup Relations. Oxford University Press, Oxford (2010)
5. Broschak, J.P., Davis-Blake, A.: Mixing standard work and nonstandard deals: the consequences of heterogeneity in employment arrangements. Acad. Manag. J. **49**(2), 371–393 (2006)
6. Chattopahyay, P., George, E.: Examining the effects of work externalization through the lens of social identity theory. J. Appl. Psychol. **86**(4), 781 (2001)
7. Choi, J., Sy, T.: Group-level organizational citizenship behavior: effects of demographic faultlines and conflict in small work groups. J. Organ. Behav. **31**(7), 1032–1054 (2010)
8. Guillaume, P., Sullivan, S.E., et al.: Are there major differences in the attitudes and service quality of standard and seasonal employees? An empirical examination and implications for practice. Hum. Resour. Manag. (2018)
9. Ho, H.: Blending nonstandard and standard employment relations. In: Academy of Management Proceedings. Academy of Management Briarcliff Manor, NY 10510, vol 2016, p 18007 (2016)
10. Liu, X., Li, X.: Employment-status-based faultlines in diverse employment work groups and their activating factors: a grounded theory exploration. Chin. J. Manag. **7**, 009 (2015)
11. Mullen, B., Brown, R., Smith, C.: Ingroup bias as a function of salience, relevance, and status: an integration. Eur. J. Soc. Psychol. **22**(2), 103–122 (1992)
12. Ou, A.Y., Tsui, A.S., et al.: Humble chief executive officers' connections to top management team integration and middle managers' responses. Adm. Sci. Q. **59**(1), 34–72 (2014)

13. Owens, B.P., Hekman, D.R.: How does leader humility influence team performance? Exploring the mechanisms of contagion and collective promotion focus. Acad. Manag. J. **59**(3), 1088–1111 (2016)
14. Owens, B.P., Johnson, M.D., Mitchell, T.R.: Expressed humility in organizations: implications for performance, teams, and leadership. Organ. Sci. **24**(5), 1517–1538 (2013)
15. Owens, B.P., Wallace, A.S., Waldman, D.A.: Leader narcissism and follower outcomes: the counterbalancing effect of leader humility. J. Appl. Psychol. **100**(4), 1203 (2015)
16. Rego, A., Simpson, A.V., et al.: The perceived impact of leaders' humility on team effectiveness: an empirical study. J. Bus. Ethics **148**(1), 205–218 (2018)
17. Sobel, M.E.: Asymptotic confidence intervals for indirect effects in structural equation models. Sociol. Methodol. **13**, 290–312 (1982)
18. Tajfel, H., Billig, M.G., et al.: Social categorization and intergroup behaviour. Eur. J. Soc. Psychol. **1**(2), 149–178 (1971)
19. Terry, D.J., Callan, V.J.: In-group bias in response to an organizational merger. Group Dyn. Theory Res. Pract. **2**(2), 67 (1998)
20. Turner, J.C., Hogg, M.A., et al.: Rediscovering the Social Group: A Self-categorization Theory. Basil Blackwell, Oxford (1987)
21. Haines III, V.Y., Doray-Demers, P., Martin, V.: Good, bad, and not so sad part-time employment. J. Vocat. Behav. **104**, 128–140 (2018)
22. Williams, L.J., Anderson, S.E.: Job satisfaction and organizational commitment as predictors of organizational citizenship and in-role behaviors. J. Manag. **17**(3), 601–617 (1991)

Administrative Resilience and Adaptive Capacity of Administrative System: A Critical Conceptual Review

Md Nazirul Islam Sarker$^{(\boxtimes)}$, Min Wu, Roger C. Shouse, and Chenwei Ma

School of Public Administration, Sichuan University, Chengdu 610065,
People's Republic of China
sarker.scu@yahoo.com

Abstract. Resilience development is viewed as a practical and effective approach to overcome the dynamic and uncertain conditions of an administrative system. This article explores factors of public administration that can enhance the adaptive capacity of administrative systems to achieve resilience. A conceptual model is developed, based on an extensive literature review, to interpret administrative resilience across several indicators. In addition, the case of Bangladesh riverine island (char) areas served as a case study validating the conceptual model. The study reveals that administrative resilience is the ability of the administrative system to provide appropriate measures to uncertainties and bounce back to previous conditions after facing risks, shocks, and disasters, and other threats to organizational stability. It also argues that administrative resilience is far better than the conventional administrative approaches in terms of organizational flexibility, quick response to uncertain and rapidly changing conditions, and strengthening the socio-ecological system for tackling vulnerability. Case study analysis confirms the conceptual model and reveals a great need to improve administrative resilience practices in riverine islands.

Keywords: Resilience · Administration · Risk management ·
Governance · Emergency management

1 Introduction

Resilience is a holistic term which is gained a popularity in the researcher community across natural and social science disciplines. The term refers to the ability of individuals or systems to withstand against the shocks and stress of environmental disturbance and uncertainty [4]. The concept is useful for evaluating the performance of a system in the face of adverse events and has been integrated into fields such as medicine, psychology, engineering, nursing, business, zoology, and urban management. The term resilience comes from the Latin word "resiliens" which means "to bounce back" or "to rebound". In psychology and medical sciences, it means to recovery from depression or illness. Generally, it

© Springer Nature Switzerland AG 2020
J. Xu et al. (Eds.): ICMSEM2019 2019, AISC 1002, pp. 717–729, 2020.
https://doi.org/10.1007/978-3-030-21255-1_55

focuses on the capacity to regain or bounce back to the original state after facing adverse events. Resilience is a tool to recover the original state after facing vulnerability [3].

Resilience is gaining popularity for enhancing the capacity of a vulnerable community. Such capacities may be categorized as adaptive, absorptive, and transformative. Though some studies exist on resilience in various disciplinary perspectives (e.g., psychological resilience, livelihood resilience, climate change resilience), research is still lacking with respect to administrative resilience to climate vulnerability in geographically isolated areas. We cannot deny the potential of administrative resilience because of its robustness, potential and the fact that people are living in places where natural hazards are common phenomena. Because public administrators are required to deal with such situations, organizational preparedness is necessary to control, monitor and adapt to the situation. Considering the importance of the issue, this study intends to address the research gap in public administration through exploring the potential of administrative resilience in the context of the vulnerability of natural hazards and provide a workable framework for better understanding. The rest of the paper divided into six sections. The second and third section deals with methodology and conceptual analysis of administrative resilience. The fourth and fifth section mainly focus on discussion emphasizing the administrative resilience concept, characteristics, dimensions, and a case study on riverine island areas of Bangladesh. The final section concludes the paper.

2   Methodology

2.1   Research Design

A mixed method approach consisting of a systematic literature review and case study has been conducted for obtaining research objectives. An extensive desk literature review has been done for obtaining recent literature (from 2005 to 2018). Concept analysis was conducted to clarify and validate the resilience concept through filed data. A case study (as part of the author's PhD research) also has been done to validate the conceptual framework by exploring the administrative resilience status in a disaster-prone vulnerable community of Bangladesh.

2.2   Search Strategy

This study emphasizes the resilience concept in the multidisciplinary aspects. An extensive literature search is done on some renowned databases like web of science, engineering village, Scopus and google scholar databases by using keywords such as "resilience, community resilience, resilient, ecology, vulnerability, adaptive, absorptive, transformative capacity". The following search strategies have been followed (Fig. 1).

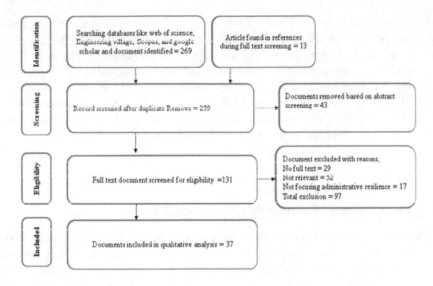

Fig. 1. PRISMA selection for identifying qualitative study

Web of science   : TS = (resilien * AND vul); TS = (resilien * AND admin)
Engineering village : resilience AND administration
Scopus   : resilience AND vulnerability; administration AND capacity
Google scholar   : Resilience, resilient, vulnerability, community, adaptive, ab−
                              sorptive, transformative capacity, administration, governance.

2.3 Inclusion Criteria

The literature search is guided by the criteria that the study must focus vulnerability and resilience in a climate vulnerable context. Resilience focused human population-based peer reviewed journal articles are considered for the time from 2005 to 2018. The focal points of the study are to identify the important definitive research on resilience, the trend of the modification and integration of the concept and meaning of resilience to other disciplines, and potential implications for society.

2.4 Exclusive Criteria

This study excludes journal articles lacking full text and those published in languages other than English. It also excludes articles unrelated to the concept of resilience in climate change vulnerability contexts.

3 Review Results

Preferred Reporting Items for Systematic Review and Meta-Analysis (PRISMA) guidelines has been followed in a systematic literature review [22]. The review search has been done by following several steps; first, 269 documents are identified with 19 from references. After removing the duplicates, 239 documents have been selected by abstract screening. Then 97 documents have been excluded due to lack of full text, irrelevant, not focusing on administrative resilience. Finally, thirty-seven documents have been selected from journal articles, books, book chapter, and working papers. The checklist of the Reporting of Observational Studies in Epidemiology (STROBE) is also followed for qualitative document selection [12]. This review has conducted from October to November 2018. A conceptual model has been developed for administrative resilience.

4 Discussion

4.1 The Concept of Administration Resilience

Resilience is considered as the most effective approach to solve administrative problems emerging from dynamic socio-ecological systems. Resilience generally focuses on the ability of a system, organization or society to withstand or recover from the effect of natural hazards. It also reveals the organization's capacity to survive an uncertain and risky situation gaining experience from the past and solve the future. Resilience is now gaining popularity in policy goals as a metaphor for the expected condition [18]. Scholars and practitioners of public administration are now turning attention to administrative resilience as a systemic way for organizations to cope with shocks, risks, vulnerability and adversity of natural hazards [18]. It provides an opportunity to deal with uncertain situations in a better way of preparation, dealing and adaptation. Since the 1980s, the concept of resilience has been used in emergency preparedness and management [5] and natural resource management [15]. Hood [7] induced the concept of resilience in public administration in the 1990s as a neglected concept which conflicted with "efficiency" to the previous administrative interest. But in 2000, resilience gained popularity in public administration to address the vulnerability, shocks, risks, and crisis. As a new domain, administrative resilience was used in governance system to address uncertainty and complexity [26]. Generally, two considerations of administrative systems are related to resilience; first, the level of concept, plan, and preparation of administrative system to prevent, mitigate and adapt to the adversity; second, the level of response to uncertainty and complexity through innovation, learning and adaptation strategies. The ultimate goal of administrative resilience is to enable people, community, system or organization to "bounce back" to a previous normal condition. Administrative resilience promotes the ability of social systems to remain stable during and after times of adversity. Some of the researchers emphasize the resilience concept for the administrative system (Table 1).

The problem of an administrative system is to develop capacity aimed at reducing costs and damage resulting from rapidly changing and uncertain conditions [11]. Ostrom and Janssen [23] mentioned that uncertainty is a condition in which options are reduced and in which outcomes are significantly influenced by others. Uncertainty stimulates barriers in administrative system which reduces

Table 1. Major concepts of administrative resilience thinking

Researchers	Summary of concepts
Comfort et al. [8]	Social, environmental or technological disasters can be mitigated and adapted by enhancing anticipate and adaptive capacity or administrative resilience
Ostrom [20]	Administrative resilience focuses on the usual level of analysis and relationship of the element of a system for managing unexpected situations
Haase [15]	Administrative resilience is identified as a key way to overcome the problems related administrative system
Toonen [25]	Resilience in public administration focuses on the enhancement of self-governing capacities of socio-ecological systems to withstand and recover from external and internal risks, shocks and disasters
Armitage [10]	Administrative law can help to build, enhance or maintaining resilience in public administration
Garschagen et al. [13]	Organizational practices can be promoted through adopting resilience thinking to manage socio-ecological systems efficiently
Milley and Jiwani [21]	The weakness of the government strategies can be solved by enhancing anticipative and adaptive capacity i.e. administrative resilience
Boyd et al. [6]	Anticipatory practices are tools of adaptive governance which can enhance administrative resilience
Duit [11]	The concept of resilience in public administration is basically focused on crisis management

system ability to maintain an operational environment. In this situation, administrators cannot predict the uncertainty of the system and ultimately fail to achieve desired outcomes. Generally, two things happen in uncertain situation [15]; first, one cannot predict, based on prior knowledge, whether a specific action can bring a specific outcome; second, changes in the operational environment reduce the effectiveness of previously employed standard problem solving procedures. In this situation, administrative resilience can respond to uncertainty, crisis, and risks through minimizing administrative constraints and ensuring better operational environment. Access to information is a major key by which the administrative system can improve its capacity to deal with the uncertainty. According to Goldsmith et al. [14], polycentric governance systems can handle the adversity in an effective way due to the heterogenous network. It is now established that the network can play a vital role in the public governance system. The heterogenous organizational network is able to enhance the administrative capacity to tackle and adapt the situation by informing decision makers.

722   M. N. I. Sarker et al.

4.2 Framework of Administrative Resilience

In response to disaster risk or disruptive events, administrative resilience shows heterogeneous nature. According to Haase [15], seven specific conditions control to the administrative resilience: (i) proper understanding of operational environment; (ii) social and ecological interaction; (iii) access to information; (iv) comparison of the process of action; (v) quick decision-making ability; (vi) capacity to make adaptation strategies; (vii) technological integration for decision making. Systems administrators can make a decision, adapt strategies and organize themselves for disaster management by adopting these seven conditions. Decision makers can also enable the adaptive, absorptive and transformative capacity of the administrative system for managing vulnerable contexts by following these conditions. According to Ostrom [20], the framework of administrative resilience focuses on the usual level of analysis and relationship of the element of a system for organizing diagnostic and specific inquiry.

A conceptual model has been developed in this study for interpreting the administrative resilience for ensuring livelihood resilience in the disaster vulnerable context. The framework (Fig. 2) focuses on the elements and dimensions of administrative resilience by identifying the relationship among them for facilitating adaptation and system's self-organization. The framework explores a relationship among the components for investigation of the status of administrative resilience without testing the casual hypothesis (Table 2).

Some studies already have been done on administrative response system [15]; adaptive governance [25], institutional response [5], institutional policy [16] in the context of disasters. Comfort et al. [8] conducted a study on administrative responses system for the earthquake in nine countries and reported that mainly four components were usually managing the disruption of natural disasters such as technology, organizational flexibility, culture and availability of information. According to Boyd et al. [6], anticipatory governance comprising political, technological, organization and participation of people can enhance resilience in the social-ecological context.

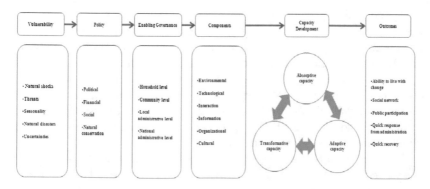

Fig. 2. A conceptual framework of administrative resilience for disaster management

Table 2. System's structure and components of administrative resilience

Administrative system	Administrative resilience components	Resilience conditions
System structure	Environmental component	Environment
	Technological component	Technology
	Interaction component	Interaction
System process	Information component	Information
	Organizational component	Rapidity, and adjustment
	Cultural component	Pattern identification

The present study basically focuses on the administrative resilience of a natural hazard prone geographically isolated riverine island (char) in Bangladesh. This study adopted the conceptual framework of comfort [17], Hasse [15], Cutter et al. [9], Johnson and Matanoski [19] and customized according to the basis of local context. The framework consists of seven steps: vulnerability, policy, enabling governance, components, capacity development and outcome which starts from vulnerability analysis and ends at outcome through an integrated way of comprising all possible steps, process and components. This framework is developed for analyzing and assessing any disaster related context so that administrators can easily take a decision about possible required intervention. The vulnerability step focuses all the vulnerable situations like natural shocks, threats, seasonality, natural disasters and uncertainties. People of the developing countries are very much familiar with this situation. Step two emphasizes the policy related issues because of its dependency on administrators and focuses on political, financial, social and natural conservation related policies. Policy analysis helps the public administrator to get overall ideas of livelihood related issues and possible intervention. The third step addresses how administrative system can work for disaster vulnerable populations and it emphasizes that the administrative system should work at household, community, local administrative unit and national or state level. It will help the administrator to enable governance systems effectively for disaster management. The fourth step focuses on the components of administrative resilience like environmental, technological, interaction, information, organizational and cultural components. The details of the components are discussed in the following sessions. The fifth step emphasizes the major dimension of capacity development for administrative resilience such as adaptive, absorptive and transformative capacity (details in the following section). And the final step describes the outcome of administrative resilience. The major outcomes are the ability to live with change, social network development, public participation, quick response from public administration and quick recovery.

4.3   Components of Administrative Resilience

There are six major components of administrative resilience: environmental, technological, interaction, information, organizational and cultural. The details of these components are described based on the natural disaster context.

(1) Environmental component

An administrative system is responsible to develop, implement and monitor of rules, regulations and policy of disaster management. It is also responsible for caring for victims of natural disasters. It works to mitigate the causes of climate change and make a strategy for better adaptation. Environmental components of the administrative system also emphasize to plan, policy and awareness of the system focusing environment related issues. The major sub-components of the components are social and governmental awareness, rule, and regulations, plan for disaster management and institution [15].

(2) Technological component

Technology is a key tool to tackle the risky operational environment. The main focus of the technological component is to consider which technology is suitable to apply for effective operations. Technological components can enhance the adaptive and transformative capacity of the administrative system. It basically focuses on weather forecasting, disaster forecasting, early warning and evacuation. Technological integration is a basic task of the administrative system for effective adaptive governance especially disaster-prone period. The major sub-components are availability, and adaptability of technology, interoperability, form, integration and interoperability of technology in an administrative system [21].

(3) Interaction component

Interaction component focuses on the ability of an administrative system to interact with others such as public organizations, private organizations, voluntary organizations, donor organizations and international organizations. This component is very much responsible for exchanging information, experiences, knowledge, resources and experts to handle the situation of disasters. It works before, during and after disasters to minimize the losses. The major sub-components of the information component are resources, personnel, information, expert and evolution [16].

(4) Information component

A proper and accurate decision requires authentic information. Information is an important component of the administrative system. It helps to know the exact situation of the disaster, organizational ability, and possible intervention. Without information, no one can design, implement or monitor the governance system. So, the availability of information is better for the governance system to manage the situations. The major sub-components of information components are the availability, value, quality, authenticity and acceptance of information [15].

(5) Organizational component

Organizational component focuses on how to address the problem of a disaster context in an effective way. It helps to organize the related stakeholders to make a good and rapid solution for managing the risk immediately. Rapidity and adjustment are the major characteristics of the organizational component which help to solve the problem rapidly and adjust the situation as the previous one. Organizational component mainly emphasizes increasing the adaptive capacity of the administrative system by providing training, knowledge dissemination, planning, organizing, budgeting, and coordinating related stakeholders. It also focuses on the maximum utilization of resources for disaster management. The major sub-components of the organization components are the plan of work, authority, expert personnel, training, resource management and flexibility [25].

(6) Cultural component

Cultural component addresses how the administration can identify the similarities and differences between the present and previous problems so as to develop more effective strategies. This component emphasizes pattern-matching for solving any unique or unexpected situations raised due to natural disasters. It also considers the local adaptation strategy for disaster management. It gives importance of organizational awareness, previous experiences, the culture of the community, new approaches and trust [15].

5   Case Study on Administrative Resilience in Riverine Island Areas in Bangladesh

A case study has been done on administrative resilience of vulnerable Riverine island (char) dwellers who are living in two local administrative unit (upazila) under Gaibandha district of Bangladesh. Char is the place in riverbed which has emerged due to deposition of silt and alluvium. The main characteristics of char areas are extreme geographical isolation. Char areas are almost inaccessible and detached from mainland activities in almost all aspects of life. The unique condition of the chars mean that char-dwellers, particularly the extreme poor, are highly vulnerable. Deprivation in all basic needs of life such as food security, agricultural development, health, education, habitation, and empowerment serve to make it almost impossible for the poor to rise above the poverty cycle. The scarce earning and employment opportunities are exacerbated by natural disasters. In Bangladesh, there are around 900 numbers of chars (7200 sq. km) which are familiar with river erosion, drought, and flood with other natural disasters. Jamuna has a higher area than the others, in terms of percentage of total within-bank area covered rivers. Thus, while this figure works out to be 45% for the Jamuna, the corresponding figures for the Padma 30% [24]. However, 5–7% of the total population of Bangladesh lives in Char areas. The majority of the char dwellers (65%) are living in the riverine island (char) areas of Jamuna River which was the case study area for this study. As we discussed in the previous section, administrative resilience is a function of the adaptive, absorptive and transformative capacity of an administrative system to withstand against the adverse effect of natural disasters. The overall status of the administrative resilience in the char areas is presented in Table 3.

Table 3. Administrative resilience status in the char areas

Major components	Sub-components	The extent of administrative intervention			
		No	Low	Moderate	High
Adaptive capacity	Disaster management training to people		✓		
	Encouraging local level leadership		✓		
	Forming community level organization		✓		
	Early warning system		✓		
	Building disaster shelter			✓	
	Training for agricultural practices		✓		
	Income generating activities training			✓	
	Health related training		✓		
Absorptive capacity	Minimization of failure		✓		
	Early detection	✓			
	Flexibility	✓			
	Controllability		✓		
	Reducing the adverse effect				
Transformative capacity	Immediate action during a disaster		✓		
	Relief during and after the disaster		✓		
	Cash incentives				
	Evacuation of people from the place of disaster		✓		
	Ensuring public services during and after the disaster		✓		
	Controlling corruption in project implementation		✓		
	Engaging special task force for disaster management	✓			

Source: Based on focus group discussions, 2017

5.1 Adaptive Capacity

Adaptive capacity is the ability of an administrative system to cope with the stress and shocks of natural hazards. Adaptation is a process to enhance adaptive capacity which is necessary for a rural institution in developing countries to protect themselves from the harmful effect of natural disasters. Since char areas are highly vulnerable to climate variability and natural disasters, so enhancement of the adaptive capacity of the rural institution is a prior need to save them. According to Duit [11], the institution has a great influence to enhance the adaptive capacity of the rural people by transforming coping strategy to adaptive capacity. This study reveals that the adaptive capacity of the local government institution in the char areas is unable to manage the situations of natural disasters which ultimately causes huge damages [2]. The study explores the ways for enhancing adaptive capacity such as disaster management training to people, encouraging local level leadership, forming community level organization, early warning system, building disaster shelter, training for agricultural practices, income generating activities training, and health related training [25].

5.2   Absorptive Capacity

Absorptive capacity is the ability of an administrative system to absorb the adverse effect of uncertainty, shocks, risks and adverse effect of natural hazards. According to Adger et al. [1], social vulnerability is a condition of people who are susceptible to the exposure of natural hazards which can be minimized by enhancing absorptive capacity. This study reveals that the absorptive capacity of the administrative system is very poor and unable to tackle the situation during disasters. The absorptive capacity of riverbank and char dwellers is poor that means they are not capable to manage the adverse situation. Since, administrative system of the study area is dealt with socially vulnerable people, so absorptive capacity of char dwellers should be developed along with the ability of the administrative system. The study has identified five sub-indicators of absorptive capacity of the local administrative system in the study areas such as minimization of failure, early detection, flexibility, controllability, and reducing the adverse effect.

5.3   Transformative Capacity

Transformative capacity is the ability of an administrative system to transform the policy to action for the protection of the people and society from the uncertain environment. Transformative capacity is a key element of the administrative system because of the system's work nature [11]. Administrative system is always transforming the rules, regulations, policy and plan into action for implementing the strategy at field level for protecting people and society [5]. The study reveals that the transformative capacity of the administrative system is not satisfactory to save the people and their livelihood from the harmful effect of natural disasters. This study also explores some indicators of adaptive capacity so that administrator can apply it at field level for protecting people and nature such as immediate action during disaster, relief during and after disaster, cash incentives, evacuation of people from the place of disaster, ensuring public services during and after disaster, controlling corruption in project implementation, and engaging special task force for disaster management.

6   Conclusion

Administrative vulnerability is a key tool to stimulate an administrative system to protect, save and reduce livelihood resilience in the face of natural hazards. The study explores some key indicators to administrative resilience which will be helpful for measuring any adverse effect and conditions related to socio-ecological vulnerability. It provides a workable definition of administrative resilience in the context of various dimensions of vulnerability. Administrative resilience always considers public values, social norms and values, the technical capacity of public administration, environmental issues and administrative reforms in the context of shocks, risks, and other vulnerabilities. The case study reveals that the status of administrative resilience in char areas is very poor and unable to develop the system due to administrative unwillingness, ignoring char areas, huge corruption in implementing the project, avoiding local people and adaptation strategies. The conditions of various capacities of the administrative system are also poor due to ignoring the disaster management policies, related laws and regulations, avoiding local people's participation in decision making, low resources, lack of expertise, lack of information and geographical isolation. Administrative resilience can be enhanced through developing a disaster information-based database to support quick decision making, strengthening policies, accountability and transparency of the administrative unit, enabling warning system and physical structure of the community, promoting international collaboration for gaining experiences from success events and disaster management funding.

Acknowledgements. This article is funded by Sichuan University Innovation Spark Project (No.201 8hhs-21), Sichuan University Central University Basic Scientific Research Project (No.skqx201501), and National Social Science Fund Youth Project (18CGL040).

References

1. Adger, W.N.: Vulnerability. Glob. Environ. Chang. **16**(3), 268–281 (2006)
2. Adger, W.N., Vincent, K.: Uncertainty in adaptive capacity. Comptes Rendus Geosci. **337**(4), 399–410 (2005)
3. Alfani, F., Dabalen, A., Fisker, P., Molini, V.: Can we measure resilience? A proposed method and evidence from countries in the Sahel, pp. 1–28. Policy Research Working Paper, World Bank (2015)
4. Alinovi, L., Mane, E., Romano, D.: Measuring household resilience to food insecurity: application to palestinian households. In Agricultural Survey Methods, pp. 341–368. Wiley, Chichester, UK (2010)
5. Baker, D., Refsgaard, K.: Institutional development and scale matching in disaster response management. Ecol. Econ. **63**(2), 331–343 (2007)
6. Boyd, E., Nykvist, B., Borgström, S., Stacewicz, I.A.: Anticipatory governance for social-ecological resilience. Ambio **44**(1), 149–161 (2015)
7. Hood, C.: A public management for all seasons? Public Adm. **69**(1), 3–19 (1991)
8. Comfort, L.K., Sungu, Y., Johnson, D., Dunn, M.: Complex systems in crisis: anticipation and resilience in dynamic environments. J. Contingencies Cris. Manag. **9**(3), 144–158 (2010)
9. Cutter, S.L., Ahearn, J.A., Amadei, B., Crawford, P., Eide, E.A., Galloway, G.E., Goodchild, M.F., Kunreuther, H.C., Li-Vollmer, M., Schoch-Spana, M.: Disaster resilience: a national imperative. Environ. Sci. Policy Sustain. Dev. **55**(2), 25–29 (2012)
10. Armitage, D.: Resilience and administrative law. Ecol. Soc. **18**(2), 10–11 (2013)
11. Duit, A.: Resilience thinking: lessons for public administration: resilience thinking: lessons for public administration. Public Adm. **94**(2), 364–380 (2016)
12. Vandenbroucke, J.P., von Elm, E., Altman, D.G., Gøtzsche, P.C., Mulrow, C.D., et al.: Strengthening the reporting of observational studies in epidemiology (STROBE): explanation and elaboration. PLOS Med. **4**(10), e297 (2007)

13. Garschagen, M.: Resilience and organisational institutionalism from a cross-cultural perspective: an exploration based on urban climate change adaptation in vietnam. Nat. Hazards **67**(1), 25–46 (2013)
14. Goldsmith, S., Eggers, W.D.: Governing by Network: The New Shape of the Public Sector. Brookings Institution Press, Washington DC (2005)
15. Haase, T.W.: Administrative resilience: evaluating the adaptive capacity of administrative systems that operate in dynamic and uncertain conditions. Ph.D. Dissertation, University of Pittsburg, Pennsylvania, USA (2009)
16. Hegger, D.L.T., Driessen, P.P.J., Bakker, M.H.N.: Evaluations of flood risk governance in terms of resilience, efficiency and legitimacy. Flood Risk Management Strategies and Governance, pp. 55–61 (2018)
17. Kendra, J.M.: Shared risk: complex systems in seismic response (1st ed.). Environ. Hazards **2**(3), 129–130 (2001)
18. Manyena, S.B.: Rural local authorities and disaster resilience in zimbabwe. Disaster Prev. Manag. **15**(5), 810–820 (2006)
19. Johnson, E.S., Matanoski, G.M.: Disaster management: enabling resilience. In: Masys, A. (ed.) Medicina del Lavoro, vol. 78. Springer International Publishing, Cham
20. Ostrom, E.: Understanding Institutional Diversity. Princeton University Press, United States of America (2005)
21. Milley, P., Jiwani, F.: Resilience and public administration: implications for the new political governance in Canada. Second World Congress on Resilience: From Person to Society, pp. 811–816 (2014)
22. Moher, D., Liberati, A., et al.: Preferred reporting items for systematic reviews and meta-analyses: the PRISMA statement. PLoS Med. **6**(7), e1000097 (2009)
23. Ostrom, E., Janssen, M.A.: Multi-level governance and resilience of social-ecological systems. Globalisation, Poverty and Conflict, pp. 239–259. Kluwer Academic Publishers, Dordrecht (2005)
24. BBS: Statistical yearbook of Bangladesh. Dhaka, Bangladesh (2012)
25. Toonen, T.: Resilience in public administration: the work of Elinor and Vincent Ostrom from a public administration perspective. Public Adm. Rev. **70**(1), 193–202 (2010)
26. Vandenabeele, W.: Toward a public administration theory of public service motivation. Public Manag. Rev. **9**(4), 545–556 (2007)

Can Manager's Environmentally Specific Transformational Leadership Improve Environmental Performance?

Xuhong Liu[1] and Xiaowen Jie[2(✉)]

[1] Department of Police Management, Sichuan Police College, Luzhou 646000,
People's Republic of China
[2] Business School, Sichuan University, Chengdu 610064, People's Republic of China
jiexw@vip.163.com

Abstract. Drawing on the natural-resource-based view (NRBV), this research conducts an in-depth study on the relationship between manager's environmentally specific transformational leadership (ESTL) and environmental performance (EP), and explores the mediating role of proactive environmental strategy (PES) and collective organizational citizenship behaviors toward the environment (collective OCBEs), and further exploring the moderating role of environmental concern in this process. By analyzing 304 questionnaires, these results show that: (1) Managers' ESTL is significantly positively correlated with EP. (2) PES and collective OCBEs act as a mediator between ESTL and EP. (3) Environmental concern can enhance the positive relationship between manager's ESTL and two types of management control systems. The above research conclusions provide management highlights for enterprise to improve EP.

Keywords: Environmentally specific transformational leadership ·
Environmental strategy · Organizational citizenship behaviors toward
the environment · Environmental performance

1 Introduction

With the acceleration of the modern industrial revolution, resource and environmental problems have become increasingly prominent, which not only affect the development of enterprise (Hart [13]; Wang et al. [29]), but also become the main factor constraining China's high-quality economic development. The extensive development model and poor management of enterprise are considered as the principal restricting element. The construction of ecological civilization has been upgraded to a national strategy in "Made in China 2025" and "Green Industrial Development Plan (2016–2020)". Besides, in the face of increasingly strengthened environmental regulation and public opinion supervision, the inevitable trend of green development urges enterprise to carry out environmental revolution.

© Springer Nature Switzerland AG 2020
J. Xu et al. (Eds.): ICMSEM2019 2019, AISC 1002, pp. 730–742, 2020.
https://doi.org/10.1007/978-3-030-21255-1_56

Scholars discussed the positive influence of enterprise managers on EP, but ignored the factor of manager's leadership. Based on the NRBV, leadership is one of the important internal resources and capabilities of enterprise environmental management. Managers who value transformational, concerned about the environmental problems and have ESTL are constantly leading the organization in green revolution, focusing leadership activities on encouraging environmental initiatives (Boiral et al. [7]; Remus and Steger [24]). Thus, how does the manager's ESTL affect the enterprise environmental behavior, and further affect the EP? Under different levels of environmental concern, is there a difference in the impact of managers' ESTL on the two relationships? This study proposes a more complete analytical framework, considering the impact of ESTL on EP under the joint action of enterprise formal management control system (PES) and informal management control system (collective OCBEs).

2 Literature Review and Hypotheses

2.1 ESTL and EP

Resource-based view holds that unique resources and capabilities (valuable, scarce, hard to copy and replace) are the root for enterprise to gain competitive advantages and improve performance. Leadership is a unique resource that can influence an organization to achieve goals and gain competitive advantages. NRBV proposed that the natural environmental elements have been neglected in the past, it must be taken into account by enterprise and rise to the height of strategic planning (Hart [13]). Environmental leadership can affect employees and organizations to practice por-environmental behaviors, reduce the cost of pollution reduction, thereby improving enterprise EP (Tapurica and Ispasoiu [2]). However, previous studies have mostly focused on the demographic characteristics, the significant relationship between environmental leadership and EP has not been empirically tested. What is more, the role of promoting the sustainable development of enterprise mainly from the power and responsibility of managers. Especially, environmental leaders are more likely to practice ESTL than environmental transactional leadership or laissez-faire leadership style (Egri and Herman [11]). Therefore, the role of ESTL in strategic choice and EP needs to be further explored.

Current literatures have shown that environmental leadership has a positive impact on the performance of green innovation (Pan and Tian [10]). However, Zeng et al. [30] had different opinion and founded that enterprise' own motivation had no significant influence on EP. This study argues that managers are the middleman of dynamic management of resources and capabilities, and different leaderships are important factors to affect EP. When managers have the ESTL, it is easier for them to influence the environmental behavior and EP. Environmental leaders are actually transformational leaders, who can promote the concept of transformational leadership to environmental issues, convey a clear and coherent vision for the field of environmental responsibility (Boiral et al. [7]). They demonstrate commitment and action by sharing environmental values and discussing the importance of sustainability (Jang et al. [15]). Based on this, the study makes the following assumptions:

Hypothesis 1. ESTL has a significant positive impact on EP.

2.2 The Mediating Role of PES

PES not only complies with environmental regulations, but also includes important proactive environmental behaviors. Research found companies with voluntarily participate in exceeding environmental regulations show higher EP (Judge and Douglas [17]). But other scholar held different opinions, argued that environmental technologies, even the most advanced technologies under PES, may cause more pollution than the most popular and typical technologies (Brechet and Jouvet [8]). This study considers that PES can improve the quantity and quality of environmental sustainability information disclosure (Helfaya and Moussa [14]), can effectively promote technological ecological innovation and organizational capabilities, and better transform knowledge for EP (Ryszko [27]). In addition, enterprises' cognition and thinking mode of green behavior also have an important influence on decision-making results (Wei et al. [31]), managers with ESTL can base on their internal and external resources, identifying and utilizing them to promote the adoption of PES. Change to inspire the common vision of the organization. The effective deployment of PES may depend on the ability of managers with ESTL to inspire a shared vision of the organization, this is a key condition to promote greening through PES. It can convey a clear environmental vision in the field of corporate environmental responsibility, reflect the value and importance of environmental management, and be more cohesive and appealing in information sharing, thus incorporating environmental issues into corporate strategic planning. Thus, ESTL embodies the characteristics of managers' concern for environmental protection and sustainable development, so as to promote enterprise to adopt PES. So, the following hypotheses are proposed in this study:

Hypothesis 2. ESTL has a significant positive impact on PES.

Hypothesis 3. PES plays the mediating role between ESTL and EP.

2.3 The Mediating Role of the Collective OCBEs

Given both the complexity and diversity of environmental issues, it difficult for formal control systems to consider all possible behaviors that could minimize environmental impacts (Jiang and Bansal [16]). Employees OCBEs are an extra-role behavior, which are carried out on the basis of voluntary and are not controlled by formal management (Alt and Spitzeck [1]; Zhang and Liu [6]). Studies have shown that there is a positive relationship between ESTL (except occasional rewards and active management) and workplace environmental behavior (Graves et al. [12]). The individualized consideration and intellectual stimulation provided by ESTL can help employees acquire professional knowledge and technology related to environmental management. When managers "lead by example" practice environmental commitment in daily work, they will play a role model effect (Bass [4]), which conveys important information about expectation and priority behavior. Employees can learn the behaviors of leaders through the social learning mechanism, thus shape their own environmental behavior.

Employees' environmental initiative is generally considered as one of the success factors to achieve environmental goals, the reason may be that EP is related to a series of environmental daily measures, such as pollution prevention and control, environmental management system, waste minimization and recycling activities. The development of environment-friendly products or processes by employees who exceed their responsibilities will contribute to environmental benefits (Paille et al. [20]). Likewise, Environmental helping behavior can help colleagues care more about the environment, and reduce internal resource consumption. Meanwhile, through open discussion, the sharing of environmental knowledge and various views, can form an environment-friendly atmosphere, it conducive to the formation of a collective result (Kim et al. [18]). And the high level of participation by the collective OCBEs also means that they play a key role in developing lean and green practices, thus can help improve production operations and EP. Therefore, we state the following hypothesis:

Hypothesis 4. ESTL has a significant positive impact on the collective OCBEs.

Hypothesis 5. The collective OCBEs play the mediating role between ESTL and EP.

2.4 The Moderating Role of Environmental Concern

Under the new environmental paradigm, managers have different levels of inherent values and beliefs, which have discrepancy on the choice of enterprise environmental strategy. Environmental concern refers to pay attention and has full understanding of environmental problems, and support to solve these and/or show their willingness to contribute to the solution (Dunlap and Jones [25]). Under the theoretical framework of value-oriented behavior, the relationship between managers' environmental concern and enterprise environmental response was significant (Potocan et al. [22]). In the case of high environmental concern, managers attach importance to analysis and utilize resources, and adopt the internal environment orientation, which related to the environmental strategy and reflects the degree of environmental commitment, as well as its willingness to integrate environmental protection into business strategy. Accordingly, the following hypotheses are proposed in the proposed report:

Hypothesis 6. Environmental concern positively moderates the impact of ESTL on PES.

Environmental concern has been identified as an important predictor of employee green behavior in the workplace (Bissing-Olson et al. [5]). Those who concerned about the natural environment are more likely to engage in environmental behavior at work, the intensity of this relationship is also influenced by managers environmental concern (Stern and Dietz [28]). However, environmental concern may not lead to pro-environmental behaviors due to various reasons, the lack of support from managers was identified as the main obstacle (Govindarajulu and Daily [28]). In the workplace, employees will engage in behaviors supported by their superiors due to some factors, such as empowering, pleasing the supervisor or reflecting individual values. In this case, managers attitude towards environmental concern will affect employees and help them pay extra environmental behaviors. Similarly, leaders create a shared feeling among followers by expressing their common concern, which can strengthen the relationship between the working group and leaders, thus making the green advocacy of the working group a collective result (Kim [18]). Accordingly, the following hypotheses are proposed in this study (Fig. 1):

Hypothesis 7. Environmental concern positively moderates the impact of ESTL on the collective OCBEs.

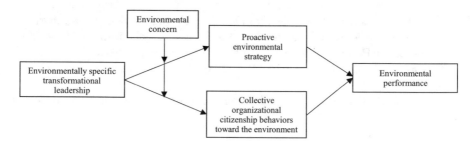

Fig. 1. Research model

3   Research Design

3.1   Sample Selection and Data Collection

Questionnaire survey was adopted in this study, enterprise with greater environmental impact in the industrial industry were selected. The research objects are middle and top managers (chairman, general manager), department managers (R&D, production, marketing manager, etc.) and persons in charge of environment and safety affairs of manufacturing, mining, construction, production and supply of electricity/heat/gas and water, water conservancy/environment/public facilities management. Taking into account the feasibility of data acquisition, the questionnaires was mainly distributed through the methods of target sampling and snowball sampling in non-probabilistic sampling. Through the small sample test, the items with factor load less than 0.5 were eliminated. A total of 377 samples were obtained after the recovery of large sample survey, after eliminating unmatched and/or missing cases, 304 valid samples were obtained with the effective questionnaire recovery rate of 80.63%. The response rates were 26.64% for top managers, 39.14% for middle managers, and 34.21% for first-line managers. Among the property rights of enterprise, state-owned enterprise account for 25.99%, while non-state-owned enterprise account for 74.01%. Among the enterprise scale, 23.36% are under 100 employees, 31.25% are between

101-500 employees, 11.84% are between 501 to 1000 employees, and 33.55% are above 1000 employees. Among the enterprises, manufacturing accounts for 75%, water conservancy/environment/public facilities management accounts for 10.83%, and other industries account for 14.17%.

3.2   Variable Measurement

In this study, foreign maturity scale with good reliability and validity developed was selected as the measurement scale. After being translated into Chinese and also translated back by two doctors major in English, the final scale was formed after discussion by academic expert groups and enterprise practice managers. Likert scale was used for all scales, with 1 representing "strongly disagree" and 5 representing "strongly agree".

ESTL were assessed using eight items adapted from Robertson [26], with a total of 8 items, item example: "Enterprise managers make a commitment to improve enterprise EP" (Cronbach's a = 0.957). PES was assessed using ten items adapted from Buysse and Verbeke [9], item example: "Enterprise shall make internal environmental reports and external environmental disclosure" (Cronbach's a = 0.963). Collective OCBEs were assessed using eight items adapted from Pinzone et al. [21], item example: "Employees take environmental protection actions that contribute positively to the enterprise image" (Cronbach's a = 0.966). Environmental concern was assessed using four items adapted from Kim [19], item example: "The earth is approaching the limit of the number of people who can bear it" (Cronbach's a = 0.701). EP were assessed using four items adapted from Judge and Douglas [17], item example: "Enterprise prevent and resolve environmental crisis" (Cronbach's a = 0.862). The scale and the nature of the enterprise have an important impact on EP. So, we have controlled them in the regression analysis.

3.3   Common Method Variance

After the data collection, the Harman's single-factor test was conducted in this study. In the case of no rotation, a total of 5 factors with eigenvalues greater than 1 were extracted to explain 67.978% of the total variation. The obtained first principal component accounted for 32.652% of the total load, and there was no significant load, the homologous error is not significant. In addition, the highest VIF value was 2.629, which was less than the rigorous threshold of 3.3, indicating that the multicollinearity effect was not serious.

3.4   Reliability and Validity Test

In this study, AMOS17.0 software was used for confirmatory factor analysis to test the discriminant validity and convergent validity of the scale. The analysis results showed that the five-factor model was significantly better than other models. Fit indices for the five-factor model was: $x^2/df = 2.353$, $RMSEA = 0.067, TLI = 0.911, CFI = 0.918$, and the AVE values of each factor loading were all greater than 0.5, indicating that the whole measurement tool had discriminative validity and convergent validity. The descriptive statistical results and correlation coefficients of variables are listed in Table 1.

Table 1. Descriptive statistics and correlation coefficients

Variable	1	2	3	4	5	6	7
1. Firm sale							
2. Firm ownership	−0.071						
3. ESTL	−0.016	0.145*					
4. EC	0.069	0.039	0.132*				
5. PES	0.119*	0.083	0.676**	0.295**			
6. Collective OCBEs	0.114*	0.135*	0.687**	0.118*	0.541**		
7. EP	0.122*	0.101	0.616**	0.160**	0.386**	0.636**	
8. Mean	2.52	1.91	3.382	4.049	3.267	3.342	3.748
9. S.d.	1.216	0.669	1.05	0.801	0.937	1.061	0.94

Note: $*p < 0.05, **p < 0.01, ***p < 0.001$

4   Result and Disscussion

4.1   Mediating Effect

Hierarchical regression method is used to test the main effect in this study. The results of hierarchical regression are shown in Table 2. ESTL has a significant positive impact on EP ($\beta = 0.615, P < 0.001$), **hypothesis 1** is verified. Baron and Kenny [3] are usually adopted to verify the mediating effect through three steps. First, it is assumed that ESTL has a significant positive impact on EP. Second, ESTL has a significant positive impact on PES ($\beta = 0.781, P < 0.001$), and **hypothesis 2** is verified. Finally, after the addition of PES, the influence of ESTL on EP decreased significantly ($\beta = 0.322, P < 0.001$), the adjusted R2 increased by 0.053, while the PES still had a significant positive impact on EP ($\beta = 0.375, P < 0.001$). Therefore, PES plays a partial mediating role in the ESTL and EP, and **hypothesis 3** is verified.

Similarly, ESTL has a significant positive impact on the collective OCBEs ($\beta = 0.727, P < 0.001$). **hypothesis 4** is verified. After joining the collective OCBEs, the influence of ESTL on EP decreased significantly ($\beta = 0.351, P < 0.001$), the adjusted R^2 increased by 0.059, while collective OCBEs still had a significant positive impact on EP ($\beta = 0.363, P < 0.001$). Therefore, the collective OCBEs play a partial mediating role in the ESTL and EP, and **hypothesis 5** is verified.

Due to the limitations of the mediation effect inspection standards of Baron and Kenny [3], this study further verifies the mediating effect according to the bootstrapping analysis by Preacher and Hayes [23]. In the specific operation, the PROCESS plug-in was adopted, the number of repeated samples was 5000, and

Table 2. Regression analysis results of mediating role

	PES		Collective OCBEs		EP			
	Model 1	Model 2	Model 3	Model 4	Model 5	Model 6	Model 7	Model 8
Firm sale	0.126*	0.130***	0.129**	0.133***	0.130*	0.133**	0.084	0.085
Firm ownership	0.092	−0.022	0.135**	0.03	0.11	0.021	0.029	0.011
ESTL		0.781***		0.727***		0.615***	0.322***	0.351***
PES							0.375***	0.363***
Collective OCBEs								0.456
R^2	0.023	0.62	0.032	0.55	0.027	0.397	0.45	0.059
ΔR^2	0.023	0.597	0.032	0.518	0.027	0.37	0.053	32.559***
ΔF	3.473**	470.941***	5.049***	345.498***	4.184*	184.091***	29.077***	0.085

Note: $*p < 0.05, **p < 0.01, ***p < 0.001$

the deviation correction bootstrapping 95% confidence interval was obtained. If there is no zero in the confidence interval, it indicates the mediating effect is significant. The total impact, direct impact and indirect impact of mediation are shown in Table 3. Among them, the total impact of ESTL on EP is 0.551, and the direct impact is 0.165. ESTL on overall EP of the indirect influence of 0.386, the ESTL type through PES on EP of the indirect effects of 0.197, the ESTL through collective OCBEs for EP of the indirect influence of 0.189. So, the mediating role of PES and collective OCBEs are verified.

Table 3. Environmental concern on the role of ESTL and PES

	Effect value	Standard error	Deviation correction confidence interval(Boot95% CI)	
			LLCI	ULCI
Total impact of EFTL on EP	0.551	0.041	0.213	0.463
Direct effect of EFTL on EP	0.165	0.064	0.14	0.359
Total indirect effect of EFTL on EP	0.386	0.063	0.262	0.511
Indirect effect of EFTL on EP(PES)	0.197	0.072	0.075	0.353
Indirect effect of EFTL on EP(Collective OCBEs)	0.189	0.057	0.073	0.299

4.2   Moderating Effect of Environmental Concern

In order to eliminate collinearity, the independent variable and moderator variable are standardized respectively when constructing the product term of independent variable and moderator variable. The test results are shown in Table 4. The product of ESTL and environmental concern has a significant

738   X. Liu and X. Jie

positive impact on PES ($\beta = 0.066, P < 0.05$), and the adjusted R^2 has also increased by 0.004, indicating that environmental concern is positively regulating ESTL and PES. The relationship of environmental strategy, **hypothesis 6** is verified.

The product of ESTL and environmental concern significantly positively affected the collective OCBEs ($\beta = 0.143, P < 0.01$), and the adjusted R^2 increased by 0.017, indicating that environmental concern is positively regulating the relationship between ESTL and collective OCBEs, **hypothesis 7** is verified.

Table 4. Environmental concern on the role of ESTL and PES

	PES		Collective OCBEs		EP			
	Model 1	Model 2	Model 3	Model 4	Model 9	Model 10	Model 11	Model 12
Firm sale	0.126*	0.130***	0.129**	0.133***	0.131*	0.125***	0.132***	0.119**
Firm ownership	0.092	−0.022	0.135**	0.03	−0.021	−0.027	0.029	0.018
ESTL		0.781***		0.727***	0.783***	0.758***	0.725***	0.670***
EC					0.016	0.015	0.015	0.017*
ESTLEC						0.066*		0.143**
R^2	0.023	0.62	0.032	0.55	0.621	0.665	0.552	0.567
ΔR^2	0.023	0.597	0.032	0.518	0.001	0.004	0.002	0.017
ΔF	3.473**	470.941***	5.049***	345.498***	0.202	2.890*	1.146	11.677**

Note: $*p < 0.05, **p < 0.01, ***p < 0.001$

In order to more intuitively understand the mediation role of environmental concern, this study used the simple slope analysis method, took one standard deviation of each around the mean value of environmental concern, and divided the sample data into the high environmental concern group and the low environmental concern group, and made regression respectively in the two groups.

It can be seen intuitively that, the difference in the impact of EFTL on the PES and the collective OCBEs under different levels of environmental concern, as shown in Figs. 2 and 3. At the level of high environmental concern, managers ESTL has a greater impact on PES and collective OCBEs. At the low level of environmental concern, the impact of managers ESTL on PES and collective OCBEs is relatively small.

5   Conclusions and Suggestions

This study discusses the impact of ESTL on EP and its mechanism of action on the basis of NRBV, transformational leadership theory and social learning theory, analyzes the role of PES and collective OCBEs in the above relationship, and verifies the role of environmental concern in the overall model. The study results show that: (1) ESTL is positively affecting EP; (2) PES and collective OCBEs have the mediating role between ESTL and EP; (3) environmental concern can enhance the positive relationship between ESTL and PES/collective OCBEs.

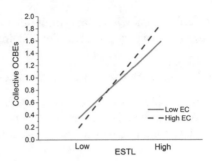

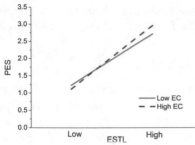

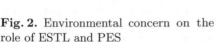

Fig. 2. Environmental concern on the role of ESTL and PES

Fig. 3. Environmental concern on the role of ESTL and collective OCBEs

5.1 Theoretical Contributions

This study expands the research scope of transformational leadership from the perspective of natural resource basics. Previous studies have focused on the role of enterprise managers in environmental values, environmental commitments, and personal traits in environmental management, while neglecting ESTL is an important internal resource and capability. Moreover, the only few researches on ESTL mainly focuse on descriptive theories, but not to teste the impact of ESTL on enterprise EP. The study results clarify that ESTL is an antecedent variable of enterprise EP.

Unveils the "black box" of the influence mechanism of ESTL on EP. On the one hand, enterprise choose PES means that it has environmental revolution in the organization, it requires managers to have ESTL. Through building a shared vision in organizational transformation, defining clear and coherent priorities, and applying the rights and responsibilities to coordinate and communicate across organizations or departments, to promote a proactive strategy to improve organizational EP and sustainability. On the other hand, it studies the impact of ESTL on the collective OCBEs in the organizational level. It supplements the research of the OCBEs in general to promote the effective operation in organization.

5.2 Management Implications

In transitional economy, the importance and necessity of environmental management of Chinese enterprise has become increasingly prominent. However, in the context of increasingly complex environmental issues, environmental management should not only stay at the end of the governance process, the effectiveness of enterprise managers ESTL should also be brought into play.

The study results provide important inspiration to managers: to enhance the ESTL with a focus on the front of the decision-making of strategy, and to promote the active construction of environmental strategic; in practice, the guidance of collective OCBEs should be pay more attention, for example, encourage ecological innovations stemming from personal suggestions or ideas, carry out skills training of knowledge and methods to solve environmental problems, and inspire employees active OCBEs. Thus, it increases the frequency of desired behavior.

5.3   Limitations and Future Research Directions

The limitations of this study are mainly as follows: (1) Applicability of this study to other industries. Selecting industrial enterprise that are greatly affected by the environment has certain representativeness, however, it does not mean that the ESTL of other industries managers has the same impact. Future studies may explore applicability to other industries, such as services industry. (2) This study used homologous cross-sectional data, this can lead to results that are affected by common method biases, and data from different sources should be considered in the future studies. (3) Chinese cultural background of the research situation. Although this study adopts the maturity western scale and the conclusions consistent with theoretical derivation, but China has more emphasis on the cultural context of collectivism, and high right distance. Therefore, in the future research, the ESTL measurement questionnaire in the Chinese cultural context can be developed to mine more appropriate and representative scales.

Acknowledgements. This work was supported by the Key Program (Grant No. 16AGL003) of the National Social Science Foundation of China.

References

1. Alt, E., Spitzeck, H.: Improving environmental performance through unit-level organizational citizenship behaviors for the environment: a capability perspective. J. Environ. Manag. **182**, 48–58 (2016)
2. Tapurica, O.C., Ispuaoiu, C.E.: Analyzing the influence of environmental leadership on pollution abatement costs. Young Econ. J. Revista Tinerilor Economisti **10**(20), (2013)
3. Baron, R.M., Kenny, D.A.: The moderator-mediator variable distinction in social psychological research: conceptual, strategic, and statistical considerations. J. Pers. Soc. Psychol. **51**(6), 1173 (1986)
4. Bass, B.M.: Leadership and Performance Beyond Expectations. Collier Macmillan, New York (1985)
5. Bissing-Olson, M.J., Iyer, A., Fielding, K.S., Zacher, H.: Relationships between daily affect and pro-environmental behavior at work: the moderating role of pro-environmental attitude. J. Organ. Behav. **34**(2), 156–175 (2013)
6. Zhang, J.L., Liu, J.: Organizational citizenship behaviour for the environment: measurement and validation. J. Bus. Ethics **109**(4), 431–445 (2012)
7. Boiral, O., Talbot, D., Paille, P.: Leading by example: a model of organizational citizenship behavior for the environment. Bus. Strat. Environ. **24**(6), 532–550 (2015)
8. Brechet, T., Jouvet, P.A.: Environmental innovation and the cost of pollution abatement revisited. Ecol. Econ. **65**(2), 262–265 (2008)
9. Buysse, K., Verbeke, A.: Proactive environmental strategies: a stakeholder management perspective. Strat. Manag. J. **24**(5), 453–470 (2003)
10. Pan, C.L., Tian, H.: Stakeholder pressure, corporate environmental ethics and proactive environmental strategy. J. Manag. Sci. **29**(3), 38–48 (2016). (in Chinese)

11. Egri, C.P., Herman, S.: Leadership in the north american environmental sector: Values, leadership styles, and contexts of environmental leaders and their organizations. Acad. Manag. J. **43**(4), 571–604 (2000)
12. Graves, L.M., Sarkis, J., Zhu, Q.H.: How transformational leadership and employee motivation combine to predict employee proenvironmental behaviors in china. J. Environ. Psychol. **35**, 81–91 (2013)
13. Hart, S.L.: A natural-resource-based view of the firm. Acad. Manag. Rev. **20**(4), 986–1014 (1995)
14. Helfaya, A., Moussa, T.: Do board's corporate social responsibility strategy and orientation influence environmental sustainability disclosure? UK evidence. Bus. Strat. Environ. **26**(8), 1061–1077 (2017)
15. Jang, Y.J., Zheng, T., Bosselman, R.: Top managers environmental values, leadership, and stakeholder engagement in promoting environmental sustainability in the restaurant industry. Int. J. Hosp. Manag. **63**, 101–111 (2017)
16. Jiang, R.J., Bansal, P.: Seeing the need for ISO 14001. J. Manag. Stud. **40**(4), 1047–1067 (2003)
17. Judge, W.Q., Douglas, T.J.: Performance implications of incorporating natural environmental issues into the strategic planning process: an empirical assessment. J. Manag. Stud. **35**(2), 241–262 (1998)
18. Kim, A., Kim, Y., Han, K., Jackson, S.E., Ployhart, R.E.: Multilevel influences on voluntary workplace green behavior: individual differences, leader behavior, and coworker advocacy. J. Manag. **43**(5), 1335–1358 (2017)
19. Kim, S.H., Kim, M., Han, H.S., Holland, S.: The determinants of hospitality employees pro-environmental behaviors: the moderating role of generational differences. Int. J. Hosp. Manag. **52**, 56–67 (2016)
20. Paille, P., Chen, Y., Boiral, O., Jin, J.: The impact of human resource management on environmental performance: An employee-level study. J. Bus. Ethics **121**(3), 451–466 (2014)
21. Pinzone, M., Guerci, M., Lettieri, E., Redman, T.: Progressing in the change journey towards sustainability in healthcare: the role of greenhrm. J. Clean. Prod. **122**, 201–211 (2016)
22. Potocan, V., Nedelko, Z., Peleckiene, V., Peleckis, K.: Values, environmental concern and economic concern as predictors of enterprise environmental responsiveness. J. Bus. Econ. Manag. **17**(5), 685–700 (2016)
23. Preacher, K.J., Hayes, A.F.: Asymptotic and resampling strategies for assessing and comparing indirect effects in multiple mediator models. Behav. Res. Methods **40**(3), 879–891 (2008)
24. Ramus, C.A., Steger, U.: The roles of supervisory support behaviors and environmental policy in employee "ecoinitiatives" at leading-edge european companies. Acad. Manag. J. **43**(4), 605–626 (2000)
25. Dunlap, R.E., Robert, E.J.: Environmental Concern: Conceptual and Measurement Issues. Greenwood Press, Westport, CT (1985)
26. Robertson, J.L., Barling, J.: Contrasting the nature and effects of environmentally specific and general transformational leadership. Leadersh. Organ. Dev. J. **38**(1), 22–41 (2017)
27. Ryszko, A.: Interorganizational cooperation, knowledge sharing, and technological eco-innovation: the role of proactive environmental strategy-empirical evidence from Poland. Pol. J. Environ. Stud. **25**(2), 753–764 (2016)
28. Stern, P.C., Dietz, T.: The value basis of environmental concern. J. Soc. Issues **50**(3), 65–84 (1994)

29. Wang, R., Wijen, F., Heugens, P.P.: Government's green grip: multifaceted state influence on corporate environmental actions in China. Strat. Manag. J. **39**(2), 403–428 (2018)

30. Zeng, S., Meng, X., Zeng, R., Tam, C.M., Tam, V.W., Jin, T.: How environmental management driving forces affect environmental and economic performance of smes: a study in the Northern China district. J. Clean. Prod. **19**(13), 1426–1437 (2011)

31. Wei, Z.K., Yang, Y., Wei, Z.I.: Paradox cognition institutional environment and green performance. Manag. Rev. **30**(11), 76–85 (2018). (in Chinese)

Financial Indicators and Stock Price Movements: The Evidence from the Finance of China

Qiang Jiang[1], Xin Wang[1], Yi Li[1], Dong Wang[2], and Qing Huang[1(✉)]

[1] Business School, Sichuan University, Chengdu, People's Republic of China
490004245@qq.com
[2] Victoria Energy Policy Centre, Victoria Institute of Strategic Economic Studies,
Victoria University, Melbourne, Australia

Abstract. As a representative of the capital market, the stock market has become a barometer of the national economy. The stock prices fluctuation plays as a benchmark role in leading public and private enterprises. This paper studied the relationship between the quarterly financial indicators and stock price movements in Chinas financial industry. Through the methods of literature research and empirical data statistical analysis, regression models are established to explore the relationship between these variables. The results indicated that total asset turnover rate has the largest impact on the stock price increase in financial industry companies among those selected financial indicators. Besides, we compared the results of banks and other financial enterprises to further research. It turns out that there is a big difference between Banks and non-bank financial enterprises. As for banks, there is a significant positive correlation between the multiple of cash dividend protection and stock price movements.

Keywords: Financial industry · Listed companies · Financial indicators · Stock price · Capital market

1 Introduction

Nowadays, the rapid development of Internet finance in China has made tremendous impact on the traditional commercial banking business. At the same time, many capital market investors are more bullish about non-bank financial companies in terms of their growth prospects or the product diversity of securities companies. With the acceleration of marketization of interest rates in China, traditional banking businesses need to be adjusted correctly to the context of market competition. Efforts should be made to develop off-balance-sheet business, but the impact of business factors in the table can not be ignored. For investors, due to the scarcity of channels in getting the off-balance-sheet business information, analysis of the company's financial position and the overall

© Springer Nature Switzerland AG 2020
J. Xu et al. (Eds.): ICMSEM2019 2019, AISC 1002, pp. 743–758, 2020.
https://doi.org/10.1007/978-3-030-21255-1_57

744   Q. Jiang et al.

value of the company from the financial indicators through the financial statements of listed companies remains the same important as before. Enterprises
need to consider that essential part when they make stock investments.

The financial indicators usually refer to the relative indicators for enterprises
to summarize or evaluate their financial status and operating results. According to the "General Rules for Corporate Finance" of the PRC, the financial
indicators for the enterprises can be divided into three types. They are the solvency targets, operational capability indicators, profitability indicators. "Rise"
(or increase) refers to the rate of increase of a stock, it can clearly reflect the
fluctuations in stock price. Stock price can represent the value of a company is
in a certain extent. One of the main sources of profit for investors investing in
equities is the dividend distributed by the company. Another one could be the
bid-ask premium resulting from the change in stock price. Whether the firm is
an investor or a financier, it is essential to study these two factors. Exploring
the relationship between the financial indicators of the financial industry and
the changes of the stock price really benefits us a lot.

At present, scholars have made great achievements in the research on the
relationship between financial indicators of listed companies and stock price
changes. Models and effective methods are also widely used in investment practice and theoretical analysis. However, with the passage of time, China's capital
market continues to develop. Has the relationship between financial indicators
and stock prices ever changed these days? After studying the previous literature,
this paper found that there are few articles of the relationship between financial indicators and stock price changes from the financial industry in the past
two years. Previous studies tend to focus on exploring the relationship between
enterprise performance and stock price changes, and believe that the indicators
reflecting profitability are the most important factors of the financial indicators
of a company. Most scholars thought it is not easy to study those characteristics
such as large changes in cash flow of enterprises. So they avoided talking about it.

Different from other articles, the comparative analysis between Banks and
non-bank financial enterprises is a major feature of this paper. By the means
of comparing the banking and non-banking financial institutions according to
the sample data, hoping to play a development role in relevant researches. In
addition, this paper applies the impact of financial indicators on the company's
stock price changes to the special field of the financial industry for analysis.
This paper uses the latest data during 2014–2017 of listed companies, on the
perspective of company's overall financial indicators, to explore the relationship
between financial indicators and stock price changes based on the whole vision
on finance industry.

This study can be divided into six parts. The first part is the introduction.
The second part is a review of the relevant research status. Then put forward
hypotheses. The third part is the research design. The fourth part shows the
empirical results and the statistical analysis process. The fifth part shows the
summary and discussion of the research results. Finally, the sixth part lists the
references.

2   Literature Review and Hypothesis Development

2.1   Literature Review

(1) Relevant research methods

Some scholars have explored and innovated the effectiveness of relevant research methods to analyze the influence of financial indicators on stock price changes. Beaver et al. [4] established a regression model by using the percentage change of price return for grouped dependent variables and found that reverse regression is a more effective method to test the incremental explanatory power of price percentage change relative to the lag value of accounting income. Floros et al. [9] compared several prediction technologies based on symmetry error statistics under the static and dynamic methods. They used ECM - GARCH model to study the dividend yields effect on market performance. It turned out that the p/e ratio can predict stock total revenue.

Some researchers focus on studying and forecasting short-term stock prices. And over time, the technology of prediction is improving. In 1988, Campbell, J. Y. and Shiller, R. J. found long historical averages of real earnings help forecast present values of future real dividends [6]. Bin and Waldyn [26] tried to develop a financial expert system that incorporates these features to predict short term stock prices. The expert system is comprised of two main modules: a knowledge base and an artificial intelligence platform.

(2) Financial indicators and stock price changes

Many factors affect the stock price changes of listed companies. As Reddy and Parab [20] said, the importance of stock returns and Economic Value-Added have attracted various research scholars over the past several years. Maskun and Ali [17] studied the effect of current ratio, return on equity, return on asset and earning per share to the price of stock in indonesian stock exchange. Some scholars are keen to study the relationship between EVA (economic value-added) and stock returns. According to these researches, there is a significant positive correlation between EVA and stock returns, but they are not mutually causal. Baybordi et al. [3] found a significant and positive relationship between economic value-added and stock return. However, Ray [19] found little support to the fact that high-EVA firms lead to higher stock market performance and shareholders value creation.

Some scholars believe that future earnings expectation and cash flow may be the important reasons affecting stock price changes, and have carried out relevant studies. Through the experiment, it is found that the expected income has a significant impact on the stock price change, but the utility of cash flow is not obvious. There is some evidence to support this claim. Sloan [24] found that stock price behaviors reflected that investors considered more about future earnings expectations when making investment decisions, but failed to directly reflect the impact of information contained in the accrual basis of current earnings and cash flow components on investors' decisions. O'Hara [18] found that the dividend per share doesn't seem to predict the performance of the stock. Cash flow turned out to have no significant impact on share prices.

However, some scholars hold different opinions and believe that dividend distribution is an important factor affecting stock price changes. Capstaff [7] tested the signaling theory of dividends by investigating the stock price reaction to dividend announcements on the Oslo Stock Exchange. The results indicate that the stock market reaction is most pronounced for large, positive dividend announcements that are followed by permanent cash flow increases. Kvamvold and Lindset [15] also examined price impacts from dividend flows and reached similar conclusions. De Cesari and Huang-Meier [8] highlight private information in stock prices as an important determinant of dividend policy and contributes to the literature on the real effects of financial markets.

(3) Related researches on other objective factors

In recent years, with the development of financial computerization, scholars around the world are more and more concerned about the stock price's reaction to news and the influence of other factors on the stock price. As Bin et al. [26] said, Information emanating from these platforms such as: Google, Wikipedia and the like can significantly affect, or be affected by, changes in stock market. In turn, these platforms influence changes in the stock market. Bernardo et al. [5] investigate if Twitter data can be used to predict or describe stock market prices by using sentiment polarity. They found companies with a high number of tweets show a weaker relationship between the two variables. Grabellus et al. [10] introduced a new information density indicator to provided a more comprehensive understanding of price reactions to news.

Some researchers have found that other objective factors besides financial indicators can also have a significant impact on stock price changes. Hearn et al. [12] studied the asset pricing implications arising from imperfect investor protection using a new governance measure which captures the impact of controlling block holders. Jiang [14] Behavioral theories suggest that investor misperceptions and market mispricing will be correlated across firms. Scharfstein [22] mentioned that ignoring private information would lead to herd behavior of some investment managers, which would have some impact on investors decision-making level and stock price changes. Gilchrist [11] believed that the increase of some indicators dispersion would lead to the increase of stock price, like Tobin Q and actual investment. In addition, there are some macro factors that can affect stock price changes. Such as exchange rate. Rutledge et al. [21] provided implications for Chinese monetary policy makers and global investors through researching the relationship between Chinese RMB exchange rates and Chinese stock prices. Bastianin and Manera [2] studied how stock market volatility responds to oil price shocks and came to some valid conclusions.

(4) Further development and application

Since the relationship between financial indicators and stock price changes has been studied to a certain extent, some experts and scholars try to extend stock prices to the network and analyze the trend of financial computerization. Combined with computer network to predict stock price, this will be one of the future development directions of this paper. Through the performance analysis, Lee et al. [16] believe that network indicators can be regarded as important supplementary indicators in predicting the global stock market and regional relative directions especially during market crisis periods. Heiberger [13], Schinckus [23], and Sui [25] are focusing on the increasing computerization of the financial markets and the consequences of such process on our ability to collect information about financial prices.

Researches on the relationship between stock price and financial index are quite a lot both in domestic and abroad. Besides, the research scope is wide. Previous studies always focus on industry to study the general commonality of the industry but ignore the commonness and difference between enterprises in the same industry. For example, in the same financial service industry, banking enterprises are quite different from other financial companies. Many of these reasons are difficult to fully explain, but we cannot deny that such differences exist objectively whether in preferential policies or other aspects.

2.2 Hypothesis Development

The total asset turnover rate represents the operating ability of a corporation. And the return on equity (ROE) and the earnings per share (EPS) can reflect the profitability of the company. We propose our first hypothesis. In view of the general characteristics of high liabilities of financial enterprises, as for the asset-liability ratio index, we assume that there is no significant difference between the impact of this financial index on listed Banks and non-bank listed companies. We propose the second hypothesis. For further analysiswe assume that bank-specific indicators, such as capital adequacy ratio, have a significant effect on the share price changes of banking enterprises because the capital adequacy ratio is one of the important criteria for measuring the investment security of banking enterprises. At the same time, we should consider whether some specific indexes of financial enterprises in banks and non-financial ones will affect the ability of financial indexes to explain the price volatility of shares. We propose our final hypotheses.

H1 The total asset turnover rate, ROE and EPS all have a significant relationship with the stock price changes of financial enterprises, and have a positive impact on the stock price.

H2 The asset-liability ratio has no obvious influence on the stock price changes of financial enterprises.

H3 The capital adequacy ratio has a significant effect on the share price changes of banking enterprises.

3   Research Design

3.1   Variable Selection

Dependent variable: In this paper, the dependent variable refers to the fluctuation degree of stock price, represented by the quarterly rise of stock price(PCT-CHG). Because the stock prices of different companies vary too much, it is easy to generate large error results by using the absolute value measurement. Therefore, this paper selects the quarterly stock price rise of each company as the proxy variable to measure the stock price changes for analysis, which makes the sample more effective for research.

Independent variables: Based on the literature research and the consideration of the criteria and habits, and to reflect the majority of investors' stock investment choice as far as possible, this paper selects these four indicators as the main explanatory variables. They are total asset turnover rate, earnings per share, the average return on equity and asset-liability ratio.

Control variables: In order to make the research model better fitting the overall distribution of the data and to enhance the fitting effect, we use some control variables in our research. In this paper, these variables involve the company's solvency, profitability and growth ability three aspects, concluding financing, investment and business activities generated cash flow, net profit after the deduction of the abnormal profit and loss account for net income and cash dividend cover, the year-on-year growth rate of cash flows, the basic earnings per share, net profit year-on-year growth rate, net assets year-on-year growth indicators and so on.

3.2   Empirical Model

The general multiple regression model of this study is as follows.

$$Y = \beta_0 + \beta_1 * ASSETSTURN1 + \beta_2 * ROE\_AVG + \beta_3 * EPS\_TTM \qquad (1)$$
$$+ \beta_4 * DEBTTOASSETS + \varepsilon$$

(In this model, $Y = PCT\_CHG$ is the quarterly rise in share prices, $ASSETSTURN1$ is the total asset turnover rate, EPS_TTM represents the earnings per share, ROE_AVG is the average return on equity, $DEBTTOASSETS$ represents the asset-liability ratio, is a random disturbance item) where,

$$PCT - CHG = [(closing\ price\ of\ the\ last\ trading\ day\ of\ the\ interval \qquad (2)$$
$$- closing\ price\ before\ the\ first\ trading\ day\ of\ the\ interval)$$
$$/closing\ price\ before\ the\ first\ trading\ day\ of\ the\ interval] * 100\%$$

$$ASSETSTURN1 = [net\ operating\ income/\ average\ total\ assets] * 100\% \quad (3)$$

$EPS_TTM =$ *A weighted average of current net profits attributable to common* (4)

shareholders/outstanding common sharesissued
during the current period

$$ROE_AVG = [net\ profit/average\ net\ assets] * 100\% \qquad (5)$$

$$DEBTTOASSETS = [total\ liabilities/total\ assets] * 100\% \qquad (6)$$

3.3   Data

The data of this paper comes from Wind financial database system. The research sample selects all A-share listed companies quarterly financial indicators in the financial industry classified by the CSRC in Shanghai and Shenzhen stock exchanges. The sample number of companies is 77 that concludes 25 banking financial enterprises and 52 non-bank financial enterprises. The sample ranges from January 2014 to December 2017, as shown in Table 1:

Table 1. Sample selection distribution

Shanghai and Shenzhen Stock Exchange A-share	Banking financial enterprises	Non-bank financial enterprises	sample total
Number of listed companies	25	52	77
Number of samples	25	52	77

4   Empirical Analysis and Results

4.1   Descriptive Statistics

We made a statistical description of the stock price changes of listed enterprises in the financial industry and the main financial indicators. The results are shown in Table 2. After excluding cases by list, a total of 715 sample list states were obtained. It can be seen that :(1) for the quarterly rise index, its maximum value is 368.2035%, and its minimum value is −60.9995%. The wide range means that the quarterly rise of stock prices of financial industry enterprises vary greatly from different enterprises. The average quarterly gain was 7.625759%. It shows that on average, the stock price trend of financial enterprises is rising. (2) the standard deviation of the total asset turnover rate is only 0.1900115, indicating that there is little difference in the total asset turnover rate among enterprises in the financial industry. (3) the maximum value of quarterly average earnings per share is 5.2991, the minimum value is −0.8977, and the standard deviation is only 0.76. It shows that the company's earnings per share volatility is not large, the overall performance is good. (4) the average of level the asset-liability ratio index is 75.038541%, both greater than 50%, and the standard deviation is not large, indicating that financial industry enterprises generally show the characteristics of high leverage (Table 3).

Table 2. Descriptive statistics table of independent variables[a]

Variables	Min	Max	Mean	Std. dev	Skewness	Kurtosis
PCT_CHG	−60.9995	368.2035	7.625759	33.7207336	3.246	19.954
ASSETSTURN1	0.0039	2.4904	0.093417	0.1900115	6.140	51.802
DEBTTOASSETS	−0.8268	98.1570	75.038541	21.1050659	−1.405	1.373
EPS_TTM	−0.8977	5.2991	0.814990	0.7562103	1.460	3.074
ROE_AVG	−153.3160	105.5322	7.512368	11.9656616	−3.673	71.329

[a]OCFPS_TTM represents business activities generated cash flow per share, FCFTOCF represents financing activities generated cash flow, ICFTOCF represents investment activities generated cash flow, INVESTINCOMETOEBT represents gains on the changes in the value of total profit ratio, OCFTOOR represents business activities generated cash flow, OPERATEINCOMETOEBT represents business accounted for the proportion of the total profit, net earnings, ROE_BASIC represents weighted return on net assets, YOYCF represents year-on-year growth rate of cash flows, YOYDEBT represents total debt year-on-year growth rate, YOYEPS_BASIC represents the year-on-year growth rate of basic earnings per share, YOYPROFIT represents the year-on-year growth rate of net profit, and YOY_EQUITY represents the year-on-year growth rate of net assets.

Table 3. Descriptive statistics table of control variables

Variables	Min	Max	Mean	Std. dev	Skewness	Kurtosis
DEDUCTEDPRO FITTOPROFIT	−3742.4109	272.5482	78.818898	182.7815175	−18.639	370.182
OPERATEIN COMETOEBT	−13634.0703	427.0325	11.619489	489.6021908	−25.110	692.063
OCFPS_TTM	−68.2013	242.1906	3.056519	15.5028446	8.412	101.463
FCFTOCF	−39291.5000	16455.2374	−124.535700	2.6033517E3	−9.966	140.484
ICFTOCF	−9754.1093	44025.1667	112.834859	2.2030845E3	12.994	236.516
INVESTINCOME TOEBT	−327.0927	13732.1715	76.697245	482.1362304	26.738	755.233
ROE_BASIC	−2228.3700	105.5300	5.232672	74.8949321	−29.217	872.078
YOYCF	−12252.6287	3.0200E6	4294.175376	1.0306879E5	29.228	856.125
YOYDEBT	−101.4855	40896.5556	302.749770	2.2563372E3	13.935	226.579
YOYEPS_BASIC	−2800.0000	36427.1186	117.345449	1.4800027E3	21.047	476.308
YOYPROFIT	−14553.7405	64419.4462	142.146417	2.3620987E3	22.411	610.908
YOY_EQUITY	−44.3097	18558.6357	109.290718	863.1902990	16.462	311.835

4.2   Correlation Test

After considering the influence of control variables, the partial correlation analysis results were shown in Table 4. As for the variable ASSETSTURN1, p

Table 4. Correlation

Control variables			PCT_CHG
DEDUCTEDPROFITTOPR OFIT & OCFPS_TTM & FCFTOCF & ICFTOCF & INVESTINCOMETOEBT & OCFTOOR & OPERATEINCOMETOEBT & ROE_BASIC & YOYCF & YOYDEBT & YOYEPS_BASIC& YOYPROFIT & YOY_ASSETS & YOY_EQUITY	PCT_CHG	Correlation	1.000
		Significance (two-tailed)	
		df	0
	EPS_TTM	Correlation	−.095
		Significance (two-tailed)	.012
		df	699
	ROE_AVG	Correlation	.006
		Significance (two-tailed)	.871
		df	699
	ASSETSTURN1	Correlation	.311
		Significance (two-tailed)	.000
		df	699
	DEBTTOASSETS	Correlation	−.068
		Significance (two-tailed)	.070
		df	699

$= 0 < 0.05$ rejects the null hypothesis, it can be seen that the quarterly increase in PCT_CHG is significantly correlated with the total asset turnover rate ASSETSTURN1, and the correlation is 0.311. Therefore, we can preliminarily judge that the quarterly rise of listed financial companies is closely related to the total asset turnover rate and earnings per share and the quarterly rise is positively correlated with the total asset turnover rate and negatively correlated with earnings per share. For variable EPS_TTM, $p = 0.012 < 0.05$. It can be considered that quarterly rise PCT_CHG and EPS_TTM have a negative correlation.

4.3   Regression Analysis

(1) All sample regression

When all control variables are considered to enter the model, regression analysis is conducted on the total asset turnover rate, average return on equity, earnings per share, asset-liability ratio and the rise of stock price, which are the main explanatory variables in this paper. It can be seen from Table 6 that the model passed the F test, indicating that the model has certain rationality. However, as can be seen from Table 5, the tolerance of OPERATEINCOME-TOEBT, ROE_AVG, ROE_BASIC, YOYDEBT, YOY_ASSETS, YOY_EQUITY and INVESTINCOMETOEBT are less than 0.1 in the co-linear diagnosis. There may be serious problems of co-linearity between explanatory variables in this model. Therefore, the regression model should be modified (Table 6). We use stepwise regression[1] method to modify the previous regression model, and the results are shown as follows.

---

[1] The stepwise regression analysis was conducted according to the method of F test (criterion: probability of f-to-enter $<= 0.05$, probability of f-to-remove $>=0.1$).

Table 5. Coefficients

Model		Unstandardized coefficients		Standardized coefficients	t	Sig.	Collinearity statistics	
		B	Std. error	Beta			Tolerance	VIF
1.	(Constant)	−15.638	6.444		−2.427	.015		
	ASSETSTURN1	79.696	9.279	.371	8.589	.000	.658	1.520
	DEBTTOASSETS	.160	.081	.095	1.909	.049	.534	1.874
	DEDUCTEDPROFITTOPROFIT	−.014	.026	−.056	−.535	.593	.111	8.997
	EPS_TTM	−4.505	1.930	−.101	−2.334	.020	.661	1.512
	OCFPS_TTM	−.067	.086	−.032	−.780	.436	.747	1.339
	FCFTOCF	.000	.001	.034	.766	.444	.636	1.572
	ICFTOCF	.001	.001	.039	.886	.376	.626	1.598
	INVESTINCOMETOEBT	.082	.033	1.254	2.446	.015	.005	213.851
	OCFTOOR	5.722E-5	.000	.030	.797	.426	.872	1.147
	OPERATEINCOMETOEBT	.082	.033	1.289	2.461	.014	.004	223.098
	ROE_AVG	−.480	1.181	−.120	−.407	.684	.014	71.213
	ROE_BASIC	.960	1.177	.240	.816	.415	.014	70.659
	YOYCF	1.520E-6	.000	.005	.141	.888	.983	1.017
	YOYDEBT	.000	.002	.027	.213	.831	.076	13.204
	YOYEPS_BASIC	.001	.001	.033	.921	.357	.939	1.065
	YOYPROFIT	.001	.000	.042	1.157	.248	.930	1.075
	YOY_ASSETS	.002	.004	.145	.671	.503	.026	37.977
	YOY_EQUITY	−.008	.005	−.209	−1.383	.167	.054	18.554

a. Dependent variable: PCT_CHG

Table 6. Model summary

Model	R	R square	Adjusted R square	Std. error of the estimate	Change statistics					
					R square change	F change	df1	df2	Sig. F change	
1	.380a	.144	.122	32.1911035	.144	6.515	18	696	.000	

Table 7 shows that all three models have passed the T test. In this paper, model 4 with the maximum goodness of fit is selected to establish the sample regression equation. (where, DEDUCTEDPROFITTOPROFIT = [net profit/net profit excluding non-recurring profit and loss] *100%) F test results indicate that the overall regression effect of the model is good. That is to say, there is a significant linear correlation between the total asset turnover rate and the rise rate of financial industry enterprises in general. Weighted return on equity (ROE_BASIC) is positively correlated with the increase of stock price. But the linear relationship between the increase and earnings per share is not significant.

For different industries, due to differences in their industries and the particularity of a certain type of industry, we judge the value of the investment differently. In one industrythere are also some differences between those companies. So we need continue to do the following analysis.

Table 7. Coefficients

Model		Unstandardized coefficients		Standardized Beta	t	Sig.	Collinearity statistics	
		B	Std. error				Tolerance	VIF
2	(Constant)	4.034	1.368		2.948	.003		
	ASSETSTURN1	61.143	7.705	.285	7.935	.000	1.000	1.000
3	(Constant)	−1.347	1.755		−.768	.443		
	ASSETSTURN1	80.370	8.585	.374	9.361	.000	.781	1.280
	DEDUCTEDPR OFITTO- PROFIT	.047	.010	.192	4.790	.000	.781	1.280
4	(Constant)	−4.259	2.017		−2.112	.035		
	ASSETSTURN1	75.434	8.711	.351	8.659	.000	.751	1.331
	DEDUCTEDPR OFITTO- PROFIT	.044	.010	.176	4.395	.000	.768	1.302
	ROE_BASIC	.414	.143	.104	2.884	.004	.960	1.042

Table 8. Model summary

Model	R	R Square	Adjusted R square	Std. error of the estimate	Change statistics				
					R square change	F change	df1	df2	Sig. F change
2	.285a	.081	.080	32.9558732	.081	62.967	1	713	.000m
3	.331b	.110	.107	32.4600828	.029	22.947	1	712	.000
4	.347c	.120	.116	32.2945944	.010	8.316	1	711	.004
a. Predictors: (Constant), ASSETSTURN1									
b. Predictors: (Constant), ASSETSTURN1, DEDUCTEDPROFITTOPROFIT									
c. Predictors: (Constant), ASSETSTURN1, DEDUCTEDPROFITTOPROFIT, ROE_BASIC									

(2) Regression of banking financial enterprises

After testing the normal distribution, we did the correlation analysis between the indexes and the dependent variable. We found that the total asset turnover, cash dividend cover, weighted average return on net assets, return on equity, asset-liability ratio and cost income these six indicators are significantly correlated PCT_CHG (Table 8). Then we added independent variables to do regression analysis. As can be seen from Table 9, R squared after adjustment =0.659>0.5, indicating that the model has a strong overall fitting effect on sample data. In Table 10, Sig.=0<0.01, indicating that ROE_AVG has a significant positive correlation with the rise. At this time, only one major explanatory variable ROE_AVG is retained in the model. For banking financial enterprises, the average return on equity has a significant positive impact on the quarterly rise of stock prices. Then we excluded the missing values according to the list, and the stepwise regression results are as follows (Tables 11, 12, 13, 14).

Table 9. Model summary

Model	R	R square	Adjusted R square	Std. error of the estimate	Change statistics					
					R square change	F change	df1	df2	Sig. F change	
5	.816a	.666	.659	13.5770139	.666	95.860	1	48	.000	
a. Predictors: (Constant), ROE_AVG										

Table 10. Coefficients

Model	Unstandardized coefficients		Standardized beta	t	Sig.	Collinearity statistics	
	B	Std. error				Tolerance	VIF
5 (Constant)	−32.280	4.785		−6.746	.000		
ROE_AVG	4.049	.414	.816	9.791	.000	1.000	1.000
a. Dependent variable: PCT_CHG							

We added some special indicators like STMNOTE_BANK_129 to further analysis. It can be seen that the Model 6 Passed the F test. In the regression model, the specific index cost to income ratio (STMNOTE_BANK_129) did not pass the test. Other special indicators such as capital adequacy ratio have not been seen to have a significant correlation with the rise. However, OCFTODIV-IDEND has passed the T test. (OCFTODIVIDEND = [net operating cash flow per share/cash dividend per share] *100%.) This model shows that for banking enterprises, the positive dividend distribution policy has a significant positive impact on the quarterly rise of their stock prices, while the impact of specific banking indicators on stock price changes is not significant. And the results of stepwise regression analysis are as follows.

Table 11. Model summary

Model	R	R square	Adjusted R square	Std. error of the estimate	Change statistics					
					R square change	F change	df1	df2	Sig. F change	
6	.341a	.116	.098	27.1175445	.116	6.325	1	48	.015	
a. Predictors: (Constant), OCFTODIVIDEND										

(3) Regression of non-bank financial enterprises

For securities companies, insurance and other non-bank financial enterprises, through simple correlation and regression analysis, we found that the total asset turnover rate, current asset turnover rate and non-current asset turnover rate are correlated with the growth rate at the significance level of 0.01. The broker insurance special index to their stock price rise role is not significant. The stepwise

Table 12. Coefficients

Model		Unstandardized coefficients		Standardized beta	t	Sig.	Collinearity statistics	
		B	Std. error				Tolerance	VIF
6	(Constant)	1.764	4.269		.413	.681		
	OCFTODIVIDEND	.255	.102	.341	2.515	.015	1.000	1.000

b. Dependent variable: PCT_CHG

Table 13. Model summary

Model	R	R square	Adjusted R square	Std. error of the estimate	Change Statistics					
					R square change	F change	df1	df2	Sig. F Change	
7	.291a	.085	.080	37.497	.085	19.251	1	208	.000	

a. Predictors: (Constant), ASSETSTURN1

Table 14. Coefficients

Model		Unstandardized coefficients		Standardized beta	t	Sig.	Collinearity statistics	
		B	Std. error				Tolerance	VIF
7	(Constant)	3.171	3.164		1.002	.317		
	ASSETSTURN1	33.306	7.591	.291	4.388	.000	1.000	1.000

b. Dependent variable: PCT_CHG

regression results are shown in the following tables when the asset-liability ratio, earnings per share and average return on equity are included for comprehensive analysis. As can be seen from model 7, for non-bank financial enterprises, there is no statistically significant relationship between average return on equity, asset-liability ratio and eps and quarterly stock price rise. The correlation between current assets turnover and non-current assets turnover is not obvious. But the total asset turnover rate has a more prominent impact on the rise. Although the fitting effect of the final model is not good, it can still explain the positive correlation between the total asset turnover rate and the quarterly rise of stock price to some extent.

5   Conclusion and Prospect

5.1   Conclusion

(1) The importance of the total asset turnover rate

The total asset turnover rate can reflect the enterprise's asset management efficiency. Among the financial indicators selected as the main explanatory variables, the total asset turnover rate has the largest positive impact on the overall stock price rise of financial enterprises. For the whole financial industry, it can be considered that quarterly stock price rise is significantly correlated with the total asset turnover rate in this paper. There is still a significant relationship between the total asset turnover rate and the rise of the whole financial industry when all variables are taken into account. These results fully show that the asset utilization efficiencys importance to stock prices. The total asset turnover rate is a manifestation of the utilization efficiency of the company's assets, so for all financial companies, efforts should be made to improve the product's market share, and the more efficient utilization of assets is needed to enhance the positive impact on stock prices changes.

(2) The impact of dividend polices in a company

Dividend distribution policy is an important part of the company's financial policy. The cash dividend guarantee multiple is the ability of the net cash flow generated by the normal business activities of an enterprise to pay dividends. Robert D. Arnott [1] found if dividends are high today, future returns tend to be higher. The dividend yield is another subjective measure of financial success. Obviously, investors will be more willing to invest in a company when they see that it is inclined to pay dividends. In 2000, H. Thomas O'Hara and several other authors used these three ratios to build a successful model, which took five years into consideration [18]. For banking financial enterprises, we should pay attention to the positive influence of cash dividend guarantee multiple on the stock price change of the company. Cash dividend security multiple has a significant positive correlation with the growth rate in the research. The implementation of a positive dividend policy is beneficial to the improvement of the company's value. This is of great theoretical and practical significance to further study in this field.

(3) Different influences in different companies

For the same indicators, due to the differences in various aspects within the industry, with the existing influence of other factors such as organizational structure, the degree of effects on stock price rise between banking financial enterprises and non-bank financial enterprises can be different. While investing in financial listed companies, investors should consider the difference between the influence of such common but important financial indicators on the stock price changes of Banks and non-bank financial enterprises. In this paper, for non-bank financial enterprises, the total asset turnover rate explains most strongly among the explanatory variables studied to the increase of the stock price but it plays an insignificant role in stock price increase of bank enterprises. For all financial enterprises, weighted return on equity is positively correlated with the increase of stock price. However, the linear relationship between the increase and the average return on equity and earnings per share is not significant in this paper. The effect of average return on equity and asset-liability ratio on the quarterly stock price rise of financial enterprises is not significant in the statistical results of this paper. And the average return on equity only has some significant impact on a few banking financial enterprises. Therefore, the positive correlation between ROE and stock price changes has not been verified before. Besides, there is no obvious correlation between the capital adequacy ratio and the rise of the company's stock price. According to the research data of this paper, the special index has no significant impact on the change of the stock price, that is, the previous hypothesis of this paper is not valid. It can be seen that the specific indicators have no significant impact on the stock price changes of listed companies although they can reflect the characteristics of the industry and the company standards and differences within the industry. That might be the reason that Chinas banking sector is heavily regulated and supported by the government.

5.2   Research Prospect

In this paper, there is no trend analysis on the time series. Therefore, it may not be able to clearly and intuitively reflect the impact of the previous financial indicators on the current period of growth. That is also what we need to research in the future. The sample data in this paper are limited. If the data are more sufficient and the explanatory variables are more abundant, the fitting degree of the model will be better. Since the empirical model does not explain the overall sample data sufficiently, this paper suspects that there may be other factors besides indicators such as the total asset turnover rate that influences the stock price rise. For example, the hype has a significant positive impact on financial services and communications. In addition, the relationship between some indexes and dependent variables may not be satisfied with a simple linear relationship which has turned out to be a poor fitting effect. The results and conclusions of this paper are subject to time and practice.

Acknowledgements. This paper is supported by an MOE (Ministry of Education in China) Youth Foundation Project of Humanities and Social SciencesProject number:14YJC790053)Basic research foundation of Sichuan University(peoject number: skyb201402, xyzx1506, skzx2015-sb68, skzx2015-zx04, skqy201622).

References

1. Arnott, R.D., Asness, C.S.: Surprise! higher dividends= higher earnings growth. Financial Analysts Journal **59**(1), 70–87 (2003)
2. Bastianin, A., Manera, M.: How does stock market volatility react to oil price shocks? Macroecon. Dyn. **22**(3), 666–682 (2018)
3. Baybordi, A., Nejad, K.G., Kargar, E.F.: Evaluating the relationship between economic value-added and stock return in companies listed at tehran stock exchange. Manag. Adm. Sci. Rev. **4**(1), 215–221 (2015)
4. Beaver, W.H., Lambert, R.A., Ryan, S.G.: The information content of security prices: A second look. J. Account. Econ. **9**(2), 139–157 (1987)
5. Bernardo, I., Henriques, R., Lobo, V.: Social market: Stock market and twitter correlation. In: International Conference on Intelligent Decision Technologies. Springer (2017), pp. 341–356

6. Campbell, J.Y., Shiller, R.J.: Stock prices, earnings, and expected dividends. J. Financ. **43**(3), 661–676 (1988)
7. Capstaff, J., Klaeboe, A., Marshall, A.P.: Share price reaction to dividend announcements: empirical evidence on the signaling model from the oslo stock exchange. Multinatl. Financ. J. **8**(1/2), 115–139 (2004)
8. De Cesari, A., Huang-Meier, W.: Dividend changes and stock price informativeness. J. Corp. Financ. **35**, 1–17 (2015)
9. Floros, C., Jaffry, S., Ghulam, Y.: Predicting returns with financial ratios: evidence from greece. Int. J. Financ. Econ. Econ. **14**(1), 31–44 (2009)
10. Füss, R., Grabellus, M., et al.: Something in the air: Information density, news surprises, and price jumps. J. Int. Financ. Mark. Inst. Money **53**, 50–75 (2018)
11. Gilchrist, S., Himmelberg, C.P., Huberman, G.: Do stock price bubbles influence corporate investment? J. Monet. Econ. **52**(4), 805–827 (2005)
12. Hearn, B., Phylaktis, K., Piesse, J.: Expropriation risk by block holders, institutional quality and expected stock returns. J. Corp. Financ. **45**, 122–149 (2017)
13. Heiberger, R.H.: Predicting economic growth with stock networks. Phys. A: Stat. Mech. Appl. **489**, 102–111 (2018)
14. Hirshleifer, D., Jiang, D.: A financing-based misvaluation factor and the cross-section of expected returns. Rev. Financ. Stud. **23**(9), 3401–3436 (2010)
15. Kvamvold, J.L.S.: Do dividend flows affect stock returns? J. Financ. Res. **41**(1), 149–174 (2018)
16. Lee, T.K., Cho, J.H., et al.: Global stock market investment strategies based on financial network indicators using machine learning techniques. Expert. Syst. Appl. **117**, 228–242 (2019)
17. Maskun, A.: The effect of current ratio, return on equity, return on asset, earning per share to the price of stock of go-public food and beverages company in Indonesian stock exchange. Int. J. Acad. Res. **4**(6), (2012)
18. O'Hara, H.T., Lazdowski, C., et al.: Financial indicators of stock price performance. Am. Bus. Rev. **18**(1), 90 (2000)
19. Ray, K.: Eva as a financial metric: the relationship between eva and stock market performance. Eur. J. Bus. Manag. **6**(11), 105–114 (2014)
20. Reddy, Y., Narayan, P.: The impact of eva and traditional accounting performance measures on stock returns: evidence from india. IUP J. Account. Res. Audit. Pract. **16**(1), 25 (2017)
21. Rutledge, R.W., Karim, K.E., Li, C.: A study of the relationship between renminbi exchange rates and chinese stock prices. Int. Econ. J. **28**(3), 381–403 (2014)
22. Scharfstein, D.S., Stein, J.C., et al.: Herd behavior and investment. Am. Econ. Rev. **80**(3), 465–479 (1990)
23. Schinckus, C.: An essay on financial information in the era of computerization. J. Inf. Technol. **33**(1), 9–18 (2018)
24. Sloan, R.G.: Do stock prices fully reflect information in accruals and cash flows about future earnings? Account. Rev. **71**, 289–315 (1996)
25. Sui, G., Li, H., et al.: Correlations of stock price fluctuations under multi-scale and multi-threshold scenarios. Phys. A: Stat. Mech. Appl. **490**, 1501–1512 (2018)
26. Weng, B., Lu, L., et al.: Predicting short-term stock prices using ensemble methods and online data sources. Expert. Syst. Appl. **112**, 258–273 (2018)

Empirically Analyzing the Future Intentions of Pakistani Students to Stay or Leave: Evidence from China

Kashif Iqbal[1]([✉]), Hui Peng[1], Muhammad Hafeez[1], Khurshaid[2], and Israr Khan[3]

[1] School of Economics and Management, Beijing University of Posts and Telecommunications, Beijing 100876, People's Republic of China
kashii42@yahoo.com
[2] University of Peshawar, Peshawar, KPK, Pakistan
[3] School of Information and Communication Engineering, Beijing University of Posts and Telecommunications, Beijing 100876, People's Republic of China

Abstract. The rise of China as a new destination for international students has been widely reported in both the domestic and international mass media, but academic research into the phenomenon and its theoretical implications are lacking. The aim of this study is to investigate empirically the factors that influence the decision-making process of Pakistani students to stay or leave China after completing their studies. For this purpose, we undertook a survey of 189 Pakistani students recently studying in different universities of Beijing, China, through a structured questionnaire technique and apply a binary logit model. Based on the survey results, the majority of Pakistani students are willing to stay and work in the future in China. While a large number of the survey respondents are thinking that wages in China are high, the working environment is better and the lifestyle is better in China as compared to Pakistan. In addition, a group of considerable survey respondents is cynic about limited job opportunities for foreigners in China.

Keywords: Brain-gain migration · Logit model · Random utility theory · China

1 Introduction

The last few decades witnessed striking growth in international migration flows. In particular, the international movement of the highly educated experienced an impressive surge [25]. There is a different form of migration, some forms have to do with "brain drain" (e.g. specialist, educated and good brain leave etc.) and other types are more comprehensive (e.g. refugees, deportations etc.) [20]. The concept of brain drain' was introduced in the late 1960s and 1970s, and soon became well rooted in the economics literature. "Brain drain" is defined differently by different authors, e.g. by Bein, Docquier, and Rapoport as the

© Springer Nature Switzerland AG 2020
J. Xu et al. (Eds.): ICMSEM2019 2019, AISC 1002, pp. 759–769, 2020.
https://doi.org/10.1007/978-3-030-21255-1_58

"proportion of working-age individuals" (aged 25 and over) with at least tertiary education, born in a given country but living elsewhere [4]. There are two different schools of thought discussing the brain drain phenomena.One school of thought argues that the brain drain is beneficial for sending countries [7,10,13]. While the other one argues that the brain drain is harmful to sending countries [1,12,21].

In a world of expanding global corporate collaborations and transnational social networks, the appeal for internationally-educated professionals has dramatically increased in the last thirty years. In 2016, there were nearly 5.1 million internationally mobile students (i.e. 2.3% of all tertiary students), up from 2.1 million in 2000 [10]. Recently, China is hosting a huge number of international students and encouraging them to participate in the labour market. According to the ministry of foreign affairs (MOFA), over 440,000 foreigner students in China in 2016, marking a 35% increase from 2012. China attracts more international students than any other Asian countries and ranks third globally, behind the United States and the United Kingdom [23]. Chinese government's ultimate aim is to host 500,000 international students, to become the biggest host country for international students in Asia, and a major study destination in the world [19]. The inflow of international students to China has increased from 1236 in 1978 to 328,330 from 189 countries in 2013 [26]. Over the past five years, international students studied in China increased to a large extent, hitting 440,000 in 2016, among the international students in China, 60% were from Asia, 18% Europe and 11% Africa [17]. Like other Asian countries, China has become the most popular destination for Pakistani students. The number of Pakistani students has risen from 5,000 to 23, 197 during the last five years (Wen, Hu,and Hao, 2017). The recent patterns of international students are reported in Table 1.

Table 1. Number, percentage, region, and country of international students in China (2016)

Region	Number	Percentage	Country	Number	Percentage
Asia	264,976	59.84%	South Korea	70,540	15.93%
Europe	71,319	16.11%	United Kingdom	41,975	9.38%
America	61,591	13.91%	USA	32,975	7.20%
Africa	38,077	8.60%	Thailand	26,694	5.21%
Oceania	6,807	1.54%	Pakistan	23,197	4.96%

Source: Chinese Ministry of Education (http://en.moe.gov.cn/News/Top_News/201604/t20160420_239196.html)

The president of China Xi Jinping announced the Belt and Road Initiative (BRI) in 2013 to strengthen China's connectivity with the world. From last one decade, there has been a significant increase of international students coming from Belt and Road Initiative (BRI) countries such as Kazakhstan, Thailand, India, Vietnam, Pakistan, Mongolia, and Malaysia. It has been observed from the existing literature that there have been a few attempts to look at brain-drain migration in Chinabut this is the first attempt to study the Pakistani student's intentions to stay or leave after graduation. Given the above background and knowing the importance of the international student's attraction to China motivates us to empirically analyze the factors that compel Pakistani students to stay or leave China after graduation and to provide the new insight for policymakers. The rest of the study is organized in a manner that; Sect. 2 provides a detailed literature review, Sect. 3 explains the methodology and data collection procedure. Section 4, discusses and analyzes the data, and finally, the paper is concluded in Sect. 5.

2   Theoretical Background and Literature Review

This section reviews some of the theoretical explanations for skilled human capital migration. A different theoretical background has been proposed to know why international migration happens. Here, a comprehensive theoretical background is used to explain why human capital migration happens. According to the neo-classical model of [22], the wage difference in both sending and receiving countries play a key role to migrate from one country or region to another country or region. While,on the other hand the classical school of thought considers that migration is a collective decision of the individual and his family to migrate from one country to another country [18].

The prior literature identified the push-pull model approach, according to push-pull model, there is a set of factors that compels the individual to move from one country or region to another country or region for a better lifestyle, better opportunities, better education, etc. [8]. Mainly, push set of factors includes the factors that push skilled human capital from home country to host countries, such as lack of opportunities, lack of basic facilities, poor working environment, corruption, and nepotism etc. [13]. While, the pull set of factors includes the factors that pull skilled human capital from the host countries to home countries, such as more opportunities, better education, better lifestyle and better working condition etc. [5]. Except for push-pull approach, there are also other socio-economic factors that compel individual to migrate from less developed to developed countries, like social attachments, culture, religion and political stability etc. [9].

To stay or leave the host country after graduation is a tough decision for students to take. There are different factors that influence the student's decisions to stay or leave the host countries. The most influencing factor is their attachment with family and friends in their home country [14]. Moreover, the return intentions linked with their expectations of life in origin countries as compared with their expectation in destination countries.To get great respect, better job opportunities and better education in the host countries individuals are compelled to stay in the host countries after their graduation [24]. Besides, good opportunities, social ties with their friends and family, and patriotism compel individuals to go back to their home country after graduation [16].

3   Research Methodology

3.1   Survey Participant and Data Collection Procedure

The survey is conducted in Beijing, the capital of China, as a case study. Beijing is the top destination for international students in China, attracting nearly 24% of the total number of international students in China. The data is collected from 189 Pakistani students that are studying in different full-time degree programmes such as bachelor, master, and Ph.D., in different universities in Beijing, China. These students are majoring in natural sciences, social sciences, physical sciences, information technologies, and management sciences. We adopted a structured questionnaire technique to collect the data set based on closed-ended questions that are indented to elicit the factors that influence the intention of Pakistani students recently studying in China. This research model based on random utility theory. According to the random utility theory, an individual is capable of evaluating the utility associated with a set of viable alternatives and subsequently selecting that he perceives will yield maximum utility [2].

For data collection, we generated our own micro data based and collect the data online through survey monkey (https://www.surveymonkey.com/). We employ three different sets of explanatory variables to explain the Pakistani students' intentions to stay after their graduation. The set of demographic variables (gender, marital status, and age), the set of explanatory variables (socio-economic background) and the perception-related variables (wages, opportunities, working environment etc.).

3.2   Econometric Strategy

Statistical computational tools, Statistics, and data (Stata), and Statistical Package for Social Sciences (SPSS) is used to evaluate and analyze the primary data. The respondents were asked to provide their demographic information, including gender, marital status, the field of study, level of study, work experience, etc. The earlier research considered that some of these qualifications and occupations may have important effects on respondent's intention toward Pakistani students to stay or return to their home country after graduation in China. The Correlation matrix applied to analyze the sample data, identifying the relationships between independent variables and the dependent variable. Logit regression analysis was used to examine the predictive power of all factors on the overall Pakistani students to stay or leave after graduation in China.

3.3   Model Specification

This paper uses the binary logit model estimated using the maximum likelihood technique. This model addresses the dichotomy of whether or not a student intends to return home and identifies the determinants of such intentions. Logistic regression model can be written as follow:

$$\Pr ob(FI = 1) = \frac{ez}{1 + ez} \tag{1}$$

In Eq. 1, e = natural logarithm base,

$$FI = future\,intention = \begin{cases} FIS = 1, & if\,future\,intention\,to\,stay\,in\,China \\ FIL = 0, & if\,future\,intention\,to\,\,leave\,China \end{cases} \tag{2}$$

Based on the introduction and literature review, this paper dedicates to compute empirically future (intention) brain gain for China. The functional form of models is as follows:

$$FI = f(LCHH, EC, FCL, FCS) \tag{3}$$

$$FI = \alpha_0 + \alpha_1 LCHH + \alpha_2 EC + \alpha_3 FCS + \alpha_4 FCL + \varepsilon_{it} \tag{4}$$

where, FI (future intention) is the dependent variable, while LCHH (Life comparison of home country with the host country), EC (education environment in China), FCS (factors that compel to leave China, and FCL (factors that compel to stay in China) are independent variables in the model. ε_{it} is an error term measuring the extent to which the model cannot fully explain future intentions.

4 Results and Discussion

4.1 Descriptive Analysis

For primary data analysis, we use the Statistics and Data (Stata) and Statistical Package for Social Sciences (SPSS). A total of 189 Pakistani students participate in the survey. Table 2 demonstrates the descriptive statistics of the participants. The number of males (83%) is more than female (17%) participants, while the number of singles (57%) is more than married (43%) participants. Moreover, education-wise, the majority of the participants are doing masters (52%), Ph.D., (25%) and only (16%) of the students are bachelor students. By age-wise, most of the participants are young. Participants with age 25–30 are 45%, while the participants with age 30–35 are 33% of the sampling population.

4.2 Correlation Matrix

A student's plans after graduation provided the strongest prediction for whether a student will desire to stay or leave the host country. The matrix correlations in Table 3 demonstrate that there is a negative correlation between individual future intentions (FI), LCHH (Life comparison of home country with the host country), EC (education environment in China), FCS (factors that compel to leave China, and FCL (factors that compel to stay in China). It indicates that improved standard of living in home countries attract an individual to return back to their home countries [13]. Moreover, the better education in home country attracts an individual to leave the host country and return back to their home

Table 2. Descriptive statistics of future intentions (N = 189)

Variables	Number	Frequency	Percentage
Gender	Male	157	83
	Female	30	17
Marital status	Single	107	57
	Married	77	41
Education	Undergraduate	31	16
	Master	99	52
	Ph.D.	48	25
	Languages	5	2
	Other	4	2
Age	18-25	23	12
	25–30	46	45
	30–35	66	33
	35–40	15	8
	40 or above	3	1

country [3]. Factors compel to stay (FCS) in China negatively associated with compelling factors to leave China (FCL). It infers that a better life standard in host country increases the brain gain chances for the home country. Moreover, due to a better learning environment and factors that compel the individuals to stay (FCS) in China respectively [6,14,15].

Table 3. Matrix of correlations

Variables	FI	LCHH	EC	FCL	FCS
FI	1				
LCHH	−0.178	1			
EC	−0.193	0.335	1		
FCL	−0.096	−0.112	0.03	1	
FCS	−0.215	0.18	0.198	0.181	1

4.3 Logistic Regression

The impact of future intentions is quantified through Logistics modeling due to the nature of the dichotomous dependent variable. The empirical results estimates are reported in Table 4. The outcomes from logistics modeling signify that better living standard in China compels the individual to stay in China have a statistically insignificant negative impact on future intentions. It is enlightening that better life standard in the home country and addition in features to stay in China are more likely to be brain gain for China, the resulted are supported by the work of [5]. The better learning environment in China compels the individual to stay in China, have a negative impact on future intentions but it is statistically insignificant.

The goodness of fit (GOF) from logistics regression is statistically significant at the 1% level. To overcome the functional form and omitted variable problem, Link test has been applied. The prediction square (hatsq) of link test, is statistically insignificant which revealed that econometric model of study has not functional form and omitted variable problem. Another problem of primary data analysis, there is a chance of multi-co-linearity. The estimated magnitudes become unstable and wildly inflated the standard errors in the presence of multi-co-linearity. To counter this, Variance inflating factor (VIF) and tolerance (I/VIF) have been computed to identify the multi-co-linearity level in the model. As a rule of thumb, VIF and tolerance values are greater than 10 and 0.1 respectively, may need to be further investigated. Mean VIF is 1.01, 1.00 and 1.03 in push factor model, pull factor model and both factors model respectively which is acceptable (see Table 4).

4.4 Model Accuracy

To measure the estimated model validity and accuracy, sensitivity and specificity analysis are conducted to pinpoint the model accuracy. The results of sensitivity and specificity analysis are demonstrated in Table 5. True positive Pr $(+|D)$ and true negative Pr$(-|\sim D)$ depict the accuracy of the model. Sensitivity measures the proportion of observed positives that were predicted to be positive while specificity measures the proportion of observed negatives that were predicted to be negatives [11]. Ideally, the test will result in both being high, but usually, there is a tradeoff. The geometrical illustration of sensitivity and specificity is depicted in Fig 1. It infers that model accuracy (correctly classified) is 85.64%.

4.5 Receiver Operating Characteristics Curve (ROC) Analysis

For the purpose of estimated model validation, we also applied the additional goodness of fit test i.e. ROC. ROC summary index is famous and recently used to measure the goodness of fit of an estimated model. It is also known as the area under the ROC curve (AUROC). The AUROC is presented in Fig. 2. This curve illustrates the pattern of hit rate cut-off value (HRC) and false rate cut-off value (FRC). A higher magnitude of AUROC depicts the better-estimated model. For an appropriate model, the AUROC value must lie between 0.5 and 1.0 [11]. The current study estimated model has 0.6885 value of AUROC which infers that model is 68.34% correctly predicted.

Table 4. Matrix of correlations

Logistics regression model (Dependent variable = FI), N = 189				
	dy/dx	Std. Err.	Z-stats	Prob.
LCHH	−0.0368	0. 268	−1.37	0.17
EC	−0.03	0.0223	−1.34	0.179
FCL	−0.0117	0.0116	−1.01	0.315
FCS	−0.0176**	0.0092	−1.91	0.056
Diagnostics statistics				
Logistic GOF	110.46*			
Pearson chi2	0			
Link Test				
_hat	2.5096*	0.947	2.65	0.008
_hatsq	−0.4815	0.2812	−1.71	0.087
_cons	0.8185	0.7127	−1.15	0.251
VIF statistics				
	VIF	TOL		
LCHH	1.17	0.853		
EC	1.15	0.8666		
FCL	1.1	0.9092		
FCS	1.06	0.9431		
Mean VIF	1.12			

Note: * and ** indicate the level of significance at 1% and 5% respectively.

Table 5. Sensitivity and Specificity analysis

Cross-validation analysis					
Sensitivity	Pr(+| D)	98.77%	False − rate for true D	Pr(−|D)	1.23%
Specificity	Pr(−|~D)	3.85%	False + rate for true ~D	Pr(+|~D)	96.15%
Positive predictive value	Pr(+| D)	86.49%	False + rate for classified	Pr(~ D|+)	13.51%
Negative predictive value	Pr(−|~D)	33.33%	False − rate for classified	Pr(D|−)	66.67%
Correctly classified	85.64%				

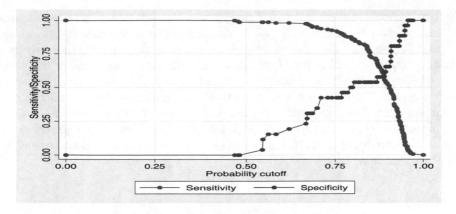

Fig. 1. Model validity and accuracy analysis

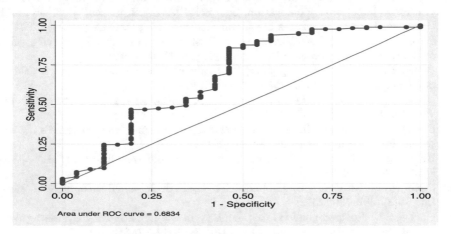

Fig. 2. Receiver operating curve (ORC) analysis

5   Conclusion and Policy Implications

(1) Main results

This article investigates the return and non-return intentions of Pakistani students after graduation in China. Based on our survey results, the majority of Pakistani students are willing to stay and work in the future in China. A large number of the survey respondents are thinking that wages in China are high, the working environment and lifestyle is also better as compared to Pakistan. However, a group of considerable survey respondents is cynic about limited job opportunities for foreigners in China.

(2) Policy implications

The empirical results of the survey should be a high concern for the policy makers of China. This foreign (Pakistani) students' potential can be turned into an economic advantage (brain gain) if the Chinese government makes sure to implement a labour friendly policy, making it one of the national priorities. Thus, Chinese law enforcement agencies need to formulate and implement a strategy that: (a) China needs to implement suitable labour policy to utilize this foreign (Pakistani) students' potential efficiently. (b) China needs an active migration monitoring system that focuses on and monitors the migration movement. (c) China needs to revisit migration policies and create more employment opportunities by promoting economic development, resolving employment problems, encouraging and guiding non-state-owned businesses, and increasing its employment capacity.

References

1. Agrawal, A., Kapur, D., et al.: Brain drain or brain bank? the impact of skilled emigration on poor-country innovation. Soc. Sci. Electron. Publ. **69**(1), 43–55 (2011)
2. Horowitz, J., et al.: Advances in random utility models. MPRA (53026) (2014)
3. Batista, C., Lacuesta, A., Vicente, P.C.: Testing the brain gain'hypothesis: micro evidence from cape verde. J. Dev. Econ. **97**(1), 32–45 (2012)
4. Beine, M., Docquier, F., Rapoport, H.: Brain drain and economic growth: theory and evidence. J. Dev. Econ. **64**(1), 275–289 (2001)
5. Beine, M., Docquier, F., Oden-Defoort, C.: A panel data analysis of the brain gain. World Dev. **39**(4), 523–532 (2011)
6. Boncea, I.: Turning brain drain into brain gain: evidence from romania's medical sector. Proc. Econ. Financ. **20**, 80–87 (2015)
7. Byra, L.: Rethinking the brain drain: dynamics and transition. Int. Rev. Econ. Financ. **28**(4), 19–25 (2013)
8. Docquier, F., Rapoport, H.: Globalization, brain drain, and development. J. Econ. Lit. **50**(3), 681–730 (2012)
9. Dorius, S.F.: Socioeconomic status mobility in the modern world system: growth and allocation effects. Soc. Indic. Res. **138**(3), 1049–1074 (2018)
10. Fransen, S., Ruiz, I., Vargas-Silva, C.: Return migration and economic outcomes in the conflict context. World Dev. **95**, 196–210 (2017)
11. Han, X., Stocking, G., et al.: Will they stay or will they go? International graduate students and their decisions to stay or leave the us upon graduation. PloS One **10**(3), 1–18 (2015)
12. Hatton, T.J.: The economics of international migration: a short history of the debate. Labour Econ. **30**, 43–50 (2014)
13. Hussain, S.M.: Reversing the brain drain: Is it beneficial? World Dev. **67**, 310–322 (2015)
14. Ifanti, A.A., Argyriou, A.A., et al.: Physicians' brain drain in greece: a perspective on the reasons why and how to address it. Health Policy **117**(2), 210–215 (2014)
15. Li, H., Ma, Y., et al.: Skill complementarities and returns to higher education: evidence from college enrollment expansion in china. China Econ. Rev. **46**, 10–26 (2017)
16. Lim, S.S.: Aspirations of migrants and returns to human capital investment. Soc. Indic. Res. **138**(1), 317–334 (2018)
17. Lu, Z., Li, W., et al.: Destination china: international students in chengdu. Int. Migr. (2018)

18. Massey, D.S., Arango, J., et al.: Theories of international migration: a review and appraisal. Popul. Dev. Rev. **19**(3), 431–466 (1993)
19. Saddozai, S.K., Peng, H., et al.: Investigation of talent, talent management, its policies and its impact on working environment. Chin. Manag. Stud. **11**(4), 538–554 (2017)
20. Stark, O., Byra, L.: A back-door brain drain. Econ. Lett. **116**(3), 273–276 (2012)
21. Su, Y., Tesfazion, P., Zhong, Z.: Where are migrants from? inter- vs. intra-provincial rural-urban migration in china. China Econ. Rev. **47**, 142–155 (2017)
22. Todaro, M.: Internal migration in developing countries: a survey. In: Population and economic change in developing countries, pp. 361–402 (1980)
23. Wei, H., Yi, J., Zhang, J.: Brain drain, brain gain, and economic growth in china. China Econ. Rev. **38**(HDRP-2009-37), 322–337 (2016)
24. Wu, W., Yu, B., Spender, J.C.: Domains and opportunities in knowledge and aerospace management in china: an integrative perspective. Chin. Manag. Stud. **9**(4), 473–481 (2015)
25. Zakharenko, R.: Human capital acquisition and international migration in a model of educational market. Reg. Sci. Urban Econ. **42**(5), 808–816 (2012)
26. Zheng, M., Yeung, W.J.J.: For money or for a life: a mixed-method study on migration and time use in china. Soc. Indic. Res. **139**(4), 1–33 (2017)

Study on Fluctuation and Regulation of Potato Market Price in China: Based on the View of Stable Crop for the Potato

Qianyou Zhang[1], Xinxin Xu[1(✉)], Yuanling Zhang[2], Yuerong Zheng[1],
and Jinqiu Tian[1]

[1] Business School, Chengdu University, Chengdu, Sichuan, People's Republic of
China
xuxinxin@cdu.edu.cn

[2] The Strategic Research Center on Stable Crop for the Potato, Xichang University,
Xichang 615013, People's Republic of China

Abstract. By utilizing the monthly potato market price data from 2011
to 2015 and time series model such as X11 seasonal adjustment method
and H-P (Hodrick–Prescott) filtering method, this paper analyzes the
period and rule of potato market price fluctuation. The study found that
the peak of the National Potato Price Index appears around every May,
the lowest point appears around every October, the highest and lowest
points appear alternately each year. In the past 5 years, the potato price
fluctuation in China is large, the fluctuation frequency is high, on the
average, the drop of peak and valley variation rate is 42.18%, the average
rising month for each cycle is 7.25 months, with an average cycle length
of 14.25 months.

Keywords: Potato · Market price · Fluctuation cycle · Empirical
study

1 Introduction

At the beginning of 2015, by promoting potato as a staple food, the National
Ministry of Agriculture proposed potato as the fourth major grain except wheat,
corn and rice in our country which can guarantee the food security in China. The
NO.1 Document issued by the Central Committee in 2016 proposed to promote
structural reform on the agricultural supply side and establish a large food concept by promoting the development of potato as a staple food. In February 2016,
the Ministry of Agriculture issued "The Guideline on Promoting Potato Industry Development" and announced to develop potato as a staple food product. It
suggests that by 2020, the potato planting area will be expanded to more than
66 billion square meters, the average production per square meter will increase
to 19 kg, and the total output will reach 130 million tons. However, the domestic
potato market price is unstable and fluctuates drastically in recent years. The

J. Xu et al. (Eds.): ICMSEM2019 2019, AISC 1002, pp. 770–779, 2020.
https://doi.org/10.1007/978-3-030-21255-1_59

low price affects the farmers and the high price affects the consumers. With the unpredictable potato price, the potato producers, operators, consumers, and supervision systems are unable to make timely and effective decisions, which will impede the development of potato to be a staple food. Therefore, it is useful to stabilize and control the potato market price fluctuations by studying the fluctuation cycle and rule of potato market price. It has important practical and theoretical value for promoting the strategy.

2   Literature Review

The sharp fluctuation of potato prices not only damage potato farmers interests, but also greatly affect consumers interests. It is harmful to the development of potato market and industry. Hence, the study of potato market price change has become one of the important parts of researchers' attention. Liu and Luo [9] used monthly market price data of 592 wholesale markets from 1996 to 2010 in Chinese Agriculture Infomation net and studied the long-term trend of Chinese potato wholesale market price change, the influence of seasonal factors, weekly factors, and forecasted the price change. Cai [3] analyzed the potato price fluctuation which caused from the market supply and demand, production cost, emergencies, and information asymmetry, the disconnection of industrial structure and the dispersion of production and operation. Liu and He [8] studied the potato price changes characteristics. However, due to the inconsistent data sources, the above research results are not the same, and no conclusion has been reached. Cong [2] analyzed the conduction effect and long-term cointegration of potato wholesale market price index and field purchase price index. Other researchers have studied the medium and long term supply and demand change and sustainable development of potato [11,12]. At the same time,many researchers have studied the price fluctuate of agriculture products [4,5,13–15].

Domestic research on potato price fluctuations has both qualitative and quantitative research, but mainly focuses on empirical aspects, especially explaining the reasons for potato price fluctuations. Throughout the domestic and foreign scholars to treat potatoes as vegetables, paying attention to the characteristics of its price changes, but treat potatoes as a staple food (belonging to quasi-public goods), research on its production and price rules, price support policies, and so on, are still relatively lacking. Based on this, this paper on the basis of absorbing the previous research results, by utilizing the monthly price data of potato market from 2011 to 2015, and some time series model such as X11 seasonal adjustment method and H-P (Hodrick–Prescott) filtering method, to analyze the cycle and regularity of price fluctuation in potato market.

3   Methods and Data

3.1   The Methods

Moderate fluctuation is the natural attribute of market price. Under the action of competition law and supply and demand law, price fluctuates with value, especially the price of agricultural products. Agricultural products price fluctuates frequently due to the characteristics of dispersibility, long cycle and slow turn of agricultural production, constraints of natural conditions, and unbalanced supply and demand. The potato price is one kind of agricultural products price, which has the general characteristics of agricultural products price and its particularity.

This paper uses X11 seasonal change method to study the patterns and characteristics of potato price fluctuation. The method decomposes time series $\{Y_t\}$ into four components: trend component T_t (Trend), which is the change caused by the basic factors of general and long-term effects in each period; seasonal component S_t (Seasonal Fluctuation), which is the cyclical change caused by natural seasonal changes and social conventions; periodic components P_t (Periodicity), which refers to the regular rise and fall of time series; irregular components I_t (Irregular Variations), there are non-trend, non-periodic random changes caused by occasional and temporary factors except for the former three components [6]. The time series decomposition is designed to separate the trend component T_t, the seasonal component S_t, and the irregular component I_t of the economic time series, and then analyze the statistical characteristics of the periodic components P_t. Generally, the symmetric moving average method and the high-order moving average method are used, and the trend component T_t, the seasonal component S_t, and the irregular component I_t of the original time dynamic series are finally separated by multiple iterations, and the adjusted series without the seasonal component is obtained. As the trend component and the periodic component are regarded as one component during the seasonal adjustment, it is necessary to separate the trend component from the periodic component. H-P (Hodrick–Prescott) filtering method is the most commonly used method to remove trend components. Let $\{Y_t\}$ be an economic time series that contains only trend and periodic components, $\{Y_t^d\}$ is the trend component, $\{Y_t^c\}$ is the periodic component, i.e. $Y_t = Y_t^d + Y_t^c$ [10].

The H-P filtering is designed to separate $\{Y_t^d\}$ from $\{Y_t\}$, a time estimation sequence $\{Y_t^d\}$ is selected to minimize the actual value and the sample value. $\{Y_t^d\}$ is often defined as the solution to the minimization problem of the following loss functions:

$$\min \sum_{t=1}^{T} \left\{ \left(Y_t - Y_t^d\right)^2 + \lambda \sum_{t=1}^{T} \left[\left(Y_{t-1}^d - Y_t^d\right) - \left(Y_t^d - Y_{t-1}^d\right)\right]^2 \right\} \tag{1}$$

This paper uses EVIEWS6.0 system default to filter the potato price index series by H-P, calculate the variation rate (Ratio of variation), i.e. which is used as the division basis for the potato market price period [7].

3.2   Data Source

In this study, the national potato market price index from January 1, 2011 to December 31, 2015 is provided by China Potato Website [1]. When the time series is decomposed, the sample data of 1825 days are divided into groups and the daily potato price index is converted into monthly data. This study is based on the monthly data.

4   Empirical Analysis

4.1   Long-Term Volatility Trend

According to the national historical data, the national potato price fluctuates frequently during the five-year period. In the 1825 sample data, the minimum value (Minimum) is 1.36, which appeared on November 6, 2011, the maximum value (Maximum) is 2.90, which appeared on May 20, 2013. The range (Range) is 1.54, the mean (Mean) is 2.1045, the standard deviation (Std. Deviation) is 0.37036, and the Variance (Variance) is 0.137. The statistics are shown in Table 1, and the national potato price index is shown in Fig. 1.

Table 1. Description statistics of national potato price index

N	1823
Range	1.54
Minimum	1.36
Maximum	2.90
Mean	2.1045
Std. Deviation	0.37036
Variance	0.137

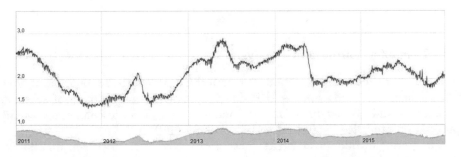

Fig. 1. Trend of the national potato price index

4.2   Analysis of Seasonal Component

Before empirical analysis, this paper tests whether the potato price index series is stationary, Because only the non-stationary time series can be decomposed by H-P filter method and X11 seasonal adjustment method. In this paper, the most commonly used DF-ADF unit root test is used to test the stability of the time series. The results show that the time series of potato price index is non-stationary, it includes random component, and the intercept term is non-zero, which indicates that the trend component is also included in the time series. Therefore, the trend cycle can be separated from the time series.

Through adjusting the nation potato seasonal price, the seasonal trend component is obtained as shown in Fig. 2. In the past 5 years, the seasonal component of potato price index has shown obvious regularity, and the fluctuation amplitude of seasonal period shows a relatively stable trend with the change of year. In every May, potato price index reaches its highest point. And the price index falls between June and October, with October being the lowest point. Price index rises again from November to May. This is a complete vitality cycle. Although there are slight changes during the period, the underlying trend does not change.

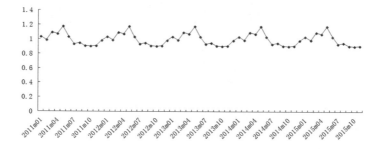

Fig. 2. Seasonal factors of the national potato price index

4.3   Periodic Component Analysis

In order to quantitatively obtain the periodic characteristics of potato price fluctuations, this study applies H-P filtering analysis on the basis of seasonal adjustment and obtained the trend component and periodic component of the national potato price index time series, as shown in Fig. 3.

The periodic component is the fluctuation value of the reaction fluctuation after the long-term trend is removed. The mutation rate is calculated by dividing the periodic value by the long-term trend, and it is an important relative index of fluctuation period. After calculation, the national potato price variability rate is shown in Table 2.

Through the analysis, it is found that the potato market price fluctuation shows obvious periodicity, and there is a decreasing trend of peak drop. To

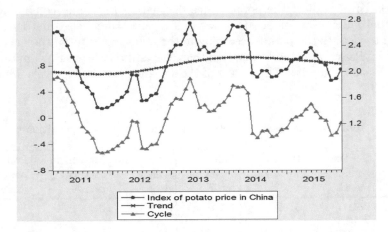

Fig. 3. Decomposition of national potato price index using H-P filtering method (=14400)

Table 2. The national potato price variability rate

Time	Variability rate	Time	Variability rate	Time	Variability rate
2011.1	30.1%	2012.9	−19.9%	2014.5	16.9%
2011.2	31.4%	2012.10	−19.2%	2014.6	−11.0%
2011.3	28.8%	2012.11	−10.3%	2014.7	−13.7%
2011.4	20.9%	2012.12	−0.5%	2014.8	−9.2%
2011.5	12.4%	2013.1	10.1%	2014.9	−8.9%
2011.6	4.7%	2013.2	14.1%	2014.10	−13.3%
2011.7	−6.8%	2013.3	13.5%	2014.11	−12.5%
2011.8	−10.7%	2013.4	20.6%	2014.12	−8.3%
2011.9	−15.5%	2013.5	27.8%	2015.1	−7.4%
2011.10	−26.1%	2013.6	18.4%	2015.2	−1.6%
2011.11	−27.0%	2013.7	7.3%	2015.3	0.3%
2011.12	−26.2%	2013.8	9.0%	2015.4	1.7%
2012.1	−24.3%	2013.9	4.5%	2015.5	5.8%
2012.2	−21.4%	2013.10	5.0%	2015.6	9.4%
2012.3	−18.8%	2013.11	8.7%	2015.7	4.2%
2012.4	−14.9%	2013.12	10.7%	2015.8	−0.5%
2012.5	−2.2%	2014.1	15.3%	2015.9	−2.0%
2012.6	−3.2%	2014.2	22.4%	2015.10	−12.8%
2012.7	−22.9%	2014.3	20.9%	2015.11	−11.2%
2012.8	−23.1%	2014.4	21.2%	2015.12	−4.0%

comprehensively reflect the potato price fluctuation in China, combining with the actual fluctuation of potato price index, this study uses the peak-valley method to identify the period according to the following criteria: the fluctuation rate "from Peak to Bottom" is greater than 5%; and the "from Bottom to Peak" drop is greater than 3%. The results of the specific partitioning cycle are shown in Table 3.

Table 3. The results of potato price fluctuation cycle partition

Serial number	Starting and ending month	Peak variation rate%	Valley variation rate%	Peak and valley drop	Months in ascending phase	Long period (months)
1	2011.1–2012.5	31.4	−27.0	58.4	6	17
2	2012.5–2013.5	27.8	−23.1	50.9	8	13
3	2013.5–2014.2	27.8	4.5	23.3	5	10
4	2014.2–2015.6	22.4	−13.7	36.1	10	17
5	2015.6–	9.4	−12.8	–	–	–
Average	–	–	–	42.18	7.25	14.25

Note: - indicates that the period is incomplete or does not meet the criteria of a cycle division and does not calculate the corresponding index.

The Chinese potato price fluctuates sharply, and the fluctuation frequency is high. From January 2011 to December 2015, potato prices experienced four complete cycle fluctuations. A new round of upswing has begun since June 2015. The average variation rate of peak and valley is 42.18%. Among them, there are two large fluctuations of peak and valley drop over 50%, the average period is 14.25 months, and the average monthly increase price index in each cycle is 7.25 months. The trend of change in price index variability is shown in Fig. 4.

5   Policy Suggestions

Through the quantitative analysis of potato market price index, it is found that potato price shows obvious cyclical fluctuation. To stabilize the fluctuation of potato production and to arouse the enthusiasm of farmers to produce potatoes, promote the implementation of potato staple food strategy, the government should establish and improve the potato market price control mechanism in the following aspects.

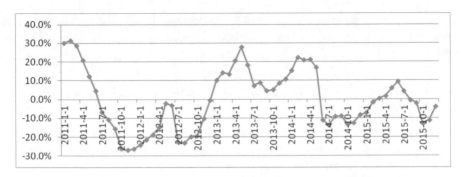

Fig. 4. Trend of potato price index variability in China

5.1   Establish Potato Price Monitoring and Warning Mechanism

At present, potato growers can only passively accept prices, and the price transmission mechanism is unreasonable. The market participants' expectations for
potato production and market prices are not clear, and the potato production is
ambiguous. Therefore, the monitoring and warning mechanism of potato price
should be set up, and the mechanism of sending and issuing potato production
information should be perfected, which focuses on the potato planting area, total
yield, market price, scale management and other indexes in the last year, and
strengthens the price monitoring and price warning ability in the main potato
producing areas. To solve the problem of asymmetric information, strengthen
the transparency of the market and make the potato industry compete fairly,
the price information publishing platform covering the main potato production
areas in China is established.

5.2   Establish and Improve Potato Storage System

At present, the construction of the potato storage system lags behind, especially
in the vast mountainous areas, the bad potato rate remains high. As a result,
potato farmers are eager to cash out potatoes after harvest, which can easily
cause market prices fluctuate intensely. To realize the staggered peak of potato
market and the balanced listing in the whole year, it is suggested to increase the
subsidy of potato storage circulation system, to establish farmers' self-storage,
and set up a specialized cooperative for potato storage. The large-scale storage
storehouse of enterprise produces the potato storage system of "three in one"
storage.

5.3   Establish Potato Insurance System

Potato market prices are highly volatile under the influence of uncontrollable
factors such as climate, environments and pests. Although potato sales revenue is
an important source of income for farmers, the sharp fluctuation of market prices

makes the potato farmers income and life uncertain. In view of this situation, it is suggested to introduce potato insurance system to protect the benefits of the potato farmers.

5.4 Cultivate New Type of Potato Production and Operation Subject

It is suggested to promote moderate scale operation in the main potato producing areas, form a relatively stable chain of industrial interests through professional large households, family farms, specialized cooperatives, leading potato processing enterprises, enhance its ability to resist market risks, and ensure the stability of production in this region, thus providing a stable supply for the main potato market area. In the main potato selling area, the circulation management should be strengthened, the circulation condition should be improved, the circulation efficiency should be optimized, and the stability of potato price should be maintained.

Acknowledgments. This research is supported by The Strategic Research Center on Stable Crop for the Potato(Grant NO.MLS1801).

References

1. China Potato Network Data Center [EB/OL] http://sjzx.tudouec.com (2018)
2. Cong, Z.: Study on price conduction effect of potato production and marketing. Ph.D. thesis, Chinese Academy of Agricultural Sciences, Chengdu (in Chinese) (2014)
3. Cai, H.: A probe into the causes of potato price fluctuation in china. Price Theory Pract. **9**, 64–65 (2013)
4. La Via, G., Nucifora, A., Cucuzza, G.: Short term forecasting of vegetables prices in sicily. In: VI International Symposium on Protected Cultivation in Mild Winter Climate: Product and Process Innovation, vol. 614, pp. 857–862 (2002)
5. Li, G., Xu, S., Li, Z.: Short-term price forecasting for agro-products using artificial neural networks. Agric. Agric. Sci. Proc. **1**, 278–287 (2010)
6. Li, J., Q, X.: Principles of statistical science. Master's thesis, Fudan University Press, Shanghai (in Chinese) (2004)
7. Liu, F., Guo, R., Wang, C.: Study on the fluctuation law of pig production in china. Agric. Outlook **4**, 28–31 (2011)
8. Liu, F., He, Z.: Study on price fluctuation and formation mechanism of fresh fruit and vegetable products in China. Master's thesis, China Agricultural Press, Shanghai (in Chinese) (2012)
9. Liu, Y., Luo, q.: Study on volatility of potato wholesale market price in china. Chin. Veg. **7**(170), 14–19 (2011)
10. Mao, X., Zeng, Y.: Price cycle recognition of pigs based on time series decomposition. China's Rural. Econ. **12**, 4–13 (2008)
11. Mi, J., Gao, Q., et al.: A supply and demand equilibrium forecast of potato in medium and long term. Chin. J. Agric. Resour. Reg. Plan. **6**, 27–34 (2015)
12. Qu, D., Xie, K.: 2015 beijing world potato congress paper excerpt-sustainable development of china's potato industry. Agric. Mark. Wkly. **29**, 26–26 (2015)

13. Ricci, U.: Die, synthetische ökonomie von henry ludwell moore. J. Econ. **1**(5), 649–668 (1930)
14. Schultz, H.: The standard error of a forecast from a curve. J. Am. Stat. Assoc. **25**(170), 139–185 (1930)
15. Tinbergen, J.: Bestimmung und deutung von angebotskurven ein beispiel. Zeitschrift für Nationalökonomie **1**(5), 669–679 (1930)

Author Index

A

Ahmed, Shariq, 641, 694
Ain, Qurat ul, 523
Akhtar, Yasmeen, 523
Akram, Muhammad Umair, 390
Akram, Umair, 574, 586
Ali, Sheeraz, 537
Amjad, Fiza, 390
Arshad, Farhan, 694

B

Baig, Sajjad Ahmad, 390

C

Cen, Hongyi, 557
Chen, ChuanBo, 105
Chen, Dong, 149
Chen, Jingdong, 119, 654, 668
Chen, Mo, 654
Chiu, Yung-ho, 557
Chuluunsukh, Anudari, 469

D

Dai, Jingqi, 270
Deng, Yanfei, 279
Diao, Xinyi, 221
Duan, Jiahui, 13
Duca, Gheorghe, 325

F

Fan, Jiaqi, 162
Fang, Qian, 61
Feng, Chun, 234

G

Gan, Shengdao, 135
Gang, Jun, 234
Gen, Mitsuo, 469
Gladchi, Viorica, 325
Gu, Xin, 75
Guo, Chunxiang, 310

H

Hafeez, Muhammad, 759
Hashim, Muhammad, 348, 390
He, Dongmei, 492
He, Rongrong, 616
He, Yue, 600
Hou, Shuhua, 450, 483
Huang, Chao, 162
Huang, Ling, 201
Huang, Qing, 743
Huang, Yong, 335
Hui, Peng, 574

I

Iqbal, Aimon, 641
Iqbal, Kashif, 759

J

Jiang, Qiang, 743
Jiang, Wen, 600
Jiang, Xianglan, 402
Jiang, Yufan, 376
Jiao, Jian, 557
Jie, Xiaowen, 730
Jin, Zelin, 557

K
Kamran, Asif, 537
Ke, Ge, 49
Khan, Abdullah, 641, 694
Khan, Israr, 759
Khurshaid, 759
Kumar, Nitin, 510

L
Lei, Yu, 93
Li, Dan, 40
Li, Jun, 295
Li, Kun, 93
Li, Liping, 40
Li, Qiang, 75
Li, Shan, 49
Li, Shuaifeng, 616
Li, Shuang, 105
Li, Shuijin, 270
Li, Wei, 49
Li, Yi, 743
Li, Ying, 557
Liang, Dianjie, 628
Liao, Chengcheng, 376
Liu, Haiyue, 201
Liu, Linlin, 243
Liu, Shiyi, 201
Liu, Shuangji, 683
Liu, Tingting, 270
Liu, Xinyun, 135
Liu, Xuhong, 730
Liu, Yunqiang, 415
Liu, Zhen, 335
Lu, Jingjing, 177
Lu, Yi, 243, 460
Luo, Jiarong, 402

M
Ma, Chenwei, 717
Ma, Sheng, 105
Ma, Zixin, 430
Majeed, Abdul, 574
Mancl, Karen, 279
Maqbool, Asif, 348
Mei, Hongchang, 177, 255
Meng, Zhiyi, 221
Mu, Yinping, 402

N
Nazam, Muhammad, 348, 390
Nedealcov, Maria, 325

P
Peng, Hui, 759

Q
Qian, Xiaoye, 705
Qiao, Jingwen, 119
Qiu, Rui, 450, 483
Quan, Quan, 415

R
Randhawa, Mahmood Ahmad, 348

S
Sarker, Md Nazirul Islam, 717
Shafi, Mohsin, 189, 683
Shah, Sita, 510
Sheng, Yi, 61
Shoukat, Farhana, 537
Shouse, Roger C., 717
Shu, Wei, 234
Shuai, Malian, 600
Sohaib, Muhammad, 574, 586
Song, Xiaoting, 189
Su, Rongjia, 628
Sun, Rongwei, 270, 483
Syed, Nadeem A., 537

T
Tanveer, Yasir, 586
Tariq, Anum, 574, 586
Tian, Jinqiu, 770
Travin, Serghei, 325
Tu, Mingzhou, 243

W
Wan, Ling, 26
Wang, Changfeng, 586
Wang, Dong, 743
Wang, Hong, 49, 135
Wang, Lishuai, 295
Wang, Min, 450
Wang, Qian, 376
Wang, Qinyun, 430
Wang, Rui, 105
Wang, Xin, 743
Wang, Xinhui, 335
Wang, Xuan, 705
Wang, Yile, 201
Wang, Yongyi, 335
Wang, Yujie, 243
Wei, Jiayi, 365

Wei, Lai, 460
Wei, Yuzhu, 255
Wu, Jing, 162
Wu, Min, 717
Wu, Wenjing, 310

X
Xiong, Wenyu, 600
Xu, Chuanpeng, 600
Xu, Dong, 600
Xu, Jiancheng, 13
Xu, Jiuping, 1
Xu, Lei, 279
Xu, Sichen, 415
Xu, Wenxin, 668
Xu, Xinxin, 770
Xue, Weixian, 654

Y
Yan, Jinjiang, 335
Yan, Qinqin, 93
Yang, Ruo, 189
Yang, Wei, 135
Yang, Yang, 376
Yang, Yongzhong, 189, 683
Yi, Sheng, 149
You, Ming, 415

Yousaf, Tahir, 523
Yu, Weiping, 600
Yu, Xiaozhong, 26
Yuan, Yuan, 279
Yun, YoungSu, 469

Z
Zhan, Chengyan, 243
Zhang, Dan, 600
Zhang, Hongjiang, 243
Zhang, Hua, 49
Zhang, Jinsong, 430
Zhang, Lei, 492
Zhang, Qianyou, 770
Zhang, Xiaoqian, 430
Zhang, Yanyan, 546
Zhang, Yuanling, 770
Zheng, Jianguo, 13
Zheng, Yanglinfeng, 705
Zheng, Yuerong, 770
Zhong, Jihui, 162
Zhou, Shoujiang, 49
Zhou, Zihan, 492
Zhu, Jialing, 415
Zhu, Mengyuan, 270
Zu, Xu, 221

Printed in the United States
By Bookmasters